Sperm Competition and the Evolution
of Animal Mating Systems

CONTRIBUTORS

Steven N. Austad
David J. Bruggers
Kimberly M. Cheng
George D. Constantz
Michael C. Devine
Donald A. Dewsbury
Boyce A. Drummond III
M. Brock Fenton
Donald C. Gilbert
Simon R. Greenwell
Mark H. Gromko
Darryl T. Gwynne
T. R. Halliday
A. H. Harcourt
Paul H. Harvey
Nancy Knowlton
Frank McKinney
G. A. Parker
Rollin C. Richmond
John Sivinski
Robert L. Smith
Christopher K. Starr
Richard H. Thomas
Randy Thornhill
P. A. Verrell
Jonathan K. Waage
David W. Zeh

Sperm Competition and the Evolution of Animal Mating Systems

Edited by **ROBERT L. SMITH**

Department of Entomology
University of Arizona
Tucson, Arizona

1984

ACADEMIC PRESS, INC.

(Harcourt Brace Jovanovich, Publishers)
Orlando   San Diego   New York   London
Toronto   Montreal   Sydney   Tokyo

Academic Press Rapid Manuscript Reproduction

ACADEMIC PRESS, INC.
Orlando, Florida 32887

United Kingdom Edition published by
ACADEMIC PRESS, INC. (LONDON) LTD.
24/28 Oval Road, London NW1 7DX

Library of Congress Cataloging in Publication Data

Main entry under title:

Sperm competition and the evolution of animal mating
systems.

Includes bibliographies and index.
1. Spermatozoa. 2. Competition (Biology) 3. Sexual
behavior in animals. 4. Behavior evolution. I. Smith,
Robert L. (Robert Lloyd)
QP255.S64 1984 591.56 84-45674
ISBN 0-12-652570-6 (alk. paper)

PRINTED IN THE UNITED STATES OF AMERICA

84 85 86 87 9 8 7 6 5 4 3 2 1

During one year as a graduate student at Arizona State University, I had the privilege of attending lectures in evolutionary ecology delivered by Professor Donald Tinkle. Don was an inspired scholar who inspired scholarship. He gave me a point of view.

A friend and graduate student peer who benefited from Professor Tinkle's lingering influence in the Department of Zoology was Don Pyle. The second Don later became interested in sperm competition in Drosophila, and produced important work on the subject.

Both Dons died while this volume was being assembled. It is to their memory and their work that the book is dedicated.

Contents

1. Sperm Competition and the Evolution of Animal Mating Strategies
G. A. Parker

2. Male Sperm Competition Avoidance Mechanisms: The Influence of Female Interests
Nancy Knowlton and Simon R. Greenwell

3.   Sperm in Competition
John Sivinski

4.   Male Mating Effort, Confidence of Paternity, and Insect Sperm Competition
Darryl T. Gwynne

5.   Alternative Hypotheses for Traits Believed to Have Evolved by Sperm Competition
Randy Thornhill

6. **Sperm Transfer and Utilization Strategies in Arachnids:
   Ecological and Morphological Constraints**
   Richard H. Thomas and David W. Zeh

7. **Evolution of Sperm Priority Patterns in Spiders**
   Steven N. Austad

8. **Sperm Competition and the Evolution of Odonate Mating Systems**
   Jonathan K. Waage

13. Sperm Competition in Amphibians
T. R. Halliday and P. A. Verrell

14. Potential for Sperm Competition in Reptiles: Behavioral and Physiological Consequences
Michael C. Devine

15. Sperm Competition in Apparently Monogamous Birds
Frank McKinney, Kimberly M. Cheng, and David J. Bruggers

16. Sperm Competition in Muroid Rodents
Donald A. Dewsbury

17. Sperm Competition? The Case of Vespertilionid and Rhinolophid Bats
M. Brock Fenton

18. Sperm Competition, Testes Size, and Breeding Systems in Primates
Paul H. Harvey and A. H. Harcourt

19. Human Sperm Competition
Robert L. Smith

List of Contributors

Steven N. Austad[1], Department of Biological Science, Purdue University, West Lafayette, IN 47909, U.S.A.

David J. Bruggers, James Ford Bell Museum of Natural History, Department of Ecology and Behavioral Biology, University of Minnesota, Minneapolis, MN 55455, U.S.A..

Kimberly M. Cheng, Department of Poultry Science, University of British Columbia, Vancouver, B. C., V6T2A2, Canada.

George D. Constantz, Department of Limnology, Academy of Natural Science, 19th and The Parkway, Philadelphia, PA 19103, U.S.A.

Michael C. Devine, Biological Sciences, Fairleigh Dickinson University, 100 River Road, Teaneck, NJ 07666, U.S.A.

Donald A. Dewsbury, Department of Psychology, University of Florida, Gainesville, FL 32611, U.S.A.

Boyce A. Drummond III, Department of Biological Sciences, Felmley Hall 106, Illinois State University, Normal, IL 61761, U.S.A.

M. Brock Fenton, Department of Biology, Carleton University, Ottawa, K1S5B6, Canada.

Donald C. Gilbert, Department of Biology, Jordan Hall 138, Indiana University, Bloomington, IN 47401, U.S.A.

Simon R. Greenwell, Department of Biology, Yale University, New Haven, CT 06511, U.S.A.

Mark H. Gromko, Department of Biology, Bowling Green University, Bowling Green, OH 43402, U.S.A.

---

[1]Present address: Department of Biology, University of New Mexico, Albuquerque, New Mexico, 87131.

Darryl T. Gwynne, Department of Biology, University of New Mexico, Albuquerque, NM 87106, U.S.A.

T. R. Halliday, Department of Biology, The Open University, Walton Hall, Milton Keynes, MK76AA, England.

A. H. Harcourt, Department of Applied Biology, University of Cambridge, Pembroke Street, Cambridge, CB23DX, England.

Paul H. Harvey, School of Biology, University of Sussex, Brighton, Sussex, BN19QG, England.

Nancy Knowlton, Department of Biology, Yale University, New Haven, CT 06511, U.S.A.

Frank McKinney, James Ford Bell Museum of Natural History, Department of Ecology and Behavioral Biology, University of Minnesota, Minneapolis, MN 55455, U.S.A.

G. A. Parker, Department of Zoology, University of Liverpool, Liverpool, L693BX England.

Rollin C. Richmond, Department of Biology, Indiana University, Bloomington, IN 47405, U.S.A.

John Sivinski, Department of Entomology and Nematology, University of Florida, Gainesville, FL 32611, U.S.A.

Robert L. Smith, Department of Entomology, University of Arizona, Tucson, AZ 85721, U.S.A.

Christopher K. Starr, Department of Plant Protection, Visayas State College of Agriculture, Baybay, Leyte 7127, Philippines.

Richard H. Thomas, Department of Ecology and Evolutionary Biology, University of Arizona, Tucson, AZ 85721, U.S.A..

Randy Thornhill, Department of Biology, University of New Mexico, Albuquerque, NM 87106, U.S.A.

P. A. Verrell, Animal Behaviour Research Group, Department of Biology, The Open University, Milton Keynes, MK76AA, England.

Jonathan K. Waage, Department of Biology and Medicine, Brown University, Providence, RI 02912, U.S.A.

David W. Zeh, Department of Ecology and Evolutionary Biology, University of Arizona, Tucson, AZ 85721, U.S.A.

Preface

This book is about why females mate with more than one male and how the resultant competition among spermatozoa has created selection on a range of attributes from gamete morphology to species' mating systems.

Antoni van Leeuwenhoek observed his own sperm and in 1677 (reticently) communicated what he had seen in a letter to The Royal Society. A little less than 200 years later Charles Darwin produced his second major treatise on organic evolution. Paradoxically, Darwin, the elucidator of sexual selection (*The Descent of Man and Selection in Relation to Sex,* 1871) may have contributed to delay in recognition of this volume's subject. It is clear that Darwin believed females to be generally monogamous: "It is shown by various facts, given hereafter, and by the results fairly attributable to sexual selection, that the female, though comparatively passive, generally exerts some choice and accepts **one** male in preference to the others." He thus denied the requisite precondition for sperm to compete, namely, multiple mating by females. This view (though a popular and comforting one among human males) is wrong for most species. A variety of circumstances compel females of most species to (at least occasionally) mate with several males during a single reproductive cycle such that sperm from two or more ejaculates may contest the fertilization of relatively few ova.

G. A. Parker was first to clearly conceptualize sperm competition as a subset of sexual selection and its implications for the insects (1970, *Biol. Rev.* **45:** 525-567) when he recognized the oppositional forces that it creates. He proposed that selection would simultaneously favor adaptations to facilitate the preemption of previous ejaculates, and adaptations that would resist preemption.

Appropriately, Geoff Parker introduces the topic here with a general review, an overview of recent developments, and a substantial expansion of his original theories. Parker's chapter is followed by three generic ones: Nancy Knowlton and Simon Greenwell consider sperm competition from the female perspective; John Sivinski treats inter- and intraejaculate competition, emphasizing evolutionary causation at the level of the individual male gamete; and Darryl Gwynne asks if high paternal investment and sperm precedence positively correlate in the insects.

Chapter 5 is a caveat. Randy Thornhill discovers alternative contexts for characters initially believed to have evolved by sperm competition; his contribution cautions the reader and instructs the prospective investigator in the field.

The remaining chapters are arranged phylogenetically, and bring the topic of sperm competition to diverse animal taxa. Some chapters were recruited because the subject taxon is popular and hence rich in data as, for example, the chapters on *Drosophila* and Lepidoptera. Others were solicited for some unique and intriguing problem they could address such as the evolution of direct vs. indirect sperm transfer among the arachnids or the problem for kinship theory presented by multiple mating and sperm competition in the Hymenoptera. Brock Fenton considers the remarkable potential for sperm competition among certain temperate bat species whose females store (and in some cases nurture) sperm through winter hibernation, and George Constantz reports mixed strategies and male-caused female genital trauma as possible sperm competition adaptations in poeciliid fishes. Several chapters are exceptional in that they contain original data collected by their authors for the express purpose of addressing questions on sperm competition. These include the contribution by Jon Waage on the dragonflies, Mark Gromko *et al.* on *Drosophila*, Don Dewsbury on rodents, and Frank McKinney *et al.* on "apparently monogamous birds." The *Drosophila* chapter and others (those on spiders, Lepidoptera, amphibians, and reptiles) represent exhaustive compilations of relevant literature that will survive as definitive references for anyone entering this field. The book concludes with Paul Harvey and A. H. Harcourt's test of predictions concerning testes size and mating systems in the primates, and I have taken some license in exploring the possibility that sperm competition has caused selection on humans.

Throughout the project, I have encouraged authors to speculate on the importance of sperm competition as an evolutionary force. I did this in the belief that conjecture (in the absence of sufficient data to infer scientific conclusions) is currently desirable for this subject. Speculation piques interest, stirs controversy, and, most importantly for a young topic, inspires data gathering. If the book is to be faulted on this account, the criticism should be directed to me, for I sometimes pressed authors to the limits of their comfort in order that interesting possibilities might be clearly exposed.

I believe this book is unique among those of its kind in that it is more than a simple collection of papers. With few exceptions, authors have had the opportunity to view the products of other contributors, to comment on them, and to cross reference highly relevant sections. This has, in my opinion, yielded substantial benefits for the reader. I had initially undertaken to homogenize nomenclature, but abandoned this because the imposition of standard syntax seemed capricious and premature. Consequently, authors have defined terms within the context of their subjects. This may be problematical for some readers, but I think it will permit a more natural shakeout with conservation of the most useful terminology.

The current effort was needed to synthesize a diffuse literature, on arthropods especially, and to provide a point of departure for inevitable expansion of the study into the vertebrate taxa. The work covers metazoans from arachnids to man, and because of its taxonomic scope should enjoy a wide readership. Its potential audience will similarly include a diversity of interests in biological phenomena from reproductive physiology to ethology. The book assembles and reviews literature, generates new theory, proposes testable predictions, and may chart the course for future research. It at least suggests many points of departure.

RLS

Acknowledgments

I first wish to recognize all of the fine scientists who contributed to this volume. I thank them for their chapters, for their reviews of each others' papers, and most of all for their tolerance, patience, and sustained encouragement throughout the project. A number of persons (in addition to authors) assisted the editor with advice or by reviewing manuscripts. Their names follow: R. Abugov, J. Alcock, S. A. Altmann, S. A. Arnold, H. J. Brockmann, E. L. Cockrum, R. H. Crozier, V. Delesalle, W. J. Gertsch, W. D. Hamilton, W. B. Heed, B. J. Kaston, A. Kodric-Brown, T. Markow, R. E. Michod, P. A. Racey, K. G. Ross, R. Rutowski, R. L. Trivers, L. J. Vitt, K. J. Wells, M. J. West-Eberhard, W. A. Wimsatt, R. B. Zimmerman. I thank them all. No less worthy of my appreciation are the many others who made my job much easier by reviewing manuscripts for individual authors prior to my receiving them. Finally, I thank Jill Smith, who spent many hours typesetting the text, drafting figures, and generally attending to details of the book's production.

1

Sperm Competition and the Evolution of Animal Mating Strategies

G. A. PARKER

I. INTRODUCTION

The reproductive success of a male individual depends mainly on how many of his sperm are successful in fertilizing eggs. Thus sexual selection will not end with adaptations that relate to mating with many females. Many male adaptations— behavioral, morphological, and physiological—relate to enhancing the success of self's sperm against rival sperm. Much of sperm structure and organization can be interpreted in terms of gamete competition (Sivinski 1980; this volume). Sessile males with external fertilization may simply maximize gamete production; they may have few alternative ways of allocating reproductive effort. However, for species with internal fertilization, there exists a rich variety of sexually-selected adaptive pathways that appear to relate to enhancing the success of self's ejaculate relative to rival ejaculates, *i.e.*, to "assure paternity" (Smith 1979a).

I earlier attempted to review the main lines of adaptation in insects to this form of sexual selection (Parker 1970a). Potential competition within the female tract between the ejaculates from rival males is likely to be intense in insects; sperm are provisioned within the sperm stores and are extremely long lived, giving much capacity for the "overlapping" of ejaculates. During the past decade it has become evident that many other groups display many and varied adaptations to sperm competition (competition between ejaculates from two or more different males over the fertilization of ova); the present book is a testament to the ubiquity of sperm competition throughout the animal kingdom. In addition, there have been many important theoretical developments, including the population genetics analysis of Prout and Bundgaard (1977). Further it has become evident that the degree of multipaternity in sibships will be of vital importance in kin selected interactions between sibs (see Hamilton 1964; Starr, this volume). The main aims of the present paper are to attempt a general synthesis of the influence of sperm competition in evolutionary adaptation, and also to examine some further theoretical problems that are posed by sperm competition.

One of the difficulties in being allowed to revisit a topic after a number of years (and many developments) have elapsed is the temptation to seek evidence in favor of one's former speculations. There are certainly alternative explanations that are sometimes more plausible than those I suggested in 1970 (Thornhill, this volume). Possibly my earlier undoubted understatement of the role of the female in the evolutionary dynamics of sperm competition is a case for scrutiny; in the present paper I have tended to argue that the degree of multipaternity of offspring will be of rather minor selective importance to the female compared with its importance to males. Alternative views may be found elsewhere (*e.g.*, Pease 1968; Walker 1980; Knowlton and Greenwell, this volume). However, few would disagree that there will generally exist a vast asymmetry in selection intensity between the two sexes. Males stand to be affected by far the most, since usually the comparision is whether or not the female obtains the highest fitness payoff from her brood of offspring. In the former, the currency is offspring number, in the latter it is usually offspring quality, derived from variance in the father's genetic quality. This variance in paternal genetic quality may well be low (*e.g.,* Maynard Smith 1978a). But although selection intensity may usually be much more intense on males, it may often be relatively easy and inexpensive for a female to prevent a mating, and difficult and costly for a male to inflict a mating upon her. This makes it hard to predict the resolution of sexual conflict over mating, and the issue is still controversial (*e.g.,* Knowlton and Greenwell, this volume).

How does sperm competition generate conflicts of interest between individuals? Firstly, as noted above, it can produce conflicts between males and females over whether or not mating should take place. Secondly, sperm competition generates a series of adaptive battles between males in which the payoffs depend on the strategies played by other males in the population. For instance, if very few males attempt to mate with fertilized females, the value of a paternity assurance mechanism such as a copulatory plug will be low, and may easily be outweighed by its cost. Alternatively, the reverse is likely if fertilized females are generally remated.

The ESS (evolutionarily stable strategy) concept of Maynard Smith (1974) provides a method for analyzing games of conflict in which payoffs depend on the strategies played by other members of the population. Note that an ESS is not the same as simple optimization in which an animal "plays" against a fixed environment, and where payoffs are not dependent on the current frequencies of strategies in the population. A strategy is an ESS if, when adopted by most individuals, it cannot be bettered by any alternative strategy. ESS's can either be **pure** strategies (*e.g.,* when in condition *a*, play strategy *A*) or mixed strategies (*e.g.,* when in condition *a*, play strategy X with probability P_X, Y with probability P_Y, Z with probability P_Z, etc.). Many ESS's are in fact types of **Nash equilibrium.** In a Nash equilibrium, if either player deviates unilaterally from the equilibrium, his payoff is reduced (see Maynard Smith 1978b). Examples of Nash equilibria are common in conflict theory (Maynard Smith 1978b, Parker and Macnair 1979). Suppose that both players play against each other using the same type of strategy (say X, which is some investment chosen from a continuous distribution), then the equilibrium may be for both to choose the pure strategy $X*$. If each player plays a different strategy against its opponent (say X against Y), the Nash ESS may consist of the pair of pure strategies: $X*$, $Y*$. This sort of ESS will relate to contests such as mating conflicts, where the female must necessarily play a different strategy (rejection) from the male (persistence).

In the present paper, I suggest that sperm competition generates a series of conflicts (subgames) that are dynamically interrelated; the global ESS is suggested to consist of a set of "balancing" Nash equilibria, one for each subgame. Ultimately, all the adaptations of a given species (and even sets of interdependent species) may perhaps be expanded into hierarchies of interrelated Nash ESS's.

II. THE EVOLUTION OF THE TWO SEXES

Gamete competition, and later sperm competition, may account for the origin of the two sexes in multicellular organisms (Parker *et al.* 1972). A sex is best defined in terms of the size of gamete that it produces. In a primitive animal with external fertilization that sheds its gametes into sea water, the more gametes shed the greater the number of fusions obtained. On a fixed energy budget, drive to produce more gametes must be compensated by a decrease in the size of each gamete. Drive to produce smaller and smaller gametes in increasing numbers (*i.e.,* to produce males) might be countered because the zygotes produced have less provisioning and are consequently less viable and/or take longer to build up to adult size. There seem intuitively to be a number of possible solutions. Gamete size might stabilize at a unique optimum that compromises the conflicting demands of productivity on the one hand and provisioning on the other. In fact, the ESS solution to this problem is the stable coexistence of two morphs for gamete size (males and females) provided that provisioning exerts an important effect upon the survival of the zygote (Parker *et al.* 1972, Bell 1978, Charlesworth 1978, Maynard Smith 1978a, Parker 1978a, Hoekstra 1980).

Parker *et al.* (1972) used a computer simulation based on the following assumptions to demonstrate that selection would lead to anisogamy (two sexes). Imagine that there is an ancestral isogamete-producing population in which individuals differ in the size of gamete they produce (Fig. 1). Some adults will produce rather few large gametes with high prospects of survivorship as zygotes, others will produce many smaller gametes with poorer prospects of survival. Most are intermediate. Fusion between gametes is random. Gamete size is controlled by a hierarchy of alleles at a "gamete-size" locus. A parent that produces n_i gametes is assumed to produce gametes of size R/n_i, where R is the fixed total energy budget for gamete production. A zygote produced by fusion of gametes from an i parent with those of a j parent is assumed to have a viability that relates to its size as:

$$\text{Viability}_{ij} = (R/n_i + R/n_j)^x$$

where power x simply scales the relative importance of size in subsequent survival. Provided that x is sufficiently high (above 1; Parker 1978a, Charlesworth 1978), alleles for intermediate gamete sizes are lost by selection and the system stabilizes with mainly two genotypes (males and females) in a 1:1 ratio. The third genotype tends not to be represented either because it arises as the rather inviable product of the fusion of two "proto-sperm," or because it is formed by the fusion of two "proto-ova," depending on dominance. This latter fusion is rare because most gametes fuse with small gametes because of their numerical predominance. After several generations, the population is essentially anisogamous (Fig. 1) as a result of

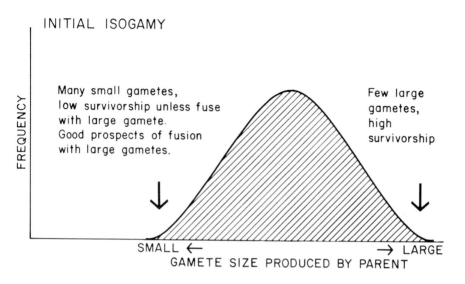

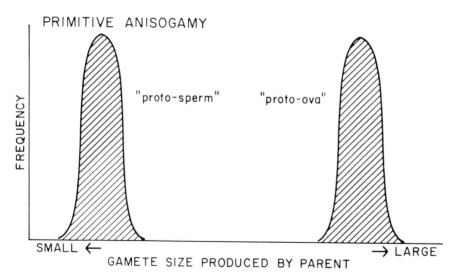

Fig. 1. Evolution of anisogamy (two sexes) from a hypothetical ancestral population with isogamy (one sex) with a range of gamete size.

disruptive selection. Analytical confirmation of this model is given by Bell (1978), Charlesworth (1978), Maynard Smith (1978a), and Hoekstra (1980).

Although alternative suggestions exist that anisogamy could have an origin dating back to prokaryotes (Baker and Parker 1973, Alexander and Borgia 1979), it is most likely that disruptive selection for two sexes would have been operative mainly during the evolution of multicellularity. Gametes would probably have been small originally (protistan) and drive to increase provisioning (to produce ova) secondary, as provisioning became more important in developing multicellular forms. Trends toward anisogamy from isogamy have been found in series of unicellular to multicellular algae (Knowlton 1974, Bell 1978). However, whatever the starting point (large to small isogametes), the ESS is likely to be anisogamy if zygote provisioning exerts an important influence on survival. Female gametes would optimize to a size that gives a maximum number of surviving offspring (Parker *et al.* 1972, Smith and Fretwell 1974), whereas sperm would be reduced in size to a minimum, bound only by considerations of optimal survival before fusion, and abilities at competitive fusion. There are obvious reasons to regard microgamete production (maleness) as the ancestral state (Maynard Smith 1978a).

Random fusion of gametes, independent of their size, is unlikely to persist for long. Females might do better to produce ova that refuse sperm fusions and only fuse with other ova. However, this sort of modifier for selective fusion cannot invade if the system starts with random fusion and proceeds to a reasonable degree of anisogamy before the modifier arises (Parker 1978a, 1979). A mutation that causes ova to accept only other ova must compete with sperm for fusions with the random-fusing ova. Because of the high numbers of sperm, such an ovum fares badly and must rely on selfing to survive. Hence ova may do better to lose their capacities to penetrate and fuse with other gametes if this allows extra expenditure on viability. There are also other reasons why the female cannot win this primordial conflict (see Parker *et al.* 1972). In contrast, a modifier that causes sperm to fuse only with ova will be highly favorable.

Sperm producers therefore survive by parasitizing the investment of ovum producers; the ESS solution (stable anisogamy) is essentially driven by sperm competition (Parker 1982b). Sperm competition acts to maintain sperm size small in order to maximize sperm productivity. Random fusion of gametes will quickly be lost in favor of disassortative fusion between eggs and sperm (Parker 1978a, 1979). By Fisher's (1930) principle, the sex ratio will approach unity.

Note that this model allows only for one sexual strategy: gamete production. It allows males to expend their reproductive energy reserve R only on gametes; sperm competition is rampant because all ejaculates must compete in the same external medium for fusions with ova. Large numbers of small sperm is the inevitable outcome. Where there is external fertilization and adults are sessile. we should expect approximately equal total gametic expenditures in the two sexes, but a vast disparity in gamete size. This is in general the case.

III. GAMETE INVESTMENT
VERSUS OTHER REPRODUCTIVE EFFORT

In animals, mobility allows for a variety of reproductive alternatives to profligate expenditure on gametes. Increased mobility in males may have arisen in response to the sexual selective pressure of being able to shed sperm as close as possible to a female who is about to release her ova. This would enhance the probability of obtaining fusions relative to less mobile males. Increased male mobility should proceed to the level where the advantage of trading one more sperm for enhanced mobility exactly balances the advantage, through sperm competition, of producing the sperm itself (Parker 1978b).

Much will depend on synchrony (communality) of spawning. If females tend to release eggs synchronously, and in close proximity, the competitive advantage to males of being able to search out specific females would be reduced and high sperm expenditure may be more favorable than increased male mobility. In species with high degrees of spatial and temporal synchrony of spawning, the sexes show similar gametic expenditure in terms of gametic mass (certain echinoderms, certain polychaetes, and certain fish), and rather similar mobility patterns. In species where there is no communal spawning, males often spend somewhat less on gametes than the female.

One effect of spatial and temporal dispersion of females during spawning is that the effect of sperm competition may be reduced. If males are mobile and tend to sequester females close to spawning, or tend to expend energy in guarding territories in which the female spawns, then often the sperm from only one male will be used to fertilize each batch of ova. The higher the proportion of occasions in which sperm competition is not prevalent, alternative male reproductive allocations such as enhanced mobility, male guarding, territoriality, and paternal care, will be favored at the expense of gametic expenditure.

IV. THE EVOLUTION OF INTERNAL FERTILIZATION

It has been argued that internal fertilization itself arose primarily by sexual selection via sperm competition (Parker 1970a). Males that could locate their sperm actually within the female would suffer much reduced chance of sperm competition and much greater chance that their sperm are used than if sperm are shed externally. However, in some groups (*e.g.,* insects) internal fertilization probably preceded copulation; primitively, females may have inseminated themselves by picking up spermatophores deposited around them by males (Alexander 1964).

Copulation could have arisen by progressive steps each fueled by sexual selection, from indirect spermatophore transferring acts. Thus the evolution both of internal fertilization, and especially copulation, can be interpreted in terms of paternity assurance.

With the onset of copulation and internal fertilization, it is generally assumed that sperm competition would become dramatically reduced in its importance as an intrasexual selective force. A vast economy in gametic expenditure is therefore expected in males that copulate, with compensatory increases in expenditure on other reproductive efforts such as female-guarding and male searching. In general, this expectation seems justified, although some species with external fertilization often show remarkable degrees of female guarding and territoriality, and probably achieve high levels of paternity assurance.

However, as many of the papers in this volume show, sperm competition is seldom eliminated even in species with internal fertilization. It can still exert a powerful influence on mating strategy. Ejaculates from different males commonly compete within the female reproductive tract, and this selective force has favored a variety of adaptations. Species in which there is sperm storage within the female (notably insects) sometimes pose different evolutionary problems from those in which there is usually little special provisioning for the sperm within the female tract (notably mammals; but see Fenton, this volume).

V. SPERM COMPETITION
AND CONFLICTING EVOLUTIONARY FORCES

There appear to be two major conflicting selective forces related to paternity assurance (Parker 1970a):

1. Selection favoring mechanisms for preemption of stored sperm. Mechanical displacement of the previously stored sperm from the sperm stores appears to be achieved in many species with sperm storage, though (with rare exceptions, see below) rather little is known about how sperm displacement is achieved. Direct displacement may not be equally feasible in groups without sperm storage (though it certainly cannot be ruled out). Selection may here act to favor increased ejaculate volume (or multiple ejaculation; see Lanier *et al.* 1979) to assure numerical predominance of self's sperm in the ensuing lottery.

2. Counter selection favoring anti-preemption mechanisms. In short, counter selection will act to favor being able to prevent future males from reducing the effectiveness of self's sperm.

Clearly these two adaptations are ultimately in conflict, and will form an evolutionary "arms race" (Dawkins and Krebs 1979). The dynamics of such arms races are interesting. If males were able, at some cost, to prevent any other male from subsequently introducing sperm, then selection would quickly favor not attempting to mate with females that are already mated; there is no point in paying the cost of hopeless persistence. As male persistence declines in the population, the paternity assurance mechanism would constitute wasted expenditure, and would in turn crash in favor of zero investment in paternity assurance. The result of this simple "2 x 2" strategy game (investment in PA/no investment; attempting to remate/not attempting) would be an endlessly oscillating limit cycle. Can we stabilize the model by allowing continuous strategy sets? One opponent in the game—say the first male to mate—can "choose" any level of investment in some method of preventing subsequent matings. His opponent—a subsequent male—can "choose" any level of investment to expend in overcoming the first male's paternity assurance mechanism. Roles (first male or subsequent male) occur randomly. Males must "choose" two strategies. One strategy is a choice of investment cost x on the machinery of remating prevention. The other is a choice of expenditure y on armament to overcome this paternity assurance adaptation. We can define a "balance of arms" in the sense that if $x = y$, the ability of each male to win is equal. Calling the value of winning V_1 and V_2 for the two males, K_1 and K_2 constants that convert armanent expenditures x and y into fitness costs, we can define the rules of the game as:

	Payoff to First Male	Payoff to Second Male
If $x > y$	$V_1 - K_1 x$	$- K_2 y$
If $x < y$	$- K_1 x$	$V_2 - K_2 y$
If $x = y$	$V_1/2 - K_1 x$	$V_2/2 - K_2 y$

It is clear that this game is the "opponent-independent costs" game, due originally to Parker (1979; see also Rose 1978). It again has no ESS. However, suppose that an individual's "choice" of strategies (its genes for x or y) are modified by environmental effects. Thus if an animal has say strategy x_A, it has a probability of obtaining armanent R with distribution $P_A(R)$ and mean R_A. Similarly if it has strategy Y_B, it achieves R with probability $P_B(R)$ and mean R_B. This modification—armament strategy is imperfectly heritable—can generate an ESS. It consists of a Nash equilibrium pair strategies for the two stable levels of investment in armament $x*$ and $y*$. This sort of ESS appears to be a fairly general solution to the evolutionary arms race problem (Parker 1983).

What does this conclusion mean for paternity assurance arms races? We would expect to see some cases in which a second male is able to overcome a first male's investment in remating prevention, and some cases where he fails to remate with mated females. As we shall see, this is often the case.

The details of this model will need altering appropriately for each particular sperm competition system. Often, males just mate and then depart from the female without any investment in future paternity assurance. This could be advantageous where the female becomes unreceptive to further matings during copulation, or where she is rather unlikely to meet another male before completing oviposition (*e.g.,* oviposition site away from the area where the sexes meet).

A. Sperm Storage and Sperm Displacement (The Insect Model)

For clarity, I shall discuss separately those systems in which there are specialized, long term sperm stores in the female (notably insects), and those where the sperm stores are much less clearly differentiated and sperm are short-lived (notably vertebrates, with certain exceptions—*e.g.,* see Fenton, Devine, this volume). It is easier to see how sperm displacement can occur in species with specialized sperm storage organs in the female, since these offer a site of large concentration of foreign sperm against which the male can direct some mechanical attack. However, in some groups with sperm storage, sperm displacement may not be possible anatomically (Devine, this volume). Further, the possibility that some sperm displacement may occur in species lacking sperm stores is certainly not a remote one (see following section). Thus the terms "insect model" and "vertebrate model" must not be taken too literally.

What evidence is there that males displace previously stored sperm? There have been numerous experiments in insects to ascertain the paternity of progeny produced after multiple matings of the same female. Sperm from a given male can be "labeled" either with genetic markers or by large doses of irradiation, which induce so high a level of dominant lethals that most eggs from the irradiated sperm fail to hatch. Generally, but by no means always, it is the last male to mate that fertilizes most offspring (Parker 1970a). Fig. 2 shows the number of cases examined in which the last male to mate fertilized 0 to 0.25, 0.25 to 0.5, 0.5 to 0.75, and 0.75 to 1.0 of the eggs laid after the last mating (data from Parker 1970a; Boorman and Parker 1976; Gwynne, this volume). In the overwhelming majority of cases, the last male to mate does best and most commonly he gains 0.75-1.0 of the subsequent progeny.

It was Lefevre and Jonsson (1962) who first proposed, following their experiments on *Drosophila melanogaster,* that sperm were displaced from the female's sperm stores on a volumetric basis, though they were unclear as to the mechanism involved in the process. Sperm were present in equal numbers in the sperm stores whether the female had been mated only once, or by two males. The second male to mate gained most fertilizations (see also Boorman and Parker 1976).

In the dung fly, *Scatophaga,* the pattern of sperm competition following multiple matings has been established by the irradiated male technique (Parker 1970b). Eggs are laid in batches and a mating precedes the laying of each batch. The last male to mate gains approximately 0.8 of the eggs of the batch about to be laid, and the sperm from previous males gain the remaining 0.2 of the batch **in the same**

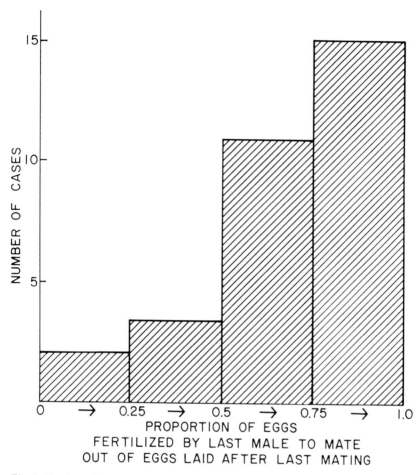

Fig. 2. Number of fertilizations obtained by the last male to mate with the female. Sperm from the last male are those most usually successful in insects.

proportions as they did at the previous oviposition (see Fig. 3). This suggests that 80% of previous sperm are displaced and that the sperm mix randomly within the sperm stores before fertilization. The longer a male dung fly copulates, the more sperm he is able to displace, though gains through increased displacement obey the law of diminishing returns with time spent copulating. Extra time spent copulating is costly to males, because it means missed opportunities to mate with new females. Using the "marginal value theorem" (Charnov 1976, Parker and Stuart 1976) it is possible to show that the observed mean copula duration (35.5 min) is close to the predicted optimum of 41 min (Parker 1970b, Parker and Stuart 1976). The discrepancy may relate to the fact that ejaculates were assumed

PATTERN OF SPERM COMPETITION IN *Scatophaga*

	First Male	Second Male	Third Male	Fourth Male
Gains at First Batch	1.0	–	–	–
Gains at Second Batch	.2	.8	–	–
Gains at Third Batch	$.2^2$	.8 x .2	.8	–
Gains at Fourth Batch	$.2^3$	.8 x $.2^2$	.8 x .2	.8

etc.

Fig. 3. Pattern of sperm competition in the dung fly, *Scatophaga stercoraria* L. The last male gets 0.8 of the batch; previous males get 0.2 x their share of the previous batch. This suggests 80% sperm displacement, then random utilization of sperm at fertilization.

to cost nothing. If the female has completed oviposition and is mated again before she leaves the oviposition site (a rather infrequent event in nature), the predicted optimum copula duration is reduced to 17.5 min because a further male will mate before any more eggs are laid. The observed copula duration with such females is 15.1 min, which suggests that males can assess whether or not the female is gravid, and behave accordingly.

It is possible to speculate upon the mechanism behind sperm displacement for some insect species. In the giant water bug, *Abedus*, the male mates continually during the female's oviposition period, and this may repeatedly displace previously stored sperm away from the site of fertilization (Smith 1979b). The *Scatophaga* data suggest that sperm mixing may be rather rapid; it is therefore possible that *Abedus* males prevent random delivery of sperm by mating repeatedly. The last male thus obtains a very high proportion of eggs (99.7%; Smith 1979b).

In the migratory locust, *Locusta migratoria migratorioides,* a direct "sperm flushing" process may account for the high degree of displacement (Parker and Smith 1975). Females lay several successive batches of eggs, and (apart from the first oviposition) usually mate once per cycle immediately after oviposition. With this pattern, some 90% of eggs at a given batch are fertilized by the last male to mate. The spermatophore body remains within the male, but the spermatophore tube everts and elongates into the female tract during copulation. It extends right up the spermathecal duct and ends in the apex of the spermatheca itself, where the sperm are then released by contractions of the male abdomen on the spermatophore body (Gregory 1956). This could well serve to displace the contents of the sperma-theca volumetrically. After copulation, part of the spermatophore tube remains within the spermathecal duct. The distal end of the tube dissolves, but the proximal end appears to act as a plug by blocking the spermathecal duct. A second mating, before the proximal part of the tube is ejected by the female some 2-3 days after the first copulation, yields only about 40% (with a large variance) of the progeny at the next oviposition (Parker and Smith 1975).

The mechanics of sperm displacement are best established for the damselfly, *Calopteryx maculata* (Waage 1979, this volume). The male mechanically removes the sperm stored from a previous male with his penis before transferring his own sperm. The penis anatomy appears to be specially modified for this function.

It is clear that mechanisms of sperm displacement often evolved independently in different groups. A major question concerns the variance in the extent of displacement. Results from *Scatophaga* (Parker 1970b, Parker and Stuart 1976) suggest that the degree of sperm displacement effort can be optimized. But why should these optima differ so widely across species (see Fig. 2)? Some considerations of this question are given by Gwynne (this volume) and also page 26 of the present paper. Some species in which there is very low displacement have plugs (*e.g.,* mosquitoes; see Parker 1970a). It may be difficult and costly to transfer sperm against mechanical barriers such as plugs, though we should expect some degree of failure of plugs in order that they be maintained (see earlier). Some failure of plugs does appear to occur in nature (*e.g.,* Bryk 1930).

In many species, constraints on the female may cause termination of copulation earlier than may otherwise be optimal for the male (Thornhill 1976, Walker 1980). For instance, in the scorpionfly, *Bittacus apicalis,* the male donates a prey item to the female during courtship. If the prey is small the female terminates copulation and the male transfers fewer sperm (Thornhill 1976). With large prey, copulation is male terminated.

In insects, a few Hymenoptera show the unique feature of requiring several ejaculates to fill the sperm stores (see Page and Metcalf 1982; Starr, this volume). Demands on the female for a vast supply of sperm can be great; she may live many years after her nuptial flights and may produce vast numbers of fertilized eggs. Why doesn't the male produce enough sperm to fill the female's sperm stores? Because of the intense male-male competition in the nuptial chase—essentially a race between drones to capture the queen—it may be better for a male to sacrifice sperm production in the interests of flight speed and agility (other possibilities are discussed by Starr, this volume). The tradeoff may be between producing less progeny if successful, but increasing the chances of actually being successful.

B. Free Sperm and Other Competitive Strategies (The Vertebrate Model)

The mechanics of copulation clearly allow previous sperm to be displaced from the female's sperm stores in insects. In many vertebrates, it is perhaps less likely that direct displacement could occur because the ejaculate typically disperses within the female tract. Some mechanical displacement might occur, however. The penis itself might have evolved in response to sperm competition, males able to insert sperm further into the female tract and therefore closer to the ova being favored. Aspects of penis morphology and copulatory behavior may be related to sperm displacement and/or plug removal. For instance, it seems possible that repetitive copulatory thrusts coupled with raised annuli such as the foreskin and frenulum may sometimes serve to displace recent prior ejaculates back down the vagina. Increased penis length may have been favored so that sperm can be ejaculated into the far end of the vagina or even into the uterus (see also Smith, this volume). It has been suggested that penile spines (Milligan 1979) and intromissions (Hart and Odell 1981) function in the removal of plugs in rodents, and Dewsbury (1981) has made a very plausible general case for the significance of the rodent copulatory pattern in terms of adaptation to sperm competition.

If there is reduced possibility of direct sperm displacement in animals without special sperm storage organs, perhaps the most immediate solution is the re-emphasis of numerical productivity of sperm. It would be pointless for a male insect to produce a vast ejaculate containing, say, a billion sperm, when perhaps less than a thousand would fill the sperm stores. Indeed, they don't; female *Drosophila* store some 650 sperm (Kaplan *et al.* 1962), and the male typically transfers some 4,000 sperm. The difference may be attributable to achieving a high level of displacement. However, the sperm competition advantage of vast sperm numbers would be great in an animal lacking sperm storage organs: in a raffle, chances of success are directly proportional to the number of tickets bought. If tickets are cheap and the prize valuable, the solution is obvious. Paradoxically, selective pressures again resemble those prevalent in external fertilization. We shall return to the problem of "how many sperm" later (p. 44).

There are other important consequences of lacking sperm storage organs. Perhaps the most notable of these will be that sperm will live for a much shorter time in the female tract; they are not specially provisioned by the female, and often even appear to be phagocytosed within the female tract. It is unusual for sperm to survive much more than 5-6 days in mammals (usually much less; though in bats sperm survival is unusually prolonged, see Fenton, this volume), and 12-13 days in birds. In contrast, insect sperm can survive periods of several years. Short sperm life must tend to moderate the potential for sperm competition in vertebrates. However, vertebrates are generally highly mobile and have such startling visual and chemical cues of estrus that it is doubtful if the actual extent of sperm competition is drastically reduced. The fact that male vertebrates often show adaptations which appear to prevent second males from mating lends support to this view.

A third difference between insects and vertebrates may be evident. In insects, the timing of fertilization is sometimes highly predictable, and expected gains relatively deterministic through time. Fertilization is likely to begin fairly soon after mating **(relative to the life of the sperm)** and may continue gradually throughout the life of the female (*e.g.,* Parker 1970b). However, the male mammal meeting a female in estrus may not be able to predict the timing of ovulation and hence of fertilization. Although the timing of fertilization can often be referenced with respect to the onset of estrus (*e.g.,* Laing 1945), this is often imprecise and males may anyhow lack the appropriate information. Scaled with respect to the relative longevities of the sperm within the female tract, this may constitute an important difference between insect and mammal systems.

In summary, groups without sperm storage may have (1) less opportunity for mechanical replacement, (2) shorter sperm longevity, and (3) timing of fertilization often less predictable relative to sperm longevity.

C. Counter Selection to Prevent Sperm Competition

There appear to be many adaptations to prevent another male's sperm competing with self's sperm (Parker 1970a). These adaptations are both morphological and behavioral. As we shall see, some sperm competition prevention mechanisms may not be in female interests and may represent cases of sexual conflict (Trivers 1972, Parker 1979).

1. Plugs

In certain acanthocephalan worms (Abele and Gilchrist 1977), insects (see Parker 1970a), spiders (Austad, this volume), mammals (*e.g.,* see Martan and Shepherd 1976), and snakes (Devine 1975, 1977, this volume), the male secretes a plug after mating which serves to block the female tract for some time and act as a "chastity belt." These were originally interpreted as various aids to fertilization, but there is little evidence for this (Martan and Shepherd 1976). Plugs are generally not 100% effective in preventing further matings (see Fenton, this volume, for bats; Parker and Smith 1975, for the migratory locust; review of Parker 1970a, for Lepidoptera; Mosig and Dewsbury 1970, and Matthews and Adler 1977, for rats) as the model described earlier would predict. In insects, the female must generally dissolve or eject the plug before she can oviposit. However, in ditrysian lepidoptera, this is not necessary because there are two separate genital openings: one for copulation and one for oviposition (see Drummond, this volume). A rather similar arrangement occurs in spiders (Austad, this volume).

Although there is the possibility that the female might obtain certain nutrients from the dissolving plug, it seems likely that the plug may often be of direct disadvantage to the female for a variety of reasons. For instance, Dr. C. Wiklund, Zoology Department, Stockholm University (pers. comm.) has pointed out that the large sphragis of the ditrysian butterfly *Parnassius apollo* may render oviposition more difficult and less efficient for the female. Female *P. apollo* oviposit simply by dropping their eggs onto the food plant; it appears to be impossible for a plugged female to test the substrate carefully with her ovipositor and position and cement the eggs on the substrate as do many insects. The female migratory locust must dissolve and finally eject part of the male's spermatophore tube, which must at least exert minor energetic costs.

2. Prolonged Copulation

Sometimes, insect copulation can be very prolonged (*e.g.,* one week in Brimstone butterflies; Labitte 1919). It is a notoriously variable entity, ranging from a few seconds in some species up to a day or more in others. It is tempting to suggest that in some species the male may himself be acting as a mechanical plug (Parker 1970a). In many species, agents in the male seminal fluid act to suppress the female's receptivity to further mating attempts. It would certainly pay males to remain in copula with a female for sufficient time to allow these agents to act. Such mechanisms may not be restricted to insects. Lanier *et al.* (1979) point out that in rats prolonged copulatory stimulation reduces female receptivity and increases active resistance. Male rats typically show repeated copulation with the same female. The extended "copulatory locks" of canids could have a function related to sperm competition (Parker 1974). However, Voss (1979) suggests that locking in rodents may relate to situations of monogamy rather than sperm competition.

In mammals, prolonged copulation could allow "self's" sperm to move closer to the ova and hence have a better chance of fertilization than those sperm introduced later, though such effects do not appear to have been established (Lanier *et al.* 1979). In insects, eggs are usually fertilized as they move down the oviduct past the duct of the spermatheca. Hence prolonged copulation could here serve only to reduce the probability that the female remated before ovipositing. However, coupled with the reduction of unreceptivity, it could have a role in reducing the probability of sperm competition.

It seems unlikely that sexual conflict could be present in the induction of unreceptivity. A behavioral response to the mechanical act of copulation, or to chemical agents in the seminal fluid, could hardly evolve if it is directly disadvantageous to the female that performs it. Thus the fact that female insects often do become unreceptive after copulation (*N.B.,* many don't) must be taken as evidence that, for them, it pays not to mate with further males. The specialization of the stimulus necessary to induce unreceptivity, and the unreceptivity itself, would evolve mutualistically in the two sexes and without conflict.

3. Postcopulatory Guarding

Males of many species guard reproductive females for some time after copulation has occurred. This usually appears to function to reduce the probability of sperm competition and hence to assure paternity. In insects, postcopulatory guarding has been studied in the dung fly *Scatophaga* (Parker 1970c), in Odonata (*e.g.,* Jacobs 1955; Waage, this volume), and in the cactus fly *Odontoloxozus* (Mangan 1979). The male may remain attached to the female but without genital coupling ("contact guarding" or "passive phase"), or may actively guard the female but without attachment to her ("non-contact guarding") as in many Odonata. After copulation, the guarding stage usually continues until all the eggs are laid and a high degree of paternity is assured.

The evolution of postcopulatory guarding poses certain problems, depending on the ancestral behavior of the females (Parker 1970c). The same sort of problems apply in general to most paternity assurance adaptations. Suppose originally that females were unreceptive to further males after their first mating. Then a postcopulatory guarding trait in males cannot spread because paternity is already assured by the female's unreceptivity—a mutant "guarding" male would invest valuable time that could be devoted to mate-searching. So even if guarding is advantageous to the female in assisting her to lay more efficiently without harassment from courting males (as is the case in *Scatophaga*; Parker 1970d), guarding behavior cannot spread unless the benefits to the male via increased oviposition efficiently affecting his progeny) exceeds his losses in terms of competitive mate searching.

For *Scatophaga,* this problem has been analyzed as follows (Parker 1970c). Fig. 4a shows the fertilization rate (rate at which a male accrues reproductive success) that would be achieved by a mutant male that showed guarding behavior in an ancestral population of non-guarders, in relation to the proportion of females that can be mated again after an initial mating (*i.e.,* a measure of the effectiveness of female rejection abilities). Provided that more than about 10% of females can be remated, the guarding mutation will spread to fixation in the population, replacing non-guarders. Its advantage increases as it spreads, and if female unreceptivity declines. A decline in unreceptivity could occur if the time cost of a copulation to a female is less than the potential costs of harassment during oviposition, as appears to be the case for *Scatophaga* (see Parker 1970d). In the present population with full female receptivity, a rare individual lacking the guarding habit would experience a marked selective disadvantage at all normal conditions of density of competitors around the oviposition site (Fig. 4b). This arises because his sperm would be displaced by another male. It would pay to abandon the guarding phase only when there are no other competitors present. Males always show guarding after copulating with quiescent females in the presence of fresh dung, but not in the absence of fresh dung (Parker 1970d). Normally there are several male competitors present at the droppings. It would probably never pay to leave a female

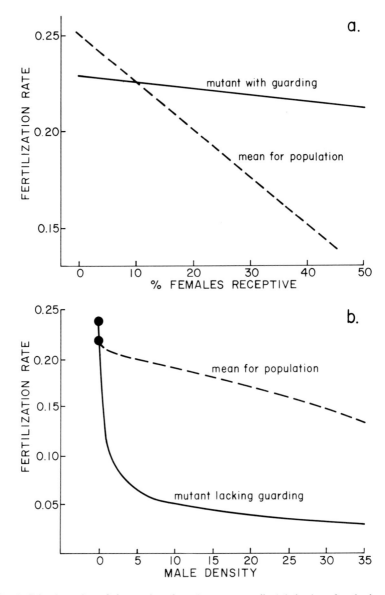

Fig. 4. Selective value of the passive phase (contact guarding) behavior of male dung flies. a. compares the fertilization rate of a mutant with the guarding pattern in an (ancestral) population of nonguarders with mean population fitness, in relation to varying proportions of females that can be remated. b. compares the fertilization rate of a mutant lacking the guarding pattern with mean population fitness in the present population of guarders, at ranging levels of competition density (density of searching males). After Parker (1970c).

immediately after copulation even when there are no observable competitors, because the odds are that a competitor would arrive some time during oviposition. For matings away from the droppings, this would not be so, and the absence of guarding in such cases appears to be adaptive.

This example serves to illustrate precisely how important can be the interaction between male behavior and female behavior in the evolution of paternity assurance mechanisms. A more detailed consideration of this sort of interaction is given in section VII.

There are probably more arthropod examples of precopulatory guarding than postcopulatory guarding, but rather few show both (see Parker 1970c). The selective advantage of precopulatory guarding appears to be in monopolizing a female until she is ready to mate, in contrast to the paternity assurance function of postcopulatory guarding. In contrast, vertebrates that show guarding typically guard the female both before and after copulation. In many mammals, it becomes rather difficult to separate the two guarding components because the male performs multiple ejaculation during estrus, and the distinction between pre- and postcopulatory guarding is probably much less important than for arthropods.

Will there be sexual conflict over postcopulatory guarding? As argued above for *Scatophaga*, it seems likely that guarding may often be beneficial to the female in terms of avoidance of repeated sexual harassment. Hence provided the guarding trait can begin to spread, it is likely to accelerate toward fixation as female receptivity begins to increase. Disadvantages to the female are sometimes conceivable. These may include energetic and "inconvenience" costs of pulling males around during oviposition. For postcopulatory guarding these would not seem to be extensive compared with the benefits. In contrast, costs to the female of precopulatory guarding could be significant. In the amphipod *Gammarus* (*e.g.,* Birkhead and Clarkson 1980) and the isopod *Asellus* (*e.g.,* Ridley and Thompson 1979), females may be attached to males for several days on end, which may have considerable energetic consequences. In a few instances precopulatory guarding may become permanent (or nearly so); it is then often associated with male dwarfing (*e.g.,* the echiuroid worms of the genus *Bonellia;* Baltzer 1931), a phenomenon that has oddly attracted little attention (but see the excellent review by Ghiselin 1974).

4. Take-over Avoidance

It seems certain that selection would act to favor increased fighting, grasping and guarding abilities so as to reduce the probability of take-over during copulation or postcopulatory guarding. Take-overs usually result in copulation by the new male and consequent sperm displacement. The benefits of increased efficiency in avoiding take-overs may be temporary, because of counter selection to increase abilities to achieve take-over. Certain forms of grasping apparatus appear to be related to enforcing copulation rather than avoidance of take-over (Thornhill, this volume).

There appear to be various simple behavioral adaptations for avoidance of take-over. One of the commonest concerns removal during copulation to areas of low risk; *i.e.,* areas where competitor density is low. For example, copulating pairs typically fall out of swarms of Nematocera to copulate hidden in the grass beneath the swarm (*e.g.,* Thornhill 1980b). In dung flies, males that meet females on the dropping (where the density of competitor is highest) typically fly their females to the surrounding grass downwind of the dropping to copulate. This is where the competitor density is lowest. However, copulating in the grass takes longer than it does on the dropping because the dropping surface is hotter. Thus at low competitor densities it pays males to copulate on the dropping to take advantage of the shorter copula duration; at higher male densities it pays them to emigrate to the downwind surrounding grass because of the lower risk of take-over. They appear to obey these predictions (Parker 1971).

The habit of removal to an area of low competitor density to avoid take-over can be stable only so long as once fixed, it will not pay unpaired males to search in the removal site. In dung flies, it would not pay unpaired males to search for females mating in the downwind surrounding grass because: (1) take-overs of copulating pairs are relatively difficult to achieve, and (2) few, if any, newly-arriving females can be captured in the downwind surrounding grass. Searching in the downwind surrounding grass is therefore relatively unprofitable despite the occurrence there of several copulating pairs. A similar argument may explain the stability of the habit of dropping out of swarms in Nematocera.

Usually, there will be no sexual conflict involved with mechanisms to avoid take-over. Although the costs of take-over to the female will generally be small compared with those of the male, they will be present. At best, a female that suffers take-over incurs the time waste involved in an extra copulation, provided that the benefits to the female of multiple paternity and/or male-male competition are insignificant (see section VI). Often the act of take-over will itself be physically harmful to the female (*e.g.,* dung flies; Hammer 1941). Hence it will generally be in female interests to avoid take-over whenever possible; male and female interests may therefore coincide. A possible exception could be the case of a female that is being courted by a male of very low guarding ability. If such a male is likely to be ousted by a bigger male during copulation (when damage to the female could be extensive), then it may pay her to precipitate a take-over at an early and less costly stage. This seems a plausible explanation for the behavior of female elephant seals, where the female is much more likely to emit a cry in response to copulatory attempts of subordinates than those of dominant bulls (Cox and LeBoeuf 1976). The cry attracts the dominant bull and the subordinate is dispelled. Cox and LeBoeuf (following Hingston 1933) interpret this in terms of female incitation of male-male competition, the idea being that the cry ensures that the female gains "good genes" for her sons. The problem with this argument is that of maintaining genetic variance in males (see *e.g.,* Williams 1975; Maynard Smith 1978a; Thornhill 1980a; Parker 1982a); as selection proceeds and genetic variance in male guarding and fighting ability diminishes toward zero, females would be precipitating competition without receiving any benefit (but see Lande 1980, 1981; Hamilton and Zuk 1983).

In summary, it seems unlikely that there will generally be sexual conflict in the evolution of male measures to ensure privacy and to avoid take-over, although in certain cases (*e.g.,* elephant seal) it may pay females to precipitate an early take-over if a later one is likely, and which may be more costly to her.

VI. FEMALE INTERESTS
(MATE CHOICE, DIVERSITY OF PROGENY, SPERM DEPLETION, PATERNAL INVESTMENT)

Because it is the subject of a later chapter (Knowlton and Greenwell, this volume), only a brief discussion is given here of what may be termed "the female perspective" in sperm competition. Throughout the preceding sections I have already outlined the extent to which paternity assurance mechanisms might be beneficial or costly to females. If beneficial, the main line of adaptation would proceed simply through selection on males, though as we have seen, the historic constraints of female behavior may still be vitally important. If male adaptations are costly to females, we must consider ESS solutions that include the female as a "player."

Walker (1980) has argued that the typical sperm displacement patterns observed in insects result mainly from selection on females, and that selection on male adaptation is constrained mainly by female interests. My own view (see next section) is that the eventual outcome of selection will often be a Nash equilibrium compromise between the interests of the two sexes, when there is sexual conflict. Conflicts over whether or not to mate perhaps resolve most commonly toward female interests, whereas perhaps the opposite is true of conflicts about the degree of sperm displacement or plugging after mating has begun. Contrary cases abound, however; *e.g.,* in Mecoptera enforced copulation is rife, but the female often determines the degree of sperm displacement (Thornhill 1976, this volume). Sexual conflict in sperm competition systems is discussed more fully in the next section, and by Knowlton and Greenwell (this volume) who present some alternative views derived from a discrete (2 versus 2) strategy analysis, rather than allowing each "player" to choose from a continuous strategy set (Parker 1983).

Walker (1980) suggests that monogamy (prolonged female unreceptivity after a first mating) may result from an advantage to the female in using sperm from the first male encountered. Reasons why the first male encountered may be best were envisaged to be that they may carry good genes (*e.g.*, genes for rapid development). This poses the problem that the good genes are likely to fixate, at which point the female strategy (being unreceptive after the first mating) is likely to crash if the alternative strategy (receptivity) is less costly (Parker 1982a).

Although the issue of female choice of males with good genes is still highly controversial, there is no doubt that the female mating pattern can be modified in relation to nongenetic benefits from males. For instance, unreceptivity may be favorable when time and risk involved in additional copulations exceeds that in rejecting males, and receptivity may be favorable when the reverse applies (*e.g.*, see Parker 1970a). Polyandry might be beneficial to females for reasons of indirect paternal investment that might be quite important in some species (Walker 1980) and remating will always be favored when the female's sperm supply is beginning to deplete so that fertility begins to drop (Parker 1970a) or when the first mating has been inadequate (see Walker 1980).

One possible genetic benefit for females arising through polyandry has been claimed to relate to increased diversity of progeny (Murray 1964, Pease 1968, Richmond and Ehrman 1974, Caldwell and Rankin 1974, Walker 1980). There is little doubt that the "sib-competition plus unpredictable environments" model for the evolution of sexual reproduction versus sexuality (Williams 1975, Maynard Smith 1978a) should generate an advantage to females in multiple mating over monogamy (Barton and Post, 1983), provided there is not complete sperm displacement. Sib-competition in unpredictable environments could be important in many insects (especially colonizers; see Caldwell and Rankin 1974), but I cannot envisage that even this selective force will be as powerful as that acting on males to assure paternity.

A second reason why multiple mating might have an advantage over single mating to females relates to the fact that if two strategies have the same mean payoff, but one (multiple mating) a lower variance, the strategy with the lower variance can sometimes be favored (Gillespie 1977; Rubenstein 1982; Knowlton and Greenwell, this volume). For the multiple mating case, this argument can work for small populations or founder effects; however, benefits of lower variance to females are likely to be small compared with the benefits of paternity assurance to males. Unless some special argument (such as sib-competition or reduced variance) is made, it will not pay females to produce varied offspring so as to "hedge their bets" over male genotypes. Suppose there exist genotypes A, B, C, etc. in males with frequencies p_A, p_B, p_C, etc. If the male genotype favored in the next generation is not predictable, a strategy for female monogamy will obtain exactly the same fitness as one for multiple mating. The strategy for monogamy, in a large population, will ensure that each female carrying it mates once with males in their population frequencies, so that the fitness of the strategy will (very approximately) be:

$$p_A(1-a) + p_B(1-b) + p_C(1-c) \ldots \ldots \text{etc.}$$

in which $(1-a)$, $(1-b)$ etc. are selection coefficients relating to the progeny of type A, B, C males in the next generation. Exactly the same expected fitness accrues to the strategy for multiple mating. The proportions of each type of sperm stored in multiple-mating females will be the same as for single-mating females; the mean fitness of each strategy will therefore be identical.

A further problem could seem to be that if recurrent mutants occur rarely through time, an unpredictable environment should usually result (in the absence of frequency-dependent selection) in the fixation of alleles and a loss of heterozygosity. The opportunity for action of both this "genetic diversity" effect and the "sib-competition" effect will therefore be greatly restricted.

In summary, it is argued that the female's **mating pattern** is much less likely to arise for reasons of "good genes" or "genetic diversity" than from the everyday enivornmental pressures on females due to copulation time waste on the one hand, and energy or time waste in harassment by courting males on the the other. In contrast, unless it is the female that terminates copulation (*e.g.,* Thornhill 1976), the **sperm displacement pattern** is perhaps adjusted more by the relatively intense intrasexual selection on the male (see section VIII, and Gwynne, this volume) than by selection on the female. Both patterns may ultimately be compromises between male and female interests (next section).

VII. RESOLUTION OF SEXUAL CONFLICT: EVOLUTIONARY DYNAMICS OF SPERM COMPETITION SYSTEMS

Cases where male and female evolutionary interests differ (sexual conflict, see Trivers 1972; Parker 1979) pose interesting evolutionary problems. As we have seen, certain lines of male adaptation to sperm competition may be of disadvantage to females. Imagine a paternity assurance mechanism such as a plug, that is costly to the female. Any cost felt by the female is also felt at least in part by the male if he will be the father of some of the progeny. But since he may not be the father of all her progeny, and may mate with other females, costs need not be symmetric for male and female. Benefits will also be asymmetric. The male may achieve many more total progeny by having the plug, even though it may reduce the expected offspring number produced by each female he mates with. The only way that the female can experience a benefit is via her sons, who experience higher reproductive success by forming plugs (the "sons effect" of Fisher 1930). This benefit via sons is always much diluted compared with the direct benefit of the characteristic to the male (Parker 1979). A character that confers a mating advantage to males can therefore often be of direct disadvantage to females (Parker 1979).

What would we expect the outcome of this sort of conflict to be? Once again the game may resemble a true arms race, in which males invest in armament to achieve matings and females invest in means of preventing extra matings. Alternatively, armament may relate to the nature and extent of plug production by males, and to the expenditure by females in dissolving or removing the plug. If strategies of investment are determined strictly by genes, the game becomes the "opponent-independent costs" model to which there is no ESS (see section V). However, if there is environmental variation in the armament strategies (*i.e.,* the genetic strategy prescribes a mean or expected armament level, about which there is random variation due to environmental factors), the ESS is likely to be a Nash equilibrium for the expected armament levels in the two sexes. The female will expend a cost, say $x*$, on armament, and the male will spend $y*$, but because of environmental effects there will be a distribution of arms levels in each sex, and the outcome of contests will vary (Parker 1983). The equilibrium is likely to follow biological intuition. For example, if it is relatively very costly for males to invest, say, in plugs, compared to female investment in plug removal, females will usually be seen to "win" by removing plugs against male interests. Equally, if the value of winning to males is very high compared to females, then males will tend to win against female interests.

However, there are at least three separate "games" of this sort present in sperm competition systems (see Fig. 5). Firstly, there is the fundamental inter-male conflict favoring paternity assurance adaptations on the one hand, and "anti-paternity assurance" on the other (1, Fig. 5.). Then there is sexual conflict if male paternity assurance mechanisms are costly to the female (2, Fig. 5). Thirdly, sexual conflict will also operate if it would pay female to avoid costs of male persistence and other "anti-paternity assurance" adaptations (3, Fig. 5). If it pays females to perform multiple mating, then game 3 in Fig. 5 becomes not a conflict, but a mutual benefit, and this will have the effect of pushing the "global solution" to the 3 games over toward "anti-paternity assurance." Similarly, if it pays females to have a guarding male or a plug (*e.g.,* to reduce harassment from searching males), then game 2 can become mutualistic rather than conflicting. This will tend to swing the global solution toward paternity assurance. This solution is likely to be associated with poor abilities to reject males, and with high costs of being unreceptive.

Clearly all three aspects of the game are interdependent. For instance, if females can be highly effective at being unreceptive at little cost, this will reduce the degree to which it pays males to attempt to mate with mated (unreceptive) females. In turn this will reduce the benefit of expenditure on paternity assurance, and the degree to which females invest in reducing the costs of paternity assurance. Thus (as discussed in section V, C, 3), high levels of female unreceptivity should be

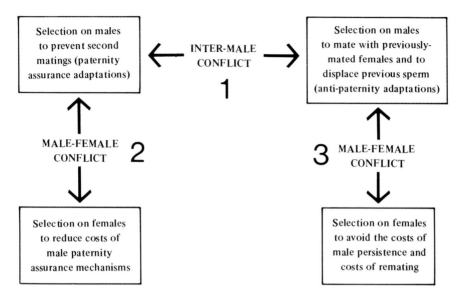

Fig. 5. Three subgames involved in the evolutionary dynamics of sperm competition systems. See text for further details.

associated with a reduction of the entire adaptive complex normally associated with sperm competition.

The sperm competition system will hence have complex dynamics involving direct (within-game) and indirect (from outside a game) evolutionary feedbacks. Since adaptation in all aspects will have costs, the global ESS is unlikely to be the one in which certain aspects of the game are **never** played (*e.g.,* perfect un-receptivity, with no male tendency to persist could not be stable since it would pay females to economize expenditure on unreceptivity to the level where it will pay males to show some persistence). The ESS is more likely to consist of a set of "homeostatic" Nash equilibria in which all components of investment are represented to some degree, depending on the cost constraints of each armament and the benefits of winning. As stressed earlier, general predictions about the outcomes of sexual conflicts are difficult to make. Although the advantage to males of obtaining a solution closer to their interests is likely to be higher (asymmetric benefits), it may be relatively easier for females, morphologically and behaviorally, to prevent males from achieving their objective (asymmetric costs). These asymmetries act in a contradictory fashion in the sense of Parker and Rubenstein (1981).

VIII. EXPENDITURE ON PATERNITY ASSURANCE

Recently, several authors (Maynard Smith 1978a, Grafen 1980, Werren *et al.*
1980) have pointed out that a high certainly of paternity does not obviously favor
the evolution of male parental care, or vice versa. For example, Grafen (1980)
claims that "if a male cannot mate with other females while caring for the young
of his present mate but can desert his mate and her offspring at any time to search
for another mate, then certainty of paternity will not affect the male's optimal
degree of parental care." To support his assertion, he shows that if the benefits
of male parental investment are devalued by a constant factor r (relating to the
probability of paternity of the offsping being cared for), then the male's optimal
expenditure on parental investment is entirely independent of r (see Fig. 6). The
argument assumes continuous breeding, and that males act to maximize their rate
of benefit via progeny. This argument is correct mathematically (see Parker and
Stuart 1976, for proof). Grafen points out that for decreased paternity to generate
decreased male parental care, the time between broods must also decrease, but
felt that "there seems to be no general reason to expect this." At first sight this
statement seems odd, since males cannot achieve increased paternity without some
cost in terms of time, energy, or sperm. Thus decreasing paternity assurance might
intuitively be expected to decrease time between broods for males. However, at

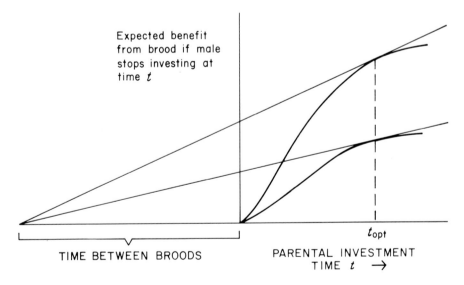

Expected benefit
from brood if male
stops investing at
time t

t_{opt}

TIME BETWEEN BROODS

PARENTAL INVESTMENT
TIME t $\rightarrow$

Fig. 6. Optimal time investment (t_{opt}) in male parental investments. Modified from Grafen
(1980).

the population level, the time between broods depends mainly on female physiology, not on how males allocate their reproductive efforts.

In the present section it is concluded that the optimal investment in paternity assurance can be increased somewhat by high paternal investment, because this can increase the inter-brood interval, especially in species where it is rather difficult for females to find males. When females can be mated as soon as they become receptive (and hence without any time delay), then increased paternal investment should not affect the optimal expenditure on paternity assurance. Other authors (Werren *et al.* 1980; Knowlton 1982; Knowlton and Greenwell, this volume) have proposed alternative suggestions and models to explain the observed association between high certainty of paternity and high male parental care.

What will determine the male's expenditure on paternity assurance? Let us assume for simplicity that adaptation will proceed to an optimum for the male (*e.g.*, Parker 1970b), *i.e.*, female interests are independent of degree of sperm displacement. Imagine a species that lives in high population density so that receptive females can be found instantaneously by males. It is clear that if the adult sex ratio is unity, the interval between matings for a male must equal the interval between mating for the female. Thus suppose that one female per adult male become available every T time units. Then if a male spends investment time I ($I < T$) with each female, the mean search time per male (= time taken for a male to find a female) must be $T - I$. At the population level there will be a perfect "compensation" for the effect of any increase in investment time by a decrease in male search time.

This "compensation effect" will, however, not be felt by a mutant individual that has, say, a slightly higher parental investment (I_m) than the rest of the population (I_p); he will have a cycle time of $[T - I_p + I_m]$ compared to the rest of the male population, which will have cycle time T. Thus extra male parental investment will not spread provided that:

$$\frac{\text{benefits of investing } I_m}{T - I_p + I_m} < \frac{\text{benefits of investing } I_p}{T} \tag{1}$$

Exactly the same condition applies if we substitute investment in paternity assurance (PA) for male parental investment (PI).

With perfect compensation between male investment and search time so that cycle duration T remains constant, it is clear that PA and PI will have independent stable optima as summarized in Fig. 7. Proof of the graphical representation is given in Appendix A.

For many species, *e.g.*, those in which unpaired males wait for receptive females in leks or other encounter sites, or where receptive females can quickly attract males by visual or chemical cues, it seems likely that females will easily find males. Thus for a wide range of operational sex ratio, the "perfect compensation" rule will be a reasonable approximation.

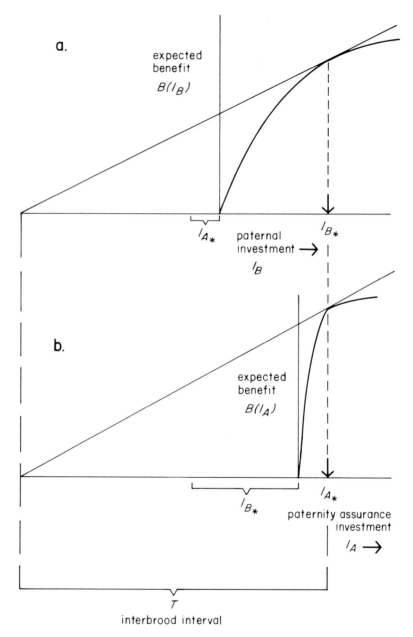

Fig. 7. Optimal time investments in paternal care and paternity assurance for cases where the interbrood interval stays constant (independent of *PA* or *PI*). a. shows optimal male parental investment, I_{B*}, when paternity assurance is at its optimal level. b. shows optimal investment I_{A*} in paternity assurance when male parental investment is at its optimum.

Now consider a slightly different model (Appendix B), but which uses the same variables. Consider a species where encounter of the sexes is more difficult. The female now sustains a significant "search delay" while looking for a male. This delay will be inversely proportional to the density of searching males, which will in turn be an inverse function of the investment time of males with females. The point is that the cycle time will not here be constant, it will increase as the density of searching males decreases. The density of searching males will be decreased if males increase their time investment with each female. Thus the "compensation" rule will not apply, and if it pays to increase male *PI* it will also pay to increase *PA* to some extent. (From Fig. 7 it can be seen that if *T* increases, the optimal *PA* will also increase since the tangent will touch *B[I]* at a higher value for *I*.) Proof that the optimal *PA* is dependent on the optimal *PI* when there is no compensation is given in Appendix B.

In fact, it turns out (Appendix B) that even if it becomes advantageous to increase *PI* considerably, this can exert only a relatively small effect on the optimal *PA* expenditure unless the ability of the sexes to encounter each other is low.

However, where male *PI* is so extensive that females may suffer considerable delay in their rate of production of broods because of the difficulty of finding an "unencumbered" male, the optimal expenditure on paternity assurance may well be high. An obvious candidate for such an effect would seem to be the giant water bug, *Abedus herberti* (Smith 1979a). Females of this species lay their eggs on the backs of males, which show extensive brooding behavior (Smith 1976a, 1976b, 1979a). It seems highly likely that the rate at which females can produce broods is limited by their capacity to find nonbrooding males. Thus the interval between successive broods (*T*) may be much longer than that of an ancestral form lacking any parental care of the fertilized eggs. In consequence, we would expect that the male's expenditure on *PA* would be increased over the ancestral form. Indeed, it turns out that *Abedus* males manage to achieve the highest degree of paternity assurance yet recorded for any insect (Smith 1979b).

In conclusion:

1. When the female can find a mate easily, the degree male parental investment will not exert any significant effect on the male's expenditure on paternity assurance.

2. When it is difficult for females to find males, male expenditure on paternity assurance should increase with expenditure on paternal care. However, unless the female's capacity to find males is very limited this effect may not be great.

3. Ultimately, the optimal expenditure on paternity assurance is likely to be much more sensitive to chances in the relationship between benefit and increasing investment (function *B[I_A]*, see Appendix B) than on the degree of other male investments such as *PI*.

It is stressed that these conclusions rely on the assumption that as male *PI* increases, the time between broods increases because it takes a female longer to find a male when she needs one. Suppose, however, that an increase in male *PI* brings about a compensatory decrease in female *PI*, because (say) the offspring need a fixed total provisioning before they become independent. The increase in time it takes a female to find a male will then be partly offset by the reduction in female *PI*. Thus the time between broods will not be increased as much as in the above model, and the dependency between *PA* and male *PI* will be even weaker than suggested by the analysis in Appendix B).

An extensive review of male expenditure on paternity assurance and other mating investments with the same female is given by Gwynne (this volume).

IX. MULTIPLE VERSUS SINGLE EJACULATION

In several species, the male copulates (and ejaculates) several times with the same female. Though less common in invertebrates than vertebrates, several insects show multiple genital contacts during the time the male is paired to the female. In the fly *Sepsis cynipsea* (Parker 1972) and the desert locust *Locusta migratoria* (Parker and Smith 1975) it seems unlikely that sperm is transferred during the repeated copulatory attempts since these usually culminate in a much longer copulation. In both these insects, guarding males typically mate **after** oviposition, and fertilize eggs at the next oviposition. The copulatory attempts have been interpreted as a means by which the male tests continually for female receptivity. The pattern is quite different in the water bug, *Abedus*. Here the male mates repeatedly both **before** and **during** oviposition (Smith 1979a). The male broods the eggs on his back and copulates on average once for every 1.4 eggs oviposited. It seems highly likely that sperm are transferred at each copulation and that this pattern might have arisen as a result of sperm competition (Smith 1979b).

In birds and mammals, multiple mating and multiple ejaculation are commonplace. The extensive work of D. A. Dewsbury and his co-workers has illuminated a rich diversity in mating patterns in rodents, most of which show multiple ejaculation (*e.g.,* Dewsbury 1975, this volume). Multiple ejaculation is also especially common in felids, though again the pattern is very variable (see Eaton 1978). For instance, adult male lions may copulate some 100 times per day throughout an estrous period of 6-7 days; whereas bobcats copulate some five times per day throughout 5 days, and Geoffroy's cat typically mates once during the one day of estrus (see references cited in Eaton 1978).

There have been rather few adaptive interpretations of multiple ejaculation patterns. Bertram (1975) has argued that if the female solicits many matings during estrus she may ultimately increase the survivorship of her offspring. If there are many copulations, the value of each mating to a male becomes very small. Hence the value of fighting another male within the pride for a given copulation becomes negligible. This reduces intermale damage, which reduces the chances of take-over by outsiders, which in turn reduces the degree of infanticide. This ingenious explanation would be plausible only if (1) the cost of damage in several, low value fights is less than that for a single, high value fight; and (2) males initially possessed some capacity for reducing their fighting in accordance with the known number of copulations of a given female. Otherwise, a mutant female with enhanced solicitation would sustain the costs (time and energy) of the extra matings without any benefit. Indeed, they may cause their males to fight more often and to sustain more damage than normal females.

Eaton (1978) suggests that high copulation frequencies in felids may have two advantages to the female. It might increase the chance that she is mated most by a "vigorous" male; low sexual vigor may correlate with a high probability of take-over and infanticide. Alternatively, it might allow the female to achieve a higher reproductive success than her competitors; by sexually exhausting a local male this could prevent the male from fertilizing other females, hence delaying conception. This second suggestion seems perhaps unlikely in view of counter selection on males to mate with other females.

A direct, intermale advantage for multiple ejaculation has been proposed by Lanier et al. (1979) following their experiments on the laboratory rat. The rat shows several copulatory series; each consisting of several mounts, with and without vaginal penetration, and eventually culminating in a mount with ejaculation. About seven such series will be completed (Beach and Jordan 1956). This number vastly exceeds that necessary to initiate female physiological responses leading to preparation of the uterus for implantation (see Lanier et al. 1979). Lanier et al. conducted experiments in which female rats were mated to two males in succession. When each male completed just one ejaculatory series, there was little effect of mating order on the number of progeny sired by each male. But when the first male completed five series, he achieved a significantly higher proportion of offspring than if he completed only one series against a second male allowed to copulate for five series or to satiety (whichever occurred first). Lanier et al. propose that the multiple ejaculation pattern has evolved in response to sperm competition; a male gains a higher probability of fertilization by having a greater number of sperm in the female tract. Similar effects were obtained by Dewsbury and Hartung (1980) in experiments in which two males of different genotypes were permitted access to an estrous female.

The idea that multiple ejaculation has evolved for this function works best if ejaculate volume is constrained at a fixed quantity, so that the only way to increase sperm contribution to the female tract is by repeated copulation. It is a less plausible solution if there is little cost of varying the amount of sperm in each ejaculation, since we must then ask why males do not simply increase ejaculate volume and copulate only once. That ejaculates have a cost to males has recently been stressed by Dewsbury (1982; see also Nakatsuru and Kramer 1982).

Should a male discharge a large amount of sperm as a single dose into the female tract, or should he discharge into the female many much smaller ejaculates, through time?

One feature of multiple ejaculation patterns is that, with the exception of rodents (*e.g.*, hamster, see below), they tend to be spread out temporally; often they spread through the whole period during which estrous cues are emitted by the female. Another feature of many mammalian reproductive cycles is that the exact timing of ovulation during estrus is to some extent unpredictable, hence it may be difficult for a male to "predict" a best time to mate during the period of the estrous cue. A third feature of mammalian reproduction is that (aside from certain specialized groups such as bats) sperm tend to have a very high death rate within the female tract (*e.g.*, Salisbury and vanDenmark 1961) compared, say, to insects and other groups with specialized sperm storage organs in the female. The following model is a simplistic attempt to incorporate these features and to investigate conditions under which a multiple ejaculation strategy would be evolutionarily stable against a single ejaculation strategy.

A. The Model

Assume that during the estrous period *t,* the timing of conception is unpredictable. In other words there is no "best time" to mate within *t* since the probability of ova becoming vailable is constant. Imagine two possible strategies: *S* (single ejaculation) is for ejaculating once at the start of *t; M* (multiple ejaculation) is for ejaculating *n* times at equal intervals during *t,* and only $1/n$ sperm are inseminated on each occasion. Multiple ejaculation is assumed for present purposes to have no greater cost than single ejaculation. Assume also that proportion *d* of all (*M* and *S*) sperm die in the female tract in each interval between successive ejaculations made by an *M* strategist; *d* may well be under female control.

So far the model has the properties of an unpredictable estrus with sperm death. Now consider the following sperm competition game. Two males are available throughout time *t* and both have access to the female for matings. Let us assume that the chances of fertilization for a given male at a given step during estrus are proportional to:

$$\frac{\text{number of his surviving sperm}}{\text{total surviving sperm}}$$

Support for this model of sperm competition comes from the work of Martin *et al.* (1974). However, we are interested only in the effects of sperm competition on mating strategy. We therefore assume that whenever mating takes place, there will be sufficient sperm to ensure fertilization. It is obvious that an *S* strategist should always mate at the start of time t if the probability that ova become available remains constant through t. While he delays mating, he has zero probability of a fertilization. This is why the analysis is restricted to the case where S mates at the start of t, and M mates at the start and at equal intervals throughout t. We ask the question: "Which strategy, S or M, will be favored by selection?"

Mathematical details of the model are outlined in Appendix C. The conclusion is that the multiple ejaculation strategy M is evolutionarily stable against S only when there is a rather high rate of sperm death during estrus. The actual rate of sperm death required to favor M depends on the number of ejaculations delivered by M during estrus, but only weakly so (see Fig. 8). Roughly speaking, multiple ejaculation will be the ESS if over 90-95% of sperm delivered at the start of estrus (or at the start of the competition) will be dead by the end of estrus (the end of the competition). When conditions are reversed, S is stable. In fact, if only two ejaculations are possible for the M strategy, the probability of death of a sperm during the entire contest must be greater than 89% for M to be stable; if M delivers 20 ejaculations, the probability of death must exceed 95%. It seems likely that within the range that M is stable, the number of ejaculations would be set by some competitive equilibrium; such a model has not been investigated.

With a constant death rate, a given ejaculate shows a negative exponential decline in numbers during estrus. An alternative (possibly more plausible) model of sperm death would assert an increasing rate of sperm death through time. For the same proportion of mortality of sperm during estrus, this second model would yield more pessimistic conditions for M to be an ESS than generated by the "constant death rate" model.

The hypothesis we obtain is that when two males have full access to the same female throughout estrus, multiple ejaculation will be an ESS if almost all the sperm inseminated at the start of estrus will be dead by the end of estrus. It is easy to extend the model to the case where one dominant male guards the female during estrus, but a second male may be able to achieve a "sneak" mating (as can occur in many species with defensive guarding, see review of Dunbar 1981; Rubenstein 1980). In this case we would expect a "sneak" to obtain only one opportunity to mate; he should therefore put in as large an ejaculate as possible (S strategy). The dominant male can, in contrast adopt the M strategy, and should do so if the sperm death rate is high.

B. Data

From the above analysis, we assume that the mating pattern will be a reflection of how long sperm survive in the female tract. Does mammalian sperm show the sort of death rate that could favor multiple rather than single mating strategies? We can deduce from the model that, roughly, if almost all sperm introduced at the

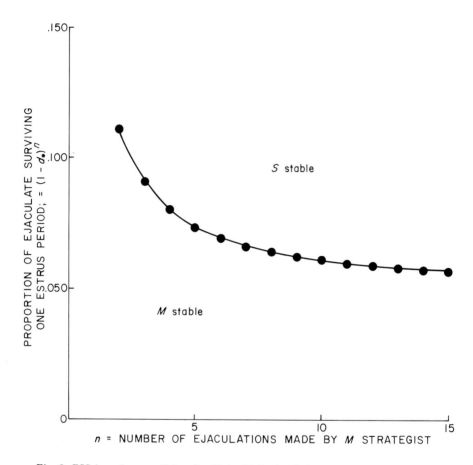

Fig. 8. ESS boundary condition for M (multiple ejaculation) versus S (single ejaculation) strategies. d is the probability of mortality of sperm between two successive ejaculations. With n ejaculations by an M male, during estrus, $(1 - d)^n$ sperm would survive one full estrus of those inseminated at the start of estrus. $(1 - d*)^n$ is the ESS boundary condition between S and M strategies. The risk of death during estrus must be very high if M is to be stable.

start of estrus would be dead by the end of estrus, M would be advantageous through sperm competition. Alternatively, if an ejaculate can survive much longer than the duration of estrus, S would be the better strategy.

Table I shows the results of a survey for 10 mammals for which details of sperm longevity and ejaculatory pattern could be obtained. Cases of true copulation-induced ovulation (*e.g.,* cat, mustelids, rabbit) were omitted since they require a rather different model from the one just investigated (the timing of ovulation can be influenced by copulation in other species, *e.g.,* pigs; see Pitkjanen 1959).

Unfortunately, because of the wide ranges of estrus and estimates of sperm life, many cases turn out to be borderline candidates for M or S. A clear fit between predicted and observed patterns appears to occur with pig, guinea pig, hamster (but see below), and dog; and a clear discrepancy is evident with mouse and cow. "Optimistic" evaluations of inconclusive cases can lead to better support for the theory (see Table I), but until further and more accurate data are available I cannot claim there there is general support (or lack of it) for the model. More extensive data on natural mating patterns would also be desirable, since the species listed in Table I are domestic and patterns may be somewhat aberrant. Eaton (1978) found that those wild felids with polygamy and a tendency for intense internal competition were those with the most notable capacity for multiple ejaculation. Unfortunately, sperm longevity is not known for these species.

Perhaps the most important criticism of the above way of testing the model is as follows. The model assumes that there will **always** be enough surviving sperm to achieve fertilization, because a male should always remate with the female when this is not so. Hence we are really defining multiple mating as remating by the same male within the fertilizing life of the previous ejaculate. The data on fertilizing life of the sperm in Table I give averages or ranges for ejaculate lives. We assume that it will pay to be M if the duration of estrus is slightly shorter than this (a guess for 95% sperm mortality). But there is real danger that before this stage it would pay a male to remate anyhow because of reducing prospects of fertilization. This effect makes it very difficult to interpret cases such as the mouse (Table I); if sperm last only 6-12 h, it is remarkable that males do not mate at least twice during the 12-24 h estrus.

A further difficulty (D. Dewsbury, pers. comm.) is that in rodents, M strategists often do not spread their ejaculations equally throughout estrus. For instance, male hamsters deliver some 10 ejaculations over some 30-40 min in all. The estrus period spans some 27 h; the hamster is therefore closer to S than M in the sense of the model (that it delivers all its sperm during a relatively short period within estrus). Carnivores, primates, and ungulates spread their ejaculations much more widely throughout estrus.

A fascinating byproduct of this analysis is that it reveals a highly significant correlation between the duration of estrus and the fertilizing life of the sperm (Fig. 9). The slope of the regression approximates to 1, so that the mean survival time of the sperm is approximately equal to the estrus duration.

Why should this be so? Presumably, there has been selection for appropriate (optimal) longevity. The extent of provisioning should be fined-tuned in relation to an expected period of potential fertilization. There seem to be two possibilities:

1. The male provisions the sperm (*e.g.*, by energy within each sperm or by agents in the seminal fluid) so that they can survive to the end of estrus. For many species, fertilization occurs within estrus, so that sperm deposited at the start of estrus would be capable of fertilization at any time during estrus (*e.g.*, sheep). Extra provisioning

TABLE I

Data on Estrus Duration, Fertilizing Capacity, and Mating Pattern in Some Animals

Species	Duration of Estrus (h)	Authority	Fertilizing Life of Sperm (h)	Authority
Horse	Mean 120 Mean usually 96-144 depending on breed Range 1-37 days	Cupps *et al.* 1969 Fraser 1980 Hafez 1969	144 144	Day 1942 Burkhardt 1949
Cow	6-30 Mean 17 9-28 12-24 13-27 Mean 19	Hammond 1927 Hafez 1969 Fraser 1980 Cupps *et al.* 1969	$> 50(?)$ 26 26.5-5.32	Vandeplassche and Paredis 1948 Jarosz 1962 Laing 1945
Sheep	Usually 24, up to 72 Mean 36 Mean 38 (Merino) Range 36-48	Fraser 1980 Hafez 1969 Cupps *et al.* 1969	+48 (Shropshire) 30-48	Green 1947 Bishop 1961
Pig	24-48(?) 40-80 Mean 59 48-72	Fraser 1980 Hafez 1969 Cupps *et al.* 1969 Ensminger 1970	24-48	Ensminger 1970
Guinea Pig	Mean 3-5 Mean 8-9 Range 1-42	Boling *et al.* 1939 Young *et al.* 1937	22	Soderwall and Young 1940
Rat	Mean 13.7 Range 1-28	Young 1941	14	Soderwall and Blandau 1941
Mouse	12-24 (usually toward 12)	Snell *et al.* 1940	6 12	Merton 1939 McGaughey *et al.* 1968 (critical of Merton
Hamster	Mean 27.4	Asdell 1964	13	Miyamoto and Chang 1972
Dog	Mean 216 Range 168-312	Cupps *et al.* 1969	Mean 82(?) Up to 264 264	Griffiths and Amoroso 1939 Doak *et al.* 1967 Thibault 1973
Man	No estrus, mating throughout cycle	–	28-48 Max. 120	Bishop 1961 Thibault 1973

TABLE I

**Data on Estrus Duration, Fertilizing Capacity, and Mating Pattern
in Some Animals (*Continued*)**

Mating Pattern	Authority	Predicted Pattern	Fit
Probably usually 5-10 matings by stallion per estrous female (*M*)	Fraser 1980	Borderline *S*	−?
3-10 matings by bull per estrous female (*M*)	Fraser 1980	*S*	−
Sometimes ram mates several times. Some mature rams will mate each ewe only once (*M/S*)	Fraser 1980	Borderline	+?
3-7 matings by boar per estrous female 7-11 matings 4-8 matings, mean 6.6 (*M*)	Fraser 1980 Burger 1952 Hafez 1969	*M*	+
Apart from Kunkel and Kunkel (1964) most data suggests typically only one ejaculation per estrus (*S*)	Young and Grunt 1951 Grunt and Young 1952 Rood 1972	*S*	+
Usually about 7 ejaculatory scores per estrus (*M*)	Beach and Jordan 1956	Borderline *M*	+?
Not usually repeated mating (*S*)	McGill 1962	*M*	−
Usually about 10 ejaculations per estrus (*M*)	Beach and Rabedeau 1959 Bunnell *et al.* 1977	*M*	+
Single ejaculation (*S*)	Eaton 1978	*S* following (Doak *et al.* 1967)	+?
Mating throughout cycle, sometimes within life of previous ejaculate (*M?*)	—	*M*	+?

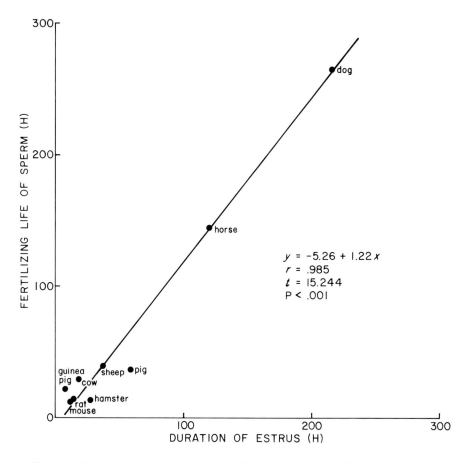

Fig. 9. Relationship between the mean fertilizing life of sperm and the mean duration of estrus, based on nine mammal species (those in Table I, excluding humans). The regression is still very significant if $\log_{10}$ of data (x, y values) is taken to reduce the effects of horse and dog, but not if horse and dog are excluded.

would represent wasted expenditure. However, the hypothesis works less well for cattle; ovulation occurs several hours **after** the end of estrus and inseminations earlier than about 16 h before the end of heat are not fertile (Laing 1945).

2. The female provisions the sperm so that they "cover" the normal period of ovulation. Even in mammals, sperm tend to become stored in various parts of the female tract (see Thibault 1973). Though perhaps less plausible than (1), it could pay the female to provision in this way if for some reason there is a chance that she will receive no further matings after the start of estrus. For most species examined, a single mating at any time during estrus would usually be adequate to ensure full fertility.

If an ejaculate is provisioned to last for roughly the duration of estrus without any sperm death, then it would always pay to be S rather than M. If, however, full fertility can be achieved by only a tiny fraction of an ejaculate, then pro-visioning could be such as to ensure that mortality does not reduce the ejaculate below this fraction before the end of estrus. This could theoretically permit an M strategy. For artificial insemination, although bull semen can be diluted 50-200 times, most other species can be diluted only by around 2-5 times to ensure full fertility (Foote 1969), which would not permit an M strategy.

Multiple ejaculation is clearly a common mating strategy in mammals. If the simplest way to increase a given male's sperm load within the female tract is by repeated ejaculation, then the strategy is perhaps most readily explained by the theory of Lanier et al. (1979); i.e., in a competitive mating situation it pays to increase sperm contribution by multiple ejaculation. When the male has the "evolutionary option" to increase ejaculate volume, it may, however, often pay to introduce a single large ejaculate rather than many small ones, unless sperm death rates are very high. The theory of Lanier et al. (1979) seems a better ex-planation of multiple ejaculation if males facultatively alter their mating pattern in accordance with the risk of sperm competition. Thus a good strategy may be to have a fixed morphological and physiological regime for ejaculate production (rather than a capacity for two volume levels), and to use the behavioral mechanism of multiple mating to increase sperm load when another male is in competition for the same female. It is probably easier and less costly to vary copulation pattern than to vary ejaculate volume.

The evolutionary analysis of strategies of mating and sperm production in male mammals in relation to sperm competition and multiple mating is in its infancy. The present discussion must be regarded as a prospective survey; the adaptive dynamics of the system are complex and would repay further study.

X. SPERM COMPETITION AND MALE-MALE DISPUTES

Can knowledge about prior mating be used to settle a dispute between two males over a female, assuming that one male (the "resident") has already mated and the other (the "interloper") hasn't? Ethologists have known for decades that asym-metries (such as size difference) can be very important in determining the outcome of a contest. Recently, the role of such asymmetries in the evolution of animal contests has been investigated theoretically (Parker 1974, Maynard Smith and Parker 1976, Hammerstein 1981, Parker and Rubenstein 1981, Hammerstein and Parker 1982). Where animals can regulate costs continuously in a dispute, it appears that solutions to contests should be related in a "commonsense" fashion to asymmetries in payoff. That is, the winner should generally have higher fighting ability, or should have more to gain from winning. Battles should be settled by the asymmetric cue without appreciable escalation.

Consider a dispute between two males over a receptive female. The resident has already mated with the female; the interloper has just arrived. Can there be an asymmetry in the value of the female to the two males so that we can predict the convention for settlement of the dispute? If the value of winning is greater for the resident, he should occupy the conventional "winning role." and vice versa, assuming that the resident and interloper have similar fighting abilities.

Packer (1979) claims that the female may be worth more to the resident. He argues as follows. Consider a point in estrus by which time n males have copulated with the female. The nth male has chance $1/n$ of siring the offspring, assuming that the chances are equal for all ejaculates. the $(n + 1)$th male would have chance $1/(n + 1)$, and since $1/n > 1/(n + 1)$, the resident would have more to gain from winning than the interloper.

The logic of this argument is seductive; I earlier (1975) proposed the same effect in correspondence with J. Maynard Smith, but later retracted it, for the following reason. The value of a resource to a given opponent is not its absolute value, but the difference in expected payoff between winning and losing, excluding any contest costs. Viewed in this way, we can get a quite different asymmetry.

First, imagine the simple case where there is likely to be just one dispute during guarding of the female (see Table II). If the resident wins, he obtains all the offspring. Call this value G fitness units. If he loses, he gains $G(1 - p)$ where $p =$ the probability that the last male's sperm will be used for fertilization. Thus the value of winning for the resident is $pG = G - [G(1 - p)]$ (Table II). Note that we can discount the cost of the owner's copulation in our calculation, since prior investment is irrelevant (Dawkins and Carlisle 1976, Parker 1974). However, at the time of the dispute, the interloper has not yet copulated, and so we must count the cost $(- c$ fitness units) of copulating in estimating the resource value for the interloper. Thus if he wins, he gains $Gp - c;$ if he loses he gets 0. The value of

TABLE II

Resource Value Asymmetry With Just One Dispute During Guarding

	(1) Gain If Wins	(2) Gain If Loses	Resource Value (1) − (2)
Resident's values	G	$G(1 - p)$	Gp
Interloper's values	$Gp - c$	0	$Gp - c$

p = probability that last male's sperm will be used at fertilization
$- c$ = cost of copulation to interloper
G = number of offspring to be produced

winning is therefore $Gp - c$ (Table II). Both males would have to guard the female (equally) until the end of estrus, so guarding costs are asymmetric. In other words, there is no asymmetry arising from the effect of sperm competition; the only asymmetry concerns the cost $(- c)$ of copulation. Thus the magnitude of the copulation cost should set the asymmetry in resource value in favor of the resident, assuming there is usually only one dispute during guarding.

Now consider the case where typically two disputes would occur during guarding. First. assume that all ejaculates fare equally in competition for ova, irrespective of their order, as may be an approximation for many vertebrates. We can calculate very simply the value of winning for resident and interloper, following the same logic as in Table II. This time, however, there are two disputes to consider instead of one (see Table III). Consider the second dispute. Let us assume that copulation costs nothing, since we wish to find asymmetries due to sperm competition. When the present resident won the first dispute, there is no asymmetry in the value of winning between resident and interloper. Similarly, if we consider the first dispute, no asymmetry will occur if there is no change in ownership at the second. For these two cases, it is obvious that gains will be ordered as if there was only one dispute,

TABLE III

Asymmetries in Resource Value Between Resident and Interloper
When There are Typically Two Disputes During the Time a Female is Guarded,
And Where All Ejaculates Have Equal Chances at Fertilization
("Vertebrate" Model)

	(1) Gain If Wins	(2) Gain If Loses	Resource Value (1) − (2)
SECOND DISPUTE			
A. Resident won first dispute			
Resident's values	G	$G/2$	$G/2$
Interloper's values	$G/2 - c$	0	$G/2 - c$
B. Original resident lost first dispute			
Resident's values	$G/2$	$G/3$	$G/6$
Interloper's values	$G/3 - c$	0	$G/3 - c$
FIRST DISPUTE			
A. No change in male at second dispute			
Resident's values	G	$G/2$	$G/2$
Interloper's values	$G/2 - c$	0	$G/2 - c$
B. Change in male at second dispute			
Resident's values	$G/2$	$G/3$	$G/6$
Interloper's values	$G/3 - c$	0	$G/3 - c$

p, c, and G as in Table II

which we already know to be symmetric apart from copulation costs. But if there is a change of resident (take-over) at one or other dispute, the situation changes so as to be asymmetric in favor of the interloper. The possibility of having three ejaculates in competition causes the interloper's gain from winning to exceed that of the resident.

It is easy to see why the asymmetry favors the interloper. Consider the nth resident. If he wins the last dispute before the end of estrus, he gains G/n; if he loses he gains $G/(n + 1)$. The interloper would gain $G/(n + 1)$ by winning, and 0 by losing. Thus the value of winning is greater for the interloper if:

$$\frac{1}{n + 1} > \frac{1}{n} - \frac{1}{n + 1}$$

which is always true if $n > 1$ (i.e., if there are two or more disputes per estrus guarding). It is easy to show from the above equation that at the last dispute:

$$(\text{interloper's resource value}) - (\text{resident's resource value}) = \frac{G(n - 1)}{n(n + 1)}$$

i.e., for $n \geqslant 2$, the asymmetry will decrease sharply with n, the number of changes in residency.

Table IV shows the parallel case where there is sperm displacement following the *Scatophaga* pattern (the "insect model"). Cases where there is no change of resident at the dispute not currently under consideration are omitted because they are covered in Table II, and are symmetric. Remember that in *Scatophaga* the last male to mate gains p = 80% of the eggs; other males gain from the remaining 20% in the proportions they would have attained previous to the last mating (see Fig. 3). It turns out that there will be no asymmetry in the first dispute when there is a take-over at the second (assuming copulation costs are insignificant so that $c = 0$). The only case that generates an asymmetry in the insect model is a second dispute after the previous resident lost the first dispute. The value of winning is Gp^2 for the resident and Gp for the interloper. If there is virtually complete displacement ($p \to 1$), or lack of it ($p \to 0$), there will be no asymmetry. The difference:

$$(\text{resource value for interloper}) - (\text{resource value for resident}) = Gp(1 - p)$$

i.e., the asymmetry is maximized at intermediate degrees of sperm displacement ($p = 0.5$).

We can make the following conclusions, on the assumption that copulation costs are insignificant. In both the "vertebrate" model (all ejaculates have equal chances) and the "insect" model (sperm is displaced), sperm competition can create an imbalance in resource value in favor of the interloper when there is a possibility

TABLE IV

Asymmetries in Resource Value Between Resident and Interloper When There is Sperm Displacement of the *Scatophaga* Pattern, and Where Typically Two Disputes Occur During Guarding[a]

	(1) Gain If Wins	(2) Gain If Loses	Resource Value (1) − (2)
A. Second dispute, original resident lost first dispute			
Resident's values	Gp	$Gp(1-p)$	Gp^2
Interloper's values	$Gp - c$	0	$Gp - c$
B. First dispute, male changes at second dispute			
Resident's values	$G(1-p)$	$G(1-p)^2$	$Gp(1-p)$
Interloper's values	$Gp(1-p) - c$	0	$Gp(1-p) - c$

[a]Only cases where there is a change of male at the other dispute are considered, since other cases follow Table I.
$p, c,$ and G as in Table II.

that more than two ejaculates can compete for fertilizations. The asymmetry between resident and interloper decreases sharply as the number of different ejaculates in competition increases above two. In the "insect" model, the asymmetry is increased by tendencies for intemediate levels of displacement; complete displacement (or lack of it) will make the dispute symmetric. (The case of a complete lack of displacement can, in fact, be discounted from the analysis, since postcopulatory female guarding and battles for ownership would here constitute a maladaptive alternative to searching for new females.)

However, there can of course be a resource value asymmetry in favor of the resident if the costs of copulation are significant. This is certainly the case in *Scatophaga* (Parker and Thompson 1980), where the cost of a given copulation is high, and can be estimated in terms of missed opportunities to mate with new females. There is a high level of sperm displacement, which tends to reduce asymmetry in resource value due to sperm competition effects. Hence overall, *Scatophaga* shows a significant resource value asymmetry in favor of the resident male. This may also apply in other insect species. Residents are often conventional winners of disputes, possibly indicating that copulation costs may be more significant than sperm competition effects. Perhaps the most general reason for the observation that residents are commonly seen to win disputes concerns fighting ability. Asymmetries in fighting ability are likely to be highly important in determining the outcomes of contests; resources will therefore tend to be held by individuals of high fighting ability.

If having mated does correlate with a higher resource value because of copulation costs, then it is tempting to speculate over which visual cues might be specialized as a result. A bizarre possibility is that the conspicuous genital displays and colorations used in threat by many male primates (see Eibl-Eibesfeldt 1970) originated as an "I have just copulated" cue. Perhaps Wickler's (1967) suggestion that genital displays originated from urinary marking is a more plasible solution!

Finally, consider the fact that the asymmetry in resource value due to sperm competition will decrease with the number of competing ejaculates. If copulation costs maintain a constant asymmetric component of resource value in favor of the resident, while sperm competition generates an asymmetric component in favor of the interloper that reduces as the number of ejaculates increases, it seems likely that a net asymmetry in favor of the interloper would be found only at low numbers (*e.g.,* 2 or 3) or competing ejaculates (Fig. 10). Clearly, if interlopers were to have a resource value asymmetry in their favor that subsequently become used to settle contests, owners would leave their females immediately after copulating rather than wasting time guarding. Guarding must be associated with a tendency to win if it is to be maintained.

This outlines a problem that would arise in a more rigorous treatment of the problem. We are attempting to ascertain what convention settlement will arise as a result of increasing ejaculate competition; unfortunately the degree of ejaculate competition is dependent on the form of conventional settlement. There is no problem if the asymmetry in resource value always favors the resident (or interloper) whatever the number of matings. But if it favors interlopers at relatively low numbers of matings, and residents at high numbers of matings, there may well be no ESS to the problem, for the following reason. If high numbers of matings occur, a "resident wins" convention will appear. This will reduce the number of matings, favoring an "interloper wins" convention, which in turn will increase the number of matings. This must remain a subject for more rigorous future contemplation, but perhaps the best hope for an ESS occurs when take-overs are relatively infrequent so that there is usually likely to be only one ejaculate present at the time of the dispute. Here (Fig. 10) the net asymmetry will favor a "resident wins" solution, which should then be stable.

XI. SPERM SIZE AND NUMBER WITH INTERNAL FERTILIZATION

Cohen (1973) has suggested that males produce so many sperm because most of them are defective. He argues that errors in chiasmata formation during meiosis could be so prevalent that only a tiny proportion of sperm in each ejaculate are suitable for fertilization. As evidence, Cohen found a highly significant correlation between mean chiasmata number and what is termed "sperm redundancy" (= number of sperm ejaculated/number of sperm used in fertilization).

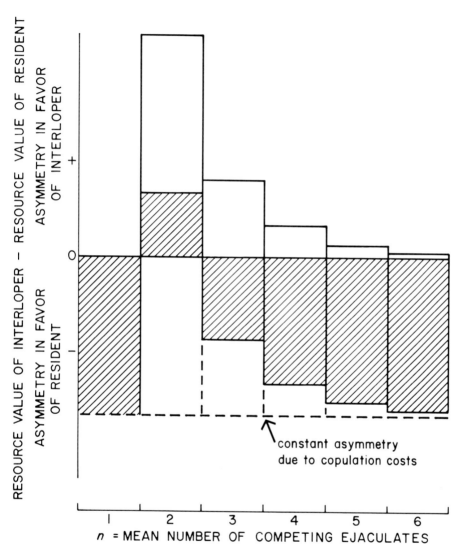

Fig. 10. Possible pattern of changes in resource value asymmetry between residents and interlopers when sperm competition follows the "raffle principle," in relation to n, the mean number of competing ejaculates that will typically be present within the female when disputes occur. The asymmetry due to copulation costs is constant in favor of the resident male, but the asymmetry due to sperm competition is zero for $n = 1$, and in favor of the interloper, but decreasing, for $n \geqslant 2$. The net asymmetry (shaded) may vary from favor of the resident, to the interloper, and back to resident, as n increases from 1. See text for further explanation.

This theory, though ingenious, fails to answer the question "why so many small sperm?" Female gametes are, of course, also produced by meiosis, and should therefore suffer similarly. Cohen argues that they indeed might, and suggests as evidence the very high level of degeneration found in populations of mammalian oocytes, only a small number of which are eventually left for ovulation (but see Bernstein *et al*. 1981). The difference in gamete production is attributed to the fact that it would be costly for the female to provision defective gametes. Thus whether or not Cohen is correct about the high frequency of genetic defects arising from meiosis, the question of why sperm are small, unprovisioned and numerous compared with ova is not actually covered by Cohen's theory. Rather it is an assumption that **because** sperm are small and uncostly, it is left to the female to sort out the suitable ones.

I began this paper by outlining the disruptive selection theory for the evolution of gamete dimorphism, stressing that it was essentially sperm competition that keeps sperm small and allows the differentiation of the two sexes. With ancestral sessile forms with external fertilization in sea water, sperm from many males must compete simultaneously for fusions with ova, and there is no difficulty in seeing why males would produce so many small gametes.

We have also seen that sperm competition is not extinguished by internal fertilization. Is it in fact sperm competition that maintains the two sexes, and keeps sperm size small in species with internal fertilization? The following is a summary of Parker (1982b).

The problem resolves into two questions. Firstly, why do sperm stay small with internal fertilization, and secondly, why does the male produce so many of them?

The first question is hardest to answer. Sperm competition can indeed be a powerful force in maintaining the sperm small. As the number of competing ejaculates reduces, the effectiveness of sperm competition in keeping sperm provisioning minimal also reduces. Thus if usually there is only one ejaculate within the female, but on rather rare occasions there are just two competing ejaculates, the effects of sperm competition will be lowest. However, such low levels of sperm competition still may be adequate to keep sperm small. Consider the benefits that might accrue to a male by putting more cytoplasmic reserve into sperm as an aid to the viability of the zygote after fusion. A unit of extra provisioning is diluted n-fold when allocated among n sperm; hence the benefits of the extra cytoplasmic provisioning are similarly reduced if only one sperm is used for fertilization. In contrast, when two ejaculates are in competition, sperm **numbers** may be the major feature that affects the probability of winning the fertilization. Thus it pays to have more sperm, rather than the same number of bigger ones. Roughly speaking, sperm will stay small if the probability of having two ejaculates in competition within the female tract is greater than four times the ratio of sperm size/ovum size. This ratio is clearly minute in most species, so the probability of sperm competition need be only tiny to maintain the strategy of producing small sperm.

It is easier to answer the question, why are there so many sperm with internal fertilization? Although internally-fertilizing species typically show less sperm investment than their sessile, externally fertilizing counterparts, they still produce vast numbers of sperm, and more than can reasonably be accounted for in terms of difficulties in getting to the ova after ejaculation (Cohen 1973). Imagine a mammal in which double mating (competition between ejaculates from two males) occurs only rarely, with frequency p. When ejaculates compete, let us assume that male A's sperm have a chance of fertilizing = (total of A's sperm)/(total A + total B sperm), following Martin *et al.* (1974). We now ask: What fraction k of this total mating effort should a male spend on the ejaculate? Assume simplistically that if he expends proportion k of his total mating expenditure on sperm, he can spend only $(1 - k)$ on mate-searching, etc. Thus sperm have energetic costs that affect alternative reproductive efforts (see Dewsbury 1982, Nakatsuru and Kramer 1982). The number of new mates he encounters will be directly proportional to his mate-searching effort (total mobility). Thus if other males spend k on sperm, a mutant expending k_m gets (relatively $(1 - k_m)/(1 - k)$ matings. The ESS amount of expenditure, $k*$, turns out to be: $k* = p/(4 - p)$, *i.e.*, approximately $p/4$ if $p \to 0$ (Parker 1982b). Thus a male should spend on an ejaculate a proportion of his total mating effort that is roughly equivalent to one quarter of the probability of double mating. Of course, this figure is intended only to indicate the order of magnitude of expenditure on sperm, since forms of inter-male competition other than active searching may yield different relative numbers of matings; *i.e.*, deviations from $(1 - k_m)/(1 - k)$. Suppose the chances of double mating are as low as 4%. Bearing in mind the immense expenditure of most male mammals on mating effort (*i.e.*, to achieve a mating with a female), we would expect at 1% expenditure on the ejaculate to constitute a significant volume. Thus given that each sperm will be minute, sperm competition can readily explain the numbers of sperm found in species with internal fertilization.

XII. SUMMARY

The evolution of anisogamy (two sexes) is driven by competition between the gametes of different parents, a primitive form of ejaculate competition. Initially, sperm competition may have acted to maintain sperm size small in order to maximize sperm productivity; it would also have favored increased competitiveness (against rival sperm from different ejaculates) in terms of morphological features (Sivinski 1979). Internal fertilization may have evolved in response to sperm competition: males able to release sperm closer to females or actually within the female reproductive tract gaining numerical predominance in paternity of offspring. Increased male mobility (mate searching effort) and internal fertilization may both act to reduce the extent of sperm competition, and may allow enhanced expenditure on alternative male reproductive efforts (mate searching, guarding, parental investment) at the expense of sperm production. However, sperm competition appears to remain an important selective pressure in most animal groups.

There are two major selective pressures related to paternity assurance: selection favoring mechanisms for preemption of stored sperm, and counter selection to prevent preemption (anti-preemption). This conflict may generate an arms race which may result in a solution in which in a species, a second male sometimes overcomes the first male's paternity assurance adaptation, and sometimes not. Sperm displacement is more likely to occur in groups with (1) sperm storage organs in the female, and (2) male genitalia that are capable of acting mechanically against the stored sperm. Many male adaptations, both morphological and behavioral, can be interpreted in terms of preemption and anti-preemption of stored sperm.

Certain lines of male adaptation to sperm competition may conflict with female interests. These can be classified into two groups: (1) cases where the female benefits from several matings—conflict is between female interests and male adaptation to prevent remating, and (2) cases where the female benefits from mating once only—conflict is between female interests and male courtship persistence and the degree of sperm displacement. Male-female conflict, and the conflict between male adaptations of preemption and anti-preemption, act as dynamically interrelated arms races; the global ESS (evolutionarily stable strategy) may be a set of homeostatic Nash equilibria.

The link between paternity assurance and male parental investment is examined in a model that assumes continuous breeding and in which females sustain a time delay between broods due to the difficulties of finding a male. It is concluded that when a female can find a mate easily, the degree of male parental investment will not be correlated with expenditure on paternity assurance (following Grafen 1980, and others). However, if it is difficult for females to find males, paternity assurance effort and male parental investment should be positively correlated, though the correlation will be weak unless it is extremely difficult for the sexes to meet.

Males of many species show a pattern of multiple insemination of a given female, rather than transferring sperm in a single ejaculation. Lanier *et al.* (1979) interpret this is as a means by which a male increases his total sperm load within a female, and as an adaptation to sperm competition pressures. It is probably more flexible (*i.e.,* allows facultative switches) and also less costly anatomically to change the sperm load in this way than by adopting the alternative strategy of simply increasing ejaculate volume and retaining single ejaculation.

A model of such alternative options was examined which included the two strategies: S (single, large ejaculation at the start of estrus) versus M (several small ejaculations spread at equal intervals throughout estrus). S always wins unless sperm mortality is very high indeed (say around 90-95% or more mortality of sperm during estrus). In mammals, the sperm life appears to track quite closely the duration of estrus, so that many species lie around the borderline condition between M and S. This effect may itself be due to optimal provisioning of sperm for a period of longevity that matches the period of potential fertilization.

Can sperm competition generate an asymmetry in the value of a guarded female between the guarding male (who has already mated with the female) and an interloper male (who has not mated with the female)? Such asymmetrics in resource value between opponents are likely to be important in determining the outcome of contests. Contrary to Packer (1979), it turns out that in consideration of **sperm competition**, the female can either be more valuable to the interloper, or equally valuable to each opponent, depending on how many take-overs occur typically during the guarding of a given female. However, in consideration of the **cost of copulation and ejaculation**, there can be a resource value asymmetry in favor of the guarder.

Finally, sperm competition may be a major selective force for keeping sperm small and numerous. It helps to explain why anisogamy is maintained with internal fertilization, as well as how it originated under external fertilization.

ACKNOWLEDGMENTS

I am most grateful to all those who have offered many helpful suggestions for improving the first draft of this paper, and especially to Bob Smith and Don Dewsbury. It is a very great pleasure to thank Bob and Jill Smith for making my visit to Tucson in 1980 so extremely pleasant and memorable; I am extremely indebted to them both for their hospitality. I also thank Miss J. Farrell for typing.

APPENDIX A

Proof That the Tangent Method Gives the ESS Investment
for PA or PI When There is Perfect "Compensation" in Search Time

Assume continuous breeding. Let $I*$ be the ESS investment time for PA or PI. Let I_m be a mutant individual that has a different investment time. Suppose benefits of investing I are summarized by the benefit function $B(I)$. From equation (1), page 27, it is necessary that the rate of progeny production by a mutant is less than that of the ESS strategies, $i.e.$:

$$\frac{B(I_m)}{T - I* + I_m} < \frac{B(I*)}{T}$$

for $I*$ to be an ESS. If we subtract mean population fitness rate from the mutant's fitness rate, the result must always be negative if $I*$ is an ESS. The gradient of this difference with respect to I_m will be positive for $I_m < I*$, and negative for $I_m > I*$. At $I_m = I*$, the gradient will be zero. So we can write:

$$\frac{d}{dI_m} \left[\frac{B(I_m)}{T - I* + I_m} - \frac{B(I*)}{T} \right]_{I*} = 0, \text{ at } I_m = I*$$

$$\therefore \frac{B(I_m) - B'(I_m)(T - I* + I_m)}{(T - I* + I_m)^2} = 0, \text{ at } I_m = I*$$

and setting $I_m = I*$ gives the familiar marginal value theorem result (Charnov 1976, Parker and Stuart 1976) that

$$\frac{B(I*)}{T} = B'(I*)$$

but in which T includes investment time $I*$. This equation shows that at the ESS, the gradient of $B(I)$ must equal the gradient of the tangent drawn from a point $T - I*$ to the left of the origin of $B(I)$.

APPENDIX B

Optimal *PI* and *PA* When Females Experience a "Search Delay" Inversely Proportional to the Density of Searching Males

Assume continuous breeding and a 1:1 adult sex ratio. Suppose it takes time K for a female to be found after becoming receptive when there is one searching male to each searching female. K is a constant that summarizes the "aptitude for encounter." Then if there are n males per female, the "search delay" will be reduced from K to K/n.

Let time t = the time taken for a female to mature eggs, etc. and to become receptive. Cycle time T now becomes:

$$T = t + K/n$$

Suppose each male invests a total of time I on *PI* plus *PA*. We first ask how this will affect cycle time T.

The number of males that will become available for pairing during one cycle time T must be $T/I = n$. Hence the female's "search delay" time S_f is

$$S_f = K/n = \frac{KI}{t + S_f}$$

and so $S_f^2 + tS_f - KI = 0$

The solution of this quadratic that we require is:

$$S_f = \frac{-t + \sqrt{t^2 + 4KI}}{2}$$

Standardizing time $t = 1$, so that I is a measure of male investment time relative to female investment time gives

$$S_f = \frac{\sqrt{4KI + 1}}{2} - \frac{1}{2}$$

Clearly, if $K = 0$ (females never need to wait long for a mating) or $I = 0$ (males invest nothing so that there is a vast number of them available for pairing as each female becomes receptive) then $S_f = 0$ and the cycle time will be constant at T (following the "compensation" rule as in Appendix A.

If KI (the product of time taken for one male to find one female and the ratio of male investment to female investment) is high, then S_f can be significant and can therefore add to the cycle duration T. For instance if $KI = 1$, then $S_f = 0.6$ and $T = 1.6$; if $KI = 2$, then $S_f = 0.8$ and $T = 1.8$; if $KI = 4$, $S_f = 1.5$ and $T = 2.5$. Thus as male investment increases, so does T, but not markedly so since a unit increase in KI promotes a change in T related to $\sqrt{KI}$.

This means that if for some reason it became favorable to increase (say) *PI* because of some change in the *PI* benefit function, it would also be favorable to increase *PA* investment. However, because of the relationship between S_f and I, the extent of the adjustment may not be very great unless K is high (low "aptitude for encounter").

To establish the optima for male *PA* and *PI*, call investment in $PA = I_A$, and investment in $PI = I_B$. Since total benefit is degree of paternity assurance multipled by the number of surviving offspring, mean population fitness is measured by the reproductive rate:

$$\frac{B(I_A) \cdot B(I_B)}{T}$$

where T = the cycle duration. For a male we can call the cycle time:

$$T = S_m + I_A + I_B$$

in which S_m is the time taken for a male to find a female. By the procedure adopted in Appendix A, it is easy to show that the ESS investment in PA (= I_{A*}) again obeys marginal value theorem and has, holding I_B constant,

$$B'(I_{A*}) = \frac{B(I_{A*})}{S_m + I_{A*} + I_B} \tag{1A}$$

Since with a 1:1 adult sex ratio the male cycle duration must equal that of the female, then

$$S_m + I_{A*} + I_B = S_f + t$$

$$\therefore \text{If } t = 1, S_m = \frac{\sqrt{4K(I_{A*} + I_B)}}{2} + \frac{1}{2} - (I_{A*} + I_B)$$

An exactly parallel set of equations can be established for the optimal investment I_{B*} when I_A is held constant.

It is obvious that I_{A*} and I_{B*} are now interdependent. The male should expend on each in accordance with the modified version of marginal value theorem given in equation (1A).

It is obvious that if male expenditure on PA is already high so that the degree of PA is also high, then even large increases in PI will not much affect the optimal expenditure on PA. The optimal PA expenditure is therefore likely to be much more sensitive to changes in the function $B(I_A)$ than to changes in PI.

APPENDIX C

Sperm Competition and Repeated Matings

Two males compete during an estrous period t. Both can copulate and stay with the female throughout time t. Assume:

1. The timing of conception is unpredictable over time t, so that sperm present at any given time have equal chances of fertilization as those present at any other time.

2. Males can "play" either strategy:

   S = single ejaculation at **start** of time t

   M = multiple ejaculation; n times at equal intervals throughout t and $1/n$ amount of sperm in each ejaculate

3. Proportion d of sperm die in each interval (each "step") between successive ejaculations of M.

4. Chances of fertilization for a given male at a given step proportional to:

$$\frac{\text{number of his surviving sperm}}{\text{total surviving sperm}}$$

5. Multiple ejaculation has no greater cost than single ejaculation.

We seek the evolutionarily stable strategy ESS at a given value d (sperm death rate) and n (number of ejaculations that would be performed during estrus by an M male). For M to be an ESS against S, the payoff of M against M must be greater than the payoff of S against M, i.e.,

$$E(M,M) > E(S,M)$$

Now $E(M,M) = 1/2$. Both males will have equal chances if they play the same strategy. What is $E(S, M)$? This =

$$\frac{1}{n} \times \frac{1}{1 + \frac{1}{n}}$$

conceptions at the first step before any sperm have died.

$$+ \frac{1}{n} \times \frac{(1-d)}{(1-d) + \frac{1}{n} + (\frac{1-d}{n})}$$

at the second step when proportion d of sperm from first step have died.

$$+ \frac{1}{n} \times \frac{(1-d)^2}{(1-d)^2 + \frac{1}{n} + \frac{(1-d)}{n} + \frac{(1-d)^2}{n}}$$

at the third step.

$$\frac{1}{n} \times \frac{(1-d)^{n-1}}{(1-d)^{(n-1)} + \frac{1}{n} + \frac{(1-d)}{n} + \frac{(1-d)^2}{n} \cdots \frac{(1-d)^2}{n}}$$

at the nth (last) step. At the ith step, it is clear that the gain

$$= \frac{(1-d)^i - 1}{n(1-d)^i - 1 + 1 + (1-d) + (1-d)^2 \ldots + (1-d)^i - 1}$$

and since the progression

$$1 + (1-d) + (1-d)^2 \ldots (1-d)^i - 1 = \frac{1 - (1-d)^i}{1 - (1-d)}$$

$$= \frac{1 - (1-d)^i}{d}$$

We can see that M will be an ESS against S if

$$\frac{1}{2} > \sum_{i=1}^{n} \frac{d(1-d)^i - 1}{nd(1-d)^i - 1 + 1 - (1-d)^i}$$

It is easy to iterate by computer a threshold value for d that exactly balances the above equation at any given n; i.e., at this value a mutant S would be selectively neutral.

For S to be stable against M we require that

$$E(S,S) > E(M,S)$$

Again

$$E(S,S) = 1/2$$

$$E(M,S) = \frac{1}{n} \cdot \frac{\frac{1}{n}}{1 + \frac{1}{n}}$$

at the first step,

$$+ \frac{1}{n} \cdot \frac{\frac{1 + (1-d)}{n}}{(1-d) + \frac{1}{n} + \frac{1-d}{n}}$$

at the second step and so on. By similar reasoning, we get the condition for S to be an ESS is

$$\frac{1}{2} > \sum_{i=1}^{n} \frac{\frac{1}{n}[1 - (1-d)^i]}{nd(1-d)^i - 1 + 1 - (1-d)^i}$$

This equation generates exactly the same threshold value for d as the equation above for M to be an ESS against S. Thus it turns out that if $d >$ the threshold value, M will be an ESS; if $d <$ the threshold value, S will be an ESS. High sperm mortality favors M.

For Fig. 8 we plot threshold value for mortality in terms of the proportion of sperm that would be surviving by the end of estrus out of an ejaculate transferred at the start of estrus, i.e., $(1-d)^n$. Sperm mortality must be very extensive during estrus to favor M.

REFERENCES

Abele, L., and S. Gilchrist. 1977. Homosexual rape and sexual selection in acanthocephalan worms. *Science* 197:81-83.

Alexander, R. D. 1964. The evolution of mating behavior in arthropods. In *Insect Reproduction*, K. C. Highnam (ed.), pp. 78-94. Symp. R. Entomol. Soc. Lond. No. 2.

Alexander, R. D., and G. Borgia. 1979. On the origin and basis of the male-female phenomenon. In *Sexual Selection and Reproductive Competition in Insects*, M. S. Blum and N. A. Blums(eds.), pp. 417-440. Academic Press, New York.

Asdell, S. A. 1964. *Patterns of Mammalian Reproduction*. Cornell University Press, Ithaca, NY.

Baker, R. R., and G. A. Parker. 1973. The origin and evolution of sexual reproduction up to the evolution of the male-female phenomenon. *Acta Biotheor.* 22:49-77.

Baltzer, F. 1931. Echiurida. *Handb. Zool.* 2:62-168.

Barton, N., and R. Post. 1983. Competition between siblings and advantage of mixed families. *J. Theor. Biol.*, in press.

Bell, G. 1978. The evolution of anisogamy. *J. Theor. Biol.* 73:247-270.

Bernstein, H., G. S. Byers, and R. Michod. 1981. Evolution of sexual reproduction: Importance of DNA repair, complementation, and variation. *Am. Nat.* 117:537-549.

Bertram, B. R. 1975. Social factors influencing reproduction in wild lions. *J. Zool. Lond.* 177: 463-482.

Beach, F. A., and L. Jordan. 1956. Sexual exhaustion and recovery in the male rat. *Q. J. Exp. Psychol.* 121:121-133.

Beach, F. A., and R. G. Rabedeau. 1959. Sexual exhaustion and recovery in the male hamster. *J. Comp. Physiol. Psychol.* 52:56-66.

Birkhead, T. R., and K. Clarkson. 1980. Mate selection and precopulatory guarding in *Gammarus pulex*. *Z. Tierpsychol.* 52:365-380.

Bishop, D. W. 1961. Biology of spermatozoa. In *Sex and Internal Secretions Vol. II, 3rd. ed.* Balliere, Tindall & Cox, London.

Boling, J. L., R. J. Blandau, J. C. Wilson, and W. C. Young. 1939. Postparturitional heat responses of newborn and adult guinea pigs: Data on parturition. *Proc. Soc. Exp. Biol. Med.* 42:128-132.

Boorman, E., and G. A. Parker. 1976. Sperm (ejaculate) competition in *Drosophila melanogaster*, and the reproductive value of females to males in relation to female age and mating status. *Ecol. Entomol.* 1:145-155.

Bryk, F. 1930. Monogamie einrichtungen bei schmetterlingsweibchen. *Arch. Frauenk. Konst-Forsch.* 16:308-313.

Bunnell, B. N., B. D. Boland, and D. A. Dewsbury. 1977. Copulatory behaviour of golden hamsters (*Mesocricetus auratus*). *Behaviour* 61:180-106.

Burger, J. F. 1952. Sex physiology of pigs. *Onderstepoort J. Vet. Res. Suppl.* 2:1-218.

Burkhardt, J. 1949. Sperm survival in the genital tract of the mare. *J. Agric. Sci.* 39:201-203.

Caldwell, R. L., and M. J. Rankin. 1974. Separation of migratory from feeding and reproductive behavior in *Oncopeltus fasciatus*. *J. Comp. Physiol.* 88:383-394.

Charnov, E. L. 1976. Optimal foraging: The marginal value theorem. *Theor. Popul. Biol.* 9:129-136.

Charlesworth, B. 1978. The population genetics of anisogamy. *J. Theor. Biol.* 73:347-357.

Cohen, J. 1973. Crossovers, sperm redundancy, and their close association. *Heredity* 31:408-413.

Cox, C. R., and B. J. LeBoeuf. 1977. Female incitation of male competition: A mechanism in sexual selection. *Am. Nat.* 111:317-335.

Cupps, P. T., L. L. Anderson, and H. H. Cole. 1969. The estrus cycle. In *Reproduction in Domestic Animals*, H. H. Cole and P. T. Cupps (eds.), pp. 217-250. Academic Press, New York.

Dawkins, R., and T. R. Carlisle. 1976. Parental investment, mate desertion and a fallacy. *Nature (Lond.)* 262:131-133.

Dawkins, R., and J. R. Krebs. 1979. Arms races between and within species. *Proc. R. Soc. Lond. B. Biol. Sci.* 205:489-511.

Day, F. T. 1942. Survival of spermatozoa in the genital tract of the mare. *J. Agric. Sci.* **32**: 108-111.

Devine, M. C. 1975. Copulatory plugs in snakes: Enforced chastity. *Science* 187:844-845.

Devine, M. C. 1977. Copulatory plugs, restricted mating opportunities and reproductive competition among male garter snakes. *Nature (Lond.)* 267:345-346.

Dewsbury, D. A. 1975. Diversity and adaptation in rodent copulatory behavior. *Science* 190: 945-954.

Dewsbury, D. A. 1981. On the function of the multiple intromission, multiple ejaculation copulatory patterns of rodents. *Bull. Psychon. Soc.* 18:21-23.

Dewsbury, D. A. 1982. Ejaculate cost and mate choice. *Am. Nat.* 119:601-610.

Dewsbury, D. A., and T. G. Hartung. 1980. Copulatory behavior and differential reproduction of laboratory rats in a two-male, one-female competitive situation. *Anim. Behav.* 28:95-102.

Doak, R. L., A. Hall, and H. E. Dale. 1965. Longevity of spermatozoa in the reproductive tract of the bitch. *J. Reprod. Fertil.* 13:51-58.

Dunbar, R. I. M. 1983. Intraspecific variations in mating strategy. In *Perspectives in Ethology, Vol. 5.* P. Bateson and P. Klopfer (eds.), in press.

Eaton, R. K. 1978. Why some felids copulate so much: A model for the evolution of copulation frequency. *Carnivore* 1:42-51.

Eibl-Eibesfeldt, I. 1970. *Ethology: The Biology of Behavior.* Holt, Rinehart & Winston, New York.

Ensminger, M. E. 1970. *Swine Science, 4th ed.* Interstate Printers, Danville, IL.

Fisher, R. A. 1930. *The Genetical Theory of Natural Selection.* Clarendon Press, Oxford.

Foote, R. H. 1969. Physiological aspects of artificial insemination. In *Reproduction in Domestic Animals, 2nd ed.,* H. H. Cole and P. T. Cupps (eds.), pp. 315-353. Academic Press, New York.

Fraser, A. F. 1980. *Farm Animal Behaviour, 2nd ed.* Balliere, Tindall & Cassell, London.

Ghiselin, M. 1974. *The Economy of Nature and the Evolution of Sex.* University of California Press, Berkeley, CA.

Gillespie, J. H. 1977. Natural selection for variance in offspring numbers: A new evolutionary principle. *Am. Nat.* 111:1010-1014.

Grafen, A. 1980. Opportunity cost, benefit and the degree of relatedness. *Anim. Behav.* 28: 967-968.

Gregory, G. E. 1965. The formation and fate of the spermatophore in the African migratory locust, *Locusta migratoria migratorioides* Reiche and Fairemaire. *Trans. R. Entomol. Soc. Lond.* 117:33-66.

Green, W. W. 1947. Duration of sperm fertility in the ewe. *Am. J. Vet. Res.* 8:299.

Griffiths, W. E. B., and E. Amoroso. 1939. Prooestrus, oestrus, ovulation, and mating in the greyhound bitch. *Vet. Rec.* 51:1279-1284.

Grunt, J. A., and W. C. Young. 1952. Psychological modification of fatigue following orgasm (ejaculation) in male guinea pigs. *J. Comp. Physiol. Psychol.* 45:508-510.

Hafez, E. S. E. 1969. *The Behaviour of Domestic Animals, 2nd ed.* Balliere, Tindall & Cassell, London.

Hart, B. L., and V. Odell. 1981. Elicitation of ejaculation and penile reflexes in spinal male rats by peripheral electric shock. *Physiol. Behav.* 26:623-626.

Hamilton, W. D. 1964. The genetical theory of social behaviour I and II. *J. Theor. Biol.* 7:1-52.

Hamilton, W. D., and Zuk, M. 1983. Heritable true fitness and bright birds: A role for parasites? *Science,* in press.

Hammer, O. 1941. Biological and ecological investigations on flies associated with pasturing cattle and their excrement. *Vidensk. Medd. Dan. Naturhist. Foren.* 105:1-257.

Hammerstein, P. 1981. The role of asymmetries in animal contests. *Anim. Behav.* 29:193-205.

Hammerstein, P., and G. A. Parker. 1982. The assymetric war of attrition. *J. Theor. Biol.* 96:647-682.

Hammond, J. 1927. *The Physiology of Reproduction in the Cow.* Cambridge University Press, Cambridge.

Hingston, R. W. G. 1933. *The Meaning of Animal Colouration and Adornment.* Edward Arnold, London.

Hoekstra, R. F. 1980. Why do organisms produce gametes of only two different sizes? Some theoretical aspects of the evolution of anisogamy. *J. Theor. Biol.* 87:785-793.

Jacobs, M. E. 1955. Studies on territorialism and sexual selection in dragonflies. *Ecology* 36: 566-586.

Jarosz, S. 1962. Obtaining fertilized and unfertilized ova following spontaneous and induced ovulation in cows. *Anim. Breed. Abstr.* 30:187-188.

Kaplan, W. D., V. E. Tinderholt, and D. H. Gugler. 1962. The number of sperm present in the reproductive tracts of *Drosophila melanogaster* females. *Drosophila Inf. Serv.* 36:82.

Knowlton, N. 1974. A note on the evolution of gamete dimorphism. *J. Theor. Biol.* 46:283-285.

Knowlton, N. 1982. Parental care and sex role reversal. In *Current Problems in Sociobiology,* King's College Sociobiology Group (eds.), pp. 203-222. Cambridge University Press, Cambridge.

Kunkel, P., and I. Kunkel. 1964. Beitrage zur ethologie des Hansmeerschuseinchens *Cavia aperea f. porcellus* (L.) *Z. Tierpsychol.* 21:610-641.

Labitte, A. 1919. Observations sur *Rhodocera rhamni. Bull. Mus. Hist. Nat. Paris* 25:624-625.

Laing, J. A. 1945. Observations on the survival time of the spermatozoa in the genital tract of the cow and its relation to fertility. *J. Agric. Sci., Camb.* 35:72-83.

Lande, R. 1980. Sexual dimorphism, sexual selection and adaptation in polygenic characters. *Evolution* 34:292-305.

Lande, R. 1981. Models of speciation by sexual selection on polygenic traits. *Proc. Natl. Acad. Sci.* 78:3731-3735.

Lanier, D. L., D. Q. Estep, and D. A. Dewsbury. 1979. Role of prolonged copulatory behavior in facilitating reproductive success in a competitive mating situation in laboratory rats. *J. Comp. Physiol. Psychol.* 93:781-792.

Lefevre, G., and U. B. Jonsson. 1962. Sperm transfer, storage, displacement and utilization in *Drosophila melanogaster. Genetics* 47:1719-1736.

Mangan, R. L. 1979. Reproductive behavior of the cactus fly, *Odontoloxozus longicornis,* male territoriality and female guarding as adaptive strategies. *Behav. Ecol. Sociobiol.* 4: 265-278.

Martin, P. A., T. J. Reimers, J. R. Lodge., and P. J. Dziuk. 1974. The effect of ratios and numbers of spermatozoa mixed from two males on proportions of offspring. *J. Reprod. Fertil.* 39:251-258.

Martan, J., and B. A. Shepherd. 1976. The role of the copulatory plug in reproduction of the guinea pig. *J. Exp. Zool.* 196:79-84.

Matthews, M., and N. T. Adler. 1977. Facilitative and inhibitory influences of reproductive behavior on sperm transport in rats. *J. Comp. Physiol. Psychol.* 92:727-741.

Maynard Smith, J. 1974. The theory of games and the evolution of animal conflicts. *J. Theor. Biol.* 47:209-221.

Maynard Smith, J. 1978a. *The Evolution of Sex.* Cambridge University Press, Cambridge.

Maynard Smith, J. 1978b. Optimization theory in evolution. *Annu. Rev. Ecol. Syst.* 9:31-56.

Maynard Smith, J., and G. A. Parker. 1976. the logic of asymmetric contests. *Anim. Behav.* 24:159-175.

McGaughey, R. W., J. H. Marston, and M. C. Chang. 1968. Fertilizing life of mouse spermatozoa in the female tract. *J. Reprod. Fertil.* 16:147-150.

McGill, T. E. 1962. Sexual behavior in three inbred strains of mice. *Behaviour* 19:341-350.

Merton, H. 1939. Studies on reproduction in the albino mouse: III. The duration of life of spermatozoa in the female reproductive tract. *Proc. R. Soc. Edinb. Sect B (Biol. Sci.)* 59:207.

Milligan, S. R. 1979. The copulatory pattern of the bank vole (*Clethrionomys glareolus*) and speculation on the role of penile spines. *J. Zool.* 188:279-283.

Miyamoto, H., and M. C. Chang. 1972. Fertilizing life of golden hamster spermatozoa in the female tract. *J. Reprod. Fertil.* 31:131-134.

Mosig, D. W., and D. A. Dewsbury. 1970. Plug fate in the copulatory behavior of rats. *Psychon. Sci.* 10:315-316.

Murray, J. D. 1964. Multiple mating and effective population size in *Cepea nemoralis*. *Evolution* 18:283-291.

Nakatsuru, K., and D. L. Kramer. 1982. Is sperm cheap? Limited male fertility and female choice in the lemon tetra (Pisces, Characidae). *Science* 216:753-755.

Packer, C. 1979. Male dominance and reproductive activity in *Papio anubis*. *Anim. Behav.* 27: 37-45.

Page, R. E., and R. A. Metcalf. 1982. Multiple mating, sperm utilization, and social evolution. *Am. Nat.* 119:263-281.

Parker, G. A. 1970a. Sperm competition and its evolutionary consequences in the insects. *Biol. Rev.* 45:525-567.

Parker, G. A. 1970b. Sperm competition and its evolutionary effect on copula duration in the fly, *Scatophaga sterocoraria*. *J. Insect Physiol.* 16:1301-1328.

Parker, G. A. 1970c. The reproductive behavior and the nature of sexual selection in *Scatophaga stercoraria* L. (Diptera:Scatophagidae): VII. The origin and evolution of the passive phase. *Evolution* 24:774-788.

Parker, G. A. 1970d. The reproductive behaviour and the nature of sexual selection in *Scatophaga stercoraria* L. (Diptera:Scatophagidae): V. The female's behaviour at the oviposition site. *Behaviour* 37:140-168.

Parker, G. A. 1971. The reproductive behavior and the nature of sexual selection in *Scatophaga stercoraria* L. (Diptera:Scatophagidae): VI. The adaptive significance of emigration from the oviposition site during the phase of genital contact. *J. Anim. Ecol.* 40:215-233.

Parker, G. A. 1972. Reproductive behaviour of *Sepsis cynipsea* (L.) (Diptera:Sepsidae): I. A preliminary analysis of the reproductive strategy and its associated behaviour patterns. *Behaviour* 41:172-206.

Parker, G. A. 1974. Assessment strategy and the evolution of fighting behaviour. *J. Theor. Biol.* 47:223-243.

Parker, G. A. 1978a. Selection on non-random fusion of gametes during the evolution of anisogamy. *J. Theor. Biol.* 73:1-28.

Parker, G. A. 1978b. The evolution of competitive mate-searching. *Ann. Rev. Entomol.* 23: 173-196.

Parker, G. A. 1979. Sexual selection and sexual conflict. In *Sexual Selection and Reproductive Competition in Insects,* M. S. Blum and N. A. Blum (eds.), pp. 123-166. Academic Press, New York.

Parker, G. A. 1982a. Phenotype limited evolutionarily stable strategies. In *Current Problems In Sociobiology,* King's College Sociobiology Group (eds.), pp. 173-201. Cambridge University Press, Cambridge.

Parker, G. A. 1982b. Why so many tiny sperm? The maintenance of two sexes with internal fertilization. *J. Theor. Biol.* 96:281-294.

Parker, G. A. 1983. Arms races in evolution: An ESS to the opponent-independent costs game. *J. Theor. Biol.,* in press.

Parker, G. A., R. R. Baker, and V. G. F. Smith. 1972. The origin and evolution of gamete dimorphism and the male-female phenomenon. *J. Theor. Biol.* 36:529-553.

Parker, G. A., and M. R. Macnair. 1979. Models of parent-offspring conflict: IV. Suppression: Evolutionary retaliation by the parent. *Anim. Behav.* 27:1210-1235.

Parker, G. A., and D. I. Rubenstein. 1981. Role assessment, reserve strategy and the requisition of information in asymmetric animal contests. *Anim. Behav.* 29:221-240.

Parker, G. A., and J. L. Smith. 1975. Sperm competition and the evolution of the precopulatory passive phase behavior in *Locusta migratoria migratorioides. J. Entomol. Ser. A* 49: 155-171.

Parker, G. A., and R. A. Stuart. 1976. Animal behavior as a strategy optimizer: Evolution of resources assessment strategies and optimal emigration thresholds. *Am. Nat.* 110:1055-1076.

Parker, G. A., and E. A. Thompson. 1980. Dung fly struggles: A test of the war of attrition. *Behav. Ecol. Sociobiol.* 7:37-44.

Pease, R. W. 1968. The evolution and biological significance of multiple pairing in Lepidoptera. *J. Lepid. Soc.* 22:197-209.

Pitkjanen, I. G. 1959. The time of ovulation in sows. *Anim. Breed. Abstr.* **27**:212.

Prout, T., and J. Bundgaard. 1977. The population genetics of sperm displacement. *Genetics* **85**:95-124.

Richmond, R. C., and J. Ehrman. 1974. Incidence of repeated mating in the superspecies *Drosophila paulistorum. Experientia* **30**:489-490.

Ridley, M., and D. J. Thompson. 1979. Size and mating in *Asellus aquaticus. Z. Tierpsychol.* **51**:380-397.

Rood, J. P. 1972. Ecological and behavioural comparisons of three genera of Argentine cavies. *Anim. Behav. Monogr.* **5**:1-82.

Rose, M. R. 1978. Cheating in evolutionary games. *J. Theor. Biol.* **75**:21-34.

Rubenstein, D. I. 1980. On the evolution of alternative mating strategies. In *Limits to Action,* J. R. Staddon (ed.)., pp. 65-100. Academic Press, New York.

Rubenstein, D. I. 1982. Risk, uncertainty and evolutionary strategies. In *Current Problems in Sociobiology,* King's College Sociobiology Group (eds.), pp. 91-111. Cambridge University Press, Cambridge.

Salisbury, G. W., and van Denmark, N. L. 1961. *Physiology of Reproduction and Artificial Insemination in Cattle.* W. H. Freeman, London.

Sivinski, J. 1980. Sexual selection and insect sperm. *Fla. Entomol.* **63**:99-111.

Smith, C. C., and S. D. Fretwell. 1974. The optimal balance between size and number of offspring. *Am. Nat.* **108**:499-506.

Smith, R. L. 1976a. Brooding behavior of a male water bug *Belostoma flumineum. J. Kans. Entomol. Soc.* **49**:333-343.

Smith, R. L. 1976b. Male brooding behavior of the water bug *Abedus herberti* (Hemiptera: Belostomatidae). *Ann. Entomol. Soc. Am.* **69**:740-747.

Smith, R. L. 1979a. Paternity assurance and altered roles in the mating behaviour of a giant water bug, *Abedus herberti* (Heteroptera, Belostomatidae). *Anim. Behav.* **27**:716-725.

Smith, R. L. 1979b. Repeated copulation and sperm precedence: Paternity assurance for a male brooding water bug. *Science* **205**:1029-1031.

Snell, G. E., E. Fekete, K. P. Hummel, and L. W. Law. 1940. The relation of mating, ovulation and the estrous smear in the house mouse to the time of day. *Anat. Rec.* **76**:39-54.

Soderwall, A. L., and R. J. Blandau. 1941. The duration of the fertilizing capacity of spermatozoa in the genital tract of the rat. *J. Exp. Zool.* **88**:55-63.

Soderwall, A. L., and W. C. Young. 1940. The effect of aging in the female genital tract of the fertilizing capacity of guinea pig spermatozoa. *Anat. Rec.* **78**:19-29.

Thibault, C. 1973. Sperm transport and storage in vertebrates. *J. Reprod. Fertil. Suppl.* **18**: 39-53.

Thornhill, R. 1976. Sexual selection and nuptial feeding behavior in *Bittacus apicalis* (Diptera: Mecoptera). *Am. Nat.* **110**:153-163.

Thornhill, R. 1980a. Competitive, charming males and choosy females: Was Darwin correct? *Fla. Entomol.* **63**:5-30.

Thornhill, R. 1080b. Sexual selection within mating swarms of the lovebug *Plecia nearctica* (Diptera:Bibionidae). *Anim. Behav.* **28**:405-412.

Trivers, R. L. 1972. Parental investment and sexual selection. In *Sexual Selection and the Descent of Man, 1871-1971,* B. Campbell (ed.), pp. 136-179. Aldine-Atherton, Chicago.

Vandeplassche, M., and R. Paredis. 1948. Preservation of the fertilizing capacity of bull semen in the genital tract of the cow. *Nature (Lond.)* **162**:813.

Voss, R. Male accessory glands and the evolution of copulatory plugs in rodents. *Occas. Pap. Mus. Zool. Univ. Mich.* **689**:1-27.

Waage, J. K. 1979. Dual function of the damselfly penis: Sperm removal and transfer. *Science* **203**:227-232.

Walker, W. F. 1980. Sperm utilization strategies in non-social insects. *Am. Nat.* **115**:780-799.

Werren, J. H., M. R. Gross, and R. Shine. 1980. Paternity and the evolution of male parental care. *J. Theor. Biol.* **82**:619-631.

Wickler, W. 1967. Socio-sexual signals and their intraspecific imitation among primates. In *Primate Ethology,* D. Morris (ed.), pp. 69-174. Weidenfeld and Nicholson, London.

Williams. G. C. 1975. *Sex and Evolution.* Princeton University Press, Princeton, NJ.

Young, W. C. 1941. Observations and experiments on mating behavior in female mammals. *Q. Rev. Biol.* **16**:135-156, 311-335.

Young, W. C., E. W. Dempsey, C. W. Hagquist, and J. L. Boling. 1937. The determination of heat in the guinea pig. *J. Rab. Clin. Med.* **23**:300-302.

Young, W. C. E., and J. A. Grunt. 1951. The pattern and measurement of sexual behavior in in the male guinea pig. *J. Comp. Physiol. Psychol.* **44**:492-500.

2

Male Sperm Competition Avoidance Mechanisms: The Influence of Female Interests

NANCY KNOWLTON

SIMON R. GREENWELL

61

I. INTRODUCTION

Sperm competition (Parker 1970a), like other aspects of reproductive biology described by the theory of sexual selection, can be viewed within the traditional context of interactions among males and between males and females (Darwin 1871, Huxley 1938). Recent work has focused mainly on the more obvious behavioral and morphological features related to direct competition among males. For example, multiple copulations (Smith 1979), testis size (Harcourt *et al.* 1981), penis morphology and size (Waage 1979, this volume; Smith, this volume), prolonged copulation or guarding of females by males (Parker 1970a), and sperm plugs (Parker 1970a) may be viewed as adaptations which reduce the likelihood of a male's sperm being preempted by sperm from prior or subsequent matings.

The ability of females to influence whether or not sperm competition avoidance mechanisms evolve has received less emphasis (but see especially Walker 1980). This neglect is curious, since females in many species have considerable physical control over copulation and the fertilization of their eggs and thus some potential for evolutionary "manipulation." Although the outcome of a selective conflict between the sexes (Parker 1979) can be difficult to predict (Maynard Smith 1977, Parker 1979, Schuster and Sigmund 1981, Knowlton 1982), such situations should not be ignored. In this paper we will (1) review how mechanisms in males that reduce sperm competition can be costly to females, with special emphasis on the relationship between multiple mating and female fitness; (2) analyze the likely evolutionary outcomes given such costs; and (3) interpret some patterns observed in nature (particularly the relationship between certainty of paternity and male parental care) in light of these theoretical considerations. Our main concern is to determine the conditions under which selection on females is likely to counteract selection on males for avoiding sperm competition.

II. COSTS TO FEMALES

It should be stated at the outset that sperm competition avoidance mechanisms need not be costly to females (*e.g.,* Parker, Thornhill, this volume). Parker (1970b) and Borgia (1981), for example, present data showing that female dung flies are less likely to be damaged and can oviposit more rapidly when they are effectively protected by their mates from other males. In other species, excess sperm, spermatophores, and sperm plugs can provide nutrients for females (Thornhill 1976; Sivinski, this volume). However, since our purpose is to determine whether the interests of females can inhibit development of sperm competition avoidance mechanisms, we will concentrate on potential costs which may arise directly as a consequence of the actual methods used or indirectly through their effectiveness in reducing sperm competition. Four major categories of costs are presented below, in order of increasing conceptual complexity. Although we cite possible examples where appropriate, quantitative documentations of these costs are not currently available.

A. Costs Stemming from Methods Used

The methods used by males to reduce sperm competition may impose direct costs on females independent of their sperm competition consequences. Although such costs may often be quite trivial (*e.g.*, sperm displacement typically takes little time and does not injure the female), there are a number of probable exceptions. For example, mechanisms aimed at preventing female rematings could be costly if they resulted in tissue traumatization (as described for poeciliid fishes by Constantz, this volume) or reduced efficiency of oviposition (as suggested for several insect sperm plugs; see Parker, this volume). Among the mechanisms which function to prevent females from using previously acquired sperm, multiple copulations (*e.g.*, as described for giant water bugs by Smith 1979) are particularly likely to be costly because they may interfere with other activities or increase the risk of predation (see Trivers 1976, Daly 1978). Insemination through the body wall (see Hinton 1964), which may have arisen in the context of competition with previously deposited sperm (if sperm injected directly into the female were more likely to fertilize the eggs; Lloyd 1979) probably at least initially had negative effects on females.

B. Nutrient Losses

Males in some species donate nutrients to females while mating, for example through food offerings or via materials received with the sperm (*e.g.*, for insects see Thornhill 1976; Boggs and Gilbert 1979; Gwynne, this volume). Therefore, if sperm competition avoidance mechanisms such as sperm plugs make it more difficult for females to mate again, there could be energetic costs to females. The potential evolutionary significance of this cost may be limited, however. As male contributions become more substantial, male coyness (*e.g.*, Rutowski 1980) with previously mated females would be a more likely evolutionary response (Walker 1980, Dewsbury 1982), if refusing to mate with a recently mated female were less costly than preventing the female from remating.

C. Reduction in Amount of Sperm Available

Since it is in the male's interests for the female to use all his sperm before soliciting other matings, sperm competition avoidance mechanisms that prevent a female remating could result in her having insufficient numbers of sperm for fertilization of all her eggs. Generally, however, one would expect males to contribute a modest excess of sperm so that all potentially laid eggs could be fertilized. Thus in most species females are unlikely to suffer such a cost. A possible exception may be found in honey bees, in which the sperm supplies of a single male cannot meet the needs of a queen for her entire lifespan. (This limitation in sperm numbers may be the result of sexual selection for another character, high maneuverability in flight, necessary for a male to have any chance of obtaining a copulation [see Parker, this volume].) Queens must remove male genital parts from their reproductive tracts in order to copulate again, and wild queens that lay only unfertilized eggs (drone producers) are sometimes found (T. Seeley, pers. comm.).

D. Reduction in Sperm Diversity

Effective sperm competition avoidance mechanisms reduce either the number of males which contribute sperm to a female or the evenness of their effective contributions. Although the average sperm quality available to singly mated females is not expected to differ from the average sperm quality of multiply mated females (see Parker, this volume), several theoretical analyses suggest that more diverse sperm supplies gained by multiple matings could increase the fitness of multiply mated females over singly mated females. Multiple matings have the effect of increasing the variance in sperm types within a single female and correspondingly reducing the variance in the number of sperm types among females or among broods of the same female.

1. Variance in Offspring Numbers

Gillespie has modeled the populational consequences of both between-generation variance (1973, 1977) and within-generation variance (1974, 1977) in the number of offspring produced (see Fig. 1). The basic idea behind these models is that under certain circumstances there can exist a relationship of decreasing returns between the number of viable offspring produced and parental fitness. This leads to the conclusion that if two strategies produce the same mean number of surviving offspring but have different variances, then the strategy with the lower variance will be favored. In relation to sperm competition, these models indicate that females that don't fertilize all their eggs with sperm from a single male may be fitter because the variance in their reproductive success decreases as a result.

A. WITHIN GENERATION VARIANCE

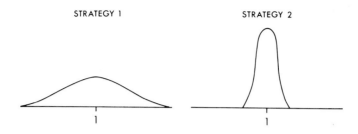

B. BETWEEN GENERATION VARIANCE

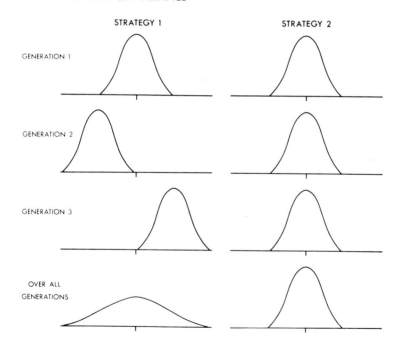

NUMBER OF SURVIVING FEMALE OFFSPRING PER FEMALE

Fig. 1. Probability distributions for the number of female offspring per female given within (A) or between (B) generation variance. Gillespie (1977) predicts that in both cases strategy 2 will have the higher fitness. In (A) the strength of this effect depends on population stability and size, as it is mediated through reduced probability of extinction. Within any generation we would predict that multiply mating females would show less variance in the expected number of surviving offspring. However, for between generation variance (B) we see no justification for distinguishing singly and multiply mated females.

The least restrictive of Gillespie's models (1973) predicts that strategies which vary little from one generation to the next in their success can be favored over strategies whose success fluctuates more widely through time. We can see no obvious reason why the amount of between-generation variance in the number of offspring produced by multiply mated individuals should be less than the between-generation variance for single maters, however (Fig. 1B). It is much easier to envision the possibility of more variance in reproductive success among females that mate singly than among females that mate multiply in any single generation (Fig. 1A). This stems from the fact that some singly mated females might do very well and others might do very poorly, while the reproductive success of multiply mated females would be averaged out through fertilizations involving a mixture of sperm qualities.

One of the clearest indications that multiple mating can reduce variance in reproductive success is found in the work of Page and Metcalf (1982). Theoretically and empirically, they have shown in the honey bee (which has a multi-allelic sex-determination locus for which homozygosity is lethal) that variance in brood viability decreases with increases in the number of matings by the queen (average brood success is unchanged). Although most species lack this peculiar genetic feature, males that successfully court and carry recessive genetic defects will be found at low frequencies in most populations. The likelihood of genetic incompatibility between two potential mates will be even greater when there is strong overdominance (*e.g.*, as in sickle cell anemia) or when there are moderate amounts of inbreeding.

The strength of the advantage of low variance within generations depends on overall population stability (average number of surviving daughters per female approximating one) and decreases with increasing population size, however. This makes it difficult to interpret the selective importance of even unambiguous differences in variance such as were described by Page and Metcalf (1982). Thus the costs to females of sperm competition avoidance mechanisms arising from the considerations outlined by Gillespie may be of importance only when population sizes are small and stable, and/or costs to females in achieving multiple matings are comparatively low (*i.e.*, mean numbers of offspring produced by singly and multiply mated females are nearly equal).

Rubenstein (1982) took the populational approach of Gillespie and applied it to individuals. Rather than comparing strategies producing **equal** average numbers of offspring per individual but with different populational variances, he asked whether a strategy producing a lower variance in expected success in resource acquisition when used over the lifetime of an individual would result in a **higher** average number of offspring for that individual. In this case, lower variance should produce higher average fitness if there is an ecological relationship of diminishing returns between resources available to the individual and numbers of offspring produced. Although relating this idea to single versus multiple matings is not as straightforward, a loosely parallel situation would be one in which females producing a mixture of high and low quality offspring or using a mixture of high and low quality sperm would do nearly as well as females with only high quality offspring or sperm and much better than females with only low quality offspring or sperm (see below).

2. Variance in Offspring Quality

Formal similarities between benefits of multiple matings and benefits of sexual reproduction suggest that given sibling competition and unpredictable environments, multiply mating females might on average do better than singly mating females (Parker, this volume). There is certainly no question that "environments," defined broadly, are unpredictable. The optimum genotype for an offspring will depend on when and where it is, and females vary genetically in many ways that are distinct from whatever genetic factors influence mating behavior. Even if females "knew" their own genotype and had perfect information about the environment, an accurate assessment of the genotypes of all potential mates would not be possible.

The effect of sperm diversity on variance in offspring quality is illustrated in Fig. 2. As indicated, the combination of sib-competition and soft selection (Wallace 1975) yields an advantage to multiply mated females. In Maynard Smith's (1976) model, a female's offspring compete with both siblings and offspring from other females, and only a limited number of individuals that are among those best adapted to prevailing conditions (which are unpredictable from generation to generation) survive. Alternatively, sib-competition could occur while the offspring were still being cared for by the parent. In this case, if each female were only able to rear successfully some most viable fraction of the clutch, the mean fitness of the surviving offspring produced by multiply mated females would exceed that of singly mated females.

3. Female Incitation of Male-male Sperm Competition

Many authors have suggested that females should act so as to fertilize their eggs with sperm bearing the "best" genes. Cox and LeBoeuf (1977) went on to suggest that the probability of this could be increased if females encouraged aggressive interactions among potential mates, the argument being that they would then be more likely to produce effectively aggressive, reproductively successful sons. A parallel argument could be made with respect to sperm; by mixing the sperm of several males, females could insure that their eggs were fertilized by the most competitively successful sperm, increasing the likelihood that their sons would also have competitive sperm. Alternatively, however, females could benefit through the production of sons that effectively avoid sperm competition. Given the

SINGLY MATED   MULTIPLY MATED

FEMALES   FEMALES

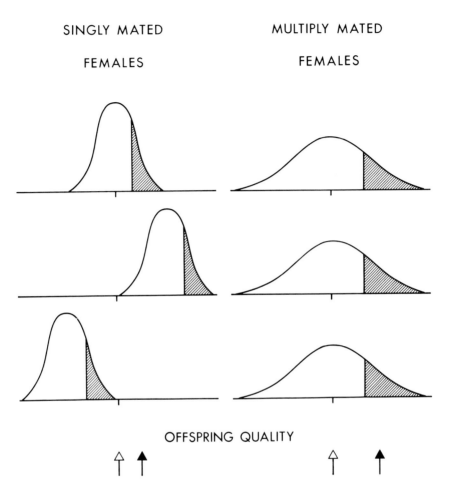

OFFSPRING QUALITY

Fig. 2. The differential effect of "soft" selection acting on sperm received or offspring produced by singly mated and multiply mated females. Both types of female receive sperm that is of the same mean fitness when averaged over many females (white arrows). After "soft" selection has occurred, the average fitness of multiply mated females' offspring is higher (black arrows). The total area under each curve represents the number of siblings or sperm before the operation of "soft" selection, while the hatched areas indicate those individuals that survived "soft" selection.

potential for these two opposing forces operating via sexual selection and the success of sons, the significance of the argument that females benefit by encouraging competition among the sperm of different males remains unclear.

Females would benefit more broadly if there were a correlation between sperm competitiveness and the viability of the individual bearing competitive sperm genes. Such a correlation is unlikely, however, except in the case of gross genetic abnormalities (see Sivinski, this volume). Since genetic abnormalities of this magnitude would typically be apparent in courting males, incitation of sperm competition for this purpose would seem to be unnecessary.

4. Magnitude of Importance of Sperm Diversity

In the above discussion we have assumed that multiply mated females receive a more varied array of sperm within a brood than singly mated females. But because even a single mating produces substantial genetic variability (Williams 1975), it could be argued that the additional benefits associated with producing more variable offspring from multiple matings will be comparatively slight. Related to this is the more controversial point (see Jarvi et al. 1982) that there will be little genetic variance in traits most closely associated with fitness. In considering whether additional matings should ever be beneficial, however, it must be remembered that costs of additional matings may also be low, and certainly often less than the potential two-fold cost of sexual reproduction itself.

It should also be noted that the influence of multiple mating on female fitness depends to some extent on the temporal patterning of mating and egg production that characterizes a population. When potentially highly fecund females use sperm received during a short period to fertilize their entire production of eggs, reduction in sperm variability from sperm competition avoidance mechanisms will be at its maximum. In species which have less temporally restricted mating opportunities, many males may father a female's offspring over a lifetime, although any one clutch would be more likely to have a single father when sperm competition avoidance mechanisms are well developed. In such cases, the potentially substantial benefits of mixed paternity associated with sibling competition among offspring in a single brood would be minimized. Some sibling competition between members of different broods might be possible, although females must then devote more resources to offspring that may ultimately fail. On the other hand, this arrangement could theoretically reduce the level of parent-offspring conflict (Parker and Macnair 1979).

III. THE RESOLUTION OF CONFLICTING MALE AND FEMALE INTERESTS

In general, two arguments are invoked to explain or predict the outcome of an evolutionary conflict between two parties (Parker 1979). The first is to assume that the party experiencing the greater selection pressure will win the conflict (e.g., the evolution of anisogamy; Parker 1979). The second argument, which is based on the concept of phylogenetic inertia, predicts that the party that is better able to "manipulate" the other, based on existing adaptations, will win the evolutionary confrontation (*e.g.,* parent-offspring conflict; Parker and Macnair 1979).

However, in trying to explain the evolution of male sperm competition avoidance mechanisms that are costly to females, only the first argument, based on relative selection pressures, has been seriously considered (but see Walker 1980). Consequently, since the selection pressures acting on males are generally much greater than those acting on females (quantity versus quality of offspring respectively), it is usually concluded that females will not greatly influence the evolution of sperm competition avoidance mechanisms in males. But will this conclusion still hold if the second argument mentioned above is also taken into account?

As Parker (this volume) has pointed out, if females are able to decide when copulation is terminated, the female sex is in a potentially strong position to influence the evolution of sperm competition avoidance mechanisms. Starting from this premise we have utilized the principles of games theory in order to construct a model to determine the evolutionary dynamics of associated male and female mating strategies and hence the evolution of sperm competition avoidance mechanisms. We have assumed that males may or may not have sperm competition avoidance mechanisms, while females have the prerogative to determine when copulation ends. The latter seems reasonable since in many species females must cooperate with males if insemination is to occur (*e.g.,* many insects, fishes, etc.). In particular we have examined the consequences of females terminating mating when males that are attempting to reduce sperm competition are detected. Given the widespread phenomenon of female coyness (*e.g.,* Darwin 1871), termination of mating in response to novel male behavior or morphology would seem to be probable enough to deserve theoretical investigation. The model has been kept simple in order to provide qualitative predictions; the building of a very precise (and complicated) model seems inappropriate since the estimation of parameter values will typically be difficult, especially for the evolutionary antecedents of present-day situations.

A. The Model

Assume that males either do or do not have some behavioral or morphological trait causing the active prevention of sperm competition (*e.g.,* multiple copulations, sperm plugs, etc.). "Tolerant" females do not distinguish between "active" and "passive" males, while "intolerant" females do distinguish, and terminate the mating when they detect that their mate is attempting to limit their ability to use previously acquired sperm or to obtain future mates. When an active male/ intolerant female mating is interrupted, both individuals achieve some fractional success from this mating and have some probability of finding another mate within the time period during which they would have been unavailable for remating had the first mating continued. These time periods need not be the same for males and females (*e.g.,* a female may spend extra time ovipositing), since the relative successes of strategies used by members of one sex are not defined by the success of strategies in the other sex. Individuals that remate can achieve, at most, the same total reproductive success as individuals that have a complete single mating. Note that it is not necessary to assume that there is a direct relationship between the time spent in an initial, interrupted mating and the success achieved from it, because these parameters can be varied independently.

Active males have a higher certainty of paternity than passive males for both complete and interrupted matings because it is assumed that premature termination of mating occurs after the active male has implemented his sperm competition avoidance mechanism. Although this assumption may be invalid for certain organisms, the qualitative predictions of the model are unaffected by changing this assumption. Tolerant females may experience a cost or a benefit associated with sperm competition avoidance mechanisms.

The payoffs to active and passive males, and tolerant and intolerant females, resulting from the four possible mating combinations are shown in Fig. 3. For males, the relevant parameters are:

p = the proportion of males in the population which actively attempt to prevent sperm competition.

r = the certainty of paternity of a passive male (one which makes no attempt to prevent sperm competition) relative to an active male (which does). It is assumed that $r \leqslant 1$ (*i.e.,* that the certainty of paternity of active males is at least as high as that of passive males, even if the cost of the trait that is responsible for reducing sperm competition is incorporated into r).

f_m = the number of offspring that an active male can father as a result of an initial mating with an intolerant female, relative to the number of offspring that a male can father in a mating that is not prematurely terminated ($0 \leqslant f_m \leqslant 1$).

a_m = the probability of an active male mating again (within the time period during which he would have been unavailable for remating had the initial mating continued) following the termination of an initial, incomplete mating with an intolerant female.

For females, there is a parallel set of parameters:

q = the proportion of tolerant females in the population.

s = the reproductive success of a female that tolerates a mating with a male that actively prevents sperm competition, relative to the reproductive success of a female that mates with a passive male ($s < 1$ when the sperm competition avoidance mechanism is costly to the female).

f_f = the number of offspring produced by an intolerant female from an initial mating with an active male, relative to the number of offspring that are produced when a mating is not prematurely terminated ($0 \leqslant f_f \leqslant 1$).

a_f = the probability of an intolerant female mating again (within the time period during which she would have been unavailable for remating had the initial mating continued) following the termination of a mating with an active male.

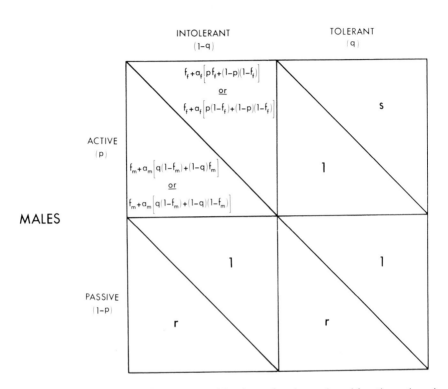

Fig. 3. The payoff matrix for tolerant and intolerant females mating with active and passive males. For each of the four sectors, payoffs to males appear below the diagonal; payoffs to females appear above the diagonal. For active male/intolerant female matings, the expression for the payoff depends on whether the first, interrupted mating was less than (upper expression) or greater than (lower expression) one-half complete with respect to the expected success from uninterrupted matings. We assume only one additional mating is possible after an interrupted mating ends, in the time period required for one complete, initial mating.

B. Analysis

The first step is to determine the equations that underlie the dynamics of the male and female strategies as shown in Fig. 4. These can be obtained by setting expressions for the payoffs to active and passive males equal to one another and solving for the frequency of tolerant females (q) at which this occurs. The favored male strategies either side of this equilibrium can subsequently be determined. This procedure is then repeated for tolerant and intolerant female payoffs to determine the dynamics of the female strategies. For each sex, the position of

MALE DYNAMICS

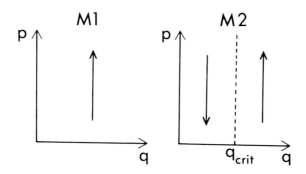

FEMALE DYNAMICS

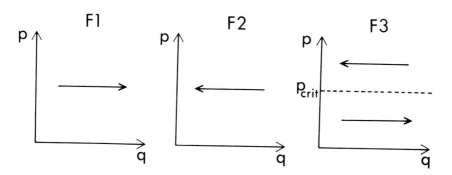

Fig. 4. Direction of change in the frequencies of male and female mating strategies as a function of the frequency of mating strategies in the opposite sex. The differences between M1 and M2, and between F1 and F2 are determined by the parameter values. In (F3), intolerance carries a cost that is independent of matings with active males (not indicated in Fig. 3), but otherwise the conditions are identical to those for F2.

the critical value and the strategies favored either side of it are determined by the values of the parameters. A fuller derivation that illustrates the dynamics more precisely is included in the Appendix.

For males, it is clear that if tolerant females are sufficiently common, active males will always increase in frequency because they gain the benefits of increased certainty of paternity without frequently suffering from interrupted matings with intolerant females (see Fig. 4, top). This result is obtained from Fig. 3 by comparing the payoff to active and passive males when encountering a tolerant female (active payoff = 1; passive payoff = r; $1 > r$). But it is not necessarily true that active males will be more successful than passive males for all frequencies of tolerant females. As the second graph for males (M2) of Fig. 4 illustrates, there may be some critical frequency of tolerant females, q_{crit}, below which passive males do better than active males. This situation will occur when the payoff to an active male mating with an intolerant female is less than the payoff to a passive male. Values of q_{crit} for various values of the other parameters are shown in Table I. They indicate that the evolution of active prevention mechanisms is favored (*i.e.*, is less likely to be blocked by the presence of intolerant females) when the probability of a male mating again (a_m) is high, when the relative number of intolerant female's offspring that an active male can potentially father (f_m) is high, and when the relative certainty of paternity of passive males (r) is low.

For females, in this simple analysis, there can be no critical frequency of active males which yields an advantage to tolerant females on one side of the critical frequency and a disadvantage on the other side. This follows from the equal payoffs accorded to tolerant and intolerant females when mating with passive males. Tolerant females will therefore always decrease or always increase in frequency when there are active males in the population, depending upon which strategy does better in interactions with active males (see F1, F2 of Fig. 4). In general, tolerance is favored when the reproductive success of tolerant females mated with active males (s) is high, when the probability of a female mating again (a_f) is low, and when the relative reproductive success of an intolerant female mating with an active male (f_f) is low.

If, however, we assume that intolerance carries a cost independent of interactions with active males (which is not indicated in Fig. 3) such that the payoffs to tolerant and intolerant females mating with passive males are now unequal, then the possibility exists, in situations in which tolerance would otherwise always decrease, of a critical value of p, p_{crit}, above which tolerance would decrease and below which tolerance would increase (F3 of Fig. 4). For example, the ability of an intolerant female to terminate a mating might depend upon some morphological character that would have to be maintained. This would give tolerant females an advantage over intolerant females when mating with passive males since the structure could impose a cost without any benefit being derived from it.

The male and female dynamics illustrated in Fig. 4 can now be combined in all the possible male/female pairwise combinations to determine in each case the resultant coevolutionary path of the male and female strategies (see Fig. 5). Widely varying outcomes are possible. At one end of the spectrum, tolerant females and active males have a selective advantage for all values of p (F1 of Fig. 4) and q (M1 of Fig. 4), which invariably favors the evolution of sperm competition

TABLE I

Values of q_{crit}[a] for Seven Values of a_m[b], Three Values of r [c], and Five Values of f_m[d,e]

a_m	r	.01	.10	.25	.50	.75
0.01	.25	.24	.16	<0	<0	<0
	.50	.49	.44	.33	<0	<0
	.75	.75	.72	.66	.49	<0
0.10	.25	.22	.15	<0	<0	<0
	.50	.47	.42	.30	<0	<0
	.75	.73	.70	.64	.44	<0
0.25	.25	.20	.12	<0	<0	<0
	.50	.43	.38	.24	<0	<0
	.75	.69	.66	.59	.33	<0
0.50	.25	.17	.08	<0	<0	<0
	.50	.38	.31	.15	<0	<0
	.75	.63	.59	.50	.00	<0
0.75	.25	.14	.05	<0	<0	<0
	.50	.33	.26	.07	<0	<0
	.75	.56	.52	.40	<0	<0
0.90	.25	.13	.04	<0	<0	<0
	.50	.30	.23	.03	<0	<0
	.75	.52	.47	.33	<0	<0
0.99	.25	.13	.03	<0	<0	<0
	.50	.29	.21	.00	<0	<0
	.75	.50	.44	.30	<0	<0

The column group header spanning .01–.75 is f_m.

[a]Minimum frequency of tolerant females required for active males to increase in frequency.
[b]Probability of males remating.
[c]Relative certainty of paternity of passive males.
[d]The number of intolerant female's offspring that an active male can father relative to the number for tolerant females.
[e]When f_m is .75 or larger, q_{crit} is always less than 0 (no minimum frequency of tolerant females) for all values of a_m and r indicated.

avoidance mechanisms in males (Fig. 5A). At the other end of the spectrum, active prevention of sperm competition by males is only favored when tolerant females are common (M2 of Fig. 4), a situation which cannot be stably maintained because tolerant females are always at a disadvantage (F2 of Fig. 4). In this case we would not expect to find traits in males related to the avoidance of sperm competition (Fig. 5E). Other combinations of selective pressures produce less predictable outcomes. Fig. 5F, for example, suggests the possibility of oscillating frequencies, or more likely (based on Parker 1983) a mixture of the four male

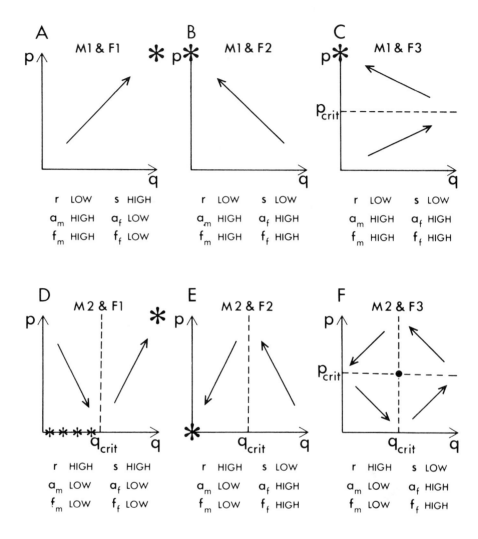

Fig. 5. Evolutionary dynamics of male (vertical component) and female (horizontal component) mating strategies combined. Asterisks indicate likely outcomes resulting from these dynamics. (C) and (F) are identical to (B) and (E) respectively, except that intolerant females suffer some cost independent of their encounters with active males. Parameter values favoring each dynamic are indicated, although these values need not all hold for the dynamic to be achieved.

and female strategies. Other apparently predictable outcomes are not likely to be stable over long periods of time; although Figs. 5B and 5C indicate an end point of active males and intolerant females, this would seem to be the starting point for sexual conflict of the form described by Parker (1979). The eventual outcome would depend on the strength of selective pressures on males and females and the efficacy of mutations affecting these strategies (Parker 1979). Finally, some combinations (Fig. 5D) suggest that reaching the ultimately stable outcome of active males and tolerant females might require considerable time; the frequency of tolerant females changes very slowly when active males are rare (because tolerant and intolerant females can be equally successful with passive males), making it difficult to reach the high critical frequency of tolerant females, q_{crit}, required for active males to increase in frequency. Thus, in this case, at any one time (*e.g.,* the present), it is unlikely that all species will have reached the equilibrium of tolerant females and active males.

IV. NATURAL HISTORY PATTERNS AND MODEL PARAMETERS

A. The Timing of Detection by Females of Active Male Strategies

One influential parameter in our model is f_m, the relative number of intolerant female's offspring that an active male can potentially father, which will depend on the type of mechanism used by males to reduce sperm competition. Some mechanisms, for example sperm plugs and postcopulatory guarding, can only be detected by the female after sperm transfer has been completed. In such cases, f_m can approach 1 and no critical frequency of tolerant females is required in order for active males to increase in frequency (Table I). Other male sperm competition avoidance mechanisms, such as the removal of sperm from previous matings before any new sperm are transferred (described by Waage 1979, for a damselfly as requiring 90% or more of the total time spent in copula) are detectable before the male achieves any reproductive success at all, and therefore $f_m = 0$. Note that the probability of remating for males (a_m) will not necessarily be high when f_m is low, because a_m will often be largely determined by the sex ratio among potentially reproductive individuals (the operational sex ratio; Emlen and Oring 1977). In such cases, high starting frequencies of tolerant females may be required, decreasing the likelihood that such a strategy can evolve (see Table I). Thus, the model predicts that strategies like that described by Waage for damselflies should be rarer than sperm competition avoidance mechanisms that are implemented after sperm transfer. Although it is not clear how common preliminary sperm removal strategies are throughout the animal kingdom (*e.g.,* the mammalian penis may function in a comparable fashion; Parker, this volume), at least in insects pre-insemination sperm removal does not appear to be common, despite the fact that the morphological and behavioral specializations do not seem to be extraordinarily complex.

B. Sperm Competition, Certainty of Paternity, and Paternal Investment

Recently, there has been considerable interest in the relationship between certainty of paternity and paternal investment (see Werren *et al.* 1980). On one side, several authors have pointed out that in species in which male parental care is well developed, certainty of paternity is also quite high (*e.g.,* Loiselle and Barlow 1978; Ridley 1978; Alexander and Borgia 1979; Gwynne, this volume). On the other, theoreticians have argued that there is no simple reason for high certainty of paternity to favor the evolution of male parental care or vice versa (Maynard Smith 1978, Grafen 1980, Werren *et al.* 1980). The existence of this pattern and the lack of an obvious explanation has prompted theoreticians to consider other models with more complex assumptions (Werren *et al.* 1980; Knowlton 1982; Parker, this volume).

The model presented here predicts a potentially strong association between certainty of paternity and male parental care for two reasons. First, any substantial pre- or postcopulatory investment by the male will affect probabilities of remating for males and females (a_m, a_f); as male parental investment increases relative to female investment, a_m will increase and a_f decrease because of the shift in operational sex ratio. These changes favor the evolution of sperm competition avoidance mechanisms by making dynamics such as those shown in Fig. 5A more likely (see also Borgia 1979).

Second, certainty of paternity of passive males relative to active males (r) should also be directly influenced by particular types of male parental investment. For example, as male nutrient contributions increase, the selection pressures acting on females to achieve multiple matings also increase (because the amount of paternal nutrient investment a female receives will be determined by the number of matings that a female achieves). As the number of males that a female mates with increases, the relative certainty of paternity of passive males (r) decreases. In our model when r is low, the starting frequency of tolerant females required for the spread of active males is also low or zero (Table I), making the evolution of sperm competition avoidance mechanisms (and hence the evolution of high certainty of paternity for active males) more likely. Although the importance of nutrients to females will also make male prevention of multiple matings more costly to females

(s low), this can at most lead to sexual conflict (Fig. 5B), which may often be resolved in favor of the male's interests (Parker, this volume).

In summary, a correlation between male parental care and effective sperm competition avoidance mechanisms (high certainty of paternity) is to be expected. This correlation is achieved via the influence of paternal investment on the evolution of certainty of paternity and not vice versa; *i.e.,* our model does not predict that increased certainty of paternity should itself select for further increases in male parental care (in contrast to many previous models). Our reasoning also shows that paternal investment cannot be viewed as a single parameter to be plugged into a model; the form of the investment (*e.g.,* nutrients to female while mating vs. nutrients to or care of offspring) can influence the extent of its impact on the evolution of sperm competition avoidance mechanisms.

V. SUMMARY

Sperm competition avoidance mechanisms can be costly to females. These costs arise directly from the methods used by males or through the resulting reduction in the ability of the female to mate effectively more than once. The most widely experienced cost is probably reduction in offspring diversity which, like sexual reproduction, is advantageous when environments are unpredictable and siblings compete with each other. Female incitation of competition among sperm from several males is unlikely to be of great importance.

Overall, costs to females associated with sperm competition avoidance mechanisms will typically be less than the potential benefits of the mechanisms to males. Despite this fact, these costs may often be great enough to favor the implementation of a strategy in females whose evolutionary consequence is to reverse the direction of selection acting on males. Our model suggests that in some situations this may enable females to prevent the evolution of male sperm competition avoidance mechanisms (that are costly to females) by "manipulating" the fitnesses of male mating strategies. This evolutionary manipulation can be achieved through the ability of females in many species to determine when copulation is terminated. By terminating copulation prematurely whenever a male attempts to implement a sperm competition avoidance mechanism, the female can differentially decrease the fitness of active males relative to passive males.

For certain sets of parameter values, passive males will be fitter than active males and the evolution of sperm competition avoidance mechanisms will be prevented. For other sets of parameter values, the costs to females of sperm competition avoidance mechanisms will be less than the costs of prematurely terminating copulation, and the potential for males to increase their reproductive success by increasing certainty of paternity will be high, thus favoring the evolution of sperm competition avoidance mechanisms. This latter condition is likely to apply whenever males invest heavily in the offspring, which probably helps to explain the documented association between high certainty of paternity and high paternal investment. In still other situations the model predicts selection for sperm competition avoidance mechanisms in males and selection for resistance to such mechanisms in females. The evolutionary outcome of such a sexual conflict can only be decided by the strengths in each sex of the opposing selection pressures and by the ability of one sex to manipulate the other (Parker 1979).

Unfortunately for scientists, these two aspects of sexual conflict over sperm competition are likely to work in opposite directions (selection more intense on males but females having more physical control over their bodies), potentially balancing each other and thus making it difficult to predict evolutionary outcomes. By examining the consequences of the ability of females to determine when copulation ends, our model has reduced the domain over which these inconclusive arguments need be applied and has shown that under certain circumstances females can influence the evolution of sperm competition avoidance mechanisms.

ACKNOWLEDGMENTS

We thank V. Delesalle, M. Hatziolos, B. Keller, G. Parker, R. Smith, D. Zeh, and R. Zimmerman, who made many helpful suggestions on the manuscript, and B. Keller for his help with the figures. We especially appreciate the patience of the editor.

APPENDIX

Using the payoff matrix in Fig. 3 and following Maynard Smith (1977) and Parker (1979), active prevention of sperm competition by males will increase in frequency whenever (for $f_m < 0.5$)

$$q + (1 - q)[f_m + a_m(q[1 - f_m] + [1 - q]f_m)] > qr + (1 - q)r \qquad (1)$$

or (for $f_m \geqslant 0.5$)

$$q + (1 - q)[f_m + a_m(q[1 - f_m] + [1 - q][1 - f_m])] > qr + (1 - q)r \qquad (2)$$

—that is, when the payoff to active males exceeds the payoff to passive males. These inequalities simplify to:

$$q^2(a_m[2f_m - 1]) + q(1 + a_m - 3a_m f_m - f_m) + (f_m[1 + a_m] - r) > 0 \qquad (3)$$

and

$$q([1 - f_m][1 - a_m]) + (f_m[1 - a_m] + a_m - r) > 0 \qquad (4)$$

respectively. Similarly, female toleration will increase in frequency whenever (for $f_f < 0.5$)

$$ps + (1 - p) > p(f_f + a_f[pf_f + (1 - p)(1 - f_f)]) + (1 - p) \qquad (5)$$

or (for $f_f \geqslant 0.5$)

$$ps + (1 - p) > p(f_f + a_f[p(1 - f_f) + (1 - p)(1 - f_f)]) + (1 - p) \qquad (6)$$

—that is, when the payoff to tolerant females exceeds the payoff to intolerant females. These inequalities simplify to:

$$p^2(a_f[1 - 2f_f]) + p(s - f_f - a_f[1 - f_f]) > 0 \qquad (7)$$

and

$$p(s - f_f - a_f[1 - f_f]) > 0 \qquad (8)$$

respectively.

The expressions on the left hand side of inequalities 3, 4 and 7, 8 can be plotted against Δp and Δq respectively in order to determine the rate of change in the frequency of the male and female strategies as a function of the frequency of the strategies present in the opposite sex (see Figs. 6, 7). For each inequality, the constraints imposed by the range of values that the parameters can take (mostly between 0 and 1) limit the number of general classes of positions that each curve can take to two. The positions of the curves, which depend upon the specific parameter values, determine the evolutionary dynamics of male and female strategies. The distance of the curve from the "x-axis," at any one point, is a measure of the rate at which the mating strategies of that sex are changing in frequency (for that particular frequency of mating strategies in the opposite sex). When the curve lies above the "x-axis" strategy "Y" is increasing while when the curve lies below the "x-axis" strategy "Y" is decreasing. The points at which the curves cross the "x-axis" are therefore critical since they represent positions either side of which different mating strategies are favored. By determining for each class of curves which mating strategy is favored for each frequency of mating strategies in the opposite sex, the dynamics of the male and female strategies can be expressed as trajectories in state-space (Figs. 4,

5) and the equilibrium positions discovered. At this stage, the distinction between the f_m, $f_f \geq$ 0.5 and f_m, $f_f < 0.5$ inequalities disappears since Figs. 4 and 5 only illustrate the direction of evolution; they do not incorporate the rate of evolution of the mating strategies.

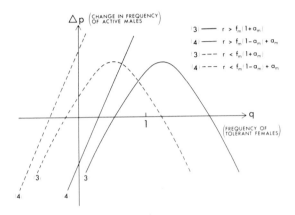

Fig. 6. The change in p (frequency of active males) as a function of q (frequency of tolerant females). For all values of q where a curve falls above the "x-axis," active males will increase in frequency ($\Delta p > 0$). The parenthetical number next to each curve indicates the Appendix inequality from which it was derived, while the key shows the parameter relationships which distinguish the pair of curves associated with each inequality. The shapes of the curves are dictated by the fact that the coefficient of q^2 in (3) must be negative and the coefficient of q in (4) must be positive. The critical values of q (q_{crit}) occur where the curves cross the "x-axis" in the range $0 < q < 1$. The sign and absolute value of q_{crit} depend upon the parameters a_m, f_m, and r.

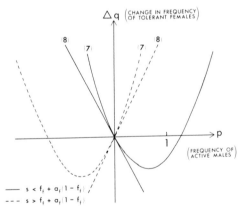

Fig. 7. The change in q (frequency of tolerant females) as a function of p (frequency of active males). For values of p for which a curve falls above the "x-axis," tolerant females will increase in frequency. The parenthetical number next to each curve indicates the Appendix inequality from which it was derived, and the key shows the parameter relationships which distinguish the pair of curves associated with each inequality. All the equations must cross the "x-axis" at $p = 0$. Equation (7) additionally crosses the "x-axis" at a biologically irrelevant point; its U-shape is required because the coefficient of p^2 must be positive.

REFERENCES

Alexander, R. D., and G. Borgia. 1979. On the origin and basis of the male-female phenomenon. In *Sexual Selection and Reproductive Competition in Insects,* M. S. Blum and N. A. Blum (eds.), pp. 417-440. Academic Press, New York.

Boggs, C. L., and L. E. Gilbert. 1979. Male contribution to egg production in butterflies: Evidence for transfer of nutrients at mating. *Science* 206:83-84.

Borgia, G. 1979. Sexual selection and the evolution of mating systems. In *Sexual Selection and Reproductive Competition in Insects,* M. S. Blum and N. A. Blum (eds.), pp. 19-80. Academic Press, New York.

Borgia, G. 1981. Mate selection in the fly *Scatophaga stercoraria:* Female choice in a male-controlled system. *Anim. Behav.* 29:71-80.

Cox, C. R., and B. J. LeBoeuf. 1977. Female incitation of male competition: A mechanism in sexual selection. *Am. Nat.* 111:317-335.

Daly, M. 1978. The cost of mating. *Am. Nat.* 112:771-774.

Darwin, C. 1871. *The Descent of Man and Selection in Relation to Sex.* John Murray, London.

Dewsbury, D. A. 1982. Ejaculate cost and male choice. *Am. Nat.* 119:601-610.

Emlen, S. T., and L. W. Oring. 1977. Ecology, sexual selection, and the evolution of mating systems. *Science* 197:215-223.

Gillespie, J. 1973. Polymorphism in random environments. *Theor. Popul. Biol.* 4:193-195.

Gillespie, J. H. 1974. Natural selection for within-generation variance in offspring number. *Genetics* 76:601-606.

Gillespie, J. H. 1977. Natural selection for variance in offspring numbers: A new evolutionary principle. *Am. Nat.* 111:1010-1014.

Grafen, A. 1980. Opportunity cost, benefit and degree of relatedness. *Anim. Behav.* 28:967-968.

Harcourt, A. H., P. H. Harvey, S. G. Larson, and R. V. Short. 1981. Testis weight, body weight and breeding system in primates. *Nature* 293:55-57.

Hinton, H. E. 1964. Sperm transfer in insects and the evolution of haemocoelic insemination. In *Insect Reproduction,* K. C. Highnam (ed.), pp. 95-107. Symp. R. Entomol. Soc. Lond. No. 2, Royal Entomological Society, London.

Huxley, J. S. 1938. The present standing of the theory of sexual selection. In *Evolution: Essays on Aspects of Evolutionary Biology,* G. R. deBeer (ed.), pp. 11-42. Clarendon Press, Oxford.

Jarvi, T., E. Roskaft, and T. Slagsvold. 1982. The conflict between male polygamy and female monogamy: Some comments on the "cheating hypothesis." *Am. Nat.* 120:689-691.

Knowlton, N. 1982. Parental care and sex role reversal. In *Current Problems in Sociobiology,* Kings College Sociobiology Group (eds.), pp. 203-222. Cambridge University Press, Cambridge.

Lloyd, J. E. 1979. Mating behavior and natural selection. *Fla. Entomol.* 62:17-34.

Loiselle, P. V., and G. W. Barlow. 1978. Do fishes lek like birds? In *Contrasts in Behavior,* E. S. Reese and F. J. Lighter (eds.), pp. 31-75. Wiley, New York.

Maynard Smith, J. 1976. A short-term advantage for sex and recombination through sib-competition. *J. Theor. Biol.* 63:245-258.

Maynard Smith, J. 1977. Parental investment: A prospective analysis. *Anim. Behav.* 25:1-9.

Maynard Smith, J. 1978. *The Evolution of Sex.* Cambridge University Press, Cambridge.

Page, R. E., Jr., and R. A. Metcalf. 1982. Multiple mating, sperm utilization, and social evolution. *Am. Nat.* 119:263-281.

Parker, G. A. 1970a. Sperm competition and its evolutionary consequences in the insects. *Biol. Rev.* 45:525-567.

Parker, G. A. 1970b. The reproductive biology and the nature of sexual selection in *Scatophaga stercoraria* L. (Diptera:Scatophagidae): V. The female's behaviour at the oviposition site. *Behaviour* 37:140-168.

Parker, G. A. 1979. Sexual selection and sexual conflict. In *Sexual Selection and Reproductive Competition in Insects,* M. S. Blum and N. A. Blum (eds.), pp. 123-166. Academic Press, New York.

Parker, G. A. 1983. Arms races in evolution: An ESS to the opponent-independent costs game. *J. Theor. Biol.*

Parker, G. A., and M. R. Macnair. 1979. Models of parent-offspring conflict. IV. Suppression: Evolutionary retaliation by the parent. *Anim. Behav.* 27:1210-1235.

Ridley, M. 1978. Paternal care. *Anim. Behav.* 26:904-932.

Rubenstein, D. I. 1982. Risk, uncertainty and evolutionary strategies. In *Current Problems in Sociobiology,* Kings College Sociobiology Group (eds.), pp. 91-111. Cambridge University Press, Cambridge.

Rutowski, R. L. 1980. Courtship solicitation by females of the checkered white butterfly, *Pieris protodice. Behav. Ecol. Sociobiol.* 7:113-117.

Schuster, P., and K. Sigmund. 1981. Coyness, philandering and stable strategies. *Anim. Behav.* 29:186-192.

Smith, R. L. 1979. Repeated copulation and sperm precedence: Paternity assurance for a male brooding water bug. *Science* 205:1029-1031.

Thornhill, R. 1976. Sexual selection and paternal investment in insects. *Am. Nat.* 110:153-163.

Trivers, R. L. 1976. Sexual selection and resource-accruing abilities in *Anolis garmani. Evolution* 30:253-269.

Waage, J. K. 1979. Dual function of the damselfly penis: Sperm removal and transfer. *Science* 203:916-918.

Walker, W. F. 1980. Sperm utilization strategies in nonsocial insects. *Am. Nat.* 115:780-799.

Wallace, B. 1975. Hard and soft selection revisited. *Evolution* 29:465-473.

Werren, J. H., M. R. Gross, and R. Shine. 1980. Paternity and the evolution of male parental care. *J. Theor. Biol.* 82:619-631.

Williams, G. C. 1975. *Sex and Evolution.* Princeton University Press, Princeton, NJ.

3

Sperm in Competition

JOHN SIVINSKI

Sperm Competition and the Evolution
of Animal Mating Systems

85

I. INTRODUCTION

The sperm of external fertilizers are usually simple in design and largely similar (Franzen 1956, 1970). By contrast, internal fertilizers produce an often bizarre gametic fauna that includes giants extending many times the length of the male, immobile dwarfs, active sperm lacking genomes, and polyflagellated projectiles. An attractive explanation for this diversity is the relatively greater variance among internal female genitalic environments as compared with the homogeneity of open water. Additionally, sperm deposited internally may have to be equipped for various stays in the female tract and structured to penetrate eggs whose surfaces and chemistries are adapted to a variety of oviposition habitats (in part, Baccetti 1972, Cohen 1977).

Such elemental diversities of time, terrain, and proximate mission, however, may be insufficient to explain the range of sperm morphology and behavior. For example, Cohen (1977) found a surprising inverse relationship between the complexity of flagellar fiber arrangements and the distance the gamete must swim. Head and midpiece structures are frequently most elaborate where the journey is the least difficult, and no particularly clear correlation exists between the relative complexities of sperm and the egg membranes they penetrate (Cohen notes that the very complex sperm of gastropods penetrate eggs that are "effectively naked").

What other factors are responsible for evolution of diversity in sperm? Competition for fertilization is a likely context. Consider again the problem of gamete morphology in external and internal fertilizers. Certain antagonisms are likely to be more acute among internal fertilizers. Competition between sperm concentrated in a storage organ or other niche of the female system could be aggressively resolved by physical or chemical means unlikely to be effective in the more diffuse cloud of an external ejaculate, and the short time between gamete release and external syngamy restricts the scope of competition between ejaculates. Whether fertilization is internal or external there are circumstances where females will be confronted with multiple ejaculates. When females retain their eggs they are more able to choose the paternity of their offspring through sperm manipulation. If so, internal fertilization places a premium on sperm that circumvent such choice by force or "deceit." The more abundant opportunities for internally fertilizing adults or their sperm to influence sperm competition may be partially responsible for the adaptive radiation of their gametes.

While opportunities for competition may be greater in internal fertilizers, the motives are universal. There are at least five ways competition could have influenced gametic evolution and these occur in or between three levels of organization: competition among ejaculates; within ejaculate conflict; conflict between the male parent and his gametes; competition between sperm and choosing females;

and competition between nuclear programming and organelles located in sperm. The following is an attempt to discern the mark of such conflicts on the forms of gametes, the structures of their populations, and the architecture of the organs that deal with them.

II. INTRAEJACULATE COMPETITION AND MALE/GAMETE CONFLICT

A. Competition Among Sibling Sperm and the Loss of Gamete Individuality

Sperm have a dual nature. As cells of a male's body, they serve as tools of his reproductive interests. At the same time, they are microorganisms typically with one of what can be quite a number of possible genomes. In the latter role, they encounter competition for ova from tens to tens of millions of siblings.

Selection at the gametic level may not be harmonious with selection on the diploid adult. A male's interest is to obtain the greatest possible number of fertilizations. Competition among his sperm might initially be a matter of indifference, but means of resolving competition could evolve that would lower the quality of the ejaculate as a whole. A gene expressed in a gamete that poisoned, injured, or stole resources from sibling sperm can increase in frequency (carnivorous protozoa and antibiotics hint at possible aggressive forms).[1] Because the sperm of an ejaculate have a 0.5 probability of sharing the identical "violent" allele, damage to the ejaculate could increase until the cost balanced the benefits of less competition due to fewer gametes (similar to sibling conflicts described by Trivers 1974). Where violent sperm are immune to their own weapons or "recognize" and preferentially destroy nonviolent sperm, ejaculate efficiency could sink to half that of sperm sibships not containing a "violent" allele. When more successful alleles are sex-linked, the number of zygotes per ejaculate need not be reduced for diploid interests to be challenged. Offspring sex ratios will be unadaptively distorted when one sex chromosome gives an advantage to its bearer over the vessel of its homolog (Hamilton 1967, Maynard Smith 1978).

Conflicts between diploid and haploid generations in animals and higher plants might have a predictable outcome. An individual that provides resources to another

---

[1] Rothstein (1979) argues that inhibitory traits should be rare, since their value is directly correlated with their frequency. That is, when inhibitory alleles arise, their ability to influence the average performance of large numbers of competing genes is apt to be low compared with the cost involved in inhibiting. This argument lacks force when applied to gametes. Among sperm, a mutation that survives one generation will automatically occur in 50% of the next population, *i.e.*, half the sperm in an ejaculate of a male heterozygous for inhibition will be inhibitors.

possesses a means of manipulating the recipient (Alexander 1974). A sperm's body consists of cytoplasm obtained directly from the male. There would be abundant opportunities for an adult to adjust the cellular machinery of its sperm to muzzle gametic gene expression.

Male suppression is a possible explanation for the apparent inability of the sperm genome to influence its phenotype (Sivinski 1980; Crow 1979, who argues that "Whatever the evolutionary reasons for the nonfunctioning of genes in sperm cells may be . . ." expression of sperm genes would result in more opportunities for adaptations to arise that damage siblings; Dawkins 1982). The evidence for the absence of haploid effects, *i.e.* transcription of genes within sperm, is considerable (*e.g.,* citations in Beatty 1975a, b). McCloskey (1966) and Lindsley and Grell (1969) in continuation of Muller and Settles' 1927 work have shown that *Drosophila melanogaster* sperm nearly devoid of chromosomes can differentiate and function. When matched with eggs that have a corresponding excess of chromosomes, normal individuals can be produced. Mouse sperm with whole chromosome duplications and deletions are functional and capable of fertilizations (Ford 1972). Variances of morphological characters of sperm in inbred (homozygous) and outbred (heterozygous) lines of mice are not significantly different, demonstrating that genotypic diversity need not result in phenotypic variety (Beatty 1971, 1975; Pant 1971).

There are suggestions, however, that sperm genes may not be completely mute. Whatever results in the greater than average success of *t*-allele-bearing sperm in heterozygous rodents appears to take place sometime between meiosis and fertilization (Braden 1958, 1960, 1972; Yanagisawa *et al.* 1961). Differences in *t* and + sperm head antigens have been found (Yanagisawa *et al.* 1974). DNA-dependent RNA polymerase has been discovered in mouse late spermatids (Moore 1972; see also section IV). Perhaps during evolutionary conflicts resulting from different and conflicting directions of selection, gametic genes break through the diploid's blockade and enjoy temporary expression (see the discussion of "arms races" in Dawkins and Krebs 1979).

Microgametophytes (pollen) are, in many ways, functional equivalents of sperm. It is puzzling, then, that haploid effects are common in pollen (Mulcahy 1975, 1979; Mulcahy and Kaplan 1979; see section II). Differences in the population structure of the two forms of "gametes" might account for the disparity in haploid expression. Sperm are typically in close proximity to large numbers of siblings, whereas pollen can be more diffusely spread. "Ejaculate" mixture in wind pollinating species seems likely. Even the dense masses of pollen on bees often come from several plants, increasing the likelihood that after deposition on a style, nearby competitors are unrelated. The opportunity for pollen to affect siblings unfavorably may be considerably less than for sperm (Sivinski 1980; a notable exception occurs in plants where large numbers of sibling gametophytes are packaged together in pollen dispersal units: see Wilson 1979). The mechanics of insemination make plants more vulnerable than animals to self-fertilization. Expression and recognition of the haploid genome can prevent inbreeding (see Heslop-Harrison 1975, also Bremermann 1980).

Selective asymmetries between levels of organization will influence the frequency and form of adaptations. In sperm, because of the opportunities for "parental" manipulation, we might find more "altruism" and less aggression than predicted by the coefficient of relatedness between gametes.

B. Sperm in Groups: Cooperation, Manipulation, and Evidence for Diploid-haploid Conflict

A modest-size metazoan's abilities exceed the summed capacities of half a billion conspecific protozoa. With this in mind, Cohen (1975) asked why redundant sperm, supposedly produced as a shotgunlike means of hitting upon an ovum, have not been selected to form multicellular coalitions better at fertilization (in the case of man, ". . .a planarianlike organism exquisitely suited for finding the egg."). While not attempting to answer Cohen's query (see Sivinski 1980, and citations; Parker, this volume), it should be pointed out that sperm occasionally do function in groups, and that castes of specialized gametes form conglomerates of some, if less than metazoan, complexity (see Fig. 1).

Under certain circumstances, ejaculate efficiency can be enhanced by grouping sperm into multicellular units. For participating gametes, however, group existence bears a price, the closer proximity of competitors (see Alexander 1974, discussion of benefits and drawbacks to sociality). Males and sperm might disagree as to the extent of cooperation (again, see Trivers' discussion of parent/offspring conflict, 1974). Where some sperm, by performing a specialized role within the group, actually lower their own chances of fertilizing an egg, the potential for conflict between haploid and diploid generations is particularly great. Adaptations that appear to be resolutions of such conflicts suggest a history of haploid expression and diploid suppression.

There are several multi-sperm formations that seem to be aggregates of equals. All potentially increase their probability of penetrating an egg by their association. Some of this grouping is clearly produced by the metazoan. Sperm, particularly insect sperm, are commonly packaged in bundles that break apart in the vas deferens or female tract. These packets are sometimes mobile and have been mistaken for single sperm (Nur 1962). Bundled gametes are often amassed under proteinaceous caps that may serve as sources of nutrition or concentrations of enzymes that break down genital tract secretions for assimilation (citations in Mackie and Walker 1973). A cluster consists of descendants from a single spermatocyte that have remained in proximity through development. Their numbers depend on how many mitotic divisions occur prior to meiosis (e.g., in the grasshopper

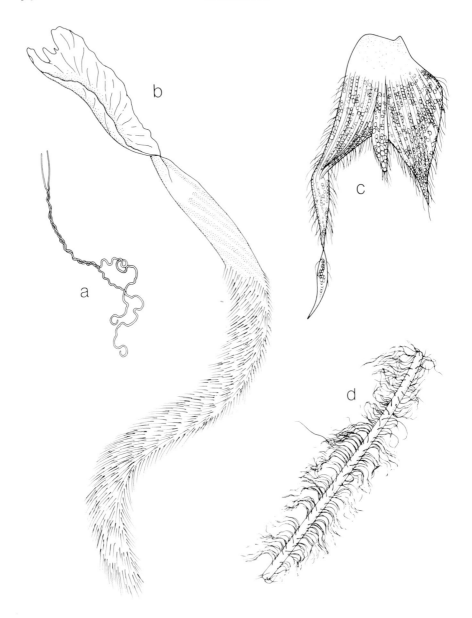

Fig. 1. Sperm in groups (not to scale).   A. Paired sperm of the silverfish, *Thermobia domestica* (redrawn from Bawa 1964). B. The apyrene sperm of the mollusk, *Clathrus clathrus,* and its fertile eupyrene passengers (redrawn from Fretter 1953). C. Another atypical molluscan sperm (*Cinctescala eusculpta*) with typical sperm attached (redrawn from Nishiwaki 1974). D. Spermatostyle and associated spermatozoa of a gyrinid beetle, *Dineutus* sp. (redrawn from Breland and Simmons 1970). Drawing by Susan Wineriter.

Melanoplus sp., seven premeiotic divisions result in 512 sperm. See White 1955, Virkki 1969, Kurokawa and Hihara 1976, for the peculiar relationship between bundle number and phylogenetic position; also Sivinski 1980a).

Other relationships occur where free, mature gametes join together in the ducts of the male reproductive system. Sperm of some whirligig beetles (Gyrinidae) become attached in groups of 100 or more to rodlike objects, spermatostyles, as they pass through the vas deferens (see Fig. 1d). Partially disintegrated rods have been recovered from spermathecae. Spermatostyles are moved by the efforts of their crew, but loose sperm are also motile (Breland and Simmons 1970). The function of these aggregates is obscure. The sperm may benefit from material in the spermatostyle, or perhaps they deliver the rod as a parental investment to the site of female absorption (see also section III). The rods might even be useful in competition with rival ejaculates, perhaps as a means of allowing large groups of siblings to force themselves toward and then occupy privileged positions in the spermatheca.

Paired spermatozoa are a rare, but widely distributed, phenomenon. Pairing may occur in the testes, *e.g.,* the silverfish *Thermobia domestica* (Thysanura), or during passage down the vas deferens, as in the water beetle *Dytiscus marginalis* (Bawa 1975). In *T. domestica,* sperm are only motile when anterior thirds of a pair are entwined (Bawa 1964; see Fig. 1a; similar union occurs in another silverfish *Lepisma saccharina:* Werner 1964). *Dytiscus marginalis* sperm are motile as individuals but have adaptations for paired life, including spurs that mesh at points of attachment (Mackie and Walker 1974; Bawa 1975). Other gametes typically joined at the head with tails and midpieces free are found in the millipede *Polydesmus* sp. (Reger and Cooper 1968), the mollusk *Turritella* (Idelman 1970), woolly opossum (Phillips 1970), and opossum (Holstein 1965; for the numerologist, marsupials have two-channeled vaginas, bifurcated penises, and some have sperm that swim in pairs). In the beetle *Colymbetes,* groups of three or more are formed (Mackie and Walker 1974).

Pairing probably represents some sort of cooperation, but to what purpose is unclear. Two flagella might generate more motive power (Afzelius 1970). The sperm of the millipede *Polydesmus,* however, are odd, crescent-shaped objects without flagella, and apparently unmotile (Reger and Cooper 1968). Other suggested functions include protection of each others' acrosomes (the tip of the sperm, important in fertilization) or enhanced ability to penetrate an ovum (Mackie and Walker 1974).

Assuming probably unrealistically high levels of haploid expression, partnerships might become increasingly uneasy as fusion of egg and sperm pronuclei approaches. Given sperm are not genetically identical products of the second spermatocyte, the cost of having an attached competitor may, at some point, outweigh the advantage of its cooperation. Both sperm are equally related to the male, who should program gametes to remain together until all benefits of the coupling are realized.

A potential conflict could arise over the point where siblings are to be jettisoned. The presence of a common rival, another male's ejaculate, might modify gametic selfishness and select for extended unions. It is better to aid a "sibling" gamete within range of fertilization than to compete as an individual, on an approximately equal basis, with unrelated gametes.

Among haploid-expressing sperm, then, the length of pairing might be directly related to the degree of sperm competition between or among males. While hardly compelling as evidence, the two known pairing durations of insect sperm are worth comparing. *Dytiscus marginalis* sperm are paired in the male tract, but are mostly single in the spermatheca (Mackie and Walker 1974, Bawa 1975). *Thermobia domestica,* however, are found together in female storage organs (Bawa 1975). While multiple matings are typical of female silverfish, the potential for sperm competition is not necessarily high. Females must replenish sperm supplies shed at each adult ecdysis (Sweetman 1938, Hinton 1981). Male *D. marginalis,* during days-long couplings, fill their mates' vaginas with bright white cement that hardens on exposure to air. Females are "corked shut" from fall to winter, suggesting very low levels of sperm competition (Balduf 1935).

Unlike cooperating equals there are a number of specialized sperm morphs that, by nature of their role, have little probability of obtaining fertilizations and constitute a sort of "worker caste." The enormous transport cells of some mollusks are cases of extreme dimorphism. They have literally become "mobile penises," carrying thousands of brother gametes upon their tails. These mammoth vessels swim up the female tract as far as the ovaries, liberating smaller sperm as they go (Fig. 1b and c; Fretter 1953). Large, multiflagellated sperm of certain snails entangle masses of normal gametes, either concentrating or helping to propel siblings during copulation. Once in the female, their midpieces degenerate, perhaps to provide nutrition for siblings (Woodard 1940, Hanson *et al.* 1952; extensive reviews in Nishiwaki 1964, and Fain-Maurel 1966). Other sperm morphs in mollusks are fusiform objects of little motility, which may serve solely as food sources for ejaculate mates (Pelseneer 1935, Hanson *et al.* 1952). Among Lepidoptera, a class of atypical sperm probably "displace or inactivate eupyrene sperm from previous matings or delay further mating by the female" (R. Silbergleid *et al.,* pers. comm.).

Specialized sperm serving peculiar, nonfertilizing functions are also genetically unusual. They either lack genomes altogether (**apyrene** sperm) or contain only part of the normal chromosomal complement (**oligopyrene** sperm; gametes with complete haploid genomes are **eupyrene**).

Is it possible that incomplete nuclei result from parental pruning of rebellious gametic genomes? Modification of a sperm for a supportive role results in a lower probability that it will participate in a zygote. A "gene" that allowed a "worker" spermatocyte to complete normal fertile development at the cost of ejaculate as a whole might increase in frequency if the greater success of helped "siblings" did not offset the cost of sterility; that is, among the gametes of diploids the probability of sharing a gene is typically 0.5. Such a gene would be an outlaw, resistant to "parental" manipulation that forces cells into forms with less than average chances of obtaining fertilizations. A male could squash gametic rebellion by removing damaging genes from his sterile caste; that is, take the genome from the cell.

An absence of chromosomes could indicate a history of haploid gene expression acting in the context of parent-offspring conflict. A test of the argument lies in nuclear constituency of specialized sperm in haplo-diploid species. Since sperm of a haploid male are genetically identical, there should be no selection for gametic rebellion or genetic disenfranchisement (see also Starr, this volume). I know of only one relevant case.[2] Male rotifers are thought to be haploid and practice a sort of traumatic insemination. The first objects to enter the female are rigid rods produced by atypical, immobile spermatozoa. These sperm degenerate after extruding the rod and never encounter ova (Koehler and Birky 1966; the projectiles may be useful in breaching female tissue: Koehler 1965). Atypical sperm have "intact nuclei."

C. Haploidy and Competition in Ova

Eggs are another class of gametes that could provide comparative evidence for male/sperm conflict. Ova rarely compete for fertilization and so might, by previous argument, escape suppression by the diploid genome. Do haploid effects, perhaps related to choice of sperm, occur in eggs? An answer of "not likely" is indicated by the lack of time in which a haploid ovum genome could express itself. In animals as diverse as pigeons, dogs, and roundworms, expulsion of even the first polar body takes place after penetration by the sperm. Only rarely does the second division

---

[2] One type of specialization, extra chromosomes, automatically renders sperm infertile. Polyploids could provide additional amounts of nucleic acid constituents and associated proteins to zygotes formed by eupyrene ejaculate mates (see Fain-Maurel 1966; and citations). Because males passing such sperm run the risk of losing offspring by producing polyploid zygotes, this form of investment might arise only when ova discriminate against sperm with chromosomal abnormalities (see Ito and Leuchtenberger 1955, for evidence of such discrimination in a clam; chromosomal abnormalities may be very widespread: Cohen 1975). Once such morphs are established, they might evolve further specializations as "helpers." Since gene expression is not necessarily curbed, it is difficult to imagine polyploidy as a means of enforcing parental dictation of a helper role.

A number of invertebrates produce diploid, tetraploid, or higher ploid sperm (sperm with excess chromosomes are referred to as hyperpyrene; see Fain-Maurel 1966). They are usually larger versions of eupyrene sperm, leading to suspicions that they result from meiotic error. Some, however, are made in species-specific numbers and are occasionally very abundant. In the annelid *Allolobosphora caliginosa* they ". . .smother the typical line with their number" (Chatton and Tuzet 1943). Others bear morphological peculiarities suggestive of specialized roles. Certain carabid beetles, for instance, have hyperpyrene sperm with multiple flagella (Bouix 1963). There is nothing to prevent parental investment via hyperpyrene sperm in haplo-diploid species. Several wasps and ants have giant polyploid gametes (citation in Fain-Maurel 1966; see section III).

and haploidy precede fertilization (Cohen 1971, 1977). This last-minute finale to oogenesis suggests that (1) ovular choice or other possible haploid expressions are unimportant and that reduction is selected to occur only with the imminent arrival of the sperm pronucleus, or (2) haploidy is disadvantageous and selection favors retention of the diploid genome.

Among the costs of reduction is the loss of nuclear programming. The "lampbrush" configurations of chromosomes during oocyte growth and organization suggests active RNA transcription, a poor time for discarding genetic material. Chromosome repair may only be possible in the presence of a homolog making haploids vulnerable to lesion (Bernstein *et al.* 1981). Perhaps less plausible but in keeping with the spirit of this discussion are certain risks inherent in haploidy prior to the completion of cell provisioning. A gene appropriating excess resources could leave other eggs with an alternative allele undervictualed, decreasing the fecundity of the parent. The diploid can preclude female-egg conflict by limiting the range of an "outlaw" gene's expression, *i.e.* by keeping all mutations in the company of cooperative, probably dominant alleles. Reduction to haploidy almost always takes place outside of the ovary, the oocyte's source of supply.

A possibility of post-oocyte growth competition for resources exists as well, and might aid in explaining the rarity of sexual dimorphism in zygotes, *i.e.* offspring at the end of parental investment in the absence of parental care. When males determine sex, the danger of the wrong "sexed" sperm penetrating an inappropriate investment may force homogametic females into equal distribution of resources among her ova (Sivinski 1980; assuming eggs/females can't effectively choose sperm on the basis of the sex chromosome they contain; see section III). On the other hand, heterogametic females might be vulnerable to intragenomic competition, where genes on sex chromosomes might strive to acquire a richer cytoplasm than is in the parents' interest. In the two unambiguous instances of egg sexual dimorphism I'm aware of, gender is determined maternally but not by simple heterogametic sex chromosomes. Female eggs of the marine worm *Dinophilus apatris* are much larger than male eggs. Sex determination occurs prior to meiosis, perhaps due to the balance of multiple factors in the maternal genome (Bacci 1965). In the peach scale *Pseudolacaspis pentagone,* male eggs are white and female eggs coral pink, indicating some difference in the nature of parental investment. Sex is determined cytologically (Brown and Bennett 1957). A case resembling a predicted conflict occurs in the common grackle, where heterogametic females tend to produce larger male eggs but the relationship between size and sex is not invariable (Howe 1976, 1977).

Oddly enough, considering its breathless finish, ovarian meiosis often begins early in female development and then pauses shortly after the onset of the diplotene stage of prophase I, the orientation of homologous chromosomes into tetrads (Austin and Short 1972). In humans, for example, meiosis starts in the female embryo but is completed following ovulation, a hiatus of up to 50 years. Stop and go meiosis can be seen either as a brake to impending haploidy (at loci not affected by crossovers) or as an early chromosomal assortment driven by competition to occupy that portion of the oocyte destined to become the ovum and avoid the genetic death of inclusion in the first polar body. This latter view requires predetermination of chromosomal fate early in meiosis. The first chromosome to occupy the favorable location in the cell would have won the first round in its struggle for perpetuation to the certain cost of its homologue. Jockeying for position could force such a critical stage of meiosis father and farther back into development (it is tempting to consider chromosome crossovers being manipulated by a polar body-bound allele to attach itself to a more fortunate chromatid and consign its competitor to the void).

III. INTEREJACULATE COMPETITION

Phenotypic characteristics of sperm dictated by the male's diploid genotype are referred to as diploid effects. In the absence of haploid expression, sperm competitions become contests between extensions of adult genomes. As might be expected, different males produce gametes of different abilities. When female chickens, mice, rabbits, and cattle are artificially inseminated with similar numbers of sperm from two or three males, sperm from particular animals are consistently more effective in obtaining fertilizations (Beatty 1975a, b; and citations: "It is an exception for equal numbers of sperm to produce equal numbers of offspring").[3]

For diploid effects to evolve by interejaculate competition, adults must potentially place unrelated sperm together. Variance in the characteristics of the sperm themselves might provide the motive for multiple inseminations. Ejaculate mixture could be "deliberate." Discriminating females might actively arrange competition between gametes. Proximity leading to conflict could also be merely the effect of selection acting in other contexts.

Note that unusual sperm are occasionally found in taxa with unusual opportunities for ejaculate mixture, e.g. the termite *Mastotermes darwinensis* whose colonies contain multiple reproductives (see Sivinski 1981 and citations), and birds of the snipe family Scolopacidae whose females produce multiple clutches over relatively short periods of time, each preceded by mating with a different male (J. van Rhijn, pers. comm.; see McFarlane 1963).

---

[3] In the wasp *Dahlbominus fuscipennis,* gametes from males bearing a mutant have an advantage over wild type competitors (Wilkes 1966). Diploid effect is an inappropriate term, since males are haploid, but this is another example of male genotype correlated with the competitive ability of sperm.

A. Female Encouragement of Sperm Competition

Gametophyte (pollen) fitness can be related to the fitness of the sporophyte it cooperates in producing. Genes that are transcribed in both haploid and diploid genomes have an "overlapping" pattern of expression (Mulcahy 1979). To the extent that selection acts similarly in the two contexts, fitness of an overlapping haploid is a predictor of fitness in the diploid and, presumably, vice versa. In corn, for instance, the speed of pollen tube growth is positively correlated with seedling weight, ear weight, and kernel weight of the resultant plant (Ottaviano *et al.* 1980).

Females could benefit in such instances by passively accepting fertilizations by victors of gametic competition. Statistically, better fathers could be obtained for offspring by pitting together large numbers of competitors. Paternal filtering through intense competition may be partially responsible for success of angiosperms which receive massive, simultaneous doses of pollen from insect symbionts (Mulcahy 1979).[4] Selection of gametes could be intensified if females produce a challenging genital environment in which only the most "vigorous" could reach an egg (again, see Mulcahy 1979, for a discussion of the importance of stylar tissue as an "impediment" to fertilization; see also section III for discussion of female choice of compatible sperm genomes).

An equivalent form of selection could occur in animals if sperm fitness predicts offspring quality. Females might solicit ejaculates from numerous males without any precopulatory regard for their quality (this supposes heritable differences in sperm competitiveness perhaps due to recurrent mutation: see Borgia 1979).

Gamete fitness need not guarantee its genetic program's success in a diploid genome. Alleles might have favorable expression in gametophyte or sperm but bear disastrous pleiotropic effects into the diploid. Evening primrose demes are thought to contain genes that result in inferior winter survival but that are maintained by their prowess when part of pollen grain genomes (Henbert Nilson 1923, in Haldane 1932). Mouse sperm carrying *t*-alleles are more likely to win fertilizations, but resultant animals bear effects ranging from short tails to homozygotic sterility (Lewontin 1970; see Alexander and Borgia 1978).

Genes expressed in the haploid generation may not be expressed in the diploid (nonoverlapping, in terminology of Mulcahy 1979). It would seem that the greater the difference between the environments of the haploid and diploid generations,

---

[4]Cohen (1967, 1971, 1975, 1977) has long championed a related argument, that dangers inherent in crossovers leave only a few sperm with intact chromosomes in each ejaculate, and that the female can recognize and/or filter these individuals through haploid effects. He supports his position with a significant correlation between the number of sperm per egg and the number of chiasmata per genome (see, however, Wallace 1974; Parker, this volume). If haploid effects are common, the correlation could also be consistent with males providing undamaged haploid genome samples to choosing females. The greater the potential genetic variance (*i.e.,* the recombination rate) within an ejaculate, the larger the ejaculate would need to be to contain the best possible sperm for a particular egg or female (Sivinski 1980a).

the more likely it would be for distinct sets of genes to evolve for the production of organism and gamete. Sperm in the internal genitalia of a female probably have less in common with the organism that deposited them than male microgametophytes and sporophytes of higher plants whose open air and multicellular worlds overlap. If so, the portion of the diploid genome reflected in the sperm phenotype may predict very little about the nature of the genes that produced the diploid phenotypes.

Where all correlations between sperm and offspring quality are equally possible, females could select mates at both individual and gametic levels. Even occasional exposure of ova to the sperm of two or more males would generate selection for females to produce sons not only with the ability to obtain mates, but also able to pass a competitive ejaculate. Information on an adult male's quality could probably be more easily obtained than insights into the uncorrelated qualities of his ejaculate. Females might discriminate at the two levels by choosing several of the best available adult phenotypes and allowing their sperm to compete.

Some patterns of mating are consistent with expectations of competition encouragement. Mated short-tailed crickets, *Anurogryllus arroreus*, respond to calling males and, presumably, mate more than once before any eggs are laid (T. Walker 1980). Other copulatory strategies suggest gametic level selection is unimportant. *Drosophila melanogaster* remate only after ca. 78% of the first ejaculate is exhausted (Gromko and Pyle 1978). Subsequent fertilizations are accomplished by sperm that have had the opportunity to compete, but the late introduction of the second ejaculate is hardly indicative of an arranged sperm confrontation.

Where females accept the results of competition between sperm or encourage it by gathering antagonists and providing arenas, the paternity of offspring should be independent of the order of mating. In insects and other taxa, the last, sometimes the first, males to copulate obtain the majority of subsequent fertilizations. Predictable paternity on the basis of order demonstrates adult influence over fertilization (Parker 1970, males protecting paternity generate order effects; Walker's 1980 argument for female control of ordering not relevant here, since females would abandon ordering to maximize competition). Order effects do not prove that competition between sperm is absent. Adult stratagems could obscure the smaller struggle within. Low variance, *i.e.*, high predictability, in last or first male fertilization rates, however, would refute the contention that competition between sperm is important in determining paternity (note that high variance could result from differences in competitive ability of adults, not necessarily differences between sperm). Measurements of variance in natural populations of insects have apparently not been made (see Walker 1980). Among rodents, variance in order effects seem to be considerable (*e.g.*, Levine 1967; Dewsbury, this volume).

B. Incidental Proximity of Ejaculates

Females may mate with more than one male for reasons other than arranging competition or replenishing completely exhausted sperm stores. An effect of such behavior is placing together unrelated ejaculates. A result of proximity would be selection for sperm to obtain fertilizations at the cost of rivals. Walker (1980) has reviewed the adaptive significance of multiple mating. Advantages include increasing genetic diversity, mating with males of increasing genetic superiority, increased predation protection for females (see Sivinski 1980b, for evidence of greater survival in mating female stick insects), and transfer of nutrients. This last benefit is of particular interest in considering some curious aspects of sperm morphology. Multiple copulations in the context of female mate choice, as opposed to filtering mates through sperm competition, are considered in section IV.

C. Paternal Investment Via Sperm:
Implications for Sperm Competition

The number of sperm accepted by a female usually exceeds the number of ova she will produce. One possible reason for the redundancy is the use of additional sperm as a parental investment (see Gwynne, this volume). A female, for example, might store and release multiple sperm as a nutritional boost to the zygote (Afzelius 1970). Polyspermy, more than one sperm penetrating the ovum, occurs in selachians (Ruckert 1899), urodeles (Jordan 1893), reptiles (Oppel 1892), bryozoans (Bonnevie 1907), birds (Blount 1909), and insects (Richards and Miller 1937; citations in Mackie and Walker 1974; see, however, Lefevre and Jonsson 1962).

Insect sperm often contain large mitochondria whose configuration is radically changed to a crystalline form during spermiogenesis. Peculiarities of the crystal are consistent with a paternal investment. They do not possess the biochemical activity of mitochondria, are structurally stable during the life of the sperm, and are metabolically inactive in terms of sperm activity. Many contain large amounts of protein and are completely absorbed by the zygote (Perotti 1973, Bacetti et al. 1977).

Some male animals, primarily arthropods, produce huge gametes that might contribute resources to zygotes regardless of the selective reasons for their gigantism (see following subsection on sperm polymorphism). In the featherwinged beetle, *Ptinella aptera,* they are twice as long as the adult (Taylor 1981, 1982). Other ptiliids have sperm that are only slightly less enormous relative to body size. *Bambara invisibilis* females can store only ca. 28 sperm (Dybas and Dybas 1981). At 14 mm some sperm of the house centipede, *Scutigera forceps,* are longer than the testes and nearly as long as the animal itself (Ansley 1954). Painters frog,

Discoglossus pictus, has 2.3 mm sperm; bundles of sperm are visible with the naked eye (Favard 1955). Ostracod spermatozoa can be relative titans, 10 mm long, many times the male length (Bauer 1940). *Drosophila melanogaster* sperm average 1.7 mm; in other *Drosophila* they reach ca. 15 mm (Beatty and Burgoyne 1971, Perotti 1975). Mitochondrial derivatives compose 50% or more of their volume (Perotti 1973). The club-shaped sperm of ticks range from 1 to 7 mm (Rothschild 1961, Austin 1965). Heavy-bodied sperm of the clerid beetle *Divales bipustulatus* stretch 10 mm. Most of the tail is composed of derivatives (Mazzini 1976). Sperm of the backswimmer *Notonecta glauca* are 15 mm long, with a volume of 38,000 μ^3. Mitochondrial crystals occupy 90% of these sluggish giants (Afzelius *et al.* 1976).

Peculiar nonfertilizing gametes may be forms of paternal investment. Oligopyrene sperm in some molluscs may penetrate trophic eggs, ova that are later consumed by sibling larvae (Dupouy 1964); they might do so to enhance the eggs' value as food. Instances of pseudogamy (parthenogenesis "stimulated" by sperm) are also suggestive of paternal investment through gametes. The sperm makes no genetic contribution but must penetrate the egg for development to proceed. Perhaps eggs pirate material from male gametes (see Lloyd 1979, Sivinski 1980a). The relative size of the egg and sperm differs across pseudogamous taxa. As dimorphism increases, the less likely it is that sperm can contribute significantly to the zygote (assuming no polyspermy). The entire ejaculate, however, could still be nutritionally useful to the female.

There are advantages to males who invest directly in the zygote with enriched sperm rather than indirectly, through substances absorbed by the female. Indirect investments increase future fecundity, but may provide for ova a male does not fertilize due to subsequent inseminations by other males. A shortened time between resource transfer and deposition in a zygote minimizes the probability of a female dying before producing the investing male's offspring.

Storage difficulties inherent in small animals maintaining large, perhaps resource-laden, sperm might select for frequent copulations and the mixing of rival ejaculates. *Drosophila melanogaster* sperm are relatively big and copulatory patterns indicate females remate to replenish diminished sperm stores (*e.g.,* Gromko and Pyle 1978).

Sperm and ejaculatory fluids might be used by a male as a nuptial gift to a mate. Rather than benefiting zygotes, resources would serve as investments in the fecundity and good health of his offspring's mother (see Thornhill 1976, 1980; Gwynne, this volume, for discussions of paternal investment). Bedbugs probably use the massive ejaculates they receive as food (Hinton 1974). In the purple martin ectoparasite *Hesperocimex sonorensis,* females inseminated by well-fed males produced three times as many eggs as those paired with starving males (Ryckman 1958). Further evidence of ejaculate value might be inferred from the homosexual relationships in Cimicidae. Males of the bat bedbug *Afrocimex* are transvestites, with pseudofemale paragenital structures that apparently invite homosexual attentions.

Hinton (1964) felt these sodomies were a means of food sharing, an inappropriate argument since it proposes the feeding of sexual rivals. Aid in stealing an ejaculate for consumption might account for the mimicry (see Lloyd 1979, who argues that inviting male attentions reduces the fertilizing capabilities of rivals). Some insect sperm storage organs suspiciously resemble stomachs. The spermathecae of morabine grasshoppers are capable of holding 800 times the ejaculate of similar-sized relatives. Much of the proteinaceous material held by the spermatheca is destroyed by a range of enzymes (Blackith 1973).

Where females benefit directly from male investment, and zygotes only in-directly, males become particularly vulnerable to exploitation. A female collecting seminal material for her own nutrition might harbor the remains of several un-related ejaculates competing for fertilizations (see, however, Gwynne, this volume, whose correlation between paternal investment and assurance of paternity would indicate lower than expected levels of competition).

D. Strategies of Competition: Getting There

Cohen (1977) has remarked on a puzzling negative correlation between flagellar complexity in sperm and the distance of their average journey. The epitome of his paradox are the globular spermatozoa of cycads. Covered with a fur of flagella, they travel about half the length of their body. (Arthropod sperm are sometimes much longer than the journeys they undertake.)

When interejaculate competition is possible, emphasis on distance is misleading. No matter how short the race, the object is to win. Among competitors starting together, *i.e.* at random, distance is immaterial compared with relative velocity. Even co-occurrence of short distance and high speed need not be unexpected. Parker (1970) has argued that competition to be nearer ova has been important in the evolution of intromission devices. Greater intrasexual competition, resulting in deeper penetration toward the site of egg production might simultaneously produce more effective gametic propulsion.

A gamete can get there first not only through speed, but by leaving first. Sperm in the storage organs of insects often show a sort of perpetual agitation or excite-ment prior to a fertilization (*e.g.*, Lefevre and Jonsson 1962, Taylor 1982). It is possible that the disturbance is the collective result of competitors jockeying for position.

Competition in time might account for the extraordinary reproduction of the fish ectoparasite *Gyrodactylus* (Trematoda). Unequally developed products of a polyembryonic partitioning of the ovum lodge inside siblings. A parent contains an embryo which has another embryo within it, which holds yet another embryo bearing a tiny fourth, and final, embryo (Dawes 1968). In hermaphroditic matings reminiscent of couplings between Russian dolls, sperm introduced into the adult reach all the nested embryos (Cohen 1977). It is possible that sperm attempting to get a jump on future rivals could have played a role in the evolution of this hyperviviparity. The initial stage might have resembled the occasional cases of mother-daughter insemination in the livebearing guppy *Poecilia reticulata.* Sperm reaching the hollow offspring-containing ovary inseminate both generations (Spurway 1953, Cohen 1977). If mates are rare, small internalized siblings that permit insemination increase their chances for sexual reproduction.

Sperm might succeed not only by leaving first or proceeding more rapidly, but also by leaving more often. One such case occurs in the bedbugs. Insemination in bedbugs and their relatives is traumatic. Sperm are injected through the body wall and homosexual injections occur in the Cimicidae, Plokiophilidae, and Anthocoridae (Carayon 1974). Among male *Xylocoris maculipennis* (Anthocoridae), sperm of the "mounter" migrate to the "mountee's" seminal vesicles, expanded portions of the vas deferens that serve as holding pens for sperm. Here they mingle with the victim's sperm and are found in his subsequent ejaculates (Carayon 1974). Parasitization of a rival's genitals multiplies a male's ability to deliver sperm (Lloyd 1979). Males would be expected to resist "rape" and sperm might have adaptations to infiltrate defenses. Phagocytosis of sperm occurs in both sexes of some related bedbugs (Carayon 1966). In *X. maculipennis,* the sperm are not attacked but do wait a curiously long 24 h before moving toward the reproductive tract. Rapist *X. maculipennis* have apparently countered any defenses raised to date. Sperm of traumatic inseminations, in general, tend to have enlarged motor organelles, presumably to overcome the mechanical resistance inherent in travel through a body (Baccetti and Afzelius 1976). In other species of homosexual parasites, should they occur, perhaps the converging defenses of an exploited male intensifies selection for power.

The potential for genital parasitization by mobile cells exists in certain vertebrates as well. Primordial germ cells migrate to regions of gonadal development through the bloodstream in some mammalian embryos. In cases where placental circulations become fused prior to the cells' arrival at gonadal primordia, cells from one embryo can reach developing siblings. This results in gonads which contain the genotypes of both individuals (Austin and Short 1972: exchange of germ cells has been considered an accidental occurrence).

E. Strategies of Competition: Eupyrene Sperm Polymorphism

Sperm within an ejaculate, all with complete genetic complements and apparently capable of fertilization, sometimes display striking polymorphisms. Sperm classes in the stinkbug *Arvelius albopunctatus* have nuclear volumes of 200, 400, and 1600 μ^3 (Schrader and Leuchtenberger 1950). Differences are due to nuclear proteins and RNA. (Several genera of tropical pentatomids have apyrene sperm formed in the "harlequin" lobe of the testes, Schrader 1960.) Certain *Drosophila* species have two, three, and, in a single case, perhaps four sperm size classes ranging from 46 μ to 430 μ (Beatty and Sidhu 1969, Beatty and Burgoyne 1971).

Can these morphs result from sperm competition? Big gametes might swim faster, block exits, or force smaller gametes out of storage organs. Featherwinged beetles have relatively gigantic sperm. A large number of females collected by Dybas and Dybas (1981) had "a single large spermatozoon protruding outside the body from the vagina," suggesting that gametes themselves might sometimes block reinsemination. The expense of large, tough sperm, useful in experienced females carrying rival ejaculates, coupled with the advantages of providing great numbers of sperm to virgins, might result in dimorphisms (Sivinski 1980a). Producing numerous small gametes might also increase representation in a mixed ejaculate and raise the probability of fertilization. If so, large and small gametes could lie on opposite ends of a competitive spectrum. Males which combine the alternatives would be reminiscent of sporting weapons that anticipate any game with a rifle mounted over a shotgun (a strategy perhaps parallel to the postulated mother-determined distribution of digger and patroller haploid male bees, Alcock 1979; and fighting and pacific haploid male parasitic Hymenoptera in figs, Hamilton 1979).

The distribution of the different gamete sizes within the female is sometimes suggestive of divergent fates. In *Drosophila* species, all morphs are passed to the female, but are not randomly distributed among storage organs. Only the largest occupy the ventral receptacle (Beatty and Burgoyne 1971), a storage organ distinct from the spermatheca and whose contents are typically the first to be used for fertilization (Fowler 1973). The occupancy of the best site by the largest morph could be indicative of an ability to move more rapidly than small competitors whose efforts would presumably be comparable to tiny siblings. It could also be a tactic on the level of the male parent who directs the size classes to different locations in the female tract, with large ones being sent to the front where competition is greatest. Sperm-trimorphic *Drosophila pseudoobscura* are an exception to the rule; the ventral receptacle and spermathecae are drained at about the same rate (Patterson 1947). Equal utilization, as it occurs in at least this one species, is counter to the above explanations.

Animals with eupyrene sperm polymorphisms have female storage organs. Storage can increase the probability of ejaculate mixing and, hence, competition. Natural populations of sperm-trimorphic *Drosophila pseudoobscura* are known to have high rates of multiple inseminations (Cobb 1977, Levine *et al.* 1980). Females of the sperm-dimorphic *D. subobscura,* however, rarely mate more than once (Maynard Smith 1956). If size classes in this latter species are not an anachronism from a polyandrous past, polymorphism might serve other than a competitive purpose.

Schrader (1960) has suggested that the largest sperms in pentatomids provide resources for the zygote with their disproportionate amounts of nuclear RNA and proteins. Why, then, are there masses of impoverished sperm? The benefit to a male of differential investment in offspring is not entirely clear. In some *Drosophila* species, the largest sperm may participate in forming the first zygotes. Males might provide extravagantly for their firstborn simply because the probability that a female will survive to produce subsequent offspring is low. Small gametes could merely be cheap insurance that a rarely encountered female Methuselah will continue to bear a male's offspring. Relatively precipitous drops would be expected in the natural survivorship schedules of polymorphic species. Interestingly, the short morphs of the sperm-trimorphic *Drosophila obscura* are rare in storage organs, although they are known to be common in the uterus after copulation (Beatty and Sidhu 1969). Small sperm may serve as material reserves for larger sperm, or as a paternal investment.

Yet another alternative is that large sperm last longer in storage organs, bulky food reserves being translated into longevity. This is doubly unlikely in *Drosophila* species, as large sperm are the first used or are used at a similar rate as smaller siblings, and the large protein stores of the mitochondrial crystals are not metabolized by the sperm (see subsection on paternal investment).

IV. COMPETITION BETWEEN CHOOSING FEMALES AND SPERM

A. Postcopulatory Female Choice

Females are thought to enhance their reproductive success by the choice of mates (Borgia 1979, Thornhill 1980). When choice is limited to discrimination between adults, the further from the end of parental investment a female chooses between mates, the more she could benefit from her selection (for the implications of gametic level filtering see preceding section). To decide before fertilization avoids gamete wastage (Mecham 1961). A decision prior to copulation precludes the hazards, time loss, and energetic expense of extraneous mating (see Parker 1974, Sivinski 1980b). To discriminate between mates at a distance aids in forestalling rape or the persistent, perhaps dangerous, attentions of unwanted males.

It might be expected, then, that postcopulatory choosing between males by favoring particular sperm would be uncommon. There are circumstances, however, where females might be forced, or prefer, to choose between mates by identifying and manipulating their sperm. Ejaculates compelled into mated females by genetically inferior rapists might be discarded (see Thornhill 1980b). A female wishing to enhance the genetic diversity of offspring might try to suppress the sperm of previous mates (see Walker 1980). When a female signals to attract mates, as in the bulk of pheromone-emitting insects, she surrenders the ability to discriminate as a distance. Here, and in other systems where males are difficult to avoid or simultaneously compare, it might prove less expensive to engage in multiple copulations and then choose a mate by sorting sperm. Courtships that attract predators or cost females time would make it convenient to internally discriminate between cellular models of male genotypes (see Daly 1978, for some phenotypic costs of sexual behavior).

The machinery of choice could be either morphological or chemical. Chemical complexity of the female and ovum offers a number of potential means of choosing (discussion of female/sperm interaction in Fowler 1973, Cohen 1977; see Yanagimachi 1977). The elaborate genitalia and multiple sperm storage organs of some insects might perform ejaculate sorting. The female tract and associated muscles are a major factor in the movement of many sperm (Hinton 1964; Blackith 1973 argues for spermathecal filtering; citations in Walker 1980).

Males and their sperm should attempt to subvert female discrimination. Penile complexity could be due to selection for "little openers, snipers, levers, and syringes" that circumvent female choice (Lloyd 1979; see Waage 1979, this volume). Sperm might resist manipulation by being too large to imprison in various nooks of the female tract or carry structures that lodge in membranes and resist transportation. Barbs on grasshopper sperm might give the female greater traction for moving the cell (Afzelius 1970), but barbs could also be a means of maintaining position. A number of arthropod sperm are aflagellate and immobile, a surrender that could evolve under absolute female control (see Baccetti and Afzelius 1976; also Dallai 1979).

Choosing females might seek additional information about male quality in the nature of their ejaculates. Searching for signs of fitness in cells could result in miniature versions of the advertisements, deceits, and scrutinies that characterize the intersexually selected macroscopic world (see Trivers 1972). In this vein, ejaculate size could provide females with a means of judging male vigor, gametic exuberance being a display of an ability to accrue resources (Wilson 1979, discussing excess pollen per ovule). Not all sperm redundancy ("excess" sperm per egg) is consistent with this hypothesis. In insects, while females often expel or digest a portion of the ejaculate, more sperm are stored and maintained than are strictly necessary for fertilization (see Sivinski 1980a). There seems to be little gain in materially supporting an extraneous portion of a no longer informative display. Enlarged sperm, like enlarged ejaculates, might demonstrate male success. There is evidence against this proposal in the negative correlation between sperm size and number among related insects (White 1954, 1973; Virkki 1969; Kurokawa and Hihara 1976; see Sivinski 1980a). If males invest in fewer, larger sperm, they are not revealing any greater resource accrual.

Titanic sperm, as sometimes encountered in arthropods, are sufficiently bizarre to suggest intersexually selected ornamentation, peacocks' tails on the cellular level. However, the cost to females of accommodating such monstrosities would be expected to put an early finish to runaway selection for elaborateness (see Fisher 1930 for runaway sexual selection; also Thornhill 1979). It is conceivable that females could read the constituents of sperm as a chemical abstract of its manufacturer's quality. Males might eventually load their sperm with a library of informative compounds forcing enlargement of the gametes (see Thornhill 1980; and citations for discussion of display information content).

B. Choosing Gender Through Sperm Preference

Females might wish to choose the gender of their offspring because of local mate competition (Hamilton 1967, Borgia 1980), local resource competition (Clark 1978), the correlation between parental physical condition and offspring reproductive success (Trivers and Willard 1973; see, however, Williams 1979), the attractiveness of the father and presumably his sons (Burley 1981), and local scarcity of one sex (Verner 1965, Werren and Charnov 1978). In the common circumstance of male determination of offspring gender, this can be accomplished prior to fertilization only by biased production of gametes by "cooperating" males or female discrimination between sperm. Manipulation of gender by mothers requires a haploid effect in sperm or a phenotypic label left over from an earlier stage of spermatogenesis to recognize gamete "sex" (a possible exception could occur in the body louse *Pediculus humanus,* where some males sire sons, others daughters; Hindle 1919).

Where sex chromosomes differ in size, there would likely be at least a tiny phenotypic difference in weight or nuclear dimension between X and Y gametes (the nuclear ultrastructure of X and O sperm differs in the homopteran *Dalbulus maidus;* Kitajima and Da Cruz-Landim 1972). A number of artificial separation techniques, however, have suffered a history of unrepeatability (Beatty 1975; see however Pinkel *et al.* 1982). Some suggestive evidence of gender expression in sperm does exist. Delayed fertilization in some frogs, flies, mealybugs, butterflies, and copepods results in male-biased sex ratios (Werren and Charnov [1978] argue that delayed fertilization is indicative of a local male scarcity that females capitalize on by producing sons). Among Werren and Charnov's examples are male heterogametic organisms, which in the absence of differential mortality of zygotes are apparently choosing gender at the gametic level. A maternally inherited cytoplasmic factor in humans seems to recognize and destroy Y sperm, thereby assuring its own reproduction (Leinhart and Vermelin 1946, in Grun 1976; discussed by Eberhard 1980). Sex ratios of *Drosophila melanogaster* offspring sometimes vary with the age of the father. Sons predominate in the eggs fertilized by young males, daughters in those sired by older males. Mange (1970) has argued that the difference is due to preferential use of sperm by females. There are several *Drosophila* examples of nonrandom utilization associated with various disomic or attached forms of X and Y chromosomes (reviewed by Fowler 1973). It is not clear whether these nonrandom recoveries are due to competition or preference, but they are suggestive of gender expression in the sperm phenotype. In haploid-expressing pollen, the greater success, under certain conditions, of X-bearing gametophytes is thought to be due to choice exerted by the style rather than competition between grains (Lewis 1942; as discussed by Hamilton 1967).

Homogametic females choosing the gender of their offspring are in conflict with the discarded "sexed" sperm, but not necessarily the male who provided the ejaculate. Males lose fertilizations to competitors, however, where sperm redundancy is low and additional copulations are required to provide a sufficient quantity of correctly sexed gametes (*e.g., Drosophila melanogaster,* where stored sperm-to-egg ratios approach 1:1, Lefevre and Jonsson 1962). When sex ratios are biased to minimize competition between siblings, multiple paternity generates yet another conflict (see Hamilton 1967 for the effects of inbreeding on within-brood competition and sex ratio). From the father's, but not the mother's, point of view the sibship contains unrelated individuals and males would prefer to produce the minority sex. Borgia (1980) has suggested that male sperm might mimic female sperm under conditions of local mate competition and so produce extra sons by subterfuge. One characteristic available for recognition would be the presence of sex chromosome antigens. It is worth noting that, in mammals, **all** sperm carry Y antigens (McLaren 1965, Katsch and Katsch 1965, Cohen 1971). For ubiquitous Y antigens to be an evolved mimicry, biased sex ratios in mammals must usually favor males. Local resource competition resulting in male biases occurs in at least one mammal, the prosimian *Galago crassicaudatus* (Clark 1978), but requires female control of resources, a stable saturated environment, and inflexible territory boundaries, a peculiar set of phenomena.

Males in haplo-diploid species are related only to the diploid of the following generation and might strive to produce extra daughters (J. Brockman, unpubl. ms). A peculiar case of sperm polymorphism in a haplo-diploid species was historically interpreted as sex ratio control by sperm. In the parasitic hymenopteran *Dahlbominus fuscipennis,* there are five sperm morphs; two morphs characterized respectively by sinistral (left-handed) and dextral (right-handed) coilings reach the spermatheca (Lee and Wilkes 1965, Wilkes 1965). Sinistral sperm were thought to be unable to penetrate the micropyle, thereby capping unfertilized eggs and producing males (Wilkes and Lee 1965). It is not clear why a female would abandon control of sex ratio to a potential antagonist. One possible benefit, male investment via infertile gametes in a mother's sons, fails to find support in the incomplete penetrance by sperm. It is even less obvious why males would produce a sperm that resulted in the development of an individual containing no genes of direct paternal descent. A male could benefit only under conditions of local mate competition where he would initiate the production of mates for his daughters (the sexes might not disagree as to sex ratio here). Male progeny, however, are known to increase with adult female density, demonstrating female control of offspring gender in response to changes in local mate competition (Victorov and Kochetova 1973). The evidence of outbreeding casts doubt on male production by left-handed sperm.

V. COMPETITION AMONG SYMBIOTIC GENOMES

A number of replicating entities coexist alongside the chromosomes. Mito-chondria, plastids, plasmons, bacteriods, fungi, virus, and various other bits of programming pass down the generations via gametes. Conflicts of reproductive interest between replicators might lead to selfish manipulation of their gametic vehicles (see Eberhard 1980, for a review of organelle competition).

Gamete volume could be a limiting factor for replicators inherited through the cytoplasm. Mitochondrial DNA (= mtDNA), for example, is enormously more abundant in ova than sperm: 10^6:1 genomes in clawed frogs (Dawid and Blackler 1972). This results in a serious competitive disadvantage for any mtDNA that happens to be located in a male, assuming that the proportion of paternal mtDNA in the zygote foretells its relative abundance in the ova and sperm of the following generation. The disadvantage might be tempered by larger sperm carrying increased cargos of mtDNA. The bulk of the gargantuan sperm of certain insects consists of mitochondrial crystals; mtDNA may be abundant in these derivatives (Perotti 1973). Such giant cells are similar to what might be the expected production of a mitochondrial genome able to influence sperm size (Sivinski 1980a). Organelles with primarily maternal inheritance might destroy "male" sperm. The resulting sex ratio bias benefits organelles but runs counter to the interests of both parents (Eberhard 1980).

A manipulative organelle would come into conflict with a nuclear genome programmed to allocate a certain amount of resources to a particular number of sperm. From the mitochondrial point of view, male (*i.e.* nuclear) fecundity could drop precipitously before selection disfavored further gamete inflation (recall the 10^6:1 abundance ratio). Victory would be expected to go to the nucleus, sheer size and diversity offering possibilities of control that would exhaust the genetic repertoire of an organelle. Beck *et al.* (1971) identified 63 nuclear genes affecting mitochondrial functions in yeast, while yeast mitochondria are believed to contain only 60-70 cistrons (Kroon and Saccone 1976).

A sperm-borne organelle's fate after fertilization is not always clear (Gillham 1978). In many cases, replicators appear to degenerate (citations in Eberhard 1980). Either maternal organelles somehow take advantage of numbers to remove competitors or the nucleus attempts to preclude inoculation by parasites, *i.e.* organelles less controllable than those already present. Where sperm inflation of organelle DNA in sperm has been attempted, organelle death could represent an ovum's means of protecting its future son's virility. Grun (1976) suggests that ova kill male line replicators to prevent sex between the maternal and paternal organelle populations, precluding dangerous recombinants (a domesticating adaptation that might be described as spaying the Red Queen; see Van Valen 1973, for Red Queen theory; see, however, Eberhard 1980).

VI. DISCUSSION AND CONCLUSIONS

There are at least five arenas of conflict occurring in and between three selection levels that could influence the evolution of sperm. It is a measure of ignorance that I have argued that some aspects of sperm biology could evolve in nearly all of these contexts. Gigantic sperm, for example, are considered as a means of paternal investment, agents of aggressive displacement, displays of male quality, and the monstrous constructions of meddling organelles. The evidence that would permit choosing among or discarding all these alternatives is, to my knowledge, inconclusive or unavailable.

While admitting ignorance of the particulars of gametic evolution, it seems almost inevitable that gametic diversity is, in part, the result of conflict. The opportunities available to sperm to experience and resolve competition are enormous. They are, after all, the last pre-synaptic clue to male quality, the farthest cellular extension of the male body into the sometimes contrary world of female reproductive interests, and, beside the egg, the closest complex objects to the fusion of pronuclei, the final resolution of most male contention.

VII. SUMMARY

The variety of sperm morphology and behavior is not easily explained by adaption to genital or egg topology. Competitions and conflicts in situations such as the following may influence the evolution of spermatozoa.

(1) Sperm of a male, bearing a coefficient of relatedness of 0.5 to many or most ejaculate mates, could compete in ways that lower the effectiveness of the ejaculate as a whole. The present absence of phenotypic expression (haploid effect) by genes in sperm may be an adaptation to suppress such "outlaws." Peculiar "worker" sperm who aid ejaculate mates but are not suited for fertilizing often

lack genomes altogether. Chromosomes may have been removed as the final resolution of male/sperm conflicts.

(2) Males produce sperm of different fertilizing abilities. If these differences are heritable and if ejaculates occasionally mix then females wishing competitive sons might encourage competition between sperm through multiple matings. Such an argument could be refuted through low variance in order effects (which male in a series of mates obtains what proportion of fertilizations).

(3) Sperm from different males may be put in competition as an effect of unrelated selection for multiple matings. One advantage to repeated insemination may be harvesting of resources present in sperm themselves. Giant protein-rich gametes, in particular, may be specialized bearers of paternal investment.

Once ejaculate mixture becomes sufficiently commonplace, sperm may evolve to defeat rivals spatially, *e.g.* through structures that allow more rapid movement or by novel directions of approach, *e.g.* through the body cavity in cases of traumatic insemination. Occasionally gametes overcome competitors temporally as in cases of hyperviviparity, the fertilization of embryos. The evolution of sperm polymorphisms may be due to differences in the competitive terrain of the female reproductive tract.

(4) Females might sometimes exercise postcopulatory mate choice by manipulation of sperm. Gametes that resist such choice would be at an advantage. In theory a common context for female sperm sorting would be attempts to control offspring gender. Such activities would be hampered by the lack of gene expression in sperm.

(5) A number of non-nuclear nucleic acids occur in gametes. There are potential conflicts between these and nuclear genes, *e.g.* organelles may be best represented in the next metazoan by being present in large numbers in inflated sperm. Competitions are most likely to be won by nuclear genes given their greater number and range.

ACKNOWLEDGMENTS

I have relied heavily on the researches and reviews of a number of sperm scholars. Interested readers are particularly urged to see the works of J. Cohen, R. A. Beatty, and D. L. Mulcahy. J. E. Lloyd, G. Parker, R. Smith, R. Thornhill, and T. J. Walker made many helpful criticisms. Remaining errors are due to my obduracy. P. Sivinski translated papers and B. Hollien prepared the manuscript.

REFERENCES

Afzelius, B. A. 1970. Thoughts on comparative spermatology. In *Comparative Spermatology,* B. Baccetti (ed.), pp. 565-571. Academic Press, New York.

Afzelius, B. A. (ed.) 1975. *The Functional Anatomy of the Spermatozoon.* Pergamon Press, Oxford.

Afzelius, B. A., B. Baccetti, and R. Dallai. 1976. The giant spermatozoon of *Notonecta. J. Submicrosc. Cytol.* 8:149-161.

Alcock, J. 1979. The evolution of intraspecific diversity in male reproductive strategies in some bees and wasps. In *Sexual Selection and Reproductive Competition in Insects,* M. S. Blum and N. A. Blum (eds.), pp. 381-402. Academic Press, New York.

Alexander, R. D. The evolution of social behavior. *Annu. Rev. Ecol. Syst.* 5:325-383.

Alexander, R. D., and G. Borgia. 1978. Group selection, altruism, and the levels of organization of life. *Annu. Rev. Ecol. Syst.* 9:449-583.

Ansley, H. R. 1954. A cytological and cytophotometric study of alternative pathways of meiosis in the house centipede (*Scutigera forceps,* Rafinesque). *Chromosoma* 6:656-695.

Austin, C. R. 1965. *Fertilization.* Prentice-Hall, Englewood Cliffs, NJ.

Austin, C. R., and R. V. Short. 1972. *Germ Cells and Fertilization.* Cambridge University Press, Cambridge.

Baccetti, B. 1972. Insect sperm cells. *Adv. Insect Physiol.* 9:315-395.

Baccetti, B. 1979. Ultrastructure of sperm and its bearing on arthropod phylogeny. In *Arthropod Phylogeny,* A. P. Gupta (ed.), pp. 609-644. Van Nostrand Reinhold, New York.

Baccetti, B., and B. A. Afzelius. 1976. *The Biology of the Sperm Cell.* S. Karger, Basel.

Baccetti, B., and R. Dallai. 1978. The spermatozoon of Arthropoda. III. The multiflagellate spermatozoon in the termite *Mastrotermes darwiniensis. J. Cell Biol.* 76:569-576.

Baccetti, B., R. Dallai, and B. Fratello. 1973. The spermatozoon of Arthropoda. XXII. The 12 + 0, 14 + 0 or aflagellate sperm of Protura. *J. Cell Sci.* 13:321-335.

Baccetti, B., R. Dallai, V. Palline, F. Rosati, and B. A. Afzelius. 1977. Protein of insect sperm and mitochondrial crystals. *J. Cell Biol.* 73:594-600.

Bacci, G. 1965. *Sex Determination.* Pergamon Press, Oxford.

Balduf, W. V. 1935. *The Bionomics of Entomophagous Coleoptera.* E. W. Classey, Hamton, England.

Bauer, H. 1940. Uber die chromosomen der bisexuallen und der parthenogenetischen rasse des ostracoden *Heterocyris incongruens* Ramd. *Chromosoma* 1:620-637.

Bawa, S. R. 1964. Electron microscope study of spermiogenesis in a firebrat insect *Thermobia domestica* Pack. *J. Cell Biol.* 23:431-444.

Bawa, S. R. 1975. Joined spermatozoa. In *The Functional Anatomy of the Spermatozoon,* G. A. Afzelius (ed.), pp. 259-266. Pergamon Press, Oxford.

Beatty, R. A. 1971. The genetics of size and shape of spermatozoon organelles. In *Proc. Int. Symp. Genetics of the Spermatozoon.* Organizer, New York.

Beatty, R. A. 1975a. Genetics of animal spermatozoa. In *Gamete Competition in Plants and Animals,* D. L. Mulcahy (ed.), pp. 61-68. No. Holland Publ. Co., Amsterdam.

Beatty, R. A. 1975b. Sperm diversity within the species. In *The Functional Anatomy of the Spermatozoon,* B. A. Afzelius (ed.), pp. 319-327. Pergamon Press, Oxford.

Beatty, R. A., and P. S. Burgoyne. 1971. Size classes of the head and flagellum of *Drosophila* spermatozoa. *Cytogenetics* 10:177-189.

Beatty, R. A., and N. S. Sidhu. 1969. Polymegaly of spermatozoan length and its genetic control in *Drosophila* species. *Proc. R. Soc. Edinb. Sect. B (Biol. Sci.)* 71:14.

Beck, J. C., J. H. Parker, W. X. Balcauagg, and J. R. Matoon. 1971. Mendelian genes affecting development and function of yeast mitochondria. In *Autonomy and Biogenesis of Mitochondria and Chloroplasts,* N. K. Boardman, A. W. Linnane, and R. M. Smillie (eds.), pp. 194-204. No. Holland Publ. Co., Amsterdam.

Bernstein, H., G. S. Byers, and R. E. Michod. 1981. Evolution of sexual reproduction: Importance of DNA repair, complementation, and variation. *Am. Nat.* 117:537-549.

Blackith, R. E. 1973. Clues to the Mesozoic evolution of the Eumastacidae. *Acria* 2:5-28.

Blount, M. 1909. The early development of the pigeon's egg, with especial reference to polyspermy on the origin of the periblast nuclei. *J. Morphol.* 20:1-64.

Bonnevie, E. 1907. Untersuchungen uber keimzellen. II. Physiologische polyspermie bei bryozoen. *Z. Naturwiss. Jena* 41:567-598.

Borgia, G. A. 1979. Sexual selection and the evolution of mating systems. In *Sexual Selection and Reproductive Competition in Insects,* M. S. Blum and N. A. Blum (eds.), pp. 19-80. Academic Press, New York.

Borgia, G. A. 1980. Evolution of haploidy: Models for inbred and outbred systems. *Theor. Popul. Biol.* 17:103-128.

Bouix, G. 1963. Sur la spermatogenes des *Carabus:* Modalite et frequence de la spermiogenese atypique.

Braden, A. W. H. 1958. Influence of time of mating on the segregation ratio of alleles at the *T*-locus in the house mouse. *Nature* 181:786-787.

Braden, A. W. H. 1960. Genetic influences on the morphology and functions of the gametes. *J. Cell. Comp. Physiol.* 56:17-29.

Braden, A. W. H. 1972. *T*-locus in mice; segregation distortion and sterility in the male. In *The Genetics of the Spermatozoon,* R. A. Beatty and S. Gluecksohn-Waelsch (eds.), pp. 289-305. Organizers, New York.

Breland, O. P., and E. Simmons. 1970. Preliminary studies of the spermatozoa and the male reproductive system of some whirligig beetles (Coleoptera:Gyrinidae). *Entomol. News* 81:101-110.

Bremermann, H. J. 1980. Sex and polymorphism as strategies in host-pathogen interactions. *J. Theor. Biol.* 87:671-702.

Brown, S. W., and F. D. Bennett. 1957. On sex determination in the diaspine scale *Pseudolacaspis pentagona* (Targ.) (Coccoidea). *Genetics* 42:510-523.

Burley, N. 1981. Sex ratio manipulation and selection for attractiveness. *Science* 211:721-722.

Carayon, J. 1966. Traumatic insemination and the paragenital system. In *Monograph of Cimicidae,* R. L. Uniager (ed.), pp. 81-166. Hornshafter, Baltimore, MD.

Carayon, J. 1974. Insemination traumatique heterosexuelle et homosexuelle chez *Xylocoris maculipennis* (Hem. Anthocoridae). *C. R. Acad. Sci. Paris D* 278:2803-2806.

Chatton, E., and O. Tuzet. 1943. Sur la formation des gonies polyvalentes et des spermies geantes chez deux lombriciens.*C. R. Acad. Sci. Paris D* 77:710-712.

Clark, A. B. 1978. Sex ratio and local resource competition in a prosimian primate. *Science* 201:163-165.

Cobb, G. 1977. Multiple insemination and male sexual selection in natural populations of *Drosophila pseudoobscura. Am. Nat.* 111:641-656.

Cohen, J. 1967. Correlation between sperm 'redundancy' and chiasma frequency. *Nature* 215:862-863.

Cohen, J. 1971. Comparative physiology of gamete populations. *Adv. Comp. Physiol. Biochem.* 4:268-380.

Cohen, J. 1973. Crossovers, sperm redundancy and their close association. *Heredity* 31:408-413.

Cohen, J. 1975a. Gamete redundancy—wastage or selection? In *Gamete Competition in Plants and Animals,* D. L. Mulcahy (ed.), pp. 99-144. No. Holland Publ. Co., Amsterdam.

Cohen, J. 1975b. Gametic diversity within an ejaculate. In *The Functional Anatomy of the Spermatozoon,* B. A. Afzelius (ed.), pp. 329-339. Pergamon Press, Oxford.

Cohen, J. 1977. *Reproduction.* Butterworths, London.

Crow, J. F. 1979. Genes that violate Mendel's rules. *Sci. Am.* 240:134-143.

Daily, M. 1978. The cost of mating. *Am. Nat.* 112:771-774.

Dallai, R. 1979. An overview of atypical spermatozoa in insects. In *The Spermatozoon,* D. W. Fawcett and J. M. Bedford (eds.), pp. 253-265. Urban and Schwarzengerg, Baltimore, MD.

Dawes, B. 1968. *The Trematoda.* Cambridge University Press, Cambridge.

Dawid, I. B., and A. W. Blackler. 1972. Maternal and cytoplasmic inheritance of mitochondrial DNA in *Xenopus. Dev. Biol.* 29:152-161.

Dawkins, R. 1982. *The Extended Phenotypes.* W. H. Freeman, Oxford.

Dawkins, R., and J. R. Krebs. 1979. Arms races between and within species. In *The Evolution of Adaptation by Natural Selection,* J. Maynard Smith and R. Holliday (eds.), pp. 487-511. The Royal Society, London.

Dupouy, J. 1964. La teratogenese germinale male des gastropods et ses rapports avec l'oogenese atypique et la formation des oeufs nourriciers. *Asch. Zool. Exp. Gen.* 103:217-368.

Dybas, L. K., and H. S. Dybas. 1981. Coadaptation and taxonomic differentiation of sperm and spermathecae in featherwinged beetles. *Evolution* 35:168-174.

Eberhard, W. G. 1980. The evolutionary consequences of intracellular organelle competition. *Q. Rev. Biol.* 55:231-249.

Fain-Maurel, M. 1966. Acquisitions recents sur les spermatogeneses atypiques. *Annee Biol.* 513-564.

Favard, P. 1955. Spermatogenese de *Discoglossus pictus* Otth. etude cytologique de maturation du spermatozoid. *Ann. Sci. Nat. Zool.* 17:370.

Fisher, R. A. 1940. *The Genetical Theory of Natural Selection.* Clarendon Press, Oxford.

Ford, C. E. 1972. Gross genome unbalance in mouse spermatozoa, does it influence capacity to fertilize? In *The Genetics of the Spermatozoon,* R. A. Beatty and S. Gluecksohn-Waelsch (eds.), pp. 359-369. Organizers, New York.

Fowler, G. L. 1973. Some aspects of the reproductive biology of *Drosophila;* Sperm transfer, sperm storage and sperm utilization. *Adv. Genet.* 17:293-260.

Frazen, A. 1956. On spermiogenesis, morphology of the spermatozoon, and biology of fertilization among invertebrates. *Zool. Bidr. Upps.* 31:355-482.

Frazen, A. 1970. Phylogenetic aspects of the morphology of spermatozoa and spermiogenesis. In *Comparative Spermatology,* B. Baccetti (ed.), pp. 29-46. Academic Press, New York.

Fretter, V. 1953. The transference of sperm from male to female prosobranch, with reference, also, to the pyramidellids. *Proc. Linn. Soc. Lond.* 164:217-224.

Gillham, N. W. 1978. *Organelle Heredity.* Raven Press, New York.

Gromko, M. H., and D. W. Pyle. 1978. Sperm competition, male fitness and repeated mating by female *Drosophila melanogaster. Evolution* 32:588-593.

Grun, P. 1976. *Cytoplasmic Genetics and Evolution.* Columbia University Press, New York.

Haldane, J. B. S. 1932. *The Causes of Evolution.* Longmans, Green, and Co., London.

Hamilton, W. D. 1967. Extraordinary sex ratios. *Science* 156:477-488.

Hamilton, W. D. 1979. Wingless and fighting males in fig wasps and other insects. In *Sexual Selection and Reproductive Competition in Insects,* M. S. Blum and N. A. Blum (eds.), pp. 167-220. Academic Press, New York.

Hanson, J., J. T. Randall, and S. T. Bayley. 1952. The microstructures of the spermatozoa of the snail *Viviparus. Exp. Cell Res.* 3:65-78.

Heslop-Harrison, J. 1975. Male gametophyte selection and the pollen-stigma interaction. In *Gamete Competition in Plants and Animals,* D. L. Mulcahy (ed.), pp. 177-190. No. Holland Publ. Co., Amsterdam.

Hindle, E. 1919. Sex inheritance in *Pediculus humanus* var. *corporis. J. Genet.* 8:167-277.

Hinton, H. E. 1964. Sperm transfer in insects and the evolution of haemocoelic insemination. In *Insect Reproduction Symposium No. 2,* K. C. Highnam (ed.), pp. 95-107. R. Entomol. Soc., London.

Hinton, H. E. 1974. Symposium on reproduction of arthopods of medical and veterinary importance. III. Accessory function of seminal fluid. *J. Med. Entomol.* 11:19-25.

Hinton, H. E. 1981. *Biology of Insect Eggs. Vol. 1.* Pergamon Press, Oxford.

Holstein, A. F. 1965. Elektronmikroskopische untersuchungen am spermatozoon des opossum (*Didelphys virginiana* Kerr). *Z. Zellforsch. Mikrosk. Anat.* 65:905-914.

Howe, H. F. 1976. Egg size, hatching asynchrony, sex and brood reduction in the common grackle. *Ecology* 57:1195-1207.

Howe, H. F. 1977. Sex ratio adjustment in the common grackle. *Science* 198:744-746.

Idelman, S. 1960. Evolution de la spermatogenese chez un mollusque prosobranch *Turritella communis. Proc. Eur. Reg. Conf. Electron Microsc.* 2:942-946.

Ito, S., and C. Leuchtenberger. 1955. The possible role of the activation process of the clar *Spisula solidissima. Chromosoma* 7:328-329.

Jordan, E. O. 1893. The habits and development of the newt (*Diemyctylus viridescens*). *J. Morphol.* 8:269-366.

Katsch, S., and G. F. Katsch. 1965. Perspectives in immunological control of reproduction. *Pac. Med. Surg.* 73:28-43.

Kitajima, E. W., and C. DaCruz Landim. 1972. An electron microscopic study of the process of differentiation during spermiogenesis in the corn leaf hopper *Dalbulus maidis* Del. and W. *Rev. Biol.* 8:5-19.

Koehler, J. K. 1965. An electron microscope study of the dimorphic spermatozoa of *Asplanchna* (Rotifera). I. The adult testes. *Z. Zellforsch. Mikrosk. Anat.* 67:57-76.

Koehler, J. K., and C. W. Birky. 1966. An electron microscope study of the dimorphic spermatozoa of *Asplanchna* (Rotifera). II. The development of "atypical spermatozoa." *Z. Zellforsch. Mikrosk. Anat.* 70:303-321.

Kroon, A. M., and C. Saccone. 1976. Concluding remarks. In *The Genetic Function of Mitochondrial DNA,* C. Saccone and A. M. Kroon (eds.), pp. 343-347. No. Holland Publ. Co., Amsterdam.

Kurokawa, H., and F. Hihara. 1976. Number of first spermatocytes in relation to phylogeny of *Drosophila* (Diptera:Drosophilidae). *Int. J. Insect Morphol. Embryol.* 5:51-63.

Lee, P. E., and A. Wilkes. 1965. Polymorphic spermatozoa in the hymenopterous wasp *Dahlbominus. Science* 147:1445-1446.

Lefevre, G., and V. B. Jonsson. 1962. Sperm transfer, storage, displacement and utilization in *Drosophila melanogaster. Genetics* 47:1719-1736.

Levine, L. 1967. Sexual selection in mice. IV. Experimental demonstration of selective fertilization. *Am. Nat.* 101:289-294.

Levine, L., M. Asmussen, O. Olvera, J. R. Powell, M. E. de la Rose, V. M. Salceda, M. I. Baso, J. Guzman, and W. W. Anderson. 1980. Population genetics of Mexican *Drosophila.* V. A high rate of multiple insemination in a natural population of *Drosophila pseudoobscura. Am. Nat.* 116:493-503.

Lewis, D. 1942. The evolution of sex in flowering plants. *Biol. Rev.* 17:46-67.

Lewontin, R. C. 1970. The units of selection. *Annu. Rev. Ecol. Syst.* 1:1-18.

Lindsley, D. L., and E. H. Grell. 1969. Spermiogenesis without chromosomes in *Drosophila melanogaster. Genetics Suppl.* 61:46-67.

Lloyd, J. E. 1979. Mating behavior and natural selection. *Fla. Entomol.* 62:17-34.

Mackie, J. B., and M. H. Walker. 1974. A study of the conjugate sperm of the dytiscid water beetles *Dytiscus marginalis* and *Colymbetes fuscus. Cell Tissue Res.* 143:505-519.

Mange, A. P. 1970. Possible non-random utilization of X and Y bearing sperm in *Drosophila melanogaster. Genetics* 65:95-106.

Mazzini, M. 1976. Giant spermatozoa in *Divales bipustulatus* F. (Coleoptera:Cleridae). *Int. J. Insect Morphol. Embryol.* 5:107-115.

Maynard Smith, J. 1956. Fertility, mating behavior and sexual selection in *Drosophila subobscura. Genetics* 54:261-279.

Maynard Smith, J. 1978. *The Evolution of Sex.* Cambridge University Press, Cambridge.

McFarlane, R. W. 1963. Taxonomic significance of avian sperm. *XIII Int. Ornithol. Congr. Ithaca* 91-102.

McLaren, A. 1965. Growth of male young in mothers immunized against Y antigen. *Transplantation* 3:28-38.

McCloskey, J. D. 1966. The problem of gene activity in the sperm of *Drosophila melanogaster. Am. Nat.* 100:211.

Mecham, J. S. 1961. Isolating mechanisms in anuran amphibians. In *Vertebrate Speciation,* W. F. Blair (ed.), pp. 24-61. Univ. Texas Press, Austin, TX.

Moore, G. P. M. 1972. A cytological demonstration of the DNA-transcription enzyme RNA polymerase during mammalian spermatogenesis. In *The Genetics of the Spermatozoon,* R. A. Beatty and S. Gluecksohn-Waelsch (eds.), pp. 90-96. Edinburgh.

Mulcahy, D. L. (ed.). 1975. *Gamete Competition in Plants and Animals.* No. Holland Publ. Co., Amsterdam.

Mulcahy, D. L. 1979. The rise of the angiosperms: A genecological factor. *Science* 206:20-23.

Mulcahy, D. L., and S. M. Kaplan. 1979. Mendelian ratios despite nonrandom fertilization? *Am. Nat.* 113:419-425.

Muller, H. J., and F. Settles. 1927. The non-functioning of genes in spermatozoa. *Z. Indukt. Abstammungs-Vererbungsl.* 43:285-312.

Nishiwaki, S. 1964. Phylogenetic study on the type of the dimorphic spermatozoa in *Prosobranchia. Sci. Rep. Tokyo Kyoiku Daigaku Sect. B* 11:237-275.

Nur, U. 1962. Sperms, sperm bundles and fertilization in a mealy bug *Pseudococcus obscura* Essig (Homoptera, Coccoidea). *J. Morphol.* 111:173-183.

Oppel, A. 1892. Die befruchtung des reptilieneies. *Arch. Mikrosk. Anat.* 39:215-290.

Ottaviano, E., M. Sari-Gorla, and D. L. Mulcahy. 1980. Pollen tube growth rates in *Zea mays:* Implications for genetic improvement of crops. *Science* 210:437-438.

Pant, K. P. 1971. Patterns of inheritance in the midpiece length of mouse spermatozoa. In *The Genetics of the Spermatozoon,* R. A. Beatty and S. Gluecksohn-Waelsch (eds.), pp. 116-119. Organizers, New York.

Parker, G. A. 1970. Sperm competition and its evolutionary consequences in the insects. *Biol. Rev.* 45:525-567.

Parker, G. A. 1974. Courtship persistence and female guarding as male time investment strategies. 48:157-184.

Patterson, J. T. 1947. The insemination reaction and its bearing on the problem of speciation in the *Mulleri* subgroups. *Texas Univ. Publ.* 4720:37-41.

Pelseneer, P. 1935. *Essai d'Ethologie Zoologique.* Brussels.

Perotti, M. E. 1973. The mitochondrial derivative of the spermatozoon of *Drosophila* before and after fertilization. *J. Ultrastruct. Res.* 44:181-198.

Perotti, M. E. 1975. Ultrastructural aspects of fertilization in *Drosophila.* In *The Functional Anatomy of the Spermatozoon,* B. A. Afzelius (ed.), pp. 57-68. Pergamon Press, Oxford.

Phillips, D. M. 1970. Ultrastructure of spermatozoa of the woolly opossum *Caluromys philander. J. Ultrastruct. Res.* 33:381-397.

Pinkel, D., B. L. Gledhill, S. Lake, D. Stephanson, and M. A. Van Dilla. 1982. Sex preselection in mammals? Separation of sperm bearing X and "O" chromosomes in the vole *Microtus oregoni. Science* 218:904-906.

Reger, J. F., and D. P. Cooper. 1968. Studies on the fine structure of spermatids and spermatozoa from the millipede *Polydesmus* sp. *J. Ultrastruct. Res.* 23:60-70.

Richards, A. G., and A. Miller. 1937. Insect development analysed by experimental methods, a review, I: Embryonic stages. *J. N. Y. Entomol. Soc.* 45:1-60.

Rothschild, Lord. 1961. Structure and movements of tick spermatozoa (Arachnida, Acarii). *Q. J. Microsc. Sci.* 102:239-247.

Rothstein, S. I. 1979. Gene frequencies and selection for inhibitory traits with special emphasis on the adaptiveness of territoriality. *Am. Nat.* 113:317-331.

Ruckert, J.: Festschr. 1899. *Zum 70.* Geburtstag von Carl von Kupffer, Jena.

Ryckman, R. E. 1958. Description and biology of *Hesperocimex sonorensis,* new species, an ectoparasite of the purple martin (Hemiptera:Cimicidae). *Ann. Entomol. Soc. Am.* 51:33-47.

Schrader, R. 1960. Evolutionary aspects of aberrant meiosis in some Pentatomidae (Heteroptera). *Evolution* 14:498-508.

Schrader, R., and C. Leuchtenberger. 1950. A cytochemical analysis of the functional interrelationships of various cell structures in *Arevelius albopunctatus* (De Geer). *Exp. Cell Res.* 1:421-452.

Sivinski, J. 1980a. Sexual selection and insect sperm. *Fla. Entomol.* 63:99-111.

Sivinski, J. 1980b. The effects of mating on predation in the stick insect *Diaphomera veliei. Ann. Entomol. Soc. Am.* 75:553-556.

Spurway, H. 1953. Spontaneous parthenogenesis in a fish. *Nature* 171:437.

Sweetman, H. L. 1938. Physical ecology of the firebrat *Thermobia domestica* (Packard). *Ecol. Monogr.* 8:285-311.

Taylor, V. A. 1981. The adaptive and evolutionary significance of wing polymorphism and parthenogenesis in *Ptinella* Motschusky (Coleoptera:Ptidiidae). *Ecol. Entomol.* 6:89-98.

Taylor, V. A. 1982. The giant sperm of a minute beetle. *Tissue & Cell* 14:113-123.

Thornhill, R. 1980a. Competitive, charming males and choosy females: Was Darwin correct? *Fla. Entomol.* 63:5-30.

Thornhill, R. 1980b. Rape in *Panorpa* scorpionflies and a general rape theory. *Anim. Behav.* 23:52-59.

Trivers, R. L. 1972. Parental investment and sexual selection. In *Sexual Selection and the Descent of Man, 1871-1971,* B. Campbell (ed.), pp. 136-179. Aldine-Atherton, Chicago.

Trivers, R. L., and D. E. Willard. 1973. Natural selection of parental ability to vary the sex ratio of offspring. *Science* 179:90-92.

Van Valen, L. 1973. A new evolutionary law. *Evol. Theory* 1:1-30.

Verner, S. 1965. Selection for sex ratio. *Am. Nat.* **99**:419-421.

Victorov, G. A., and N. I. Kochetova. 1973. On the regulation of sex ratio in *Dahlbominus fuscipennis* Zett. *Entomol. Rev.* **52**:434-438.

Virkki, N. 1969. Sperm bundles and phylogenesis. *Z. Zellforsch. Mikrosk. Anat.* **101**:13-27.

Waage, J. K. 1979. Dual function of the damselfly penis: Sperm removal and transfer. *Science* **201**:916-918.

Walker, T. J. 1980. Reproductive behavior and mating success of male short-tailed crickets: Differences between and within demes. *Evol. Biol.* **13**:219-260.

Walker, W. F. 1980. Sperm utilization strategies in nonsocial insects. *Am. Nat.* **115**:780-799.

Wallace, H. 1974. Chiasmata have no effect on fertility. *Heredity* **33**:423-429.

Werner, G. 1964. Untersuchungen uber die spermiogeneses beim silberfischen, *Lipisma saccharina* L. *Z. Z. Zellforsch. Mikrosk. Anat.* **63**:880-912.

Werren, J. H., and E. L. Charnov. 1978. Facultative sex ratios and population dynamics. *Nature* **227**:349-350.

White, M. J. D. 1954. Patterns of spermatogenesis in grasshoppers. *Aust. J. Zool.* **3**:222-226.

White, M. J. D. 1973. *Animal Cytology and Evolution, 3rd ed.* Cambridge University Press, London.

Wilkes, A. 1965. Sperm transfer and utilization by the arrhenotokous wasp *Dahlbominus fuscipennis* (Zett) (Hymenoptera:Eulophidae). *Can. Entomol.* **97**:647-657.

Wilkes, A. 1966. Sperm utilization following multiple inseminations in the wasp *Dahlbominus fuscipennis. Can. J. Genet. Cytol.* **8**:451-461.

Wilkes, A., and P. E. Lee. 1965. The ultrastructure of dimorphic spermatozoa in the hymenopteran *Dahlbominus fuscipennis* (Zett.) (Eulophidae). *Can. J. Genet. Cytol.* **7**:609-619.

Williams, G. C. 1979. The question of adaptive sex ratio in outcrossed vertebrates. In *The Evolution of Adaptation by Natural Selection,* J. Maynard Smith and R. Holliday (eds.), pp. 567-580. R. Soc., London.

Wilson, M. F. 1979. Sexual selection in plants. *Am. Nat.* **113**:777-790.

Woodard, T. M. 1940. The function of the apyrene spermatozoa of *Goniogasis laqueata* (Say). *J. Exp. Zool.* **85**:103-123.

Yanagimachi, R. 1977. Specificity of sperm-egg interaction. In *Immunobiology of Gametes,* M. Edidin and M. H. Johnson (eds.), pp. 255-285. Cambridge University Press, Cambridge.

Yanagisawa, K., L. Dunn, and D. Bennett. 1961. On the mechanism of abnormal transmission ratios at the *t*-locus in the house mouse. *Genetics* **46**:1635-1644.

Yanagisawa, K., D. R. Pollard, D. Bennett, L. C. Dunn, and E. A. Boyse. 1974. Transmission ratio distortion at the *t*-locus: Serological identification of two sperm populations in *t*-heterozygotes. *Immunogenetics* **1**:91-96.

4

Male Mating Effort, Confidence of Paternity, and Insect Sperm Competition

DARRYL T. GWYNNE

I. INTRODUCTION

Trivers (1972) laid the groundwork for understanding how relative parental investment by males and females influences the operation of sexual selection and leads to the evolution of the basic differences between the sexes. Parental investment (PI) is defined as investment (risk taken and energy expended) in a single offspring at the cost of the parent's ability to invest in other offspring. There is

Copyright © 1984 by Academic Press, Inc.
All rights of reproduction in any form reserved.
ISBN 0-12-652570-6

a basic inequality in the investment in each gamete by the sexes, an inequality often magnified by greater female investment in gestation, parental care, etc. The theory predicts certain fundamental sexual differences; individuals of the sex which invests less per offspring (usually males) should compete for fertilization of members of the opposite sex. Typical male characteristics should, therefore, function in competition for mates. The sex investing more in each offspring (usually females) is not expected to exhibit strong sexual competitiveness but instead characteristics associated with discrimination of mates.

In addition to a difference in investment by the sexes in the offspring there is a concomitant difference in the assurance of parentage (Alexander and Borgia 1979, Alexander *et al.* 1979). A female not only invests more in each offspring but also has a higher assurance that (1) investment in a gamete will pay off reproductively, and (2) each offspring is her progeny. Males are much less able to predict which gametes will become offspring and after fertilization have a much lower confidence of parentage (see Maynard Smith 1978).

Alexander *et al.* (1979) have stated that this difference between the sexes can also account for the differences between male and female attributes, *i.e.,* because a female has absolute assurance of parentage she is in a better position to invest in offspring and, thus, to be choosy about who fathers them. Conversely, the uncertainty of male paternity can in itself lead to selection for traits associated with competition among males for fertilizations.

Thus, when males invest in young, one of the typical "feminine" traits they are expected to evolve is a higher confidence of parentage (see Trivers 1972, Borgia 1979). This conclusion is supported by models of Knowlton and Greenwell (this volume) which predict selection for increased confidence of parentage by males when there is an increase in paternal investment. However, these models do not show that increasing certainty of paternity should itself select for escalated paternal investment.

In insects male parental care is rare, being restricted to a few species in the order Hemiptera (Smith 1980). Most cases of male insects investing in offspring involve indirect efforts such as feeding a female with a "nuptial gift" of a prey item or glandular product (Thornhill 1976a, Thornhill and Alcock 1983). A main thesis of this chapter is that insects in which males invest by providing benefits to females and/or offspring should show adaptations to ensure paternity.

Smith (1979b) showed that the above prediction holds true in a species of water bug with paternal care. By examining competition between the ejaculates of two different males he revealed that the last male to mate fertilizes virtually all of the eggs he subsequently cares for. In this chapter the test is extended by considering all insect groups with male investment in offspring, especially those with male contributions of nutrients to the female. These prezygotic male contributions are not true parental investment but are instead a form of mating effort since they function to attract females and secure matings (Alexander and Borgia 1979). In the first part of this chapter I define more clearly the kind of reproductive effort that these male prezygotic nutrient investments represent. Even though these efforts are mating effort, they are likely to be important to female reproduction and thus may represent a limiting resource, ultimately influencing the evolution of sexual differences in a similar way as paternal care. The second part of the chapter reviews insect sperm competition literature and examines the hypothesis that sperm precedence levels for the last male to mate with a female are higher in those species where the male provides nutrient or other benefits to the female and/ or her offspring (see Thornhill 1979).

II. MATING EFFORT, PARENTAL EFFORT, AND THE EVOLUTION OF SEXUAL DIFFERENCES

Mating effort (ME) and parental effort (PE) were defined by Low (1978) as the two subsets of an organisms's total reproductive effort (RE). PE is the sum of all PI and ME is the sum of the risk and energy involved in obtaining fertilizations.

It is the basic difference between the sexes in their control over gametes which led Alexander and Borgia (1979) to conclude that males cannot invest parentally before fertilization. They argue that since PE is put into offspring and ME secures matings (Low 1978), male prezygotic effort is more correctly identified as ME. Selection should, therefore, favor males not to provide prezygotic PE because they simply do not "know" which of or what percentage of their gametes will become offspring. On the other hand, essentially all female prezygotic effort is PE because she is in a position to invest parentally, before fertilization, due to her ability to direct resources to individual gametes. Therefore, Alexander and Borgia (1979) argue that male prezygotic efforts which provide benefits to the female and/or her offspring, and which have been considered as PE (*e.g.,* Trivers 1972, Thornhill 1976a, Gladstone 1979), are more correctly identified as ME.

Trivers (1972) did exclude "effort expended in finding a member of the opposite sex, or in subduing members of one's own sex" from his definition of PI. Even though this is ME (as defined by Low [1978]), the separation between the excluded effort and PI was not clear in Trivers' paper. Although he included male prezygotic efforts (*e.g.,* providing the female with food at mating and defending a place for her to feed or nest) in his definition of PI, these efforts would also fall into his above cited exclusion in that they often play a critical role in obtaining copulations. Although these efforts, included by Trivers as PI, should be considered ME, they are a different category of ME in that immediate benefits are provided to the female and/or her offspring. The importance of this kind of male ME is that, like male PE (such as paternal care), it can result in a reduction in the different operation of sexual selection on the sexes because of a decrease in the variation in male reproductive success (see Borgia 1979). Males, because of what they have to offer, can limit female reproduction. When the total time, energy, and risk incurment by males in "female benefitting" ME or PE exceeds that (mostly PE) of the female, females will compete for males as mates (and will thus allocate some effort to ME) and males will discriminate among females as mates, likely using cues which indicate maternal and other reproductive abilities (*e.g.,* Gwynne 1981). A male's efforts benefit females and reduce his opportunities to mate with other females. This results in the evolution of mechanisms to increase the confidence of paternity for the fewer matings secured (see Knowlton and Greenwell, this volume).

Alexander and Borgia (1979) classified ME into (1) competitive interactions with the same sex; (2) transfer of benefits to members of the other sex as part of securing mates; and (3) evidence of commitment to PE to be directed at the offspring resulting from the mating. A more important division of ME, however, may be into one that provides material or risk-lowering benefits to the female and/or her offspring and decreases the probability of a male locating and inseminating other females, and a second that includes efforts expended to acquire fertilizations which do not provide these benefits. The former might best be termed "nonpromiscuous mating effort" in contrast to the latter, "promiscuous mating effort." (The word promiscuous here refers to a strategy in which a male maximizes the number of females inseminated and is not intended to imply random mating or any lack of choice by males; see Selander 1972.)

III. EVOLUTION OF NONPROMISCUOUS MATING EFFORT AND PARENTAL EFFORT

Paternal effort is derived from energies formerly used in mating (*i.e.,* ME). Paternal effort results from selection favoring a shift of RE from ME to PE. Parker *et al.* (1972), Parker (1978a, this volume), and Alexander and Borgia (1979) have discussed the origin of gamete dimorphism and ultimately the origin of the sexes. "Maleness" evolved as a result of the diversion of PE into ME concomitant with anisogamy and some gametes becoming motile because of selection associated with locating and fertilizing larger, more sessile gametes (Alexander and Borgia 1979). The "primitive" male, therefore, apportioned almost all his RE into ME by producing spermatozoa and attempting to place them in areas where they were likely to fertilize the maximum number of eggs. Several steps are likely to have been involved in the evolution of male PE or nonpromiscuous ME from promiscuous ME:

1. The "primitive male" system involved no male PE and all ME directed toward locating a mate and competing with other males for access to ova (*i.e.*, promiscuous ME).

2. Conditions arise that permit males to incidentally contribute benefits to the female and/or offspring as a result of promiscuous mating effort. At this point, however, there are no costs, in the form of diminished opportunities to mate other females, for the contributing males (*e.g.*, males that defend resources [food, oviposition sites] to which females are attracted, and incidentally provide a benefit by preventing other males from interfering with female utilization of the resource). (See also Werren *et al.* 1980.)

3. Nonpromiscuous ME directs efforts toward female/offspring benefits and consequently the male decreases his opportunites to mate other females. With increases in a male's efforts directed to nonpromiscuous ME tactics, there is a reduction in the differential operation of selection on the sexes as described above, and it is here that an increase in the male's confidence of paternity should be selected so that he acquires the maximum number of fertilizations associated with the fewer females he will mate.

4. Finally, some aspect of the male's ME may now become true PE as true paternal care directed toward eggs and/or young.

Some examples will serve to illustrate the progression from promiscuous ME to nonpromiscuous ME and/or PE. Information available on the comparative reproductive behavior of dragonflies and damselflies (order Odonata) suggests an evolutionary progression of ME to PE as outlined above. Males of many species exhibit some kind of territorial mating strategy (territory is here defined as any defended area). In some species of dragonflies, males leave the female after mating (Corbet 1962; Waage, this volume) and thus invest energies and take risk only in competing with other males (*e.g.*, territory defense, etc.) and by maximizing fertilizations by inseminating as many receptive females as possible (*i.e.*, promiscuous ME). Males of many species guard the female against other males while she oviposits. This includes "non-contact guarding" species, where the male flies around the ovipositing females, and species in which males remain "in tandem" (by continuing to hold the female's thorax with the apical abdominal appendages while she oviposits) after mating (see Waage, this volume). Some portions of effort expended by "contact guarding" males involve promiscuous ME in the form of behavioral and morphological adaptations to prevent sperm competition from other males (Waage 1979a, b, this volume) and, in non-contact guarding species, for locating and copulating with other females while guarding previously-mated ovipositing females as well (Alcock 1979; Waage 1979a, this volume). Part of the effort, however, is nonpromiscuous ME because males contribute while decreasing the probability of contact with other females during the guarding of mated females. The advantages to both sexes of postcopulatory guarding are increased oviposition as a result of decreased interference by other males (Waage 1978, 1979a, this volume) and diminished predation on females. An increase in nonpromiscuous ME is especially obvious in males that show tandem guarding. This activity precludes opportunities to secure other females for its duration.

Similar trends have been suggested for the evolution of paternal care in insects (Smith 1980) and fish (Williams 1975, Loiselle 1978). These authors suggest that male parental care may have evolved in species in which males utilized territorial mate-acquisition strategies.

Both male PE and nonpromiscuous ME benefit the female and/or her offspring and, at the same time, reduce the number of fertilizations a male can obtain. Only these kinds of investments by males can decrease the differences in the operation of sexual selection on the sexes by reducing the variance in male reproductive success and causing males to become a resource required by females to maximize their own reproductive success.

How does increased male PE or nonpromiscuous ME evolve? The selection pressures that have brought about these kinds of male investments are poorly understood. Trivers (1972) and Thornhill (1976a) have suggested that female choice of males who provide immediate benefits could drive the system. An alternative explanation, however, is that the transfer of benefits to a female advance the male's as well as the female's genetic interests. Males who invest parentally must have, at least initially, gained a reproductive advantage through the increased survival, and ultimately reproduction, of their mates (see Alexander and Borgia 1979). The general scenario for the evolution of male PE or nonpromiscuous ME involves some change in the social or physical environment that increases male reproductive success via investment. Males that invest in this way, and females that choose these males, therefore gain. In the example of the evolution of male guarding in odonates, a presumed change in the environment, *e.g.,* an increase in the numbers of males interfering with the ovipositing female (Waage 1979b), would have strongly selected for postcopulatory guarding.

IV. CAN MATING EFFORT AND PARENTAL EFFORT BE DISTINGUISHED?

The preceding discussion has applied the interpretation of Low (1978) and Alexander and Borgia (1979) in distinguishing certain male reproductive effort as PE or ME. To recapitulate: ME is different from PE in that ME is effort in acquiring fertilizations whereas PE is investment in offspring. Promiscuous ME can be separated from PE because an increase in promiscuous ME expended by males should not decrease differences between the sexes. However, nonpromiscuous ME and PE lead to the same predictions. Hence, the two are indistinguishable. Alexander and Borgia (1979) have argued that male prezygotic effort such as courtship feeding must be considered ME and not PE because it is effort expended to acquire fertilizations. However, even postzygotic male investment such as paternal care of eggs and/or young may be difficult to classify as ME or PE. This can be illustrated with some examples.

Males of many species of fish care for eggs exclusive of any maternal care. In most of these groups the male defends a nest site and cares for the eggs of several females (Ridley 1978). Evidence from several well-documented studies suggests that nest sites are limited (Cunningham 1979, Downhower and Brown 1981). Therefore, males are investing postzygotically by guarding their offspring—this investment is apparently PE. However, as long as there is a chance that other females will lay eggs in his nest, egg guarding by a male could function as ME with the male investing prezygotically to acquire more matings. In sticklebacks, females actually prefer to mate with a male with eggs in his nest (Ridley and Rechten 1981). Indeed, males will steal eggs from other nests and transfer them to their own (Rohwer 1978). Thus, postzygotic parental care could function simultaneously both as PI and as prezygotic ME to acquire additional matings. Paternal guarding of eggs can be demonstrated to be PE exclusively only if the male copulated once and guards only one female's eggs, forfeiting additional opportunities to mate.

The same reasoning can be applied to Smith's (1976a, b 1979a, b) studies of brooding by male water bugs (Hemiptera:Belostomatidae). In these species several females eventually oviposit on a male's back. As long as there is space available for eggs on a male's back (there is some evidence that male backs are a limiting resource for females; Smith 1979a), the investment can be considered ME involved in advertising the male's parental abilities that may aid in the acquisition of matings with other egg-laden females. However, females reject males as mates if they have no "egg room" left (Smith 1979b). Since completely encumbered males do continue to brood their eggs (Smith 1979b), the investment could then be considered exclusive PE.

V. PATERNAL INVESTMENT AND CONFIDENCE OF PATERNITY IN INSECTS

Male PE and nonpromiscuous ME are predicted to have similar effects on the operation of sexual selection and the evolution of sexual differences. This is because both kinds of male effort benefit the female and/or offspring and result in a decrease in the maximum number of fertilizations a male can obtain. Because of the decrease in the nurturant male's opportunities for mating, selection should perfect mechanisms to assure confidence of paternity.

Models developed by Werren *et al.* (1980) show that increased confidence of paternity should be strongly linked with male parental investment where the male's opportunities for additional matings are decreased. This is because level of paternity can reflect the number of opportunities for promiscuous mating; a higher level of paternity results in fewer opportunities for promiscuous fertilizations and thus a greater advantage in being parental. The paper goes on to argue against the theme prevalent in the literature that confidence of paternity was an important pre-requisite for the evolution of male parental care (*e.g.,* Ridley 1978, Blumer 1979). Reviewing male parental care (mainly birds and fishes), Werren *et al.* (1980) state that since promiscuous matings are common in these groups there should be little selective advantage to increased male paternity.

Although male PE is rare in insects (Smith 1980), there are many examples of other ways males may invest in a female and/or offspring by feeding her or guarding her while she oviposits. Because of the nature of most of these male contributions, the opportunities for promiscuous matings (*i.e.,* those without the female-benefitting investment) are greatly reduced. In katydids and scorpionflies, for example, males provide food to the female at mating (see below), and insem-ination is dependent on the amount of food provided by the male (Thornhill 1976b, Gwynne *et al.* 1984). When males provide prezygotic nutritious gifts or parental care (Smith 1979a, b), it would appear to be difficult to obtain fertiliza-tions without the concomitant investment. Therefore, in these species, increased confidence of paternity is expected to have evolved (see Thornhill 1979). In species where females are likely to mate more than once, paternity assurance mech-anisms have to minimize fertilizations by sperm from competing males or at least delay inseminations so that the current nurturant male obtains a maximum number of fertilizations for his fewer matings.

Mechanisms to achieve this include mate guarding, sperm plugs, and sperm displacement mechanisms (see Parker 1970a). Sperm displacement occurs when the sperm of the last male to mate with a female fertilizes most of her eggs. Sperm competition results when the ejaculates from two or more males overlap and com-pete for fertilizations of a single female's eggs. Parker (1970a) was the first to dis-cuss the evolutionary consequences of sperm competition and argued that insects are preadapted for high levels of this potent form of male-male competition because female insects often mate many times before eggs are fertilized and because sperm are usually stored and remain viable until the female's death.

Many insect species have now been examined for levels of sperm competition (reviewed by Parker 1970a, and Boorman and Parker 1976). These reviews have revealed that some degree of sperm competition almost invariably occurs after a second insemination and that **usually** the ejaculate of the last male to mate com-petes with the previous males' ejaculates to such a degree that he acquires more than 50% of the fertilizations. However, a great deal of variability does exist in the amount of sperm precedence an ejaculate from the last male is able to achieve.

Differences in sperm precedence levels exist within species depending on such variables as time between matings and number of matings. There are also great differences in levels of precedence among species. Despite this fact, these dif-ferences are commonly glossed over in studies of insect mating systems and it is assumed that the last male to mate obtains most of the fertilizations.

Parker (1970a, b) interpreted the sperm precedence to reflect the fact that each male is maximizing his total number of egg fertilizations rather than maximizing egg fertilizations per female. The argument presented below is that additional variability for sperm precedence levels among species may be explained by considering whether males of a particular species invest or are likely to invest in nonpromiscuous ME or PE. An increase in either should therefore result in increased sperm precedence.

VI. INSECT SPERM COMPETITION STUDIES

Most studies of insect sperm competition use one of two methods to determine paternity in experiments where more than one male is mated to a single female; the offspring are identified by genetic markers or one of the fathers is sterilized so that nonviable eggs are assumed to be his offspring. A problem inherent in both methods is the bias that may result from differences in the competitiveness of sperm types. Sterile male studies are particularly sensitive because sterile sperm may not be as vigorous as normal sperm. A second problem with sterile male studies is that a certain proportion of eggs fertilized by a normal male do not hatch and a certain proportion of eggs fertilized by a "sterile" male may hatch; thus paternity may be assigned incorrectly in some cases. The first problem is ameliorated somewhat by reciprocal matings, *i.e.*, the level of sperm displacement is calculated by taking the mean of two values, one with the "marked male" (mutant or sterile) mating last and the reciprocal, the normal male last. Boorman and Parker (1976) published a method to overcome the second problem; they developed a formula that incorporates values from "control" matings (*i.e.*, with normal males and with just sterile males) along with values from the reciprocal matings to determine a much less biased sperm precedence value. Their actual precedence value, termed the P_2 value, is the proportion of the offspring fathered by the last male to mate. Thus, 0.5 indicates complete mixing, and values between 0.5 and 1.0 indicate some level of sperm precedence by the last male.

Table I presents P_2 values from all available sperm competition studies. Boorman and Parker (1976) calculated many of these values (indicated in Table I), but P_2 values are included that were calculated from studies postdating Boorman and Parker. In many cases where control values and P_2 values for reciprocal matings were not given, they were calculated from figures or tables presented by the authors of the respective papers. If several experimental regimes were employed (*e.g.*, differing times between matings) and if information was available from the literature, I attempted to use the most biologically meaningful data (*i.e.*, data from laboratory matings approximating normal field conditions). For details refer to each species in Table I.

TABLE I

P_2 Values and Male Investment or Possible Investment

Species (Family)	Reference	P_2 Value	Male Investment	Notes
ORDER ODONATA				
Calopteryx maculata (Calopterygidae)	Waage 1979a	very high	Male postcopulatory guarding of female	Male's penis removes almost all of the sperm from previous matings.
ORDER ORTHOPTERA				
Schistocerca gregaria (Acrididae)	Hunter-Jones 1960	.99		
Locusta migratoria (Acrididae)	Parker and Smith 1975	.86		Only data from one mating per oviposition cycle used as females under natural conditions are reluctant to mate more frequently (Mika 1959 in Parker and Smith 1975).
Paratettix texanus (Tetrigidae)	Nabours 1927	.72	Large proteinaceous spermatophore	
Blattella germanica (Blattidae)	Cochran 1979	.43	Male uric acid contribution	
ORDER HEMIPTERA				
Abedus herberti (Belostomatidae)	Smith 1979b	.99	Postcopulatory brooding of eggs by male	

Oncopeltus fasciatus (Lygaeidae)	Economopoulus and Gordon 1972	.72	
Dysdercus koenigii (Pyrrocoreidae)	Harwalker and Rahalkar 1973	.66	Females mate only once between clutches so only data on first clutch after mating was used.
Nezara viridula (Pentatomidae)	McLain 1981	.75	Possible nutritional benefits to female from mating
ORDER COLEOPTERA			
Tribolium confusum (Tenebrionidae)	Vardell and Brower 1978	.82	Data from time after second male removed.
Tribolium castaneum (Tenebrionidae)	Schlager 1960	$\cong .99^*$	
Trogoderma inclusum (Dermestidae)	Vick et al. 1972	.52	
Anthonomus grandis (Curculionidae)	Bartlett et al. 1968	.59	Females mate more than once per day in the field (Mitchell 1967); thus only data for one day or less used. (Previous studies [see Boorman and Parker 1976] ignored this.)
Conotrachelus nenuphar (Curculionidae)	Huettel et al. 1976	.53	
Epilachna varivestis (Coccinellidae)	Webb and Smith 1968	.70	
ORDER LEPIDOPTERA			
Spodoptera frugiperda (Noctuidae)	Snow et al. 1970	$.72^*$	

127

TABLE I

P_2 Values and Male Investment or Possible Investment
(Continued)

Species (Family)	P_2 Value	Reference	Male Investment	Notes
Spodoptera litura (Noctuidae)	1.00	Etman and Hooper 1979		P_2 value determined from the fact that second mating flushes all previous sperm.
Trichoplusia ni (Noctuidae)	.92	North and Holt 1968		
Heliothis virescens (Noctuidae)	.99	Flint and Kressin 1968		"All or none" displacement, suggesting a sperm plug, P_2 values therefore calculated only from data where displacement occurred.
Heliothis virescens (Noctuidae)	.80	Pair *et al.* 1977	Transfer of zinc by males	"Sterile males" in experiments resulted from 4-8th generation backcrosses with *H. subflexa*.
Choristoneura fumiferana (Tortricidae)	0	Retnakaran 1974		Females call and mate at most once every 4 h (in field). In cases where females mated twice in one day (8% of sample), displacement was essentially all or none.
Laspeyresia pomonella (Olethreutidae)	.58*	Proverbs and Newton 1962		
Plodia interpunctella (Pyralidae)	.68	Brower 1975	Transfer of amino acids by male (Greenfield 1982)	

Species	P_2	Reference	Notes
Bombyx mori (Bombycidae)	.34(?)*	Omura 1939	
Colias eurytheme (Pieridae)	1.00	Boggs and Watt 1981	Transfer of amino acids and lipids by males (Table II)
Euphydryas editha (Nymphalidae)	.919	Labine 1966	Calculated (from Labine 1966). Data taken after second mating to the point where natural infertility starts (5 days).
Papilio spp. (Papilionidae)	1.00*	Ae 1962, Clarke and Sheppard 1962	
ORDER DIPTERA			
Drosophila melanogaster (Drosophilidae)	>.85	Gromko et al, this volume	Evidence that P_2 increases with increasing interval between matings (1-14 days). Females under natural conditions typically do not remate for 4-6 days (Gromko and Pyle 1978).
Dacus oleae (Tephritidae)	.60*	Cavalloro and Delrio 1974	
Ceratitis capitata (Tephritidae)	.63*	Katiyar and Ramirez 1970	
Rhagoletis pomonella (Tephritidae)	.76	Myers et al. 1976	
Culicoides mellitus (Ceratopogonidae)	.28	Linley 1975	

129

TABLE I

P_2 Values and Male Investment or Possible Investment
(Continued)

Species (Family)	Reference	P_2 Value	Male Investment	Notes
Glossina austeni (Glossinidae)	Curtis 1968	.25*		Female larviposits.
Aedes aegypti (Culicidae)	George 1967	.07		Culicids studied have an effective sperm plug. Data not available to calculate P_2 values but they appear to be low (see Parker 1970a).
Scatophaga stercoraria (Scatophagidae)	Parker 1970	.81*	Postcopulatory guarding of female	
ORDER HYMENOPTERA				
Dahlbominus fuscipennis (Eulophidae)	Wilkes 1966	.38(if remating occurs within 5 min) .20(with interval of 10-12 days)		Many wasps emerge from one host. Females mate soon after emergence and usually do not remate (*D. fusci-pennis*, Baldwin *et al* 1964; *N. vitri-pennis*, Holmes 1974).
Nasonia vitripennis (Pteromalidae)	Holmes 1974	very low		Data presented does not allow determination of P_2.

*from Boorman and Parker 1976.

VII. PATERNAL INVESTMENT AND SPERM PRECEDENCE

There are several mechanisms by which male insects increase nonpromiscuous ME: (1) In some species males provide nourishment to females during mating. The forms of nourishment transfer were reviewed by Thornhill (1976a) and include food captured or collected by males and fed to females during copulation (*e.g.*, nuptial prey items of some scorpionflies), nourishment from male glandular products including salivary secretions (some flies [Diptera] and scorpionflies), secretions from glands in the integument (dorsal and tibial glands of some crickets) and products passed with the sperm (*e.g.*, spermatophores of katydids which are eaten by the female; secretions in or with the spermatophore and absorbed by females of some grasshoppers and Lepidoptera); and female cannibalism of the male soma following mating. (2) Males may invest time and energy guarding the female, commonly while she oviposits. This tactic is thought to be most advantageous when the last male to mate obtains most of the fertilizations. (3) The male may guard or brood the eggs laid by the female after mating (rare in insects and restricted to a few families in the order Hemiptera; Smith 1976a, b, 1980).

There are sperm competition data available for only a few species in which males are known to invest more than promiscuous ME. These are species with more obvious forms of nonpromiscuous ME such as postcopulatory guarding of mates or eggs. More subtle investments are found in some species of Lepidoptera and Orthoptera in which males transfer proteins and other potentially nutritious substances into the female's reproductive tract at mating. Little information on levels of sperm precedence is available for species in which males are known to transfer nutrients. This information is supplemented with data available for related species with similar life histories. Levels of sperm precedence of species in which males invest by nonpromiscuous ME and "probable" nonpromiscuous ME species are examined below.

A. Male Provides Nourishment

1. Orthoptera

Female cockroaches are known to feed on uric acid provided by males from specialized storage glands. Uric acid is either voided onto spermatophores subsequently eaten by females or is consumed directly by females feeding on exposed uricose glands of males following mating (Schal and Bell 1982). The translocation of labeled nutrients from male uricose glands to eggs in the female has been demonstrated for both *Blattella germanica* (Mullins and Keil 1980) and *Xestoblatta hamata* (Schal and Bell 1982). These papers presented evidence that male-donated nutrients are a significant contribution to the nutrition of females and suggested that these nutrients represented paternal investment. However, as uric acid is a common waste product of insects (Chapman 1971), males may store urates and use mating as a means of excretion (Roth and Dateo 1964) in order to donate their waste products to fernale reproduction. Thus urate production may not represent a large cost to males (although production of special glands do cost something). This might explain the low P_2 value for *B. germanica* (Table I) which indicates complete mixing of ejaculates.

The acridid grasshoppers is another group of insects in which males pass protein-aceous products into the female's genital tract at mating. A thorough study by Friedel and Gillott (1977) showed that specialized proteins, produced in the accessory glands of male *Melanoplus sanguinipes,* are transferred with the spermatophore into the female at mating. These proteins are not broken down but are incorporated intact into the ovaries within 24 h. Sperm displacement information from the acridids *Schistocerca gregaria* and *Locusta migratoria* (Table I) shows a very high level of displacement. Basic reproductive biology is similar for the three species. Females mate more than once (Table II) and each copulation precedes an oviposition bout. Parker and Smith (1975) were able to obtain two matings within one cycle with *L. migratoria,* and this decreased the P_2 to 0.36. However, Mika (1979) indicated that females in the field are reluctant to remate within one mating-oviposition cycle. High male investment may likely explain the extremely high P_2 values for *L. migratoria* and *S. gregaria.*

Male grouse locusts (Tetrigidae) produce a very large proteinaceous spermatophore, some one-fifth of adult body length, which is apparently digested internally by the female (Farrow 1963). Male *Paratettix texanus* show a high level of sperm precedence by the second male (Table I).

Male nutrient investment at mating is known for several other families of Orthoptera (see Gwynne 1983). In tree crickets (Gryllidae) it has been shown that female feeding on male glandular secretions increases the number of eggs females lay (Bell 1979).

Males of most species of katydids (Tettigoniidae) produce a large mass of proteinaceous material attached to the spermatophore. The female eats this after mating (Gwynne 1983). Katydid spermatophores are important to female reproduction; nutrients are translocated to maturing eggs (Bowen *et al.* 1984) and additional spermatophores eaten by females increase both the number and weight of eggs they produce (Gwynne, unpubl. data). In two species with a very large spermatophore, females compete for access to males capable of producing a spermatophore (Gwynne 1981, 1983), indicating that female reproduction may be limited by spermatophore nutrients. There are no sperm competition data on katydids, but their large male investment and the consequent prediction of a high level of sperm displacement make this group a prime candidate for future study.

2. Lepidoptera

Males of several species of butterflies and most moths transfer spermatophores and other accessory gland secretions to the female at mating. Studies that involved labeled proteinaceous components of these secretions have shown that they are absorbed into the female's genital tract and used in part to produce eggs (Table II). In addition to proteins, male products known to be absorbed by females are lipids (Marshall 1980) and trace elements; Engebretson and Mason (1980) showed that male tobacco budworm moths, *Heliothis virescens,* transfer over a third of their zinc reserves to the female at mating. Much of this is translocated to eggs (Table I). Prezygotic nutrient investments by lepidopterans appear to represent a substantial cost to the male as they inhibit his ability to copulate again (Rutowski 1979). In addition, females of two species of butterflies with large spermatophores are known to solicit courtship from males; this suggests that they actively seek spermatophore nutrients (Rutowski, 1981, Rutowski *et al.* 1981).

Sperm competition information is available for only three species in which transfer of male nutrients is known. In two species the P_2 values are high; for the butterfly *Colias eurytheme* there is complete sperm precedence by the second male and in the moth *Heliothis virescens* two estimates of the P_2 place it between 0.80 and 0.99 (Table I). For the moth *Plodia interpunctella* proteinaceous substances are transferred at mating but sperm precedence by the male is lower (P_2 = 0.68). However, Greenfield (1981) presents evidence that male contributions may add little to female fecundity in this species.

These studies of male-contributed nutrients involve three families of butterflies and two families of moths, so similar processes may likely be found in related species with similar life history characteristics. There are sperm competition data available for a number of other lepidopteran species (Table I). Here I will review several life history characteristics that should be expected in species with male contributed nutrients (Table II), and examine the P_2 values for the Lepidoptera (Table I) to see if species with these life histories have higher levels of last male sperm precedence.

These species of lepidopterans have long-lived adult females, most of which mate more than once (Table II). Several authors have argued that multiple mating by adult females may function to obtain nutritional resources from males (Thornhill 1976a, Byers 1978, Alcock *et al.* 1978) and females of the species in which males contribute nutrients do mate more than once (Table II). Byers (1978) stated that a paternal nutrient investment is most likely to be found in those species of Lepidoptera that are long-lived as adults and that mature eggs after eclosion. Studies by Gilbert (1972) showed that some long-lived adult female *Heliconius* species feed on a mixture of both pollen and nectar. This group of butterflies uses additional factors (possibly amino acids) to mature eggs later in adult life, after eclosion, and acquires these nutrients from pollen (Dunlap-Pianka *et al.* 1977) and

TABLE II

Some Life History Characteristics of Orthoptera and Lepidoptera in Table I

Species (Family)	Type of Male-Produced Nutrition	Female Mating Frequency	Adult Stage
ORTHOPTERA			
Melanoplus sanguinipes (Acrididae)	Accessory gland protein contribution (Friedel and Gillott 1977).	Mean of 2.1 per day, but as many as 5 times per day (Pickford and Gillott 1972).	
Schistocerca gregaria (Acrididae)		Many times, copulation appears to stimulate egg laying (Norris 1954, in Hunter-Jones 1960).	
Locusta migratoria (Acrididae)		Female reluctant to remate within one oviposition cycle but many egg pods are laid (Mika 1959, in Parker and Smith 1975).	
Blattella germanica (Blattidae)	Uric acid contribution to female and eggs (Mullins and Keil 1980).	Mating twice between egg cases rare; about 20% of females remate after each egg case (Cochran 1979).	
LEPIDOPTERA			
Lymire edwardsii (Ctenuchidae)	Amino acid contribution to female and eggs (Goss 1977).		
Colias eurytheme (Pieridae)	Amino acid (Boggs and Watt 1981) and lipid (Marshall 1980) contribution.	Up to 3 times (field spermatophore counts) (Rutowski et al. 1981).	

Species			
Danaus plexippus (Danaidae)	Amino acid contribution to female and eggs (Boggs and Gilbert 1979).	Up to 8 times, $\bar{X}$ = 2.51 (field spermatophore counts) (Pliske 1973).	Long-lived, adults migrate.
Heliconius erato (Nymphalidae)	Amino acid contribution to female and eggs (Boggs and Gilbert 1979).	Up to 3 times, $\bar{X}$ = 1.07 (field spermatophore counts) (Pliske 1973).	*Heliconius* spp. are long-lived as adults and mature many eggs after emergence. Some species use amino acids from pollen (Dunlap-Pianka *et al.* 1977).
Heliconius hecale (Nymphalidae)	Amino acid contribution to female and eggs (Boggs and Gilbert 1979).	Three times "on the average" (Boggs and Gilbert 1979).	
Heliconius charitonius (Nymphalidae)	Amino acid contribution to female and eggs (Boggs 1981).	Once (Ehrlich and Ehrlich 1978).	
Dryas julia (Nymphalidae)	Amino acid contribution to female and eggs (Boggs 1981).	Up to four times (Ehrlich and Ehrlich 1978).	
Euphydryas editha (Nymphalidae)		Up to 2 times, $\bar{X}$ = 1.27. Some evidence that females absorb spermatophores (field spermatophore counts) (Ehrlich and Ehrlich 1978).	This species does not mature eggs after adult molt (Labine 1968).
Papilio spp. (Papilionidae)		Up to 5 times, $\bar{X}$ = 1.73 (field spermatophore counts) (Burns 1968).	Some species very long-lived.
Spodoptera frugiperda (Noctuidae)		Up to 4 times, $\bar{X}$ = 1.73 (field spermatophore counts) (Miyashita and Fuwa 1972).	Long-lived, migrates (Swann and Papp 1972).
Trichoplusia ni (Noctuidae)			Long-lived, migrates (Swann and Papp 1972).

TABLE II

Some Life History Characteristics of Orthoptera and Lepidoptera in Table I
(Continued)

Species (Family)	Type of Male-Produced Nutrition	Female Mating Frequency	Adult Stage
Heliothis virescens (Noctuidae)	Zinc contribution to female and eggs (Engebretson and Mason 1980).	Up to 6 times, $\overline{X} = 1.33$ (field spermatophore counts) (Hendricks *et al.* 1970).	Long-lived.
Choristoneura fumiferana (Tortricidae)		Three or more times in nature (Outram 1968).	Adult females lay 1 or 2 egg masses before flying, but fecund females are known to disperse (McKnight 1968).
Plodia interpunctella (Pyralidae)	Amino acid contribution (Greenfield 1982).		Short lived.
Laspeyresia pomonella (Olethreutidae)		Up to 10 times (in laboratory) (Proverbs and Newton 1962).	Short-lived.
Bombyx mori (Bombycidae)			Very short-lived (Metcalf *et al.* 1951).

from males during mating (Boggs and Gilbert 1979). Most lepidopterans do not actively feed on pollen and, even though amino acids are present in many butterfly nectars (Gilbert and Singer 1975), adequate nutrients for egg production are probably limiting (Marshall 1982). Byers (1978) cites evidence that the quality of lepidopteran eggs can decrease as nutrients become less available over the oviposition period.

Long-lived species of Lepidoptera with multiple mating females are thus the most likely candidates for increased male nutrient investment. An example is the monarch butterfly, *Danaus plexippus;* one species examined by Boggs and Gilbert (1979) has male-transferred nutrients. The monarch, feeding exclusively on nectar, is a long-lived migratory species in which females mate up to eight times (Table II).

Table I lists nine species of Lepidoptera for which sperm competition data are available. The two butterflies, swallowtails (*Papilio* spp.) and a checkerspot (*Euphydryas editha*) have very high P_2 values. Multiple mating by females occurs in both species (Table II). Both also have long-lived adult stages; some swallowtails survive up to six months (Tyler 1975).

Values of P_2 are also presented for four species of noctuid moths (Table I). In one of these (*Heliothis virescens*) the male provides nutrients to females at mating (see preceding). Multiple mating by females is the rule for members of this family which are also characteristically long-lived as adults (Byers 1978). Many species migrate as adults, moving great distances from areas where they emerged (as adults) to spend extended periods of time feeding before ovipositing (Byers 1978). At least two of the four species of noctuids in Table I migrate (Swann and Papp 1972), all are long-lived and strong fliers, and typically mate many times (Table II). Sperm displacement is almost complete for these species (mean P_2 for the four species = 0.88, range 0.72-1.00; Table I), and is achieved in *Spodoptera litura* by flushing of all sperm from previous matings (Etman and Hooper 1979; Table I).

Female spruce budworms, *Choristoneura fumiferana,* emerge, mate, and quickly lay a large number of eggs (Sanders and Lucuik 1975). Although females live a realtively short time, they commonly migrate from their emergence area and mate two or three times (Outram 1968, 1971). Retnakaran (1974) provides sperm competition data for this species. Females mated twice within a 24-h period showed essentially either total or no sperm precedence by the second male. Repeat matings probably do not occur in nature before 24 h, because females call (*i.e.,* release pheromone) at most only once within a 24-h period (Outram 1971). In fact, only 8% of females given two males in one day actually mated twice. With 48 h between matings there is no sperm precedence by the second male, suggesting that first males somehow block further inseminations.

Sperm competition studies are available for two other moth species, the codling moth, *Laspeyresia pomonella,* and the silk moth, *Bombyx mori* (see Table I). Although females in both species mate more than once (from laboratory data;

Table I), they are relatively weak fliers and short-lived as adults. Adults of both species deposit their eggs and die within a few days after emergence (List and Newton 1921, Metcalf *et al.* 1951). Although ejaculates from second males do exhibit some sperm precedence there is much sperm mixing. The mean P_2 value is substantially lower than those reported for other moth species.

3. Other Orders

A recent paper by McLain (1981) showed that stinkbugs (*Nezara viridula*) have a reasonably high level of last male sperm precedence (P_2 = 0.75). Females of this species mate frequently and McLain (1981) suggests that they may obtain nutritional benefits from the males during copulation, citing evidence that starved nonvirgins live one-third longer than starved virgins (Mitchell and Mau 1969). Eggs are deposited in masses; the 0.75 value conceals the fact that the P_2 value is low for the first egg mass (0.141) but increases rapidly so that the third and subsequent masses show almost complete last male precedence.

Other species of insects with high levels of sperm precedence may have male nutritional investment. Evidence cited above suggests that these kinds of investments are often not obvious to the observer as they are in the form of secretions passed into the female's genital tract. Leopold (1976) reviews evidence of this sort for several other species of the above-mentioned orders as well as species in the Coleoptera and Neuroptera.

There are no available data on sperm competition for insect species in which males provide the female with a nuptial prey item (see Thornhill 1976a). Prey contributed by male scorpionflies contributes to female fitness both in terms of the number of eggs produced and in reduced risk of predation on females that acquire nuptial prey from males (Thornhill 1976b). Thornhill (1979) predicted that scorpionflies should have a high level of sperm precedence because of the male nutrient investment. He pointed out that the length of the male penis may allow flushing of previously stored sperm from the female storage organ.

B. Postcopulatory Guarding

When a male guards his mate after copulation there are time and energy costs as well as costs associated with a decrease in the rate at which other females are encountered (Parker 1974). This kind of male behavior has typically been interpreted as functioning to prevent sperm competition from other males (*i.e.,* promiscuous ME; *e.g.,* Parker 1970a; Waage, this volume). Another interpretion, however, is that the evolved function of guarding is not to exclude other males but to provide benefits to the female and ultimately the offspring (*i.e.,* nonpromiscuous ME; see above, and Thornhill, this volume). In the damselfly *Calopteryx maculata,* the male hovers around the ovipositing female, keeping away conspecific males.

Guarded females benefit from longer periods of undisturbed oviposition (Waage 1979b). In dung flies. *Scatophaga stercoraria,* the male clings to the female dorsum while she oviposits (Parker 1978a). There are several advantages that female flies obtain from male guarding, especially if a large male is involved. Female dung flies select larger males as mates (Borgia 1981) and large males provide protection from physical damage caused by male-male interactions, allow for more rapid oviposition (see also Parker 1970c), and provide quicker escape for the coupled pair in that larger males are better able to fly while coupled, thus better equipped to escape predation or other mishaps (Borgia 1981).

Data on sperm precedence are available for the above-mentioned two species (Table I). The dung fly P_2 value is higher than almost all other Diptera. Exact values are not known for *C. maculata,* but Waage (1979a, b) has shown that the penis of the male is able to remove 88-100% of sperm from previous matings. In this volume, Waage shows that penile sperm removal is likely to occur in other odonate species with postcopulatory guarding.

The high P_2 values in the Acrididae (Table I) may not only be the result of the possible male investment in nutrition as discussed above, but also the fact that both species may provide benefits to the female by remaining on her back while she oviposits (Parker 1970a).

Both hypotheses for the evolved function of postcopulatory guarding predict high levels of sperm precedence. The "female benefitting" hypothesis, however, predicts that increased sperm precedence (= increased confidence of paternity) would evolve at some point after the male starts to provide benefits (see Werren *et al.* 1980). The alternate hypothesis suggests that guarding should be a result of selection to "protect" the male's ejaculate because of a high potential for sperm displacement without gaurding. A problem separating the two hypothesis is that with an increase in any male contribution to the female (PI or nonpromiscuous ME) a high level of sperm precedence is predicted along with mechanisms to prevent female remating. One mechanism is postcopulatory guarding by males (see below). In contrast to the sperm competition hypothesis, the hypothesis suggested here always predicts some benefit to the female or offspring resulting from the male's guarding behavior.

C. Paternal Care of Eggs

Smith (1979b) has shown that males of the water bug *Abedus herberti* have a very high confidence of paternity with P_2 values averaging 0.99 (Table I). He discussed this result in the context of the paternal investment expended by males as a result of brooding eggs adhered to their wing covers by females with which they had mated. The P_2 values of this species are greater than the three other hemipterans that have been studied (Table I).

D. Other Species With High P_2 Values

Most insect species have P_2 values that indicate sperm mixing with some second male advantage (Table I). Both *Drosophila melanogaster* and two species of *Tribolium* show high levels of precedence yet are not known to have paternal investment.

Male *Drosophila* transfer a number of proteins, lipids, and other substances from their accessory glands to females at mating (Gromko *et al.,* this volume) and some of these may contribute to the nutritional requirements of the female. Multiple copulations by male *Drosophila* reduce their fertility due to a reduction of accessory gland secretions and not sperm supplies, suggesting that the secretions are costly (Lefevre and Jonsson 1962). Female *D. melanogaster* commonly remate in nature and increase fecundity as a result (Gromko *et al.,* this volume). Although remating by females prevents depletion of spermathecal sperm (Pyle and Gromko 1978), it may in addition serve to transfer accessory gland products important to female reproduction.

Male *Tribolium* transfer a spermatophore to the female along with a fluid that solidifies on contact with air (Sokoloff 1974). In *Tribolium* there appears to be no blockage to male sperm as a result of mating (Schlager 1960, Vardell and Brower 1978). Thus, products introduced by males of both *Drosophila* and *Tribolium* may represent important nutrients and, therefore, may explain the high P_2 values for these genera.

VIII. PATERNITY-ENSURING MECHANISMS

High levels of sperm precedence would not result in high confidence of paternity if females remate soon after a given mating. Males should, therefore, evolve mechanisms that prevent females from remating for a critical period of time. Various types of "sexually selected adaptations that reduce sperm competition" were discussed by Parker (1970a). These kinds of adaptatons should be especially prevalent in those species with males that provide benefits to females.

Sperm plugs as barriers to further inseminations were discussed by Labine (1964) and Ehrlich and Ehrlich (1978) for various species of butterflies. Sperm plugs are not known for species in which nutrients are transferred to females but are known in closely related species (genus *Danaus;* Ehrlich and Ehrlich 1978) and in other species with high levels of sperm precedence (*Euphydryas;* Labine 1964). Female postmating nonreceptive periods are common in Lepidoptera (Ehrlich and Ehrlich 1978) and appear to be a result of male substances introduced into the bursa copulatrix (in moths; Riddiford and Ashenhurst 1973) or pressure from the physical presence of the spermatophore in the bursa (in butterflies; Obara *et al.* 1975). In *Heliconius erato,* a species with proven nutrient transfer to females, an "antiaphrodisiac" substance is transferred from the male to the female. This substance repels males for up to several weeks (Gilbert 1976).

The internal spermatophore can act as a sperm plug in acridid grasshoppers (Parker and Smith 1975) as well as possibly providing benefits to the female (discussed previously). A similar situation would occur in odonates (Waage 1979a, b), dung flies and other insects with male postcopulatory guarding (Parker 1974). In katydids, where males make a substantial contribution in large spermatophores, females are known to have a refractory period of at least a week (Morris *et al.* 1976).

IX. DISCUSSION AND CONCLUSIONS

Parker (1970a) regarded patterns of ejaculate competition in insects as results of intermale competition. He showed that in most species males mating last achieved some measure of sperm precedence but that the amount of precedence varied and was rarely complete (Table I). He interpreted this as a result of selection maximizing the overall fertilization rate of males rather than maximizing the number of eggs fertilized per female (see Parker 1970b). If males were able to acquire many matings it is conceivable that overall fertilization rate could be maximized by mating with many females with little sperm precedence (*i.e.,* P_2 about one-half). This appears to be the case in many species (Table I). By this argument, if the cost of individual matings increases, males should obtain fewer copulations and, as a result, they should be selected to increase P_2 for each female they fertilize as well as to perfect mechanisms to prevent or delay additional matings by the female (Parker 1970a). This chapter has argued that a cost of mating for males in certain species involved parental investment such as paternal care of offspring, or what I have termed the "nonpromiscuous" component of mating effort—risks incurred and energy expended by the male in order to obtain fertilizations which benefit the female and/or her offspring. Smith (1979b) addressed this point in his explanation of why male water bugs, which invest parentally by carrying eggs, have almost perfect confidence of paternity due to a high level of sperm precedence. "Male water bugs lack the opportunity to optimize their overall fertilization rate because egg-covered males are rejected as mates by females." There is, therefore, a "limitation on the absolute number of eggs a male water bug is allowed to fertilize"

Large costs to the male for individual matings can also be derived from promiscuous mating effort, that which functions in obtaining fertilizations but does not provide immediate benefits to the female or offspring. This sort of effort includes the time taken to search for (Parker 1978b) or guard fertilizable females, or the energetic costs of providing sperm or ejaculatory materials (Dewsbury 1982). The argument presented here is that levels of sperm precedence are always expected to be higher when males expend parental investment or nonpromiscuous mating effort but not when males expend only promiscuous effort.

As stressed by Walker (1980) and Knowlton and Greenwell (this volume), most discussions of sperm competition have not considered the evolutionary interests of the female. Although males gain by displacing the sperm of previous males, additional copulations may represent either a net cost or a net benefit to the female (Knowlton and Greenwell, this volume). There are both time and energy costs in mating as well as the possibility of more severe costs such as increased probability of predation or injury (Daly 1978). If remating represents an overall cost to the female it should be in her interests to reduce the amount of precedence by the last mating male, thus reducing the value to males of copulations with nonvirgins. Indeed, low P_2 values are characteristic of several species in whch females usually mate but once (Walker 1980).

Walker (1980) argues that a high level of sperm precedence by the second male should occur in species in which females acquire benefits from males, since high levels of sperm displacement would represent a reward for males. There are several possibilities for benefits derived by females from multiple matings (see Alcock *et al.* 1978; Walker 1980; Sakaluk and Cade 1983; Knowlton and Greenwell, this volume). These include (1) insurance of an adequate sperm supply; (2) minimizing the loss of time and energy required to resist male advances; (3) acquiring genetic benefits; and (4) acquiring immediate benefits from the male. This last benefit (4) only, represents an expenditure of parental or nonpromiscuous mating effort by the male.

1. If females of a particular species use additional matings to replenish sperm supplies, a high P_2 is expected, but is less a "reward" than a simple mechanical outcome of the sperm storage organ of the female containing mainly sperm from the additional mating. In fact, if a female of this same species obtains no other benefit from mating, a low P_2 (as a "nonreward" to males; see above) is predicted if she is forced to remate before her sperm supplies ave been used up.

2. Mated females should not provide a high P_2 "reward" to males when the benefit is a relative one due to the time and energy saved in allowing persistent males to mate rather than repelling them. Indeed, a decrease in P_2 is once again expected under these circumstances.

3. There is some debate over whether females can theoretically obtain genetic benefits from remating. One suggested function of multiple mating by females is to increase the genetic diversity of offspring. Although Williams (1975) argues that multiple mating would add little to the genetic variability in offspring obtained from a single mating, Parker (this volume) indicates that in unpredictable environments this strategy by females might, on average, be more successful. Also presented is the argument that females may remate in order to compensate for a previous mating with a genetically inferior male. Parker (this volume) predicts that multiple mating under these conditions will

not, on average, be more successful (but see chapter by Knowlton and Greenwell, this volume, which suggests that benefits of multiple mating may come from a reduction in the variance of female reproductive success).

4. In species where males provide benefits to their mates and/or offspring, females are expected to multiply mate and to reward males with a high confidence of paternity via high levels of sperm displacement in combination with mechanisms to prevent ejaculate displacement (Walker 1980; Knowlton and Greenwell, this volume). There is good evidence that certain types of male mating or parental effort do provide benefits by increasing fecundity, via supplied nutrition in crickets, katydids, and scorpionflies; by preventing interference with oviposition in damselflies; by decreasing risk of predation in scorpionflies; and by providing care of eggs in giant water bugs (for references, see preceding). Indeed, for water bugs (Smith 1979a) and certain katydids (Gwynne 1981, 1983), the services males supply to females have apparently become limiting on female reproduction; females of these species appear to be limited in their total number of matings by sexual competition with other females for male donations (Gwynne, unpubl. data).

Two contributions to the present volume have outlined theoretical outcomes to sexual conflict over the degree of sperm competition. Parker (this volume) suggests that male interests should usually outweigh the costs to females. However, as Knowlton and Greenwell (this volume) point out, selection may be more intense on males, but females have physical control over their bodies; as a result they are in a direct position to manipulate ejaculates (see also Lloyd 1979). Thus, although it is usually in the female's interest to lower P_2 (even if remating represents a low cost to her—see preceding), if relative benefits to the male in achieving a higher P_2 are much greater than the benefit to the female of decreasing P_2, males are expected to evolve adaptations to achieve higher levels of sperm precedence. As stated above, there should be strong selection on males to achieve high levels of sperm precedence when the costs of individual matings are high. If this cost is due to promiscuous mating effort only, the degree of precedence by the last male to mate should be determined by the cost to the female of the additional mating. However, if this cost to males imparts important material benefits to his mate or her offspring, P_2 levels are always expected to be high. The available information on sperm competition and these sorts of male donations in insects supports this prediction.

X. SUMMARY

Although male parental investment is rare in insects, males of a number of species from different orders provide benefits to the female by guarding her or giving her food during mating. These prezygotic male donations are mating effort rather than parental effort as they apparently function to secure matings and are not directed at individual offspring. However, as these male efforts are likely to be costly, they are expected to influence the operation of sexual selection on the sexes in a similar way to parental investment. Thus, they represent a "non-promiscuous" form of mating effort in contrast to "promiscuous" mating effort which functions to secure matings but does not provide benefits to females or offspring.

Males that invest via nonpromiscuous mating effort or parental effort are expected to have a high confidence of paternity. This hypothesis is examined for insects by comparing the measured levels of last male sperm precedence in a number of different species. The data available support the hypothesis: Species in which males provide or are likely to provide direct benefits to their mates show the highest levels of last male sperm precedence.

ACKNOWLEDGMENTS

I appreciate very much the valuable comments and criticisms of the following people: Ron Aiken, John Alcock, Steve Austad, Carol Boggs, Mike Cunningham, Gary Dodson, Howard Evans, Rob Longair, Larry Marshall, Kevin O'Neill, Geoff Parker, Ron Rutowski, Scott Sakaluk, Steve Schuster, Bill Shields, Bob Smith, Jon Waage, Bruce Woodward, and Sam Zeveloff. Extra thanks go to Randy Thornhill who contributed many ideas and suggestions which have greatly improved the chapter. The manuscript was written while my research and I were supported by a grant (BNS-7912208) from the National Science Foundation (U.S.A.) and by a Queen Elizabeth II Fellowship (Australia).

REFERENCES

Ae, S. A. 1962. Some problems in hybrids between *Papilio bianor* and *P. maakii*. *Academia* 33:21-38. (Published by Nazan Univ.)

Alcock, J., E. M. Barrows, G. Gordh, L. J. Hubbard, L. Kirkendall, D. W. Pyle, T. L. Ponder, and F. G. Zalom. 1978. The ecology and evolution of male reproductive behaviour in the bees and wasps. *Zool. J. Linn. Soc.* 64:293-326.

Alcock, J. 1979. Multiple mating in *Calopteryx maculata* (Odonata:Calopterygidae) and the advantage of non-contact guarding by males. *J. Nat. Hist.* 13:439-446.

Alexander, R. D., and G. Borgia. 1979. On the origin and basis of the male-female phenomenon. In *Sexual Selection and Reproductive Competition in Insects*, M. S. Blum and N. A. Blum (eds.), pp. 417-440. Academic Press, New York.

Alexander, R. D., J. L. Hoogland, R. D. Howard, K. M. Noonan, and P. W. Sherman. 1979. Sexual dimorphisms and breeding systems in pinnipeds, ungulates, primates and humans. In *Evolutionary Biology and Human Social Behavior: An Anthropological Perspective*, N. A. Chagnon and W. Irons (eds.), pp. 402-435. Duxbury Press, North Scituate, MA.

Baldwin, W. F., E. Shaver, and A. Wilkes. 1964. Mutants of the parasitic wasp *Dahlbominus fuscipennis* (Zett.) (Hymenoptera:Eulophidae). *Can. J. Genet. Cytol.* 6:453-466.

Bartlett, A. C., E. B. Mattix, and N. M. Wilson. 1968. Multiple matings and use of sperm in the boll weevil, *Anthonomus grandis*. *Ann. Entomol. Soc. Am.* 61:1148-1155.

Bell, P. 1979. Mate choice and mating behavior in the black horned tree cricket, *Oecanthus nigricornis* (Walker). M.S. thesis, Univ. Toronto.

Blumer, L. S. 1979. Male parental care in the bony fishes. *Q. Rev. Biol.* 54:149-161.
Boggs, C. L. 1981. Selection pressures affecting male nutrient investment at mating in Heliconine butterflies. *Evolution* 35:931-940.
Boggs, C. L. and L. E. Gilbert. 1979. Male contribution to egg production in butterflies: Evidence for transfer of nutrients at mating. *Science* 206:83-84.
Boggs, C. L., and W. B. Watt. 1981. Population structure of Pierid butterflies. IV. Genetic and physiological investment in offspring by male *Colias*. *Oecologia* 50:320-324.
Boorman, E., and G. A. Parker. 1976. Sperm (ejaculate) competition in *Drosophila melanogaster*, and the reproductive value of females to males in relation to female age and mating status. *Ecol. Entomol.* 1:145-155.
Borgia, G. 1979. Sexual selection and the evolution of mating systems. In *Sexual Selection and Reproductive Competition in Insects*, M. S. Blum and N. A. Blum (eds.), pp. 19-80. Academic Press, New York.
Borgia, G. 1981. Mate selection in the fly *Scatophaga stercoraria:* Female choice in a male-controlled system. *Anim. Behav.* 29:71-80.
Bowen, B. J., C. G. Codd, and D. T. Gwynne. 1984. The katydid spermatophore (Orthoptera: Tettigoniidae): Male nutritional investment and its fate in the mated female. *Aust. J. Zool.* in press.
Brower, J. H. 1975. Sperm precedence in the Indian meal moth, *Plodia interpunctella. Ann. Entomol Soc. Am.* 68:78-80
Burns, J. M. 1968. Mating frequency in natural populations of skippers and butterflies as determined by spermatophore counts. *Proc. Nat. Acad. Sci.* 61:852-859.
Byers, J. R. 1978. Biosystematics of the genus *Euxoa* (Lepidoptera:Noctuidae). X. Incidence and level of multiple mating in natural and laboratory populations. *Can. Entomol.* 110: 193-200.
Cavalloro, R., and G. Delrio. 1974. Mating behavior and competitiveness of gamma-irradiated olive fruit flies. *J. Econ. Entomol.* 67:253-255.
Chapman, R. F. 1971. *The Insects: Structure and Function*. Elsevier, New York.
Clarke, C. A., and P. M. Sheppard. 1962. Offspring from double matings in swallowtail butterflies. *Entomol.* 95:199-203.
Cochran, D. G. 1979. A genetic determination of insemination frequency and sperm precedence in the German cockroach. *Entomol. Exp. Appl.* 26:259-266.
Corbet, P. S. 1962. *A. Biology of Dragonflies*. Witherby, London.
Cunningham, M. A. 1979. Sexual dimorphism, paternal care and female mate preferences in bluntnose minnows, *Pimephales notatus.* M.S. thesis, Ohio State University, Columbus, OH.
Curtis, C. F. 1968. Radiation sterilization and the effect of multiple mating of females in *Glossina austeni. J. Insect Physiol.* 14:1365-1380.
Daly, M. 1978. The cost of mating. *Am. Nat.* 112:771-7744.
Dewsbury, D. A. 1982. Ejaculate cost and male choice. *Am. Nat.* 119:601-610.
Downhower, J. and L. Brown. 1981. The timing of reproduction and its behavioral consequences for mottled sculpins (*Cottus bairdi*). In *Natural Selection and Social Behavior: Recent Advances in Theory and Research*, R. D. Alexander and D. W. Tinkle (eds.), pp. 78-95. Chiron Press, Portland, OR.
Dunlap-Pianka, H., C. L. Boggs, and L. E. Gilbert. 1977. Ovarian dynamics in Heliconine butterflies: Programmed senescence versus eternal youth. *Science* 197:489-490.
Economopoulos, A. P., and H. T. Gordon. 1972. Sperm replacement and depletion in the spermatheca of the S and CS strains of *Oncopeltus fasciatus. Entomol. Exp. Appl.* 15: 1-12.
Ehrlich, A. H., and P. R. Ehrlich. 1978. Reproductive strategies in the butterflies. I. Mating frequency, plugging, and egg number. *J. Kans. Entomol. Soc.* 51:666-697.
Engebretson, J. A., and W. H. Mason. 1980. Transfer of ^{65}Zn at mating in *Heliothis virescens. Environ. Entomol.* 9:119-121.
Etman, A. A. M., and G. H. S. Hooper. 1979. Sperm precedence of the last mating in *Spodoptera litura. Ann. Entomol. Soc. Am.* 72:119-120.
Farrow, R. A. 1963. The spermatophore of *Tetrix* Latreille (Orthoptera:Tetrigidae). *Entomol. Mon. Mag.* 99:217-233.

Flint, H. M., and E. L. Kressin. 1968. Gamma irradiation of the tobacco budworm: Steriliza-tion, competitiveness, and observations on reproductive biology. *J. Econ. Entomol.* **61**: 477-483.

Friedel, T., and C. Gillott. 1977. Contribution of male-produced proteins to vitellogenesis in *Melanoplus sanguinipes*. *J. Insect Physiol.* **23**:145-151.

George, J. A. 1967. Effect of mating sequence on egg-hatch from female *Aedes aegypti* (L.) mated with irradiated and normal males. *Mosq. News* **27**:82-86.

Gilbert, L. E. 1972. Pollen feeding and reproductive biology of *Heliconius butterflies. Proc. Nat. Acad. Sci. U.S.A.* **69**:1403-1407.

Gilbert, L. E. 1976. Postmating female odor in *Heliconius* butterflies: A male-contributed aphrodisiac? *Science* **193**:419-420.

Gilbert, L. E., and M. C. Singer. 1975. Butterfly ecology. *Ann. Rev. Ecol. Syst.* **6**:365-397.

Gladstone, D. E. 1979. Promiscuity in monogamous colonial birds. *Am. Nat.* **114**:545-557.

Goss, G. J. 1977. The interaction between moths and pyrrolizidine alkaloid-containing plants including nutrient transfers via the spermatophore in *Lymire edwardsii* (Ctenuchidae). Ph.D. dissertation, University of Miami, Miami, FL.

Greenfield, M. D. 1982. The question of paternal investment in Lepidoptera: Male-contributed proteins in *Plodia interpunctella. Int. J. Invert. Reprod.* **5**:323-330.

Gromko, M. H., and D. W. Pyle. 1978. Sperm competition, male fitness, and repeated mating by female *Drosophila melanogaster. Evolution* **32**:588-593.

Gwynne, D. T. 1981. Sexual difference theory: Mormon crickets show role reversal in mate choice. *Science* **213**:799-780.

Gwynne, D. T. 1983. Male nutritional investment and the evolution of sexual differences in Tettigoniidae and other Orthoptera. In *Orthopteran Mating Systems: Sexual Competi-tion in a Diverse Group of Insects,* D. T. Gwynne and G. K. Morris (eds.), pp. 337-366. Westview Press, Boulder, CO.

Gwynne, D. T., B. J. Bowen, and C. G. Codd. 1984. The function of the katydid spermato-phore and its role in fecundity and insemination (Orthoptera:Tettigoniidae). *Aust. J. Zool.,* in press.

Harwalker, M. R., and G. W. Rahalkar. 1973. Sperm utilization in the female red cotton bug. *J. Econ. Entomol.* **66**:805-806.

Hendricks, D. E., H. M. Graham, and A. T. Fernandez. 1970. Mating of female tobacco bud-worms and bollworms collected from light traps. *J. Econ. Entomol.* **63**:1228-1231.

Holmes, H. B. 1974. Patterns of sperm competition in *Nasonia vitripennis. Can. J. Genet. Cytol.* **16**:789:795.

Huettel, M. D., C. O. Calkins, and A. J. Hill. 1976. Allozyme markers in the study of sperm precedence in the plum curculio, *Conotrachelus nenuphar. Ann. Entomol. Soc. Am.* **69**: 465-468.

Hunter-Jones, P. 1960. Fertilization of eggs of the desert locust by spermatozoa from succes-sive copulations. *Nature* **185**:336.

Katiyar, K. P., and E. Ramirez. 1970. Mating frequency and fertility of mediterranean fruit fly females alternatively mated with normal and irradiated males. *J. Econ. Entomol.* **63**: 1247-1250.

Labine, P. A. 1964. Population biology of the butterfly, *Euphydryas editha*. I. Barriers to multiple inseminations. *Evolution* **18**:335-336.

Labine, P. A. 1966. The population biology of the butterfly, *Euphydryas editha*. IV. Sperm precedence - a preliminary report. *Evolution* **20**:580-586.

Labine, P. A. 1968. The population biology of the butterfly, *Euphydryas editha.* VIII. Ovi-position and its relation to patterns of oviposition in other butterflies. *Evolution* **22**:799-805.

Lefevre, G., and U. Jonsson. 1962. Sperm transfer, storage, displacement and utilization in *Drosophila melanogaster. Genetics* **47**:1719-1736.

Leopold, R. A. 1976. The role of male accessory glands in insect reproduction. *Ann. Rev. Entomol.* **21**:199-221.

Linley, J. R. 1975. Sperm supply and its utilization in doubly inseminated flies, *Culicoides melleus. J. Insect Physiol.* **21**:1795-1788.

List, G. M., and J. H. Newton. 1921. Codling moth control for certain sections of Colorado. *Agric. Exp. Stn. Colo. Agric. Coll. Bull. 268.*

Lloyd, J. 1979. Mating behavior and natural selection. *Fla. Entomol.* **62**:17-34.

Loiselle, P. V. 1978. Prevalence of male brood care in teleosts. *Nature* **275**:98.

Low, B. S. 1978. Environmental uncertainty and the parental strategies of marsupials and placentals. *Am. Nat.* **112**:197-213.

Marshall, L. D. 1980. Paternal investment in *Colias philodice-eurytheme* butterflies (Lepidoptera:Pieridae). M.S. thesis, Arizona State University, Tempe, AZ.

Marshall, L. D. 1982. Male nutrient investment in the Lepidoptera: What nutrients should males invest? *Am. Nat.* **120**:273-279.

Maynard Smith, J. 1978. *The Evolution of Sex.* Cambridge University Press, Cambridge.

McKnight, M. E. 1968. A literature review of the spruce, western, and 2-year-cycle budworms. *U.S. Dep. Agric. For. Serv. Res. Pap. RM-44.*

McLain, D. K. 1981. Sperm precedence and prolonged copulation in the southern green stink bug *Nezara viridula. J. Georgia Entomol. Soc.* **16**:70-77.

Metcalf, C. L., W. P. Flint, and R. L. Metcalf. 1951. *Destructive and Useful Insects, Their Habits and Control.* McGraw-Hill, New York.

Mika, G. 1959. Über das Paarungsverhalten der Wanderheuschrecke *Locusta migratoria* R. und F. und deren Abhängigkeit vom Zustand der inneren Geschlechtsorgane. *Zool. Beitr.* **4**: 153-203.

Mitchell, H. C. 1967. Natural boll weevil behavior: Interrelationships of field movements, matings and oviposition. M.S. thesis, Mississippi State University, Mississippi State, MS.

Mitchell, W. C., and R. F. L. Mau. 1969. Sexual activity and longevity of the southern green stink bug *Nezara viridula. Ann. Entomol. Soc. Am.* **62**:1246-1247.

Miyashita, K., and M. Fuwa. 1972. The occurrence time, reiterative ability, and duration of mating in *Spodoptera litura* F. (Lepidoptera:Noctuidae). *Appl. Entomol. Zool.* **7**:171-173.

Morris, G. K., G. E. Kerr, and D. T. Gwynne. 1976. Ontogeny of phonotaxis in *Orchelimum gladiator* (Orthoptera:Tettigoniidae:Conocephalinae). *Can. J. Zool.* **53**:1127-1130.

Mullins, D. E., and C. B. Keil. 1980. Paternal investment of urates in cockroaches. *Nature* **283**:567-569.

Myers, H. S., B. D. Barry, J. A. Burnside, and R. H. Rhode. 1976. Sperm precedence in female apple maggots alternately mated to normal and irradiated males. *Ann. Entomol. Soc. Am.* **69**:39-41.

Nabours, R. K. 1927. Polyandry in the grouse locust *Paratettix texanus* Handcock with notes on the inheritance of acquired characters and telegony. *Am. Nat.* **61**:531-538.

Norris, M. J. 1954. Sexual maturation in the desert locust (*Schistocerca gregaria* Forskol) with special reference of the effects of grouping. *Anti-locust Bull. No. 18.*

North, G. T., and G. G. Holt. 1968. Genetic and cytogenetic basis of radiation-induced sterility in the adult male cabbage looper *Trichoplusia ni.* In *Symposium on the Use of Isotopes and Radiation in Entomology,* I.A.E.A./F.A.O., pp. 391-403. Vienna 1967.

Obara, Y, H. Tateda, and M. Kuwabara. 1975. Mating behavior of the cabbage white butterfly, *Pieria rapae curcivora* Boisduval. V. Copulatory stimuli inducing changes of female response patterns. *Zool. Mag.* **84**:71-76.

Omura, S. 1939. Selective fertilization in *Bombyx mori. Jpn. J. Genet.* **15**:29-35. (English resumé.)

Outram, I. 1968. Polyandry in spruce budworm. *Can. Dep. For. Rural Dev. Bi-mon. Res. Notes* **24**:6-7.

Outram, I. 1971. Aspects of mating in the spruce budworm *Choristoneura fumiferana* (Lepidoptera:Tortricidae). *Can. Entomol.* **103**:1121-1128.

Pair, S. D., M. L. Laster, and D. F. Martin. 1977. Hybrid sterility of the tobacco budworm: Effects of alternate sterile and normal matings on fecundity and fertility. *Ann. Entomol. Soc. Am.* **70**:952-954.

Parker, G. A. 1970a. Sperm competition and its evolutionary consequences in the insects. *Biol. Rev.* **45**:525-567.

Parker, G. A. 1970b. Sperm competition and its evolutionary effect on copula duration in the fly *Scatophaga stercoraria. J. Insect Physiol.* **16**:1301-1328.

Parker, G. A. 1970c. The reproductive behaviour and nature of sexual selection in *Scatophaga stercoraria* L. (Diptera:Scatophagidae). V. The female's behaviour at the oviposition site. *Behaviour* 37:140-168.

Parker, G. A. 1974. Courtship persistence and female-guarding as male time investment strategies. *Behaviour* 48:157-184.

Parker, G. A. 1978a. Evolution of competitive mate searching. *Ann. Rev. Entomol.* 23:173-176.

Parker, G. A. 1978b. Searching for mates. In *Behavioural Ecology: An Evolutionary Approach*, J. R. Krebs and N. B. Davies (eds.), pp. 214-244. Sinauer, Sunderland, MA.

Parker, G. A., R. R. Baker, and V. G. F. Smith. 1972. The origin and evolution of gamete dimorphism and the male-female phenomenon. *J. Theor. Biol.* 36:529-553.

Parker, G. A., and J. L. Smith 1975. Sperm competition and the evolution of the precopulatory passive phase behaviour in *Locusta migratoria migratorioides*. *J. Entomol. Ser. A. Gen. Entomol.* 49:155-171.

Pickford, R., and C. Gillott. 1972a. Courtship behavior of the migratory grasshopper *Melanoplus sanguinipes* (Orthoptera:Acrididae). *Can Entomol.* 104:715-722.

Pickford, R., and C. Gillott. 1972b. Coupling behaviour of the migratory grasshopper *Melanoplus sanguinipes* (Orthoptera:Acrididae). *Can. Entomol.* 104:873-879.

Pliske, T. E. 1973. Factors determining mating frequencies in some new world butterflies and skippers. *Ann. Entomol. Soc. Am.* 66:164-169.

Proverbs, M. D., and J. R. Newton. 1962. Some effects of gamma radiation on the reproductive potential of the codling moth, *Carpocapsa pomonella* (L.) (Lepidoptera:Olethreutidae). *Can. Entomol.* 94:1162-1170.

Pyle, D. W., and M. H. Gromko. 1978. Repeated mating by female *Drosophila melanogaster:* The adaptive importance. *Experientia* 34:449-450.

Retnakaran, A. 1974. The mechanism of sperm precedence in the spruce budworm, *Choristoneura fumiferana* (Lepidoptera:Tortricidae). *Can. Entomol.* 106:1189-1194.

Riddiford, L. M., and J. B. Ashenhurst. 1973. The switchover from virgin to mated behavior in female *Cecropia* moths: The role of the bursa copulatrix. *Biol. Bull.* 144:162-171.

Ridley, M. 1978. Paternal care. *Anim. Behav.* 26:904-932.

Ridley, M., and C. Rechten. 1981. Female sticklebacks prefer to spawn with males whose nests contain eggs. *Behaviour* 76:152-161.

Rohwer, S. 1978. Parent cannibalism of offspring and egg raiding as a courtship strategy. *Am. Nat.* 112:429-440.

Roth, L. M., and G. P. Dateo, Jr. 1964. Uric acid in the reproductive system of the cockroach *Blatella germanica*. *Science* 146:782-784.

Rutowski, R. L. 1979. The butterfly as an honest salesman. *Anim. Behav.* 27:1269-1270.

Rutowski, R. L. 1981. Courtship solicitation by females of the checkered white butterfly, *Pieris protodice*. *Behav. Ecol. Sociobiol.* 7:113-117.

Rutowski, R. L., C. E. Long, L. D. Marshall, and R. S. Vetter. 1981. Courtship solicitation by *Colias* females (Lepidoptera:Pieridae). *Am. Midl. Nat.* 105:334-340.

Sakaluk, S., and W. Cade. 1983. The adaptive significance of female multiple mating in house and field crickets. In *Orthopteran Mating Systems: Sexual Competition in a Diverse Group of Insects*, D. T. Gwynne and G. K. Morris (eds.), pp. 319-336. Westview Press, Boulder, CO.

Sanders, C. J., and G. S. Lucuik. 1975. Effects of photoperiod and size on flight activity and oviposition in the eastern spruce budworm (Lepidoptera:Tortricidae). *Can. Entomol.* 107:1287-1299.

Schal, C., and W. J. Bell. 1982. Ecological correlates of paternal investment of urates in a tropical cockroach. *Science* 218:170-172.

Schlager, G. 1960. Sperm precedence in the fertilization of eggs in *Tribolium castaneum.* *Ann. Entomol. Soc. Am.* 53:557-560.

Selander, R. K. 1972. Sexual selection and dimorphism in birds. In *Sexual Selection and the Descent of Man, 1871-1971*, B. Campbell (ed.), pp. 180-230. Aldine-Atherton, Chicago.

Smith, R. L. 1976a. Brooding behavior of a male water bug *Belostoma flumineum* (Hemiptera: Belostomatidae). *J. Kans. Entomol. Soc.* 49:333-343.

Smith, R. L. 1976b. Male brooding behavior of the water bug *Abedus herberti* (Hemiptera: Belostomatidae). *Ann. Entomol. Soc. Am.* 69:740-747.

Smith, R. L. 1979a. Paternity assurance and altered roles in the mating behaviour of a giant water bug, *Abedus herberti* (Heteroptera:Belostomatidae). *Anim. Behav.* 27:716-725.

Smith, R. L. 1979b. Repeated copulation and sperm precedence: Paternity assurance for a male brooding water bug. *Science* 205:1029-1031.

Smith, R. L. 1980. Evolution of exclusive postcopulatory paternal care in the insects. *Fla. Entomol.* 63:63-78.

Snow, J. W., J. R. Young, and R. L. Jones. 1970. Competitiveness of sperm in female fall armyworms mating with normal and chemosterilized males. *J. Econ. Entomol.* 63:1799-1802.

Sokoloff, A. 1974. *The Biology of Tribolium, Vol. 2.* Oxford University Press, Oxford.

Swann, L. A., and C. S. Papp. 1972. *The Common Insects of North America.* Harper and Row, New York.

Thornhill, R. 1976a. Sexual selection and paternal investment in insects. *Am. Nat.* 110:153-163.

Thornhill, R. 1976b. Sexual selection and nuptial feeding behavior in *Bittacus apicalis* (Insecta: Mecoptera). *Am. Nat.* 110:529-548.

Thornhill, R. 1979. Male and female sexual selection and the evolution of mating strategies in insects. In *Sexual Selection and Reproductive Competition in the Insects,* M. S. Blum and N. A. Blum (eds.), pp. 81-121. Academic Press, New York.

Thornhill, R., and J. Alcock. 1983. *The Evolution of Insect Mating Systems.* Harvard University Press, Cambridge, MA.

Trivers, R. L. 1972. Parental investment and sexual selection. In *Sexual Selection and the Descent of Man, 1871-1971,* B. Campbell (ed.), pp. 136-179. Aldine-Atherton, Chicago.

Tyler, H. A. 1975. *The Swallowtail Butterflies of North America.* Naturegraph Publ., Healdsburg, CA.

Vardell, H. H., and J. H. Brower. 1978. Sperm precedence in *Tribolium confusum* (Coleoptera: Tenebrionidae). *J. Kans. Entomol. Soc.* 51:187-190.

Vick, K. W., W. E. Burkholder, and B. J. Smittle. 1972. Duration of mating refractory period and frequency of second matings in female *Trogoderma inclusum* (Coloeptera:Dermestidae). *Ann. Entomol. Soc. Am.* 65:790-793.

Waage, J. K. 1978. Oviposition duration and egg deposition rates in *Calopteryx maculata* (P. DeBeauvois) (Zygoptera:Calopterygidae). *Odontologica* 7:77-88.

Waage, J. K. 1979a. Dual function of the damselfly penis: Sperm removal and transfer. *Science* 203:916-918.

Waage, J. K. 1979b. Adaptive significance of postcopulatory guarding of mates and nonmates by male *Calopteryx maculata* (Odonata). *Behav. Ecol. Sociobiol.* 6:147-154.

Walker, W. F. 1980. Sperm utilization strategies in nonsocial insects. *Am. Nat.* 115:780-799.

Webb, R. E., and F. F. Smith. 1968. Fertility of Mexican bean beetles mated alternately with normal and apholate-treated males. *J. Econ. Entomol.* 61:521-523.

Werren, J. H., M. R. Gross, and R. Shine. 1980. Paternity and the evolution of male parental care. *J. Theor. Biol.* 82:690-631.

Wilkes, A. 1966. Sperm utilization following multiple insemination in the wasp *Dahlbominus fuscipennis. Can. J. Genet. Cytol.* 8:541-461.

Williams, G. C. 1975. *Sex and Evolution.* Princeton University Press, Princeton, NJ.

5

Alternative Hypotheses for Traits Believed to Have Evolved by Sperm Competition

RANDY THORNHILL

I. INTRODUCTION

Charles Darwin (1871) used the term sexual selection to refer to the nonrandom differential reproduction of individuals that results from competition for access to mates. He felt that sexual selection operates in two ways: (1) through competition between members of one sex (usually males) for members of the opposite sex, and (2) through preference by members of one sex (usually females) for certain members of the opposite sex. Since 1871, the theory of sexual selection has been greatly clarified and expanded (Fisher 1930; Bateman 1948; Parker 1970, 1979; Trivers 1972; Emlen and Oring 1977; Alexander and Borgia 1979; Borgia 1979; West-Eberhard 1979).

Sperm Competition and the Evolution
of Animal Mating Systems

It is clear that sexual selection can be more subtle than envisioned by Darwin. Male-male competition may occur between the time of insemination and conception. This form of sexual selection typically involves sperm competition—the competition between ejaculates of two or more males for the fertilization of the eggs of a single female (Parker 1970). Furthermore, male-male competition may continue between the time of conception and birth (male-induced abortion; Trivers 1972, Schwagmeyer 1979) and even after the young are born. The latter circumstance may take the form of infanticide by a male other than the father (Trivers 1972, Hrdy 1979). Female choice need not cease with insemination. Lloyd (1979), Sivinski (1980a, this volume), and Walker (1980) have hypothesized that females may choose among ejaculates of different males after insemination (see also Willson 1979). Female choice potentially includes a variety of subtle mechanisms which may operate during and after mating (Thornhill and Alcock 1983; Thornhill 1983, unpubl. data).

Sexual selection can be a very powerful agent in bringing about evolutionary change (Darwin 1871; Fisher 1930; Parker 1970, 1979; West-Eberhard 1979). That sperm competition can be a potent evolutionary force leading to male morphologies and behavior was first discussed by Parker (1970). Appropriately, Parker's 1970 paper has become a classic in evolutionary biology and behavior. It has led to many studies (*e.g.*, the papers and references in this volume) of sperm competition as the selective force that has molded male reproductive characteristics. However, the selective context of sperm competition has often been used as a theory directing research without the consideration of alternative selective contexts. This has been especially true in insect reproductive biology where sperm competition has been most broadly applied. Sperm competition is the only context considered when many investigators study insect characteristics such as copulation duration, copulatory frequencies, interactions of males and females during copulation, and post- and precopulatory interactions of females and males. I too have done this (Thornhill 1974). There are alternative selective contexts to explain these characteristics, and the consideration of alternative hypotheses is as important in the study of adaptation as in other types of scientific endeavors.

In this paper I will argue that alternative hypotheses and mutually exclusive and falsifiable predictions from the alternatives should always be considered. I will examine in detail the function of a male morphological feature in scorpionflies (Mecoptera) of the genus *Panorpa* (Panorpidae) that I initially interpreted as having evolved in the context of reducing sperm competition, but in which, as subsequent studies have indicated, another selective context has been more important. Then I will briefly address a few other characteristics that I or others have viewed to have evolved via sperm competition, and suggest alternatives that appear to need consideration before the functions of these characteristics can be understood. Finally, I briefly address the general methodologies used to study evolved function—broadly the experimental and comparative methods—in terms of usefulness, precision, and validity.

In this paper I use the terms adaptation and function in the sense of Williams (1966). The working hypothesis (as opposed to the conclusion) in evolutionary biology (at least my own brand) is that characters of organisms are adaptations evolved via selection acting among individuals. The function of a trait refers to how that trait has contributed to differential reproduction of individuals (that is, contributed to Hamiltonian [1964] inclusive fitness) in evolutionary history. All traits have multiple effects. Studies of adaptation involve efforts to understanding which of many plausible effects is most likely the function (*i.e.*, describe the trait's selective history).

II. THE DORSAL CLAMP OF MALE PANORPA

A. Background Work

I began work on scorpionflies in 1971, the year following the publication of Parker's seminal paper. The males have behavior and morphological features that I initially interpreted as evolved in the context of reducing sperm competition. As my studies developed it became more and more difficult to accept this interpretation. Furthermore, alternative hypotheses had not been rigorously considered. This led to experiments beginning in 1977 designed to determine the function of the traits. The work I discuss in this section deals with the function of a clamp-like structure on the dorsum of the male's abdomen in scorpionflies of the genus *Panorpa*. The dorsal clamp is formed from parts of the dorsum of the male's third and fourth abdominal segments. The clamp holds the female's wings during mating. (See Mickoleit [1971] for details of the structure of the dorsal clamp.) Solitary males often attempt to disrupt copulating pairs and such males are occasionally successful. This led me initially to the interpretation that the clamp functions to prevent a copulating male's mate from being usurped and inseminated by an intruder, reducing the probability of the ejaculate of the usurped male being the fertilizing ejaculate. This interpretation was in following with Parker's (1970) general functional view of male-grasping morphologies as evolved to prevent "take-over" where the copulating male is displaced from the female by an intruding male who in turn inseminates the female resulting in increased sperm competition. As discussed below, the dorsal clamp appears, instead, to function in the context of forced copulation—it increases the probability that a male can inseminate an unwilling female.

I briefly summarize pertinent background information before I discuss experiments that address the function of the dorsal clamp.

In North America, the family Panorpidae contains only the genus *Panorpa*. *Panorpa* is the largest genus of North American Mecoptera with 47 described

species (Byers and Thornhill 1983). Adult *Panorpa* are medium-sized insects that
live among the herbs of moist forests. They are relatively weak fliers. When disturbed, they fly only a few feet before coming to rest on the herbs. Adults of most
North American *Panorpa* species are nocturnal or crepuscular in their mating
activities (Thornhill 1979a). They feed throughout the day and night. Adult
Panorpa scorpionflies are scavengers. Their diet consists almost entirely of dead
arthropods that they locate among the herbs by olfaction (Thornhill 1980a).

Male *Panorpa* exhibit three alternative forms of mating behavior which are present within the behavioral repertoire of each individual male. The three alternatives
appear to represent a single conditional strategy (Thornhill 1980b, 1981). Two
behavioral alternatives employed by males to obtain copulations involve nuptial
feeding, *i.e.,* the male presents a food item to the female during courtship and the
female feeds on it throughout copulation. In one case a male feeds a female a
salivary mass that he secretes. These masses are hard and typically pillar-shaped
and are attached to a leaf during their secretion. After saliva secretion, males
stand near their salivary mass and disperse distance sex pheromone from an eversible sac among the genitalia (Thornhill 1973). The pheromone is attractive to
conspecific females at distances up to 8 m (Thornhill 1979b). A female attracted
by the pheromone feeds on the saliva during copulation, which may last a few
hours in some species. Alternatively, a male may feed a female a dead arthropod
during copulation. In this case a male locates a dead arthropod, feeds on it briefly,
and then disperses sex attractant while standing adjacent to the dead arthropod.
In both alternative behavior patterns, males display with wing movements and
abdominal vibrations to females attracted to the pheromone. Also, males defend
nuptial offerings of both types from other males that attempt to usurp them
through aggression.

The third behavior employed by males is forced copulation. A forced copulation attempt involves a male without a nuptial offering (*i.e.,* dead insect or
salivary mass) rushing toward a passing female and lashing out his mobile abdomen
at her. (Males engaging in forced copulation do not release pheromone.) If such a
male successfully grasps a leg or wing of the female with his genital claspers, he then
attempts to re-position her so as to secure the anterior edge of the female's right
forewing in his dorsal clamp. When the female's wing is secured, the male attempts
to grasp the genitalia of the female with his genital claspers. The male retains hold
of the female's wing with the dorsal clamp throughout copulation (Thornhill
1979a, 1980b). Forced copulation in *Panorpa* is in no way an abnormal or
"aberrant" behavior. It is an aspect of the evolved behavioral repertoire of individual males that is widespread among species of the genus (Thornhill 1979a, 1980b,
1981).

The behavior of females toward males with and without a nuptial offering is
distinctly different. Females flee from males that approach them without a nuptial
offering; however, females approach males with nuptial offerings and exhibit coy

behavior toward them. Also, females struggle to escape from the grasp of forceful males, but females do not resist copulation with resource-providing males (Thornhill 1979a, 1980b).

The limited resource of dead arthropods has been the important determinant of the evolution of male and female reproductive behavior in *Panorpa* (Thornhill 1979a, 1980a, b, 1981). The mating system is one of resource-defense polygyny (Thornhill 1981).

I have shown in detailed laboratory and field experiments and observations involving several species of *Panorpa* that the extent of use of each of the three behavioral alternatives by males is related to the availability of dead arthropods in the habitat, which is determined by absolute abundance of arthropods and by male-male competition for the arthropods (Thornhill 1979a, 1981). Individual males prefer to adopt the three alternatives in the following sequence: dead arthropod > salivary mass > forced copulation. That is, when males are excluded from dead arthropods via male-male competition, they secrete saliva if they can (a male's ability to secrete saliva is determined by his recent history of obtaining food), and males only adopt forced copulation when the other two alternatives cannot be adopted. A male's body size influences his ability in male-male competition and thus which alternatives are adopted most frequently. Large males tend to adopt the use of dead arthropods as nuptial gifts. Medium-sized males most frequently use saliva. Forced copulation is adopted most frequently by small males.

Males switch from one alternative to another in an adaptive fashion (Thornhill 1981). The behavioral alternatives contribute differently to male fitness. The preference of alternatives employed by males is consistent with female choice and thus male mating success. Females prefer males with arthropods over males with salivary secretions and actively attempt to avoid forced copulating males. Also, the alternatives appear to be associated with different male mortality probabilities. Relative to large- and medium-sized males, small males tend to lose in the competition for food, and thus are forced to feed on dead arthropods in the webs of web-building spiders, which results in high mortality (Thornhill 1981).

Web-building spiders are the most important predators of *Panorpa* of both sexes. Scorpionflies feed to an important extent (about 25% of natural feeding observations) on dead insects enswathed in silk in spider webs (Thornhill 1975, 1978). Scorpionflies have defenses against spiders including a noxious oral effluent that is dabbed on spiders, but feeding in spider webs still results in high mortality. Sixty-five percent of predation on *Panorpa* during 1971-1974 and 1977 was by web-building spiders (Thornhill 1975, 1978).

B. Female Fitness in Relation to Male Alternatives

Behavior suggesting forced copulation has been described in a number of insect and vertebrate species (references in Thornhill 1980b; see also Farr 1980, and

McKinney *et al.,* this volume). I have argued that three criteria should be considered in order to distinguish forced copulation from the many other forms of behavior that might resemble forced copulation. These three criteria are: (1) forced copulation must be clearly distinguished from aggressive male courtship and female coyness, (2) force-copulating males must incur reproductive success associated with this behavior, and (3) female fitness should be reduced by forced copulation. These three criteria appear to apply to certain duck species (see McKinney *et al.,* this volume). The first two criteria were shown to be met in *Panorpa* in an earlier paper (Thornhill 1980b), and some evidence that criterion (3) applies to *Panorpa* was presented at that time. I now focus on female fitness associated with the three male alternative behaviors. It is critical to first demonstrate that forced copulation by *Panorpa* males results in reduced female fitness (*i.e.,* that the behavior is actually forced copulation) before I discuss experiments to determine the function of the dorsal clamp.

Laboratory tests with *P. latipennis* were conducted in Ann Arbor, MI, in 1978 and 1979 to determine if female preference for males exhibiting the behavioral alternatives is adaptive in *Panorpa.* There is as much variation in volume among eggs laid by a single female as between females in both *P. latipennis* and *P. mirabilis* (Thornhill, unpubl. data). Thus, I focused on the number of eggs laid by females as an appropriate measure of fitness. Laboratory tests were designed to answer the question: Does female fertility (number of eggs laid) vary in relation to the behavioral alternatives exhibited by a female's mate and is the direction of variation consistent with the direction of female preference?

I conducted two tests to determine the numbers of eggs laid by females that interacted only with males employing forced copulation. One test involved the placement of each of eight field-collected non-gravid females (slender females with few or no mature eggs; Thornhill 1974) with a single field-collected male that had been starved (no food provided) for four days prior to experimentation. The starving of males causes them to be incapable of secreting saliva (Thornhill 1979a). No food was supplied to the pairs during the tests. The absence of food, coupled with the starving of males results in forced copulation being the only option available to these males. Each of the eight pairs was housed in a 5-gallon glass terrarium containing a water supply, leafy branches, and a 2-in high, 4-in diameter, soil-filled glass or aluminum oviposition bowl. (See Byers 1963, Thornhill 1974 for discussion of procedures for housing *Panorpa* and for obtaining eggs.) The pairs were kept together for 8 or 10 days. Eggs are laid in batches of various sizes. The oviposition bowls were checked daily for eggs, and the soil was changed each time the bowls were checked.

In the second test I set up another eight terraria, each with a pair of *P. latipennis* in exactly the same way, but in this case I used gravid females (*i.e.,* those with heavy expanded abdomens containing mature eggs). During the 8-10 days of both tests the terraria were observed periodically to determine if any males were exhibiting salivary secretion behavior or if salivary masses had been secreted. Even when a saliva mass has been eaten by a female a faint ring of salivary material remains on the substrate. No saliva or salivary secreting behavior was observed during the two tests.

Gravid and non-gravid females who spend 8-10 days only in the presence of force-copulating males lay few eggs (16 females laid 56 eggs, mean = 0.37 per day). Number of eggs laid per day by the 8 gravid females (mean = 0.58) is almost four times greater (Mann-Whitney U, $p < 0.001$) than that laid by the 8 non-gravid females (mean = 0.15), and gravid females laid their eggs sooner (mean days before any eggs laid = 5.2) than non-gravid females (mean = 7.9 days, Mann-Whitney U, $p < 0.01$) (Table I). The differences in laying by gravid vs. non-gravid females were expected because gravid females should deposit their eggs before non-gravid females mature eggs and deposit them.

I dissected females at the end of this test to determine whether they contained viable sperm. Sperm was present in all cases and quantity appeared normal (see Thornhill 1974, 1980b for techniques). Males are active with little mortality for about 14 days without food under laboratory conditions.

A second set of two tests determined female fertility when females are kept with males only capable of secreting saliva. These tests were conducted in the same manner as the previous two experiments except each male used here was not

TABLE I

Eggs Laid by Non-gravid and Gravid Females While Housed for 8-10 Days
With Males Employing Alternative Forms of Mating Behavior

	No. Eggs Laid	Range	Mean No. Eggs/ Female	Mean No. Eggs/Day	Mean No. Days Before Any Eggs Laid
FORCED COPULATION:					
Non-gravid females	12	0-3	1.5	0.15	7.92
Gravid females	44	0-9	5.5	0.58	5.21
No. or mean both female types	56	0-9	3.5	0.37	6.57
SALIVA:					
Non-gravid females	163	4-30	20.3	2.09	7.42
Gravid females	174	6-40	21.8	2.23	4.88
No. or mean both female types	337	4-40	21.1	2.16	6.15
CRICKET:					
Non-gravid females	250	18-49	31.3	3.21	6.00
Gravid females	221	6-56	27.6	2.83	3.24
No. or mean both female types	471	6-56	29.5	3.02	4.62

starved, but instead was provided with a large amount of food for 24 h (two large crickets) prior to testing. This allows males to nourish themselves, making them capable of saliva secretion. No food was added to the test terraria. Males were frequently observed secreting and displaying around salivary masses and females were frequently observed eating saliva.

The 16 females kept only with saliva-secreting males laid 337 eggs in the 8-10 days of experimentation (mean/day = 2.2 eggs) compared with the 56 eggs laid by 16 females kept with force-copulators (mean/day = 0.37 eggs) (Table I). Means are significantly different (Mann-Whitney U, p < 0.001). Gravid (mean = 2.1 eggs/day) and non-gravid (mean = 2.2 eggs/day) females kept with saliva-secreting males laid the same number of eggs (Table I). Egg laying by females kept with saliva-secreting males (mean = 6.2 days before first eggs laid) was initiated at the same time (i.e., no significant difference) as egg laying by females kept with force-copulators (mean = 6.6 days). Egg laying by gravid females housed with saliva-secreting males began significantly (Mann-Whitney U, p < 0.001) earlier (mean = 4.9 days before egg laying) than egg laying by non-gravid females housed with males exhibiting this behavior (mean = 7.4 days before egg laying).

Another set of two tests examined the fertility of females interacting primarily with males providing dead crickets as nuptial offerings. The design of these two tests was exactly the same as the previous two, except one medium-sized cricket per day was added to each terrarium. The terraria were fequently observed for salivary masses. Evidence of salivary secretion was found in four of the eight terraria. Males in all terraria were frequently observed displaying their cricket to females and females were frequently observed feeding on a cricket during mating.

Female fertility is significantly increased by interaction with males providing crickets as nuptial offerings (Table I). The 16 females laid 471 eggs during the test (mean = 3.02 eggs/day) compared with 337 eggs (mean = 2.16 eggs/day) laid by 16 females exposed to saliva-secreting males (Mann-Whitney U, p < 0.01). The former females laid their eggs significantly sooner (mean = 4.6 days before any eggs laid) than the latter females (mean = 6.2 days Mann-Whitney U, p < 0.01). Non-gravid females (mean = 3.2 eggs) and gravid females (mean = 2.8 eggs) kept with cricket-providing males laid similar average numbers of eggs per day. However, gravid females initiated egg laying sooner (mean = 3.2 days before any eggs laid) than non-gravid females (mean = 6.0 days; Mann-Whitney U, p < 0.001). Non-gravid and gravid females began egg laying significantly sooner (Mann-Whitney U least significant level of probability = 0.01) when kept with cricket-providing males than when kept with males employing the other two behavioral alternatives.

In a final laboratory test on fecundity in *Panorpa latipennis,* 35 male-female pairs were set up as before except: (1) the test was conducted for 7-34 days dependent on death of the female or termination of the replicate, and (2) pairs were given copious food every other day (crickets, fresh fruit, fresh beef). No female laid eggs after 26 days (11 females survived to 34 days) and no female died before 7 days in a terrarium. Thus, the test was limited to fecundity under high food availability between 1-26 days. This test did not separate the fecundity of non-gravid and gravid females.

The range in numbers of eggs laid by a female during the test was 0-161, and the range in numbers of eggs per day per female was 0-9.1. The average numbers of eggs per day for all days of the experiment was 3.04, and females laid 3.46 eggs per day on average during the first 10 days of the test. These means apparently represent maximum fecundity because of no food limitation, and they are not significantly different from the mean of 3.01 eggs per day for females interacting with males providing crickets as nuptial offerings.

Following the two tests with crickets as nuptial offerings and the one test with 35 pairs placed under high food density, the right forewing length of a number of females was measured and compared with fecundity. There is a highly significant correlation between fecundity and female body size ($r = 0.84$, $N = 40$, $p < 0.001$). There is a weaker relationship between fecundity and the number of days a terrarium was included in the test involving 35 pairs ($r = 0.60$, $N = 30$, $p < 0.001$).

Collectively, these tests reveal that females choose males in an adaptive manner: Female preference of mates and female fecundity are strongly related. Females mating with resource-providing males only lay more eggs per unit time than females experiencing forced copulation only. Also, females that mate primarily (or only) with males using dead crickets achieve the greatest relative fecundity, which apparently approaches the highest fecundity that can be attained by females.

I mentioned in the previous section that the behavioral alternatives employed by males contribute differently to male mating success as a result of the pattern of female preference for the alternatives. Furthermore, as mentioned earlier, males that lose in competition for dead arthropods (the force-copulators) may experience higher mortality from spider predators. The above tests further address male reproductive success associated with the three behavioral alternatives. Thus, overall male fitness (number of matings, number of eggs laid by a male's mate per unit time, and probable male survivorship) follows the pattern: male provides dead arthropod > male provides saliva > forces copulation. Also, as discussed below, forced copulations result in insemination only 50% of the time compared with a 100% insemination rate during copulations involving resource provision by males. This further reduces the relative reproductive success of force copulators.

Results from field and laboratory studies of *Hylobittacus apicalis* (Mecoptera: Bittacidae) also reveal adaptive mate choice by females (Thornhill 1979c, 1980c, 1983). In this species, males feed females a prey arthropod during courtship and mating. Females that prefer males with large prey, the typical preference, lay more eggs per unit time than do females that mate indiscriminately. In *H. apicalis*, it also appears that the feeding of the female by the male may increase female survivorship. Web-building spiders inhabiting the herbs are the most important predators of this species. The predation on males by spiders is significantly greater than that on females. Males exhibit significantly more movement than females; this difference in movement of the sexes is associated with the male obtaining nuptial offerings. Females hunt on their own only when male density is low: they typically depend primarily on males to provide them with food. A female that prefers males with large prey probably has to feed less on her own and thus the preference may reduce female exposure to web-building spider predation.

The same scenario is relevant for understanding the adaptive value of female choice in *Panorpa*. As discussed earlier, males feed more and die more in spider webs than females, and the extent of feeding in spider webs by males is positively related to the extent of male-male competition. Females who prefer males with resources of high quantity or quality not only experience higher fertility but also may be forced to feed on their own less, including a reduction of feeding in spider webs. Thus, females preferring males with superior nuptial offerings may have enhanced fitness in terms of both fecundity per unit time and survivorship. In *Panorpa*, survivorship is clearly a relevant fitness parameter for females because it correlates with lifetime fecundity.

The likelihood that females will need to feed on their own is increased if females become nonresponsive to resource-providing males following mating. I have demonstrated a period of female sexual nonreceptivity in scorpionflies of the family Bittacidae (Thornhill 1976a, 1977). In bittacids, the male apparently transfers material via the ejaculate that turns off female receptivity (Thornhill 1976a; J. B. Johnson and Thornhill, unpubl. data). During the period of nonreceptivity bittacid females lay eggs (Thornhill 1976a). Early observations indicated that female *Panorpa* exhibit a period of sexual nonreceptivity following mating (Thornhill 1974). During this period females were observed not to respond to visual or olfactory (pheromonal) cues from resource-providing males. I predicted that selection will have favored females that return to receptivity sooner after forced copulation than after copulations involving resources.

Laboratory experiments with *P. latipennis* were conducted in 1979 and 1980 in Ann Arbor, MI, to determine the duration of the nonreceptivity period following matings with males employing each of the three behavioral alternatives. Males employing each of the three alternatives were obtained in the same way as previously described in the tests to determine female fecundity in relation to each alternative. A single male was placed with a single, individually-marked, sexually receptive, virgin female in a 5-gallon terrarium. (Virginity of females is assured by capturing recently-emerged females in nature [Thornhill 1974].) The terraria were set up just prior to dawn, and the pairs were housed as before. Following observed copulation, I tested the receptivity of females immediately and at 6- or 12-h intervals until receptivity was apparent. Variable intervals were used because *Panorpa* shows no sexual activity during the day and thus receptivity could only be tested at night, dawn, or dusk. Testing for receptivity involves placement of females in a container with males releasing pheromone and then scoring the response of a female to the male. Receptive females readily approach courting males, whereas nonreceptive females do not (see Thornhill 1974 for details of methods). During nonreceptivity, females do not interact with other males unless forced copulation occurs. Twelve terraria, each containing a male and a female, were used for each of the two alternatives involving resources provided to females (*i.e.,* dead arthropod or saliva). Numerous terraria containing a force-copulating male and a female were set up. Females from only eight of the terraria containing force-copulators became sexually unreceptive, because in only these cases were females actually inseminated. Forced copulation results in a 50% insemination rate, whereas copulations involving male-provided resources always result in insemination. Ejaculate sizes for the three alternative mating behaviors appear comparable. (See Thornhill [1980b] for data and for techniques to determine ejaculate size and whether a female has been inseminated following copulation.) Copulation duration is variable in *Panorpa.* Data on *P. penicillata,* a Mexican species, suggested that nuptial arthropod size may be positively related to copulation duration (Thornhill 1979a). However, subsequent work indicates no relationship. Forced copulations and copulations involving resources are of similar durations ranging from 30 min to 3 h under most circumstances. Not infrequently, however, copulation will extend to 5 h and durations of 12 h have been recorded (Thornhill, unpubl. data).

Although sample sizes are less than ideal, the results seem clear-cut (Fig. 1). Females that experience forced copulation with insemination return to sexual receptivity sooner than females mating with resource-providing males. Hours to return of receptivity varied from: 18-36 (mean = 27.8 h or about one day) for the eight females experiencing forced copulation with insemination, 36-48 (mean = 40 h or about 1.8 days) for 12 females mating with saliva-providing males, and 36-42 (mean = 38.5 h) for 12 females mating with arthropod-providing males. The Kruskal-Wallis nonparametric analysis of variance by ranks indicates a highly significant difference ($0.001 < p < 0.01$) in time elapsed before return to receptivity among the three categories of females. The Newman-Keuls test revealed significant differences ($p = 0.01$ in each case) when the category forced-copulated females is compared with each of the two categories of females receiving resources during mating; the latter two categories are not significantly different.

When females resume receptivity following copulation, they remain receptive until inseminated. It appears that this pattern of receptivity followed by mating and nonreceptivity occurs through the lifetime of a female *Panorpa* (Thornhill 1974, unpubl. data), and that this pattern is common in the Mecoptera in general (Thornhill 1976a, 1977, unpubl. data).

Male and female reproductive interests are in conflict with forced copulation in *Panorpa* (Thornhill 1980b; see Trivers 1972 and Parker 1979 for general discussions of conflict of reproductive interests between the sexes). Forced copulation is employed by males excluded from other alternatives via male-male competition. Male reproductive success is increased by those males that practice forced

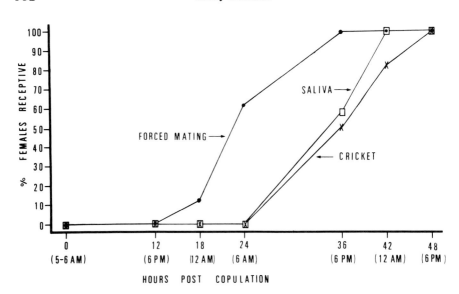

Fig. 1. The onset of sexual receptivity (see text for measurement details) in *Panorpa latipennis* females after mating with males engaging in each of the three mating alternatives. Forced mating = female mated with male with no nuptial offering and exhibiting forced copulation. Saliva = female mated with male providing saliva as nuptial gift. Cricket = female mated with male providing a dead cricket as a nuptial gift.

copulation. Force-copulators place their ejaculates inside the sperm storage organs of females in 50% of actual forced copulations. I do not know if sperm of forced copulators fertilizes eggs, but it would be impossible to explain the evolution of the behavior if force-copulators did not fertilize eggs, unless one resorted to group selectionist interpretations. Female fitness is reduced by forced copulation. This behavior clearly reduces female fecundity, and probably increases the likelihood of female mortality, because females might be forced to feed more on their own (despite a reduced nonreceptivity period) than in the case of copulations with resource-providing males. In addition, female fitness may be reduced by forced copulation because females may lose the option of mate choice—said differently, females may lose control of which male fertilizes her gametes. Given the presumed importance of female choice as a means by which females increase the probability that their offspring will possess quality paternal genes, relinquishing this may be extremely detrimental to female fitness. (See Thornhill [1980d] and Borgia [1979] for discussion of mate choice theories and so-called problems with these theories.) It is unclear at present whether female *Panorpa* choose males on the basis of relative genetic quality (as expressed in the male's morphological and behavioral phenotypes) in addition to choice on the basis of material benefit. However, it is reasonable to expect male genetic quality and a male's ability to secure a limited resource (in this case dead arthropods) to be positively correlated. In *Panorpa* contests among males are settled by aggression. Field studies reveal that a number of males quickly gather around dead arthropods placed in the habitat (Thornhill 1981). Thus, the male winning the contest is highly unlikely to win by chance alone.

Of course, the argument could be made that forced copulation enhances female fitness. If the tendency to exhibit forced copulation is sufficiently heritable, a female could gain an advantage by allowing forced copulation by the "best" forced-copulators in the context of low food availability. Females could determine the "best" in this situation by always fleeing when a force-copulator approached and by vigorously struggling to free themselves when captured by the swiftest force-copulator. I cannot directly address this argument. However, it is my impression, although detailed experiments have not been conducted, that females behave the same way toward force-copulators regardless of the female's age or feeding history. This suggests that females do not readily accept force-copulators as suitable mates under any condition.

I have suggested (Thornhill 1980b) that male *Panorpa* may have won the conflict of interests between the sexes over evolutionary time. However, this interpretation is probably incorrect. The data above indicate selection on females that has favored a return to receptivity more quickly after forced copulation than after copulation involving a resource provided by a male. A quick return to receptivity is likely to be contrary to the interests of the male that successfully forces himself on the female. Upon resumption of receptivity, the female will seek another mate whose sperm may take precedence in fertilization over the sperm of the force-copulator. Gwynne (this volume) provides comparative data from a wide variety of insects showing that in species in which males provide material benefits to females, the last male to mate has higher fertilization precedence in the next batch of eggs than the last male to mate in species in which males appear to contribute only sperm at mating. Also, the fact that females manage to free themselves 85% of the time when grasped by a male attempting forced copulation indicates strong selection on females to avoid forced copulation. Moreover, the low insemination success rate of forced copulation suggests strong selection on females to discriminate against force-copulators. Evidence reported elsewhere (Thornhill 1980b) reveals that the lower insemination rate associated with forced copulation does not result from reduced sperm quantity in force-copulating males, which suggests that females somehow cause the low insemination rate (see also below). It appears then that selection is strong on both sexes: on males to force copulate when other reproductive routes are closed, and on females to resist forced copulation, but when experiencing it to return to receptivity more quickly.

To a significant extent, a female that experiences copulation with a saliva-providing male has reduced fitness compared with a female mating with a large-arthropod-providing male. A period of nonreceptivity is initiated by a copulation involving saliva and this period extends for the same duration as when a female mates with a male providing a large arthropod. Thus, females mating with males with saliva may have to feed on their own to match the fecundity of females that mate with arthropod-providing males. However, abundance of spider predators among the herbs in *Panorpa* habitat in southeastern Michigan, where most of my work has been done, varies considerably in time and space (Thornhill, unpubl. data), which should inluence costs and benefits associated with choices by females of males providing resources of different qualities and quantities. Of course, the nature and extent of predators would also influence costs and benefits associated with the alternatives for males.

A critical remaining question is to what extent in nature is female fecundity limited and female survivorship reduced by a shortage of males providing resources of high quality to females? To an important extent this question can be addressed. My long-term study of competition for food in the form of dead arthropods among and within the nine species of *Panorpa* occurring in southeastern Michigan reveals intense competition in all years of the study (1971-1974, 1977-1980). Both sexes show interference competition (aggression) and males are more dominant than females. The competition influences survivorship (and hence fitness) negatively and has both intra- and interspecific components (Thornhill 1980a, 1981). Work in the southern Appalachian Mountains also indicates intense competition in assemblages of *Panorpa* species (Thornhill 1981). The intensity of competition in *Panorpa* indicates that males with suitable dead arthropods conferring high fitness to females only occasionally will be available. Certainly the availability of males with high quality nuptial offerings has limited female reproduction in the evolutionary history of *Panorpa*. Otherwise, male and female aggression around arthropods would not have evolved to their present levels of sophistication. Furthermore, both sexes employ elaborate adaptations for feeding on dead arthropods in spider webs (Thornhill 1975). Finally, males have evolved two behavioral alternatives in addition to using dead arthropods as nuptial offerings, and the two additional alternatives are associated with a reduction in male reproductive success compared with use of dead arthropods.

C. The Function of the Dorsal Clamp: Experiments

I will briefly summarize the methods and results of two earlier experiments (Thornhill 1980b), because they are critical for understanding the additional experiments I will describe.

The first experiment was conducted as follows. A small amount of warmed beeswax was applied to the dorsal clamps of 40 *P. latipennis* males. The males had been starved for 4 days, to prevent saliva secretion. The beeswax, when dry, fills and covers the dorsal clamp, rendering it inoperative. Warm water was applied to the dorsal clamps of 40 control males. Ten treated and 10 control males were placed in each of four 10-gallon glass terraria containing a water supply and leafy branches. Ten receptive virgin females were added to each of the four terraria. All insects used in experiments were individually marked. The experiments mimicked natural densities. Over a 4 h period, I recorded the number of males grasping females with their genital claspers, the total number of grasps made by all males, the number of males coupling the forewing of a female in the clamp, the total number of wing couplings by all males, and the number of males in forced copulations.

Treated (dorsal clamp inoperative) and untreated males grasped females with their genital claspers with similar frequency. Thirty-one treated males made 59 force-copulation attempts. Thirty-six untreated males made 55 attempts. Treated males tried to reposition females so as to secure their forewing in the dorsal clamp, but the females escaped by struggling. Twenty-two percent of untreated males that attempted forced copulation (N = 36) succeeded. Fifteen percent of actual force-copulation attempts by all untreated males (N = 55) led to copulation; however, only 50% of the actual forced copulations led to insemination. Thus, 11% of the 36 untreated males attempting forced copulation actually inseminated a female while none of the treated males did. This experiment demonstrates that forced copulation cannot be accomplished without an operative dorsal clamp.

In the second experiment (Thornhill 1980b) a different set of individually marked *P. latipennis* males with dorsal clamps covered with beeswax were placed in a large terrarium containing small dead crickets. These males began dispersing pheromone while standing adjacent to a cricket. The females that had avoided forced copulation in the previous experiment and thus were still virgin were then added approximately one hour after the end of experiment one. Most of the females were in copulation with males within an hour. All females examined (N = 20) after copulation contained an ejaculate. This experiment demonstrates that the dorsal clamp is not necessary for holding the female for insemination when males provide a small dead arthropod. It also suggests that females are more willingly inseminated by males with a resource than by force-copulators. The females did not readily mate with the males in this experiment because of increased receptivity between experiments. Receptive females appear to avoid force-copulators under all circumstances, but are attracted to resource-providing males.

In two additional experiments like the above, I used saliva-secreting males (obtained as before) instead of cricket-providing males in the second experiment. Results were similar to those in the previous two experiments. In the third experiment, 35 of 40 treated males attempted forced-copulation (none were successful) and 37 untreated males did so. Thirty percent (11 of 37) of untreated males that attempted forced copulation succeeded. Six of the 11 females (55%) experiencing forced copulation were inseminated. Sixteen percent (6 of 37) of males attempting forced copulation actually inseminated a female. In the fourth experiment, 24 of the 29 females that avoided forced copulation with males in the third experiment were in copulation with saliva-guarding males within 1.5 h after place-ment with these males. The sperm storage organs of these 24 females contained apparently normal-sized ejaculates upon examination soon after the termination of copulations.

The results of the third and fourth experiments confirm those of the first two. These experiments reveal that the dorsal clamp is essential for forced copulation and that females are more willingly inseminated by males providing a resource (dead cricket or saliva) than by force-copulators.

The experimental results are consistent with the interpretation that the dorsal clamp evolved in the context of increasing the success of forced copulation at-tempts when sexual competition forced individuals into this alternative during *Panorpa* evolutionary history. As mentioned earlier, the three alternatives are possessed by all males and apparently represent a single conditional strategy.

Yet, despite the apparent uselessness of the clamp in unforced copulation, the female's wing is placed in the clamp organ during both forced and unforced copulation. Could the dorsal clamp have evolved solely or in part in some other selective context? The experiments only superficially address this question. They were designed to test predictions from a forced copulation hypothesis. Of course, the predictions could be consistent with those stemming from an alternative evolu-tionary hypothesis(es) for the existence of the dorsal clamp. A reasonable alterna-tive was identified at the beginning of the paper. It views the dorsal clamp as evolved in the context of sperm competition—more specifically, as a structure that prevents disruption of copulating pairs and the insemination of the female of the pair by an intruding male. Aspects of the reproductive behavior of male *Panorpa* are consistent with this hypothesis—for example, pair disruptions by in-truding males are not infrequent (Thornhill 1974), disruptions sometimes result in the intruder copulating with the female (Thornhill 1974), and the clamp is used in both forced and unforced copulations. Mutually exclusive predictions from the two alternative hypotheses are easily identified. If the clamp evolved in the context of reducing the probability of the takeover and insemination of a copula-ting male's mate, one would expect treated males (dorsal clamp occluded with beeswax) to experience higher takeover rates than untreated males. If the clamp evolved in the context of forced copulation (or some other context consistent with predictions from the forced copulation hypothesis), one would expect the treated and untreated males to experience similar rates of takeover.

If one of these predictions is supported, the hypothesis that generates the other prediction is very weak—in my opinion, the hypothesis would be falsified. In this sense they are strong predictions. Numerous other less rigorous predictions (*i.e.*, not mutually exclusive and thus not providing potential for falsifiability) can be imagined but will not be addressed here.

The two mutually exclusive predictions from the two hypotheses were tested with a laboratory experiment designed as follows. Ten medium-sized crickets

12 x 3 x 3 mm) were taped to leaves in each of two 10-gallon terraria. Ten treated males were added to one of the terraria and 10 untreated males were added to the other. All males were individually marked prior to their placement in terraria. The males were allowed enough time to find crickets and begin dispersing pheromone while guarding them. I then added receptive females to the containers of males. Females were added and others removed until all males were in copulation. In some cases a copulation was terminated and another initiated by the same male before I achieved a situation of all males in copulation in both terraria. I then added a single marked male for each copulating pair in the terraria. These males served as potential intruders. The containers were observed constantly for the first 4 h after the introduction of intruding males. I recorded disruption attempts by intruders and actual takeovers. This was repeated with different males and females on two additional nights. Thus, I used a total of six terraria, three each containing 10 treated males and three each containing 10 untreated males.

The results reveal that the clamp does not reduce the probability of disruption of a copulating pair by an intruding male. Disruption attempts were equally frequent in terraria containing both treated (N = 17) and untreated males (N = 20). Two pairs in which a treated male was a member and three pairs in which an untreated male was a member were disrupted.

There are alternative hypotheses other than sperm competion that can be considered in an attempt to understand the evolution of the dorsal clamp. Felt (1895) observed female *Panorpa debilis* palpating the dorsum of the third and fourth abdominal segments of males during courtship, and later noticed a "peculiar organ" at this location. I have observed palpation of the dorsal clamp during courtship by females of several *Panorpa* species. Felt concluded that the structure probably secretes a volatile oil that attracts the female to the male. I have conducted a histological study of the dorsal clamp (Thornhill, unpubl. data; see Thornhill [1973] for methods). It has no associated glandular tissue. Felt's hypothesis is also inconsistent with data presented in the present paper. When the dorsal clamp of resource-providing males is covered with beeswax, presumably preventing any odors from being released, females readily mate with such males. Felt's hypothesis is incorrect.

The dorsal clamp could have evolved in the context of species recognition. The dorsal clamp varies in morphology across *Panorpa* species (Thornhill, unpubl. data). The examination (olfactory and apparently visual) by the female prior to copulation could reduce the probability of an interspecific mating error. This hypothesis must be examined because it has been, and still is, a widely used evolutionary explanation for species differences in courtship and mating behavior and associated morphological features. Darwin (1871) argued that sexual selection was the most important context for the evolution of sexual differences in sexual behavior and morphology, but Wallace (1889) identified species and sex recognition as the more likely contexts (see Thornhill 1980d). After Darwin and until recently, premating and mating behavior and associated morphological features have been generally viewed as functioning as reproductive isolating mechanisms—that is, they are adaptations which prevent wasted reproductive effort by individuals associated with heterospecific interactions (Alexander 1969; Thornhill and Alcock 1983; M. J. West-Eberhard, pers. comm.). However, the species' identification hypothesis should not be forgotten by investigators studying morphological and behavioral differences between species. It can serve as an alternative hypothesis in studies of traits presumed to have evolved in the context of sexual selection or reproductive competition in general. Although interspecific mating errors could potentially occur because of the co-occurrence in time of sexually active adults of several species of *Panorpa*, evidence indicates that females do not depend on cues from the dorsal clamp for species (or sex) discrimination. Male *Panorpa* produce species-specific sex pheromones that attract females from a distance and may also serve in close-range interaction (Thornhill 1979b). Males exhibit species-specific courtship actions (wing and body movements). Also, when females approach a heterospecific male they do so to obtain a meal—*i.e.*, they attempt to feed on the nuptial offering (Thornhill, unpubl. data). Females never behave coyly toward heterospecific males as they always do toward conspecifics. Females of large species attempt to usurp, via aggression, the nuptial offerings of males of small species. They are sometimes successful. Resource-holding males behave aggressively toward heterospecific females that approach them. These considerations indicate that species and sex identity are discerned prior to close-range courtship by both sexes. Finally, the experiments involving treated males are strongly inconsistent with the sex—and species—discrimination hypotheses. In these experiments the dorsal clamp was covered with opaque beeswax and thus it would apparently not emit normal visual (or olfactory) cues. Yet, females readily mated with treated males who offered nuptial gifts.

The fact that females often examine the dorsal clamp during courtship raises intriguing possibilities. What information might the female be ascertaining by this behavior? The dorsal clamp occurs in a few genera of Mecoptera other than *Panorpa*. However, in some tropical mecopteran taxa it is not clamp-like in morphology. Instead, the posterior portion (derived from the anterior dorsum of the fourth abdominal segment) has been lost evolutionarily, leaving the anterior portion only (derived from the posterior dorsum of the third abdominal segment), which has evolved into a long, posteriorly-projecting conical structure extending to the male's abdominal tip (see Crampton 1931). This structure may play a role in male-male contests for females in a fashion similar to the "horns" of beetles (see Otte and Stayman 1979, Eberhard 1979).

At this point in my research on the function of the dorsal clamp of *Panorpa*, it appears that this structure evolved in the context of forced copulation. The alternative hypotheses I have considered were either falsified (the sperm competition hypothesis and the pheromone-emission hypothesis) or very inconsistent with existing evidence (the sex identification and the species identification hypotheses). Only further experiments could falsify the sex and species identification hypotheses. The requirement that scientific hypotheses be empirically falsifiable has received much attention in the recent literature. This requirement, the so-called criterion of demarcation, is usually attributed to the philosopher K. Popper (trans. 1959). But Darwin clearly understood the importance of falsifiability in science. His work is full of statements about how the theory of evolution could be falsified (Ghiselin 1969, Alexander 1979). The idea that alternative hypotheses should be considered in scientific study is not new; it is implicit in the structure of science. I hope it is clear that by considering alternative hypotheses and mutually exclusive predictions from them, I have arrived at a better answer than if I had tested predictions from a single hypothesis without consideration of alternatives.

III. OTHER TRAITS OFTEN VIEWED AS EVOLVED BY SPERM COMPETITION

Parker (1970) listed a number of reproductive characteristics of insects which he felt evolved in the context of reducing sperm competition. In some cases he provides alternative hypotheses. However, the use of alternative hypotheses has been ignored in many studies of insect reproductive biology. It would be of value to take Parker's list and construct alternative hypotheses useful for directing future research. Voss (1979) has done this for copulatory plugs in rodents. He considers five alternatives and all available evidence bearing on them, and he generates a prediction that provides an important additional test. I will briefly address some characteristics of insect reproductive behavior that are widespread and have been interpreted almost solely as functioning to reduce sperm competition. I also discuss one example for vertebrates.

Patterns of copulation frequency and duration in insects have been widely interpreted as adaptations to reduce sperm competition (*e.g.,* Parker 1970; Thornhill 1976a, b; Alcock *et al.* 1977). Studies by Sivinski (1977, 1980b) suggest that sperm competition may have been an important factor in the evolution of lengthy copulation duration in certain stick insects (Phasmatodea). Sivinski tested experimentally two alternative hypotheses: (1) lengthy copulation evolved in the context of reduction of sperm competition, and (2) lengthy copulation evolved in the context of diminishing the probability of predation on copulating individuals. Sivinski's work has not eliminated the second hypothesis at this point (see Sivinski 1980b). Smith (1979a, b) and Smith and Smith (1976) have demonstrated that copulation frequency is important in reducing the risk of cuckoldry of paternal male water bugs (Belostomatidae). In other cases, the effects of copulation duration and frequency on sperm competition are unknown.

I originally interpreted the long copulation (2.3 days) in the "lovebug" (Bibionidae:*Plecia nearctica*) as functioning to reduce sperm competition. Sperm transfer in this insect is complete at about 12.5 h of copulation (Thornhill 1976b). Males hover in flight, in an aggregation above their emergence site, where they compete for recently emerged females that take flight through the aggregation (Thornhill 1980e). Copulating pairs fly from the emergence site. The pairs visit flowers and feed on nectar and pollen. The female is much larger than the male and she gains the best feeding site on flowers; males may feed little during the long copulation. After separation of the pair, the female lays eggs in the soil surface and litter, and then dies (Thornhill 1976b). I found in my studies in Florida that disruption of copulating pairs by intruding males occurs fairly frequently when pairs are on flowers (Thornhill 1976b). The high densities of lovebugs in Florida are associated with the recent movement of this species into the state and subsequent outbreaks. The insects are so numerous that interactions of single males with pairs are unnaturally frequent; such interactions very rarely occur under more natural densities in Mexico (Thornhill 1980e).

Sharp *et al.* (1974) found that copulating female lovebugs fly faster than do single females or males, and that the male apparently assists the female in flight. Given that the female lovebug lives only a short period of time and that the competition for females is severe in this species, a male when he obtains a mate may have been selected to remain in copulation and help his mate find food by aiding her movement from the emergence site to food and from foraging patch to foraging patch. Females that are aided by males may produce more and/or better nourished eggs, which would increase male reproductive success and could provide the selective advantage associated with the evolution of lengthy copulation. The fact that a male may fertilize all the eggs a female lays following mating may be incidental (a fortuitous effect) to the actual adaptations and without evolutionary consequences. Other alternative hypotheses to explain copulation duration in the lovebug can also be imagined based on accumulating information on this species' biology.

With regard to studies of copulatory frequency in insects (and other animals), it is likely that some of the diversity in copulatory frequency may simply represent the frequency necessary to fertilize the eggs of a female rather than representing a male's attempt to increase the competitiveness of his ejaculates relative to those of other males. I discuss this further under the last example mentioned, which involves copulatory frequency in birds.

Pre- and postcopulatory "guarding" of females by males, which may involve contact or noncontact guarding, is common in insects (*e.g.*, see Parker 1970, 1979; Alcock *et al.* 1977; Sivinski 1977, 1978; Waage 1978, this volume). Typically these behaviors are viewed as adaptations which reduce sperm competition because of their presumed role in preventing takeovers by intruding males. It is easy to see the advantage to males. Parker (1970) does point out that precopulatory guarding may function as a stimulation to the female's reproduction. Many studies (for example, those cited) report data consistent with the sperm competition hypothesis; however, in only a few cases have alternatve hypotheses been considered (see Sivinski 1977, 1978, 1980b; Parker 1978). I provide alternative hypotheses here (also see Gwynne, this volume). In some species in which mate-guarding occurs (e.g., dung flies [Parker 1978; Borgia 1979, 1981], Odonata, and perhaps many others), males control territories containing oviposition sites. Since the male owns the territory he is very familiar with variation in oviposition site quality within it. The male may not be guarding the female but instead he may be helping her locate suitable sites to deposit her eggs. This could, of course, increase male reproduction. Also, in species in which males guard females while females are in the male's territory, but oviposition occurs elsewhere, the precopulatory attendance of the male may be an aspect of courtship. Finally, Borgia's (1981) work on dung flies clearly reveals benefits to a female from being attended by her mate, and potential benefits (other than reducing sperm competition) to attending males (see Gwynne, this volume, for discussion).

I provide one final example. I hope its discussion will not only illustrate the usefulness of alternative hypotheses other than sperm competition, but will be of some use in future work on copulatory behavior of monogamous birds. I acknowledge the contributions of L. Benshoof, Department of Anthropology, University of New Mexico, to several points in the following discussion.

Birkhead (1979) has argued that a male monogamous bird that can disguise his mate's fertile period would have a selective advantage due to reduction of the probability of cuckoldry by neighboring males (i.e., reduce sperm competition), and that the fertile period of the female could be disguised by the male via a reduction in the frequency of copulation or by an extension of the length of the period over which copulation occurs. This argument led Birkhead to the following two predictions about the copulatory behavior of monogamous birds: (1) in colonial species, where inconspicuous copulation is not possible due to very close proximity of other males, the female's fertile period should be disguised by conspicuous (in terms of behavior and location) and frequent copulation over a relatively long period of time; (2) in solitary species (i.e., noncolonial and territorial), inconspicuous and infrequent copulation should evolve. Birkhead's summary of data suggests that the comparison between colonial and solitary monogamous species regarding copulatory frequency and the duration of copulatory period is met, and that male magpies (a solitary species) tend to copulate inconspicuously.

Benshoof and I (1979) have proposed a hypothesis for the evolution of concealed ovulation and the concomitant continuous receptivity of human females that may be applicable for an understanding of high copulation frequency and extended copulation period in colonial monogamous birds. We hypothesized that concealed ovulation in humans is an adaptation that disguises the female's fertile period from males (including the mate) and by so doing allows the female to gain benefits and reduce costs associated with copulation outside the pair bond. Selection is expected to favor extrabond copulation by females when benefits exceed costs to individual reproduction. Benefits to females associated with extrabond copulations may derive from obtaining resources or increased status for herself and/or her offspring, or from superior paternal genes for her offspring. Costs to a female include detection of infidelity by a mate which might result in reduction or elimination of paternal care. Concealed ovulation, as opposed to conspicuous estrus, reduces this cost because it increases the probability that a female can engage in extrabond copulation without her mate's detection. Concealed ovulation in humans is associated with continuous sexual receptivity and frequent copulation. Perhaps females of colonial monogamous birds have been selected to reduce signs of maximum fertility or extend such signals in time in the context of deception of mates and other nearby males for reasons similar to those we suggested for humans. Recently, a number of investigators have proposed other evolutionary hypotheses to explain concealed ovulation in humans. D. Symons' (1979) hypothesis is similar to our own.

The social setting of humans influences the benefits and costs of infidelity. Numerous monogamous pairs in very close proximity is a human social setting apparently not shared with other mammals; however, colonial monogamous birds nest in close proximity and thus have a social setting somewhat like humans. In many colonial bird species copulation outside the pair bond appears to be common. (See Benshoof and Thornhill 1979 for discussion and references.) Benefits that female colonial birds might gain through extrabond copulation are unclear at this time. Benshoof and I (1979) have suggested that they might gain material and/or genetic benefits. Genetic benefits involve obtaining superior paternal genes via extrabond copulation. Material benefits may be in the form of food, nest material, or protection from predators. Female Caspian Terns have been observed to receive a fish from a male just prior to an extrabond copulation with him (J. Quinn, pers. comm.). In certain herons nest material is limited (D. Mock, pers. comm.). A female could exchange an extrabond copulation for a choice piece of nest material. Also, a female by copulating with a nonmate nesting nearby and giving that male some probability of parentage might benefit if later the male was likely to warn her or her offspring of predators; this will only be likely if the female has some guarantee of future delivery of this benefit—that is, the circumstance involves a system of evolved reciprocity. Future studies of colonial pair-bonding birds could be designed so that results would refute or support this hypothesis as applied to such species. Thus, Benshoof and I agree with Birkhead that the **frequency** of copulation exhibited by colonial species is an adaptation of the male to reduce the probability of cuckoldry. However, the lengthy period in which copulation occurs in such species may best be viewed not as an adaptation by the male to prevent cuckoldry, but as an adaptation by the female to disguise her fertile period. Also, see Lumpkin (1981) for a reasonable alternative hypothesis to that provided by Birkhead.

Also, Birkhead's interpretation of reduced copulatory frequency in solitary monogamous birds as an adaptation for preventing cuckoldry must be questioned. The interpretation ignores Williams' (1966) warning with regard to adaptation as an onerous concept: It should not be invoked unnecessarily. For a reduced frequency of copulation to be considered an adaptation, one must suppose that it has been reduced from the optimal frequency necessary for fertilization of eggs. Lake (1975, cited in Birkhead 1979) found that a single insemination is usually sufficient to fertilize all the eggs of a clutch in domesticated birds. In view of these considerations, it seems unlikely that the relative infequency of copulation in solitary monogamous birds is an adaptation to prevent cuckoldry (reduce sperm competition), but rather is an adaptation to fertilize the eggs.

IV. METHODOLOGY USED TO STUDY EVOLVED FUNCTION

In the work I discussed on the function of the dorsal clamp in *Panorpa* I used experimental and observational methods only. I do not want to give the impression that these are the only, or the best, methods for studying evolved function. Among biologists in general there appears to be a dichotomy in terms of acceptable methodology. Some biologists use the comparative method successfully to shed light on the evolved function (*i.e.*, the selective history) of characteristics. Darwin (*e.g.*, 1859, 1871) employed the comparative method for an understanding of speciation, adaptation and the evolutionary history of life in general. Williams' (1975) analysis of the function of sex employed comparisons of the timing of sex in plant and animal life cycles. Alexander *et al.* (1979) and Clutton-Brock and Harvey (1977) used the comparative method to gain insight as to the selective history of patterns of sexual dimorphism in vertebrates. Other biologists feel that studying complex characteristics, which are often relatively unmodifiable within species, but vary among species, cannot be approached scientifically because at best correlational data obtained by way of species or population comparisons can be generated and correlational data do not comprise scientific evidence, *i.e.*, do not address cause and effect. Field observations supporting correlations are considered "post hoc" and thus weak or useless. Stearns (1976) has criticized the comparative method in some detail as an approach for elucidating the evolution of life history traits. His point of view is that "Only experimentation truly tests theory."

The experimental and comparative methods differ in that the comparative method involves no manipulation which is one of its strengths (see Alexander 1979). The experimental method involves some systematic variation (*i.e.*, manipulation) of the parameter of interest. Laboratory experiments typically take the form of attempts to control all parameters but the one of interest, which is manipulated. Field experiments often do not involve controls of presumably irrelevant or of confusing variables. Instead, all parameters except the one of interest are allowed to vary naturally. Typically, with field experiments, compared with laboratory experiments, one is less certain about the influence of other important variables on the result. However, even with laboratory experiments all confusing variables cannot always be controlled. The number of confusing variables that could cause a given effect is potentially infinite. The comparative method attempts to randomize the influence of confusing variables on the effect of interest (see Alexander 1979). If appropriate comparisons exist and can be recognized, the comparative method can lead to precise results, results indicative of cause and effect. Fortunately, life exhibits almost incredible diversity providing vast numbers of appropriate comparisons for almost any question of the evolved function of a characteristic. The ingenuity in use of the comparative method involves recognition of appropriate comparisons, *i.e.*, those that because of their number and nature are likely to randomize the influence of other parameters on the result.

Problems are inherent in both methods; neither is perfect. Experiments designed to ascertain cause and effect are perfected through time by investigators interested in a given cause-effect relationship. The definitive experiment cannot be done. All we can hope for is refinement; better controls and manipulations are created. The definitive analyses demonstrating cause-effect via the comparative method cannot be done. Again, those interested in a given cause-effect relationship make new and better comparisons.

Some questions cannot easily be addressed experimentally, but can be answered by appropriate comparisons. For example, it would be difficult, or impossible, to reduce variation in male reproductive success in a group of polygynous vertebrates and observe the predicted influences on the evolution of sexual dimorphism and male life history traits. Many questions can be approached from both the comparative and the experimental approach. The experiments and observations I described earlier provide considerable functional understanding of the dorsal clamp. It is unlikely that experimental removal of the presumed selection maintaining the dorsal clamp would result in observable atrophy of the organ though time. I consider the following prediction an important test for understanding the function of the dorsal clamp in *Panorpa:* The frequency in which males of a species exhibit forced copulation should be positively related to the size of the clamp (or other morphological correlates of the effectiveness of the structure as a clamp).

When both experimental and comparative methods can be used, it is reasonable to use both, or either. These two methods are based on different assumptions. Indeed, field versus laboratory experiments are based on different assumptions. Thus, the use of as many different methods as feasible to examine an hypothesis is the most rigorous approach. When results from multiple mehods all point to the same conclusion, one achieves the most confidence about cause and effect. The important issue is not which method is best, but instead we should focus on refinement and clarification of all our methods. Clutton-Brock and Harvey (1977, 1979), Maynard Smith (1978) and Thornhill and Alcock (1983) have discussed some problems in applying the comparative method that investigators employing this method might keep in mind.

V. SUMMARY

G. A. Parker's important ideas on the role of sperm competition, a form of sexual selection, in the evolution of male reproductive traits have guided much thinking and research in reproductive biology. This is especially true in insect reproductive biology where sperm competition has been applied to explain diversity of copulation duration and frequency, many interactions of males and females before, during and after mating, and sexual morphologies of males. The selective context of sperm competition has often been used as a hypothesis directing investigation without consideration of alternative hypotheses. I argue that alternative hypotheses and mutually exclusive and falsifiable predictions from the alternatives can be a profitable approach toward understanding evolved function. I use this approach to analyze in detail the evolved function of a clamp-like structure on the dorsum of the abdomen of male *Panorpa* scorpionflies which holds the female's wings during mating. I initially viewed the clamp as having evolved in the context of reducing sperm competition (*i.e.*, preventing take-over of a male's mate by an intruder), because observations in general were consistent with the hypothesis. In this paper I examine five alternative hypotheses to explain the evolution of the clamp. Two of these, including the sperm competition hypothesis, are eliminated and two others are very inconsistent with existing data. It appears that the dorsal clamp functions in the context of forced copulation—it increases the probability that a male can inseminate an unwilling female. The validity of this hypothesis rests on the existence of actual forced copulation in *Panorpa*. It is difficult to distinguish forced copulation for the many other forms of behavior that resemble forced copulation. I describe results of my study of forced copulation in *Panorpa* focusing on the influence of forced copulation on female fitness.

I examine several behavioral traits that have been interpreted as evolved in the context of sperm competition and suggest alternatives that deserve consideration before the evolved functions of the traits can be understood.

I briefly discuss general methods for studying evolved function. The various methods in evolutionary biology are based on very different approaches for elucidating cause-effect, and thus the use of multiple methods in examining a hypothesis about evolved functions provides a very strong test.

ACKNOWLEDGMENTS

I thank Gary Dodson, Darryl Gwynne, Larry Marshall, Geoff Parker, Chris Starr, and Bruce Woodward for criticizing the manuscript. This research was supported by National Science Foundation grants DEB77-01575, DEB79-10193, and BNS79-12208. Dick Alexander, Tom Moore, Bob Storer, and Don Tinkle (deceased) provided research space in the Museum of Zoology at the Univerity of Michigan during 1977-1980. F. Evans and R. Storer provided research space and housing at the Edwin S. George Reserve in 1980. Mike Brown's and Nancy Thornhill's assistance in 1980 is gratefully acknowledged.

REFERENCES

Alcock, J., G. C. Eickwort, and K. P. Eickwort. 1977. The reproductive behavior of *Anthidium maculosum* (Hymenoptera:Megachilidae) and the evolutionary significance of multiple copulations by females. *Behav. Ecol. Sociobiol.* 2:385-396.

Alexander, R. D. 1969. Comparative animal behavior and systematics. *Syst. Biol. Proc., Intl. Conf. Natl. Acad. Sci.* 494-517.

Alexander, R. D. 1979. *Darwinism and Human Affairs.* University of Washington Press, Seattle, WA.

Alexander, R. D., and G. Borgia. 1979. On the origin and basis of the male-female phenomenon. In *Sexual Selection and Reproductive Competition in Insects*, M. S. Blum and N. A. Blum (eds.), pp. 417-440. Academic Press, New York.

Alexander, R. D., J. L. Hoogland, R. D. Howard, K. M. Noonan, and P. W. Sherman. 1979. Sexual dimorphisms and breeding systems in pinnipeds, ungulates, primates, and humans. In *Evolutionary Biology and Human Social Behavior: An Anthropological Perspective*, N. A. Chagnon and W. G. Irons (eds.), pp. 402-435. Duxbury Press, North Scituate, MA.

Bateman, A. J. 1948. Intrasexual selection in *Drosophila. Heredity* 2:349-368.

Benshoof, L., and R. Thornhill. 1979. The evolution of monogamy and concealed ovulation in humans. *J. Soc. Biol. Struct.* 2:95-106.

Birkhead, T. R. 1979. Mate guarding in the magpie *Pica pica. Anim. Behav.* 27:866-874.

Borgia, G. 1979. Sexual selection and the evolution of mating systems. In *Sexual Selection and Reproductive Competition in Insects*, M. S. Blum and N. A. Blum (eds), pp. 19-80. Academic Press, New York.

Borgia, G. 1981. Mate selection in the fly *Scatophaga stercoraria:* Female choice in a male-controlled system. *Anim. Behav.* 29:71-80.

Byers, G. W. 1963. The life history of *Panorpa nuptialis* (Mecoptera:Panorpidae). *Ann. Entomol. Soc. Am.* 56:142-149.

Byers, G. W., and R. Thornhill. 1983. The biology of Mecoptera. *Annu. Rev. Entomol.* 28:203-228.

Clutton-Brock, T. H., and P. H. Harvey. 1977. Primate ecology and social organization. *J. Zool. (Lond.)* 183:1-39.

Clutton-Brock, T. H., and P. H. Harvey. 1979. Comparison and adaptation. *Proc. R. Soc. Lond. B Biol. Sci.* 205:547-565.

Crampton, G. C. 1931. The genitalia and terminal structures of the male of the archaic mecopteran, *Notiothauma reedi,* compared with related holometabola from the standpoint of phylogeny. *Psyche* 38:1-21.

Darwin, C. 1859. *On the Origin of Species.* Facsimile of the first edition, 1964. Harvard University Press, Cambridge, MA.

Darwin, C. 1871. *The Descent of Man, and Selection in Relation to Sex.* Appleton, New York.

Eberhard, W. G. 1979. The function of horns in *Podischus agenor* (Dynastinae) and other beetles. In *Sexual Selection and Reproductive Competition in Insects*, M. S. Blum and N. A. Blum (eds.), pp. 231-258. Academic Press, New York.

Emlen, S. T., and L. W. Oring. 1977. Ecology, sexual selection,and the evolution of mating systems. *Science* 197:215-223.

Farr, J. A. 1980. The effects of sexual experience and female receptivity on courtship-rape decisions in male guppies: *Poecilia reticulata* (Pisces:Poeciliidae). *Anim. Behav.* 28:1195-1202.

Felt, E. P. 1895. The scorpionflies. In *Tenth Report of the State Entomologist on the Injurious and Other Insects of the State of New York.*

Fisher, R. A. 1930. *The Genetical Theory of Natural Selection.* (2nd ed. 1958). Clarendon Press, Oxford.

Ghiselin, M. T. 1969. *The Triumph of the Darwinian Method.* University of California Press, Berkeley, CA.

Hamilton, W. D. 1964. The genetical evolution of social behavior, I, II. *J. Theor. Biol.* 7:1-52.

Hrdy, S. B. 1979. Infanticide among animals: A review, classification, and examination of the implications for the reproductive strategies of females. *Ethol. Sociobiol.* 1:13-40.

Lake, P. E. 1975. Gamete production and the fertile period with particular reference to domesticated birds. *Symp. Zool. Soc. Lond.* 35:225-244.

Lloyd, J. E. 1979. Mating behavior and natural selection. *Fla. Entomol.* 62:17-34.

Lumpkin, S. 1981. Avoidance of cuckoldry in birds: The role of the female. *Anim. Behav.* 29:303-304.

Maynard Smith, J. 1978. Optimization theory in evolution. *Annu. Rev. Ecol. Syst.* 9:31-56.

Mickoleit, G. 1971. Zur phylogenetischen und funktionellen bedentung der so genannten notaforgane de Mecoptera (Insecta, Mecoptera). *Z. Morph. Tiere* 69:1-8.

Otte, D., and K. Stayman. 1979. Beetle horns: Some patterns in functional morphology. In *Sexual Selection and Reproductive Competition in Insects,* M. S. Blum and N. A. Blum, (eds.), pp. 259-292. Academic Press, New York.

Parker, G. A. 1970. Sperm competition and its evolutionary consequences in the insects. *Biol. Rev.* 45:525-567.

Parker, G. A. 1978. Searching for mates. In *Behavioural Ecology: An Evolutionary Approach,* J. R. Krebs and N. B. Davies (eds.), pp. 214-244. Sinauer, Sunderland, MA.

Parker, G. A. 1979. Sexual selection and sexual conflict. In *Sexual Selection and Reproductive Competition in Insects,* M. S. Blum and N. A. Blum (eds.), pp. 123-166. Academic Press, New York.

Popper, K. R. 1959. *The Logic of Scientific Discovery.* Hutchinson, London.

Schwagmeyer, P. L. 1979. The Brue effect: An evolution of male/female advantages. *Am. Nat.* 114:932-938.

Sharp, J. L., N. C. Leppla, D. R. Bennett, W. K. Turner, and E. W. Hamilton. 1974. Flight ability of *Plecia nearctica* in the laboratory. *Ann. Entomol. Soc. Am.* 67:735-738.

Sivinski, J. 1977. Factors affecting mating duration in the walkingstick *Diapheromera velii* Walsh (Phasmatodea:Heteronemiidae). M.S. Thesis, University of New Mexico, Albuquerque.

Sivinski, J. 1978. Instrasexual aggression in the stick insects *Diapheromera velii* and *D. covillae* and sexual dimorphism in the Phasmatodea. *Psyche* 85:395-405.

Sivinski, J. 1980a. Sexual selection and insect sperm. *Fla. Entomol.* 63:99-111.

Sivinski, J. 1980b. The effects of mating on predation in the stick insect *Diapheromera velii* Walsh (Phasmatodea:Heteronemiidae). *Ann. Entomol. Soc. Am.* 73:553-556.

Smith, R. L. 1979a. Repeated copulation and sperm precedence: Paternity assurance for a male brooding water bug. *Science* 205:1029-1031.

Smith, R. L. 1979b. Paternity assurance and altered roles in the mating behaviour of a giant water bug, *Abedus herberti* (Heteroptera:Belostomatidae). *Anim. Behav.* 27:716-725.

Smith, R. L., and J. B. Smith. 1976. Inheritance of a naturally occurring mutation in a giant water bug. *J. Hered.* 67:182-185.

Stearns, S. C. 1976. Life-history tactics: A review of the ideas. *Q. Rev. Biol.* 51:3-47.

Symons, D. 1979. *The Evolution of Human Sexuality.* Oxford University Press, Oxford.

Thornhill, R. 1973. The morphology and histology of new sex pheromone glands in male scorpionflies, *Panorpa* and *Brachypanorpa* (Mecoptera:Panorpidae and Panorpodidae). *Great Lakes Entomol.* 6:47-55.

Thornhill, R. 1974. Evolutionary ecology of Mecoptera (Insecta). Ph.D. Dissertation, University of Michigan, Ann Arbor, MI.
Thornhill, R. 1975. Scorpionflies as kleptoparasites of web-building spiders. *Nature* **253**:709-711.
Thornhill, R. 1976a. Sexual selection and nuptial feeding behavior in *Bittacus apicalis* (Insecta: Mecoptera). *Am. Nat.* **110**:529-548.
Thornhill, R. 1976b. The reproductive behavior of the lovebug, *Plecia nearctica* Hardy (Diptera:Bibionidae). *Ann. Entomol. Soc. Am.* **69**:843-847.
Thornhill, R. 1977. The comparative predatory and sexual behavior of hangingflies (Mecoptera: Bittacidae). *Misc. Publ. Mus. Zool. Univ. Mich.* **69**:1-49.
Thornhill, R. 1978. Some arthropod predators and parasites of adult Mecoptera. *Environ. Entomol.* **7**:417-416.
Thornhill, R. 1979a. Male and female sexual selection and the evolution of mating strategies in insects. In *Sexual Selection and Reproductive Competition in Insects*, M. S. Blum and N. A. Blum (eds.), pp. 81-121. Academic Press, New York.
Thornhill, R. 1979b. Male pair-formation pheromones in *Panorpa* scorpionflies (Mecoptera: Panorpidae). *Environ. Entomol.* **8**:886-889.
Thornhill, R. 1979c. Adaptive female-mimicking behavior in a scorpionfly. *Science* **205**:412-414.
Thornhill, R. 1980a. Competition and coexistence among *Panorpa* scorpionflies (Mecoptera: Panorpidae). *Ecol. Monogr.* **50**:179-197.
Thornhill, R. 1980b. Rape in *Panorpa* scorpionflies and a general rape hypothesis. *Anim. Behav.* **28**:52-59.
Thornhill, R. 1980c. Mate choice in *Hylobittacus apicalis* (Insecta:Mecoptera) and its relation to some models of female choice. *Evolution* **34**:519-538.
Thornhill, R. 1980d. Competitive, charming males and choosy females: Was Darwin correct? *Fla. Entomol.* **63**:5-30.
Thornhill, R. 1980e. Sexual selection within mating swarms of the lovebug, *Plecia nearctica* (Diptera:Bibionidae). *Anim. Behav.* **28**:405-412.
Thornhill, R. 1981. *Panorpa* (Mecoptera:Panorpidae) scorpionflies: Systems for understanding resource-defense polygyny and the evolution of alternative male reproductive efforts. *Ann. Rev. Ecol. Syst.* **12**:355-386.
Thornhill, R. 1983. Alternative female choice tactics in the scorpionfly *Hylobittacus apicalis* (Mecoptera) and its implications. *Am. Zool.*
Thornhill, R., and J. Alcock. 1983. *The Evolution of Insect Mating Systems.* Harvard University Press, Cambridge, MA.
Trivers, R. L. 1972. Parental investment and sexual selection. In *Sexual Selection and the Descent of Man, 1871-1971*, B. Campbell (ed.), pp. 136-179. Aldine-Atherton, Chicago.
Voss, R. 1979. Male accessory glands and the evolution of copulatory plugs in rodents. *Occ. Pap. Mus. Zool. Univ. Mich.* **689**:1-27.
Waage, J. K. 1978. Oviposition duration and egg deposition rates in *Calopteryx maculata* (P. de Beauvois) (Zygoptera:Calopterygidae). *Odonatologica* **7**:77-88.
Wallace, A. R. 1889. *Darwinism.* (3rd Ed. 1923). Macmillan, London.
Walker, W. F. 1980. Sperm utilization strategies in nonsocial insects. *Am. Nat.* **115**:780-799.
West-Eberhard, M. J. 1979. Sexual selection, social competition, and evolution. *Proc. Phil. Soc. Am.* **123**:222-234.
Williams, G. C. 1966. *Adaptation and Natural Selection.* Princeton University Press, Princeton, NJ.
Williams, G. C. 1975. *Sex and Evolution.* Princeton University Press, Princeton, NJ.
Willson, M. F. 1979. Sexual selection in plants. *Am. Nat.* **113**:777-790.

6

Sperm Transfer and Utilization Strategies in Arachnids: Ecological and Morphological Constraints

RICHARD H. THOMAS

DAVID W. ZEH

Copyright © 1984 by Academic Press, Inc.
ISBN 0-12-652570-6

I. INTRODUCTION

The arachnids are a class of terrestrial or secondarily aquatic arthropods which includes such varied groups as mites and ticks, spiders, harvestmen, pseudoscorpions, and scorpions (Table I). Mating behavior and methods of sperm transfer are remarkably diverse in the class as a whole, ranging from spermatophore deposition irrespective of the presence of females ("dissociation"; Alexander 1964) to insemination with liquid sperm via copulatory organs. Comparative research has been useful in suggesting evolutionary transitions in sexual behavior (Alexander and Ewer 1957; Brinck 1958; Ghilarov 1958; Alexander 1964; Schaller 1964, 1971, 1979; Weygoldt 1966b, 1970c, 1975). One conclusion of this work is that similar forms of reproductive behavior have evolved independently in several groups and that ". . . ecological factors may have played a great part in the differentiation of certain modes of reproduction" (Schaller 1979). However, with the exception of Alexander (1964), only minor attempts have been made to explain the diversity of arachnid reproductive behavior in terms of ecological factors.

The purpose of this paper is threefold. First, we provide a comprehensive review of arachnid reproductive behavior. Second, we develop some new hypotheses concerning the relationships between ecological factors, sexual strategies, and life

TABLE I

The Orders of Arachnids Ranked Roughly According to the Estimated Number
of Species

---

  1. Acari
       Parasitiformes (primarily fluid feeders)
          Mesostigmata (trophically diverse)
          Ixodida (ticks)
       Acariformes (primarily particulate feeders)
          Prostigmata (chiggers, spider mites, dust mites)
          Oribatei (moss or "beetle" mites)
          Acaridei (scabies and slime mites)
  2. Araneae (spiders)[a]
  3. Opiliones (=Phalangida; harvestmen)
  4. Pseudoscorpiones (=Chelonethida; false scorpions)
  5. Solifugae (sun spiders or wind scorpions)
  6. Scorpiones (scorpions)
  7. Amblypygi (tailless whipscorpions)
  8. Uropygi (whipscorpions)
  9. Palpigradi (micro whipscorpions)[b]
10. Ricinulei (ricinuleids)
11. Schizomida (schizomids)

---

[a]Not treated in this chapter (see Austad, this volume).
[b]Mating behavior unknown.

history characteristics in this group. Third, where possible, we relate reproductive morphology to current controversy regarding the relative importance of male competition versus selection on females in shaping sperm utilization patterns (Walker 1980).

The array of arachnid reproductive strategies rivals any other in the animal kingdom; however, with the possible exception of the spiders (which are treated in detail in another chapter; see Austad, this volume), evolutionary biologists have overlooked these organisms. This is unfortunate since arachnids are often easy to rear, manipulate, and observe, and are therefore highly desirable subjects for experiments in sperm competition. We hope to stimulate further interest in the group by summarizing a scattered literature and suggesting avenues for future research.

Indirect sperm transfer (Schaller 1964) is the primitive mode of reproduction in arachnids (Schaller 1954, 1971; Alexander 1964; Weygoldt 1966b, 1970a, b, c). Sperm are enclosed by males in packets (spermatophores) and deposited on the substratum from which the females actively take them up. Males and females form pairs in most species that transfer sperm indirectly (Group I species; Fig. 1). A number of mite and pseudoscorpion species do not pair, however (Group II species; Fig. 1). There is disagreement over whether or not pairing represents the primitive type of mating behavior (Schaller 1979). This topic will be considered subsequently. Males in Group III species (Fig. 1) transfer sperm directly to the genital opening of the female. Depending on the taxonomic group, the sperm may or may not be enclosed in spermatophores. In some taxa, liquid sperm or spermatophores are released from the male gonopore and then transferred to the female by various appendages (e.g., chelicerae, pedipalps, walking legs). The harvestmen possess a true penis, as do some mites.

The chapter is organized as follows: We begin by presenting a simple model for the evolution of dissociation. Since we argue that dissociation is derived from pair formation, we refer to nonpairers as Group II species (Fig. 1). We also discuss the implications of dissociation for sperm competition and life history evolution. Next we discuss the nonpairing orders and compare natural history observations with predictions of the model. We also present other hypotheses relevant to sperm competition and reproductive strategies. In the third and fourth sections, we review the reproductive behavior of Group I and Group III species, respectively. In discussing mites and ticks, special emphasis is placed on female reproductive morphology and its possible effects on sperm competition.

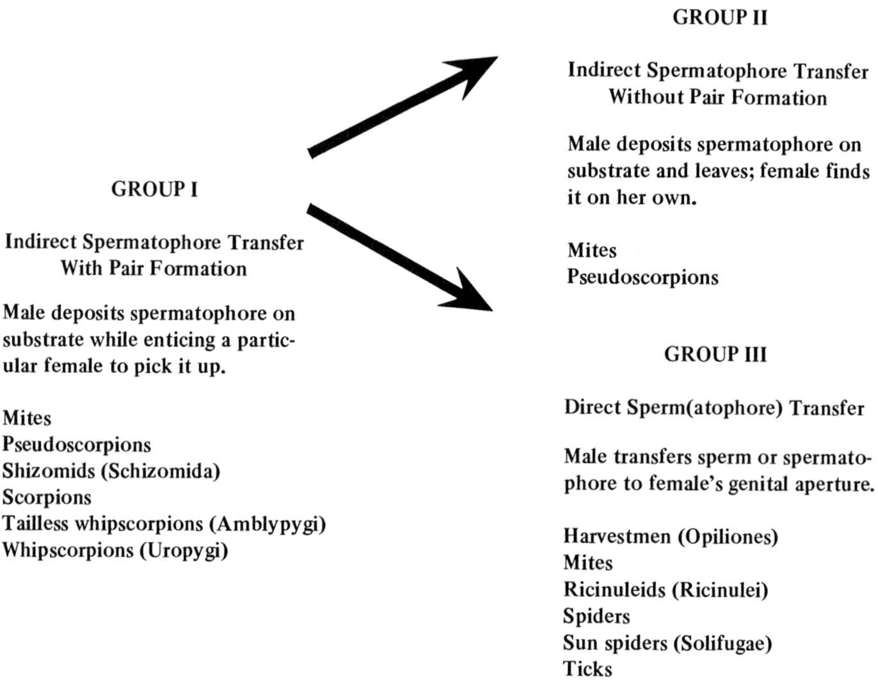

GROUP II

Indirect Spermatophore Transfer
Without Pair Formation

Male deposits spermatophore on
substrate and leaves; female finds
it on her own.

GROUP I

Indirect Spermatophore Transfer
With Pair Formation

Male deposits spermatophore on
substrate while enticing a particular female to pick it up.

Mites
Pseudoscorpions

Mites
Pseudoscorpions
Shizomids (Schizomida)
Scorpions
Tailless whipscorpions (Amblypygi)
Whipscorpions (Uropygi)

GROUP III

Direct Sperm(atophore) Transfer

Male transfers sperm or spermatophore to female's genital aperture.

Harvestmen (Opiliones)
Mites
Ricinuleids (Ricinulei)
Spiders
Sun spiders (Solifugae)
Ticks

Fig. 1. An evolutionary sequence for the development of sperm transfer mechanisms in Arachnida (modified from Alexander 1964). Arachnid taxa are categorized and appear under the appropriate group.

II. INDIRECT SPERMATOPHORE TRANSFER WITHOUT PAIR FORMATION (GROUP II)

Alexander (1964) argues that in arthropods association between the sexes during sperm transfer must have preceded evolutionarily the dissociation of the sexes. The premise is that initially the female would need her attention directed to an object, the spermatophore, before sperm uptake would be possible. Support for this hypothesis is that primitive or proto-arthropods such as *Limulus* (horseshoe crabs) or *Onychophora* exhibit pairing behavior. In contrast, Schaller (1971) and Weygoldt (1966b, 1970c) assume that sperm transfer without pair formation is the more primitive mode in arachnids. The evidence for their hypothesis comes from evolutionary trends in pseudoscorpions (Weygoldt 1966b, 1969b). As Schaller (1979) points out:

. . . increasing complexity of sexual structures and behavior patterns does fit very well with the conventional, morphologically consolidated system of pseudoscorpionid families. Both morphologically and ecologically primitive familes, Chthonidae and Neobisiidae, also demonstrate the primitive [i.e., sperm transfer without pairing] form of sexual behavior.

Schaller's hypothesis appears to hold for pseudoscorpions. However, for reasons given below, we concur with Alexander (1964) that dissociation (and direct transfer) is derived from pair formation (Fig. 1). First, we caution against equating simple with primitive. Second, the distribution of spermatophores among arthropods suggests that spermatophores evolved prior to the occupancy of terrestrial habitats (Davy 1960, Alexander 1964). If this is the case, then it is difficult to postulate selection for clumping sperm (i.e., producing spermatophores) without first evolving close association between individual males and females (i.e., pairing) (Alexander 1964). We suggest that dissociation evolved soon after the arthropods colonized terrestrial habitats. Once dissociation became established in populations, it severely restricted the kinds of suitable habitats available to these species (see model below). This habitat restriction prevented the extensive morphological diversification and adaptive radiation seen in pairing taxa. This could account for the current patterns in pseudoscorpions.

A. Cost-benefit Model for the Evolution of Dissociation

Indirect sperm transfer via a spermatophore placed on the substrate is an important precondition for the evolution of dissociation. In indirect sperm transfer, pairing does not assure fertilization because of: (1) mechanical problems of transfer (see section on scorpions); and (2) the inability of males to actively remove or displace sperm from previous matings, since females physically control sperm uptake. Therefore, indirect sperm transfer facilitates the evolution of dissociation by reducing the difference in the per spermatophore fertilization success between a pairing and a nonpairing strategy (see model below).

We now present a cost-benefit model for the evolution of dissociation from indirect sperm transfer. The essential point is that the fitness conferred by a strategy depends on the difference between the benefit and the cost incurred by that strategy and not simply on the value of the benefit. Fig. 2 examines the effect of increasing desiccation on the costs and benefits of dissociation and pair formation. The cost functions have one or two components. The cost common to both types of sperm transfer is the expense of producing and packaging sperm per spermatophore deposited. This cost is assumed to be approximately equal for the two strategies in perfectly humid environments. With decreasing ambient humidity (increasing desiccation), the packaging costs for nonpairers increase more rapidly because their spermatophores are exposed to desiccation for longer time periods.

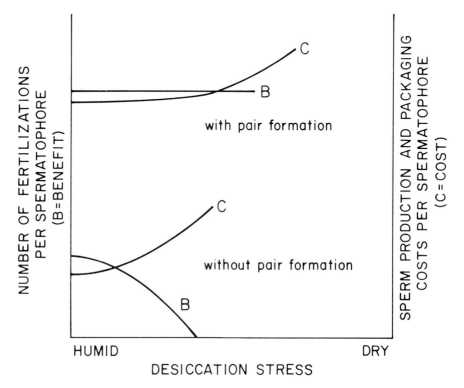

Fig. 2. Cost-benefit analysis of pair formation versus nonpairing during spermatophore deposition in arachnids. See text for further explanation.

This is represented by the more rapid increase in the overall cost curve for non-pairers in Fig. 2. A second cost component is restricted to pair formers and includes the expense of locating a female and enticing her to take up a spermatophore. This cost will include time and energy as well as mortality risks. It is equal to the cost of pairing minus the cost of dissociation at 100% humidity (or 0 desiccation stress).

The benefit curves indicate the expected or average number of sperm which fertilize eggs per spermatophore deposited. The benefit curve for the nonpairing strategy is a monotonically decreasing function of desiccation because of the inherent time delay between spermatophore deposition and uptake. In dry environments, sperm simply dry out before they are discovered by females. In contrast, when spermatophores are deposited only during pairing, the number of fertilizations per spermatophore varies only slightly with desiccation. Because of the increased uncertainty of spermatophore uptake associated with dissociation, the benefits of pairing are higher on a per spermatophore basis. However, the costs of pairing are also higher especially when desiccation stress is low. Therefore, nonpairing males can produce more spermatophores given the same total reproductive effort. In humid environments the total reproductive success (number of fertilizations per spermatophore X number of spermatophores produced) of a nonpairing strategy may equal or exceed that of a pairing strategy. Pairing should always be favored in xeric environments.

We assert that low desiccation stress is a necessary (but not sufficient) condition for males to switch from pairing to dissociation. What additional factors favor dissociation, and what are its consequences for sperm competition and life history characteristics? As our model suggests, dissociation is favored when mortality risks from interactions with mates are high. This seems likely in arachnids that are often voracious carnivores and that rely extensively on tactile sense organs (empirical support for this assertion is provided by many observations of mate cannibalism in scorpions and solifuges; consult those sections of the chapter for details). Also, we suggest two opposing hypotheses for the effect of population density on the benefits to dissociation. The first hypothesis is that dissociation evolved as a male strategy to fertilize several females, made possible by low investments of time and energy per female. Therefore we would predict dissociation will occur in species that usually exist at moderate to high densities where "access" to several females is possible. This promiscuous male strategy is most effective when the point of decreasing returns in fertilization is quickly reached with each females. Generally, this would occur where females pick up one or only a few spermatophores and then become unreceptive. The alternative hypothesis is that dissociation evolved in response to low densities. If meetings between males and females are rare events, males might increase the female-spermatophore encounter rate by depositing many spermatophores randomly in the environment. This would only be effective in extremely mesic environments with slow rates of spermatophore deterioration.

We end this section with a caveat. The model does not explicitly treat the role of female choice in the evolution of dissociation. By not physically interacting with males, females sacrifice much of their opportunity to evaluate male quality, although the spermatophore (a male product), presumably contains discernible cues to male fitness. This sacrifice may have been especially significant to the initial evolution of dissociation. Perhaps the reduction in discriminatory opportunity was counterbalanced by earlier reproduction and/or lower mortality in females that did not pair. More refined models should explicitly include the effects of female choice on the dynamics of this process. This could be accommodated using a game theory approach (see Parker, this volume; and Knowlton and Greenwell, this volume).

B. Effects of Dissociation on Sperm Competition and Life Histories

We expect that sperm competition *per se* and typical sperm competition avoidance mechanisms (*e.g.,* female guarding, sperm plugs, sperm displacement or removal) will be relatively less important in nonpairing species. First, intense sperm competition might impede the evolution of dissociation by favoring increased investment in individual females. If dissociation does evolve, numerous constraints limit male options to prevent sperm competition. Clearly, dissociation makes mate guarding impossible, while sperm plugs or sperm displacement are unlikely when females control sperm uptake. Competition among males will primarily be mediated through the quantity and timing of the spermatophores produced. Nonpairing males should invest in copious sperm production and, as we discuss below, this should have consequences for their life histories.

Recently, Warner and Harlan (1982) argued that sperm storage in females may result in differing life histories for males and females and sexual dimorphism (smaller males). By storing sperm, females can defer reproductive effort and grow larger, whereas males must invest earlier in reproduction (i.e., sperm production) at the probable cost of reduced somatic investment and lowered survivorship. We expect that selection for early reproductive effort by male arachnids will result in accelerated development to the definitive molt at the cost of smaller male size. The nonpairing arachnids (some mites and pseudoscorpions) pass through several nymphal stages before the adult instar is reached and no further molting occurs. Emergence of males before females is particularly advantageous in populations with discrete, nonoverlapping generations (Wiklund and Fagerstrom 1977). By emerging first, males can increase their probability of inseminating virgin females, which would not be the case if females were always available and generations overlapped. These arguments relating sperm storage to sexual dimorphism are not specific to the nonpairing arachnids; however, they are particularly relevant to this group since reproductive competition is mediated through the quantity of spermatophores produced, and because arachnid populations often exhibit discrete generations.

C. Mites

Reproduction without pairing is common in the Oribatei (see Schaller 1971 for discussion of this group) and has been observed in a number of families of Prostigmata, e.g., Nanorchestidae (Schuster and Schuster 1977), Ragidiidae (Ehrnsberger 1977), Tydeidae (Schuster and Schuster 1970), Hydryphantidae (Mitchell 1958), Anystidae (Schuster and Schuster 1966), and Eriophyoidea (Oldfield et al. 1972). With the exception of the superfamily Eriophyoidea, which includes many gall formers, these Prostigmata are free-living and predominantly soil and litter inhabitants.

The reproductive biology of the Eriophyoidea, a haplodiploid superfamily of economically important, host-specific plant parasites, has been investigated by Oldfield and colleagues (Oldfield et al. 1972, Oldfield 1973, Oldfield and Newell

1973). Members of this superfamily have paired spermathecae. Oldfield (1973) found sperm storage in only one of the two spermathecae in 16 of 20 species examined. In all 16 of these species, spermatheca size is nearly identical to the size of the sperm reservoir in the spermatophore. This suggests that females are only inseminated once. Oldfield and Newell (1973) supported this hypothesis with a series of elegant experiments on *Aculus cornutus.* Females inseminated with one fresh spermatophore produced predominantly females (diploid) for several days after insemination but later produced only males (haploid). These females failed to recognize and/or accept fresh spermatophores 4, 8, 12, or 16 days after insemination even though they began producing males before the ninth day. However, 5- or 10-day-old noninseminated females that previously produced only males, readily picked up sperm and became inseminated. Similar results were obtained for three other species (Oldfield 1973). Apparently, sperm competition does not occur in these species.

The other four species examined by Oldfield store sperm in both spermathecae. Also, each spermatheca is as large or larger than the sperm reservoir of the spermatophore. This indicates the potential for sperm competition since multiple insemination is possible.

Oldfield and Newell (1973) also provide quantitative data on sperm and spermatophore production in *A. cornutus.* Spermatophore deposition was monitored from emergence until death for seven males. Data obtained is in accordance with our predictions on sperm allocation patterns in dissociative species. Investment per spermatophore is low, *i.e.,* spermatophores are simple in structure and composed of few sperm. However, many spermatophores are deposited, indicating that total reproductive effort is high. There was an average of 30.4 spermatophores deposited per male per day, while one male deposited 614 spermatophores in a 17-day life span. Finally, the number of sperm per spermatophore approximates the number of eggs produced per female. There was an average of 50 sperm per spermatophore (N = 10) while females produce approximately 25 eggs.

D. Pseudoscorpions

Sperm transfer has been observed in 11 of 21 pseudoscorpion families. Reproductive behavior and habitat utilization are correlated with major taxonomic categories (Table II). The Heterosphyronida and Diplosphyronida are characterized by sperm transfer without pairing (Weygoldt 1966b, 1969b, 1970c). As would be predicted by the model of dissociation, species in these two suborders typically inhabit mesic environments such as forest litter, soil, and caves (Beier 1932; Hoff 1949, 1959; Ressl and Beier 1958; Weygoldt 1969b). In contrast, species in the other suborder, the Monosphyronida, generally form pairs and utilize a much greater diversity of habitats. At least one monosphyronid family which inhabits mesic areas, the Cheiridiidae, exhibits dissociative behavior.

TABLE II

Suborders and Families of Pseudoscorpions[a]

Suborder and Families	Mating Behavior
HETEROSPHYRONIDA	
Tridenchthoniidae	Group II
Chthoniidae	Group II
DIPLOSPHYRONIDA	
Neobisiidae	Group II
Gymnobisiidae or Vachonidae	?
Syarinidae	?
Hyidae	?
Ideoroncidae	?
Menthidae	?
Olpiidae	Group II
Garypidae	Group II
Pseudogarypidae	Group II
Synsphyronidae	?
Feallidae	?
MONOSPHYRONIDA	
Pseudocheiridiidae	Group II
Cheiridiidae	Group II
Sternophoridae	?
Myrmochernetidae	?
Chernetidae	Group I
Atemnidae	Group I
Cheliferidae	Group I
Withiidae	Group I

[a]After Weygoldt (1969b, 1970c).

Quantitative data on male life histories (*e.g.*, developmental rates relative to females) and patterns of sperm allocation (in terms of the mean and variance in the number of spermatophores deposited per male) are lacking. However, there are considerable descriptive data on spermatophore characteristics such as size, number of sperm, and the amount of sperm packaging. Nonpairing species produce the simplest spermatophores, for illustrations see Schaller (1979) or Weygoldt (1969b). In the family Chthonidae (suborder Heterosphyronida; see Table II), spermatophores consist of a straight stalk surmounted by an uncovered sperm droplet. The sperm mass is surrounded on the sides by a "collar" or by spinelike structures (Weygoldt 1966b). Neobisiids (Diplosphyronida) and cheiridiids enclose sperm in simple, globular packages.

The close matching of the number of sperm per spermatophore and the number of eggs per female is consistent with our notion that ejaculate competition is less important in nonpairing species. *Neobisium, Chthonius,* and *Larea* produce 40-100 sperm per spermatophore (Weygoldt 1969b). Female *C. tetrachelatus* produce several broods of 9-15 eggs, while *N. muscorum* females produce 15-25 eggs per brood. Males in the family Cheiridiidae allocate 6-15 sperm per spermatophore, while females produce 3-5 eggs per brood (Weygoldt 1969b). This is in marked contrast to the situation in pairing pseudoscorpions. Pairing species produce "several hundreds" of sperm per spermatophore, while females produce roughly the same numbers of eggs as do nonpairing females.

Nonpairing species reportedly deposit large numbers of spermatophores, although no precise counts are published. For example, in discussing cheiridiids, Weygoldt (1969b) states: ". . . the low number [of sperm] enables these species to produce a great many spermatophores in a short period and to literally cover the substratum with them."

Observations suggest that pheromones are important in attracting females to spermatophores. Legg (1973) describes a light oil produced by the male's lateral glands which accumulates on a thickening of the spermatophore stalk in *Chthonius ischnocheles*. In many species, females exhibit characteristic searching behaviors before actually touching the spermatophore. Finally, pheromones would explain the ability to discriminate spermatophore age in various species. Old spermatophores are generally ignored, possibly due to pheromone degradation. More recent spermatophores are often destroyed by males and replaced. It is not clear from discussions (Weygoldt 1969b) whether males recognize their own spermatophores and replace them when they dry out or if males preferentially replace spermatophores produced by other males.

Females control sperm uptake in nonpairing species, and female choice of spermatophores appears unconstrained by male behavior. Females investigate spermatophores by touching them with their palpal fingers. Receptive females then step over the spermatophore on extended legs and take up sperm by releasing a drop of fluid from their genital atrium. Cues females may use for acceping spermatophores include age, possibly signaled by pheromone concentration, size of sperm mass, and nutritive content. Legg (1973) reports the presence of nutritive substances including proteins in the seminal fluid of *C. ischnocheles*. It is not known whether these substances are utilized by the female or are used to maintain sperm during storage. Another cue to male fitness that females may use in accepting or rejecting spermatophores is the number of spermatophores produced by an individual male. This would be possible if females can distinguish the spermatophores produced by particular males.

The population dynamics and life histories of a few species have been carefully investigated. Goddard (1976) and Gabbut (1967) have both examined seasonal changes in population density and age structure of temperate, litter-dwelling species. Gabbut (1967) found remarkably high densities ($600/m^2$) of *Chthonius ischnocheles*. *Neobisium muscorum* was significantly less abundant in the same study. Goddard's estimates for *N. muscorum* and *C. orthodactylus* are significantly lower than Gabbut's for similar habitats, but again *Chthonius* was more numerous than *Neobisium*. Despite differences in absolute densities (at least partially explained by different sampling methods), several patterns were common to both studies. One generation per year was produced and adult densities peaked during late summer or early fall, when most breeding occurs. Densities appear great enough to insure that several females could encounter spermatophores produced by a single male. These data are consistent with the hypothesis that dissociation is favorable under high to moderate densities, although more systematic study is needed to rule out the low density hypothesis.

We conclude this section by discussing an intermediate form of sperm transfer exhibited by *Serianus carolinensis* (Diplosphyronida). This species occurs in low densities under debris in dry sand dunes near Beaufort, North Carolina (Weygoldt 1966a, 1969b). The male does not initiate pairing but deposits a spermatophore only after physically encountering a female. He also constructs a path of silken threads which directs the female to a single spermatophore. When males rediscover their previously deposited spermatophores, they push over the spermatophore and destroy the webs. These behaviors suggest that spermatophores are intended for particular females and that, due to rapid desiccation of sperm in this relatively xeric habitat, spermatophores are viable for a short time. According to our model, dry environments select for pairing behavior. The behavior exhibited by *Serianus* is uncommon in pseudoscorpions; we think it represents secondary evolution of a trend toward pairing in the suborder Diplosphyronida whose species normally inhabit more mesic environments and exhibit dissociation.

III. INDIRECT SPERM TRANSFER WITH PAIR FORMATION (GROUP I)

A. Mites

Indirect sperm transfer with pair formation is found in a number of Oribatei (mostly soil-dwelling saprovores and fungivores; Schuster 1962, Schaller 1971) and many terrestrial Prostigmata (*e.g.,* Trombidiidae [André 1953], Trombiculidae [chiggers; Lipovsky *et al.* 1957], and Erythraeidae [Putmann 1966]—all members of the cohort Parisitengona; see Table II). Little is known about the size of spermatophores relative to the capacity of the female's sperm storage organs. Since males have no direct contact with the female's reproductive tract, sperm displacement cannot occur. Thus females could at least have proximal control over patterns of sperm utilization, depending on the number of spermatophores they pick up. It would be a male's advantage to produce spermatophores large enough to fill a female's sperm storage organ. Where spermatophores are not this large, the explanation should be sought in terms of some cost function to males and/or discrimination against large spermatophores by females. Very little has been reported in the literature on the occurrence of multiple inseminations or on the frequency of remating in this group. However, Putmann (1966) did not observe more than one insemination in female *Balaustium* sp. (Erythraeidae). He also presents suggestive evidence that pheromones are involved in stimulating the male to deposit a spermatophore and in inducing the female to pick it up.

C. Pseudoscorpions

Pairing behavior in pseudoscorpions is restricted to the suborder Monosphyronida (Table III). Weygoldt (1966b, 1969b, 1970c) provides comprehensive reviews of mating behavior and spermatophore morphology. In the family Chernetidae, males initiate mating by grasping the female with their palpal chelae. This grasp is maintained throughout the mating "dance." Eventually the male deposits a spermatophore and pulls the female over it. The spermatophores are complex structures equipped with a drop of fluid lying beneath the sperm mass. During sperm transfer the fluid contacts the sperm mass, triggering a swelling mechanism. Swelling propels sperm directly into the spermatheca(e) of the female. In many chernetids several spermatophores are transferred during a single mating event, up to 10 in *Dinocheirus arizonensis* (Zeh, unpubl. data).

Chernetid species display a diversity of spermathecal shapes ranging from elongate and tubular to short and spherical (Fig. 3). Because of their morphological diversity, these species would provide excellent experimental organisms for testing Walker's (1980) hypotheses on spermathecal morphology and sperm precedence patterns. That is, do we see spheroid spermathecae in species with one insemination or with precedence of first insemination, in contrast to tubular spermathecae in species with precedence of sperm from the most recent insemination? Little is published on the frequency of mating in these species. Female *D. arizonensis* will mate with several males for up to a week after their first insemination (Zeh, unpubl. data). After this time they become unreceptive.

Some chernetid species are sexually dimorphic for the size and shape of the palpal chelae (Chamberlin 1931, Muchmore 1974). Male chelae are larger and more robust. The chelae are employed in aggressive struggles between males (Fig. 4). Zeh (unpubl. data) has demonstrated experimentally that chela size is positively correlated with aggressive ability in *D. arizonensis*. In competitive situations larger males (males with larger chelae) have higher mating success, especially under conditions of high density. The density effect results from the mate searching strategy of male chernetids. Males grasp or "capture" females to initiate mating, and they are usually unable to distinguish sex before attempting a capture. At high

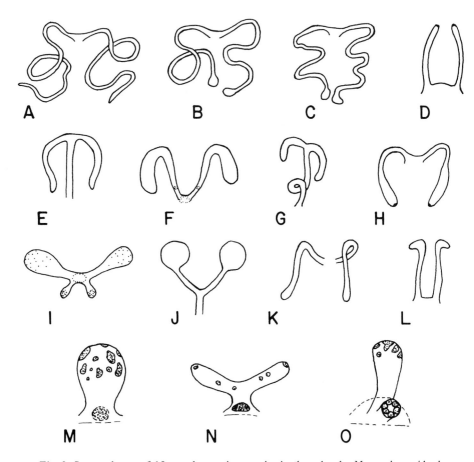

Fig. 3. Spermathecae of 15 pseudoscorpion species in the suborder Monosphyronida show-ing the diversity of shapes. A-L are from species in the family Chernetidae. M is from an atemnid, and N-O are from maratemnid pseudoscorpions. After Muchmore (1975).

densities, the number of mating attempts initiated by small males is low because of frequent encounters with large males. This density effect may explain inter-specific variation in the degree of sexual dimorphism in Chernetidae. Species that inhabit patchily distributed microhabitats are more often dimorphic than species exploiting uniformly distributed habitats such as leaf litter. Patchy microhabitats lead to local high densities and conditions favorable to large males.

The mating behavior in the family Cheliferidae is considered the most highly evolved in pseudoscorpions (Weygoldt 1969b). Males court females with body motions and displays of the ram's horn organs, erectile tubes originating near the genital plate. Usually there is no contact between males and females until after a spermatophore is formed. The male then grasps the female and assists her in

Fig. 4. Sequence of aggressive behaviors in the pseudoscorpion *Dinocheirus arizonensis.* A. Two males approach one another before an actual physical encounter. B. One of the males grasps the pedipalps of the other, after which a struggle ensues. Males use the same grasping behavior to initiate mating (see text). The body length (cephalothorax and abdomen) of each male is approximately 4 mm. The white mark on the cephalothorax of one of the males is a spot of "queen bee paint" used for identification.

taking up the spermatophore. Apparently, the modified tarsi of the male's forelegs open the female genital atrium and press the sperm mass into the spermatheca of the female (Weygoldt 1969b).

To summarize, there is an evolutionary trend toward more proficient sperm transfer mechanisms in pairing pseudoscorpions and suggestion that sperm competition is of some importance in these species. In the less specialized family Chernetidae, males position females over the spermatophore, but do not otherwise assist in sperm uptake. However, in some chernetids the males transfer several spermatophores during a single mating event (Weygoldt 1969b; Zeh, unpubl. data). This behavior may function in sperm competition. In contrast to chernetids, cheliferid males actually assist in the transfer of sperm from the deposited spermatophore to the genital opening of the female. The males press the sperm mass into the genital opening of the female. Whether this behavior simply functions to increase the efficacy of sperm transfer or to displace sperm from previous inseminations is unknown.

C. Schizomids

The order Schizomida consists of a single family (Schizomidae or Tartaridae) of small, tropical arachnids. There are three extant genera: *Schizomus, Trithyreus,* and *Stenochrus.* Mating behavior is known for *Trithyreus* (Sturm 1958, 1973). Modder (1960) details the male genital system of *Schizomus,* and Brach (1975) describes the life history of *S. floridanus.* Pairing behavior in *Trithyreus* is atypical for arachnids in that males and females face in the same direction during spermatophore uptake. To effect sperm transfer, females must cooperate by hooking their chelicerae into two dilations at the base of the male's telson. They are then pulled over the spermatophore by forward movement of the male. Because of this behavior, female control over sperm uptake is probably the most complete in the pairing arachnids. Unfortunately, little is known of sperm utilization patterns in this order.

D. Scorpions

Mating and courtship have been described for six of the seven scorpion families. Polis and Farley (1979) and Garnier and Stockmann (1972) provide extensive reviews of the literature on mating. Francke (1979) reviews spermatophore morphology. Two behaviors are common to all of the species studied. The "promenade," when the male walks back and forth while grasping the female with his chelae, and male pectine movement (Polis and Farley 1979). The pectines are comblike sensory organs underneath the most posterior pair of legs. One function of these behaviors is to locate a suitable (hard and smooth) substrate for spermatophore deposition (Alexander 1957). After the promenade, the male deposits a spermatophore and attempts to position and jerk the female so her genital operculum catches the chitinous hooks of the spermatophore. This inserts the sperm vesicle into the female's genital atrium. Evacuation of sperm from the vesicle is then controlled by the female (Alexander 1957). Positioning of the female appears to be a critical step in sperm transfer. Alexander describes unsuccessful matings in which the male apparently was unable to properly position the female. Males often exhibit an aggressive "escape display" and quickly retreat after a single spermatophore is transferred (Alexander 1958). Alexander (1958) states that a period of 3 weeks is required for a male to produce a new spermatophore.

Mate cannibalism by female scorpions has been reported for four families and 8 of 20 species reviewed by Polis and Farley (1979). Most reports are based on laboratory observations. However, the recent work of Polis and Farley (1979) and Polis (1980) demonstrates with extensive field observations that cannibalism of mature males by mature females during the breeding season accounts for a significant proportion of male mortality; 5.8% of the males survived past their first breeding season compared with 18.9% of the females (Polis and Farley 1979). This differential mortality is explained by cannibalism and other factors related to breeding. Males become vagrant during the breeding season and consequently incur greater predation, feed less, and experience greater energetic costs due to the constant need to construct temporary burrows. Male vagrancy during the breeding season appears to occur often in vaejovids and possibly is a general phenomenon in scorpions. For example, *Urodacus abruptus* (family Scorpionidae) are vagrant and adult sex ratios are skewed, 3:1, in favor of females. This skew results from differential mortality since the sex ratio of randomly collected, laboratory-reared nymphs is 1:1 (Smith 1966). These observations provide strong evidence for our assertion that mortality risks due to mate searching and interactions with females are significant selective forces in arachnids.

A particularly intriguing aspect of scorpion biology is the existence of life history related polymorphisms in Buthidae. Francke and Jones (1982) reared two broods produced by a pair of field-collected *Centruroides gracilis* females. The nymphs hatched in the laboratory and grew to adults in individual containers kept at constant temperature in environmental chambers. All nymphs were treated similarly. Females matured by the seventh instar in an average of 302 days (N = 24). Twenty percent of the males from litter 1 and approximately 50% of the males from litter 2 matured by the sixth instar in 236 days (N = 14). The rest of the males matured by the seventh instar in 282 days (N = 19). The basis of this polymorphism is unclear, although a genetic mechanism appears likely since the nymphs were subjected to similar and constant environments. The possible selective factors maintaining this polymorphism are unknown. Seventh instar males are 1.25X larger than sixth instar males for a variety of morphometrics including chela length.

Perhaps early maturation represents an alternative mating strategy (see Rubenstein 1980). The relationship between aggressive ability, mating success, and chela size may have parallels with pseudoscorpions or, alternatively, larger male size may decrease mortality from mating attempts with females. The higher mortality or lower competitiveness of small males may be compensated by rapid development and increased access to virgin females in these seasonally reproducing scorpions. This phenomenon warrants further research; these organisms are ideal subjects for testing dynamic optimization and game theory models of life history evolution (see Maynard Smith 1982).

It should be noted that Buthidae polymorphisms occur in the females, and in some cases in both sexes of other species (see Francke and Jones 1982). If fecundity is related to body size, females could trade fecundity for an earlier age of first reproduction.

We are not aware of studies which detail the number of matings per females or patterns of female receptivity in relation to previous matings. Smith (1966) states that female *Urodacus abruptus* mate once every 2 years. He also reports that a vaginal plug is secreted after insemination, but the details of this process are not given.

E. Tailless Whipscorpions (Amblypygi)

Amblypygids are a tropical and subtropical order of secretive, nocturnal carnivores that impale prey items with large, spinous pedipalps. Some variation in mating behavior exists, but the differences are not well defined along taxonomic lines (Weygoldt 1977). *Damon variegatus, Admetus barbadensis* (Alexander 1962), *Sarax sarawakensis* (Klingel 1962), and *Charinus brasilianus* (Weygoldt 1974) are similar in that males generally do not grasp females during mating. Males initiate courtship by tapping or massaging females for extended periods with antenniform (elongate and sensory) first pair of legs. Eventually, the male deposits a spermatophore and leads or beckons the female to it. The pair then separate. The male takes a more active role in *Tarantula* (*Phrynus*) spp., *Heterophrynus alces*, and *Charinus montanus* (Weygoldt 1969a, 1974, 1977). After courtship, males seize the female and pull her over the spermatophore.

Amblypygids are long-lived. *Tarantula marginemaculata* reach maturity 8-10 months and five to eight molts after hatching (Weygoldt 1970b). Mature individuals continue to molt through life and females produce eggs one to three times per year (Weygoldt 1970b, Weygoldt *et al.* 1972). Females secrete a parchmentlike membrane that attaches the eggs to the underside of her abdomen. The eggs are carried until hatching (98 days in *T. marginemaculata*) and the nymphs cling to the female's abdomen until their first molt.

Aggressive behavior frequently occurs in encounters between two amblypygids. Contests between individuals of similar size often include violent pedipalpal attack and possible fatal injury to one or both contestants (Alexander 1962). In *Trichodamon froesi* the pedipalps are greatly enlarged in males and are unusual in bearing small, apical chelae. During fighting, males grasp the tibiae of the opponent's third pair of legs and push them upward and backward (Weygoldt 1977).

Weygoldt (1977) and Weygoldt *et al.* (1972) have examined the functional morphology of amblypygid genital organs. In *Heterophrynus, Tarantula,* and *Trichodamon* the spermathecae are paired, flattened, and roughly spherical. *Charinus brasilianus* lacks spermathecae; sperm are stored in a posterior pocket of the genital atrium. There are two stages of vitellogenesis. The second stage is induced by insemination. Mating also causes the ovary to secrete a filamentous substance which is extruded at oviposition and eventually surrounds the eggs to form the egg sac. At oviposition, all sperm are utilized and stored sperm are lost with each molt. Therefore females must remate after oviposition and/or molting.

F. Whipscorpions (Uropygi)

Uropygids are large (25-75 mm) tropical and subtropical inhabitants of Asia and the Americas. Like Amblypygi, their first pair of legs are antenniform. Uropygids are also extremely long-lived. *Mastigoproctus giganteus* has four nymphal stages and a terminal adult molt (Weygoldt 1971). Maturity is reached after 4 years. Mating behavior and spermatophore morphology are similar in *M. giganteus* (Weygoldt 1970a, 1971), *M. brasilianus* (Weygoldt 1972), and *Thelphonellus amazonicus* (Weygoldt 1978). Courtship and sperm transfer are particularly striking. Males initiate mating by grasping the female's antenniform legs with chelate pedipalps. Courtship involves back and forth movement and stroking of the female with the male's antenniform legs. This can last for hours. Eventually the male turns 180° and the female then embraces the male's abdomen while he deposits a spermatophore. Next, the male pulls the female over the spermatophore and she picks up the two sperm carriers (Fig. 5). The male then turns to face the female and uses his chelae to further insert the sperm carriers into the female's gonopore (Fig. 5). This is followed by a truly remarkable behavior pattern. The male pushes and pulls the sperm carriers while keeping the tips with the ejaculatory duct openings inserted in the gonopore. This "pumping" of the sperm carriers can last for 2 h.

Sperm transfer in Uropygi is perhaps the most proficient indirect method exhibited by arachnids. It even seems probable that this system may displace sperm during the prolonged pumping of the sperm carriers by the male. Sperm competition may have been important to the evolution of this male behavior since females multiply mate even though spermatophores contain thousands of sperm.

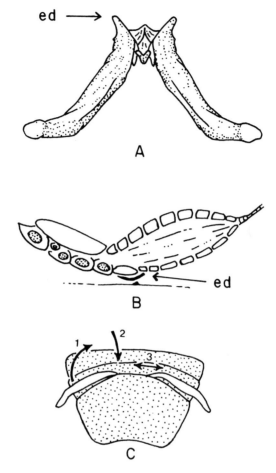

Fig. 5. Spermatophore, and sperm transfer in Uropygi. A. Dorsal view of spermatophore showing base (central structure) and the paired, elongate sperm carriers (ed = opening of the ejaculatory duct). B. Lateral view of female pressing down on the sperm carriers causing the ed ends to rotate inwards toward her genital aperture. C. Ventral view of female genital region after having picked up the sperm carriers. Arrows and numbers show how the male pushes the sperm carriers into the genital aperture of the female. After Weygoldt (1971).

IV. DIRECT SPERM TRANSFER (GROUP III)

A. Harvestmen

In number of described species, the harvestmen are surpassed among the arachnids only by spiders and the Acari. They are unusual in that males possess a protrusible penis and females an ovipositor. Harvestmen are ubiquitous in their geographical distribution, but generally prefer humid microhabitats such as forest litter, caves, and under logs. Observations on mating behavior are scattered throughout the literature (*e.g.,* Bishop 1949, Cloudsley-Thompson 1948, Edgar 1971, Forster 1954, Pabst 1953, Rodriguez and Guerrero 1976). In general, courtship is absent or of short duration. Males face females during sperm transfer and project a tubular penis into the genital aperture of the female. In some species the female uses her palpi or chelicerae to direct insertion of the penis. Mating in the family of short-legged Opiliones, the Trogulidae, is atypical. The male hangs beneath the female so that the ventral surfaces of the two are opposed (Pabst 1953).

Mating of Michigan *Leiobunum* has been observed in both the field and the laboratory (Edgar 1971). These observations have revealed some fascinating behaviors. *Leiobunum longipes* and *L. vittatum* form mating clusters near favorable oviposition sites. Clusters are composed of both sexes and range in size from a few individuals to more than 50. It is not known which sex initiates cluster formation, although males seem to be the last to abandon the clusters. The following account is for *L. longipes* (Edgar 1971). If females are present in the clusters, males fight over access to them. The palpi and chelicerae are used in struggles between males which consist of leg pulling contests. The victor then copulates with the female. During sperm transfer, the male's second pair of legs are poised to detect interruption. If a copulating male is interrupted, he may briefly leave the female to repel the intruder. Sometimes smaller males, which never directly challenge dominant males, use this opportunity to copulate with the female. When the dominant male returns, he usually again copulates with the female. In fact, repeated copulations are the rule. One male was observed to copulate 29 times with the same female in 2.5 h. After mating, the male engages in postcopulatory guarding by forming a canopy over the female with his legs. During this time the female oviposits. After the female leaves a small area defended by the male, he ignores her.

Repeated copulations and mate guarding suggest that sperm competition is rampant in harvestmen. It should be noted that the penises of harvestmen are often endowed with tubercles and terminal hooks. It is not known if these structures function to displace or remove sperm (see Waage, this volume).

The sneak strategy of small males may be related to the coexistence of two discrete male morphs in the same population. This phenomenon has been described for many species of New Zealand harvestmen (Forster 1954) and may occur in other species. The two male morphs differ in the size and structure of the chelicerae and the pedipalps, as well as in some other characters.

Because harvestmen are easy to observe and rear in the laboratory, they are well suited for experiments on sperm competition.

B. Mites and Ticks

All ticks and many mite species have direct sperm transfer. Within this grouping are both species using spermatophores and those using liquid sperm. Here we will deal with some comparatively well studied taxa with respect to reproductive biology and taxonomy (*e.g.,* Phytoseiidae, Tetranychidae, and Ixodida). There are many other mite taxa which, on the basis of morphology, probably belong in this group that remain almost unknown. The very wide range of reproductive morphologies and behaviors in the taxa discussed here could provide useful tests for a variety of hypotheses on sperm competition (see Table III).

One of the most curious adaptations occurring in the mites is the accessory genital system found in females of many Prostigmata and Mesostigmata, and all of the Acaridei. This term is unfortunate since it implies that the structures are used facultatively or secondarily; this is not the case. However, the term accessory genital system is well established in the mite literature and rather than add to an already confusing terminology we will continue to use it. In these groups sperm or spermatophores are placed into a storage organ via an external opening separate from the one through which eggs are passed when deposited. Similar situations occur in some spiders (Austad, this volume) and the ditrysian lepidoptera (Drummond, this volume). These systems have probably evolved at least three times in mites, once each in the Mesostigmata, Prostigmata, and Acaridei. Different types of accessory genital systems are illustrated in Fig. 7. Selective factors responsible for the origins of these systems are obscure and, considering their very divergent ecologies, may indeed be quite different in the various groups.

1. Taxa Without Accessory Genital Systems

Mesostigmata. Among the Mesostigmata, the primitive cohorts Sejina and Uropodina, as well as some more advanced families such as the Parasitidae (a misleading name; all species are predators), Arcticaridae, Epicriidae, and Zerconidae, do not possess an accessory genital system. Of these groups, the Uropodina are comparatively well studied because a number of species inhabit stored grain, feeding on fungi (Krantz 1961). The usual mating position is venter to venter with the male transferring a spermatophore to the ventral edge of the female's vaginal opening with his unmodified chelicerae and first pair of legs (Radinovsky 1965, Woodring and Galbraith 1976; but see Compton and Krantz 1978 for a very bizarre variation). Radinovsky (1965) observed postcopulatory guarding by males of two of the uropodine *Leiodinychus krameri.* He also observed agonistic behavior between males and presents anecdotal evidence that males prefer young virgin females.

In the Parasitidae males transfer spermatophores with their chelicerae (apparently modified for this function) to the vaginal opening of the female (Micherdzinski

1969). Little has been reported on the mating behavior of the remainder of the above-mentioned Mesostigmata.

Ixodida. Ticks are all obligate, bloodsucking, ectoparasites of vertebrates. They have been intensively studied because of their importance in disease transmission to humans and livestock. The 800 or so species are divided into two main families, the Argasidae, or soft ticks, and the Ixodidae, or hard ticks. For our purposes the hard ticks can be further subdivided into the Ixodinae (Prostriata) and the Amblyomminae (Metastriata) Generalizations on ecologies, life histories, and reproductive systems can be made for these taxa.

The argasids tend to be nest or burrow inhabiting parasites that mate off the host; the females are generally iteroparous (having multiple gonotrophic or feeding:egg-laying cycles). Total lifetime egg production for a female is generally in in the hundreds of eggs. Nymphal stages range in number from 2-7, and males tend to be nonfeeding as adults. Females store endospermatophores (produced by the ectospermatophore that the male places on the female's genital aperture) in the hypertrophied uterus (Fig. 7A), where sperm mature (Balashov 1972, Hoogstraal 1976). Thus on morphological grounds one would expect the first insemination to have priority since fertilization is known to occur in the oviduct (Balashov 1972). Prior to transfer of the ectospermatophore to the female genital opening the male inserts his mouthparts into the large and elastic female genital aperture (Feldman-Muhsam 1973). The function of this action has not been demonstrated; however, it is possible that sperm displacement or removal occurs. After withdrawing his mouthparts from the female, the male uses them to transfer an ectospermatophore to the female genital aperture. The ectospermatophore then everts the two endospermatophores (Oliver *et al.* 1974) into the female tract (Feldman-Muhsam 1973). Females multiply mate. Aeschliman and Grandjean (1973) found evidence of 40 matings on dissection of one laboratory-reared female *Ornithodorus moubata*. It is unclear whether multiple matings within a gonotrophic cycle occur in nature. Mating in each gonotrophic cycle is necessary to maintain full fecundity and egg viability, as well as the normal preoviposition period in *O. moubata* (Aeschliman and Grandjean 1973).

The Amblyomma present a marked contrast to the argasids in ecology and life history. They tend to be field ticks which mate on the host, and females are usually semelparous. Total lifetime egg production for a female generally ranges from 6-7,000, and can be as high as 20,000 (Hoogstraal 1976). A single nymphal stage is the rule and both males and females tend to require blood meals as adults. Females store endospermatophores in a thin-walled seminal receptacle which is a narrow-necked, blind outpouching of the cervical region of the vagina (Balashov 1972; Fig. 7B). From morphology, one would expect the last insemination to have priority. Prior to transfer of the ectospermatophore to the female genital opening the male inserts his chelicerae into the relatively small and inflexible female genital aperture (Feldman-Muhsam 1973). Again, the function of this action is not clear. Transfer of

TABLE III

Selected Mite Taxa and Their Mating Systems

Mating System	Taxon	Multiple Insemination	References
GROUP I	Oribatei	?	Schuster 1962, Schaller 1971
	Prostigmata		
	Trombidiidae	?	André 1953
	Trombiculidae	?	Lipovsky *et al.* 1957
	Erythraeidae	no, ?	Putmann 1966
GROUP II	Oribatei	?	Schaller 1971
	Prostigmata		
	Nanorchestidae	?	Schuster and Schuster 1977
	Rhagidiidae	?	Ehrnsberger 1977
	Tydeidae	?	Schuster and Schuster 1970
	Hydryphantidae	?	Mitchell 1958
	Anystidae	?	Schuster and Schuster 1966
	Eriophyidea		Oldfield 1973
	Eriophyidae	yes & no	
	Rhyncaphytoptidae	no	
	Sierraphytoptidae	yes	
GROUP III	Mesostigmata		
(without accessory genital system)	Sejina	?	Krantz 1978
	Uropodina	?	Radinovsky 1965, Compton and Krantz 1978
	Parasitidae	?	Micherdzinski 1969
	Ixodida	yes	Balashov 1972, Hoogstraal 1976

202

Taxon		References
Prostigmata		
Hydrachnoidea	?	Davids and Belier 1980
GROUP III (with accessory genital system)		
Mesostigmata		
Rhodacaridae	no, ?	Lee 1974 (reviews behaviors in other families also)
Macrochelidae	no	Costa 1966, 1967; Kinn and Witcosky 1977
Phytoseiidae	yes	McMurtry *et al.* 1970; Dosse 1958, 1959; Amano and Chant 1978, Schulten *et al.* 1978
Acaridei		
Acaridae	yes	Griffiths and Boczek 1977
Prostigmata		
Tetranychidae	yes	Helle 1967; Potter and Wrensch 1978, Potter 1978

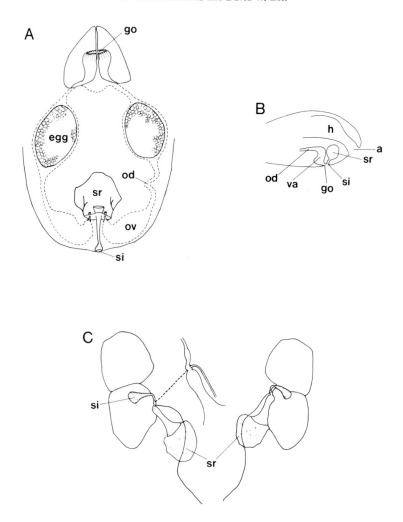

Fig. 6. Female mite accessory genital systems and male mesostigmate chelicerae. A. Schematic diagram of reproductive system of a free-living acaridid mite. B. Schematic diagram of reproductive system of a free-living tetranychid mite in median sagittal section. C. Spermathecae of the phytoseiid mite *Amblyseius* sp. D. Reproductive structures of a mesostigmate mite (*Athiasella dentata*). E. Spermatheca of the mesostigmate mite *Rhinophaga pongoicola*. F. Spermatheca of the mesostigmate mite *Tyranninyssus spinosus.* G. Male mesostigmate chelicera with spermatodactyl (*Gamasellus tragardhi*). H. Male mesostigmate chelicera with spermatodactyl (*Euepicrius filamentosus*). go = genital opening; h = hindgut; od = oviduct; ov = ovary; si = sperm (or spermatophore) induction opening; sr = seminal receptacle; sri = duct into sr; sro = duct out of sr; va = vagina; ut = uterus. A redrawn from Griffiths and Boczek (1977); B redrawn from Crooker and Cone (1979); C, E, F redrawn from Fain (1963); D, G, H redrawn from Lee (1974).

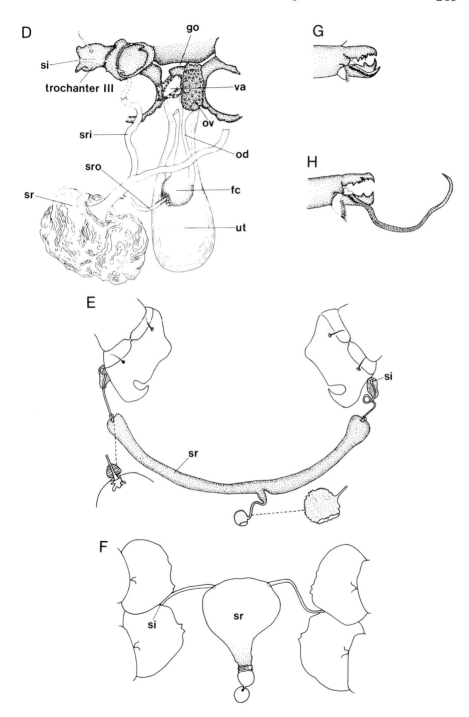

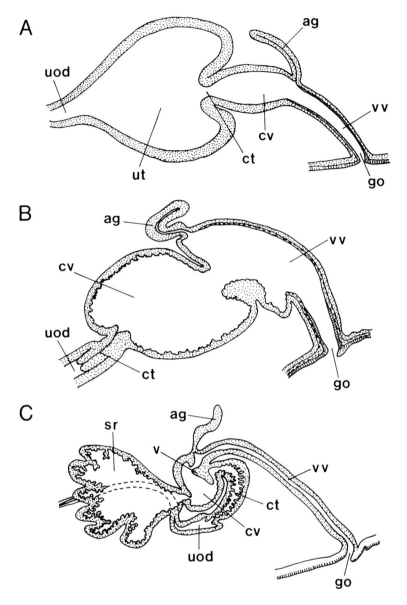

Fig. 7. Female tick reproductive system. A. Argasidae (*Ornithodoros papillipes*). B. Ixodinae (Ixodidae; *Ixodes ricinus*). C. Amblyomminae (Ixodidae; *Hyalomma asiaticum*). ag = accessory gland; ct = connecting tube; cv = cervical part of vagina; go = genital opening; sr = seminal receptacle; uod = unpaired oviduct; ut = uterus; v = valve; vv = vestibular part of vagina. Redrawn from Balashov (1972).

the ectospermatophore and insemination occur in a manner similar to that observed in the argasids, except that a single endospermatophore per ectospermatophore seems to be standard (Oliver *et al.* 1974). However, Londt and Spickett (1976) commonly found two, or multiples of two, endospermatophores in the seminal receptacles of females of *Boophilus decoloratus,* suggesting two endospermatophores per ectospermatophore. They also note that singly mated females produce as many eggs as multiply mated females (Londt and Spickett 1976). More than one mating is common (Oliver *et al.* [1974] found one to five endospermatophores in *Dermacentor occidentalis* females, and one to three in *Haemaphysalis leporispalustris* females), but it is not clear how often more than one male is involved. A number of observations indicate that after insemination males maintain their venter-to-venter mating position until the female completes engorgement and drops from the host to lay eggs, an interval ranging from a few hours to 2 days (Brown 1979, Oliver *et al.* 1974). Brown (1979) also reports on male *Dermacentor variabilis* (dog ticks) which assume the venter-to-venter mating position but feed prior to copulation, then feed following copulation until the female drops from the host. The pair remains in mating position for 4 days. In laboratory situations males are commonly observed to mate with more than one female (Gladny and Drummond 1970, Londt 1976). Oliver *et al.* (1974) observed two males competing for a female *Haemaphysalis leporispalustris* and found two ectospermatophores on her. These observations need to be expanded.

The Ixodinae (Prostriata) of the Ixodidae fit somewhere between argasids and amblyommines in their habits. They tend to be field ticks, and adults often do not require blood meals in order to reproduce. Females are semelparous and generally produce an egg batch composed of several thousand eggs (Hoogstraal 1976). A single nymphal stage and a three-host life cycle with mating occurring either on or off the host is usual (Balashov 1972, Hoogstraal 1976). Females store endospermatophores in the wide cervical part of the vagina (Fig. 7B) (Balashov 1972), apparently functionally analogous to the situation in argasids. Thus one would expect the first insemination to have priority. Prior to transfer of the ectospermatophore to the female genital opening, the male inserts his hypostome and chelicerae into the relatively small yet flexible female genital aperture (Feldman-Muhsam 1973). Balashov (1972) reports males of *Ixodes ricinus* may remain "attached" to females for several days. Thus it seems likely that postcopulatory guarding occurs here, as in the Amblyomminae.

A *caveat* should be inserted at this point; the above generalizations are based on examination of only a few taxa. Instructive divergences from the generalizations are known. For example, species of the argasid genus *Otobius,* which parasitize ungulates and hares, have a single autogenic (oviposition without feeding) gonotrophic cycle; adults usually do not feed and females die after oviposition (Balashov 1972), a remarkable convergence on the amblyommine pattern. Since much of the work on tick reproductive behavior has been conducted with the goal to control using sterile males, observations directly pertinent to questions on sperm competition have not often been made. Adult male ticks have large and complex accessory glands that secrete the seminal fluid found in endospermatophores (Balashov 1972). Balashov (1972) found accessory gland vacuole types in *Hyalomma asiaticum* containing "mucoprotein or glycoprotein, mucopolysaccharide, protein-rich disulfides, and lipoproteins." Sperm cells are known to approximately double in size in the female tract (Balashov 1972, Oliver 1974). It would be instructive to know if the female contributes to this doubling or if the materials transferred by the male is sufficient, or perhaps in excess of what is needed to maintain sperm in the female. It should be possible to determine this with labeling experiments. Pheromones are known to be important in aggregation and mating in both tick families (Sonenshine *et al.* 1979, Rechav and Whitehead 1979, Leahy 1979). It would be helpful to know, for instance, if females turn off sex attractant following mating. Brown (1979) found that male *Dermacentor variabilis* lose their ability to discriminate between sexes midway into a feeding period, perhaps suggesting a cessation of pheromone production.

Prostigmata. Among the huge number of Prostigmata without accessory genital systems, many species use spermatophores and perhaps some use liquid sperm. Unfortunately very little is known of the behaviors of these species. At least some of the secondarily aquatic Parasitengona have copulation involving a spermatophore (Davids and Belier 1980).

2. Taxa With Accessory Genital Systems

Mesostigmata. These organisms comprise a very large and diverse group ranging from soil-dwelling predators, to arthropod associates, to parasites of any animal larger than themselves (Krantz 1978). They possess one of the most curious reproductive systems found in animals. Spermatophores are introduced into one or both of the female's spermathecal pores by the male's spermatodactyls, syringelike modifications on the moveable digits of the chelicerae (Fig. 6G) (Athias-Henriot 1970, Krantz and Wernz 1979, Young 1968, Pound and Oliver 1976). These pores are found on, or around, the coxae or trochanters of the third or fourth pair of legs. They lead, via tubes, to the sperm storage organs (spermathecae).

Pound and Oliver (1976) provide a table synonomizing the many terminologies applied to the Mesostigmata accessory genital system. As can be seen in Fig. 6C, D, E, and F, there is considerable variation in accessory genital systems, both with respect to size and shape of the tubes leading to the spermatheca(e), and the spermatheca(e) itself. In all cases examined, tubes leading to spermathecae, regardless of length, are narrow and it thus seems unlikely that a male's spermatodactyls could reach the spermathecae. As Walker (1980) has suggested, this would reduce the probability of sperm displacement in multiple matings. A number of Mesostigmata have very long spermatodactyls (Fig. 6H), but in the only reported copulation no more than the tip of spermatodactyl was introduced into the spermatehcal pore (Lee 1974). This may suggest the action of epigamic selection on spermato-dactyl length, but more observations are needed to rule out the possibility of sperm displacement.

Walker (1980) has suggested a relationship betwen spermathecal shape and order of sperm precedence. He notes a trend for "spheroid spermathecae in species with monogamous [*i.e.,* one insemination] females or with precedence of the first insemination, contrasted with elongate or tubular spermathecae in species with precedence of sperm from the last insemination." One interesting consequence of the accessory genital system is that it can reverse Walker's (1980) predictions, depending on the relative positions of the tubes carrying sperm into and out of the spermathecae. When these tubes are widely separated (Fig. 6E, F), we expect that in species with tubular spermathecae the first insemination will have prece-dence, while in species with spheroid spermathecae there could be more oppor-tunity for sperm mixing and the first insemination will therefore be less likely to have precedence. When these tubes are closely spaced (Fig 6D), spermathecal shape and order of sperm precedence should be as Walker (1980) suggests since passage of sperm into and out of the spermathecae occurs in much the same way as in the insect taxa he dealt with.

One of the two groups for which detailed information exists in mating behavior does not provide support for sperm competition as a factor in the evolution of the shape and arrangement of the female accessory genital system. In the Macrochelidae (common predators in decaying material) spermathecal morphology is variable among species while mating behavior and sperm utilization are not. In *Macrocheles robustulus* the spermatheca consists of two globular chambers connected by a narrowed passage (effectively tubular) with sperm entrance and exit tubes widely separated (Costa 1966), while *M. parapisentii, M. pisentii, M. cristati,* and *M. sacerci* have globular spermathecae with sperm entrance and exit tubes widely separated (Costa 1967). In all these species we find that females mate only once (Costa 1966, 1967; Kinn and Witcosky 1977). Copulations are observed only with freshly emerged females; after the cuticle of their heavily sclerotized bodies hardens females become nonreceptive (Kinn and Witcosky 1977). There is competition among males for the opportunity to inseminate females, and in one extreme case Costa (1967) has observed older males of *M. parapisentii* regularly killing newly emerged males and even male deutonymphs (the final nymphal stage prior to the molt to the adult). This imposes strong selection for rapid development.

The final family of Mesostigmata to be discussed, the Phytoseiidae, has been intensively studied (compared with the rest of the Acari!) because of its importance in the control of plant parasitic mites (McMurtry *et al.* 1970). The Phytoseiidae is unusual in two respects. First, they are parahaplodiploid—male eggs require fer-tilization and syngamy to develop, but the paternal genome is heterochromatisized and discarded from the embryonic cells in early development (Helle *et al.* 1978,

Nelson-Rees *et al.* 1980). The rest of the Mesostigmata with accessory genital systems appear to be truly arrhenotokous (Oliver 1971, 1977). Second, they have unusual paired spermathecae (Fig. 6C), or at any rate what are called sperma-thecae—they may not be homologous to the spermathecae of the rest of the Meso-stigmata. The remainder of the internal structures of the female phytoseiid repro-ductive system remains unknown (Amano and Chant 1978). Nonetheless, some very interesting and suggestive observations have been made on members of this family. Multiple matings are commonly observed (Dosse 1959, McMurtry and Scriven 1964, McMurtry *et al.* 1970) and more than one spermatophore have been observed in a spermatheca (Dosse 1958, 1959). Amano and Chant (1978) made detailed observations on sperm transfer and mating behavior in the Phytoseiidae in their comparative study of *Phytoseiulus persimilis* and *Amblyseius andersoni.* They distinguish between the ectospermatophore, a sac which remains attached to the male's chelicera throughout copulation, and endospermatophores which are chitinized sacs, presumably containing sperm, found in the female's spermatheca. This situation is similar to that in ticks, though it is not clear in this case if the endospermatophore wall is transferred by the male or is produced by the female tract. This resistant endospermatophore disappeared from the spermatheca within 24 h in *P. persimilis,* while in *A. andersoni* it remained in reduced size for another day or two. This may indicate that the spermathecae are functioning in sperm reception rather than sperm storage, since sperm does not remain in the sperma-thecae very long relative to the duration of egg laying. The rest of the female reproductive system needs to be elucidated. Also, Amano and Chant (1978) ob-served differences in time *in copula* and in number of endospermatophores between the two species. *P. persimilis* remained *in copula* for 131 min on average, and generally one endospermatophore was transferred to one of the spermathecae, while *A. andersoni* remained *in copula* for 185 min on average and invariably one endospermatophore was transferred to each of the pair of spermathecae. By means of interrupted matings it was determined that fecundity is dependent on time *in copula.* Fecundity increased in *A. andersoni* with copulation durations up to 120 min, thereafter leveling off, suggesting the possibility that the function of the additional time *in copula* is guarding. Fecundity continued to increase throughout the period of copulation in *P. persimilis.* Dosse (1959) observed several speramto-phores in the spermathecae of *P. persimilis* and *Typhlodromus zwolferi,* and sug-gested this indicated the number of copulations. Schulten *et al.* (1978) in their study of *P. persimilis* and *Amblyseius bibens* also observed a monotonic increase in fecundity with increasing time in copula. They further observed that a single full term copulation is adequate to ensure maximum egg production in *P. persimilis,* while *A. bibens* seems to require two to three inseminations for maximum egg production. Similarly, McMurtry *et al.* (1970) reviewed a number of reports in-dicating that repeated inseminations are commonly necessary for maximum egg production in phytoseiids. These observations indicate potential for sperm competition.

An interesting aspect of Schluten *et al.*'s (1978) work is their observation that mating can be interrupted under natural conditions by conspecifics (females looking for males), their spider mite prey, and possibly by predators. The sex ratio in phytoseiids is strongly biased toward females with proportionally more males produced early in a female's reproductive life (McMurtry *et al.* 1970, Schulten *et al.* 1978). Fertilized females are known to disperse when density increases (McMurtry *et al.* 1970, Schulten *et al.* 1978). Thus, since total fecundity is dependent on time in copula, interrupted matings could have the effect of increasing the sex ratio. Changes in density of prey species and conspecifics as the resource (a plant leaf) is colonized and finally depleted may therefore influence patterns of mating and sperm utilization. Further investigation in this area should be fruitful.

Acaridei. An accessory genital system is found in all members of the suborder Acaridei (Astigmata; Fig. 6A). Multiple matings appear to be the general case for this suborder (Griffiths and Boczek 1977). Until recently it was believed that all sperm transfer occurred with liquid sperm introduced into the female by the male's aedeagus or copulatory organ. Griffiths and Boczek (1977), however, report the existence of spermatophores in three genera of stored products Acaridei (mites inhabiting foodstuffs and animal feed; see Hughes 1976) and suggest that the possession of a spermatophore is a generic character. The genus *Tyrophagus* does not possess a spermatophore, but the very closely related genus *Acarus* does (Griffiths and Boczek 1977). In *Tyrophagus* the shape of the aedeagus varies considerably and is used in species identification; in 10 species of *Acarus* there is no difference in the shape of the aedeagus (Griffiths and Boczek 1977). These two genera, therefore, may provide a system to study the origin and maintenance of spermatophores. For both these genera more than one spermatophore is required to fill the spermatheca and multiple matings are necessary for maximum egg production (Griffiths and Boczek 1977). They also note that in *Acarus siro* the spermatophore breaks down completely upon entry into the spermatheca, but in *Lardoglyphus konoi* the spermatophore retains its shape with the sperm escaping when the anterior margin breaks down. Time of spermatophore breakdown could be important to the amount of sperm mixing within the spermatheca.

Prostigmata. Of all the Prostigmata thought to have an accessory genital system, only the Tetranychidae, or spider mites, are known in any detail with respect to reproduction. The Tetranychidae are small plant parasites that feed by piercing the host tissue with their cheliceral stylets and then ingesting the tissue fluid. After the introduction and widespread use of pesticides following WWII, they became economically important agricultural pests (Jeppson *et al.* 1975). It is for this reason that they have been intensively studied. Fertilized females disperse in the aerial plankton by spinning a strand of silk and "ballooning" like many spiders (Mitchell 1970). They are arrhenotokous (Oliver 1971) and unfertilized females can produce haploid male offspring. Although they have separate copulatory and egg laying apertures, the internal connection, if any, between the spermatheca (or seminal receptacle, as it is usually called) and the ovaries is not clear (Crooker and Cone 1979; Fig. 6B). Pijnacker and Drenth-Diephuis (1973) present evidence that indicates sperm are taken up by the cells of the seminal receptacle wall and then transported to the hemolymph by which means they reach the ovaries.

The remainder of this discussion focuses on *Tetranychus urticae* because it has been the most studied species and even has available a few genetic markers (Helle 1967). Males develop more quickly than the slightly larger females, and guard quiescent female deutonymphs just prior to their molt to the sexually receptive adult form (Potter *et al.* 1976). Pheromones are used to locate females of the proper age (Cone 1979). As the mite population on a leaf increases, the functional sex ratio becomes more and more skewed toward males as females disperse and the males remain (Potter 1978). Males fight, occasionally to the death, to be the guardians of soon-to-emerge receptive females. As the functional sex ratio becomes more skewed toward males, competition for mating opportunities increases (Potter *et al.* 1976). Occasionally under natural conditions a female will mate again within a minute of the first mating (Helle 1967). Experiments using albino (a simple recessive) and normal animals have shown that the first mating has complete precedence if the mating was uninterrupted (Helle 1967, Potter and Wrensch 1978). However, if the first mating was interrupted, some eggs were fertilized by the sperm from the second mating, the number increasing the earlier the interruption occurred (Potter and Wrensch 1978). Second matings are totally ineffective in older females. The reason for this is unknown. A postcopulatory plug appears unlikely given the observation that incomplete matings of sufficient duration usually prevent successful second matings (Potter and Wrensch 1978). Sperm displacement or mixing must not occur to any obvious extent within the seminal receptacle. Frequency of effective double insemination increases from about 6% in uncrowded conditions (Helle 1967) to over 14% in crowded populations (Potter and Wrensch 1978). Potter and Wrensch (1978) suggest that the mating position (male underneath) and clavate shape of the aedeagus are partly responsible for the relatively low incidence of interrupted copulations, even under crowded conditions.

C. Ricinuleids

Ricinuleids are small (5-10 mm), heavy-bodied arachnids. Two extant genera are recognized: *Ricinoides* in Africa and *Cryptocellus* in the Americas. Ricinuleids are found in forest litter, under logs, and often abundantly in caves. They do not have eyes or trichobothria (sensory hairs). Pittard and Mitchell (1972) have described the morphology of the life stages of *C. pelaezi.* There is a six-legged "larva" similar to the larvae of mites and ticks, three nymphal stages, and an adult stage. Mating behavior is known for *R. afzelli* (Pollock 1967), *R. hanseni* (Legg 1977), and *C. lampeli* (Cooke 1967). The structure of the male's third pair of legs is a particularly noteworthy feature of the Ricinulei. Each tarsus is modified into a copulatory apparatus by the presence of cavities and fixed and movable processes (Fig. 8). Prior to copulation, male *R. hanseni* charge each copulatory apparatus with sperm from their "penis" (Legg 1977). The male then climbs on the back of a receptive female and carefully places his modified tarsus into the genital atrium of the female. The genital region is located on a pedicel which connects the cephalothorax and the abdomen. Normally, the cephalothorax and the abdomen are linked and the pedicel is not exposed. *Ricinoides hanseni* females uncouple the body regions when a courting male approaches (Legg 1977). A series of lobes near the tip of the male's copulatory apparatus fit precisely into a number (five to six) of vesicular evaginations (spermathecae) of the female's genital atrium. A single, tubular lobe of the copulatory apparatus ejaculates sperm directly into the largest spermatheca. Legg (1977) questions whether the smaller evaginations actually function in sperm transfer, and suggests they act primarily as anchor points for the male copulatory apparatus. While the smaller evaginations do permit anchoring, it is unlikely they evolved in females for this reason. Therefore, we predict they serve a significant sperm storage function. Because of direct insertion of the tarsal lobes into the female spermathecae, the potential for sperm displacement is high. Unfortunately, observations on the number of matings per female are not published.

D. Solifuges

Solifuges comprise an order of approximately 800 tropical and subtropical species which are abundant in arid habitats. They are large (to 70 mm), extremely voracious, and highly mobile carnivores. Solifuges possess huge, powerful chelicerae which in some species are as long as the rest of the cephalothorax. Mating behavior is known for the Asian and North African family Galeodidae (Amitai *et al.* 1962, Junqua 1962), and for the North American family Eremobatidae (Muma 1966c). There are some important differences in sperm transfer in the two families. Galeodid males deposit a globule of sperm on the ground and then transport it to the female genital opening with their chelicerae. In *Eremobates* spp. (Eremobatidae), sperm is emitted from the male genital opening onto the female genital opening. The males do not possess a penis; however, following sperm emission, males thrust the fixed fingers of their chelicerae into the female genital opening. This presumably forces sperm into the sperm storage sites of the female.

Muma (1966c) has subdivided the mating sequence of *Eremobates* into three phases: the attack phase (prior to sperm transfer); the contact phase (including sperm transfer); and the release phase (following sperm transfer). During the attack

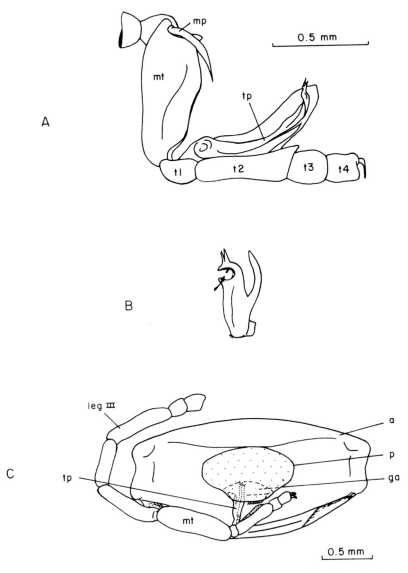

Fig. 8. Sperm transfer in Ricinulei. A. Lateral view of the third leg of a male showing the metatarsus and tarsi. The distal region of the tarsal process (tp) is inserted into the genital atrium of the female, while the metatarsal process (mp) is used to clamp onto the pedicel of the female. B. Enlarged view of the distal portion of the tarsal process. The small lobes at the top left fit into the smaller spermathecae of the female, while sperm is ejaculated into the largest spermatheca from the region indicated by the arrow. C. Posterior, end-on view of the female abdomen indicating the position of the male third leg during sperm transfer (a = abdomen; ga = genital atrium; mp = metatarsal process; mt = metatarsus; p = pedicel; tp = tarsal process; t1, t2, t3, t4 = tarsi). Modified from Legg (1977).

phase the female may retreat, become quiescent, or fight with the male. If the female is passive, or is overpowered by the male, the contact phase ensues. During contact and prior to transfer, the male turns the female on her side or back and thrusts the needlelike fixed fingers of his chelicerae into the genital opening of the female. This behavior may serve several functions: It might (1) prepare the genital tract for sperm transfer; (2) indicate to the male if the spermathecae contain sperm from previous copulations; or (3) result in the displacement and/or removal of sperm transferred by other males. There are statements supporting the second hypothesis. Muma (1966c) states that males will not complete copulations (after cheliceral insertion takes place) with recently mated females. However, they will copulate with females whose sperm supply has been exhausted. Muma does not give the sample size on these observations, and it may be premature to reject the sperm removal hypothesis outright. Precopulatory cheliceral insertion also occurs in Galeodidae. The existence of a sex-limited cheliceral flagellum in males may be related to the function of precopulatory cheliceral insertion.

One conspicuous result of Muma's study of mating is the low success rate. Only 6 of 34 mating attempts resulted in sperm transfer. In three cases, males killed females during the contact phase. These results may be an artifact of laboratory confinement, but the animals were well maintained prior to mating and other aspects of reproduction; *e.g.,* brood production was successful in the laboratory (Muma 1966a, b). It seems that pairing incurs considerable costs in these arachnids.

V. SUMMARY

This chapter provides a comprehensive review of arachnid reproductive behavior and suggests how ecological factors and morphology impinge on sperm transfer and utilization patterns. Sperm transfer in arachnids falls into three categories: indirect with pair formation (Group I); indirect without pair formation (Group II); and direct (Group III). Indirect transfer is achieved by a spermatophore deposited on the substratum. Direct transfer occurs when sperm (liquid sperm or a spermatophore) is placed directly on or into the female gonopore by the male. A model for the evolution of nonpairing (dissociation) is presented. Factors favoring dissociation include: low desiccation, high mortality resulting from interactions with mates, and mechanical difficulties of indirect sperm transfer. Dissociation should favor massive sperm production distributed among many spermatophores. Also, sperm competition and sperm competition avoidance mechanisms should be less evident in nonpairing species. Mite and pseudoscorpion taxa which exhibit dissociation are discussed. The remainder of the chapter treats taxa which form pairs. The specific mating behavior of Group I species varies considerably among the arachnid orders.

In some groups, *e.g.* schizomids, females dominate sperm uptake and the potential for sperm displacement appears minimal, while in Uropygi the potential for sperm competition and sperm displacement seem high. Although the mating behavior of many Group I species is known, observations pertinent to sperm competition, *e.g.* patterns of female receptivity, frequency of multiple insemination, etc., are often lacking. In our treatment of Group III species, special emphasis is placed on female reproductive morphology and its consequences for sperm competition. The accessory genital systems (separate passageways for sperm reception and egg fertilization) in the Acari should have important effects on sperm precedence, in some cases reversing the relations between spermathecal shape and sperm precedence predicted by Walker (1980). Finally, many Group III species exhibit behaviors indicative of serm competition, *e.g.* postcopulatory guarding and repeated copulations in harvestmen and mites. Also, precopulatory insertion of the male chelicerae into the female reproductive tract occurs in Solifugae, mites and ticks. The role of sperm competition in the evolution of these behaviors remains to be elucidated.

ACKNOWLEDGMENTS

We would like to thank V. Delesalle, A. Kodric-Brown, D. McDonald, C. Petersen, M. Zeh, and R. Zimmerman for their helpful comments on portions of the manuscript. V. Behan and S. Austad provided useful criticisms of an early draft of the mite material. We are grateful to Lynda Delph, Natacha Hennocq, and Jill Smith for their careful work on the figures. Finally, Bob Smith provided encouragement and advice throughout.

REFERENCES

Aeschliman, A., and O. Grandjean. 1973. Observations on fecundity in *Ornithodorus moubata* Murry (Ixodoidea:Argasidae): Relationships between mating and oviposition. *Acarologia* 15:206-217.

Alexander, A. J. 1957. Courtship and mating in the scorpion, *Opistophthalmus latimanus*. *Proc. Zool. Soc. Lond.* 128:529-544.

Alexander, A. J. 1958. Courtship and mating in buthid scorpions. *Proc. Zool. Soc. Lond.* 133:145-169.

Alexander, A. J. 1962. Biology and behavior of *Damon variegatus* Perty of South Africa and *Admetus barbadensis* Pocock of Trinidad, W. I. (Arachnida, Pedipalpi). *Zoologica* 47:25-37.

Alexander, A. J., and D. W. Ewer. 1957. On the origin of mating behavior in spiders. *Am. Nat.* 91:311-317.

Alexander, R. D. 1964. The evolution of mating behavior in arthropods. In *Insect Reproduction*, K. C. Highnam (ed.), pp. 78-94. Symp. R. Entomol. Soc. Lond. No. 2.

Amano, H., and D. A. Chant. 1978. Mating behavior and reproductive mechanisms of two species of predaceous mites, *Phytoseiulus persimilis* Athias-Henriot and *Amblyseius andersoni* (Chant) (Acarina:Phytoseiidae). *Acarologia* 20:196-213.

Amitai, P., G. Levy, and A. Shulov. 1962. Observations on mating in a solifugid *Galeodes sulfuripes* Roewer. *Bull. Res. Counc. Isr. Sect. B Zool.* 11:156-159.

André, M. 1953. Observations sur la fécondation chez *Allothrombium fuliginosum. Bull. Mus. Hist. Nat. Paris* 25:383-386.

Athias-Henriot, C. 1970. Obsurvations sur la morphologie externe des gamasides: Muses au point terminologiques. *Acarologia* 12:25-27.

Balashov, Yu. S. 1972. Bloodsucking ticks (Ixodoidea): Vectors of diseases of man and animals. *Misc. Publ. Entomol. Soc. Am.* 8:161-376.

Beier, M. 1932. Pseudoscorpionidea–Afterscorpione. In *Handbuch der Zoologie, Vol. 3*, Kükenthal-Krumbuch (eds.), pp. 117-192. Walter de Gruyter, Berlin.

Bishop, S. C. 1949. The function of the spur on the femur of the palpus of the male *Leiobunum calcar* (Wood) (Arachnida:Phalangida). *Entomol. News* 60:10-11.

Brach, V. 1975. Development of the whipscorpion *Schizomus floridanus*, with notes on behavior and laboratory culture. *Bull. South. Calif. Acad. Sci.* 74:97-100.

Brinck, P. 1958. Parningens uppkomst och betydelse hos insekter och närstående djurgrupper. *Entomol. Tidskr.* 78:246-264.

Brown, S. J. 1979. Further observations on the behavior of adult males of *Dermacentor variabilis* (Acari, Ixodidae) with respect to attached females. *J. Med. Entomol.* 16:262.

Chamberlin, J. C. 1931. The arachnid order Chelonethida. *Stanford Univ. Publ. Biol. Sci.* 7:1-284.

Cloudsley-Thompson, J. L. 1948. Notes on the Arachnida. 4. Courtship behaviour of the harvester *Mitopus morio. Ann. Mag. Nat. Hist.* 14:809-810.

Compton, G. L., and G. W. Krantz, 1978. Mating behavior and related morphological specialization in the Uropodine mite, *Caminella peraphora. Science* 200:1300-1301.

Cone, W. W. 1979. Pheromones of Tetranychidae. In *Recent Advances in Acarology, Vol. II*, J. G. Rodriguez (ed.), pp. 309-317. Academic Press, New York.

Cooke, J. A. L. 1967. Observations on the biology of Ricinulei (Arachnida) with description of two new species of *Cryptocellus. J. Zool. (Lond.)* 151:31-42.

Costa, M. 1966. Notes on macrochelids associated with manure and coprid beetles in Israel. I. *Macrocheles robustulus* (Berlese 1904). Development and biology. *Acarologia* 8:532-548.

Costa, M. 1967. Notes on macrochelids associated with manure and coprid beetles of Israel. II. Three new species of the *Macrocheles pisentii* complex, with notes on their biology. *Acarologia* 9:304-329.

Crooker, A. R., Jr., and W. W. Cone. 1979. Structure of the reproductive system of the adult female twospotted spider mite *Tetranychus urticae*. In *Recent Advances in Acarology, Vol. II*, J. G. Rodriguez (ed.), pp. 405-409. Academic Press, New York.

Davids, C., and R. Belier. 1980. Spermatophores and sperm transfer in the water mite *Hydrachna conjecta* Koen: Reflections on the descent of water mites from terrestrial forms. *Acarologia* 21:84-90.

Davy, K. G. 1960. The evolution of spermatophores in insects. *Proc. R. Entomol. Soc. Lond. Ser. A Gen. Entomol.* 35:107-113.

Dosse, G. 1958. Die Spermathecae, ein zusatzliches Bestimmungsmerkmal bei Raubmilben (Acari, Phytoseiidae). *Pflanzenschutzberichte* 22:125-133.

Dosse, G. 1959. Über den Kopulationsvorgang bei Raumilben aus der Gattung *Typhlodromus* (Acari, Phytoseiidae). *Pflanzenschutzberichte* 22:125-133.

Edgar, A. L. 1971. Studies on the biology and ecology of Michigan Phalangida (Opiliones). *Misc. Publ. Mus. Zool. Univ. Mich.* 144:1-64.

Ehrnsberger, R. 1977. Fortpflanzungsverhalten der Rhagidiidae (Acarina, Trombidiformes). *Acarologia* 19:67-73.

Fain, A. 1963. La spermathèque et ses canaux adducteurs chez les Acariens mesostigmatiques parasites des voies respiratoires. *Acarologia* 5:463-479.

Feldman-Muhsam, B. 1973. Copulation and spermatophore formation in soft and hard ticks. In *Proceedings of the Third International Congress of Acarology*, D. Milan and B. Rosicky (eds.), pp. 719-722. Junk, The Hague.

Forster, R. R. 1954. The New Zealand harvestmen (suborder Laniatores). *Canterbury Mus. Bull.* 2:1-329.

Francke, O. F. 1979. Spermatophores of some North American scorpions (Arachnida, Scorpiones). *J. Arachnol.* 7:19-32.

Francke, O. F., and S. K. Jones. 1982. The life history of *Centruroides gracilis* (Scorpiones, Buthidae). *J. Arachnol.* **10**:223-240.

Gabbut, P. O. 1967. Quantitative sampling of the pseudoscorpion *Chthonius ischnocheles* from beech litter. *J. Zool. (Lond.)* **151**:469-478.

Garnier, G., and R. Stockmann. 1972. Etude comparative de la pariade chez différentes espèces de scorpions et chez *Pandinus imperator. Ann. Univ. Abidjan, Ser. E (Ecologie)* **5**:475-497.

Ghilarov, M. W. 1958. Evolution of insemination character in terrestrial arthropods. *Zool. Zh.* **37**:37:707-735. (Russian with English summary.)

Gladny, W. J., and R. O. Drummond. 1970. Mating behavior and reproduction of the lone star tick, *Amblyomma americanum. Ann. Entomol. Soc. Am.* **63**:1036-1039.

Goddard, S. J. 1976. Population dynamics, distribution patterns and life cycles of *Neobisium muscorum* and *Chthonius orthodactylus. J. Zool. (Lond.)* **178**:295-304.

Griffiths, D. A., and J. Boczek. 1977. Spermatophores of some acaroid mites (Astigmata: Acari). *Entomol. Exp. Appl.* **10**:103-110.

Helle, W. 1967. Fertilization in the two-spotted spider mite (*Tetranychus urticae*:Acari). *Entomol. Exp. Appl.* **10**:103-110.

Helle, W., H. R. Bolland, R. van Arendonk, R. de Boer, G. G. M. Schulten, and V. M. Russell. 1978. Genetic evidence for biparental males in haplodiploid predator mites (Acarina: Phytoseiidae). *Genet. Agrar.* **49**:165-171.

Hoff, C. C. 1949. Pseudoscorpions of Illinois. *Bull. Ill. Nat. Hist. Surv.* **24**.

Hoff, C. C. 1959. The ecology and distribution of the pseudoscorpions of north-central New Mexico. *Univ. N. M. Publ. Biol.* **8**.

Hoogstraal, H. 1976. Biology of ticks. In *Tick-borne Diseases and Their Vectors*, J. K. H. Wilde (ed.), pp. 3-14. Edinburgh Univ. Press.

Hughes, A. M. 1976. The mites of stored food and houses (2nd ed.). *Minist. Agric. Fish. Food Tech Bull.* **9**:1-400.

Jeppson, L. R., H. R. Keifer, E. W. Baker. 1975. *Mites Injurious to Economic Plants.* Univ. Calif. Press, Berkeley, CA.

Junqua, C. 1962. Donnes sur la reproduction d'un solifuge: *Othoes saharae* Panouse. *C. R. Séanc. Acad. Sci. Ser. III Sci. Vie* **255**:2673-2673.

Kinn, D. N., and J. J. Witcosky. 1977. The life cycle and behavior of *Macrocheles bondreauxi* Krantz. *Z. Angew. Entomol.* **84**:136-144.

Klingel, H. 1962. Paarungsverhalten bei Pedipalpen (*Telyphonus caudatus* L., Holopeltidia, Uropygi und *Sarax sarawakensis* Simon). *Verh. Deut. Zool. Ges. Wien* 452-549.

Krantz, G. W. 1961. The biology and ecology of granary mites of the Pacific Northwest. I. Ecological considerations. *Ann. Entomol. Soc. Am.* **54**:169-174.

Krantz, G. W. 1978. *A Manual of Acarology, 2nd ed.* Oregon State Univ., Corvallis, OR.

Krantz, G. W., and J. G. Wernz. 1979. Sperm transfer in *Glyptholaspis americana.* In *Recent Advances in Acarology, Vol. II*, J. G. Rodriguez (ed.), pp. 441-446. Academic Press, New York.

Leahy, M. G. 1979. Pheromones of argasid ticks. In *Recent Advances in Acarology, Vol. II*, J. G. Rodriguez (ed.), pp. 297-308. Academic Press, New York.

Lee, D. C. 1974. Rhodacaridae (Acari:Mesostigmata) from near Adelaide, Australia. III. Behaviour and development. *Acarologia* **16**:21-44.

Legg, G. 1973. Spermatophore formation in the pseudoscorpion *Chthonius ischnocheles* (Chthoniidae). *J. Zool. (Lond.)* **170**:367-394.

Legg, G. 1977. Sperm transfer and mating in *Ricinoides hanseni. J. Zool. (Lond.)* **182**:51-61.

Lipovsky, L. J., G. W. Byers, and E. H. Kardos. 1957. Spermatophores—the mode of insemination of chiggers (Acarina:Trombiculidae). *J. Parasitol.* **43**:256-262.

Londt, J. G. 1976. Fertilization capacity of *Boophilus decoloratus* (Koch 1944) (Acarina: Ixodidae). *Onderstepoort J. Vet. Res.* **43**:143-145.

Londt, J. G., and A. M. Spickett. 1976. Gonad development and gametogenesis in *Boophilus decoloratus* (Koch 1844) (Acarina:Metastriata:Ixodidae). *Onderstepoort J. Vet. Res.* **43**: 79-96.

Maynard Smith, J. 1982. *Evolution and the Theory of Games.* Cambridge University Press, New York.

McMurtry, J. A., C. B. Huffaker, and M. Van de Vrie. 1970. Ecology of tetranychid mites and their natural enemies; a review. *Hilgardia* **40**:331-458.

McMurtry, J. A., and G. T. Scriven. 1964. Biology of the predaceous mite *Typhlodromus rickeri* (Acarina:Phytoseiidae). *Ann. Entomol. Soc. Am.* **57**:362-367.

Micherdzinski, W. 1969. Die Familie Parasitidae Oudemans 1901 (Acarina, Mesostigmata). *Zak. Zool. Syst. Pol. Akad. Nauk,* Krakow.

Mitchell, R. 1958. Sperm transfer in the water-mite *Hydryphantes ruber* Geer. *Am. Midl. Nat.* **60**:156-158.

Mitchell, R. 1970. An analysis of dispersal in mites. *Am. Nat.* **104**:425-431.

Modder, W. W. D. 1960. The male genital system of *Schizomus crassicaudatus*. *Ceylon J. Sci. Biol. Sci.* **3**:173-189.

Muchmore, W. B. 1974. Clarification of the genera *Hesperochernes* and *Dinocheirus* (Pseudo-scorpionida, Chernetidae). *J. Arachnol.* **2**:25-36.

Muchmore, W. B. 1975. Use of spermathecae in the taxonomy of chernetid pseudoscorpions. *Proc. 6th Int. Arachnol. Congr.,* pp. 17-20.

Muma, M. H. 1966a. Egg deposition and incubation in *Eremobates*. *Fla. Entomol.* **49**:23-31.

Muma, M. H. 1966b. Life cycles of *Eremobates*. *Fla. Entomol.* **49**:233-242.

Muma, M. H. 1966c. Mating behavior in the solpugid genus *Eremobates* Banks. *Anim. Behav.* **14**:346-350.

Nelson-Rees, W. A., M. A Hoy, and R. T. Roush. 1980. Heterochromatinization, chromatin elimination, and haploidization in the parahaploid mite *Metaseiulus occidentalis* (Acarina: Phytoseiidae). *Chromosoma* **77**:263-276.

Oldfield, G. N. 1973. Sperm storage in female Eriophyoidea (Acarina). *Ann. Entomol. Soc. Am.* **66**:160-163.

Oldfield, G. N., and I. M. Newell. 1973. The role of the spermatophore in the reproductive biology of protogynes of *Aculus cornutus* (Acarina:Eriophyidae). *Ann. Entomol. Soc. Am.* **66**:160-163.

Oldfield, G. N., I. M. Newell, and D. K. Reed. 1972. Insemination of protogynes of *Aculus cornutus* from spermatophores and description of the sperm cell. *Ann. Entomol. Soc. Am.* **65**:1080-1084.

Oliver, J. H., Jr. 1971. Parthenogenesis in mites and ticks (Arachnida:Acari). *Am. Zool.* **11**:283-299.

Oliver, J. H., Jr. 1974. Symposium on reproduction of arthropods of medical and veterinary importance. IV. Reproduction in ticks (Ixodoidea). *J. Med. Entomol.* **11**:26-34.

Oliver, J. H., Jr. 1977. Cytogenetics of mites and ticks. *Ann. Rev. Entomol.* **22**:407-429.

Oliver, J. H., Jr., Z. Al-Ahmadi, and R. L. Osburn. 1974. Reproduction in ticks (Acari: Ixodoidea). 3. Copulation in *Dermacentor occidentalis* Marx and *Haemaphysalis leporispalustris* (Packard) (Ixodidae). *J. Parasitol.* **60**:499-506.

Pabst, W. 1953. Zur Biologie der mitteleuropäischen Troguliden. *Zool. J. Linn. Soc.* **82**:1-46.

Pijnacker, L. P., and L. J. Drenth-Diephuis. 1973. Cytological investigations on the male reproductive system and the sperm track in the spider mite *Tetranychus urticae* Koch (Tetranychidae, Acarina). *Neth. J. Zool.* **23**:446-464.

Pittard, K., and R. W. Mitchell. 1972. Comparative morphology of the life stages of *Cryptocellus pelaezi* (Arachnida, Ricinulei). *Graduate Studies No. 1,* Texas Tech. Univ., Lubbock, TX.

Polis, G. A. 1980. The effect of cannibalism on the demography and activity of a natural population of desert scorpions. *Behav. Ecol. Sociobiol.* **7**:25-35.

Polis, G. A., and R. D. Farley. 1979. Behavior and ecology of mating in the cannibalistic scorpion *Paruroctonus mesaensis* Stahnke. *J. Arachnol.* **7**:33-46.

Pollock, J. 1967. Notes on the biology of Ricinulei (Arachnida). *J. W. Afr. Sci. Assoc.* **12**:19-22.

Potter, D. A. 1978. Functional sex ratio in the carmine spider mite. *Ann. Entomol. Soc. Am.* **71**:218-222.

Potter, D. A., and D. L. Wrensch. 1978. Interrupted matings and the effectiveness of second inseminations in the twospotted spider mite. *Ann. Entomol. Soc. Am.* **71**:882-885.

Potter, D. A., D. L. Wrensch, and D. E. Johnston. 1976. Guarding, aggressive behavior, and mating success in male twospotted spider mites. *Ann. Entomol. Soc. Am.* **69**:707-711.

Pound, J. M., and J. H. Oliver, Jr. 1976. Reproductive morphological spermatogenesis in *Dermanyssus gallinae* (DeGeer) (Acari:Dermanyssidae). *J. Morphol.* 150:825-842.

Putmann, W. L. 1966. Insemination in *Balaustium* sp. (Erythraeidae). *Acarologia* 8:424-426.

Radinovsky, S. 1965. The biology and ecology of granary mites of the Pacific Northwest. IV. Various aspects of the reproductive behavior of *Leiodinychus krameri* (Acarina: Uropodidae). *Ann. Entomol. Soc. Am.* 58:267-272.

Rechav, Y., and G. B. Whitehead. 1979. Male produced pheromones of Ixodidae. In *Recent Advances in Acarology, Vol. II*, J. G. Rodriguez (ed.), pp. 291-296. Academic Press, New York.

Ressl, R., and M. Beier. 1958. Zur Ökologie Biologie und Phänologie der heimischen Pseudoskorpione. *Zool. Jahrb. Abt. Syst. Oekol. Georg. Tiere* 86:1-26.

Rodriguez, T. C. A., and B. S. Guerrero. 1976. La historia natural y el comportamiento de *Zygopachylus albomarginis* (Chamberlain) (Arachnida, Opiliones:Goryleptidae). *Biotropica* 8:242-247.

Rubenstein, D. I. 1980. On the evolution of alternative mating strategies. In *Limits to Action*, J. E. R. Staddon (ed.), pp. 65-100. Academic Press, New York.

Schaller, F. 1954. Die indirekte Spermatophorenübertragung und ihre Probleme. *Forsch. Fortschr. Dtsch. Wiss.* 28:321-326.

Schaller, F. 1964. Mating behavior of lower terrestrial arthropods from the phylogenetic point of view. *Proc. 12th Int. Congr. Entomol.,* pp. 297-298.

Schaller, F. 1971. Indirect sperm transfer by soil arthropods. *Ann. Rev. Entomol.* 16:407-446.

Schaller, F. 1979. Significance of sperm transfer and formation of spermatophores in arthropod phylogeny. In *Arthropod Phylogeny*, A. P. Gupta (ed.), pp. 587-608. Van Nostrand Reinhold, New York.

Schulten, G. G. M., R. C. M. van Arendonk, V. M. Russell, and F. A. Roorda. 1978. Copulation, egg production and sex-ratio in *Phytoseiulus persimilis* and *Amblyseius bibens* (Acari: Phytoseiddae). *Entomol. Exp. Appl.* 24:145-153.

Schuster, I. J., and R. Schuster. 1970. Indirekte Spermaübertragung bei Tydeidae (Acari, Trombidiformes). *Naturwissenschaften* 57:256.

Schuster, R. 1962. Nachweis eines Paarungszeremoniells bei den Hornmilben. *Naturwissenschaften* 49:502.

Schuster, R., and I. J. Schuster. 1966. Über das Fortpflanzungsverhalten von Aystiden-Männchen (Acari, Trombidiformes). *Naturwissenschaften* 53:162.

Schuster, R., and I. J. Schuster. 1977. Ernährungs-und fortpflanzungsbiologische Studien an der Milbenfamilie Nanorchestidae (Acari, Trombidiformes). *Zool. Anz. Jena* 199:89-94.

Smith, G. T. 1966. Life history of *Urodacus abruptus* (Scorpionidae). *Aust. J. Zool.* 14:383-398.

Sonenshine, D. E., R. M. Silverstein, and P. J. Homsher. 1979. Female produced pheromones of Ixodidae. In *Recent Advances in Acarology, Vol. II,* J. G. Rodriguez (ed.), pp. 281-290. Academic Press, New York.

Stürm, H. 1958. Indirekte Spermatophoren-ubertragung bei dem Geisselskorpion *Trithyreus sturmi* Kraus (Schizomidae, Pedipalpi). *Naturwissenschaften* 45:142-143.

Stürm, H. 1973. Zur Ethologie von *Trithyreus sturmi* Kraus (Arachnida, Schizopeltidia). *Z. Tierpsychol.* 33:113-140.

Walker, W. F. 1980. Sperm utilization strategies in nonsocial insects. *Am. Nat.* 115:780-799.

Warner, R. R., and R. K. Harlan. 1982. Sperm competition and sperm storage as determinants of sexual dimorphism in the dwarf surfperch, *Micrometus minimus. Evolution* 36:44-55.

Weygoldt, P. 1966a. Spermatophore web formation in a pseudoscorpion. *Science* 153:1647-1649.

Weygoldt, P. 1966b. Vergleichende Untersuchungen zur Fortpflanzungsbiologie der Pseudoscorpione. Beobachtungen über das Verhalten, die Samenübertragungsweisen und die Spermatophoren einiger einheimischer Arten. *Z. Morphol. Oekol. Tiere* 56:39-92.

Weygoldt, P. 1969a. Beobachtungen zur Fortpflanzungsbiologie und zum Verhalten der Geisselspinne *Tarantula marginemaculata* C. L. Koch (Chelicerata, Amblypygi). *Z. Morphol. Tiere* 64:338-360.

Weygoldt, P. 1969b. *The Biology of Pseudoscorpions.* Harvard Univ. Press, Cambridge, MA.

Weygoldt, P. 1970a. Courtship behavior and sperm transfer in the giant whip scorpion, *Mastigoproctus giganteus* (Lucas) (Uropygi, Thelyphonidae). *Behaviour* 36:1-8.

Weygoldt, P. 1970b. Lebenzyklus und postembryonale Entwicklung der Geisselspinne *Tarantula marginemaculata* C. L. Koch (Chelicerata, Amblypygi) im Laboratorium. *Z. Morphol. Tiere* 67:58-85.

Weygoldt, P. 1970c. Vergleichende Untersuchungen zur Fortpflanzungsbiologie der Pseudoscorpione II. *Zool. Syst. Evolutionsforsch.* 8:241-259.

Weygoldt, P. 1971. Notes of the life history and reproductive biology of the giant whip scorpion, *Mastigoproctus giganteus* (Uropygi, Thelyphonidae) from Florida. *J. Zool. (Lond.).* 164:137-147.

Weygoldt, P. 1972. Spermatophorenbau und Samenübertragung bei Uropygen (*Mastigoproctus brasilianus* C. L. Koch) und Amblypygen (*Charinus brasilianus* Weygoldt und *Admetus pumilio* C. L. Koch) (Chelicerata, Arachnida). *Z. Morphol. Tiere* 71:23-51.

Weygoldt, P. 1974. Kampf und Paarung bei der Geibelspinne *Charinus montanus* Weygoldt (Arachnida, Amblypygi, Charontidae). *Z. Tierpsychol.* 34:217-223.

Weygoldt, P. 1975. Die indirekte Spermatophorenübertragung bei Arachniden. *Verh. Dtsch. Zool. Ges.* 1974:308-313.

Weygoldt, P. 1977. Kampf, Paarungsverhalten, Spermatophorenmorphologie und weibliche Genitalien bei neotropischen Geibelspinnen (Amblypygi, Arachnida). *Zoomorphologie* 86: 271-286.

Weygoldt, P. 1978. Paarungsverhalten und Spermatophorenmorphologie bei Geibelskorpionen: *Thelyphonellus amazonicus* Butler und *Typopeltis crucifer* Pocock (Arachnida, Uropygi). *Zoomorphologie* 89:145-156.

Weygoldt, P., A. Weisemann, and K. Weisemann. 1972. Morphologisch-histologische Undersuchungen an den Geschlechtsorganen der Amblypygi unter besonderer Beruecksichtigung von *Tarantula marginemaculata* C. L. Koch (Arachnida). *Z. Morphol. Tiere* 73:209-247.

Wiklund, C., and T. Fagerström. 1977. Why do males emerge before females? *Oecologia* 31: 153-158.

Woodring, J. P., and C. A. Galbraith. 1976. The anatomy of the adult uropodid *Fuscuropoda agitans* (Arachnida:Acari) with comparative observations on other Acari. *J. Morphol. Tiere* 150:19-58.

Young, J. H. 1968. The morphology of *Haemogamasus ambulans.* II. Reproductive system. *J. Kansas Entomol. Soc.* 41:532-542.

7

Evolution of Sperm Priority Patterns in Spiders

STEVEN N. AUSTAD

I. INTRODUCTION

Spiders, an arachnid order containing approximately 30,000 species (Levi and Levi 1968), occupy essentially every terrestrial habitat. They are well suited for detailed studies of reproductive adaptation and behavior because they can be easily located, marked, and monitored in the field, as well as maintained and manipulated in the laboratory. Their potential as experimental organisms for

Sperm Competition and the Evolution
of Animal Mating Systems

223

examining major issues in the evolution of behavior has only recently begun to be exploited (*e.g.,* Rovner 1968; Aspey 1977a, b; Reichert 1978a, b, 1979; Christenson and Goist 1979; Jackson 1980a).

Spiders are particularly well preadapted to maintain a high level of sperm competition, because females can store viable sperm for long periods, males are generally capable of mating with more than one female (although see below), and males are generally unable to monopolize access to a female for the duration of her reproductive life. Males are incapable of monopolization chiefly because the reproductive life of a female is relatively much longer than the male's. Despite these preadaptations. little is known of the pattern of sperm priority actually exhibited by spiders.

Given their large potential for sperm competition and the current scarcity of knowledge on sperm priority in the group, this paper has the following aims: (1) to summarize current knowledge on spider sperm priority patterns; (2) to identify the important parameters that might affect the evolution of these patterns; (3) to predict what patterns may be exhibited in some unstudied species; and (4) to suggest research plans that will critically examine these predictions.

II. CURRENT KNOWLEDGE ON SPIDER SPERM PRIORITY

In all spiders, the transfer of sperm from male to female is an indirect process. After maturity, males construct a special web upon which they deposit liquid sperm from their gonopores. They then transfer the sperm into bulbs within their pedipalps where it is stored until mating. The process of bulb filling is called **sperm induction**. The pedipalps are used as intromittent organs during copulation from several positions (Fig. 1).

In most spider species, sperm induction occurs shortly after the final molt, and then again after mating if the palpal sperm supply is exhausted and males mate more than once. In some of the Theridiidae and Linyphiidae, however, sperm induction occurs during the mating sequence itself (van Helsdingen 1965). This behavior facilitates sperm competition studies by allowing estimates to be made of the time required for sperm transfer. The significance of such estimates will be discussed below. Female spiders store sperm in their spermathecae until the time of egg deposition, when eggs are fertilized as they pass down the uterus.

Sperm priority has been studied in three species of spiders. Austad (1982) studied the linyphiid *Frontinella pyramitela,* Jackson (1980b) examined the salticid *Phidippus johnsoni,* and Vollrath (1980) studied *Nephila clavipes.* I will review each of these studies in some detail.

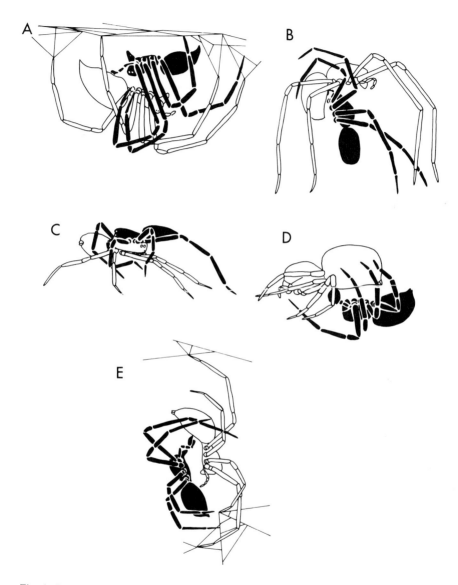

Fig. 1. Spider mating positions. Shaded spider is the male. A. Typical of the Linyphiidae. B. Typical of all mygalomorphs and many of the six-eyed spiders. C. Typical of many hunting spiders. D. From the family Thomisidae. E. From the family Clubionidae. Redrawn from Kaston (1948).

A. The Bowl and Doily Spider

Mating of *Frontinella pyramitela* takes place on this species' unusual web (Fig. 2). Females are usually receptive, except just prior to egg deposition. When females are receptive, the duration and pattern of copulation are controlled by the male. A female will occasionally bolt from a copulating male, but the male usually pursues her and continues copulation as if no interruption had occurred.

Mating typically consists of a preinsemination phase (= Phase A of van Helsdingen 1965) followed by several insemination phases. The preinsemination phase is indistinguishable from the rest of the mating sequence. The standard mating position is assumed, intromission occurs, but no sperm is transferred (Austad 1982). At the end of preinsemination, the male moves away from the female, builds a sperm web, inducts sperm, then renews copulation. The length of the preinsemination phase and occurrence of insemination phases are a function of the copulatory history of the female (Table I). Female fertility (though not fecundity) is related

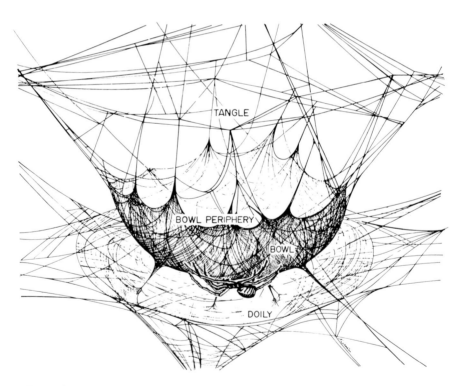

Fig. 2. Structure of the bowl and doily web. Mating occurs on the underside of the bowl. Sperm induction takes place between the bowl and the doily. Only one male at a time will occupy the bowl, although others may occupy the bowl periphery and the tangle. Males fight over access to the bowl. From Austad (1982).

TABLE I

Male Mating Behavior as a Function of Female Reproductive History in *F. pyramitela*

Female Reproductive History	N	Mean Time (min) Spent in:	
		Preinsemination	Insemination
Virgin	61	17.2 (6.0)[a]	30.7 (10.5)
Recently mated < 24 h	40	185.3 (100.5)	31.4 (7.1)[b]
Previously mated ≥ 24 h	53	25.6 (18.6)	0 (0)

[a]S.D. shown in parentheses.
[b]Only 19 out of 40 males performed insemination. These data include only those males who actually transferred sperm.

to the length of the sum of the insemination phases, as ascertained by a series of laboratory matings that were interrupted at predetermined points (Austad 1982).

Sperm priority was determined by double-mating females with x-irradiated sterile males and normal males (Table II). These experiments were repeated with the order of males reversed to establish that irradiation does not affect the competitive ability of sperm (Table II).

The first male to mate fertilizes practially all of a female's eggs if he is allowed to complete his mating. If a second mating occurs within 24 h of the first, the second male will often transfer sperm, but even so will fertilize less than 5% of the eggs on average (Tables I and II). If the second mating occurs more than 24 h after the first, copulation is terminated by the male during preinsemination, so no sperm are transferred.

Additional experiments in which the first copulation was interrupted at several points before sperm transfer was completed and a second male introduced, demonstrated that a first male will not lose previously obtained fertilizations even when replaced before finishing copulation (Austad 1982). In other words, if a first male has transferred enough sperm so that he will fertilize half of a female's eggs, he will still fertilize those eggs, even if replaced by a second male. Sperm displacement does not occur. Methodological details for all the above experiments may be found in Austad (1982).

B. A Jumping Spider

In the jumping spider, *Phidippus johnsoni*, the female controls duration of mating (Jackson 1980b). Females exhibit a complex and seemingly unpredictable pattern of receptivity. At least some females are receptive in all reproductive states, *i.e.*, virgin, recently mated, gravid, post-oviposition. Mating duration is extremely variable in all reproductive states, but is correlated with the site (inside or outside the nest) where copulation occurs. There is no sperm induction during mating.

TABLE II

Sperm Priority Experiments for *F. pyramitela*

Mating Regime	Number of Females	1st Sac	Proportion Hatch:		
			2nd Sac	3rd Sac	4th Sac
S–F[a]	20	.023 (20)[b]	.000 (7)	.000 (1)	.000 (1)
F–S	18	.935 (18)	.785 (10)	.011 (4)	.000 (3)
S–S (control)	5	.028 (5)	.000 (2)	.000 (1)	–
F–S (control)	10	.981 (10)	.931 (7)	.530 (2)	–
F only	61	.928 (61)	.864 (32)	.655 (8)	.027 (4)

[a]S = sterile male, F = fertile male. Mating regime defines the order in which the matings were performed. In these double matings the second male immediately replaced the first. Generally the intercopulatory interval was < 15 min.
[b]Numbers in parentheses indicate the number of egg sacs deposited by females of that particular mating regime. From Austad (1982).

Jackson also used X-irradiation to sterilize males, and a double mating procedure to determine sperm priority. He did not use reciprocal mating sequences (*i.e.,* sterilized male first in some experiments, second in others), it is not known whether irradiation had altered competitiveness of treated sperm related to untreated sperm. However he did examine embryos from matings involving only irradiated males to determine that they had not been rendered aspermic.

Jackson found that the first male often (54% of the time) left an externally visible mating plug covering the seminal duct openings (Fig. 3). A second copulation sometimes left the original plug intact, sometimes resulted in its disappearance, and sometimes resulted in the deposition of a new plug. He artificially plugged seminal duct openings of some females with glue to determine the effect of absolutely impenetrable plugs on male mating behavior, and found that males behaved as in perfectly normal copulations.

From 22 double mating experiments in which the sterile male always mated first, he found that 12 females deposited infertile eggs, 6 deposited a normal complement of fertile eggs, and 4 deposited some fertile eggs though fewer than normal. He concluded that although a first male may leave a mating plug which serves as an effective barrier to further insemination, sometimes a second male can remove it and completely displace the original male's sperm, and sometimes the plug is removed but only part of the original sperm is displaced.

Jackson's (1980b) study illustrates the difficulty often associated with interpreting experimental results when the normal course of sperm transfer is unknown, for his data are open to another interpretation. In examining the effect of copulation duration on egg hatching success, he found that there was tremendous variation in copulation duration, and that its relationship to hatching success was

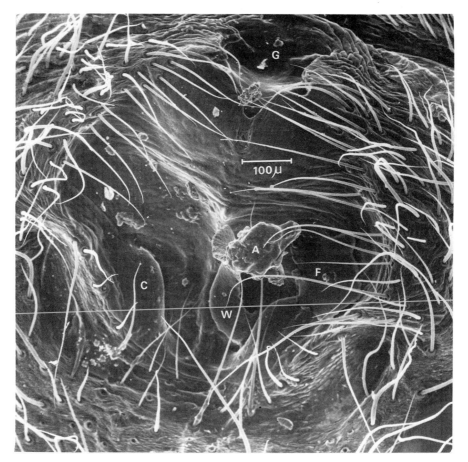

Fig. 3. Epigynum of *Phidippus johnsoni.* C = copulatory orifice; G = gonopore. Plug over one copulatory organ in three forms: A = amorphous mass; F = film; W = wedge situated in orifice. From Jackson (1980b).

complex. Specifically, if one considered only those females that deposited some fertile eggs, then copulation length was unrelated to hatching success. For instance, one female that copulated only one minute produced a full complement of fertile eggs. However, short copulations often resulted in no eggs hatching at all. Only 47% of the females that copulated for less than 30 min deposited fertile eggs, whereas 100% of those that copulated for more than 5 h did.

These results could lead to the following interpretation: The quantity of sperm needed for full female fertility can be transferred in a very short time, but the point during copulation at which sperm transfer occurs is highly variable. A conceivable reason for this variability is that spider genital morphology is often complex and successful intromission may require an intricate and precise manipulation

TABLE III

An Alternative Interpretation of Jackson's (1980b) Sperm Competition Data
for *Phidippus johnsoni*[a]

Sperm Priority Hypothesis	Infertile Eggs	Females Predicted to Have:	
		Fertile Eggs	Reduced Proportion of Fertile Eggs
First male	16.5	5.5	0
Second male	11.6	10.4	0
Random mixing	11.6	5.5	6.2
Actual experimental results	12.0	6.0	4.0

[a]I have assumed that since only 47% of singly-mated females who had short copulations actually deposited fertile eggs, and since matings were short in double-mating experiments, sperm was transferred in only 47% of the matings, regardless of a female's reproductive history. Therefore both males are assumed to have transferred sperm in only $0.47^2 = 0.22$ of the double matings; similarly neither male transferred sperm in $0.53^2 = 0.28$ of the cases; and only the first, or only the second, in $(0.53)(0.47) = 0.25$ of the matings. From these proportions, I have calculated the expected experimental results under three hypotheses of sperm priority. The actual experimental results are given for comparison.

of the pedipalp. However, regardless of the reason, the longer a male mates, the better are his chances of achieving sperm transfer.

In light of the above data I have reanalyzed Jackson's double mating results on the assumption that since all of his experimental first matings were short (average = 19 min), sperm was transferred in only 47% of the matings. Assuming that the same holds for second matings as well, I have computed the expected experimental results under three different hypotheses about sperm priority (Table III). The experimental results most closely match the prediction that sperm randomly mix.

This is not to claim that Jackson's interpretation is wrong. In fact, the existence of a mating plug which at least sometimes survives a second mating indicates that first males can succeed in monopolizing a female's eggs. I only point out the difficulty in interpretation of results without knowledge about the course of sperm transfer. Consequently the existing data, though suggestive, make definitive interpretation impossible. This should not be surprising. Even the most throughly studied arthropod, *Drosophila melanogaster,* has not had its sperm priority characteristics fully elucidated (Gromko and Pyle 1978; Gromko *et al.,* this volume).

C. The Golden Web Spider

Vollrath (1980) also used X-irradiation and double matings to examine sperm priority in the tropical orb-weaver *Nephila clavipes.* Only six females were double mated. Of these, four seem to indicate nearly complete first male sperm priority, one appears to show second male priority, and one was intermediate. Vollrath's conclusions were that although first mates generally fertilize most eggs, second mates can sometimes displace previously stored sperm—especially if the second males are large. He also noted that the relative duration of first and second matings are probably of some significance in determining priority. Interpretation is limited by the small sample size (as Vollrath acknowledges) and a lack of controls. In addition, both males were observed mating in only one of the six experiments.

Christenson et al. (1979), in a study of variation in egg hatching success in this species, found that 10-100% of egg sacs found in the field hatched no spiderlings, depending in part on the season in which the eggs were deposited. These sacs had presumably come from females inseminated by normal males. This amount of natural variation in hatching success indicates that relatively extensive controls would be needed to interpret data from double matings.

Christenson and Goist (pers. comm.) report that the palpal insertion duration is much shorter during second matings. Brief insertions are found throughout the genus and are thought not to involve sperm transfer (Robinson and Robinson 1980). However, it is difficult to imagine males would expend the risk and energy to mate with previously mated females, if they had no chance of obtaining fertilizations. Current evidence suggests that first male sperm priority exists in *N. clavipes*, but further work is clearly needed to demonstrate whether relative copulation length or body size influences this priority.

Although these three studies have been the only ones that attempts to experimentally determine the precise pattern of sperm priority in spiders with multiple mating, there is other information on the phenomenon. Martyniuk and Jaenike (1982), using electrophoretic techniques, recently found that at least four of six females of the linyphiid filmy dome spider, *Neriene radiata*, captured while in copula, produced eggs that had been fertilized by males other than the one copulating at the time of capture. Their evidence does not establish that the first mate fertilizes the majority of the eggs, but it does demonstrate that the last male does not preempt previously stored sperm.

There also exist many reports of hard coverings discovered on female genitalia, which appear to be mating plugs. These coverings have been reported in the Theridiidae (Gertsch 1979), Salticidae (Jackson 1980b), Araneidae (Abalos and Baez 1963, Levi 1975), Oxyopidae (Exline and Whitcomb 1965, Whitcomb and Eason 1965), Thomisidae (Muniappan and Chada 1970), Toxopidae (Forster 1967, cited by Jackson 1980b), and Clubionidae (Brady 1964).

In additional species the distal segment of the embolus (the intromittent portion of the male pedipalp) has been reported broken off and wedged either in the seminal duct opening or the spermatheca (Abalos and Baez 1963, Kaston 1970, Levi 1975, Gertsch 1979, Ross and Smith 1979). Males that lose this segment are often, but not always, incapable of further mating with that pedipalp (Robinson 1980), and have been called "eunuchs" by Robinson and Robinson (1980).

These processes are often much smaller than the ducts in which they are lodged, and it is not clear whether they could act as mating plugs, except as part of a co-agulatory matrix. Abalos and Baez (1963) think that they could prevent sub-sequent insemination in at least some araneids. Since more than two of these pro-cesses are sometimes in the epigynum of a single female (Kaston 1970, Gertsch 1979), they apparently do not necessarily prevent multiple mating, even though they may block effective insemination.

Even though research into this aspect of spider biology is just beginning, the incidence of first male sperm priority seems strikingly high, when compared with the insects, where it is rare (Parker 1970; Boorman and Parker 1976; Walker 1980; Gwynne, this volume).

III. PHYLOGENETIC CONSTRAINTS

Phylogenetic constraints are often overlooked in evolutionary biology. Never-theless, they may often be of paramount significance. It is important to distinguish the effect of phylogenetic constraints from other factors potentially influencing sperm priority patterns, because to the extent that these constraints are significant, the priority pattern may be nonadaptive (which is not to say maladaptive). We must therefore beware of attempting procrustean adaptive explanations of any evolved trait without at least considering alternative explanations (Gould and Lewontin 1979).

Walker (1980) has recently noted that among the insects, species with strong last male sperm priority possess elongated or ovoid spermathecae, and those with less sperm displacement possess spheroid sperm storage organs. Spermathecal mor-phology is presumably related to priority in that ovoid or elongate organs allow previously stored sperm to be pushed farther from the spermathecal duct by the last ejaculate to enter. It is this duct that sperm must again exit in order to fertilize eggs. Walker does not consider this pattern to be a reflection of the phylogenetic morphological constraints to which species are subject, but an adaptation to pro-mote an optimal pattern of sperm utilization from the female's perspective. How-ever, examination of his Table I reveals spermathecal morphology to be at least as strongly correlated with taxon as with priority pattern. Evidence to date, therefore, does not permit discrimination between phylogenetic constraint and adaptation arguments for the evolution of spermathecal morphology or sperm priority pattern.

In spiders there exists a fundamental dichotomy of spermathecal morphology that divides roughly along phylogenetic lines. This may clarify the relation between reproductive morphology and the evolution of priority patterns. I will call these two types of storage organs **cul-de-sac** and **conduit** spermathecae.

Cul-de-sac spermathecae are essentially diverticulae of the vagina. A single duct joins the lumen of the spermatheca to the vagina. Sperm must pass through this duct on its way into the storage organ, and then return by the same route when eggs are fertilized. Cul-de-sac spermathecae are found in all of the mygalomorphs (tarantulas and their allies), as well as in many families of relatively generalized araneomorphs (the so-called higher spiders). Table IV lists the spider families with this type of storage organ.

Conduit spermathecae, on the other hand, consist of an external (seminal) duct that opens near the vaginal opening and through which the male intromittent

TABLE IV

Spider Families With Cul-de-sac Spermathecae[a]

Mygalomorphae	
(atypical)	Liphistiidae
	Antrodiaetidae (folding-door trap-door spiders)
	Mecicobothriidae
	Atypidae (purse-web weavers)
(typical)	Theraphosidae ("ordinary" tarantulas)
	Paratropididae
	Pycnothelidae
	Barychelidae
	Migidae
	Dipluridae (funnel-web tarantulas)
	Ctenizidae (trap-door spiders)
	Actinopodidae
Araneomorphae	
(primitive weavers and hunters)	Sicariidae (6-eyed crab spiders)
	Scytodidae (spitting spiders)
	Loxoscelidae (brown spiders)
	Diguetidae
	Plectreuridae
	Caponiidae
	Oonopidae
	Tetrablemmidae
	Pacullidae
	Ochyroceratidae
	Leptonetidae
	Telemidae
	Textricellidae
	Dysderidae
	Segestriidae
	Hypochilidae
	Filistatidae[b]
(higher weavers)	Pholcidae (daddy-long-legs spiders)
	Tetragnathidae (long-jawed spiders)

[a]Cul-de-sac spermathecae are those in which the storage organ is a blind pouch. Seminal fluid enters and exits through the same duct. See text for explanation of the possible significance of this morphology on sperm priority pattern. From Kaston, B. J., HOW TO KNOW THE SPIDERS, 3rd ed. © 1972, 1978 Wm. C. Brown Publ., Dubuque, IA. All rights reserved. Adapted by permission.
[b]There is controversy about the position of this family.

organ dispenses seminal fluid, and an internal (fertilization) duct from which sperm issues when eggs are fertilized. Spider families in which this type of sperm storage organ occurs are listed in Table V. Although this classification is not perfect (there are some ambiguous and intermediate types) most all spiders can be accommodated within it.

The potential significance of this dichotomy in morphology is great. The cul-de-sac requires that the last sperm to enter would lie closest to the fertilization duct, and hence would be the first to exit; this assumes little mixing of ejaculates and no physical displacement of previously stored sperm out of the spermatheca itself. Conversely, the conduit would dictate that the last sperm be more distal from the fertilization duct, and therefore it would be used in later fertilizations. Spermathecal morphology, then, could be the sole determinant of the sperm priority pattern.

Spermathecal morphology also has important implications for the effectiveness of mating plugs. An effective plug in cul-de-sac females would need to block either the opening of the fertilization duct or the vagina. In either case, when first egg fertilization occurred, the plug would necessarily be dislodged. Therefore the plug's effectiveness would be limited to a single egg sac. On the other hand, plugging the seminal duct of the conduit spermatheca would not interfere with either fertilization or oviposition, so that fertilization of all eggs by the plugging male would be ensured as long as the plug, and his sperm, endured. Since most species of spiders are iteroparous, monopoly of fertilization between, as well as within, egg clutches would be expected to figure prominently in male reproductive success. Thus in a hypothetical ancestral population that contained plugging and nonplugging males, the relative reproductive rewards to each type of male might be expected to differ substantially depending upon whether the species had cul-de-sac or conduit spermathecae. These considerations alone predict that first male sperm priority and the occurrence of mating plugs would be the dominant pattern among spider

TABLE V

Spider Families With Conduit Spermathecae[a]

Araneomorphae	
(primitive weavers)	Gradungulidae
(cribellates)	Oecobiidae
	Eresidae (eresid funnel-weavers)
	Dinopidae (ogre-faced spiders)
	Uloboridae (hackled-band orbweavers)
	Dictynidae
	Amaurobiidae
	Amphinectidae
	Neolanidae
	Psechridae

TABLE V

Spider Families With Conduit Spermathecae[a] (*Continued*)

(cribellates)	Stiphiidae
	Tengellidae
	Acanthoctenidae
(higher web weavers)	Symphytognathidae
	Theridiidae (comb-footed spiders)
	Nicodamidae
	Nesticidae (cave spiders)
	Hadrotarsidae
	Linyphiidae (sheet-web weavers)
	Micryphantidae (dwarf spiders)
	Theridiosomatidae (ray spiders)
	Araneidae ("ordinary" orbweavers)
	Agelenidae (funnel-web weavers)
	Argyronetidae (water spiders)
	Desidae (marine spiders)
	Hahniidae
(3-clawed hunters)	Hersiliidae
	Urocteidae
	Mimetidae (pirate spiders)
	Archaeidae
	Mecysmaucheniidae
	Zodariidae
	Palpimanidae[b]
	Stenochilidae
	Pisauridae (nursery-web spiders)
	Lycosidae (wolf spiders)
	Oxyopidae (lynx spiders)
	Senoculidae
	Toxopidae
(2-clawed hunters)	Ammoxenidae
	Amaurobiodidae
	Aphantochilidae
	Platoridae (platorid crab spiders)
	Lyssomanidae
	Gnaphosidae (ground spiders)
	Salticidae (jumping spiders)
	Prodidomidae
	Zoridae
	Philodromidae (running crab spiders)
	Homalonychidae
	Ctenidae (running spiders)
	Cithaeronidae
	Sparassidae (giant crab spiders)
	Clubionidae (sac spiders)
	Selenopidae
	Anyphaenidae
	Thomisidae (typical crab spiders)

[a]Conduit spermathecae are those in which there are separate ducts for seminal fluid to enter (seminal duct) and leave (fertilization duct). See text for explanation of the possible significance of this morphology on sperm priority pattern. From Kaston, B. J., HOW TO KNOW THE SPIDERS, 3rd ed. ©1972, 1978 Wm. C. Brown Publ., Dubuque, IA. All rights reserved. Adapted by permission.
[b]There is controversy about the position of this family.

species with conduit spermathecae, while last male sperm priority or mixing of sperm would be most common in species with cul-de-sac spermathecae.

Several lines of inquiry seem appropriate for exploring the role of female anatomy in sperm priority pattern. An enlightening comparison would be between species that are ecologically similar, but divergent in spermathecal morphology. I would specifically suggest a comparison between some of the orbweaving tetragnathids, that have cul-de-sac organs, and the most ecologically similar araneids, that have have conduit organs. The exact species chosen for comparison will depend upon the locality. In the mideastern U.S., *Tetragnatha laboriosa* Hentz and *Leucage venusta* (Walckenaer) often occupy similar habitats. Likewise, an apt comparison might be made between the ground-dwelling tetragnathid genus *Pachygnatha* and some of the short-sighted vagrants (*e.g.*, Clubionidae, Gnaphosidae, Philodromidae). If the sperm priority patterns in these suggested comparisons were the same, then the argument that phylogenetic constraints determined priority pattern would be untenable.

The other obvious line of inquiry is to compare species that are ecologically dissimilar, but alike in spermathecal morphology. To some extent this has already been done, in that the studies of Austad (1982, 1983), Jackson (1980b), and Vollrath (1980) have examined ecologically dissimilar species that all have conduit spermathecae. However, the published results, as discussed previously, are not completely interpretable. A detailed study of sperm priority in any cul-de-sac species would be of great interest.

Although a nonadaptive explanation for sperm priority patterns in spiders is plausible, the adaptation arguments must also be considered, even if the physiological mechanisms are less than obvious. In the next two sections I consider potential adaptive advantages of differing sperm priority patterns for each sex. Then I consolidate the conclusions reached for the separate sexes.

IV. SPERM PRIORITY: THE FEMALE PERSPECTIVE

If it is assumed that a pattern of sperm priority bestows some adaptive benefit on females, then two separate but related questions must be asked. First, why should a female mate with more than one male? Second, given that females mate more than once, how might they benefit from different sperm priority patterns?

Females have been reported to mate with more than one male in a wide variety of spider species (Table VI). Among these reports are species (*e.g.*, *Nephila clavipes* and *Latrodectus hesperus*) in which the female is many times the size of the male, assuring that the observed second mating did not result from forced copulation, although there is no evidence that even equivalent size males could successfully copulate with resisting females.

Table VI includes only confirmed observations of multiple matings, and no doubt vastly underrepresents its actual incidence, since researchers are seldom deliberately looking for the phenomenon and often do not report it, even when it is observed. It is perplexing that females ever remate, if only because of the time and energy it entails (Parker 1970). Alcock *et al.* (1978) presented six possible advantages that females might derive from multiple matings. Adapting these advantages to spiders, I will discuss each briefly.

A. Sperm Supply

Females may remate to ensure an adequate sperm supply. Two general observations make this an attractive hypothesis: (1) some species remate only after oviposition (*i.e.*, after having used some of their stored sperm; see Table VI); (2) females often, though not always (see Bonnet 1935, Kaston 1970), exhibit decreasing hatching success with successive egg sacs in the laboratory (Mikulska and Jacunski 1968, Horner and Starks 1972, Jackson 1978a, Austad 1982) when they have not been allowed to mate more than once. For instance, *F. pyramitela* exhibits hatching success in the laboratory of 0.93, 0.86, 0.66, 0.02, in the first through fourth egg sac, respectively.

Unfortunately, no data exist on reproductive success of multiple- versus single-mating females in those species that remate only after egg deposition. Moreover, since many species remate before oviposition, dwindling sperm supplies cannot be the general explanation for remating. Also, even though successive egg sacs show diminished hatching success in the laboratory and this is assumed to be a result of sperm depletion, there is still no assurance that the phenomenon occurs in nature. For example, Bristowe (1958) remarked that in the English climate, the onset of autumn prevents the orbweaver *Araneus diadematus* from laying more than one egg sac, whereas it lays up to six in captivity. *Frontinella pyramitela* lays up to seven sacs in the laboratory, but the longevity of individually marked females in the field is such that it is unlikely that they ever deposit more than two sacs (Austad, unpubl. data). Thus the large decrease in egg hatching success in the third and subsequent sacs (see Table II) seemingly is not relevant to field conditions.

The above evidence suggests that sperm depletion might not be as important a selective force in nature as laboratory data seem to indicate. However the issue is by no means settled. It is plausible that certain spiders—especially those with large interindividual variance in the number of egg sacs deposited—can require sperm supplies supplemental to that stored during their first copulation. There are a number of North American species that routinely deposit multiple sacs; these might be studied to ascertain the possible importance of sperm depletion. The orbweavers *Metepeira labyrinthea* (labyrinth spider), *Mecynogea lemniscata* (basilica spider), and *Cyclosa* spp. all deposit strings of egg sacs in semipermanent webs, so

TABLE VI

Spider Species in Which Females Have Been Reported to Mate With More Than One Male

Species	When Second Mating Occurred	Source
Theraphosidae[a] (H)[b]		
Aphonopelma hentzi	before oviposition	Baerg 1963
Eurypelma californica	before oviposition	Baerg 1928
Dugesiella hentzi	before oviposition	Petrunkevitch 1911
Dipluridae (W)[c]		
Porrhothele antipodiana	?	Jackson *et al.* 1981
Dictynidae (W)		
Dictyna sublata	after oviposition	Montgomery 1903
Dictyna calcarata	before and after oviposition	Jackson 1978c
Mallos trivittatus	before and after oviposition	Jackson 1978c
Sicariidae (H)		
Sicarius spp.	before oviposition	Levi 1968
Dysderidae (H)		
Dysdera crocata	?	Jackson *et al.* 1981
Pholcidae (W)		
Pholcus phalangioides	after oviposition	Montgomery 1903
Theridiidae (W)		
Achaearanea tepidariorum	any time except when the female is very gravid	Montgomery 1903
Latrodectus spp.	?	Kaston 1970, Jackson *et al.* 1981
Linyphiidae (W)		
Frontinella pyramitela	before and after oviposition	Austad 1982
Neriene radiata	?	Martyniuk and Jaenike 1982
Micryphantidae (W)		
Ceratinopsis interpres	before oviposition	Montgomery 1910
Araneidae (W)		
Nephila clavipes	before oviposition, but only while feeding	Christenson, Goist, and Wenzl, pers. comm.
Nephila maculata	before oviposition, but only while feeding	Robinson and Robinson 1973
Argiope spp.	?	Robinson, pers. comm.
Gasteracanthus cancriformis	before oviposition	Robinson and Robinson 1980
"*Araneus*" *patagiatus*	?	Levi 1974
"*Araneus*" *cornutus*	?	Levi 1974
"*Araneus*" *umbraticus*	?	Levi 1974
"*Araneus*" *sclopetarius*	?	Levi 1974
Tetragnathidae (W & H)		
Tetragnatha spp.	before and after oviposition	McCook 1890
Pachygnatha spp.	before and after oviposition	McCook 1890
Agelenidae (W)		
Tegenaria derhami	after oviposition	Montgomery 1903
Agelenopsis aperta	after oviposition	Riechert, pers. comm.
Pisauridae (H)		
Pisaura mirabilis	before oviposition	Austad, pers. obs.
Lycosidae (H)		
Lycosa stonei	after oviposition	Montgomery 1903
Lycosa rabida	before oviposition	Montgomery 1903

TABLE VI

Spider Species in Which Females Have Been Reported to Mate With More Than One Male
(*Continued*)

Species	When Second Mating Occurred	Source
Clubionidae (H)		
Castianeira longipalpus	before oviposition	Montgomery 1910
Thomisidae (H)		
Xysticus ferox	before oviposition	Montgomery 1903
Philodromidae (H)		
Philodromus rufus	before oviposition	Dondale 1964
Salticidae (H)		
Phidippus purpuratus	before oviposition	Montgomery 1910
Phidippus johnsoni	before oviposition	Jackson 1980b
Portia fimbriata	?	Jackson *et al.* 1981
Holoplatys sp.	?	Jackson *et al.* 1981
Neon valentulus	after oviposition	Wild 1969

[a]Spiders that undergo molting after sexual maturity shed the spermathecal lining, in effect discarding all stored sperm and becoming virgins again.
[b]H = hunting spider.
[c]W = web-building spider.

that there is little doubt about sac maternity. With some careful observation of the order in which sacs are typically deposited, one could compare hatching success in serially-deposited sacs under field conditions and perhaps resolve this issue.

B. Sperm Viability

Females may remate as a bet hedge against the possibility of first mate sterility. This advantage would hold either if there is a significant chance that a female's first mate were sterile, or if there were deterioration in quality of stored sperm through time.

Neither of these arguments is compelling. As emphasized by Alcock *et al.* (1978), males would be intensely selected to produce fully competent sperm. There is no evidence for the occurrence of male sterility in spiders. Also spiders are known to be able to store fully viable sperm for long periods. For instance, in Britain the purse-web spider (*Atypus affinis*) routinely stores sperm from the autumn of one year until the following summer when first egg deposition occurs (Bristowe 1958). Similar reproductive patterns have been reported for North American tarantulas (Baerg 1958), the New Zealand salticid *Trite auricoma* (Forster and Forster 1973), and numerous other species. Thus there is no evidence that sperm deterioration within the spermathecae is a significant factor in the evolution of multiple mating behavior.

C. Storage Cost

Frequent remating might reduce the physiological costs of long term storage. However, there is no reason to suppose that these costs would exceed those of remating. Additionally, if these costs explained the incidence of multiple mating in spiders, the observed mating pattern would tend toward females copulating at regular intervals throughout their reproductive lives, whereas Table VI shows that very often they copulate several times before first oviposition.

D. Nutritional Benefits

Male-provided nutritional benefits (such as nuptial prey or nutritive secretions) have been reported in an increasing number of insects (see Thornhill 1976 and Gwynne, this volume), however the phenomenon is not common in spiders. One species, *Pisaura mirabilis,* is known in which males offer an enswathed prey item to the female as a prelude to mating. Males then copulate while the gift is being devoured. The extent to which this nuptial gift affects female reproductive success is unknown, although many authors have established a relationship between prey consumption and egg production for spiders (*e.g.,* Bonnet 1927; Turnbull 1960; Miyashita and Jacunski 1968; Kessler 1971, 1973). However, the rarity of this behavior makes it unsatisfactory as a general explanation of multiple mating.

Thornhill (1976) pointed out that is is conceivable that males might donate their own soma as a type of reproductive investment. Could spiders be doing this? Female spiders are known (indeed, are renowned) for devouring their mates. Even though the popular impression of spiders as archetypical *femmes fatales* is exaggerated, there are nonetheless reports of females at least occasionally devouring their mates in nearly every species which has been extensively studied (*e.g.,* McCook 1890; Montgomery 1903, 1910; Bristowe 1926, 1930, 1931). Many authors specify that hungry females are more dangerous to males than are well-fed females. But this argument seems weak as a general explanation also, because if females routinely preyed upon males, they would not need to actually copulate to entice them close enough to catch. They would only need to disguise their state of receptivity.

E. Genetic Diversity

A female may secure sperm from several mates to increase the genetic diversity of her progeny. This is an attractive hypothesis, especially when more obvious benefits of multiple mating are difficult to identify. However, it rests on fragile theoretical underpinning. Williams (1975) specified that a single mating provides half the within-progeny genetic variance found in the entire population. Therefore additional matings provide little additional genetic diversity. So, even if it is assumed that maximum offspring diversity is desirable, the cost of additional matings would need to be small to justify a minor gain in variability.

Moreover, the idea that within-progeny genetic diversity is desirable is a variant on the problem of the evolution and maintenance of sex, an issue for which the theoretical ground has been paced and repaced (see Maynard Smith 1978) with marginal success. The general conclusions from models developed by Williams (1975), and refined by Maynard Smith (1978), are that if genetic diversity within lineages might be selected, it would most likely be in organisms that: (1) are sessile or very sedentary; (2) exhibit high fecundity; and (3) whose offspring have a high probability of competing with one another in environments different from their parents'.

Spiders exhibit considerable variation in all of these attributes. Spider vagility ranges from the near sessility of burrowers to the relative mobility of hunters. Fecundity is correlated with female body size and longevity (Bonnet 1927, Gertsch 1979). The probability of offspring finding themselves in environments different from their parents' might plausibly be related to average dispersal distance, which in turn is probably crudely related to body size, in that smaller spiderlings are more likely to aerially disperse than are larger ones, and smaller aerial dispersers will likely disperse farther than larger aerial dispersers. Combining these factors, then, the expectation is that if genetic diversity is important as a reason for multiple mating by spider females, a trend should be observed for multiple mating species to be large and sedentary with small, aerially-dispersing young. Table VI reveals no such trend.

F. Rejection Cost

Remating by females may be less expensive than rejection of persistent males. In this section I will argue that the most plausible explanation for the existence of multiple mating by female spiders is that the cost of resistance exceeds the cost of submission. In addition, multiple mating can be expected to occur regularly throughout the taxon because it is in precisely those groups where cost of submission is highest, that the cost of resistance is also highest.

There is tremendous variation in copulation duration among spiders, both within (Jackson 1980b) and between (Montgomery 1903) species. Tarantulas mate in a matter of a very few minutes (Baerg 1958), whereas the jumping spider *Phidippus johnsoni* has been reported to mate for more than 18 h (Jackson 1980b). However mating is a mainly passive affair for females, and little or no energy expenditure occurs during mating beyond the fixed cost of resting metabolism.

Parker (1970) argued that unless a female has something to gain, the time waste from extra matings should alone be enough to selectively favor female nonreceptivity. This assumes that time spent mating is lost for other activities. At least for many of the aerial web-builders, this assumption is false. Webs still intercept prey whether there are spiders mating in them or not. *Frontinella pyramitela* females routinely break away from copulating males and successfully capture prey (Austad, unpubl. data). Several *Nephila* and *Gasteracantha* species are reported to commonly mate while feeding (Robinson and Robinson 1973, Austin and Anderson 1978). Robinson and Robinson (1980) have seen females of these genera complete an entire predatory episode—including prey capture and consumption—while in copula. For hunting spiders, which rely more on stealth to capture prey, copulation would seem to inhibit foraging effectiveness.

Submission would be costly if copulating females were more conspicuous, less vigilant, or less agile than solitary females in relation to their potential predators. Once again there is a convenient, if crude, dichotomy between aerial web-builders and hunters. In web-builders, the web is both the most conspicuous evidence of female presence and the chief predator detection mechanism (Turnbull 1973). Copulating females would not likely be much more conspicuous than solitary females, and would probably be much less conspicuous than a female that was being actively courted by a male.

For hunting spiders, the amount of predation risk associated with mating depends upon where the mating occurs. If it occurs within a burrow (many mygalomorphs, some lycosids) or silken retreat (many of the six-eyed hunters), then conspicuousness is probably not increased. However, if it occurs outside, then both the presence of two, instead of one, spiders and the unavoidable movement that mating entails, would probably increase conspicuousness. Again though, conspicuousness resulting from copulation would seem to be less than that caused by elaborate and prolonged courtship. It may be significant that in a species (*P. johnsoni*) which has been shown to copulate both inside and outside retreats, mating duration is much shorter outside (Jackson 1980b). So even though the hunting spiders apparently incur a higher cost in time and risk for mating than do web-builders, their cost of resistance seems concomitantly higher.

V. SPERM PRECEDENCE: THE MALE PERSPECTIVE

Parker (1970) has stated that intense competition between males' sperm within the female reproductive tract for access to fertilizations should lead to adaptations that: (1) allow males to achieve precedence of their sperm over previously stored sperm, and (2) protect a male's sperm from replacement by sperm from subsequent mates. These adaptations are diametrically opposed, and it is by no means clear when one type will prevail.

Male spiders (at least in web-building species about which most is known) are faced with the dual problem that each copulation is of tremendous importance in terms of eventual reproductive success, yet sperm displacement is almost impossible to avoid through behavioral means.

Each copulation is especially important to male web-building spiders because female encounter rate is low. This may not be true in some of the colonial species, which have many individuals of both sexes present in the same community web, or for some of the hunters about which very little is known. However, in species where females generally live in or close to protected areas such as webs, burrows, or retreats, males must wander to find them, and wandering is risky. Field studies of *N. clavipes*, for instance, reveal that successful travel between female webs is limited to between 4% (Christenson and Goist 1979) and 11% of males. Similarly, only about 25% of *F. pyramitela* males visit as many as two female webs (Austad, unpubl. data). If interweb travel is as risky as these data suggest, then assuring that his sperm is not displaced after copulation should be of profound significance to a male. Also, because females are so valuable to them, males can be expected to be very persistent in courtship. The unusually long interval between copulation and first sperm usage by females, and the typically long female reproductive life compared with a male's, cause behavioral paternity assurance by male spiders to be almost impossible to achieve.

The long interval between mating and first oviposition (common intervals range from about 2 weeks to 10 months) indicates a male might successfully guard a female for a considerable time, and still lose his entire reproductive investment if he finally loses an encounter with another male, or dies before the first oviposition. Because female reproductive life is so long, a male might succeed in preventing other males access to a female for his entire adult life, yet have his sperm displaced after he dies. An additional implication of long female reproductive life is that even at a low encounter rate between the sexes, females would be apt to encounter a number of males.

The inefficacy of female guarding is further suggested by the fact that the above conditions are almost identical to those proposed by Parker (1974) on theoretical grounds as preconditions to the evolution of postcopulatory mate guarding, yet it is almost unheard of in spiders. Although there are scattered reports of male spiders residing with females for long periods after mating (*e.g.*, McCook 1890, Comstock 1940), many of these have been invalidated. For example, *F. pyramitela* males have long been reported to cohabit with females for long periods (Comstock 1940, Kaston 1948). However, by studying individually marked males, I found that though males do indeed cohabit for days with soon-to-mature females, they depart soon after mating. Males may still be seen in females' webs long after maturation, but these are a succession of different individuals, each of whom remains for a relatively short time. Rovner (1968) found that similar reports of extended cohabitation were false for *Linyphia triangularis*.

Short-term postcopulatory guarding has been reported in some orbweavers (Christenson and Goist 1979, Robinson and Robinson 1980). Robinson and Robinson (1980) have seen "eunuchs" (*i.e.*, males who have lost the terminal segment of the intromittent organ, and hence are incapable of further mating) attack courting males in the orbweaver *Herennia ornatissima*; however, the duration of such female defense is unknown. In *N. clavipes* the first mate will remain on the web hub defending access to the female for several days after mating (Christenson and Goist 1979). Yet this is a far cry from the necessary effective guarding period of 4-6 weeks before the female's first oviposition (Christenson *et al.* 1979).

The absence of or rarity of postcopulatory guarding under conditions for which it might seem so adaptive suggests that males have found less obvious methods for protecting their reproductive investments. Speculations upon why counteradaptations to circumvent these methods have seemingly not evolved will be discussed in the next section.

VI. COMBINED PERSPECTIVES

I will argue in this section that the extent to which adaptations conducive to first male or last male sperm priority prevail is largely an effect of selection on the female to promote the sperm precedence pattern most favorable to her. Adaptations of this sort need not necessarily be selectively advantageous to both sexes. The fact that in some insects, such as the hymenopteran genus *Apis,* mating plugs are either removed by the female physically or are absorbed by vaginal secretions (Woyke and Ruttner 1958), hints that selection among males alone could account for some priority mechanisms. Still, it seems logical that mechanisms favorable to both sexes would be more strongly selected, other things being equal, and might also be more effective, since one sex would not be continually selected for adaptations to circumvent them.

The advantage to a female of evincing first male sperm priority is that it should both discourage subsequent males from prolonged courtship sequences (to the extent that the mechanism is foolproof and the previous mating detectable) and encourage the original mate to depart as soon as successful insemination is accomplished. If a male is assured of the fertilization of a female's eggs when he has finished mating, then he will maximize his reproductive success by departing in search of additional mates immediately. For the previously mentioned reasons, male residence in a female's proximity may be inimical to her fitness.

Last mate sperm priority would tend to encourage postcopulatory mate guarding by males. Consequently, females would have one or more males continually in their proximity and conflict would repeatedly erupt whenever another mate arrived. There is no obvious circumstance under which this pattern of male behavior might be beneficial to a female.

If first male priority were common, males might be expected to exhibit another behavior: precopulatory guarding of immature females. If interfemale searching is as risky an endeavor as it seems, and if first male priority exists, then an about-to-mature female is very valuable. Conversely, if last male priority is the rule, then there is little advantage in expending time, risk, and energy to defend an immature females.

Precopulatory guarding, or the suitor phenomenon (Robinson and Robinson 1980), is a well-known and often mentioned male behavior in spiders. In web-building species this takes the form of male conflict over the location on the web most accessible to the female (Rovner 1968, Christenson and Goist 1979, Robinson and Robinson 1980, Vollrath 1980, Austad 1983). In species that occupy nests or retreats it takes the form of cohabitation in the nest with the subadult female (references in Jackson 1978c). This behavior thus seems further indirect evidence as to the high incidence of first male sperm priority.

One might expect that if females exhibit first male sperm priority to discourage extended male presence, that they might advertise their reproductive state. Contact pheromones in female silk are known to help males locate mates (Hegdekar and Dondale 1969, Tietjen 1979, Ross and Smith 1980). It is conceivable that females could alter the silk pheromones as a function of reproductive state, therefore allowing males to distinguish virgins or immatures from mated females without having to come into direct contact with them.

Some mixture of sperm from several mates is the most common pattern of sperm priority in insects, though usually one mate or the other predominates (Parker 1979; Boorman and Parker 1976; Gwynne, this volume). This could be the realized pattern of sperm usage for one of several reasons: (1) some mixture of sperm might yield an optimal amount of genetic diversity among the female's progeny; (2) priority mechanisms are adaptive but are less than perfect; and (3) selection may simultaneously favor several competing adaptations. The first reason is unlikely to be important for the previously mentioned reasons. The second and third reasons are closely related ideas. The prevalence of sperm mixing in insects, and the fact that both Jackson's (1980b) and Vollrath's (1980) results hint at sperm mixture under at least some circumstances should keep us alive to the possibility of nonadaptive sperm mixture. However, the evidence to date suggests that first male sperm priority should be the prevalent condition among the spiders.

VII. SUMMARY

Spiders possess many predispositions for a high degree of sperm competition. However, first male sperm priority via the mechanism of mating plugs has already proven to be much more common among spiders than among insects, and there is suggestive evidence that it may be the predominant pattern of sperm usage. This may be a simple nonadaptive consequence of female reproductive anatomy, or it may be an adaptation of females to discourage prolonged courtship and mating attempts by supernumerary males. If it is nonadaptively related to female reproductive anatomy, then there are a number of spider species whose contrasting types of sperm storage organs should allow elucidation of the role of anatomy. If it is an adaptation to discourage multiple mating attempts, then there is at least one species (*viz. Pisaura mirabilis*) in which multiple mating might be encouraged and an alternate sperm priority pattern might be expected. Sperm competition studies of a few carefully chosen species should allow discrimination among the alternatives mentioned above.

Although females of many species mate with several males, the reason they do so is far from clear. The most attractive hypothesis is that the cost of female resistance exceeds the cost of submission to persistent males.

Male spiders can be expected to be very persistent in courting females because interfemale travel is risky and an individual has a low probability of encountering many females during his lifetime. Males also face a high potential for sperm displacement because of the exceptionally long interval between mating and fertilization and the long reproductive lifetime of females. The only effective method of paternity assurance then might be a mating plug.

My general prediction therefore is that because it is advantageous to females, first male sperm priority will be the rule in spiders. Exceptions are predicted in certain species, such as those with male investment in offspring beyond sperm, or those with an especially short interval between copulation and egg deposition, which are aberrant in the general scheme of spider life history characteristics.

Although absolute sperm priority would be strongly selected, some mixing may occur as a result of the imperfection of adaptations or oppositional selection on each of the sexes.

ACKNOWLEDGMENTS

I would like to thank Willis Gertsch and Vince Roth for taking time to discuss spider reproductive anatomy with me when I first became interested in sperm competition in spiders. For improving this manuscript through cogent and constructive criticism I wish to thank Terry Christenson, Winston Fulton, Darryl Gwynne, Rick Howard, Robert Jackson, B. J. Kaston, Morris Levy, Mike Robinson, Richard Thomas, and Peter Waser. I also wish to thank Geoff Parker for providing the seminal paper in this field of inquiry and for discussing sperm competition with me at length. Finally, I am grateful to Terri Werderitsh who typed the entire manuscript numerous times.

REFERENCES

Abalos, J. W., and E. C. Baez. 1963. On spermatic transmission in spiders. *Psyche* 70:197-207.

Alcock, J., E. M. Barrows, G. Gordh, L. J. Hubbard, L. Kirkendall, D. W. Pyle, T. L. Ponder, and F. G. Zalom. 1978. The ecology and evolution of male reproductive behaviour in the bees and wasps. *Zool. J. Linn. Soc.* 64:293-326.

Aspey, W. P. 1977a. Wolf spider sociobiology: I. Agonistic display and dominance subordinance relations in adult male *Schizocosa crassipes*. *Behaviour* 62:103-141.

Aspey, W. P. 1977b. Wolf spider sociobiology: II. Density parameters influencing agonistic behaviour in *Schizocosa crassipes*. *Behaviour* 62:142-163.

Austad, S. N. 1982. First male sperm priority in the bowl and doily spider (*Frontinella pyramitela*). *Evolution* 36:777-785.

Austad, S. N. 1983. A game theoretical interpretation of male combat in the bowl and doily spider (*Frontinella pyramitela*). *Anim. Behav.*

Austin, A. D., and D. T. Anderson. 1978. Reproduction and development of the spider *Nephila edulis* (Koch) (Araneidae:Araneae). *Aust. J. Zool.* 26:501-518.

Baerg, W. J. 1928. The life cycle and mating habits of the male tarantula. *Q. Rev. Biol.* 3:109-116.

Baerg, W. J. 1958. *The Tarantula*. University of Kansas Press, Lawrence, KS.

Baerg, W. J. 1963. Tarantula life history records. *J. N.Y. Entomol. Soc.* 71:233-238.

Bonnet, P. 1927. Étude et considerations sur la fécondite chez les Araneides. *Mém. Soc. Zool. Fr.* 28:1-47.

Bonnet, P. 1935. *Theridion tepidariorum* C. L. Koch. Araignée compopolite: repartition, cycle vital, moeurs. *Bull. Soc. Hist. Nat. Toulouse* 68:335-386.

Boorman, E., and G. A. Parker. 1976. Sperm (ejaculate) competition in *Drosophila melanogaster*, and the reproductive value of females to males in relation to female age and mating status. *Ecol. Entomol.* 1:145-155.

Brady, A. R. 1964. The lynx spiders of North America, north of Mexico (Araneae:Oxyopidae). *Bull. Mus. Comp. Zool.* 131:429-518.

Bristowe, W. S. 1926. The mating habits of British thomisid and sparassid spiders. *Ann. Mag. Nat. Hist. Ser 9*, 18:114-131.

Bristowe, W. S. 1930. A supplementary note on the mating habits of spiders. *Proc. Zool. Soc. Lond.* 1930:395-413.

Bristowe, W. S. 1931. The mating habits of spiders: A second supplement. *Proc. Zool. Soc. Lond.* 1931:1401-1412.

Bristowe, W. S. 1958. *The World of Spiders*. Collins, London.

Christenson, T. E., and K. C. Goist, Jr. 1979. Costs and benefits of male-male competition in the orb-weaving spider *Nephila clavipes*. *Behav. Ecol. Sociobiol.* 5:87-92.

Christenson, T. E., P. A. Wenzl, and P. Legum. 1979. Seasonal variation in egg hatching and certain egg parameters of the golden silk spider *Nephila clavipes* (Araneidae). *Psyche* 86:137-148.

Comstock, J. H. 1940. *The Spider Book*. (Revised and edited by W. J. Gertsch.) Comstock, Ithaca, NY.

Dondale, C. D. 1964. Sexual behavior and its application to a species problem in the spider genus *Philodromus* (Araneae:Thomisidae). *Can J. Zool.* 42:817-821.

Exline, H., and W. H. Whitcomb. 1965. Clarification of the mating procedure of *Peucetia viridans* (Araneida:Oxyopidae) by a microscopic examination of the epigynal plug. *Fla. Entomol.* 48:169-171.

Forster, R. R. 1967. The spiders of New Zealand: Part I. *Otago Mus. Zool. Bull.* 1:1-124.

Forster, R. R., and L. M. Forster. 1973. *New Zealand Spiders*. Collins, Auckland and London.

Gertsch, W. J. 1979. *American Spiders*, 2nd Ed. Van Nostrand Reinhold, New York.

Gould, S. J., and R. C. Lewontin. 1979. The spandrels of San Marcos and the Panglossian paradigm: A critique of the adaptationist programme. *Proc. R. Soc. Lond B. Biol. Sci.* 205:581-598.

Gromko, M. H., and D. W. Pyle. 1978. Sperm competition, male fitness and repeated mating by female *Drosophila melanogaster. Evolution* 32:588-593.

Hegdekar, B. M., and C. D. Dondale. 1969. A contact sex pheromone and some response parameters in lycosid spiders. *Can. J. Zool.* 47:1-4.

Horner, N. V., and K. J. Starks. 1972. Bionomics of the jumping spider *Metaphidippus galathea. Ann. Entomol. Soc. Am.* 65:602-607.

Jackson, R. R. 1978a. The life history of *Phidippus johnsoni* (Araneae:Salticidae). *J. Arachnol.* 6:1-29.

Jackson, R. R. 1978b. The mating strategy of *Phidippus johnsoni* (Araneae:Salticidae): I. Pursuit time and persistence. *Behav. Ecol. Sociobiol.* 4:123-132.

Jackson, R. R. 1978c. Male mating strategies of dictynid spiders with differing types of social organization. *Symp. Zool. Soc. Lond.* 42:79-88.

Jackson, R. R. 1980a. The mating strategy of *Phidippus johnsoni* (Araneae:Salticidae): III. Intermale aggression and a cost-benefit analysis. *J. Arachnol.* 8:241-250.

Jackson, R. R. 1980b. The mating strategy of *Phidippus johnsoni* (Araneae:Salticidae): II. Sperm competition and the function of copulation. *J. Arachnol.* 8:217-240.

Jackson, R. R., P. Phibbs, S. D. Pollard, and D. D. Harding. 1981. Spiders that mate more than once. *Secretary's News. Br. Arachnol. Soc.* 30:6-7.

Kaston, B. J. 1948. *Spiders of Connecticut*. State of Connecticut, Hartford, CT.

Kaston, B. J. 1970. Comparative biology of American black widow spiders. *Trans. San Diego Soc. Nat. Hist.* 16:33-82.

Kaston, B. J. 1978. *How to Know the Spiders*, 3rd Ed. Brown, Dubuque, IA.

Kessler, A. 1971. Relation between egg production and food consumption in species of the genus *Pardosa* (Lycosidae, Araneae) under experimental conditions of food abundance and food shortage. *Ecologia* 8:93-109.

Levi, H. W. 1968. Predatory and sexual behavior of the spider *Sicarius* (Araneae:Sicariidae). *Psyche* 74:320-330.

Levi, H. W. 1974. Mating behavior and the presence of the embolus cap in male Araneidae. *Proc. VI Arachnol. Congr.*, pp. 49-50.

Levi, H. W. 1975. The American orb-weaver genera *Larinia, Cercidia,* and *Mangora* north of Mexico (Araneae:Araneidae). *Bull. Mus. Comp. Zool.* 147:101-135.

Levi, H. W., and L. R. Levi. 1968. *A Guide to the Spiders and Their Kin*. Golden Press, New York.

Martyniuk, J., and J. Jaenike. 1982. Multiple mating and sperm usage patterns in natural populations of *Prolinyphia marginata* (Araneae:Linyphiidae). *Ann. Entomol. Soc. Am.* 75:516-518.

Maynard Smith, J. 1978. *The Evolution of Sex*. Cambridge University Press, Cambridge.

McCook, H. C. 1890. *American Spiders and Their Spinningwork: Vol. II. Motherhood and Babyhood: Life and Death*.

Mikulska, I., and L. Jacunski. 1968. Fecundity and reproduction activity of the spider *Tegenaria atrica* C. L. Koch. *Zool. Pol.* 18:97-106.

Montgomery, T. H. 1903. Studies on the habits of spiders, particularly those of the mating period. *Proc. Acad. Nat. Sci. Phila.* 55:59-149.

Montgomery, T. H. 1910. The significance of the courtship and secondary sexual characters of araneads. *Am. Nat.* 44:151-177.

Muniappan, R., and H. L. Chada. 1970. Biology of the crab spider, *Misumenops celer. Ann. Entomol. Soc. Am.* 63:1718-1722.

Parker, G. A. 1970. Sperm competition and its evolutionary effect on copula duration in the fly *Scatophaga stercoraria. J. Insect Physiol.* 16:1301-1328.

Parker, G. A. 1974. Courtship persistence and female guarding as male time investment strategies. *Behaviour* 48:157-184.

Petrunkevitch, A. 1911. Sense of sight, courtship and mating in *Dugesiella hentzi* (Girard), a Theraphosid spider from Texas. *Zool. Jahrbucher Abt. Syst. Oekol. Geogr. Tiere* 35: 355-376.

Riechert, S. E. 1978a. Energy-based territoriality in populations of the desert spider *Agelenopsis aperta* (Gertsch). *Symp. Zool. Soc. Lond.* 41:211-222.

Riechert, S. E. 1978b. Games spiders play: Behavioral variability in territorial disputes. *Behav. Ecol. Sociobiol.* 3:135-162.

Riechert, S. E. 1979. Games spiders play: II. Resource assessment strategies. *Behav. Ecol. Sociociol.* 6:121-128.

Robinson, M. H. 1982. Courtship and mating behavior in spiders. *Ann. Rev. Entomol.* 27:1-20.

Robinson, M. H., and B. Robinson. 1973. Ecology and behavior of the giant wood spider, *Nephila maculata* (Fabricus) in New Guinea. *Smithson. Contrib. Zool.* 149:1-76.

Robinson, M. H., and B. Robinson. 1980. Comparative studies on the courtship and mating behavior of tropical araneid spiders. *Pacific Insects Monogr.* 36:1-218.

Ross, K., and R. L. Smith. 1979. Aspects of the courtship behavior of the black widow spider, *Latrodectus hesperus* (Araneae:Theridiidae), with evidence for the existence of a contact sex pheromone. *J. Arachnol.* 7:69-77.

Rovner, J. S. 1968. Territoriality in the sheet-web spider (*Linyphia triangularis*). *Z. Tierpsychol.* 25:232-242.

Thornhill, R. 1976. Sexual selection and paternal investment patterns in insects. *Am. Nat.* 110:153-163.

Tietjin, W. J. 1979. Is the sex pheromone of *Lycosa rabida* (Araneae:Lycosidae) deposited on a substratum? *J. Arachnol.* 7:107-212.

Turnbull, A. L. 1973. Ecology of the true spiders (Araneomorphae). *Ann. Rev. Entomol.* 18: 305-348.

van Helsdingen, P. J. 1965. Sexual behaviour of *Lepthyphantes leprosus* (Ohlert) (Araneida, Linyphiidae), with notes on the function of the genital organs. *Zool. Meded. (Leiden)* 41:15-42.

Vollrath, R. 1980. Male body size and fitness in the web-building spider *Nephila clavipes. Z. Tierpsychol.* 53:61-78.

Walker, W. F. 1980. Sperm utilization strategies in nonsocial insects. *Am. Nat.* 115:780-799.

Whitcomb, W. H., and R. Eason. 1965. The mating behavior of *Peucetia viridans* (Araneida: Oxyopidae). *Fla. Entomol.* 48:163-167.

Wild, A. M. 1969. A note on the mating of *Neon valentulus* Falconer. *Bull. Br. Arachnol. Soc.* 1:62.

Williams, G. C. 1975. *Sex and Evolution.* Princeton University Press, Princeton, NJ.

Woyke, J., and F. Ruttner. 1958. An anatomical study of the mating process in the honeybee. *Bee World* 39:3-18.

8

Sperm Competition and the Evolution of Odonate Mating Systems

JONATHAN K. WAAGE

I. INTRODUCTION

In this review I present evidence for sperm displacement in damselflies and dragonflies (Odonata) and discuss the relevance of sperm competition to the evolution of odonate reproductive behavior. I will focus on patterns of genitalic morphology and behavior that suggest sperm competition has been a major determinant of much of odonate reproductive behavior (Parker 1970, 1974, this volume; Waage 1973, 1979a). Although my basic intent is to explore rather than test, I hope my development of the subject will reveal the validity and generality of my hypothesis, and the information needed to test it.

Four major sets of factors influence behavioral patterns (Fig. 1). While focusing on the evidence for sexual selection in the context of sperm competition, I do not ignore the influences of these other forces. Indeed the phylogenetic, ecological, and behavioral diversity of odonates may provide opportunities to evaluate the relative import of each. The genetic basis of behavior and morphology under discussion is virtually unknown. Thus, I approach odonate behavior in an attempt to determine if observed variations are consistent with those predicted, assuming the operation of selection in the context of sperm competition. Significant to future work are the means by which behavioral variants might be identified as either evolved functions (adaptations) or incidental effects (Williams 1966, Gould and Lewontin 1979).

II. Sexual Selection and Sperm Competition

Darwin (1859, 1871) recognized the importance of sexual selection in explaining certain morphological and behavioral traits of males and females. However, its importance has been debated and underplayed until recently (Campbell 1972, Blum and Blum 1979). Rebirth of interest in sexual selection seems to have emerged with the reaffirmation, in the late 1960s, that the focus of selection is at the level of the individual. Sexual selection's importance becomes clear when we view reproduction **not** as a populationwide cooperative venture to preserve and better the species, **but instead** as a struggle among individuals (including mates) for the representation of their genes in future generations.

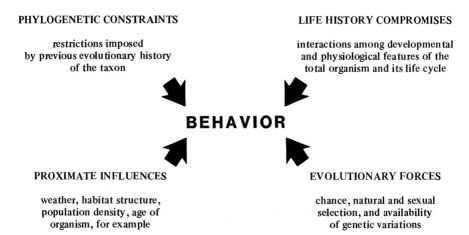

PHYLOGENETIC CONSTRAINTS

restrictions imposed
by previous evolutionary history
of the taxon

LIFE HISTORY COMPROMISES

interactions among developmental
and physiological features of the
total organism and its life cycle

BEHAVIOR

PROXIMATE INFLUENCES

weather, habitat structure,
population density, age of
organism, for example

EVOLUTIONARY FORCES

chance, natural and sexual
selection, and availability
of genetic variations

Fig. 1. Factors influencing the evolution of behavioral patterns.

Darwin (1871) pointed out that sexual selection in the context of intrasexual competition could favor devices and behaviors of males that would prevent interference from other males before and during copulation:

> So again, if the chief service rendered to the male by his prehensile organs is to prevent the escape of the female before the arrival of other males, or when assaulted by them, these organs will have been perfected through sexual selection, that is by the advantage acquired by certain individuals over their rivals.

However, Parker (1970) added a significant new dimension to sexual selection theory with the concept of sperm competition. Prior to this addition, intrasexual selection was viewed as precopulatory through the exclusion of rivals from potential mates or pair-forming sites via aggressive behavior. However, Parker (1970) proposed that intrasexual selection should also occur between insemination and fertilization when a rival's reproductive investment is reduced or nullified by gamete replacement, displacement, or dilution. He also noted that selection in this context would favor **both** the evolution of devices and behaviors that enhance an individual's ability to replace, displace, or dilute a rival's gametes, **and** those that would lessen the chances of this happening to an individual. Sperm competition thus shifts emphasis in estimating reproductive success from numbers of matings to numbers of eggs fertilized. With this comes an awareness of the kinds of morphological and behavioral traits that would enhance reproductive success.

The basic prerequisites for sperm competition are: delay between insemination and fertilization, storage of sperm in a form and location accessible to males, and possibility of multiple matings by a female prior to fertilization of some or all of her eggs. It is important to remember that while sperm competition implies a focus on intermale competition, the evolutionary perspectives of females are also critical to understanding the phenomenon and its dynamics (Knowlton and Greenwell, this volume).

Jacobs (1955) and particularly Parker (1970) predicted that much of odonate postcopulatory behavior (*e.g.*, tandem oviposition and non-contact guarding) might function to protect mates from take-overs and sperm competition by rivals during oviposition. I predicted (Waage 1973) and later verified (Waage 1978, 1979a, b) that postcopulatory male guarding of mates by the damselfly *Calopteryx maculata* functions in this respect. The discovery that *C. maculata* males use a highly specialized penis to remove the sperm of previous males (Waage 1979a) led me to examine (1) how widespread sperm displacement is among odonates, and (2) how important sperm competition might be as a major selective factor in the evolution of odonate postcopulatory and even pair-forming behaviors. Since relatively little is known about sperm displacement[1] in odonates, it seems

---

[1] I use the term displacement to denote both the removal and physical repositioning of the sperm of previous males, unless otherwise specified. Sperm competition then includes displacement, dilution, and precedence as its mechanisms.

best to begin with an examination of procedures by which it can be identified. My intent is to encourage others to apply and improve the methodology.

III. DEMONSTRATING SPERM COMPETITION IN ODONATES

The most accurate means of demonstrating relative fertilization success for two or more males inseminating the same female involves the use of genetic markers or irradiated male techniques (Boorman and Parker 1976, Smith 1979). However, most odonates lack known genetic markers, cannot be bred in the laboratory, or cannot be manipulated well enough in nature to use irradiated male techniques. Instead, an indirect assessment of sperm displacement ability and, to some extent, of the magnitude of the displacement is possible by examining the volume of sperm

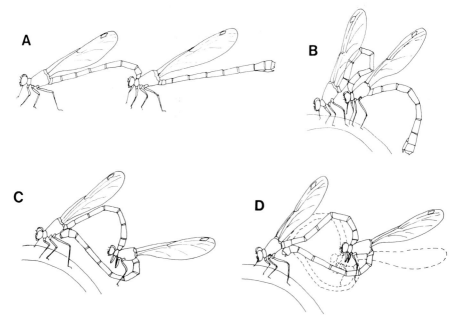

Fig. 2. Stages in the odonate copulatory cycle as illustrated for Zygoptera. A. Tandem, precopulatory stage: the male (left) holds the female's prothorax with two pairs of anal appendages. B. Sperm translocation: the male transfers sperm from his testes to a temporary storage vesicle on the venter of his second segment. C. "Wheel position" (copulation): the female brings her abdomen forward and upward to engage her genitalia with the male's. Sperm displacement and insemination occur in this position. D. Rocking or pumping movements: the male arches and depresses his abdomen, particularly the first few segments, during the sperm displacement stage. During insemination, the male's abdomen remains in the arched position.

in storage organs of females collected before, during, and after copulations (Waage 1979a).

The Odonata are well suited for this indirect method for several reasons. First, sperm transfer occurs in two distinct stages (Fig. 2B, C). Prior to mating (Fig. 2B), a male translocates sperm from his testes to a temporary storage sac (penis vesicle in Anisoptera, Fig. 6; sperm vesicle in Zygoptera, Figs. 3 and 6) located with the copulatory apparatus on the venter of his second and third abdominal segments. Dissection of this storage vesicle reveals whether or not a male has transferred his sperm to a female, and aids in the identification of whose sperm is present in females captured during or after copulations. During copulation (Fig. 2C), the male transfers his sperm from the vesicle to the female's storage organs. An estimate of the volume of sperm in full male vesicles indicates how much sperm postcopula females should carry if sperm displacement is complete. In Zygoptera, the vesicle is normally full after translocation and empty after insemination, but variations in the volumes involved may occur and should be checked.

Second, female sperm storage organs (bursa copulatrix and spermatheca, Figs. 3 and 4) are easy to remove and clear of surrounding tissues. They are transparent sacs within which the sperm mass is visible and its volume measurable.

Third, odonate sperm tends to clump into a dense, interwoven mass (Waage 1979a, Miller 1982a). Although this makes separating and counting sperm difficult, it does allow estimation of volume by measuring the external dimensions of the sperm mass. In the zygopterans I have examined (*Calopteryx, Lestes, Argia, Enallagma,* and *Ischnura*), sperm shape and size are not affected by the dilute acetic

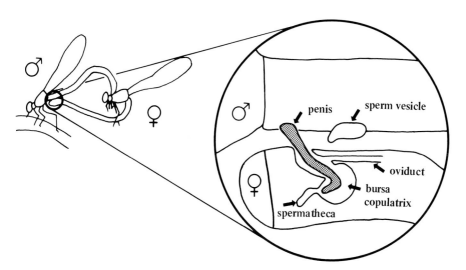

Fig. 3. Schematic of coupled genitalia in Zygoptera (major organs identified). The same orientation is maintained in subsequent detailed figures of genitalia.

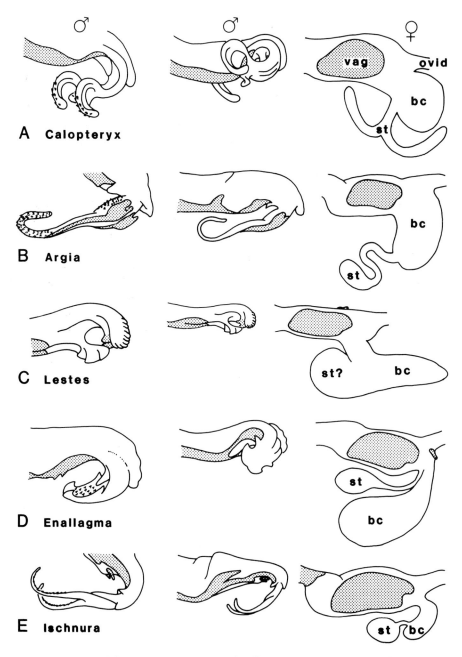

Fig. 4. Male and female genitalia representing five zygopteran genera. The distal segment of each penis (center) is shown to scale with the corresponding female genitalia (right), and enlarged to show details (left). bc = bursa copulatrix; ovid = oviduct; st = spermatheca; vag = vagina. Chitinous structures are shaded. The valve between the vagina and bursa copulatrix is not illustrated.

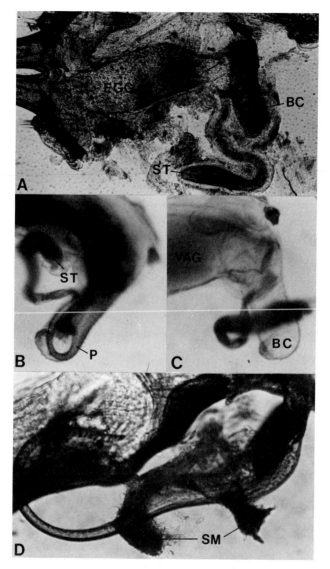

Fig. 5. Photomicrographs of *Argia fumipennis violacea* female genitalia and an in copula dissection illustrating sperm storage, fertilization and sperm removal. A. Preparation of a female in the process of oviposition. The bursa copulatrix (BC) and spermatheca (ST) are full of sperm. An egg in the vagina has its narrow end in position to receive sperm at the vaginal/bursal junction. It will be passed out between the ovipositor valves at the left. A second egg is visible at the upper right in the oviduct. B. In copula preparation showing the distal segment of the penis (P) entirely within the bursa. Sperm can be seen within the bursa and in the elongated spermatheca (ST). C. Same preparation after removal of the male's penis. No sperm remains in the bursa (BC). D. Same preparation showing the sperm mass (SM) still held on the male's distal segment by its proximally oriented spines.

acid used in removing surrounding tissues and ethanol used for preserving specimens. However, in some libellulids (*e.g.*, *Celithemis elisa*) sperm are reduced to granular structures upon exposure to acetic acid.

Finally, it is often possible to use live, tethered females to obtain matings under more controlled conditions (Waage 1979b). One use of this technique is to obtain multiple-mated females for comparison with females from single matings. If females mated twice have larger average sperm volumes than those mated once, it is likely that males simply add sperm to that from previous males. However, if average sperm volumes are the same, the implication is that the second male has removed the previous sperm prior to insemination. This can be checked by examining the male's sperm vesicle to determine that he has transferred sperm. Tethered females can also be used to obtain more mating pairs than might normally be found in nature when male-female encounter rates are low.

These approaches provide a reasonable indication of the potential for sperm displacement by comparing the volume of sperm in females collected in each of three contexts: (1) in tandem and post-sperm translocation, but precopula; (2) in copula, but prior to insemination; and (3) postcopula, but prior to oviposition. In addition, sperm volumes of females collected away from mating and oviposition areas and after they have oviposited indicate how much sperm females might carry prior to mating and how much is used for an average oviposition bout (Waage 1980). It is important to collect a sufficient sample of precopula pairs to assess how often sperm is present in females prior to mating, and how much. It cannot be assumed that mature females generally carry sperm from previous matings, even though odonates continue to mature and oviposit eggs throughout their lives (Corbet 1962).

The sperm translocation event (Fig. 2B) is a useful marker for identifying the stage in the mating cycle of a zygopteran pair discovered in tandem (Fig. 2A). If the pair enter copula (Fig. 2C) without sperm translocation they are likely to be resuming copulation after a temporary interruption rather than beginning it. Collecting both the male and female and examining the male's sperm vesicle reveal if insemination has occurred. Among different species of Anisoptera, sperm translocation may occur before or after the male encounters a female (Bick 1972, Corbet 1980) and, therefore, cannot always be used as a sequential marker. However, the penis vesicle can be checked to aid in establishing the stage of the mating cycle at capture, if it is first determined that males transfer all the sperm in the vesicle during insemination.

Pairs in copula must be collected and carefully dissected to ascertain the actual mechanism of sperm displacement and the roles played by various appendages on the distal segment of the male's penis. I have had reasonable success (30-80% of attempts) in obtaining zygopterans in copula by slowly approaching free or tethered mating pairs and, with dissecting scissors, quickly snipping the female's abdomen close to its distal end (segment 6 or 7). The pair is immediately netted, and the joined abdomens isolated and transferred to a vial of 70% ethanol. Snipping the female's abdomen first prevents the male from disengaging his genitalia.

Obtaining "in copula preparations" of anisopterans, especially libellulids, is more problematic. First, many species copulate in flight (often for less than 20 sec) or in treetops, and are extremely difficult to approach. Second, the penis of libellulids and some other Anisoptera is extensile and erectile and may partially or completely withdraw from the female when the male is killed. The male and female genitalia appear to be held together by external and perhaps internal appendages of the copulatory apparatus which also tend to release their hold when the male is killed (Miller 1981). The procedure outlined above for Zygoptera has been attempted with *Celithemis elisa,* but was unsuccessful (the genitalia disengage immediately); however, limited success was achieved with *Sympetrum rubicundulum* if the male was first decapitated. In copula pairs of both species can often be approached by slowly lowering an insect net, tent-like, over them. Alternative approaches such as freezing copulating pairs with liquid nitrogen or immobilizing them with quick-freeze aerosol sprays seem worth investigating.

Two weaknesses in the methodology described above should be noted. First, the magnitude of sperm displacement is difficult to estimate since taking specimens at various times during the displacement stage of copulation results in an underestimate of how much sperm is displaced unless the pair is collected just before insemination—the last few minutes or seconds of copulation in zygopterans (Rowe 1978, Waage 1979a, Miller and Miller 1981). If a pair is encountered in copula it may be difficult to tell how close they are to termination. One solution to this problem may result from the findings of Miller and Miller (1981) and Waage (unpubl. data) that the rocking or pumping motions of pairs during copulation (Fig. 2D) may be used to identify various stages in the copulation process.

A second weakness in my methodology concerns measuring sperm volumes in general. The assumption is made that there are no differences in the density of sperm in the storage organs of females collected in different contexts. Although it seems to have been a reasonable assumption for *Calopteryx* where sperm removal approaches 100% (Waage 1979a), it may not be so for other species where repositioning may involve compaction of the sperm mass (see discussion of *Lestes vigilax* and *Sympetrum rubicundulum* below). Therefore, while the methodology allows a general assessment of whether or not sperm displacement occurs, it may not always permit an accurate estimate of the degree of displacement.

Some damselflies can be reared and bred in laboratory cages, and there has been preliminary work with heritable phenotypic polymorphisms (Johnson 1966a, b, 1969). Where mating and oviposition can be controlled, for example in large field cages or with tethered females, genetic markers or irradiated male techniques may be feasible. Fincke (1984) has recently demonstrated sperm precedence of the last male in *Enallagma hageni* using irradiated males and field cages. Meanwhile, indirect methods and circumstantial evidence based on functional genitalic morphology must provide the major basis for evaluating the level of sperm competition in these insects.

IV. SPERM DISPLACEMENT ABILITY

A. Zygoptera: Quantitative and Comparative Evidence

Waage (1979a) showed that *Calopteryx maculata* physically remove 80-100% of the sperm of previous males from the bursa copulatrix and spermatheca of females that they will subsequently inseminate. Females carry substantial volumes of sperm both before mating and after oviposition (83-90% of the volume of postcopula females; Waage 1980). Females carrying viable sperm from recent matings are also likely to remate when they change oviposition sites (Waage 1979b).

The mechanism for sperm removal in *C. maculata* involves the morphology of the distal segment of the male's penis[2] (Waage 1979a). This segment ends in a flexible scoop-like flap, and carries two horn-like appendages (Fig. 4A). The flap is used to remove sperm from the bursa copulatrix, and the horn-like appendages remove sperm from the spermatheca. Dissections of pairs in copula have revealed that these appendages enter the spermathecal tubes. Sperm removal involves insertion and withdrawal with sperm being pulled out by the numerous proximally oriented spines on the appendages. Most in copula and interrupted copula specimens have clumps of sperm (of a previous male) under the flap or on the horn-like appendages.

The penis morphology of *C. dimidiata* is amost idential to that of *C. maculata*, differing primarily in size. The structure of the female genitalia and sperm storage organs of these species also differs only in size. As expected, male *C. dimidiata* also remove sperm of previous males (ca. 98% removal) prior to insemination (Waage, unpubl. data). The average sperm volume carried by interrupted copula females was significantly less (F = 6.70; df = 1,13; P < 0.02) than that carried by precopula females. Females of this species also carry substantial stores of sperm before mating and after oviposition.

Published illustrations of calopterygid penis morphology (Kennedy 1920) and specimens of the other North American *Calopteryx* reveal similar architecture of the distal segment, suggesting that sperm removal may be universal in this family.

---

[2] I adopt this terminology for male intromittent organs in Odonata because of its familiarity. The penis of males in this order is not homologous to reproductive organs of any other insect. Other terminology associated with the odonate penis include: ligula (Zygoptera) and vesica spermalis (Anisoptera) (Pfau 1971).

The descriptions that follow for *Argia, Enallagma, Ischnura,* and *Lestes* represent, in addition to *Calopteryx,* the major morphological variants of the zygopteran penis. Based on my investigations and the comparative studies of Schmidt (1915), Kennedy (1920), and Pfau (1971), there are four characteristic forms of the flexible distal segment of the penis among Zygoptera:

   1. A flap-like process with a pair of curved, horn- or blade-like appendages (Calopterygidae; *Calopteryx,* Fig. 4A).
   2. A flap-like process with no long appendages (many Coenagrionidae; *Enallagma,* Fig. 4D).
   3. No flap-like process but a single or forked, recurved, hook- or whip-like appendage, usually covered with proximally oriented spines (other Coenagrionidae; *Argia,* Fig. 4B, and *Ischnura,* Fig. 4E).
   4. No extensive flap- or whip-like appendage, but often a cup- or scoop-like configuration and one to several ventrally oriented ridges (Lestidae; *Lestes,* Fig. 4C).

These four morphological variants suggest widespread sperm displacement ability, since they seem to share with *Calopteryx* the scooping function of the distal flap and/or the insert/remove function of appendages bearing proximally oriented spines. What follows is a summary of current information on the sperm displacement capabilities and correspondence between male and female genitalic morphology of four representative genera of Zygoptera. This information is highly suggestive that sperm displacement is widespread among Zygoptera.

Male *Argia fumipennis violacea* appear to use the single hook-like process of the distal segment (Fig. 4B) to remove sperm from both the bursa copulatrix and spermatheca. This hook bears proximally oriented spines like the horns of *C. maculata* and it was found in the spermathecal tube of two in copula preparations (Fig. 5). All of the sperm was pulled out of the bursa copulatrix when the penis was removed and it remained attached to the spines on the hook. Dissections of a small sample (N=15) of specimens revealed that females carry stored sperm prior to mating, that this sperm is virtually all removed from the bursa copulatrix during copulations, and that postcopula females have maximal amounts of sperm. Thus sperm removal almost certainly occurs in *Argia.* Presently my data include too few post-tandem/precopula females to assess directly how much sperm a female carries prior to mating. However, interrupted copula specimens and females collected alone early in the day (assumed to be precopula) suggest there are substantial amounts.

Data for 30 pairs of *Lestes vigilax* (Fig. 4C) indicate that males remove or reposition 40-50% of the sperm of previous males (Waage 1982). The mean sperm volume carried by interrupted copula females was significantly less ($F = 9.05$; df = 1,27; $P < 0.01$) than those in pre- and postcopula females, which did not differ ($F = 0.044$, $P < 0.75$). In copula specimens indicate that some sperm is removed by the cup-like distal segment of the penis and the flanges on its ventral surface. Females carry large volumes of sperm (3+ times that of a male's sperm vesicle) in a single sac-like bursa copulatrix/spermatheca. It appears that males both remove and push aside previous sperm from the region of the bursa/spermatheca closest to the vagina. Sperm in this region are likely to have precedence since fertilization appears to occur at the junction between the vagina and bursa copulatrix (see Fig. 5A). This assumes relatively little mixing of sperm in the female's storage organs, which is probable, due to the entanglement typical of zygopteran sperm.

Of particular interest in *Lestes* is the correspondence between the absence of appendages on the distal segment of the penis and the lack of a well defined spermatheca in the female. In contrast, both *Calopteryx* and *Argia* have well defined spermathecae corresponding in number and size to the appendages on the penis (Fig. 4A,B).

Miller and Miller (1981) suggest *Enallagma cyathigerum* males can use the flap-like distal segment of the penis and the proximally oriented spines on its underside (Fig. 4D) to remove sperm from the bursa copulatrix. Like *Lestes*, *Enallagma* males lack appendages on the distal segment of the penis that might function directly to remove sperm from the single spermatheca. The volume of the *Enallagma* spermatheca is small relative to that of the bursa copulatrix, and removal of sperm from only the bursa may suffice to assure a male the majority of fertilizations even if previous sperm remain in the spermatheca.

Given the low mating frequencies of *Enallagma* species (Bick and Bick 1963, Johnson 1964a, Bick and Hornuff 1966, Parr and Palmer 1971, Fincke 1982), it is possible that males only occasionally encounter females carrying sperm. However, the costs to male reproductive success of not removing sperm, even if encountered only at low frequency, would still be great. Thus males should routinely go through the movements necessary to displace sperm while mating.

Fig. 4E shows the genitalic morphology of male and female *Ischnura verticalis*. Male penis morphology is very similar to that of *Argia f. violacea* except that the distal segment ends in a bifurcated, spiny appendage (as in *Argia sedula*). Females have a small bursa copulatrix and a rather large spermatheca joined to the bursa by a very narrow tube which seems too small for the male's distal segment to enter. Pinhey (1969) and Pfau (1971) illustrate in copula preparations of *Ischnura senegalensis* and *I. elegans,* respectively, in which the distal segment of the penis extends into the female's bursa copulatrix; and P. Miller (pers. comm.) has data suggesting sperm removal may occur in *I. elegans*. Although *Ischnura* appear to have penis morphology capable of removing sperm, at least from the bursa copulatrix, there are presently too few data to reliably assess their sperm displacement ability.

These general patterns of penis morphology and their correspondence with the structure of female storage organs, plus the demonstration of sperm displacement and its mechanism in *Calopteryx maculata, C. dimidiata, Argia f. violacea,* and *Lestes vigilax,* convincingly support the hypothesis that most male Zygoptera are capable of sperm displacement. Of course the hypothesis needs further testing and tests should be careful to distinguish between the ability or potential for sperm displacement and its actual occurrence.

B. Anisoptera: Preliminary Findings

A discussion of the potential for sperm displacement among dragonflies should begin by distinguishing two major differences between anisopteran and zygopteran morphology (Fig. 6). First, the anisopteran penis is a 4-segmented extension of the temporary sperm storage vesicle on the venter of the second abdominal segment. Sperm is translocated from the testes to an internal penis vesicle through a valve (v in Fig. 6) on the third penis segment. It is stored in the vesicle and transferred to the female through the penis (see below). Unlike sperm translocation in Zygoptera which occurs after tandem formation, in Anisoptera it can also occur prior to the male's having encountered a female. In the zygopterans examined, the entire contents of the sperm vesicle are transferred to the female during insemination (Hornuff 1968; Waage 1979a, unpubl. data). Comparable data are not yet available for anisopterans, although in specimens of the libellulids, *Celithemis elisa* and *Sympetrum rubicundulum,* penis vesicles were empty in postcopula males and contained sperm in those collected prior to copulation.

The second and more important difference between damselflies and many but not all dragonflies is that the distal segment of the anisopteran penis (epecially in the Libellulidae) is erectile and often contains a variety of chitinous and membranous processes (see Figs. 8 and 9). Except during copulation, the penis is in a relaxed state with the distal segment largely hidden. This makes it extremely difficult to determine the functions of the processes of the distal segment. In

A. ZYGOPTERA   B. ANISOPTERA

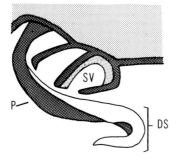 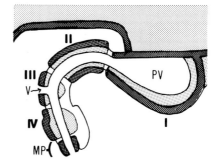

Fig. 6. Schematic showing the major morphological differences between male zygopteran (Coenagrionidae) and anisopteran (Libellulidae) genitalia. A. Zygoptera: DS = distal segment; P = penis shaft; SV = sperm vesicle. Sperm is transferred from the sperm vesicle to the female along a membranous groove on the dorsum of the penis (upper surface in figure). B. Anisoptera: I, II, III, IV = segments of the penis; MP = medial process of the distal segment; PV = penis vesicle; V = valve through which sperm is translocated to the penis vesicle prior to copulation. Dark shading represents chitinous structures; lack of shading represents flexible or membranous structures.

fresh specimens and some stored for several months in ethanol, it is possible to erect the penis and extend or inflate the processes of the distal segment. This is done by squeezing the basal segment (containing the penis vesicle) with forceps (see Miller 1981). Often sperm will be released when the penis erects. The drawings of erected penises in Figs. 8 and 9 were produced using this technique. It is not yet possible to determine if the same configurations of the processes on the distal segments occur when the penis is in the female's genital tract.

Libellulids have the most complex erectile processes (Kennedy 1922; Restifo 1972; Miller 1981, 1982b), while those of aeshnids and gomphids are fewer and less complex (Schmidt 1915, Pfau 1971). Figs. 7A and B illustrate the basic morphology of *Anax junius* (Aeshnidae) and *Gomphus submedians* (Gomphidae). Nothing is known about sperm displacement in these families and they are included here only to illustrate the range of male and female genitalic morphology in the Anisoptera. The distal segment of the *Anax junius* penis (Fig. 7A) terminates in a spongy, furrowed, bulbous pad covered with fine spines. It appears to be only slightly erectile in preserved specimens. The female has a large, spherical bursa and a pair of narrow spermathecal tubes. Unlike zygopterans and other anisopterans I have examined, the bursae of specimens preserved in ethanol are chitinous. As can be seen in Fig. 7A, the sizes of the distal segment of the penis and the bursa are comparable and one can hypothesize a removal function of the spongy pad, perhaps displacing sperm by its bulk.

Gomphus submedians (Fig. 7B) presents a unique structural feature, a long, very sharp, stiff, blade-like projection of the distal segment. This appears (from in copula dissections) to fit into a sheath-like sac in the female's bursal area. The female has a small bursa and a pair of elongated spermathecae. The region (I, Fig. 7B) above the blade-like projection of the male's distal segment may be erectile, but this has not been checked in fresh specimens. There is no obvious means by which this penis morphology would facilitate sperm removal or repositioning, unless erectile structures not yet seen are involved. However, the morphology of the spermathecae suggest the possibility of "last in, first out" sperm precedence (see Walker 1980).

Fig. 8 illustrates the stages in the erection of the penis of the libellulid, *Celithemis elisa*. Terminology for various structures on segment IV follow Kennedy (1922) and Restifo (1972). This sequence was produced by squeezing the basal penis segment (I) and may not accurately represent the positioning that occurs when the penis is in the female (see also Miller 1981). The structures on and in the segment IV are of three types: (1) heavily chitinized, rigid structures—lateral lobes, medial lobes, cornu; (2) spongy, partially inflatable, semi-rigid lobes—apical lobe; and (3) sac-like, inflatable, flexible lobes—internal lobes. In *Celithemis elisa* the lateral lobes (L), medial lobes (M), and cornu (C) are smooth, while the apical lobe (A) is covered with a dense mat of proximally oriented spines, giving it the texture of a cat's tongue. The internal lobes are smooth in *C. elisa,* but are often spiny in other libellulids (Kennedy 1922; Miller 1982b, pers. comm.).

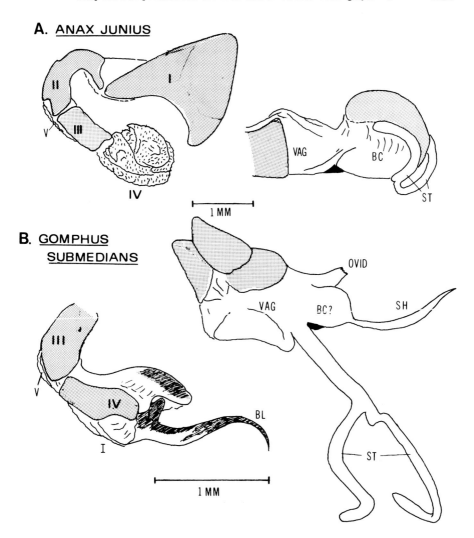

Fig. 7. Examples of anisopteran genitalic morphology. A. *Anax junius* (Aeshnidae). B. *Gomphus submedians* (Gomphidae). Male is on left and to scale with female. Orientation and terminology as in previous figures. BL = rigid, bladelike projection of the distal segment; I = possibly erectile region of the distal segment; SH = sheathlike process of the bursa; V = valve used in sperm translocation (see Fig. 6B).

The female genitalia of *C. elisa* (Fig. 9A) follow the general anisopteran format of a bursa copulatrix to which two spermathecal tubes are attached. There is considerable variation among anisopterans in the relative sizes of the bursa and spermathecae (*e.g.*, Figs. 7 and 9). Based on size measurement (in copula preparations have not been obtained for *C. elisa*) and demonstrated sperm displacement

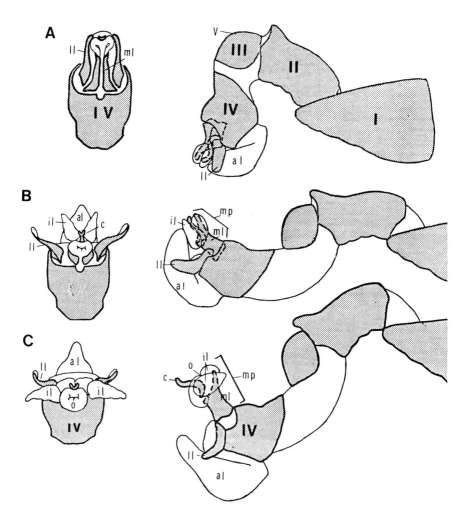

Fig. 8. Illustration of the erection of the penis of *Celithemis elisa* (Libellulidae) viewed from the side and top. A. Relaxed state. B. Extended state, distal segment not fully inflated. C. Fully erect and inflated. al = apical lobe; c = cornu; il = internal lobe; ll = lateral lobe; ml = medial lobe; mp = medial process; o = opening of sperm duct; v = valve used for sperm translocation; I-IV = segments.

ability (see below), I hypothesize that the expanded distal segment of the penis is within the bursal area during copulation. The valve between the vagina and bursa may be opened by structures (*e.g.*, lateral lobes) on the penis or by muscular movements of the female. In all libellulid females I have examined (8 spp.), there is a chitinous plate or tab (fl in Fig. 9) to which large muscles attach. These muscles may control the opening to the bursa during copulation and/or oviposition.

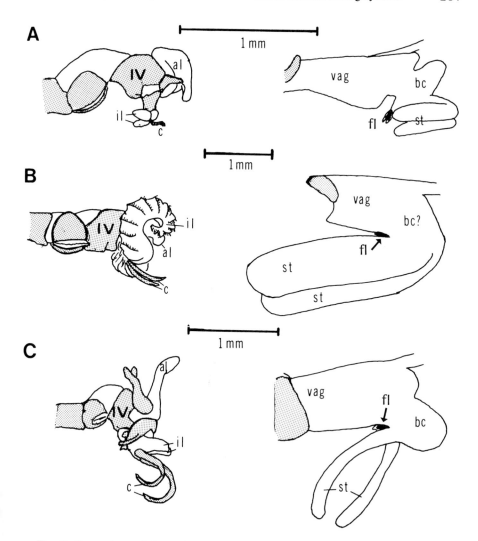

Fig. 9. Comparison of the male and female genitalia of three libellulids. A. *Celithemis elisa,* B. *Sympetrum rubicundulum.* C. *S. semicinctum.* Note the variations in the size and shape of female spermathecae (st) and the internal lobes (il) and cornu (c) of the males. Male and female genitalia to scale. Vag, bc, al, as in previous figures, fl = chitinous tab.

Sperm displacement occurs in *Celithemis elisa* (Waage, unpubl. data). The mean sperm volume for interrupted copula females was significantly less (F = 13.13; df = 1,31; P < 0.005) than those for pre- and postcopula females which did not differ (F = 4.15, P > 0.05). Precopula sperm volumes were estimated using females collected early in the day (prior to mating activity), and after oviposition. Egg counts in both these collections of females and in postcopula females revealed that early in the day females have full clutches and are likely to mate or attempt oviposition using stored sperm. Changes in sperm density within the bursa and spermathecae were also noted (although these were slight as judged by optical density of the sperm mass). Thus, it is unclear if sperm is physically removed or just packed into the back of the bursa and spermathecae; consequently, the exact magnitude of displacement cannot be determined accurately. Roughly 60% of the sperm in a postcopula female appears to belong to the last male to mate with her. Copulation lasts for mean + s.e. = 5.1 + 0.34 min (N=12), and rocking or pumping movements, similar to those of zygopterans, occur during the first 48.5 + 6.3 sec (N=11). The mechanism of sperm displacement in *Celithemis* is unknown since I have been unable to obtain in copula specimens. It is possible that the spiny apical lobe of the penis may be involved in removing or displacing sperm from the bursa. Males from postcopula, interrupted copula, and ovipositing pairs often had a thin crust of sperm on the distal segment of the penis, particularly on the ventral surface of the apical lobe or between it and the medial process. Miller (1981) has evidence for sperm displacement ability in *C. eponina* but sperm volumes have not been measured to confirm this.

Sperm displacement also occurs in the libellulid, *Erythemis* (*Lepthemis*) *simplicicollis* (Waage, unpubl. data). The copulation duration of this species is relatively brief (mean + s.e. = 19.8 + 0.6 sec, N = 273) and females in the population studied (near Lake Placid, Florida) mated several times (2.05 + 0.18 copulations/bout of oviposition, N = 63) per oviposition period. Data from 109 specimens revealed ca. 57% change in sperm volume from pre- to interrupted copula specimens and from interrupted copula to postcopula ones. The female genitalia differ from *Celithemis* in having a large bursa copulatrix and a small (< 10% of the bursa volume) spermatheca. The distal segment of the penis is correspondingly less complex.

Preliminary examinations of another libellulid, *Sympetrum rubicundulum* (Fig. 9B), suggest that it does not remove sperm but may reposition it. In this species the female's spermathecae are extremely large and the bursa small or nonexistent. There were no noticeable qualitative differences in the volumes of sperm carried by females of precopula, interrupted copula, and postcopula pairs (N=22). There are considerable variations in the density of the spermathecal sperm mass within and among these females. The volume of sperm in a typical female's spermatheca is at least 3-5 times as great as is contained in the male's penis vesicle. Thus, females appear to carry sperm accumulated from several matings. The male's penis in this species is quite unusual in that it has a pair of large inflatable internal lobes that are ringed with stiff bristles. In preparations inflated with forceps these lobes coil into a ram's horn configuration (Fig. 9B). However, I hypothesize that in the female this coiling movement is prohibited and instead the lobes expand into the spermathecal tubes as illustrated in Fig. 10. It seems possible that the function of these internal lobes and the bristles on them may be to pack existing sperm deeper

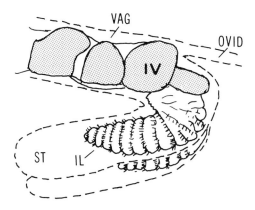

Fig. 10. Hypothesized in copula orientation of the inflated internal lobes of male *Sympetrum rubicundulum*. This schematic indicates a way in which males might use their inflatable internal lobes (IL) and the bristles on them to pack sperm of previous males into the back of the spermathecae (ST).

into the female's spermathecae. This might allow a male to compete effectively for fertilizations by moving rival sperm to the back of the spermathecae and filling the resulting space with his own.

There is considerable interspecific variation among libellulids in the size and morphology of both the distal segment of the penis and the bursa copulatrix and spermatheca (Fig. 9; see also Miller 1981, 1982b). This family is an important one for comparative studies of behavior and sperm displacement potential since there is also considerable variation within and among species in pair-forming and post-copulatory behavior (see below). Preliminary work by Miller (1981, 1982b) has revealed two major variants in libellulid penis morphology that may reflect removal, and repositioning or packing as alternatives for manipulation of sperm.

C. Alternative Functions for Odonate Penis Morphology

Sperm displacement is not the only possible function of the distal segment of the odonate penis. Natural selection has presumably shaped much of odonate penis morphology for the efficient transfer of sperm during mating. Given that the distal segment of the penis normally enters the bursa copulatrix during copulation, there seem to be four general, alternative functions for distal segment morphology (Table I).

First, the distal segment may facilitate sperm transfer to the female by opening the valve or junction between the vagina and bursa copulatrix. Although this function is consistent with some zygopteran morphology, it does not adequately explain the general presence of appendages with proximally oriented spines on the

TABLE I

Possible Functions of the Distal Segment of the Odonate Penis

A. AIDS IN SPERM TRANSFER
    1. Produces opening to bursa copulatrix or opens valve at vagina/bursa junction
    2. Produces partial vacuum in bursa to aid in drawing sperm into bursa

B. HOLDS GENITALIA TOGETHER DURING COPULATION

C. REPRODUCTIVE ISOLATION
    1. Provides a "lock and key" mechanism to prevent interspecific copulations

D. DISPLACES OR REMOVES SPERM
    1. Scoops or pulls sperm mass from bursa and spermatheca
    2. Moves or packs sperm mass further into bursa or spermatheca

distal segment or the spines on the venter of the distal flange of the penis (*e.g.,* *Enallagma,* Fig. 4D).

Some of the chitinous, non-spiny parts of the libellulid penis (*e.g.,* lateral lobes and cornu) and the blade-like process on the distal segment of the gomphid penis might serve to open the junction between the bursa and vagina. However, the erectile or semi-erectile lobes (internal lobes and apical lobe) probably do not serve an opening function since they inflate at the same time or after the chitinous structures have opened up, and they are usually covered or ringed with bristles or spines. The libellulid penis also contains an internal sperm pump and injector system in the distal segment which facilitates insemination (Pfau 1971).

The second possible function, holding genitalia together during copulation, is an improbable one for the zygopteran penis. The distal segment moves between the vagina and the bursa during the rocking or pumping movements of copulation (Fig. 2D). Thus, despite its "hook-like" appearance, the recurved distal flap and its appendages do not hold the genitalia together. The distal segment is in fact quite flexible and (in dissections of in copula pairs) is often extended distally (Miller and Miller 1981; Waage, unpubl. data). Miller and Miller (1981) found that the penis shaft is held between the female's ovipositor blades such that it is able to move between the vagina and bursa but is not easily withdrawn. This can usually be demonstrated by pulling the ovipositor blades apart, which allows the penis to slip or be pulled out easily. Clearly, the distal segment does not have a clasping function in Zygoptera.

Since some anisopterans also exhibit rocking or pumping motions during copulation, it is possible that the distal segment moves between the vagina and bursa, such that genitalic clasping is an unlikely function in these species as well. However, this needs confirmation since libellulids could hypothetically use some parts of their distal segment morphology to hold the penis within the bursa while effectively moving other parts by inflating and deflating them. At present, the best argument against the "genitalic holding" function is the evidence that this is accomplished instead by secondary structures of the male's copulatory complex (*e.g.*, hamules) that grasp the female's genital plate or ovipositor valves (Pfau 1971, Johnson 1972, Miller 1981).

A third function is reproductive isolation between sympatric species. There are two possible morphological "lock and key" barriers to interspecific copulation for odonates (Johnson 1962b, Watson 1966, Paulson 1974). The first is the fit between male anal appendages and female prothoracic and head morphology that is involved in forming and maintaining the tandem position, and the second is the fit between the structures of the male's copulatory complex on the venter of his second abdominal segment, including the penis, and the female's external and internal genitalia. Interspecific and occasionally intergeneric tandems and copulations are known for a variety of odonates (*e.g.*, Corbet 1962, Waage 1975), but there is also good evidence for species discrimination and mechanical isolation involving anal appendages (Loibl 1958, Corbet 1962, Paulson 1974, Tennessen 1975, Robertson and Patterson 1982). I know of no experimental demonstration of mechanical isolation in Odonata involving genitalia (but see Watson 1966). Indeed, comparisons of zygopteran male and female genitalic morphology indicate rather minor variations in shape among species within many genera and major differences, if present, are usually in size rather than structure. Thus, there is little evidence that reproductive isolation is a general function for zygopteran genitalia. Among Anisoptera, the "lock and key" isolating function also seems unlikely for the erectile lobes of the distal segment. However, it remains an hypothesis to be tested for the chitinous lobes and particularly for structures such as the hard, sharp blade-like processes of the gomphid penis.

It appears that the fourth possible function, sperm removal or repositioning, is the one most compatible with the basic structure of the zygopteran penis and supported by experimental evidence. The orientation of the distal flange, its appendages, and the proximally oriented spines on both are as they should be for removing sperm. Evidence that the penis moves between the vagina and bursa copulatrix during copulation, that appendages of the distal segment of the penis enters the spermatheca, and that sperm clumps are found on the penis of in copula males also support the sperm displacement function for Zygoptera.

At least some of the morphology of the distal penis segment of the libellulids and perhaps some aeshnids seems likely to function in sperm displacement. For others (*e.g.*, gomphids), if sperm competition occurs it might involve precedence by the sperm of the last male, since the elongated spermatheca could prevent mixing of sperm masses deposited by successive males. At present the only evidence for sperm displacement in Anisoptera are the data presented above for *Celithemis elisa* and *Erythemis simplicicollis*, and additional work on this suborder is needed before general statements can be made.

D. Sperm Competition and the Female Perspective

The focus to this point has been on male penis morphology and the potential for sperm displacement. Females are not neutral parties in the evolution of sperm competition among males. Selection on females is expected to influence the structure and function of sperm storage organs as well as the tendency of females to mate more than once. Unfortunately, little is known about the functional morphology of odonate female genitalia. They are encased in muscle and other tissues and are in immediate contact with the ganglion of the 8th abdominal segment. It seems likely that a female has some control over the size and shape of the sac-like bursa copulatrix and spermatheca, and should be able to control movement of sperm from the bursa to the vaginal area for fertilization of eggs during oviposition. It is possible that females can also shift the sperm mass between bursa and spermatheca and perhaps expel some or all of it (see Lloyd 1979).

Males of some species may not be able to remove sperm from the spermathecae of females due to the size, shape, or location of these organs (*e.g., Enallagma, Gomphus,* and *Sympetrum*). This raises the question of how selection for female morphology and behavior influences sperm competition among males (see also Walker 1980). Evidence of female morphology that prevents sperm removal, re-positioning, precedence, or dilution should be looked for.

A particularly relevant question is why females should mate more than once per lifetime or per oviposition event (see Walker 1980; Knowlton and Greenwell, this volume). Two important features of odonate biology must be kept in mind when considering reasons for multiple matings by females. First, all zygopteran and probably most anisopteran females cannot be forced to mate. The female must take an active role in bringing her genitalia into contact with the male's to initiate copulation. A possible exception are those libellulids that mate rapidly in mid-air. Here the male may be able to flip the female's abdomen forward and upward to enter copula by aerial maneuvering. Second, odonates continually mature eggs throughout their lives, ovipositing at several day intervals. This means multiple mating by females can occur on two time scales: between and within oviposition periods.

While some odonates may mate only once per clutch or several clutches, others are known to mate several times per clutch. For example, both *Calopteryx maculata* (Waage 1978, 1979b) and *Erythemis simplicicollis* (McVey 1981, Waage, unpubl. data) females have been observed to mate up to 5-8 times during a single visit to the water. Tethered females of these species can usually be mated several times without intervening oviposition (Waage 1979b, unpubl. data).

Those factors discussed by Walker (1980) that offer possible explanations for multiple mating by odonate females are: (1) sperm replenishment; (2) utilization of sperm or seminal fluids for nutriment; (3) increased genetic diversity of offspring; (4) genetic superiority of the last male; and (5) reduction of time and energy loss from male harassment. Two other factors can be added for odonates:

(1) mate choice with fertility assurance; and (2) access to oviposition sites controlled by males.

In all species I have studied except *Enallagma aspersum,* females collected before mating carry substantial sperm loads, even exceeding the amount of sperm a single male can provide in *Lestes vigilax* and *Sympetrum rubicundulum.* I have found that postoviposition females of *Argia sedula, Calopteryx maculata, Celithemis elisa,* and *Erythemis simplicicollis* still have substantial sperm reserves. Thus, odonate females may often arrive at oviposition sites or mating areas with substantial sperm reserves from a previous mating. In addition, there is evidence that some odonate females might be able to fertilize several successive clutches with sperm stored from a single mating. Grieve (1937) obtained five clutches of fertile eggs (N=1650) from a mated *Ischnura verticalis* female kept in a laboratory cage for 34 days. While sperm replenishment or, more importantly, replacement with fresh sperm might favor females who mate several times during their lifetime, it probably would not favor multiple matings per oviposition period.

Multiple matings and sperm storage by female odonates might also be advantageous if some of the stored sperm or its associated seminal fluids can be used as a nutritional source (Friedel and Gillott 1977; Gillott and Friedel 1977, Boggs and Gilbert 1979; Sivinski, this volume). Histological and biochemical studies of odonate spermathecae should reveal if this organ is involved in long-term sperm storage or in digestion and absorption of sperm and seminal proteins.

Females might benefit from multiple matings during the same or successive oviposition bouts by increasing the genetic diversity of their offspring. While intuitively pleasing, this benefit depends on numerous variables and is difficult to test. It also requires that sperm displacement by males be incomplete or that females oviposit part of a clutch between matings with different males.

Depending on the heritability of male courtship or pair-forming behavior, females might benefit from remating if they encounter a superior male, especially if that male displaced the sperm of previous mates. Females might also benefit from multiple mating with sperm displacement in the context of mate choice with fertility assurance (D. Fernandes, pers. comm.; see also Lloyd 1979). Mating with the first male encountered at least assures fertility. The mated female can then search for superior males (or better oviposition site) who, if and when found, will pre-empt the sperm of the less desirable male.

Whenever male density is high at or near oviposition sites, females may benefit from remating, simply to minimize harassment by these males. In addition, females may mate with males to gain access to the oviposition sites they control. Both circumstances appear to occur for *Calopteryx maculata* females (Waage 1978, 1979b). *Calopteryx* females often mate despite having nearly full sperm storage organs (Waage 1979a, b) when moving to new oviposition sites in another male's territory. They usually attempt to oviposit in a nonmate's territory without mating, but this is only possible when the male is already guarding a previous mate. The

2-3 min of courtship and copulation involved in this otherwise unnecessary mating seem to be an acceptable time and risk cost for the benefit of being guarded by the new mate for 10-15 min of oviposition (Waage 1978). Attempts by females to oviposit in a new male's territory without remating suggest at least some cost to mating, and that females are not utilizing multiple matings to increase genetic variability among offspring.

In summary, there are a number of possible reasons for multiple mating by females on both intra- and interclutch time scales. Since female behavior can directly influence the dynamics of male sperm competition, it is important that future odonate studies give greater consideration to identifying and exploring the significance of female perspectives, particularly those directly linked to reproductive success and potentially in conflict with male perspectives.

V. SPERM COMPETITION
AND ODONATE POSTCOPULATORY BEHAVIOR

This section presents an overview of odonate postcopulatory behavior and explores the link between it and sperm competition. There are two possible approaches: assume sperm displacement is universal among odonates and ask how sperm competition is lessened or avoided by postcopulatory interactions; or assume the extent of sperm displacement among odonates is unknown and ask what known patterns of postcopulatory interactions suggest about its presence or absence. Although ability to displace sperm seems widespread among odonates, especially Zygoptera, I will take the second approach since it provides greater opportunity to produce testable predictions about sperm displacement ability and the risks of sperm competition.

Two comments are necessary before proceeding. First, it is well to consider the complex array of factors besides sperm competition that could influence postcopulatory behavior (Fig. 1). Of particular importance are population density and habitat structure (Pajunen 1966, Ueda 1979). Second, even if odonate males do not remove or reposition each other's sperm, there may still be mixing or dilution effects from take-overs of mates. Thus sperm competition, at least at this minimal level, is likely to be a general selective force if males regularly mate with previously mated females.

A. Major Patterns of Odonate Postcopulatory Interactions

The following classification of odonate postcopulatory interactions is based largely on discussions by Heymer (1968), Sakagami et al. (1974) and Schmidt

(1965, 1975). Bick (1972) and Corbet (1962, 1980) have stressed the difficulty of stereotyping odonate postcopulatory behavior, due to its inherent plasticity. In fact the considerable variation within and among species for these behaviors should provide opportunities to test some of the ideas discussed below.

Table II outlines the principal categories of postcopulatory and oviposition behavior for odonates and provides an indication of their distribution among the major temperate zone families. These behaviors are illustrated in Fig. 11 for Zygoptera and Fig. 12 for Anisoptera.

Tandem oviposition and noncontact guarding involve direct actions of males which suggests they were evolved primarily in the context of sperm competition. A third category involves largely female determined behaviors that may be influenced by male activity patterns. In this situation, females are unlikely to encounter males, or possess the ability to avoid male disturbance during oviposition, with the result that potential for sperm competition is reduced. Two other explanations for behaviors in this third category, that sperm displacement does not occur, or that females are unlikely to remate, should be distinguished.

Tandem oviposition is the predominant postcopulatory behavior pattern of temperate zone zygopterans (Fig. 11), absent only in the Calopterygidae and in many species of *Ischnura*. In the Anisoptera it is common only among libellulids

TABLE II

Variations of Odonate Postcopulatory Behavior.
Major temperate zone families are included: CALOP = Calopterygidae;
LEST = Lestidae; COEN = Coenagrionidae; LIBELL = Libellulidae (including Corduliidae),
AESHN = Aeshnidae; OTHERS = Gomphidae, Cordulegasteridae, Petaluridae.
VC = very common; C = common; UC = uncommon; R = rare; -- = absent.

POSTCOPULATORY BEHAVIOR	ZYGOPTERA			ANISOPTERA		
	CALOP	LEST	COEN	LIBELL	AESHN	OTHERS
A. OVIPOSITION IN TANDEM	–	VC	VC	C	R	R?
B. NON-CONTACT GUARDING	VC	–	–	VC	--?	–?
C. FEMALE OVIPOSITS ALONE						
1. Occasional male/female encounters	–	–	–	–	VC	VC
2. Spatial or temporal separation	?	–?	*	(UC)?	C?	?
3. Effective refusal or avoidance behavior	R?	–	*	(UC)?	C?	?
4. Rapid or secretive oviposition	–	–	–	(C)	C	?
5. Submerged oviposition	UC(C)	R	(C)	–	--	–

*A number of ischnurans exhibit this behavior.
( ) Secondary behaviors that follow category A or B oviposition.

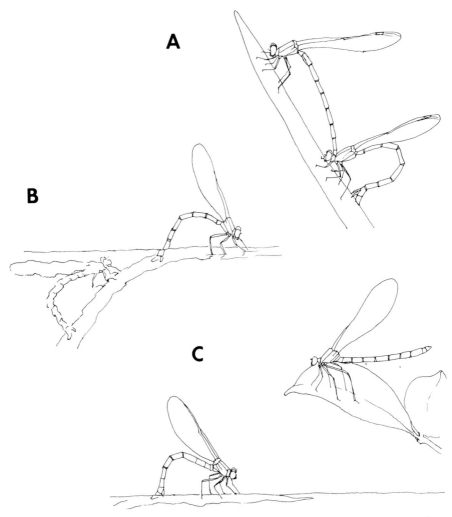

Fig. 11. Postcopulatory behavior of Zygoptera. A. Tandem oviposition at the water surface. B. Solitary oviposition by females at and below the water surface. C. Non-contact guarding. See Table II and text for discussion.

(Fig. 12C), but occurs in a few aeshnids as well (Fig. 12A). A period of oviposition in tandem may be followed by the female continuing unattended by her mate (Fig. 11B). This is particularly true of Zygoptera species in which females usually oviposit while submerged. Pairs ovipositing in tandem usually do not localize within a male's territory or the area where pair formation occurred, and several sites may be utilized during a bout of oviposition. Oviposition is often communal

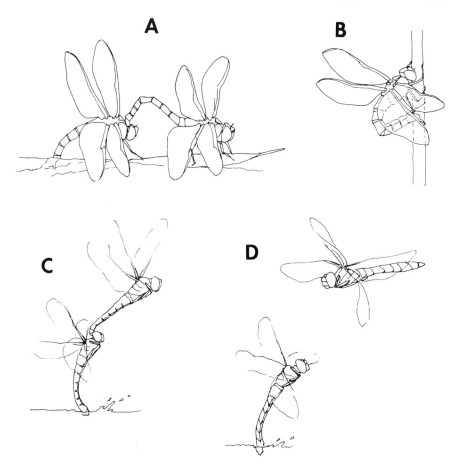

Fig. 12. Postcopulatory behavior of Anisoptera. A. Tandem oviposition (endophytic) while perched. B. Solitary oviposition (endophytic). C. Tandem oviposition (exophytic in flight. D. Non-contact guarding by hovering male. Solitary oviposition also occurs exophytically (*e.g.,* as in D, but without the male). See Table II and text for discussion.

and, among endophytically ovipositing zygopterans, may last from 10 min (*e.g., Pseudagrion,* Furtado 1972) to several hours (*e.g., Lestes,* Loibl 1958). In those exophytically ovipositing anisopterans for which data are available, tandem oviposition in flight (Fig. 12C) lasts for 3-10 min (*Diplacodes, Pantana,* and *Tramea,* Sakagami *et al.* 1974; *Sympetrum,* Ueda 1979; *Celithemis,* Waage, unpubl. data).

Non-contact guarding (Figs. 11C, 12D) occurs in the behaviorally advanced Calopterygidae (Zygoptera) and Libellulidae (Anisoptera). While guarding an ovipositing mate, the male perches nearby (Fig. 11C) or hovers above (Fig. 12D) the female and drives off approaching males. This pattern is usually associated with territorial behavior and relatively frequent, rapid matings (less than 3-5 min). Of 36 libellulids for which there are sufficient data on postcopulatory behavior, 22 exhibit non-contact guarding as their primary mode of oviposition, 7 oviposit in tandem, and in 7 the female oviposits alone. Durations of non-contact guarding range from 10-15 min in endophytically ovipositing *Calopteryx* (Waage 1978) and from 10 sec to 5 min in exophytically ovipositing libellulids (*Libellula luctuosa*, Campanella 1975; *Plathemis lydia*, Jacobs 1955; *Hemicordulia ogasawarensis*, Sakagami *et al.* 1974; *Cordulia anea amurensis*, Ubukata 1975).

In some coenagrionids, a variation of non-contact guarding occurs. Males that have released their submerged ovipositioning mates often perch on vegetation above the female. They fly at other males, and attempt tandems with any re-emerging females. O. Fincke (pers. comm.) has shown that by doing this, *Enallagma hageni* males effectively reduce the probability of take-over of re-emerging mates. This variation of guarding differs in degree from the more direct protection of mates ovipositing above water.

Oviposition by solitary females (Figs. 11B, 12B) involves one or more of four basic patterns. First, when oviposition and pair formation sites are separate or when they occur together but are widely dispersed, females may oviposit without male interference or with only occasional discovery by searching males. Many species exhibiting this behavior are riverine. These include wide-ranging and rather secretive species of Aeshnidae, Cordulegasteridae, Gomphidae, and Petaluridae (Corbet 1962, Heymer 1973). Unlike pond or lake habitats, rivers and streams allow searching males to view only a limited area of open space at one time. Males of these species wait in open, sunny areas for females moving up or downstream (*e.g., Onychogomphus forcipatus*, Kaiser 1974a) or patrol long stretches of stream searching for resting or ovipositing females. Little is known about the frequency and outcome of male-female encounters in these large, difficult-to-study species. Some knowledge of the relationship of unaccompanied egg laying to the potential for sperm competition in this primitive group may contribute to a general understanding of the evolution of anisopteran mating systems.

Second, females of some species may oviposit alone at times other than those of peak male activity. This usually involves a return of females to mating/oviposition areas late in the day (Sakagami *et al.* 1974, Schmidt 1975) but may also include females that return after the seasonal peak of mating activity, as in *Aeshna nigroflava* (Ubukata 1974); or after maturation, as in *Ischnura aurora* which mate while teneral (Rowe 1978). This temporal delay in oviposition appears to depend somewhat on male density at the oviposition site (*e.g.,* Corbet 1962, Hassan 1978) as would be expected if reduction of male interference were important. Such density-dependent changes in oviposition behavior clearly warrant further investigation in the context of potential for sperm competition.

Third, females may oviposit alone in areas of male activity if they have effective avoidance or refusal behaviors or if they can oviposit rapidly. Among Zygoptera this pattern seems to occur in the genus *Ischnura* where females of most species studied oviposit on their own and are able to repel approaching males with "wing warning" and "abdominal curving" displays (Bick 1966). Similar refusal displays have been recorded for aeshnid females that normally oviposit unattended (Corbet 1962). Refusal displays by females occur in other species as well (Pajunen 1963b, Waage 1973), but experimental evaluations of their effectiveness are needed.

Exophytic oviposition, characteristic of most Anisoptera, allows females to rapidly deposit a number of eggs. For example, estimates of egg deposition rates in Zygoptera are generally in the range of 2-10/min (Bick *et al.* 1976, Waage 1978) but are as high as 1300/min in the libellulid *Plathemis lydia* (McVey 1981). Some libellulids seem capable of rapid oviposition in areas of high male activity by furtive forays over oviposition sites and by dropping motionless to the substrate or flying temporarily away from the water when approached by males (*e.g.,* Jacobs 1955; Pajunen 1963a, 1966).

Finally, submerged oviposition, occurring only in the Zygoptera, effectively protects the female from interference. It occurs sporadically in most families, often following an initial period of tandem or guarded oviposition above water. Submerged oviposition sometimes lasts more than an hour and may incur considerable energetic costs and predation risks. Males in tandem rarely accompany their mates underwater for more than a few minutes, but may remain at the surface and attempt tandems with emerging females (*e.g.,* Bick and Bick 1963, Fincke 1982). Thus males avoid the energetic costs and risks while apparently continuing to guard submerged ovipositing mates.

It appears that all three classes of odonate postcopulatory behavior can provide some degree of protection from disturbance or take-over of ovipositing females. Tandem oviposition and non-contact guarding represent the primary postcopulatory behaviors for virtually all zygopterans and libellulids, which include the majority of odonate species. All of the odonates known to displace sperm either oviposit in tandem, or guard mates during oviposition.

B. Adaptive Significance of Tandem and Guarded Oviposition

It has generally been thought that oviposition in tandem or male guarding functions to protect females from disturbance by males (*e.g.,* Corbet 1962). This assumes that female time and energy budgets are such that delays in completion of oviposition and increased risk of death prior to completing oviposition would result in lowered fitness for unprotected females. Such time and energy budget restrictions have not been demonstrated in odonates.

More recently, the influence of sperm competition has been recognized (Parker 1970, Sakagami *et al.* 1974, Ueda 1979, Waage 1979b, Corbet 1980). This second function has been demonstrated for *Calopteryx maculata* (Waage 1978, 1979a, b). Here guarded females oviposit 5-10 times as long as unguarded ones and, except at high densities, non-contact guarding prevents take-overs.

Before attempting to distinguish between the two postulated functions for tandem and guarded oviposition, it is instructive to consider two primary disadvantages to postcopulatory male and female association. First, oviposition puts females and accompanying males at considerable risk of capture by aerial (other odonates, asilids, birds), terrestrial (hunting spiders), and aquatic (frogs, fish) predators (e.g., Corbet 1962, Jacobs 1955, Waage 1979b). This is particularly true for tandem pairs at the water surface. Second, the ability of males and females to find and mate others is reduced or eliminated during both non-contact guarding and tandem oviposition. As discussed below, the choice of oviposition sites by each sex and the male's ability to defend his territory may also be compromised by postcopulatory association, especially in tandem.

If species can be found that show non-contact guarding or tandem oviposition but are unable to displace sperm, then the proposed functions (minimizing interference, reducing sperm competition) can clearly be separated. Male disturbance of ovipositing females should also be less when sperm displacement does not occur because males have little to gain from attempting matings with already inseminated females.

Conversely, male harassment should be greater when sperm displacement occurs. Therefore, males protecting females from sperm displacement will also be protecting them from disturbance. This makes it difficult to distinguish between function and effect when sperm displacement occurs. Ecological factors such as male density and habitat structure influence the level of disturbance, making it difficult to factor out the importance of the two hypotheses by comparing levels of male harassment.

It would seem likely that male and female postcopulatory behavior should adjust to density and frequency dependent changes in population size and behaviors of others. I have not found evidence for such adjustments in manipulative studies of postcopulatory guarding by *Calopteryx maculata* (Waage 1979b). However, there are a number of studies and observations on other species suggesting that such adjustments of behavior may occur. Hassan (1978), Jacobs (1955), Krüner (1977), and Pajunen (1963a, 1966) have noted that changes in postcopulatory behavior and the risk of take-over may be associated with variations in male density and habitat structure. Ueda (1979) found two male mating tactics, "territorial" and "wandering," in *Sympetrum parvulum*. Wandering males searched for females away from the water and would either remain in tandem with a mate brought to the water or non-contact guard her during oviposition. Territorial males always exhibited non-contact guarding of their mates. Of particular significance was Ueda's observation that wandering males were more likely to oviposit in tandem at high densities, suggesting that as the risks of take-over increased, males intensified their guarding tactics. That territorial males did not also switch to tandem oviposition at high densities is intriguing. Ueda (1979) suggests they were better at repelling other males and that non-contact guarding was adequate and female encounters sufficiently frequent to permit this behavior.

Another way to explore the relative significance of the two hypotheses for evolution of postcopulatory protection of mates is to work out the phylogenetic sequences involved in the origins of sperm displacement ability, multiple mating by females, and mating systems that increase the chances of disturbance of ovipositing females. If sperm displacement is universal among odonates and occurs by homologous mechanisms, then protection from disturbance must derive largely as a by-product of protecting mates from take-overs and sperm displacement. If, however, sperm displacement occurs among scattered taxa, then postcopulatory protection probably evolved in most odonates in the context of reducing disturbance of ovipositing females (which in turn selects for paternity assurance via sperm precedence).

Phylogenetic evidence is, at this point, only suggestive. Among Zygoptera, tandem oviposition is nearly universal and non-contact guarding occurs primarily in behaviorally advanced groups, particularly the Calopterygidae. This later group shares with the libellulids, among the Anisoptera, the most complex genitalic morphology. Sakagami et al. (1974) concluded that, for libellulids, tandem oviposition and non-contact guarding have been derived from solitary oviposition. Nothing is known about the sperm displacement ability of other Anisoptera, most of which lack any postcopulatory association. Thus the possibility that sperm displacement and postcopulatory protection are nearly universal among zygopterans suggests that the sperm competition hypothesis best explains their postcopulatory behavior. For the Anisoptera, the significance of postcopulatory interaction appears more complex.

C. Adaptive Significance of Non-contact Guarding

Since tandem oviposition is generally a more effective means of protecting mates than non-contact guarding, the question arises: What are the benefits to non-contact guarding that explain its existence among a number of odonates? Three hypothetical answers are: (1) it allows males (and/or females) to take advantage of additional mating opportunities during oviposition; (2) it allows males to defend and maintain territories while guarding mates; and (3) it lowers the risk to males of predation during oviposition.

Remaining in tandem effectively prevents both the male and female from re-mating should the opportunity arise during oviposition. Alcock (1979) and Ueda (1979) have argued that non-contact guarding is advantageous for males since it frees them to take advantage of mating opportunities that may occur while a previous mate is ovipositing. The male might then be able to protect the original female and his new mate(s). This of course assumes that his original female will not be remated while he is attempting pair formation or copulating with a new female, or that the gain in eggs fertilized by mating and guarding a new female will exceed any loss of fertilized eggs via the original mate if she is taken over or disturbed in the process.

Alcock (1979) studied a low density population of *Calopteryx maculata* and found that guarding males would mate with additional females arriving at their territories and were able to protect their original mates, and ultimately the new mates. My experiments on two populations of *C. maculata* under more crowded conditions showed that males only rarely attempted additional matings while guarding mates (for an average of 15 min; Waage 1979b). In these populations, males that attempted additional matings risked losing previous mates to other males, and had a low probability of acquiring additional matings with arriving females. The differences in our results probably reflect behavioral responses to changes in population density.

Remating while guarding has only been reported for *Calopteryx maculata* (Alcock 1979, Waage 1979b) and for three libellulids: *Perithemis tenera* (Jacobs 1955), *Plathemis lydia* (Jacobs 1955; but not by Campanella and Wolf 1974), and *Sympetrum parvulum* (Ueda 1979; but not for the population he studied). These species have short copulation times, relative to other species in the same families. This makes multiple mating without loss of previous mates possible under certain male densities provided males can distinguish among mates and nonmates. Males of both *Calopteryx maculata* (Waage 1979b) and *Perithemis tenera* (Jacobs 1955) have been observed making mistakes and remating the same female or guarding nonmates.

Guarding of nonmates has also been observed in the species listed above (Jacobs 1955, Alcock 1979, Waage 1979b) as well as in *Libellula pulchella* (Pezalla 1979) and *Erythemis simplicicollis* (McVey 1981, Waage, unpubl. data). Therefore, guarding of more than one ovipositing female cannot be taken as reliable evidence of multiple mating while non-contact guarding, and indicates that non-contact guarding males do not or cannot always take advantage of multiple mating opportunities.

Since male territories represent their primary pair-forming sites (Corbet 1962, 1980; Johnson 1964a; Waage 1973) and territorial males generally have a mating advantage over males without territories (see below), non-contact guarding males might benefit from guarding both an ovipositing mate and their territory. Such territorial defense is impossible while the male is in tandem with an ovipositing female. This assumes that a male can defend his territory as well as the ovipositing female, or that defending the territory is more important to the male than defense of a single mate.

All odonates known to exhibit non-contact guarding are territorial. Males of some species defend a given territory for only part of a day (*e.g.*, *Acisoma panorpoides*, Hassan 1978; *Leucorrhinia dubia*, Pajunen 1963; *L. intacta*, E. Waltz and

L. Wolf, unpubl. data; *Pachydiplax longipennis,* K. Sherman, unpubl. data) while others return to a given area on successive days (*e.g.,* up to 8 in *Calopteryx maculata,* Waage 1973).

Greater mating success of males with territories and of males at certain territories has been demonstrated in the following non-contact guarders: *Calopteryx maculata* (Waage 1973, 1979b), *Erythemis simplicicollis* (McVey 1981), *Leucorrhinia intacta* (E. Waltz unpubl.), *Libellula pulchella* (Pezalla 1979), *Perithemis tenera* (Jacobs 1955), *Plathemis lydia* (Campanella and Wolf 1974, Jacobs 1955), and *Sympetrum parvulum* (Ueda 1979). Thus males who obtain and then lose territories are likely to suffer lower reproductive success than those who maintain them. Non-contact guarding therefore allows a male to defend two important resources: an ovipositing mate and means of future access to mates.

Males (and females) that oviposit in tandem have a risk of being predated. This risk may be high, especially in exposed habitats where escape would be hampered by tandem flight. Non-contact guarding males decrease this liability by hovering above the ovipositing female or perching above the air/water interface.

Of the three functions proposed for non-contact guarding, risk of predation is the most difficult to verify because predation events are rarely observed. Comparative and manipulative studies may be useful in testing the other hypotheses, especially with libellulid species which exhibit considerable intra- and interspecific variation of postcopulatory behavior.

VI. ODONATE MATING SYSTEMS

The discussion to this point can now be set in the broader context of odonate mating systems. Information for this comes largely from the reviews by Corbet (1962, 1980), Johnson (1964b), St. Quentin (1964), Schmidt (1965, 1975), Heymer (1968, 1973), Bick (1972), and Sakagami *et al.* (1974). Table III provides overviews of the three principal categories of odonate mating systems. These range from infrequent, basically chance encounters at widely scattered oviposition or feeding areas to frequent, localized encounters at oviposition sites within male territories. Two major determinants of the variations in odonate mating systems appear to be the dispersion and defensibility of oviposition sites, and population density. In most species, pair formation occurs at or near oviposition sites (Corbet 1962, 1980). Males aggregate in thse areas and await females ready to oviposit. The degree of male localization and agonistic behavior therefore depends on the degree to which oviposition sites are clumped and on population density.

Where oviposition sites are widely dispersed and females are unlikely to be concentrated in predictable areas, pair formation is basically opportunistic

TABLE III

Overview of Odonate Mating Systems

A. OPPORTUNISTIC ENCOUNTERS BY SEARCHING, NON-LOCALIZED MALES
1. Few, long (30-300 min) copulations per lifetime
2. No courtship, little localization, low aggression
3. Tandem oviposition (rare) or no postcopulatory association
4. Rare among Zygoptera, common in several anisopteran families

B. GENERALLY LOCALIZED ENCOUNTERS AT OR NEAR OVIPOSITION SITES
1. Several copulations (15-60 min) per lifetime or day
2. No courtship, varying degrees of localization and aggression, males perch and wait near oviposition sites or patrol
3. Oviposition in tandem or (uncommon) no postcopulatory association
4. Most common mating system for Zygoptera and Anisoptera

C. ENCOUNTERS AT TERRITORIES LOCALIZED AT OVIPOSITION SITES
1. One to many, short (5 sec-15 min) copulations per day
2. Display and courtship common, well developed territoriality
3. Non-contact guarding in male's territory, or (uncommon) tandem oviposition. Female may continue alone within or outside of male's territory
4. Limited mostly to Calopterygidae and Libellulidae

(Table IIIA). Males search for resting, feeding, or ovipositing females or wait for them in sunny areas along streams or in forests. Mating durations are long and, as expected with infrequent encounters, there is little if any postcopulatory interaction.

At the other extreme (Table IIIC), when oviposition sites are more localized and defensible, there is a tendency toward territorial behavior and frequent, short copulations. Mates are non-contact guarded in the male's territory (at least originally; see Alcock 1982). This system is exhibited by calopterygids and libellulids, and it is here that sexual selection in the context of intrasexual competition is likely to have had its strongest influence.

Intermediate between these extremes are species (Table IIIB) utilizing somewhat localized but usually not defensible oviposition sites. For example, a moderate sized pond has, for exophytically ovipositing odonates, essentially one continuous oviposition site, its surface. Here, males may localize in and occasionally defend segments of shoreline, while waiting for incoming females. Variants of this system include male patrolling, transient visitation at one or more ponds (*e.g.,* some aeshnids, Kaiser 1974b), and interception of females away from the water as they approach (*e.g., Celithemis elisa,* Waage, unpubl. data). Males of these species are often not territorial and may only defend perch sites or make agonistic approaches to other patrolling males. Matings are less frequent and longer than for species in category C (Table III). Oviposition is generally in tandem and not localized to the site of pair formation. Occasionally these species exhibit non-contact guarding and solitary oviposition. The majority of damselflies and dragonflies employ this intermediate type of mating system.

Of particular interest in the context of sperm competition is the tendency toward more frequent, short coupling and increased specialization of postcopulatory behavior (*e.g.*, non-contact guarding) as males become more localized at clumped oviposition sites. At these sites sex ratios are strongly male-biased (*e.g.*, Waage 1980), female arrivals are more predictable and opportunities for multiple matings increase. It is among these species that sperm competition is expected to be most intense. Although this has been verified in *Calopteryx maculata* (Waage 1978, 1979a, b, 1980), its generality needs further investigation. Of major importance are territorial libellulids that copulate only 3-15 sec. The general reproductive behavior of a number of these species has been described (Campanella 1975, Campanella and Wolf 1974, Hassan 1978, Jacobs 1955, Johnson 1962a, McVey 1981, Parr and Parr 1974, Pezalla 1979) but, so far no attempts have been made to determine if they displace sperm or if postcopulatory guarding is adaptive in the context of sperm competition.

Another group in need of study are damselflies of the genus *Ischnura*. These species have the longest copulations (up to 3-5 hr) of any odonates and generally lack postcopulatory association. Penis morphology and preliminary evidence suggest these species may be capable of sperm displacement, but the tendency of females to oviposit alone and effectively repel approaching males suggests otherwise. Grieve (1973) maintained a mated *Ischnura verticalis* female in the laboratory that continued to produce fertile eggs for the 34 days it survived. This and evidence from natural populations (Parr and Palmer 1971) suggest *Ischnura* females may mate only once per lifetime. Since it occupies the same habitats and is closely related to other species with category B (Table III) mating systems that oviposit in tandem, *Ischnura* is a key genus for the understanding of sperm competition and its relation to mating systems in Zygoptera.

VII. SUMMARY

This paper presents and begins to evaluate the hypothesis that sperm competition has been a major selective factor in the evolution of odonate genitalic morphology and reproductive behavior. The data at present suggest that the hypothesis is viable and that its subset of hypotheses are testable. Sperm removal or displacement can be indirectly evaluated for odonates by taking advantage of their unique reproductive behavior and morphology. This method involves the comparison of sperm volumes of females collected before, during, and after copulation; and by dissection of in copula pairs. Sperm displacement (at levels of 40-100%) has been shown to occur in five odonate species: the zygopterans, *Calopteryx maculata* and *C. dimidiata* (Calopterygidae), and *Lestes vigilax* (Lestidae); and the anisopterans, *Celithemis elisa* and *Erythemis simplicicollis* (Libellulidae). Sperm displacement probably also occurs in two other zygopteran genera, *Argia* and *Enallagma,* and in other libellulids.

In the Zygoptera, sperm removal involves use of structures on the distal segment of the penis that scoop or pull clumps of sperm from one or both of the female's sperm storage organs. Some penis morphology in other genera (*e.g., Lestes*) may function in displacing and repositioning sperm within the female's storage organs. Damselfly penis morphology appears to have been shaped through natural selection for the efficient transfer of sperm to the female, and by sexual selection, in the context of sperm competition, for the removal or displacement of rival sperm prior to insemination. Structures associated with these functions are clearly separable in most species. Comparative morphology of zygopteran genitalia suggests that the same structures used to remove or reposition sperm by the species above are found in species representing the vast majority of genera. This suggests that sperm displacement **ability** is widespread among Zygoptera. Females of most Zygoptera species and all Anisoptera species so far examined carry stored sperm from previous matings, and after ovipositing. Although sperm displacement ability may be widespread in Zygoptera, there are indications that males of some species may not always have opportunity to displace sperm. Many more species must be examined to determine the true extent of both ability and opportunity to displace sperm.

Anisopterans have erectile penises whose functional morphology is difficult to evaluate. Only a few species of libellulids have been examined for sperm displacement ability. *Celithemis elisa* and *Erythemis simplicicollis* remove sperm, but *Sympetrum rubicundulum* does not appear to. It is possible that males of this latter species push sperm of rivals into the recesses of female storage organs prior to insemination, but this needs verification. It is not presently possible to evaluate the full nature and extent of sperm displacement ability in anisopterans.

The morphology of the distal segment of the odonate penis might serve several functions other than sperm removal or displacement. These include: aiding in sperm transfer, holding genitalia together during copulation, and mechanical reproductive isolation. However, the evidence weighs heavily in favor of sperm displacement as the most likely function of the distal segment.

Odonate postcopulatory interactions fall into three categories: tandem oviposition, non-contact guarding of ovipositing females, and solitary oviposition by females. The first two categories are effective means to avoid take-overs of ovipositing females by other males. The vast majority of Zygoptera and most Libellullidae exhibit one or the other of these postcopulatory behaviors. Solitary oviposition probably occurs occasionally in most odonate species, especially following a period of tandem or guarded oviposition; however, it is the principal postcopulatory behavior only of the more primitive anisopteran families (Aeshnidae, Cordulegasteridae, Gomphidae, and Petaluridae). In general, solitary oviposition occurs in situations where male-female encounters are rare or where females effectively avoid male interference. Postcopulatory interactions (tandem and guarded oviposition) appear to function in avoiding sperm competition and/or disturbance of ovipositing females. Non-contact guarding seems to be a compromise between protection of a mate while allowing the male to either gain access to additional mates or defend his territory, or both.

Odonate mating systems range from non-localized, opportunistic encounters to highly localized pair formation at oviposition sites within male territories. Durations of copulation (several sec to several hr) and postcopulatory behaviors correlate well with the three types of mating systems. The major determinants of mating systems appear to be ecological (habitat structure and male density), but these in turn result in conditions strongly conducive to sexual (intrasexual and sperm competition) selection. A parallel exists between the calopterygid damselflies and libellulid dragonflies. Both groups exhibit frequent, short copulations, territoriality, and non-contact guarding, characteristics likely to have evolved in the context of sperm competition. Many odonates are flexible in their reproductive behavior. Habitat structure and population densities influence encounter rates and likelihood of interference with mating and oviposition. These factors in turn determine the relative benefits of guarding previous mates vs seeking new ones. Opportunities abound for comparative and manipulative studies of the interactions among various factors that influence odonate reproductive behavior. Future studies that consider the female perspective in odonate sperm competition may be especially fruitful.

ACKNOWLEDGMENTS

    I gratefully acknowledge the comments of the following colleagues during the writing and revision of this manuscript: John Alcock, George Bick, Darryl Gwynne, Peter Kareiva, James Lloyd, Peter Miller, Douglass Morse, Robert Smith, Edward Waltz, and Larry Wolf. I was fortunate to have the capable field assistance of D. Ensor, S. Dunn, D. Fernandes, E. Oseas, and D. Schoeling. Emily Oseas deserves special thanks for measuring sperm volumes and drawing the figures for this paper. I apologize to her for any artistic disasters resulting from my rearrangements. K. Sherman (unpubl.) independently suggested non-contact guarding may function to allow a male to maintain a territory while guarding an ovipositing mate. A special thanks to Richard D. Alexander and Geoffrey Parker for their respective influences on the perspective from which I view odonate behavior. This work has been generously supported by grants from NSF (DEB 77-15904 and DEB 80-04282) and the American Philosophical Society (No. 7784, 1977) to the author. A final appreciation is due to Robert L. Smith for organizing the symposium and patiently editing its product.

288 Jonathan K. Waage

REFERENCES

Alcock, J. 1979. Multiple mating in *Calopteryx maculata* (Odonata:Calopterygidae) and the advantage of non-contact guarding by males. *J. Nat. Hist.* 13:439-446.

Alcock, J. 1982. Post-copulatory mate guarding by males of the damselfly *Hetaerina vulnerata* Selys (Odonata:Calopterygidae). *Anim. Behav.* 30:99-107.

Bick, G. C. 1966. Threat display in unaccompanied females of the damselfly *Ischnura verticalis* (Say) (Odonata:Coenagrionidae). *Proc. Entomol. Soc. Wash.* 68:271.

Bick, G. C. 1972. A review of territorial and reproductive behavior in Zygoptera. *Contact-brief Nederlandse Libellenonderzoekers (Suppl.)* 10:1-14.

Bick, G. C., and J. C. Bick. 1963. Behavior and population structure of the damselfly *Enallagma civile* (Hagen). *Southwest. Nat.* 6:57-84.

Bick, G. C., and L. E. Hornuff. 1966. Reproductive behavior in the damselflies *Enallagma aspersum* (Hagen) and *E. exsulans* (Hagen). *Proc. Entomol. Soc. Wash.* 68:78-85.

Bick, G. C., J. C. Bick, and L. E. Hornuff. 1976. Behavior of *Chromagrion conditum* (Hagen) adults (Zygoptera:Coenagrionidae). *Odonatologica* 5:129-141.

Blum, M. S., and N. A. Blum (eds.). 1979. *Sexual Selection and Reproductive Competition in Insects.* Academic Press, New York.

Boggs, C. L., and L. E. Gilbert. 1979. Male contribution to egg production in butterflies: Evidence for transfer of nutrients at mating. *Science* 206:83-84.

Boorman, E., and G. A. Parker. 1976. Sperm (ejaculate) competition in *Drosophila melanogaster* and the reproductive value of females to males in relation to female age and mating status. *Ecol. Entomol.* 1:145-155.

Campanella, P. J. 1975. The evolution of mating systems in temperate zone dragonflies (Odonata:Anisoptera): II. *Libellula luctuosa* (Burmeister). *Behavior* 54:278-310.

Campanella, P. J., and L. Wolf. 1974. Temporal lek as a mating system in a temperate zone dragonfly (Odonata:Anisoptera): I. *Plathemis lydia* (Drury). *Behaviour* 51:49-87.

Campbell, B. (ed.). 1972. *Sexual Selection and the Descent of Man 1871-1971.* Aldine Publ. Co., Chicago.

Corbet, P. S. 1962. *A Biology of Dragonflies.* M. F. and G. Witherby Ltd., London.

Corbet, P. S. 1980. Biology of Odonata. *Annu. Rev. Entomol.* 25:189-217.

Darwin, C. 1859. *On the Origin of Species by Means of Natural Selection.* John Murray, London. (1964, facsimile of the 1st edition, Harvard Univ. Press).

Darwin, C. 1871. *The Descent of Man and Selection in Relation to Sex, 2nd Ed., Rev.* (1898). D. Appleton and Co., New York.

Fincke, O. M. 1982. Lifetime mating success in a natural population of the damselfly, *Enallagma hageni* (Walsh) (Odonata:Coenagrionidae). *Behav. Ecol. Sociobiol.* 10:293-302.

Friedel, T., and C. Gillott. 1977. Contribution of male-produced proteins to vitellogenesis in *Melanoplus sanguinipes. J. Insect Physiol.* 23:145-151.

Furtado, J. I. 1972. The reproductive behavior of *Ischnura senegalensis* (Rambur), *Pseudagrion microcephalum* (Rambur), and *P. perfuscatum* Lieftinck. *Malays. J. Sci.* 1(A):57-69.

Gillott, C., and T. Friedel. 1977. Fecundity-enhancing and receptivity-inhibiting substances produced by male insects: A review. *Adv. Invert. Reprod.* 1:199-218.

Gould, S. J., and R. C. Lewontin. 1979. The spandrels of San Marco and the Panglossian paradigm: A critique of the adaptationist programme. *Proc. R. Soc. Lond. B* 105:581-598.

Grieve, E. 1937. Studies on the biology of the damselfly *Ischnura verticalis* Say, with notes on certain parasites. *Entomol. Am.* 17:121-153.

Hassan, A. J. 1978. Reproductive behaviour of *Acisoma panorpoides inflatum* Selys (Anisoptera:Libellulidae). *Odonatologica* 7:237-245.

Heymer, A. 1968. Eiablagverhalten der Libellen. *Umsch. Wiss. Tech.* 21/68:665-666.

Heymer, A. 1973. Verhaltenstudien an Prachtlibellen: Beitrage zur Ethologie und Evolution der Calopterygidae Selys, 1850 (Odonata:Zygoptera). *J. Comp. Ethol. Suppl.* 11:1-100.

Hornuff, L. E., Jr. 1968. Functional morphology of the external genitalia of Nearctic damselflies. Ph.D. Dissertation, Univ. of Oklahoma.

Jacobs, M. E. 1955. Studies on territorialism and sexual selection in dragonflies. *Ecology* 36:566-586.

Johnson, C. 1962a. A study on territoriality and breeding behavior in *Pachydiplax longipennis* Burmeister (Odonata:Libellulidae). *Southwest. Nat.* 7:191-197.

Johnson, C. 1962b. Reproductive isolation in damselflies and dragonflies (Odonata). *Tex. J. Sci.* 14:297-304.

Johnson, C. 1964a. Mating expectancies and sex ratio in the damselfly *Enallagma praevarum. Southwest. Nat.* 9:297-304.

Johnson, C. 1964b. Evolution of territoriality in the Odonata. *Evolution* 18:89-92.

Johnson, C. 1966a. Improvements for colonizing damselflies in the laboratory. *Tex. J. Sci.* 18:179-183.

Johnson, C. 1966b. Genetics of female dimorphism in *Ischnura demorsa. Heredity* 21:453-459.

Johnson, C. 1969. Genetic variability in Ischnuran damselflies. *Am. Midl. Nat.* 81:39-46.

Johnson, C. 1972. Tandem linkage, sperm translocation, and copulation in the dragonfly *Hagenius brevistylus* (Odonata:Gomphidae). *Am. Midl. Nat.* 88:131-149.

Kaiser, H. 1974a. Intraspezifische Aggression und räumliche Verteilung bei der Libelle *Onychogomphus forcipatus* (Odonata). *Oecologia (Berl.)* 15:223-234.

Kaiser, H. 1974b. Verhaltensgefüge und Temporalverhalten der Libelle *Aeschna cyanea* (Odonata). *Z. Tierpsychol.* 34:398-429.

Kennedy, C. H. 1920. The phylogeny of the zygopterous dragonflies as based on the evidence of the penes. *Ohio J. Sci.* 21:19-29.

Kennedy, C. H. 1922. The morphology of the penis in the genus *Libellula. Entomol. News* 33:33-40.

Krüner, U. 1977. Revier- und Fortpflanzungsverhalten von *Orthetrum cancellatum* (L.) *Odonatologica* 6:263-270.

Lloyd, J. E. 1979. Mating behavior and natural selection. *Fla. Entomol.* 62:17-34.

Loibl, E. 1958. Zur Ethologie und Biologie der deutschen Lestiden (Odonata). *Z. Tierpsychol.* 15:54-81.

McVey, M. E. 1981. Lifetime reproductive tactics in a territorial dragonfly, *Erythemis simplicicollis.* Ph.D. Dissertation, Rockefeller Univ.

Miller, P. L. 1981. Functional morphology of the penis of *Celithemis eponia* Drury (Libellulidae; Odonata). *Odonatologica* 10:293-300.

Miller, P. L. 1982a. The occurrence and activity of sperm in mature female *Enallagma cyathigerum* (Charpentier) (Zygoptera:Coenagrionidae). *Odonatologica* 11:159-161.

Miller, P. L. 1982b. Genital structure, sperm competition and reproductive behaviour in some African libellulid dragonflies. *Adv. Odonatol.* 1:175-193.

Miller, P. L., and C. A. Miller. 1981. Field observations on copulatory behaviour in Zygoptera, with an examination of the structure and activity of the male genitalia. *Odonatologica* 10:201-218.

Pajunen, V. I. 1963a. Reproductive behavior in *Leucorrhinia dubia* v.d. Lind and *L. rubicunda* L. (Odonata:Libellulidae). *Ann. Entomol. Fenn.* 29:106-118.

Pajunen, V. I. 1963b. On the threat display of resting dragonflies (Odonata). *Ann. Entomol. Fenn.* 19:236-239.

Pajunen, V. I. 1966. The influence of population density on the territorial behaviour of *Leucorrhinia rubicunda* L. *Ann. Zool. Fenn.* 3:40-52.

Parker, G. A. 1970. Sperm competition and its evolutionary consequences in the insects. *Biol. Rev.* 45:525-567.

Parker, G. A. 1974. Courtship persistence and female-guarding as male time-investment strategies. *Behaviour* 48:157-184.

Parr, M. J., and M. Palmer, 1971. The sex ratios, mating frequencies and mating expectancies of three coenagrionids in northern England. *Entomol. Scand.* 2:191-204.

Parr, M. J., and M. Parr. 1974. Studies on the behaviour and ecology of *Nesciothemis nigerensis* Gambles (Anisoptera:Libellulidae). *Odonatologica* 3:21-47.

Paulson, D. 1974. Reproductive isolation in damselflies. *Syst. Zool.* 23:40-49.

Pezalla, V. 1979. Behavioral ecology of the dragonfly *Libellula pulchella* Drury (Odonata: Anisoptera). *Am. Midl. Nat.* 102:1-22.

Pfau, H. K. 1971. Struktur und Funktion des sekundären Kopulations apparates der Odonaten (Insecten, Palaeoptera), ihren Wandlung in der Stammesgeschichte und Bedeutung fur die adaptive Entfaltung der Ordnung. *Z. Morph. Tiere.* 70:281-371.

Pinhey, E. 1969. Copulatory factors in *Ischnura senegalensis* (Rambur). *Arnoldia* 4:1-6.

Restifo, F. A. 1972. The comparative morphology of the penis in the libellulid genera *Celithemis, Leucorrhinia* and *Libellula* (Odonata). M.S. Thesis, Ohio State Univ.

Robertson, H. M., and H. E. H. Patterson. 1982. Mate recognition and mechanical isolation in *Enallagma* damselflies (Odonata:Coenagrionidae). *Evolution* 36:243-250.

Rowe, R. J. 1978. *Ischnura aurora* (Brauner), a dragonfly with unusual mating behaviour (Zygoptera:Coenagrionidae). *Odonatologica* 7:375-383.

St. Quentin, D. 1964. Territorialität bei Libellen (Odonata). *Mitt. Muench. Entomol. Ges.* 54:162-180.

Sakagami, S., H. Ubukata, N. Iga, and M. Toda. 1974. Observations on the behavior of some Odonata in the Bonin Islands, with considerations on the evolution of reproductive behavior in Libellulidae. *J. Fac. Sci. Hokkaido Univ. Ser. VI Zool.* 19:722-757.

Schmidt, E. 1915. Vergleichende Morphologie des 2 und 3 Abdominalsegments bei männlichen Libellen. *Zool. Jahrb.* 39:87-200.

Schmidt, Eb. 1965. Zum Paarungs- und Eiablageverhalten der Libellen. *Faun. Mitt. Nordd.* 2: 313-319.

Schmidt, Eb. 1975. Zur Klassifikation des Eiablageverhaltens der Odonaten. *Odonatologica* 4: 177-184.

Smith, R. L. 1979. Repeated copulation and sperm precedence: Paternity assurance for a male brooding water bug. *Science* 205:1029-1031.

Tennessen, K. J. 1975. Reproductive behavior and isolation of two sympatric coenagrionid damselflies in Florida. Ph.D. Dissertation, Univ. Florida.

Ubukata, H. 1974. Relative abundance and phenology of adult dragonflies at a distrophic pond in Usubetsu, near Sapporo. *J. Fac. Sci. Hokkaido Univ. Ser. VI Zool.* 19:758-776.

Ubukata, H. 1975. Life history and behavior of a corduliid dragonfly *Cordulia aenea amurensis* Selys: II. Reproductive period with special reference to territoriality. *J. Fac. Sci. Hokkaido Univ. Ser. VI Zool.* 19:812-833.

Ueda, T. 1979. Plasticity of the reproductive behaviour in a dragonfly, *Sympetrum parvulum* Barteneff, with reference to the social relationship of males and the density of territories. *Res. Popul. Ecol* 21:135-152.

Waage, J. K. 1973. Reproductive behaviour and its relation to territoriality in *Calopteryx maculata* (Beauvois). *Behaviour* 47:240-256.

Waage, J. K. 1975. Reproductive isolation and the potential for character displacement in the damselflies *Calopteryx maculata* and *C. aequabilis*. *Syst. Zool.* 24:24-36.

Waage, J. K. 1978. Oviposition duration and egg deposition rates in *Calopteryx maculata*. *Odonatologica* 7:77-88.

Waage, J. K. 1979a. Dual function of the damselfly penis: Sperm removal and transfer. *Science* 203:916-918.

Waage, J. K. 1979b. Adaptive significance of postcopulatory guarding of mates and nonmates by male *Calopteryx maculata* (Odonata). *Behav. Ecol. Sociobiol.* 6:147-154.

Waage, J. K. 1980. Adult sex ratios and female reproductive potential in *Calopteryx* (Zygoptera:Calopterygidae). *Odonatologica* 9:217-230.

Waage, J. K. 1982. Sperm displacement by male *Lestes vigilax* Hagen (Odonata:Zygoptera). *Odonatologica* 11:201-209.

Walker, W. F. 1980. Sperm utilization strategies in nonsocial insects. *Am. Nat.* 115:780-799.

Watson, J. A. L. 1966. Genital structure as an isolating mechanism in Odonata. *Proc. Roy. Entomol. Soc. Lond. Ser. A Gen. Entomol.* 41:171-174.

Williams, G. C. 1966. *Adaptation and Natural Selection*. Princeton Univ. Press, Princeton, NJ.

Note: Since this review was last revised, a number of important papers have appeared on odonate sperm competition: Fincke, O. M. 1984. Sperm displacement in the damselfly *Enallagma hageni* (Odonata:Coenagrionidae). *Behav. Ecol. Sociobiol.,* in press; McVey, M. E., and B. J. Smittle. 1984. Sperm precedence in the dragonfly *Erythemis simplicicollis* (Say) (Odonata:Libellulidae). *J. Insect Physiol,* in press; Miller, P. A. 1984. The structure of the genitalia and the volume of sperm stored in male and female *Nesciothemis farinosa* Foerester and *Orthetrum cancellatum* (Burm.) (Odonata:Libellulidae). *Odonatologica,* in press; and Siva-jothy, M. T. 1984. Sperm competition in the family Libellulidae (Odonata). *Adv. Odonatol.* 2:in press.

9

Multiple Mating and Sperm Competition in the Lepidoptera

BOYCE A. DRUMMMOND III

I. INTRODUCTION

Mating frequency of sexually reproducing species is of evolutionary importance because it affects the rate at which genetic factors can be shuffled in a population (Labine 1964, Burns 1968). The increased rate of recombination within a population resulting from multiple mating allows more rapid response to changing selection

Sperm Competition and the Evolution
of Animal Mating Systems

pressures. And, depending on the vagility of the species, multiple matings can have various effects on gene flow between populations (Pease 1968, Burns 1968). For example, for species in which females are wide-ranging or migratory, gene flow between populations is high if a female mates only once (thereby carrying not only her own genes but also those of her mate to a new environment), but gene flow is lower if females mate several times, pairing with local males at each new locality, because in Lepidoptera the last male to mate usually fertilizes all remaining eggs. Conversely, for species with sedentary females and mobile males, gene flow is low if females mate only once, but increases with the degree of multiple mating. The extent of multiple mating in Lepidoptera is of much interest because there exists great diversity in the degree of vagility of female butterflies and moths. For example, females of the bagworm, *Thyridopteryx ephemeraeformis* (Psychidae), are flightless (Kaufmann 1968), but females of the monarch butterfly, *Danaus plexippus* (Danainae), are migratory and may travel over a thousand miles during a lifetime, laying eggs along the way (Urquhart 1960).

Numerous published studies (reviewed here) and my own research reveal that females of many lepidopteran species do mate more than once, which provides an opportunity for the ejaculates of two or more males to be in competition within the reproductive tract of a single female. Sperm competition implies that viable sperm from more than one male are present in the female's reproductive tract simultaneously, a condition that probably occurs in most species of Lepidoptera for which multiple matings have been recorded.

Because of their importance in studies of basic ecology (Gilbert and Singer 1975), population genetics (Ehrlich and White 1980), evolution (Turner 1981), behavior (Silberglied 1977), and applied entomology (Roelefs and Carde 1977), a great deal is known about the reproductive biology of the Lepidoptera, the second largest order of insects. In reviewing this extensive and scattered literature, I hope to facilitate comparison of the Lepidoptera with other taxonomic groups discussed in this volume. The purposes of this paper are (1) to describe the anatomical features of lepidopteran reproductive systems and summarize their effects on the movement and activity of sperm between spermatogenesis and fertilization (Section II and Appendix); (2) to assess the extent and importance of multiple mating in natural populations of Lepidoptera (Section III); (3) to identify the factors that determine the pattern of paternity among offspring of multiply-mated females; and (4) to consider the degree to which sperm competition has shaped the evolution of courtship and mating behaviors of male and female Lepidoptera (Section IV).

II. REPRODUCTIVE ANATOMY AND SPERM DYNAMICS

The higher classification of the Lepidoptera reflects the occurrence in the order of two main types of female genitalia, monotrysian and ditrysian. The monotrysian type occurs in the more primitive suborders Zeugloptera, Dachnonypha, and Monotrysia. The ditrysian type occurs throughout the Ditrysia, the suborder that includes the vast majority (ca. 97%) of lepidopteran species (Common 1970, Dugdale 1974). In monotrysian Lepidoptera a single genital opening serves the purposes of both copulation and oviposition. Ditrysian Lepidoptera have paired genital openings so that the functions of mating and egg laying are physically separated. Paired female genital openings have arisen phylogenetically at least four times in the higher insect orders: Homoptera (Cicadidae), Hemiptera (Anthocoridae), Coleoptera (*Hydroporus*), and the higher Lepidoptera (Ditrysia). The evolution of paired genital openings in these four orders is clearly a case of convergence; single genital openings occur in females of Trichoptera and Mecoptera, the closest relatives of the Lepidoptera (Matsuda 1976).

A. Reproductive Anatomy

The genitalia of both sexes of most Lepidoptera provide important characters for separating species and have been used often to delineate higher taxa (see review by Common 1975). The external genitalic structures, derived from the integument of abdominal segments VII through X, are the most used in taxonomic studies of Lepidoptera, but internal genital organs, particularly of females, are also important. Comparative morphological studies of lepidopteran genitalia have been published for males (Mehta 1933, Klots 1970, Birket-Smith 1974a, b) and females (Klots 1970, Mutuura 1972, Dugdale 1974). An excellent brief introduction to the anatomy and morphology of the reproductive organs of both sexes in Lepidoptera is given in Common (1970), and a comprehensive review of the embryology, morphology, and function of all abdominal organs may be found in Matsuda (1976). The descriptions of genital anatomy given here are selective and focus on structures of relevance to the mechanics of copulation and to sperm transfer, storage, and utilization.

1. Males

The genital segment of the male abdomen (Fig. 1A), segment IX, is formed by two chitinous half-rings: a hoodlike, dorsal tegumen, from which arise various (usually) paired processes of the genitalia, and a ventral U-shaped vinculum, the upper ends of which articulate with the ventral extremities of the tegumen. Closing

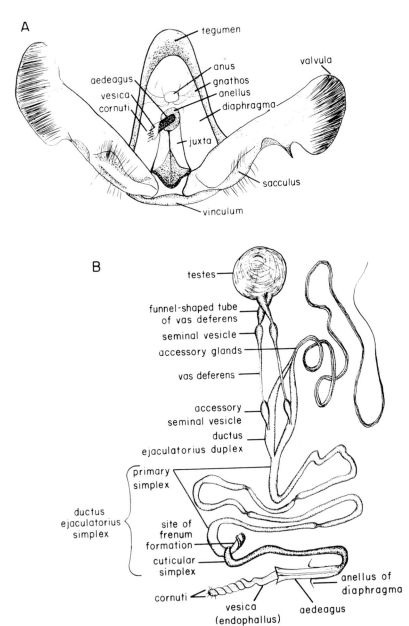

Fig. 1. Male genitalia of Lepidoptera. A. Posteroventral view of external male genitalia of the codling moth, *Laspeyresia pomonella* L. (Olethreutidae). Redrawn from Ferro and Akre (1975). B. Internal reproductive system of a male lepidopteran. Generalized, but based in part on drawings of *Heliothis zea* by Callahan (1958).

the posterior end of the abdomen is a membranous diaphragma through which the copulatory organ, the aedeagus, projects. The aedeagus is a tubular sclerotized intromittent organ or phallus (Birket-Smith 1974a), characteristic of most Ditrysia but absent in some Monotrysia. The sclerotized region of the diaphragma through which the aedeagus passes is the anellus, whose inner layer (the manica) fastens around the aedeagus. Of the various sclerotized portions of the diaphragma that support the aedeagus during copulation, a ventral plate called the juxta is characteristic of the Lepidoptera (Matsuda 1976). The distal end of the aedeagus is invaginated, forming an inner tube (endophallus), the vesica, which is everted during copulation. In many species the exterior of the everted vesica bears sclerotized areas or spines (cornuti) which may be shed in the bursa copulatrix (Common 1970). During copulation the male grasps the female's abdomen with paired clasping organs, the valvulae, which are situated laterally at the posterior end of the male's abdomen and which articulate with the tegumen and vinculum. Considerable variation in size, shape, and armature occur in the valvulae throughout the Lepidoptera (Matsuda 1976).

A generalized representation of the internal reproductive system in male Lepidoptera is shown in Fig. 1B. In most Lepidoptera the two testes are fused into a single, globular organ contained in a common scrotum, although in some groups (e.g., Zeugloptera, some Bombycoidea, some Papilionoidea) the testes remain separate, each enclosed in its own scrotum. Spermatogenesis usually begins in the later larval instars and may terminate before adulthood (Holt and North 1970) or continue throughout the life of the male (Lai-Fook 1982). Lepidoptera produce two kinds of sperm (Meves 1902, Goldschmidt 1916, Iriki 1941): eupyrene, which are nucleated and fertilize the eggs, and apyrene, which lack nuclei and are incapable of fertilizing eggs (Friedlander and Gitay 1972). Although the two sperm types are produced at different times in the life cycle of Lepidoptera (Leviatan and Friedlander 1979; Friedlander and Benz 1981, 1982; Lai-Fook 1982), both types are passed to the female during copulation and both migrate to the spermatheca (Holt and North 1970, Riemann and Thorson 1971, Friedlander and Gitay 1972, Katsuno 1977, Etman and Hooper 1979b). Several functions have been proposed for apyrene sperm (for reviews, see Thibout 1981 and Silberglied et al. 1984), but none of the proposed hypotheses have been substantiated experimentally. The possible role of apyrene sperm in sperm competition has been suggested by Silberglied et al. (1984) and is discussed in section IV. The possibility that more than one type of eupyrene sperm are produced in some species is raised by Sidhu (1972).

Arising from the posterior margin of the testis are the paired vasa deferentia, tubular secretory ducts that lead to the ductus ejaculatorius duplex. In most Lepidoptera, the vasa deferentia are conspicuously swollen for the first one-quarter of their length, forming funnel-shaped tubes (Norris 1932) that constrict slightly before dilating to form a pair of oval-shaped, sperm-storing organs, the seminal vesicles. Occasionally the constriction is more pronounced and completely separates the seminal vesicles from the funnel-shaped cephalad portion of the vasa deferentia, as in Lasiocampidae (Williams 1940). Although a single pair of seminal vesicles has been reported for most lepidopteran species studied in detail (Hewer 1932, Norris 1932, Musgrave 1937, Rakshpal 1944, Khalifa 1950, Davis 1968, Rieman and Thorson 1976, Stern and Smith 1960, Omura 1938, Callahan 1958, Callahan and Chapin 1960), some species have two pairs (Callahan and Chapin 1960, Holt and North 1970, Outram 1970, Tedders and Calcote 1967, Ferro and Akre 1975) and a few have three (Thibout 1971, Yang and Chow 1978). In each of these cases the multiple pairs of seminal vesicles are similar histologically and functionally (all store sperm bundles). Below the seminal vesicles, the vasa deferentia continue as thin, narrow tubules, sometimes called vesicular ducts (Santorini and Vassilaina-Alexopoulou 1976).

The ductus ejaculatorius duplex comprises a pair of sperm storage organs that unite caudally to form the ductus ejaculatorius simplex, a single duct leading to the aedeagus. From the cephalad ends of the duplex arise paired filamentous accessory glands, which secrete the seminal fluid. The ductus ejaculatorius duplex is a U-shaped secretory and storage gland that can dilate considerably to accommodate the secretions of the paired accessory glands and the sperm bundles and additional secretions it receives from the vasa deferentia (Callahan and Cascio 1963, Outram 1970, Ferro and Akre 1975).

The ejaculatory duct (ductus ejaculatorius simplex) is a long tube that carries sperm and accessory gland secretions from the ductus ejaculatorius duplex to the aedeagus. Callahan (1958) recognized two segments of the ejaculatory duct in *Heliothis zea*. The longer cephalad portion (45 mm) is a glandular, thin-walled tube about 1 mm in diameter, the primary simplex, which produces the precursor of the spermatophore. The shorter caudad portion (25 mm), the cuticular simplex, is a flattened tubular organ about 1.5 mm in width, in which the spermatophore precursor is molded into the resilient spermatophore prior to deposition in the female. At the junction of the primary and cuticular segments is a swollen area of twisting and coiling where the frenum of the spermatophore is formed. The caudal end of the cuticular segment of the ductus ejaculatorius simplex joins the eversible endophallus. The two-part nature of the ejaculatory duct described by Callahan (1958) has been confirmed in several other species of Noctuidae (Callahan and Cascio 1963), two species of *Laspeyresia* (Tedders and Calcote 1967, Ferro and Akre 1975), several species of *Zygaena* (Hewer 1932), and *Diatrea grandiosella* (Davis 1968). It is in this most caudal portion of the male reproductive tract that the greatest morphological variation occurs, both within and among taxa (Norris 1932, Callahan and Chapin 1960). Callahan and Chapin (1960) consider the elongation of the cuticular simplex (between the area of frenum formation and the aedeagus) to be an evolutionary advancement, as illustrated by the relatively long cuticular simplex in the higher Noctuidae (16-20 mm) compared with its near absence in the lower Phycitinae.

The most cephalad portion of the cuticular simplex is the area where the frenum (lock) is formed on the spermatophore tip. This region (Fig. 1B) consists of a thick, heavy, muscularized duct (Noctuidae) or an ejaculatory bulb (Phycitinae) associated with one or more hollow outgrowths in which the frenum is actually molded to its species-specific shape (Callahan and Chapin 1960, Norris 1932). The frenum serves to lock the collum of the spermatophore in the appropriate part of the ductus bursae of the female. The aperture of the spermatophore is situated at the base of the frenum in the Phycitinae (see Fig. 3A-D) and Noctuidae (Callahan 1958).

The 28 species of Noctuidae studied by Callahan and Chapin (1960) showed considerable variation in the relative sizes of the aedeagus and endophallus; in the number, size, and distribution of cornuti; and in the mechanism of retraction, storage, and eversion of the endophallus. The taxonomic importance of such variation has already been mentioned, but there is also a functional significance. Callahan and Chapin (1960) reported that different species of noctuids experienced different degrees of success in completing the mechanisms of copulation and concluded that the number of aberrant (unsuccessful) matings appeared to correlate directly with the complexity of spermatophore insertion. Holt and North (1970) speculated that the observed morphological variation in the endophallus, aedeagus, and cuticular simplex of males, and the considerable differences in the morphology of the bursae copulatrix of females, could explain the interspecific variability in mating success reported for agriculturally important Lepidoptera. They also stressed the potential usefulness of studying the variation in reproductive morphology of Lepidoptera to enhancing the success of biological control methods based on the sterile male technique.

2. Females

Fig. 2 presents generalized representations of the female genital morphologies of monotrysian (A) and ditrysian (B) Lepidoptera. The external genital aperture (vulva) is just below the anus on the fused segments IX-X at the posterior extremity of the abdomen. Because the eggs are extruded through the vulva, it is often called the ostium oviductus or ovipore. The vulva is usually flanked by a pair of soft hairy lobes, the papillae anales (Fig. 2B). The muscles operating the terminal segments of the female abdomen are inserted in paired sclerotized apodemes, which extend forward from segment VIII (apophyses anteriores) and from the bases of the papillae anales on segments IX-X (apophyses posteriores) (Common 1970).

The rectum and vagina become confluent distally and form a true cloaca in some species, e.g., *Micropteryx* (Mutuura 1972), *Tegeticula* and *Podurus* (Williams 1941b, c, 1947), and *Anagasta kuhniella* (Musgrave 1937). Because a cloaca occurs in some Trichoptera (*Rhyacophila*) and in the Zeugloptera, Matsuda (1976) suggests that cloacal formation by extension of the genital tract appears to be an

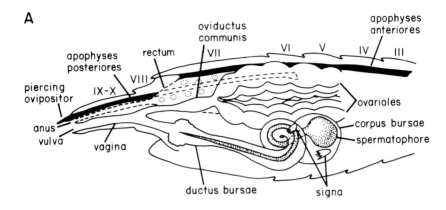

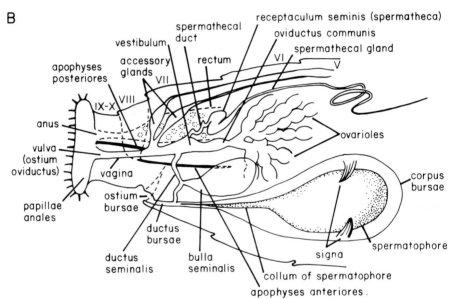

Fig. 2. Female genitalia of Lepidoptera. A. Monotrysian type, generalized, lateral view. B. Ditrysian type, generalized, lateral view.

archaic specialization that accompanied the loss of the sclerotized ovipositor in the ancestor of these groups, and that the functional awkwardness of the cloacal arrangement was later alleviated by the evolution of a special duct for copulation that opens midventrally as the ostium bursae on more anterior segments (usually on sternum VIII) in higher Lepidoptera (Ditrysia).

The largest organs in the female abdomen are the paired ovaries, each of which typically consists of four polytrophic ovarioles. Each ovary empties into a short lateral oviduct (oviductus lateralis), and these fuse into a common oviduct (oviductus communis) in the posterior area of the seventh abdominal segment. Eggs are fertilized during their passage down the common oviduct before they reach the vulva or ovipore. In monotrysian Lepidoptera (Fig. 2A) the vulva is also the copulatory opening and leads to the bursa copulatrix, an elongate sac of two parts (ductus bursae and corpus bursae) that receives the spermatophore passed to the female during copulation. Sperm from the spermatophore enter the vagina directly from the ductus bursae to fertilize the eggs in the monotrysian system (Common 1970).

The acquisition of the copulatory orifice (ostium bursae) in Ditrysia (Fig. 2B) was accompanied by the disassociation of the basal portion of the bursa copulatrix from the vagina so that the corpus bursae became connected to the new copulatory opening (ostium bursae) by the ductus bursae. At the same time, a new duct (ductus seminalis) developed to connect the bursa copulatrix with the vagina (Matsuda 1976). In the ditrysian system, therefore, sperm exit the spermatophore into the ductus seminalis, cross the oviductus communis, and travel up the spermathecal duct to the spermatheca (receptaculum seminis) where they are stored. The ductus seminalis (from the bursa copulatrix) and the spermathecal duct (from the spermatheca) join the common oviduct at approximately the same level and together mark the separation of the common oviduct from the vagina, which leads to the vulva or ovipore (ostium oviductus). In the enlarged upper part of the vagina, the vestibulum, the eggs are fertilized individually as they are held briefly with the cephalic pole (with micropyle) closely pressed against the opening of the spermathecal duct (Norris 1932, Yakhontov 1960, Callahan and Cascio 1963). Paired accessory glands open by a common duct, sometimes enlarged as a reservoir, into the vagina and their secretions are generally believed (Matsuda 1976) to be an adhesive for the eggs; accessory glands are absent in Lepidoptera which do not attach the egg to a surface during oviposition (Petersen 1907). Callahan and Cascio (1963) have suggested that the secretions of the accessory glands serve as a medium for the motile passage of the spermatozoa across the vestibulum.

The three sections of the bursa copulatorx—ostium, ductus, and corpus bursae—may be variously sclerotized. The ostium bursae is often surrounded by a sclerotized area, the sterigma (see Fig. 9A), which is sometimes formed into a projecting tube, the henia, as in Lycaenidae (Ehrlich and Ehrlich 1978). The corpus bursae is a thin-walled sac enclosed in a muscular sheath that usually inserts in one or more sclerotized plates (lamina dentatae) or spines (signa) located on the inner bursal wall (Fig. 2B). Several functions (reviewed by Hinton 1964) have been postulated for these sclerotized structures, which in most species probably serve to hold the spermatophore in place when the aedeagus is removed (Ferro and Akre 1975) or during contractions of the corpus bursae (Callahan 1958, Callahan and Cascio 1963, Thibout 1971). Secure placement of the spermatophore after mating is aided in some species by cephalad-pointing spines distributed along the interior lining of the ductus bursae (Outram 1971b). In some groups (*e.g.*, Pieridae, Sugawara 1979), there is a nonmuscular, cuticular sac (appendix bursae) of unknown function attached to the cephalad end of the corpus bursae.

B. Sperm Dynamics

As far as is known, all Lepidoptera produce spermatophores during copulation, although spermatophores of monotrysian groups are poorly known (Khalifa 1950, Hinton 1964). The size and shape of spermatophores produced by different lepidopteran groups show great diversity (Fig. 3), but are sufficiently constant within a species (Norris 1932, Drummond 1976) to be a useful taxonomic character (Petersen 1907). Of the four methods of spermatophore formation recognized by Gerber (1970), Lepidoptera conform to the more primitive of the two female-determined methods. The spermatophore is formed by sequential transfer of several secretions of the male reproductive tract, and its ultimate shape is determined in part by the interior dimensions of the bursa copulatrix of the female (bulb or body of spermatophore) and in part by the shape of the cuticular simplex and aedeagus of the male (collum and frenum).

The most detailed descriptions of copulation, spermatophore formation, and sperm transfer and storage in the Lepidoptera are those published in a series of papers on the reproductive biology of the Noctuidae by Callahan (Callahan 1958, Callahan and Chapin 1960, Callahan and Cascio 1963). Similar but less detailed descriptions have been published for other species of Noctuidae (Holt and North 1970, Chao 1981), and for Zygaenidae (Hewer 1934), Pyralidae (Phycitinae, Norris 1932; Pyraustinae, Srivastava and Srivastava 1957; Gallerinae, Khalifa 1950), Olethreutidae (Ferro and Akre 1975), Gelechiidae (Stockel 1973), Plutellidae (Thibout 1971, Yang and Chow 1978), and Bombycidae (Omura 1938). The functional morphology and myology of the copulatory apparatus in Lepidoptera have been described in detail by Arnold and Fischer (1977) for *Speyeria* (Nymphalidae) and by Steklonikov (1965) in a comparative study of several butterfly species. Neural control of spermatophore formation and sperm transfer has been studied in the silkworm, *Bombyx mori* (MacFarlane and Tsao 1974). Based on these detailed studies, I present here a brief generalized description of sperm movements during and after mating. Because mating success in ditrysian Lepidoptera is strongly influenced by the degree of variation in morphology and functional complexity of the genitalia, detailed descriptions of the events leading to insemination (copulation and spermatophore transfer) and fertilization (sperm migration and storage) are presented in the Appendix for the benefit of those who may be interested in conducting research on the reproductive biology of Lepidoptera.

During copulation, the male grasps the tip of the female's abdomen with his claspers (valvulae), inserts his aedeagus through her ostium bursae, and everts the

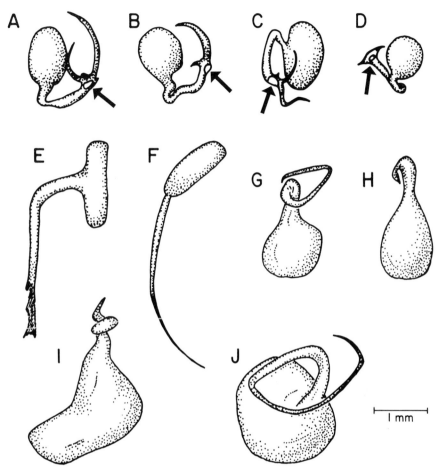

Fig. 3. Spermatophores of Lepidoptera. Pyralidae (Phycitinae): A. *Ephestia elutella* Hb.; B. *E. cautella* Wlk.; C. *Anagasta kuhniella* Z.; D. *Plodia interpunctella* Hb. Arrow shows aperture through which sperm escape. Redrawn from Norris (1932). Nymphalidae (Ithomiinae): E. *Tithorea harmonia hermias* Godman and Salvin; F. *Hypothyris mamercus mamercus* (Hewitson); G. *Godyris zavaleta amaretta* (Haensch); H. *Heterosais edessa nephele* (Bates); I. *Ceratina poecila poecila* (Bates); J. *Melinaea menophilus menophilus* (Hewitson). Original drawings. See text for discussion.

vesica, which extends to the corpus bursae. The spermatophore, molded into a tubelike structure in the cuticular simplex of the male, is forced into the bursae copulatrix via the vesica. In the corpus bursae, the body of the spermatophore becomes bulbous, while the collum is molded in the aedeagus and retained there until the sperm is transferred some 30-70 min after mating begins. After sperm transfer, the male positions the spermatophore collum so that its aperture lies opposite the opening of the ductus seminalis (Williams 1941a, Callahan 1960;

see Fig. 9B). The vesica is then retracted and the male withdraws the aedeagus and releases the female. Within the female, sperm migrate out of the spermatophore through the aperture and travel down the ductus seminalis to the common oviduct. Here the sperm cross the vestibulum and travel up the spermathecal duct to the spermatheca, the main site of sperm storage. Throughout their passage, the sperm are propelled by a combination of their own movements and peristaltic contractions of the female reproductive tract. Sperm migration from the spermatophore to the spermatheca is usually completed within a few hours after copulation begins.

III. MATING FREQUENCY IN NATURAL POPULATIONS

A. Methodology and Assumptions

Determination of mating frequency in natural populations may be obtained by direct observation of marked individuals over time, but this is a tedious and time-consuming process. Most studies documenting mating frequency of butterflies and moths come from laboratory experiments using caged adults or from indirect measurements based on dissection of wild-caught females. In the latter case, researchers capitalize on the presence of a spermatophore in the female's reproductive tract, formed *in situ* by the male during the act of copulation. Dissection of the corpus bursae (see Fig. 2) or microradiology of the female abdomen (Robin 1975) permits a count of the number of spermatophores she carries (see Fig. 9) and, by inference, of the number of times she has mated. Such a method makes three assumptions: (1) that a male transfers only a single spermatophore during each act of copulation; (2) that the passage of a spermatophore is required for insemination of the female; and (3) that the spermatophores remain recognizable in the corpus bursae regardless of age.

The validity of the first assumption has been clearly established by numerous experiments verifying the transfer of only a single spermatophore during copulation: for moths, Gelechiidae (Ouye *et al.* 1965), Olethreutidae (Dustan 1964, Proverbs and Newton 1962b), Pyralidae (Husseiny and Madsen 1964, Anwar and Feron 1971, Ashraf *et al.* 1974), Yponomeutidae (Taylor 1967), Noctuidae (Anwar et al. 1973); for skippers, Hesperiidae (Burns 1968); and for butterflies, Papilionidae (Sims 1979, Lederhouse 1981), Pieridae (Suzuki 1979, Burns 1968), Nymphalidae: Ithomiinae (Drummond, unpubl. data), Danainae (Pliske 1973), Nymphalinae (Burns 1968; Drummond, unpubl. data).

As to the second assumption, not every copulation results in the passage of a spermatophore, and failure usually is attributed to the condition of the male. Old worn males (Suzuki 1979), young males still reproductively immature, or recently-mated males (Sims 1979) sometimes fail to pass a spermatophore, even after a greatly extended period of copulation. Failure to form a spermatophore has been widely regarded as failure to inseminate, but at least one study suggests otherwise.

George and Howard (1968), in experiments on the Oriental fruit moth, *Grapholitha molesta* (Tortricidae), found that males were capable of mating two females within a 24-h period. The first mating always resulted in the passage of a large spermatophore, but subsequent matings by the same male resulted in smaller and eventually no spermatophores. Yet there was no difference in the number of fertilized eggs produced by females who received large, small, or no spermatophores. Even though mating appears to progressively deplete the accessory gland fluids to the point where no detectable spermatophores are produced, male fertility does not diminish concomitantly. The volume of accessory glands passed to the female has significant consequences for the reproductive success of both male and female (see below), but the important point is that spermatophore counts may underestimate the number of successful (inseminating) copulations, particularly in dense populations or in populations with strongly female-biased sex ratios, where males may be afforded many opportunities for mating in a short period of time.

The third assumption, that of spermatophore persistence, depends on the taxon. In some short-lived temperate zone butterflies, the spermatophore is known to persist intact in laboratory-held females for longer than the life expectancy of a female in the wild (Sims 1979, Lederhouse 1981). On the other hand, the spermatophore of the yponomeutid moth, *Porthetria dispar,* disintegrates within 2 h after forming (Taylor 1967), and rapid disintegration appears to be the rule in the lower Lepidoptera as well as in most other insect orders (Callahan and Cascio 1963). Although insect spermatophores are predominantly proteinaceous (Davey 1959), spermatophores of higher Lepidoptera contain chitin, which may contribute to the extreme durability of the spermatophore of advanced groups such as Noctuidae (Callahan 1958). The possibility that the addition of chitin to the proteinaceous spermatophore may have evolved as a male strategy to reduce courtship receptivity in females to subsequently courting males is discussed in section IV.B.

Evidence for the gradual absorption of spermatophores in some higher Lepidoptera, particularly the moth families Arctiidae (Pease 1968), Tortricidae (Outram 1968), and the butterfly subfamilies Parnassiinae (Papilionidae), Satyrinae, the heliconine Nymphalinae (Nymphalidae), and Lycaeninae (Lycaenidae) (Ehrlich and Ehrlich 1978), suggests that spermatophore counts may underestimate the number of copulations in these groups. However, the sclerotized collum of the spermatophore (Fig. 3) usually persists in the corpus bursae even if the body of the spermatophore has been completely dissolved (Pease 1968; Outram 1968; Drecktrah 1978; Drummond, unpubl. data). In fact, the progression of postcopulatory changes in spermatophores, including loss of volume (depletion), color change, and collum retraction, is sometimes so regular that it has been used as a technique for determining relative age in the females of some species, such as the European corn borer, *Ostrinia nubialis* (Pyralidae) (Showers *et al.* 1974, Elliott and Dirks 1979).

Spermatophore counts, therefore, appear to be a useful method for determining the mating frequencies of females in natural populations of Lepidoptera, but such counts may be underestimates of true mating frequency because of (1) dissolution or absorption of spermatophores in some groups, and (2) the potential for insemination without spermatophore formation under certain conditions.

Estimates of copulation frequency for males in natural populations are few because, unlike in females, there are no physical markers that accumulate to record the number of pairings. Dissection of wild-caught males, therefore, reveals nothing quantitative about their mating history, although reduction in quantity of black pigment within the ductus ejaculatorious simplex has been used to distinguish between virgin and mated status in dissected males of *Spodoptera frugiperda* (Noctuidae, Snow and Carlysle 1967). Studies of male mating frequency and male mate choice in natural populations could be facilitated if the reproductive tissues of males could be labeled such that, upon dissection of mated females, spermatophores passed by labeled males could be distinguished from those passed by unlabeled males. Although techniques for labeling spermatophores are available, to my knowledge there have been no studies involving the release of labeled male Lepidoptera into natural populations followed by capture and dissection of females to determine relative male mating frequencies.

Larval diets supplemented with low concentrations (ca. 0.05%) of Calco Oil Red dye have produced males that transfer red spermatophores clearly distinguishable to the naked eye in *Heliothis virescens* (Hendricks *et al.* 1970), *H. zea* (Burton and Snow 1970), and *Manduca sexta* (Cantelo 1973). Sparks and Cheatham (1973) tested over 60 water-soluble biological stains for their ability to stain the reproductive systems of male *M. sexta* and found four that consistently stained spermatophores so they could be recognized upon dissection of the female. Furthermore, the dye disappeared from the secretory tissues of the ductus ejaculatorius duplex in the male after mating, suggesting that dissection of recaptured treated males could indicate their mated status. Even though the amount of dye passed with the spermatophore may decrease with male age and with the number of matings (Burton and Snow 1970), dye markers appear to be a convenient tool for studying mating frequency and reproductive physiology of male Lepidoptera (Sparks and Cheatham 1973).

Radioactive tracers also have been used to label male reproductive tissues to verify spermatophore transfer, insemination, and fertilization in Lepidoptera. Amino acids labeled with ^{14}C or ^{3}H or both have been added to the diets of larvae and adults or injected into anesthetized larvae or pupae in the butterflies *Dryas julia*, *Heliconius charitonius* (Boggs 1981a), *H. erato* (Boggs and Gilbert 1979), and *Colias eurytheme* (Boggs and Watt 1981); and in the moths *Choristoneura fumiferana* (Retnakaran 1971a, 1974) and *Heliothis virescens* (Flint and Kressin 1968). Males treated in this way transfer spermatophores, sperm, and accessory fluids, the radioactivity of which can be detected after mating by autoradiography or liquid scintillation counting. In a dispersal study of the tortricid moth, *Adoxophyes orana,* Noordink and Minks (1970) supplemented larval diets with ^{22}Na or ^{32}P and released both radioactive larvae and adults. Recapture of treated individuals was verified by autoradiography, suggesting mark-and-recapture experiments using radioactive tagging as a viable method for studying mating frequency and mate choice in the field.

B. Repeat Mating by Males

1. Estimates of Copulation Frequency

Scott (1972) reviewed the mating behavior of butterflies and summarized data on the mating frequency of males, mostly derived from pairings under laboratory conditions. Of the 20 species for which data were available, eight can mate at least twice, six can mate at least four times, four can mate at least five times, one (*Heliconius*) can mate at least 10 times, and one (*Pieris protodice*) can mate at least 13 times. Most of the species that can mate more than twice are able to mate on successive days and at least three species can mate twice per day. It is doubtful that male butterflies in the wild mate more than twice in one day, a limitation established experimentally for two species of Papilionidae, in which males are known to mate several times in a lifetime (Sims 1979, Lederhouse 1981). Because many of the studies reviewed by Scott involved the continuous exposure of males to receptive females in cages, these figures may overestimate the actual mating frequency of males in natural populations. Even so, experimental studies in natural populations of butterflies have shown that males of *Poladryas minuta* (Nymphalidae, Scott 1974) and *Incisalia iroides* (Lycaenidae, Powell 1968) can average one or slightly more than one copulation per day.

Laboratory experiments with caged moths show that males are capable of inseminating several females. When one male was held with five virgin females for several days, the mean number of spermatophores produced per male was 1.9 for *Pectinophora gossypiella* (Henneberry and Leal 1979), 1.24 (maximum 3) for *Zeiraphera diniana* (Benz 1969), 1.8 (maximum 3) for *Paramyelois transitella* (Goodwin and Madsen 1964), 1.37 (maximum 10) for *Choristoneura fumiferana* (Outram 1971a), and 3.8 (maximum 5) for *Laspeyresia pomonella* (Gehring and Madsen 1963).

2. Lifetime Potential for Copulation

Several studies have attempted to determine a lifetime potential mating frequency for male Lepidoptera by providing males with continuous access to receptive virgin females for the life span of the male. For example, the mean number of matings achieved per male was 4.2 (maximum 10) for *Pectinophora gossypiella* (Gelechiidae, Ouye *et al.* 1965), 4.2 for *Earias insulana* (Noctuidae, Kehat and Gordon 1977), and 5.7 (maximum 11) for *Heliothis virescens* (Noctuidae, Flint and Kressin 1968). For *Laspeyresia pomonella* (Olethreutidae), Howell *et al.* (1978) found that the mean number of spermatophores produced per male during a lifetime was positively correlated with the degree of female bias of the sex ratio and with the number of males per cage, although overall population density *per se* had no measurable effect. By contrast, Richerson *et al.* (1976) found extremely high mating frequencies of males in dense populations of spruce budworms, *Choristoneura fumiferana.*

3. Limitations to Copulation Frequency

Male Lepidoptera are limited in the frequency with which they can mate by a number of factors. First, males of many species require a maturation period of 1-3 days after emergence in which to accumulate sufficient sperm and accessory fluids for successful copulation (Callahan and Cascio 1963). For temperate or alpine species with short flight periods, such a maturation period could severely reduce the time during which mature males are exposed to receptive females. The existence of a maturation period may be one of the factors promoting protandry in temperate zone butterflies, in which males precede females in emergence by one to a few days (see Wiklund and Fagerstrom 1977, Fagerstrom and Wiklund 1982, Lederhouse *et al.* 1982). In the genus *Papilio* (Papilionidae) for example, males of *P. zelicaon* lack normal sperm concentrations during the first adult days (Sims 1979), several species require a day or more for the hardening of the external genitalia (Lederhouse *et al.* 1982), and males of *P. polyxenes* must first establish mating territories before successful courtship can occur (Lederhouse 1982). In territorial species of Lepidoptera, inability of some males to obtain an appropriate mating territory is a further limitation to mating frequency because it reduces access to females for such males (Powell 1968, Lederhouse 1982).

Second, copulation depletes a male of reserves of sperm and accessory fluids and thus there is a refractory period of seminal replenishment during which males are less likely to initiate courtship. In *Pieris protodice,* Rutowski (1979a) found that recently mated males suffer a reduction in potency resulting from reduced amounts of ejaculate available to impregnate females and an increase in the time required for insemination of a second female. Because the courtship persistence of males dropped by a factor of 10 after copulation, Rutowski described these butterflies as "honest salesmen." This finding suggests that the prolonged court-ship of many butterfly species may function in part to allow the female to assess male potency on the basis of the strength of courtship persistence (Rutowski 1982). Furthermore, such a relationship between courtship intensity and potency may be an evolutionary product of selection on males to minimize time in court-ship and maximize time recouping when their sperm and accessory fluids are low, while at the same time not passing up opportunities to mate with highly receptive females (Rutowski 1979a). Scott (1972) suggested that species in which males produce large spermatophores or a large sphragis (copulatory plug) probably mate less often because of the increased time required to recover from the depletion of the accessory glands.

The refractory period between matings appears to increase with male age. In an experimental study of *Papilio zelicaon*, Sims (1979) found that although there was no correlation between male age and the success of the second mating, older males were less capable of remating frequently than were younger males, and older males required more time in copula to produce a spermatophore. Other studies have documented the decreased ability of older males to multiple-mate, to stim-ulate vitellogenesis and oviposition in females, or to fully inseminate their partners (*e.g.*, George and Howard 1968, Stockel 1971, Thibout and Rahn 1972).

Third, the number of successful copulations in which a male can engage during a lifetime may be limited by the total number of sperm and the volume of accessory fluids that can be produced. In many temperate zone butterfly species, spermato-genesis is completed by the time of emergence from the pupal stage (T. C. Emmel, pers. comm.), although in some species (*e.g., Papilio zelicaon*) there is evidence of continuous sperm production (Sims 1979). Long-lived tropical butterfly species probably are not limited in their cumulative mating frequency by a finite amount of sperm because T. C. Emmel (pers. comm.) has found spermatogenesis occurring in males of all ages of such groups (*e.g.*, Ithomiinae and Heliconiini). Accessory fluids from the ductus ejaculatorius duplex and the primary segment of the ductus ejaculatorius simplex are much more likely to be limiting because they are pro-duced by apocrine secretion, which results in the gradual exhaustion of the colum-nar epithelium (Callahan and Cascio 1963, Riemann and Thorson 1976). Evidence that depletion of accessory gland fluids is likely to limit successful repeat mating by males before sperm supplies are exhausted comes from previously cited experiments by George and Howard (1968) in which insemination without spermatophore transfer was achieved in the later copulations of multiple-mated males. Thus, even though some of the energetic expenditures of courtship such as search, pursuit, and display may be recouped by adult feeding, some of the fluid components passed to the female during copulation may represent irreplaceable losses.

Of critical importance here is the recent discovery that some of the nutrients in the spermatophore and seminal fluids from the accessory glands represent a male nutrient investment in reproduction, now documented for Ctenuchidae

(Goss 1977), Danainae and Heliconiini (Nymphalidae, Boggs and Gilbert 1979), and Pieridae (Boggs and Watt 1981). Although the importance of the male nutrient component to female reproduction probably varies with the resources available to populations (Boggs 1981b), it is suspected to be a general phenomenon in butterflies. Under conditions that favor substantial nutrient investment by males during mating, males may face reduced mating ability or lower survival resulting from nutrient depletion. The lack of a relationship between the quantity or timing of pollen feeding and the number of copulations in male *Heliconius cydno,* a species in which males are known to transfer nutrients during mating (Boggs 1981b), suggests that males may have limited ability to make up the loss of nutrient reserves consumed by sexual activity.

Shapiro (1982) has recently demonstrated the reduced survival of mated male *Tatochila* butterflies (Pieridae) compared with unmated males when placed in cold storage. Shapiro suggests that this reduction in survivorship in mated males may be the result of diversion of nutrients into forms transferable to the female during mating, which otherwise would have been used for maintenance under long-term cold storage. Relating the rigors of cold storage in the laboratory to actual conditions in nature is difficult (although *Tatochila* is an alpine genus), but the fact is that single-mated males suffered a reduction in life span of 45% and doubly-mated males a reduction of 60%. This suggests that even such an indirect measure as cold storage is sufficient to illustrate the sizeable cost of the male nutrient investment during copulation, regardless of whether those nutrients are available to the female after mating. Rutowski (1982) estimates that the total quantity of material passed by male butterflies during copulation may be as much as 10% of the male's body weight. Because accessory gland secretions contain a complex assortment of proteins and lipids (Marshall 1982), each copulation represents a significant energetic investment in transferred fluids alone. Amoaka-Atta (1976) has confirmed that repeat matings decrease male survivorship in a study of *Ephestia* (=*Cadra*) *cautella* (Noctuidae). Males continually exposed to excess females mated frequently (up to eight times) and had a significantly lower (P < 0.01) mean longevity (4.25 ± 1.11 days) than unmated males and mated males whose first mating was delayed by several days (7.75 ± 1.99 days). This inverse relationship between lifetime spermatophore production and survivorship, based on experiments in which caged males of the spiny bollworm, *Earias insulana* (Noctuidae), were exposed to excess virgin females (Kehat and Gordon 1977) is shown in Fig. 4.

C. Repeat Mating by Females

1. Estimates of Copulation Frequency

The increasing number of studies that report spermatophore counts of wild-caught females of Lepidoptera make possible a much more comprehensive analysis

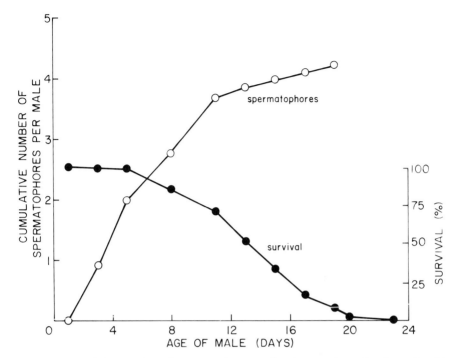

Fig. 4. Cumulative number of spermatophores produced per male, and male survivorship in the spiny bollworm, *Earias insulana* (Noctuidae). Males were kept in glass jars (25°C, 60-70% RH) with three or four receptive virgin females (replaced regularly) for the lifespan of the male, which averaged 14.7 ± 0.6 days (N = 35). Mean number of matings per male was 4.2 ± 0.3 S.E. From Kehat and Gordon (1977).

of multiple-mating in females than in males. Mating frequencies in natural populations of some 79 species of Lepidoptera based on spermatophore counts are presented in Table I. Inclusion in the table was limited to populations for which there was a minimum sample size of 20 females, although exceptions were made in a few cases where comparative data from seasonal or migrating populations were deemed more important than minimum sample size. Ehrlich and Ehrlich (1978) present spermatophore counts and other quantitative measures of female reproductive biology of butterflies, but the majority of the more than 300 species they studied are represented by only one dissected female; only 18 species are represented by 20 or more dissections. The 20-specimen minimum adopted here is a compromise; Burns (1968) calculated that a minimum sample size of 30-45 is needed for mating studies based on spermatophore counts, but adherence to this suggestion would have reduced the number of entires in Table I by about 40%. Mating frequency is expressed here as the mean number of spermatophores per **mated** female as recommended by Burns (1968) and followed by most researchers.

TABLE I

Lepidopteran Mating Frequencies in Natural Populations Based on Spermatophore Counts of Field-captured Females

SUPERFAMILY Family (Subfamily) Genus species	N^a	Mean No. Spermatophores Per Mated Female[b]	Max. No. Spermatophores Per Female	% Virgins[c]	% Mated Once[d]	% Multiply Mated[e]	Collection Locality[f] (Method)[g]	Source
TORTRICOIDEA								
Olethreutidae								
Grapholitha molesta (Busck)	629	1.25	5	2.9	78.8	18.3	Canada(B)	Dustan 1964
Tortricidae								
Choristoneura fumiferana (Clem.)	194	1.68	5	4.1	46.4	49.5	Canada(L)	Outram 1968
GELECHIODEA								
Gelechiidae								
Pectinophora gossypiella (Saunders)	2270	1.09	6	11.1	70.8	18.1	Texas(L)	Graham *et al.* 1965
PYRALOIDEA								
Pyralidae								
Cnaphalocrocis medinalis Guenee	548	-	-	70.3	-	-	Kyushu(L)	Wada *et al.* 1980
	502	-	-	76.9	-	-	Kyushu(L)	Wada *et al.* 1980
	577	-	-	70.5	-	-	Kyushu(L)	Wada *et al.* 1980
Paramyelois transitella (Walker)	230	1.19	3	5.2	79.1	15.7	Calif.(L)	Goodwin and Madsen 1964
Diatraea saccharalis (F.)	674	1.02	2	5.0	93.0	2.0	La.(L)	Perez and Long 1964
Diatraea grandiosella Dyar	3436	1.51	4	0.1	59.4	40.5	Kansas(L)	Schenck and Poston 1979
Ostrinia nubilalis (Hubner) (1st generation range of 5-year averages)	3027	1.10–1.33	4	-	71–84	15–28	Iowa(L)	Showers *et al.* 1974
(2nd and 3rd generations range of 5-year averages)		1.18–1.32						

HESPERIODEA
Hesperiidae (Pyrginae)

Taxon	N						Locality	Reference
Epargyreus clarus (Cramer)	62	1.44	3	4.8	56.5	38.7	Va.(N)	Burns 1968
Urbanus proteus (L.)	20	1.11	2	10.0	80.0	10.0	Ark.(N)	Drummond, unpubl. data
	298	-	-	78.9	-	-	Florida(M)	Walker 1978
Urbanus simplicius (Stoll)	67	1.48[2.50]	4	3.0	58.2	38.8	Trinidad(N)	Pliske 1973
	60	1.39[3.00]	4	5.0	66.7	28.3	Trinidad(N)	Pliske 1973

Hesperiidae (Hesperiinae)

Taxon	N						Locality	Reference
Atalopedes campestris (Bsd.)	55	1.07	2	1.8	90.9	7.3	Texas(N)	Burns 1968
Euphyes vestris (Bsd.)	46	1.45	3	4.3	60.9	34.8	Texas(N)	Burns 1968
Hesperia sassacus Harris	31	1.34	3	6.5	67.7	25.8	Va.(N)	Burns 1968
Hylephila phyleus (Drury)	80	1.10	2	0	90.0	10.0	Calif.(N)	Shapiro 1977
Lerema accius (Smith)	44	2.03	6	18.2	38.6	43.2	SC(N)	Burns 1968
Panoquina ocola (Edwards)	39	-	-	38.5	-	-	Florida(M)	Walker 1978
Poanes viator (Edwards)	79	1.06	2	0	93.7	6.3	Texas(N)	Burns 1968
Polites mystic (Edwards)	49	1.33	4	0	75.5	24.5	Va.(N)	Burns 1968
Thymelicus lineola (Ochs.)	54	1.09	2	0	90.7	9.3	Mich.(N)	Burns 1968
Wallengrenia otho (Smith)	171	1.13	3	1.8	87.1	11.1	Texas(N)	Burns 1968

PAPILIONOIDEA
Papilionidae

Taxon	N						Locality	Reference
Battus philenor (L.)	33	1.73	5	0	51.5	48.5	Va.(N)	Burns 1968
Papilio glaucus	220	1.16[1.23]	3	0.9	84.1	15.0	Florida(N)	Pliske 1972
	150	1.14[1.35]	3	0	86.7	13.3	Florida(N)	Pliske 1973
	84	1.75	5	0	46.4	53.6	Va.(N)	Burns 1968
	200	1.54	3	2.0	51.5	46.5	Va.(N)	Makielski 1972
	29	1.72	3	0	41.4	58.6	Md.(N)	Burns 1968
	92	1.73	3	0	37.0	63.0	Md.(N)	Burns 1968
Papilio palamedes Drury	32	1.47[2.50]	3	0	55.5	34.4	Florida(N)	Pliske 1973
Papilio polyxenes Fabr.	171	1.33[1.86]	3	2.3	67.3	30.4	NY(N)	Lederhouse 1981
Papilio troilus L.	358	1.23[2.27]	4	0.5	79.9	19.6	Florida(N)	Pliske 1973
Papilio zelicaon Lucas	97	1.23[1.37]	3	5.2	75.3	19.5	Calif.(N)	Sims 1979
	84	1.21	3	60.7	33.3	6.0	Calif.(N)	Shields 1967

Pieridae (Pierinae)

Taxon	N						Locality	Reference
Ascia monuste (L.)	99	1.20[2.00]	3	0	81.8	18.2	Florida(N)	Pliske 1973
Euchloe ausonides Lucas	43	1.74[1.00]	3	0	44.2	55.8	-(N)	Ehrlich and Ehrlich 1978

TABLE I

Lepidopteran Mating Frequencies in Natural Populations Based on Spermatophore Counts of Field-captured Females
(continued)

SUPERFAMILY Family (Subfamily) Genus species	N[a]	Mean No. Spermatophores Per Mated Female[b]	Max. No. Spermatophores Per Female	% Virgins[c]	% Mated Once[d]	% Multiply Mated[e]	Collection Locality[f] (Method)[g]	Source
Phoebis sennae (L.)	37	-	-	64.9	-	-	Florida(M)	Walker 1978
	77	1.05	2	15.6	80.5	3.9	Florida(M)	Drummond, unpubl. data
Pieris rapae crucivora Bsd.	349	1.62[2.84]	7	3.2	53.3	43.5	Japan(N)	Suzuki 1979
Pieris napi complex (L.)	27	1.42[1.67]	3	11.1	59.3	29.6	–(N)	Ehrlich and Ehrlich 1978
Pieridae (Coliadinae)								
Colias philodice Godart	26	1.15[1.40]	2	0	84.6	15.4	Calif.(N)	Stern and Smith 1960
	20	1.21[1.44]	2	5.0	75.0	20.0	Calif.(N)	Stern and Smith 1960
	25	1.12[1.43]	2	0	88.0	12.0	Calif.(N)	Stern and Smith 1960
	23	1.22[1.67]	3	0	73.9	26.1	Calif.(N)	Stern and Smith 1960
	31	1.35	3	0	67.7	32.3	Florida(N)	Rutowski 1978a
Eurema lisa Bsd. & LeConte								
Nymphalidae (Ithomiinae)								
Ithomia drymo Hbn.	27	2.91[6.00]	8	59.3	14.8	25.9	–(N)	Ehrlich and Ehrlich 1978
Nymphalidae (Danainae)								
Danaus gilippus (Cramer)	194	4.02[8.05]	15	1.0	22.7	76.3	Florida(N)	Pliske 1973
	70	2.35[5.00]	8	1.4	42.9	55.7	Trinidad(N)	Pliske 1973
	50	2.63	10	2.0	34.0	64.0	Texas(N)	Burns 1968
	91	2.23[4.33]	8	4.4	46.2	49.4	Trinidad(N)	Pliske 1973
Danaus plexippus (L.)	80	0	-	100	0	0	NY(N)	Williams *et al.* 1942
(fall migration south)								
(overwintering group)	38	1.10	2	44.7	50.0	5.3	Calif.(N)	Williams *et al.* 1942
(spring migration north)	5	3.60	5	0	0	100	Calif.(N)	Hill *et. al.* 1976
(spring migration north)	25	2.60	7	0	16.0	84.0	Calif.(N)	Ehrlich and Ehrlich 1978
Nymphalidae (Satyrinae)								
Cercyonis pegala (Fabr.)	28	1.04	2	0	96.4	3.6	Va.(N)	Burns 1968
Coenonympha tullia complex Muller	44	1.05[1.14]	2	4.5	91.0	4.5	–(N)	Ehrlich and Ehrlich 1978
Erebia epipsodea Butl.	197	1.21[1.36]	3	2.0	79.2	18.8	–(N)	Ehrlich and Ehrlich 1978

Species						Location	Reference
Euptychia hermes (Fabr.)	27	1.15[1.17]	2	3.7	81.5	14.8	–(N)	Ehrlich and Ehrlich 1978
	103	1.07[1.60]	2	1.9	91.3	6.8	Trinidad(N)	Pliske 1973
Euptychia renata Cramer	22	1.18[1.29]	2	0	81.8	18.2	–(N)	Ehrlich and Ehrlich 1978
Oeneis chryxus (Dbd.)	50	1.28[1.33]	3	6.0	76.0	18.0	–(N)	Ehrlich and Ehrlich 1978
Nymphalidae (Nymphalinae)								
Anartia amathea (L.)	104	1.18[1.00]	3	2.9	80.8	16.3	Trinidad(N)	Pliske 1973
Biblis hyperia (Cramer)	41	1.95[4.00]	7	0	41.5	58.5	Trinidad(N)	Pliske 1973
Chlosyne palla Bsd.	21	1.10[1.00]	2	4.8	85.7	9.5	–(N)	Ehrlich and Ehrlich 1978
Euphydryas anicia Dbd.	24	1.22[1.50]	2	4.2	75.0	20.8	–(N)	Ehrlich and Ehrlich 1978
Euphydryas editha Bsd.	22	1.27[1.00]	2	0	72.7	27.3	–(N)	Ehrlich and Ehrlich 1978
Phyciodes campestris Behr	22	1.10[1.00]	2	9.1	81.8	9.1	–(N)	Ehrlich and Ehrlich 1978
Poladryas minuta (Edwards)	51	1.03	-	27.4	70.6	2.0	Colo.(N)	Scott 1974
Precis coenia (Hubner)	42	-	-	35.7	-	-	Florida(M)	Walker 1978
Precis lavinia (Cramer)	101	1.50[2.00]	3	8.9	50.5	40.6	Trinidad(N)	Pliske 1973
Speyeria cybele (Fabr.)	42	1.05	2	0	95.2	4.8	Va.(N)	Burns 1968
	26	1.04	2	0	96.2	3.8	Md.(N)	Burns 1968
Nymphalidae (Nymphalinae: Heliconiini)								
Agraulis vanillae (L.) (fall migration)	102	1.16[1.70]	2	4.9	79.4	15.7	Trinidad(N)	Pliske 1973
(fall migration)	26	-	-	23.0	-	-	Florida(M)	Walker 1978
(spring migration)	20	1.05	2	0	95.0	5.0	Florida(M)	Drummond, unpubl. data
	5	-	-	0	-	-	Florida(M)	Walker 1978
Heliconius ethilla Godt.	28	1.04[1.06]	2	7.1	89.3	3.6	–(N)	Ehrlich and Ehrlich 1978
Heliconius erato L.[h]	21	(1.00)	1	9.5	(76.2)	(0)	–(N)	Ehrlich and Ehrlich 1978
	99	1.10[1.00]	3	3.0	87.9	9.1	Trinidad(N)	Pliske 1973
Nymphalidae (Acraeinae)								
Acraea egina (Cr.)	42	1.66	4	2.4	54.8	42.8	Sierra Leone-3(N)	Owen et al. 1973b
Acraea encedon (L.)	38	2.19	7	5.3	34.2	60.5	Sierra Leone-3(N)	Owen et al. 1973b
	67	1.00	1	92.5	7.5	0	Uganda-1(N)	Owen et al. 1973b
	57	1.07	2	75.4	22.8	1.8	Uganda-1(N)	Owen et al. 1973b
	22	1.00	1	90.9	9.1	0	Uganda-2(N)	Owen et al. 1973b
	29	-	0	100	0	0	Uganda-3(N)	Owen et al. 1973b
	101	1.09	2	77.2	20.8	2.0	Uganda-4(N)	Owen et al. 1973b
	58	1.10	2	82.8	15.5	1.7	Uganda-4(N)	Owen et al. 1973b

TABLE I

Lepidopteran Mating Frequencies in Natural Populations Based on Spermatophore Counts of Field-captured Females
(continued)

SUPERFAMILY Family (Subfamily) Genus species	N[a]	Mean No. Spermatophores Per Mated Female[b]	Max. No. Spermatophores Per Female	% Virgins[c]	% Mated Onced	% Multiply Mated[e]	Collection Locality[f] (Method)[g]	Source
Acraea encedon (L.)	123	1.15	4	20.3	69.9	9.8	Sierra Leone-1(N)	Owen *et al.* 1973b
	6	2.00	3	0	33.3	66.7	Sierra Leone-2(N)	Owen *et al.* 1973b
Acraea eponina (Cr.)	20	1.29	2	15.0	60.0	25.0	Sierra Leone-3(N)	Owen *et al.* 1973b
Lycaenidae (Lycaeninae)								
Agriades glandon de Prunner[h]	27	(1.05)[1.00]	2	11.1	(66.7)	(3.7)	–(N)	Ehrlich and Ehrlich 1978
Glaucopsyche lygdamus Dbd.[h]	55	(1.17)[1.38]	2	7.2	(61.8)	(12.7)	–(N)	Ehrlich and Ehrlich 1978
Plebejus icarioides Bsd.[h]	46	(1.11)[1.00]	2	13.0	(67.4)	(8.7)	–(N)	Ehrlich and Ehrlich 1978
Plebejus saepiolus Bsd.[h]	37	(1.07)[1.00]	2	5.4	(75.7)	(5.4)	–(N)	Ehrlich and Ehrlich 1978
NOCTUOIDEA								
Arctiidae								
Utethesia ornatrix bella (L.)	88	3.84[6.25]	11	9.1	13.6	77.3	Florida(N)	Pease 1968
Noctuidae								
Heliothis armigera (Hb.)	328	-	-	25.0	-	-	Botswana(L)	Roome 1975
	2433	-	-	73.7	-	-	Botswana(L)	Roome 1975
Heliothis virescens (F)	270	1.83	6	27.4	39.2	33.4	Texas(L)	Hendricks *et al.* 1970
	24	3.44	5	62.5	0	37.5	N.Miss.(L)	Stadelbacher and Pfrimmer 1973
	14	3.25	5	73.3	0	26.7	C.Miss.(L)	Stadelbacher and Pfrimmer 1973

Species	68	2.80	5	63.8	7.3	28.9	S.Miss.(L)	Stadelbacher and Pfrimmer 1973
Heliothis zea (Boddie)	1295	1.56[1.73]	5	40.1	35.1	24.8	La.(L)	Callahan 1958
	1252	1.55[2.49]	6	34.6	43.3	22.1	Texas(L)	Hendricks et al. 1970
	5062	2.02[3.03]	6	43.7	27.1	29.2	Texas(L)	Lopez et al. 1978
	1026	2.09[2.95]	6	39.8	25.8	34.4	N.Miss.(L)	Stadelbacher and Pfrimmer 1973
	797	2.21[3.19]	6	44.6	25.4	30.1	C.Miss.(L)	Stadelbacher and Pfrimmer 1973
	595	2.01[3.02]	6	46.4	24.9	28.7	S.Miss.(L)	Stadelbacher and Pfrimmer 1973
Peridroma saucia (Hbn)	239	2.19[2.36]	5	84.9	6.7	8.4	La.(L)	Callahan and Chapin 1960
Pseudaletia unipuncta (Haw)	417	1.89[2.03]	5	43.6	25.7	30.7	La.(L)	Callahan and Chapin 1960
Spodoptera litura F.	633	1.77	5	10.1	43.0	46.9	Japan(L)	Takeuchi and Miyashita 1975
	96	1.73	4	17.7	40.6	41.7	Japan(L)	Miyashita and Fuwa 1972
Euxoa perolivalis (Smith)	42	10.86	16	0	0	100	Mont.(L)	Byers 1978
Euxoa satis (Harvey)	50	5.90	11	0	14.0	86.0	Mont.(L)	Byers 1978
Euxoa atristrigata (Smith)	50	2.98	6	0	2.0	98.0	Mont.(L)	Byers 1978
Euxoa catenula (Grt.)	35	2.83	6	14.3	8.6	77.1	Oregon(L)	Byers 1978
Euxoa cicatricosa (G. + R.)	61	2.80	8	0	26.2	73.8	Calif.(L)	Byers 1978
Euxoa auxiliaris (Grote)	31	2.27	5	3.2	29.1	67.7	Wyo.(L)	Byers 1978
	100	1.66	4	1.0	49.0	50.0	Oregon(L)	Byers 1978
Euxoa plagigera (Morr.)	30	1.65	3	3.4	53.3	43.3	Wyo.(L)	Byers 1978
Euxoa andera Smith	24	1.74	3	4.2	37.5	58.3	Mont.(L)	Byers 1978
	40	1.65	3	0	47.5	52.5	Oregon(L)	Byers 1978

TABLE I

Lepidopteran Mating Frequencies in Natural Populations Based on Spermatophore Counts of Field-captured Females
(continued)

aNumber of females dissected. With few exceptions (see text), only studies reporting 20 or more dissections per population are included in the table.

bMain entries represent mean number of spermatophores per mated female for entire sample. Figures in brackets represent mean number of spermatophores per mated female for the oldest age class in the sample.

cPercent of females with no spermatophores in the bursa copulatrix.

dPercent of females with one spermatophore in the bursa copulatrix. Studies by Roome (1975), Walker (1978), and Wada *et al.* (1980) did not distinguish between singly-mated and multiply-mated females.

ePercent of females with two or more spermatophores or spermatophore colli in the bursa copulatrix.

fNo localities were reported for the species studied by Ehrlich and Ehrlich (1978).

gB = bait trap; L = blacklight trap; M = Malaise trap (two-directional); N = aerial hand net.

hThese species absorb spermatophores. Percentages for virgins are accurate (having been confirmed by egg counts, extent of fat bodies, and lack of scale loss in genital areas), but some nonvirgins without spermatophores could not be assigned to the singly-mated or multiply-mated groups (so that percentages do not total 100). It is also likely that some species, scored as singly-mated, were actually multiply-mated.

Mating frequencies for many of the populations in Table I represent minimum values for several reasons. First, as previously explained, certain groups tend to absorb spermatophores, which tends to underestimate mating frequency. Second, wild-caught females have had their reproductive lives terminated early in the interests of science; unless they were post-reproductive, it is possible that they might still have mated. Ideally, of course, a sample of females that purports to reflect mating frequency in a natural population should be composed only of post-reproductive individuals, but this is rarely the case (Scott 1972). At the very least, mating frequency should be recorded by age class, because in multiply-mated females it is necessarily true that older females will have mated more often than younger ones. Techniques for measuring the relative age of butterflies and moths (progressive wing-scale loss, ovariole condition, and fat body depletion) have been described by a number of researchers (e.g., Callahan 1958, Pease 1968, Hendricks et al. 1970, Pliske 1973, Stadelbacher and Pfrimmer 1973, Drummond 1976, Lopez et al. 1978, Ehrlich and Ehrlich 1978, Sims 1979) whose data are entered in Table I. However, to make all entries as comparable as possible, the mating frequencies are not broken down by age but represent averages for the populations studied. Where data permitted, mating frequency of the oldest female age class in each population was calculated and included in Table I for comparison with the population mean.

It is clear from these data that although the majority of females in most populations have mated once, none of the species is exclusively monogamous. Considerable variation exists among species, and even among populations of the same species, in the relative proportions of singly-mated, multiply-mated, and virgin females. Interpretation of this variation requires consideration of population density, age structure, and sex ratio; effects of seasonality, migration, and weather; and the availability and distribution of adult and larval food sources.

Although the morphology of spermatophores (Fig. 3) can be of taxonomic utility (Norris 1932, Drummond 1976), it is not likely that female mating frequencies constitute "a distinct ethologic taxonomic character" (Burns 1968). As Pliske (1973) pointed out, mating frequencies in female Lepidoptera show great intraspecific variation (see, for example entries for populations of Epargyreus clarus, Urbanus proteus, Papilio glaucus, Phoebis sennae, Danaus plexippus, and Acraea encedon in Table I). There are, however, general trends among higher taxa that may have some usefulness to taxonomists, but that are undoubtedly of much more interest to ecologists and ethologists. For example, some species show a consistently high proportion of multiply-mated females (Utethesia ornatrix, Danaus plexippus, D. gilippus, Papilio glaucus, Euchloe ausonides, Biblis hyperia, and some species of Acraea), while other species show high proportions of virgins in at least some populations (Cnapholocrocis medinalis, Papilio zelicaon, Ithomia drymo, most Noctuidae, and some populations of Acraea encedon). Despite variations in female mating frequency among populations and between years in several species of *Euxoa* (see Table I), Byers (1978) argues that intrageneric variation in mating frequencies in *Euxoa* reflects species-specific differences in biology.

2. Lifetime Potential for Copulation

A correlation between female age and the number of spermatophores carried has been documented for moths (Callahan 1958, Callahan and Chapin 1960, Outram 1968, Pease 1968, Hendricks *et al.* 1970, Stadelbacher and Pfrimmer 1973, Lopez *et al.* 1978), skippers (Pliske 1973), and butterflies (Pliske 1972, 1973; Ehrlich and Ehrlich 1978; Sims 1979; Suzuki 1979; Lederhouse 1981). Thus, some indication of lifetime mating frequencies for females can be gained by examining only the oldest, and preferably post-reproductive, females in the population. This should be a more accurate estimate of lifetime mating potential than experiments involving caged animals, which tend to overestimate the number of potential matings for females just as they do for males. Females exposed to unlimited eager males in the confines of a small cage are more likely to accept the persistent advances of males than would females in the wild, who can more easily limit male-female encounters by a variety of courtship refusal behaviors involving flight or escape (see Silbergleid 1977, Rutowski 1982, for reviews). Furthermore, caged females in mating experiments are usually provided with an oviposition substrate so that time-consuming searches for larval foodplants or other oviposition sites are eliminated. In most species, unrestricted oviposition should minimize a female's refractory period between matings by accelerating the loss of the stimulus that causes courtship unreceptivity (*e.g.*, presence of sperm in the spermatheca or an inflated spermatophore in the bursa copulatrix; see section IV). However, Byers (1978) found no significant difference in mating frequencies between samples of female noctuids (*Euxoa*) collected at light traps and those held in the laboratory with access to males. The light trap samples consisted predominantly of older females, which suggests that laboratory experiments may approximate mating frequency in the oldest age class.

Comparison of the mean mating frequency of all females in a sample with the mean mating frequency of only the oldest age class (see Table I, column 2) reveals that older females almost always carry more spermatophores than the average female in the population, sometimes considerably more. The difference in the two means is greatest for species that have high maximum spermatophore counts, but even in these species the means rarely differ by more than a factor of two.

The data in Table I represent mating frequencies for populations studied independently. In Fig. 5 I present information on a community of tropical butterflies from a lowland rain forest area in eastern Ecuador (Drummond 1976) that illustrates a clear correlation between relative age (as measured by progressive wing-scale loss) and the number of spermatophores present in the corpus bursae for females of the nymphalid subfamily Ithomiinae. These are relatively long-lived

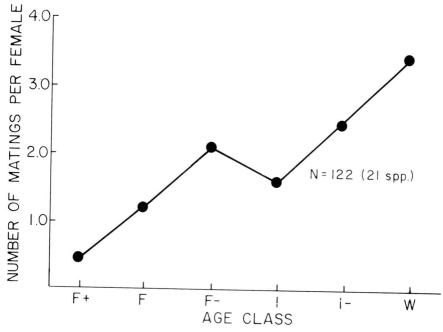

Fig. 5. Relationship between number of spermatophores per female (including virgins) and relative age in 21 species of Ithomiinae (Nymphalidae) from Limoncocha, Ecuador. Mean number of matings per mated female was 1.94; number of virgins was 17 (= 14%). Age classes based on progressive wingscale loss (F+ = teneral, F = fresh, I = intermediate, W = worn). From Drummond (1976, unpubl. data).

species (3-6 months or more) in which both males and females are reproductively active throughout life (Drummond 1976). The oldest age class (W) probably represents individuals 4-6 months old.

Another way to consider lifetime mating potential in female Lepidoptera is to focus on the maximum number of spermatophores recorded for wild-caught females. Such an analysis is presented in Fig. 6, which is based on the information in Table I, additional information from Ehrlich and Ehrlich (1978), and my own unpublished studies. All species represented by five or more female dissections are included, which suggests several taxonomic trends. First, there are several groups in which all species studied to date are known to multiply mate; all moth families, skippers (Hesperiidae), the butterfly family Papilionidae, and the nymphalid subfamilies Danainae and Acraeinae. Second, there are no groups in which the majority of species are monogamous, although nearly half of the Lycaenidae species examined contained no more than one spermatophore. In the Lycaenidae, Pieridae, Nymphalinae, and Satyrinae, the great majority of species mate no more than twice. Third, extremely large number of spermatophores have been recorded for *Utethesia ornatrix* (11) in the Arctiidae, for the genus *Euxoa* (up to 16) in the

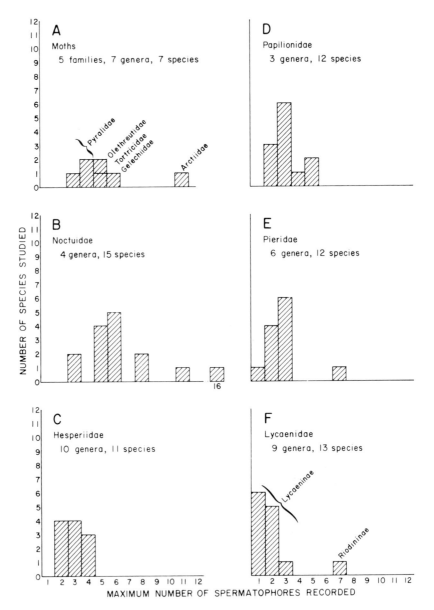

Fig. 6. Maximum number of spermatophores per female recorded for 128 species of moths (A-B), skippers (C), and butterflies (D-L), grouped by family and subfamily (Nymphalidae). Inclusion in figure is limited to species represented by a minimum of five dissections. Data compiled from sources in Table I in this chapter, Ehrlich and Ehrlich (1978), and Drummond (unpubl. data).

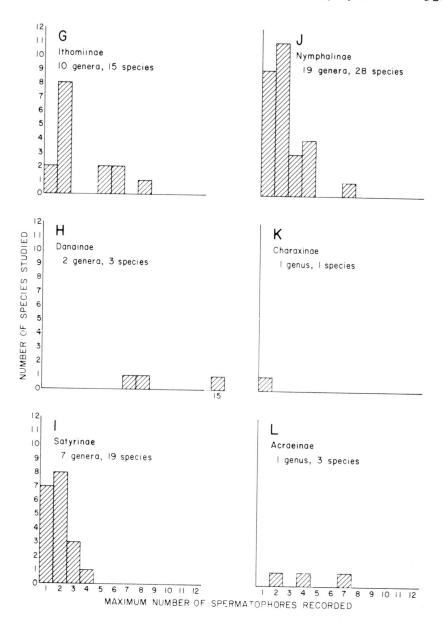

MAXIMUM NUMBER OF SPERMATOPHORES RECORDED

Noctuidae, and for members of the nymphalid subfamily Danainae (up to 15). Fourth, the large sample sizes available for three species of Pyralidae and 15 species of Noctuidae suggest that the greater number of spermatophores recorded for Noctuidae (up to 16, with a mode of 5-6 compared with only 3-4 for Pyralidae) reflects an important difference in the selection pressures shaping the mating behaviors of the two groups, which differ significantly in the degree of complexity of their genital morphology (Callahan and Chapin 1960).

Finally, two nymphalid subfamilies, Ithomiinae and Acraeinae, show considerable within-group variance in the maximum number of spermatophores recorded. Some of the long-lived Ithomiinae species with low spermatophore counts may be inadequately represented by older individuals (see Ehrlich and Ehrlich 1978), such that increased sampling of older females would more accurately reflect maximum mating frequencies and at the same time reduce the variance for the subfamily. But mating frequency in these butterflies has been correlated with oviposition mode (Fig. 7 and below), suggesting that a portion of the variance in mating frequencies reflects a divergence of reproductive strategies in the group. The strongly skewed sex ratios of some populations of *Acraea encedon* (Owen *et al.* 1973 a, b; see below), the formation of sphraga, and the absence of extended courtship before copulation in this genus (Eltringham 1912) may all contribute to the variability in mating frequency observed in the Acraeinae.

3. Factors Affecting Mating Frequency

Some of the various factors affecting mating in natural populations of Lepidoptera have been discussed by Burns (1968), Pease (1968), Pliske (1973), Scott (1972), Ehrlich and Ehrlich (1978), and Lederhouse (1981). Because mating frequency is dependent on the age structure of the population, species that exhibit strong seasonality in time of emergence and adult activity, such as most temperate zone Lepidoptera, experience seasonal fluctuations in mating frequency. Generally, the proportion of virgins decreases while the proportion of multiply-mated females increases during the flight season, although samples of agricultural pests reveal that mating frequencies are strongly affected by patterns of crop maturity in adjacent fields, timing and placement of traps (blacklight, bait, or pheromone), and voltinism (Callahan 1958, Dustan 1964, Gehring and Madsen 1973, Goodwin and Madsen 1964, Hendricks *et al.* 1970, Stadelbacher and Pfrimmer 1973, Roome 1975, Lopez *et al.* 1978, and Wada *et al.* 1980).

Seasonality in mating frequencies also occurs in migratory species. The high proportion of virgins (78.9%) reported by Walker (1978) for the skipper *Urbanus proteus* came from a sample flying southward caught in a directional Malaise trap. These individuals probably had just eclosed and begun moving to overwintering sites before mating occurred. By contrast, a nonmigratory population of this species sampled by Pliske (1978) in Trinidad contained only 3% virgins. The variability

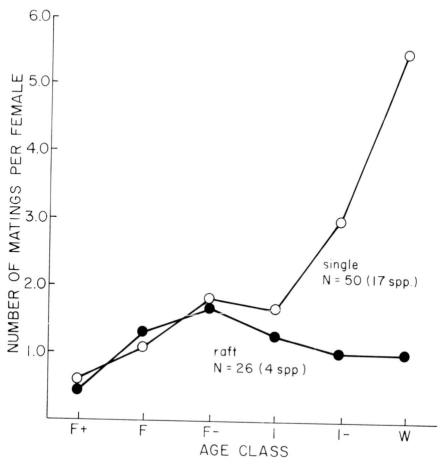

Fig. 7. Relationship between number of spermatophores per female (including virgins) and mode of oviposition expressed as a function of relative age for 21 species of Ithomiinae (Nymphalidae) from Limoncocha, Ecuador. Single = females of species that lay eggs singly on leaves of larval foodplants, mostly in deep forest understory (mean matings/mated female = 1.7). Raft = females of species that lay eggs in clusters of 4-75 on leaves of larval foodplants, mostly in disturbed or sunny areas (mean matings/mated female = 1.2). From Drummond (1976, unpubl. data).

in virginity rates and mating frequency for the Monarch butterfly, *Danaus plexippus,* is likewise affected by migration and its accompanying cycles of reproductive diapause. Nonmigratory populations in Trinidad (Pliske 1973) showed low virginity (4.4%) and a high proportion (about 50%) of multiply-mated females. Fall migrations of the Monarch heading south through New York (Pliske 1973) consisted of females in reproductive diapause and all 80 individuals dissected were virgins. Two samples representing returning spring migrations of Monarchs in

California (Williams *et al.* 1942, Hill *et al.* 1976) contained no virgins and almost all females had mated more than once. At least some of the high proportions of virgins reported for Noctuidae may be the result of sampling during prereproductive migration periods that immediately follow emergence. Northward migrations prior to mating in early spring occur in southern U.S. populations of *Peridroma saucia* (Callahan and Chapin 1960). In the army cutworm, *Euxoa* (=*Chorizagrotis*) *auxiliaris*, adults emerge in the spring on the Great Plains but migrate to and pass the summer in the Rocky Mountains where they accumulate nutritional reserves before returning in autumn to the Great Plains where they mate and oviposit (Pruess 1967).

Based on his spermatophore counts of butterflies and skippers, Burns (1968) suggested that an inverse correlation exists between mating frequency and population density. Pliske (1973) claimed that this generalization was invalid because it was based on a comparison of several species, not on a comparison of populations of different densities of the same species. Positive correlations between mating frequency and population density have been found in many blacklight collections of moths (*e.g.*, Goodwin and Madsen 1964, Graham *et al.* 1965), and the increase in mating frequency during the flight season documented earlier may be in part a function of increasing population densities over time (Gehring and Madsen 1963, Dustan 1964). Mating frequency in the butterfly *Danaus gilippus* also appears to be positively influenced by increasing density; in more dense Florida populations (200 females collected per day), Pliske (1973) found a greater mating frequency in all age classes than in less dense populations in Trinidad (15 females collected per day), and than in the widely dispersed populations studied by Burns (1968) in Texas.

Female butterflies of most species can mate on the day of emergence (Scott 1972), and sometimes before emergence! (in some *Heliconius;* see Gilbert 1976, Ehrlich and Ehrlich 1978), and thus the proportion of virgins in natural populations is generally low (Table I; see also Shields 1967, Burns 1968, Pliske 1973, Ehrlich and Ehrlich 1978). But low population densities could result in lower encounter rates between males and females, so that males become more persistent in courtship when a female is encountered, regardless of her mated status. Under such conditions, the proportions of both virgins and multiply-mated females should be greater than when encounter rates are high (Shapiro 1977). Data from skipper populations (Burns 1968, Shapiro 1977) support such an inverse correlation between population density and variance in mating frequency, which is perhaps a more accurate way of restating Burns' (1968) original generalization.

Variance in mating frequencies may be affected by territoriality in groups like *Papilio* butterflies. Because territoriality tends to decrease the effective population size by limiting the number of males that contribute genetically to the next generation, it also increases variance in male reproductive success, because males without territories rarely mate (Sims 1979). On the other hand, mating frequency of females may be independent of the absolute density in territorial species, because the availability of males (on hilltop territories) should be independent of the total number of males in the population (Lederhouse 1981). In *Papilio polyxenes,* female mating frequency was consistent from brood to brood and from year to year, despite considerable fluctuations in male densities in the mating areas (Lederhouse 1981).

Male mate choice (Rutowski 1982) may affect mating frequencies among females who differ in reproductive or survival characteristics that can be detected by courting males. In some species with polymorphic females, spermatophore counts have been used to measure mating attractiveness to males of the different female forms in natural populations. In *Colias* butterflies, Graham *et al.* (1980) demonstrated that "alba" females enjoy certain physiological advantages over their normally pigmented counterparts, but that alba females are less attractive to males than normal females and contain significantly fewer spermatophores at any given age. In the Eastern Tiger Swallowtail butterfly, *Papilio glaucus,* females are dimorphic and spermatophore counts have been used as evidence to argue that preferential mating by males with the yellow (nonmimetic) morph offsets the survival advantage of the black morph, which mimics the unpalatable papilionid butterfly, *Battus philenor* (Burns 1966). The ensuing debate over the validity of this suggestion (Prout 1967, Pliske 1972, Makielski 1972, Levin 1973) generated so many new hypotheses that David A. West (pers. comm.) has likened the whole problem to a black hole: capable of completely absorbing any hypothesis that comes near it.

Statistical analyses of female mating frequencies in the pink bollworm, *Pectinophora gossypiella* (Graham *et al.* 1965) and the nymphalid butterfly *Acraea encedon* (Owen *et al.* 1973b) suggest that the rate at which first matings occur is density-dependent, but that rates at which subsequent matings occur are density-independent. This is consistent with observations that males prefer to mate with virgins, that mated females often reject courtship advances, and that the occurrence of multiple mating in females has more to do with sperm longevity, oviposition strategy, and nutritional requirements for reproduction than with male abundance (see below).

Most of the studies that documented the importance of sex ratio in influencing mating frequency of caged males (discussed above) also demonstrated that female mating frequency increases with an increasing male bias in the sex ratio (*e.g.,* Goodwin and Madsen 1964). In the Noctuidae, an increasing male bias in the sex ratio in natural populations is often correlated with a decrease in the proportion of virgin females and an increase in the proportion of multiply-mated females, although the proportion of singly-mated females remains roughly the same (Callahan and Chapin 1960, Stadelbacher and Pfrimmer 1973).

Influence of sex ratio on mating frequencies of females has also been demonstrated in butterflies. Walker (1978) found that in southward-flying fall populations

of several migratory species the proportion of mated females was positively correlated with the degree of male bias in the sex ratio. A more dramatic example comes
from the predominantly female populations of the African butterfly, *Acraea
encedon*. In four populations studied in Uganda (UGANDA 1-4 in Table I), the
sex ratio was only 0.02-0.063, or 2-6 males per 100 females, and most females
(75-100%) were found to be unmated (Owen *et al.* 1973b), The high percentage
of virgins is understandable because males can only mate once a day and only live
2 weeks (Owen *et al.* 1973a). Two populations studied in Sierra Leone (SL-1 and
2, Table I) for comparison showed that even a slight increase in the sex ratio (to
0.16) resulted in 80% of the females being mated, of which 10% had mated more
than once. Six females from a population in eastern Sierra Leone (locality 2)
with a normal (0.5) sex ratio were all mated. Samples of other species of *Acraea*
in which the sex ratio is normal (*A. egina* and *A. eponina*, Table I) show that few
females are unmated when collected and that many have mated more than once
(Owen *et al.* 1973b). The predominantly female populations of *A. encedon* consist
of two kinds of females, those producing females only and those producing males
and females in equal abundance. Both types require fertilization of the eggs for
offspring to be produced (Chanter and Owen 1972), and the persistence of these
unusual populations, which in some cases are known to have survived for at least
50 breeding generations, apparently results from some (as yet undemonstrated)
form of frequency-dependent selective mating such that females producing both
males and females are at an advantage (Owen *et al.* 1973b).

In addition to the effects of various population parameters on mating frequency,
a number of environmental factors are known to influence the mating behavior of
Lepidoptera and, therefore, mating frequencies. High ambient temperatures can
lower population mating frequencies by inhibiting flight activity (Hendricks *et al.*
1970, Stadelbacher *et al.* 1972, Stern and Smith 1960) or by decreasing adult
longevity (Lopez *et al.* 1978). Low ambient temperatures also reduce mating
success (Henneberry and Leal 1979), and can lower mating frequencies by reducing
flight activity (Stern and Smith 1960) or by increasing the duration of copulation
(David and Gardiner 1961, Burns 1970, Wakamura 1979), which reduces mating
opportunities for species capable of mating more than once a day.

To the extent that general activity is governed by light intensity (for diurnal
and crepuscular species), humidity (for tropical species), and ambient temperatures,
mating activity will also be affected. Goodwin and Madsen (1964) report that
mating activity of the navel orangeworm (*Paramyelois transitella*) is stimulated by
twilight and is affected by slight changes in temperature and relative humidity.
For nocturnal Lepidoptera, experimental evidence confirms that mating activity is
generally inhibited by high light intensities (Wago 1977a, Henneberry and Leal
1979), but that the inhibition varies with the spectral component and that some
wavelengths at low intensities (particularly UV and sometimes red) actually stimulate mating activity (Leppla and Turner 1975, Turner *et al.* 1977). Many butterfly

species become less active in courtship and mating behavior under cloudy conditions, even if ambient temperatures are adequate (pers. obs.).

Adult and larval food sources also can affect mating frequencies significantly. Experiments with caged moths (*Cnaphalocrocis medinalis*) demonstrated that the proportion of mated females was much higher for females fed sucrose than for females fed water (Wada *et al.* 1980). Onukogu *et al.* (1980) have demonstrated significant differences in mating frequencies among cultures of the European corn borer *Ostrinia nubilalis*, reared on different larval diets. When larval resources are inadequate to finance egg production, females may rely on nutritional input from males through copulation (Boggs 1981b), thereby collecting spermatophores and accessory fluids to augment their own nutritional investment in reproduction. Finally, males of certain groups of butterflies (*e.g.*, Ithomiinae, Danainae) are completely dependent upon external sources (flower nectar and decaying vegetative tissues of Boraginaceae and Eupatoriae) of chemicals (pyrrolizidine alkaloids) required to synthesize pheromones that are essential to successful courtship and mating (Pliske 1975c, Edgar *et al.* 1976, Pliske *et al.* 1976, Schneider *et al.* 1975). The distribution and phenology of the plants containing the requisite pheromone precursor can profoundly affect the population and community structure of these butterflies, including daily patterns of courtship and mating (Drummond 1976).

D. Multiple Mating and Natural Selection: Why Do Females Mate More Than Once?

According to sexual selection theory (Trivers 1972), male reproductive advantage is maximized by mating as many times as possible. In most groups of animals, male investment in reproduction is relatively small compared to that of a female, so that a male confronted with unlimited virgins should mate with as many of them as is physiologically possible. This disparity between males and females in the degree of energetic investment in reproduction holds true for Lepidoptera, in which males have no possibility of contributing to the success of their offspring beyond the act of copulation. But the long-held view that the cost of copulation to male Lepidoptera is trivial has been revised in light of recent documentation that the spermatophore and accessory fluids may provide nourishment to the female (Boggs and Gilbert 1979, Boggs and Watt 1981, Boggs 1981a). Because selection may favor the production of larger spermatophores and greater volume of accessory fluids (Boggs 1981a), males may benefit from an ability to discriminate among females of different fecundities (Rutowski 1982). Confronted with nonvirgin females, the evolutionary decision for the male of whether or not to attempt copulation depends very much on the nature of sperm storage and patterns of sperm utilization within multiply-mated females. I examine the constraints on male mating behavior imposed by sperm dynamics in the female in section IV.

For females of most higher animals, each bout of offspring production is pre-ceded by an act of copulation so that fertilization can occur. In most arthropods, however, a male can transfer enough sperm during a single copulation to fertilize all of the eggs a female may produce during her lifetime; long-term storage of viable sperm has been reliably documented for spiders, isopods, ticks, and many insects, including Lepidoptera (Davey 1965, Parker 1970, Wigglesworth 1972). If then, in most arthropods, a single copulation supplies enough sperm to fertilize all of a female's eggs, why do females mate more than once? This question has been addressed by Alcock *et al.* (1978) and Page and Metcalf (1982) for Hymenoptera, by Walker (1980) for nonsocial insects, and by Austad (this volume) for spiders. Here I discuss potential answers to this question for Lepidoptera.

There are a number of potential disadvantages for females who engage in super-fluous matings. First, the act of copulation greatly reduces the mobility of a female (and her mate), thereby increasing her risk of predation (Sims 1979) or her vulner-ability to harsh weather, such as heavy wind and rain (Drummond 1976). Second, the long mating duration of most butterflies and moths, 1-3 h average (see Scott 1972, Shields and Emmel 1973) and up to 24 h in some groups (Hewer 1934), certainly reduces the time available to a female for obtaining nutrients or for locating larval foodplants (Wiklund 1977). Finally, additional matings sometimes result in infertility of the female, even if the first mating was successful. In most such cases (Etman and Hooper 1979a, Labine 1966, Taylor 1967), the presence of prior spermatophores in the bursa copulatrix probably prevents correct place-ment of the collum and aperture of the new spermatophore so that sperm are un-able to enter the ductus seminalis. In *Colias* (Boggs and Watt 1981) the infertility following the second mating is temporary and limited only to those eggs laid immediately after copulation. Given these disadvantages to superfluous matings, it is not surprising that females of most species of Lepidoptera have the behavioral ability to determine copulation frequency by controlling male/female encounter rates or by influencing the outcome of courtship attempts.

It is clear from the data on spermatophore counts of females in natural popula-tions that multiple mating is widespread in female Lepidoptera (Table I). The importance of "monogamy" as a mating system in the Lepidoptera has been over-emphasized (*e.g.,* Wiklund 1977, Walker 1980), in part because the short life span of many (heavily-studied) temperate-zone species means that the majority of females in such species die before a second mating occurs (Scott 1974) and because most studies reporting mean spermatophore counts are uncorrected for age-class bias (see Table I). Possible advantages to multiple mating in female Lepidoptera are discussed below within categories established by Alcock *et al.* (1978) and Walker (1980).

1. To Achieve an Adequate Sperm Supply

Repeat mating would be an advantage to a female if (1) the first mating failed to inseminate the female; (2) the first mating provided insufficient sperm to fertilize all the female's eggs; or (3) if the sperm received from the first mating, even though adequate in number initially, deteriorated in quality before all the female's eggs could be laid.

Laboratory experiments on a variety of agriculturally important moth species suggest that unsuccessful matings (failure to inseminate) occur regularly at the rate of about 5-30% of all copulations (Callahan and Chapin 1960, Callahan and Cascio 1963, Taylor 1967). Even if these experimental studies did not overestimate rates of mating failure under natural conditions (which they almost certainly do), this factor alone could account for only about half of the observed multiple matings recorded for natural populations (Table I). Pair *et al.* (1977) reported a 36% failure rate of first matings in *Heliothis,* although the proportion of unfertilized females remaining after a second copulation dropped to 19%. But many laboratory cultures of Lepidoptera exhibit poor reproductive performance, as pointed out by Snow *et al.* (1970), whose studies on the fall armyworm, *Spodoptera frugiperda,* revealed that 8% of caged females did not mate, that 24% of mated females did not oviposit, and that 30% of ovipositing females produced eggs of low viability.

Based on the absence of sperm in the spermatheca, mating failure rates between 20% and 30% have been reported for *Rhyacionia buoliana* (Olethreutidae, Pointing 1961), *Choristoneura* spp. (Tortricidae, Campbell 1961), *Atteva punctella* (Yponomeutidae, Taylor 1967), *Trichoplusia ni* (Noctuidae, North and Holt 1968), and *Heliothis virescens* (Noctuidae, Flint and Kressin 1968). By contrast, rates of insemination failure reported for *Laspeyresia pomonella* (Oleuthreutidae) are less than 3% (Robinson 1973, 1974), those for *Pectinophora gossypiella* (Gelechiidae) are less than 6% (Flint and Merkle 1980), and those for *Spodoptera litura* (Noctuidae) are about 12% (Etman and Hooper 1979a). At least some of this variation in mating success may be explained by variation among species in the complexity of copulation. In the Noctuidae, for example, Callahan and Chapin (1960) and Holt and North (1970) have shown that the percentage of males that failed to achieve insemination increased with increasing complexity of the genital armature and mechanics of copulation. Blockage of the seminal duct, by eggs or chorions of eggs, may prevent migration of sperm from the spermatophore to the spermatheca. Based on the number of eggs and dead sperm in the seminal duct of multiply-mated females of *Atteva punctella,* Taylor (1967) suggested that blockage of the genital tract could lead to multiple mating in females of this species.

Although insemination may occur rarely without the passage of a spermatophore to the female (George and Howard 1968), failure to form a spermatophore is widely reported as failure to inseminate. Many laboratory studies have shown that copulation does not always lead to spermatophore formation, with failure rates of 17% reported for *Papilio zelicaon* (Papilionidae, Sims 1978), 9% for *Atteva punctella* (Yponomeutidae, Taylor 1967), 6% for *P. polyxenes* (Lederhouse 1981), and less than 5% for *Spodoptera littoralis* (Noctuidae, Navon and Marcus 1982). Under natural conditions, Pliske (1973) found that in 15% of copulations of *Danaus gilippus* (Nymphalidae:Danainae) the male failed to pass a spermatophore. Rates of mating failure also have been inferred from rates of hatching failure of eggs laid by mated females, although hatching failure is not necessarily the result of mating failure. In cultured strains of the almond moth (*Ephestia cautella*), mating failure rates, based on hatching failure of eggs laid by mated females, range from 7% (Brower 1979) to 13-14% (Calderon and Gonen 1971, Ahmed *et al.* 1972) and up to 17% (Cogburn *et al.* 1973).

Although selection should be severe on males to produce fully competent sperm and to copulate successfully (Alcock *et al.* 1978), Taylor (1967) estimates that at least 75% of all mating failures in *Atteva punctella* are due to male inadequacies. Although Taylor found no correlation between male infertility and male age, Sims (1979) has shown that in *Papilio zelicaon* very young males (who lack suf-- ficient sperm in the ductus ejaculatorius duplex) have 10 times the insemination failure rate (45%) of mature males (4%). Male mating success also may be reduced by naturally-occurring genetic mutations that increase the rate of failure to pass a spermatophore or that decrease the number of eggs fertilized (Cotter 1963, Sims 1978, North and Holt 1968) or by reproductive cytoplasmic incompatibility caused by maternally-inherited endosymbionts (Kellen *et al.* 1980).

Insufficient sperm from the first mating as a cause for multiple mating in the Lepidoptera has not been demonstrated conclusively, although some evidence exists to support this suggestion. A single insemination has been reported as suf- ficient to fertilize all eggs produced by the butterflies *Papilio glaucus* (Levin 1973), *P. zelicaon* (Sims 1979), *P. polyxenes* (Lederhouse 1981), *Pieris rapae crucivora* (Suzuki 1979), and of some *Heliconius* (*erato* and *charitonius,* Ehrlich and Ehrlich 1978); and in the moths *Utethesia ornatrix* (Pease 1968), *Spodoptera frugiperda* (Snow *et al.* 1970), and *Earias insulana* (Kehat and Gordon 1977). In the butter- fly *Euphydryas editha* (Nymphalidae), however, Labine (1966) found that of 16 singly-mated females, 56% exhibited high fertility, 19% low fertility, and 25% produced viable offspring soon after mating, after which fertility dropped sharply. Because sperm remain viable in female *E. editha* for 7-10 days and the female can lay over 1,100 eggs in this period of time, Ehrlich and Ehrlich (1978) suggested that one mating may not always be sufficient to fertilize all eggs produced in this species. Presumably, the 25% of females in Labine's (1966) study that showed a quick drop in fertility after early success were fertilized by males that failed to transfer sufficient sperm.

In the moth *Anagasta kuhniella,* Norris (1933) found that those females that continued calling (releasing sex attractant pheromone) after mating were those with less than the normal quantity of sperm in the spermatheca. This suggests that female Lepidoptera can assess the adequacy of insemination such that inadequately inseminated females are more likely to remate than those that receive large quantities of sperm. Supporting this suggestion are experiments with *Atteva punctella* (Taylor 1967) and several species of *Heliothis* (Proshold and LaChance 1974, Raulston et al. 1975) in which females with no or few sperm in the spermatheca after mating were more likely to remate than were mated females whose spermathecae contained large amounts of sperm.

In Taylor's (1967) study, however, the insufficient sperm supply may have resulted in part from poor sperm viability or from inadequate transfer of sperm within the female after mating. The continued receptivity to courtship of mated females with inadequate sperm supplies suggests that control of receptivity is mediated by an interaction between the sperm or accessory fluids and the spermatheca rather than by the pressure exerted by the spermatophore within the bursa copulatrix (see section IV.B.). The greater accuracy of the former stimulus in indicating fertility appears to have led to its selection as the primary stimulus for initiating oviposition (Taylor 1967, Benz 1969, North and Holt 1968, Deseo 1972).

Replacement of deteriorated sperm might be a reason for repeat mating for species in which females oviposit over a long period of time. In the black swallowtail, *Papilio polyxenes,* sufficient sperm is transferred in one successful mating to fertilize all the eggs produced by a female, but the sperm deteriorates with time (beginning about 10 days after mating) and must be replenished by an additional mating to restore female fertility (Lederhouse 1981). Estimation of survivorship of female black swallowtails predicts that less than 35% would be expected to live longer than the 10 days of uniform fertility, a figure which agrees well with the proportion (30.4%) of wild-caught females that had mated more than once (Lederhouse 1981). Further evidence of sperm deterioration in this species comes from the finding that female fertility and egg hatchability in the laboratory did not decrease with the number of eggs laid but did decrease with time since mating (Lederhouse 1981). Among wild-caught, singly-mated females, Lederhouse found that fresh females had significantly higher egg hatchabilities than did females of intermediate or advanced age. But mating an additional time restored the egg hatchability of old females to that of fresh, singly-mated females. Not surprisingly, Lederhouse found that three-quarters of the females in the oldest age class (worn) had mated more than once.

The correlation between oviposition mode and mating frequency in female ithomiine butterflies mentioned previously suggests that sperm deterioration after mating may occur in these butterflies. Females that lay eggs singly had a much higher mating frequency in the oldest age class than did females that deposit rafts of up to 70-80 eggs at a time (Fig. 7). Presumably, the higher mating frequencies found among single-egg layers are a response to sperm deterioration over a longer period of oviposition activity. One could argue that single-egg layers are collecting spermatophores for nutritional reasons to finance egg production (see below), but in none of the ithomiine species studied do females eclose with matured eggs. Adult females of all species must, therefore, seek out nitrogenous sources (*e.g.,* bird droppings, decomposing animals) to permit egg development and oviposition (Drummond 1967). Thus, although the nutritional contribution that accrues to the female during copulation may be important, it does not explain the difference in mating frequencies between single- and multiple-egg layers.

2. To Minimize Physiological Costs of Long-term Sperm Storage

If the costs of the sperm-storage organ (spermatheca) and the nutrients required to maintain sperm viability exceed the costs (somatic risk and time loss) of multiple mating, then selection should favor a reduced sperm storage capability and a reduction or elimination of the female refractory period.

As Austad (this volume) points out for spiders, there is no reason to suppose that sperm storage costs are exorbitant, especially because sperm utilization appears to be an extremely efficient process in insects (Davey 1965, Parker 1970). The number of sperm stored in the spermatheca of *Papilio zelicaon* is estimated to be 55,000 to 386,000; the average number of ova is 148 (maximum 345), resulting in a sperm/ovum ratio of about 400-1200:1 (Sims 1979). This is a higher ratio than reported for other insects but is still four to five orders of magnitude less than that reported for mammals (Davey 1965). There is no doubt that all lepidopteran females store sperm; the question is, does long-term storage cost significantly more than short-term storage? Because some groups (*e.g., Heliconius*) contain species in which females store sperm from a single mating for several months as well as species that mate repeatedly during a shorter interval (Ehrlich and Ehrlich 1978), this does not appear to be a promising hypothesis. Furthermore, the diversity of mate-rejection behaviors associated with the almost universal female postmating refractory period of several days is inconsistent with this hypothesis.

3. To Enhance Female Survivorship or Reproduction Through Indirect Paternal Investment

Selection should favor repeat matings by females if by engaging in additional copulations a female enhances her survival or reproductive output. Repeat mating could enhance female survival by increasing the degree of protection from predators (Walker (1980) or by supplementing the nutrient base needed for maintenance (Thornhill 1976) and it could increase reproductive output by providing nutrients for egg production and oviposition (Thornhill 1976).

Walker (1980) presents five ways in which multiple mating could increase predator protection of females, but only three of these apply to Lepidoptera. First, the greater aposematic effect of pairs *in copula* is unlikely to encourage multiple matings among warningly-colored species, although it may reduce the cost of an otherwise advantageous pairing. Overwintering clusters of monarch butterflies, *Danaus plexippus*, in California and Mexico take advantage of such an augmented aposematic effect without resorting to copulation. Second, an increase in predator protection in aposematic species resulting from a greater concentration of released defensive compounds also is achieved easily without copulations, as illustrated by the powerful odor at the nocturnal communal roosts of pheromone-releasing *Heliconius* butterflies (Poulton 1931).

The third possibility, transfer of defensive compounds from the male to the female, as documented for cantharidin in Spanish flies, *Lytta vesicatoria* (Sierra *et al.* 1976), provides an intriguing suggestion for study in some groups of Lepidoptera. For example, in the nymphalid butterfly subfamily Ithomiinae, recent biochemical studies (Brown, in press) suggest that adults are protected not by the alkaloids contained in their larval foodplants (Solanaceae) as widely assumed (*e.g.*, Ehrlich and Raven 1964) but by pyrrolizidine alkaloids (PA's) derived through adult feeding at nectars and decomposing tissues of PA-containing plants (*e.g.*, Boraginaceae, Asteraceae:Eupatoreae). Because such PA sources are visited almost exclusively by males (Pliske 1975a, b; Drummond 1976), but both male and female ithomiines contain dehydropyrrolizidine alkaloid monoesters (Brown, in press), the possibility exists that female ithomiines obtain their defensive chemicals from the males via copulation. Differences between male and female ithomiines in PA chemistry are consistent with the mating transfer hypothesis. Males contain both PA N-oxides and free PA's but females contain almost exclusively PA N-oxides (Brown, in press). The free PA's are used by males as a precursor for pheromone synthesis (Edgar *et al.* 1976), but in this form could also be transferred to the female during mating. In the one species (*Mechanitis polymnia*) whose eggs where analyzed chemically, PA's were found to constitute almost 1% of the fresh weight (Brown, in press), which shows that the PA's in the female, whatever their source, are incorporated into reproductive tissue.

The possibility that transferred nutrients from the male could increase female longevity is supported by findings that radioactively labeled protein fed to adult male *Heliconius hecale* (Nymphalidae:Heliconiinae, Boggs and Gilbert 1979) or injected into male pupae of *Colias eurytheme* (Pieridae:Coliadinae, Boggs and Watt 1981) became distributed throughout the body of their female partners within a week of copulation. Investment in reproduction by male insects in the form of nutrients passed to the female during copulation has been reviewed by Thornhill (1976) and Gwynne (this volume). Stressing the diversity of compounds present in the accessory gland secretions passed to the female during copulation, Marshall (1982) explored the relative importance of larval and adult diets in determining the availability and utilization of male contributions to prezygotic reproductive investment in Lepidoptera. Radiotracer experiments have demonstrated that amino acids present in the spermatophore and accessory gland secretions of males are incorporated into unfertilized eggs in the female in four families of Lepidoptera, including Nymphalidae, Heliconiini (*Heliconius erato, H. hecale,* Boggs and Gilbert 1979; *Heliconius charitonius,* Boggs 1981a), Danainae (*Danaus plexippus,* Boggs and Gilbert 1979), Pieridae:Coliadinae (*Colias eurytheme,* Boggs and Watt 1981), and Ctenuchidae (*Lymire eduardsii,* Goss 1977). In addition, zinc transferred to females during copulation by males of *Heliothis zea* (Noctuidae) is incorporated into eggs (Engbretson and Mason 1980) and the diversity of accessory gland constituents (proteins, hydrocarbons, triglycerides, diglycerides, sterols, phospholipids) recently identified in *Colias philodice-eurytheme* by Marshall (1982) suggests that several other male-derived substances will be found to contribute directly to egg production in female Lepidoptera. It seems likely that male nutrient investment enhances egg production, especially in species that mature many or all of their eggs after eclosion. Male-contributed nutrients also may help maintain egg quality (Byers 1978) because, in at least some Lepidoptera, egg quality decreases progressively over the oviposition period as a result of increasing interoocyte competition for egg substrate (Campbell 1962, Wellington 1965).

Male nutrient contribution to reproduction is now believed to be widespread in the Lepidoptera (Shapiro 1982) and, depending on the balance between nutrient availability and the nutritional requirements of the female (Boggs 1981a, b; Boggs and Watt 1981, Marshall 1982), this could lead to strong selection on males to produce large nutrient-rich spermatophores (Thornhill 1976, Boggs 1981a) and to strong selection on females to collect spermatophores (and their attendant secretions) to finance egg production or enhance survivorship (Byers 1978, Boggs and Gilbert 1979, Boggs and Watt 1981).

4. To Increase Genetic Diversity Within a Female's Progeny

For species that experience variable and unpredictable environments across generations, it seems reasonable to propose that a mixed male parentage to a female's offspring, by increasing the genetic diversity within her progeny, should increase the probability of some offspring being able to survive regardless of the conditions experienced by the next generation. The vagility of many butterflies and moths makes this an attractive explanation for multiple mating in Lepidoptera.

Williams (1975), however, has explained that a single mating usually provides half the within-progeny genetic variance in the population, so that the genetic advantage obtained by additional matings is slight compared with the accompanying costs or risks to the females (Walker 1980; Austad, this volume). Furthermore, maximal genetic diversity of a female's offspring would be obtained only if sperm from all matings blended equally in the spermatheca. In other words, within-progeny genetic variance of multiply-mated females is maximized if there is no sperm precedence (see section IV.A.3.). However, last male sperm precedence could increase genetic diversity of offspring of females who mate regularly and oviposit between matings so that sperm from different males is used sequentially. Last male sperm precedence is the rule in Lepidoptera as shown by the generally high P_2 values (percent of eggs fertilized by last male to mate) presented by Gwynne (this volume) for the 11 species of Lepidoptera for which quantitative data are available.

The correlation between mating frequency and oviposition mode in ithomiine butterflies shown in Fig. 7 provides little support for the genetic diversity hypothesis. Compared with species that oviposit singly, species that oviposit in clusters (rafts) are found more often in disturbed forest and sunny areas, are more likely to be polyphagous in larval foodplant use, and use larval foodplants that exhibit greater variance in distribution and density (Drummond 1976). Based on their more stressful (sunny) environment (most ithomiines are extremely susceptible to desiccation) and the greater variation in availability of their larval foodplants, one could predict that the raft species would benefit most from increased within-progeny genetic diversity. The greater mating frequency of the single-egg layers contradicts this prediction and underscores the need to avoid oversimplification and to consider other potentially relevant factors; in this case, the probable need for sperm replenishment or male nutrient investment in species that distribute their reproductive effort over a long period of time.

5. To Minimize Loss of Time and Energy Required to Resist Insistent Males

Unless there are distinct advantages to multiple copulations, the loss of time incurred by extra matings has been advanced as a reason sufficient to select for nonreceptivity in mated females (Parker 1970).

In most Lepidoptera, females attract males either visually by wing displays (butterflies and some moths) or chemically by pheromones (most moths). Mated females can reduce the male/female encounter rate by ceasing to emit the sex attractant. In butterflies, females can reject male courtship advances in a number of ways (see Scott 1972), including the use of signals (Wago 1977b), postures (David and Gardiner 1961, Obara 1964, Suzuki et al. 1977), or evasive flights (Stride 1958; Wiklund 1977; Rutowski 1971a, b, 1979b; Lederhouse 1981). In those few groups of Lepidoptera in which males attract females, some mated females fail to respond to male attractants. For example, mated females of Papilio butterflies are less likely than virgins to fly to hilltops where territorial males form leks (Taylor 1967, Lederhouse 1982) and recently mated female ithomiine butterflies do not respond to the aphrodisiac pheromone released in courtship territories by males (Pliske 1975c, Drummond 1976).

Lederhouse (1981) concluded that mated females of the black swallowtail, Papilio polyxenes, would be unlikely to mate a second time rather than refuse or evade a courting male because the evasive flights are short compared with the duration of copulation. For a female mated to a virgin male, the ratio of mean copulation duration (44.8 min) to the mean length of evasive flight (99.8 sec) was 26.9; for a female mated to an already-mated male, the ratio increased to 154.

In at least some species, mated females may be denied the opportunity to reject male courtship advances. In species in which a large sphragis (mating plug; Fig. 8) is formed by the male during copulation, patterned courtship sequences are replaced by aerial capture and takedown of the female by the male (Scott 1972). In *Parnassius* (Papilionidae, Scott 1972, 1973), *Methona* (Nymphalidae:Ithomiinae, Pliske 1975c), *Acraea* and *Planema* (Nymphalidae:Acraeinae, Eltringham 1912), and occasionally in *Danaus plexippus* (Nymphalidae:Danainae, Pliske 1975d, Hill *et al.* 1976), the male captures the female by pouncing on her dorsum in midflight, grasping her head and thorax with his legs, and taking her quickly to the ground. At least in *Parnassius* (Scott 1972), and *Methona* (Pliske 1975c), the male must maneuver the female into a venter-to-venter position to effect copulation and form the sphragis, which may explain the need for aerial capture and takedown (Scott 1972). It is unclear whether males who capture females release previously mated females without mating with them. Scott (1972) suggests that a male could readily tell if a captured female is a virgin or not by physically detecting the sphragis, but because the last male to mate usually fertilizes all of a female's remaining eggs (Gwynne, this volume), a male could benefit by mating with nonvirgins. Although we do not know what determines the male's decision in such a case, it appears that the female does not have a choice in the matter.

Population density and sex ratio may affect the efficacy of female mate-refusal behaviors, as suggested by their effects on mating frequencies (discussed earlier). High population densities in butterflies can lead to a high incidence of unsuccessful courtships because of interference among competing males (Brower *et al.* 1965, Scott 1974, Shields 1967, Shapiro 1977). In *Pieris protodice*, high frequency of male courtships results in emigration of females and subsequent colonization of new habitats (Shapiro 1970). Labine (1966) reports that males of *Euphydryas editha* (Nymphalidae) continually harrass mated females, apparently disregarding female signals of nonreceptivity, and attempt to mate by capture. On the other hand, low population densities, which have been correlated with high proportions of both virgins and multiply-mated females (Burns 1968), result in fewer male/ female encounters. Consequently, males may be more persistent in courtship with any female (Shapiro 1977) and continued resistance to male courtship advances may be more costly to the female than an additional mating.

Although Lederhouse (1981) argues that the cost of additional matings should be small for females of species with short copulatory durations, many Lepidoptera species exhibit characteristics that make them more vulnerable during mating and that, therefore, should result in selection for reduced copulation frequency or duration: Copulating pairs are less mobile than individuals, are more conspicuous to predators than individuals, and are less wary of predators than individuals;

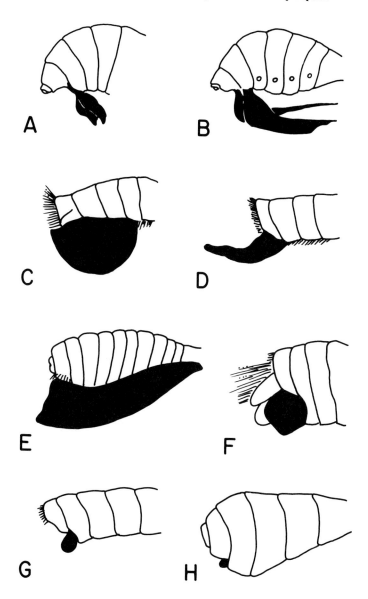

Fig. 8. Mating plugs or sphraga in female butterflies (blackened portion). Lateral views, posterior ends of abdomens at left. Not drawn to same scale. Papilionidae: A. *Eurycus cressida* Swains; B. *Euryades duponcheli* Burm.; C. *Kailasius charltonius* Gray; D. *Parnassius sikkimensis* (Elwes); E. *Parnassius mnemosyne* L. Nymphalidae: F. *Argynnis paphia* L. (Nymphalinae); G. *Actinote ozomene gabrielae* Rebel (Acraenae); H. *Dircenna loreta loreta* Haensch (Ithomiinae). A-E redrawn from Bryk (1918), F redrawn from Bryk (1919), G-H original drawings of specimens collected at Tandape (G) and Limoncocha (H), Ecuador.

copulation durations are long; and copulating pairs cannot disengage rapidly (Walker 1980). Mating butterflies tend to remain nearly motionless so that females expend little energy during copulation unless they carry the male in flight if the pair is disturbed, as in most Satyrinae (Nymphalidae) and Hesperiidae (Miller and Clench 1968). But loss of time for other activities (Parker 1970) and increased vulnerability while *in copula* (Richards 1927) should be sufficient to discourage superfluous multiple matings, as suggested by the evolution of the various mate-rejection behaviors in female Lepidoptera discussed above.

IV. SPERM PRIORITY PATTERNS AND EVOLUTION OF MATING BEHAVIOR

Several explanations have been advanced to account for the evolution of sperm priority patterns in insects (see reviews by Parker 1970; Boorman and Parker 1976; Walker 1980; Gwynne, this volume). Parker (1970) argues that intrasexual selection among males should result in stabilization of the degree of sperm displacement at an amount that yields the maximum overall fertilization rate to the male rather than that which yields the maximum possible egg gain to a male from a given mating. Gwynne (this volume) also focuses on males to interpret interspecific variability in sperm priority patterns, but suggests that the variablity results primarily from differences in the degree to which males invest in nonpromiscuous mating effort (ME) and/or male parental effort (PE) in reproduction. By contrast, Walker (1980) argues that the basic sperm priority patterns of most insect species have resulted primarily from selection acɪg on sperm utilization strategies of females. None of these hypotheses is mutually exclusive. Indeed, Walker (1980) embraces Parker's (1970) explanation as determining the optimal male strategy, but claims that the influences of female behavior and of the structure and function of the female reproductive tract largely outweigh intramale selection in determining sperm priority patterns. Furthermore, Gwynne (this volume), by focusing on the relative degrees of male nutrient investment in reproduction, and Walker (1980), by focusing on multiple mating in females as a strategy to acquire nutrients, are considering alternate sides of the same evolutionary coin. As mentioned earlier, the transfer of nutrients from male to female during mating appears to be widespread in the Lepidoptera and of considerable importance to long-lived species, so that the influence of male investment in nonpromiscuous ME and PE on the evolution of sperm priority patterns and mating strategies in Lepidoptera may be considerable.

A. Sperm Precedence in Lepidoptera:
Sperm Competition or Ejaculate Competition?

In contrast to the wealth of information about mating frequencies in natural and laboratory populations of Lepidoptera, little is known about sperm priority patterns in multiply-mated females. What we do know comes largely as a byproduct of biological control studies in which experimental comparisons of mating competitiveness are made between sterilized and normal males in economically important species (Proverbs and Newton 1962a; Labine 1966; Flint and Kressin 1968; North and Holt 1968; Snow et al. 1970; Retnakaran 1971a, b; Etman and Hooper 1969a; Sokolowski and Suski 1980). Gwynne (this volume) has discussed the difficulties in interpreting results of such studies, especially those that do not include reciprocal crosses to control for the order of mating. Other studies have investigated sperm priority patterns by using sterile males obtained from interspecific hybrids (Pair et al. 1977) or males carrying alternate alleles for a physical trait (Ae 1962, Clarke and Sheppard 1962, North and Holt 1968, Retnakaran 1974, Brower 1975, Sims 1979) or for enzyme activities measured by electrophoresis (Boggs and Watt 1981). Very few researchers have addressed the problem of sperm competition in Lepidoptera directly, so that much of the evidence is circumstantial. Scrutiny of the available data permits tentative conclusions to be drawn, but raises many more questions than can be answered given our current understanding of the reproductive biology of Lepidoptera.

Given that multiply-mated females are commonly encountered in natural populations of Lepidoptera (Table I), what outcomes are possible with regard to the utilization of sperm from two males? If sperm from the first male are no longer present or viable in the female when she mates for the second time, then, of course, no sperm competition takes place. This is simply sequential mating in which the first male fertilizes all eggs laid prior to the exhaustion of his sperm. If ejaculates from two males overlap in the female's genital tract, the pattern of their relative use describes the nature and extent of sperm competition. If sperm from the two ejaculates mix together in the spermatheca such that they fertilize offspring in proportion to their relative abundance, scramble competition is taking place. Most experimental studies on Lepidoptera have shown, however, that in twice-mated females one male or the other (usually the second) fertilizes most or all of the eggs. This is the phenomenon of sperm precedence.

Gwynne (this volume) presents the most complete analysis to date of sperm priority patterns in Lepidoptera, citing information on 11 species in the form of P_2 values (Boorman and Parker 1976). To Gwynne's summary may be added (1) a study of Laspeyresia pomonella (Olethreutidae, Sokolowski and Suski 1980), from which I calculated a P_2 of 0.65, which agrees well with the value of 0.58 calculated by Boorman and Parker (1976) from data in Proverbs and Newton (1962a); (2) additional papers by Retnakaran (1971a, b) on the spruce budworm,

Choristoneura fumiferana (Tortricidae), in which complete sperm precedence of the first male to mate (P_2 = 0) is reported; and (3) observations on a single female of *Papilio zelicaon* (Papilionidae, Sims 1979) that suggest sperm mixing and show a shift in the proportion of offspring fathered by the second male from 0.33 to 0.70 between day 1 and days 4-5 following the second copulation. Finally, Boggs (1979, cited in Boggs 1981a) states that *Dryas julia* (Nymphalinae, Heliconiini) has incomplete sperm precedence, Shapiro (unpubl., cited in Shapiro 1970) states that last male sperm precedence occurs in *Pieris rapae* (Pieridae:Pierinae), and Ehrlich and Ehrlich (1978) cite Ae (1966) to claim last male sperm precedence in *Pieris brassicae.*

1. Sperm Priority Patterns in Multiply-mated Females

The generalization that, in Lepidoptera, sperm from the last male to mate usually gain precedence over sperm stored in the female from previous matings is somewhat of an oversimplification. The P_2 values developed by Boorman and Parker (1976) for doubly-mated females range from 0 (all offspring fathered by the first male to male), through 0.5 (equal numbers of offspring fertilized by both males, *i.e.*, complete sperm mixing), to 1 (all offspring fathered by the last male to mate). Of the 11 species listed by Gwynne (this volume) only six have P_2 values above 0.9. These include three noctuid moths: *Spodoptera litura*, P_2 = 1 (Etman and Hooper 1979a), *Trichoplusia ni*, P_2 = 0.92 (North and Holt 1968), and *Heliothis virescens*, P_2 = 0.99 (Flint and Kressin 1968); and three butterflies: *Colias eurytheme*, apparent complete last male precedence (Boggs and Watt 1981), *Euphydryas editha*, P_2 = 0.919 (Labine 1966), and several species of *Papilio*, especilly *P. dardanus*, P_2 = 1 (Clark and Sheppard 1962, Ae 1962). (I do not include *Choristoneura fumiferana;* see below.) To this group could be added *Pieris rapae* (Shapiro 1970) and *P. brassicae* (Ae 1966, cited in Ehrlich and Ehrlich 1978), apparent complete last male precedence, although no quantitative data are available. In two species, *Bombyx mori* (Omura 1939) and *Choristoneura fumiferana* (Retnakaran 1972a, b, 1974), first male sperm precedence occurs, with P_2 values of 0.34 and ca. 0, respectively. The remaining species show incomplete last male precedence or extensive mixing of sperm: three moths, *Laspeyresia pomonella*, P_2 = 0.58 (Proverbs and Newton 1962a), P_2 = 0.65 (Sokolowksi and Suski 1980); *Plodia interpunctella*, P_2 = 0.68 (Brower 1975); *Spodoptera frugiperda*, P_2 = 0.72 (Snow *et al.* 1970); and the butterfly *Dryas julia*, incomplete precedence (Boggs 1981a). An additional study of *Heliothis virescens* employing sterile hybrid males (from crosses with *H. zea*) yields a P_2 value of 0.8 (Pair *et al.* 1977), considerably lower than the value of 0.99 calculated by Gwynne from Flint and Kressin's (1968) study that employed males sterilized by irradiation. Sims' (1979) preliminary results with *Papilio zelicaon* also indicate considerable sperm mixing.

Although P_2 values facilitate comparisons among species, reducing sperm priority patterns to a single number obscures intraspecific variability in the degree of sperm precedence. Several studies (Flint and Kressin 1968, Labine 1966, Retnakaran 1974, Sims 1979) have reported considerable variability in the outcome of double matings, although much of this may result from unusually short time intervals between successive matings. For example, in the spruce budworm, *Choristoneura fumiferana,* Retnakaran (1971a, b) demonstrated first male sperm precedence in females mated sequentially on successive days (as might occur in the field), but later (Retnakaran 1974) reported a shift from first male to last male sperm precedence in females mated twice in the same day. Of 15 females mated twice within a 24-h period, six showed first male precedence, six showed last male precedence, and three showed extensive sperm mixing. Likewise, in one of Labine's (1966) experiments, three females of *Euphydryas editha* were mated twice in quick succession, which resulted in one female showing second male precedence and the other two females showing first male precedence. By contrast, oviposition records of twice-mated females of *E. editha* paired on successive days showed a more consistent pattern of last male sperm precedence. In *Papilio zelicaon,* a female mated twice within 5 h showed, over a 5-day oviposition period, a shift in the proportion of eggs fathered by the second male from 33% to 70% (Sims 1979).

For *Heliothis virescens,* Flint and Kressin (1968) reported that sperm precedence occurred in almost every double mating as an "all or none phenomenon," with first male precedence in the majority of trials (78% if untreated males were first to mate, 54% if sterile males were first to mate). Thus, there is considerable variability in the outcome of double matings in *H. virescens,* even though Gwynne (this volume) calculated a P_2 value of 0.99 for this species from Flint and Kressin's data by considering only those double matings in which displacement occurred. Gwynne suggested that the incidence of first male precedence might have resulted from a sperm plug left by the first male. An alternative explanation is that the outcome of sperm competition is variable and depends on the intermating interval and relative competitiveness of the two ejaculates.

What can we conclude from these studies? First, complete mixing and proportionate utilization of sperm from more than one mating are rare in Lepidoptera. The majority of species show nearly complete sperm precedence, in most cases favoring the last male to mate, although in some species sperm precedence is not complete and may favor either the first or second male. Second, repeat matings occurring in quick succession produce more variable results than matings that approximate field conditions, separated by a day or longer.

Retnakaran (1974) has explained the importance of the time interval between successive matings in determining sperm precedence as a simple consequence of female genital architecture and sperm migration (see Fig. 2B). If a mated female remates before the sperm from the first mating are transferred to the spermatheca (which requires about 4 h in the spruce budworm), the first spermatophore will be pushed away from the ductus bursae to the rear of the corpus bursae and the second spermatophore would be in position (Fig. 9B) to release sperm that would fill the spermatheca (last male sperm precedence). On the other hand, if a mated female remates after the sperm from the first mating have completed the migration to the spermatheca, it would be impossible for the sperm from the second mating to enter the already-filled spermatheca (first male sperm precedence).

Sperm mixing can occur when the second mating occurs after only a portion of the sperm from the first mating have been transferred to the spermatheca. In this case, the second spermatophore would replace the partially-emptied first spermatophore and there would be room in the spermatheca for some of the sperm from the second spermatophore (incomplete sperm precedence). If a female's first mating is to an older male or to a male just mated, she is likely to receive a smaller than normal first spermatophore (Srivastava and Srivastava 1957, George and Howard 1968, Stockel 1971, Sims 1979, Boggs 1981a), which may contain insufficient sperm and seminal fluid to fill the spermatheca. Under such circumstances sperm mixing might result if a second mating occurred after sperm from the first male had reached the spermatheca.

2. Mechanisms Controlling Sperm Priority Patterns

Given that disproportionate use of sperm for egg fertilizations occurs in multiply-mated females, what mechanisms might account for the observed sperm priority patterns? Potential determining factors can be classified as passive or active, the latter category consisting of both male-determined and female-determined factors. Passive factors include (1) the basic genital anatomy of the female, which in ditrysian Lepidoptera requires that sperm be transferred from the spermatophore in the corpus bursae to the spermatheca before fertilization can occur (section II.B.); and (2) the sperm storage capacity of the female, which is influenced by the size and shape of the spermatheca. Active factors may be (1) female-determined, such as muscular control of sperm movement and storage within the female reproductive tract or inactivation of stored sperm by hormonal influences or nutritional deprivation; or (2) male-determined, such as physical displacement of sperm stored from a previous mating, or incapacipation of sperm stored from a previous mating. Because the time interval between matings affects all of these potential mechanisms, selection on both sexes should favor mechanisms to control female mating frequency by affecting the balance between courtship receptivity of mated females and courtship persistence of males.

The passive factors described above are "passive" only in an ecological time frame; they are certainly subject to selection over evolutionary time because of their great potential for affecting reproductive success. In fact, an attempt (Walker 1980) has been made to correlate the morphology of the sperm storage organ of the female (the spermatheca in ditrysian Lepidoptera) with sperm utilization patterns.

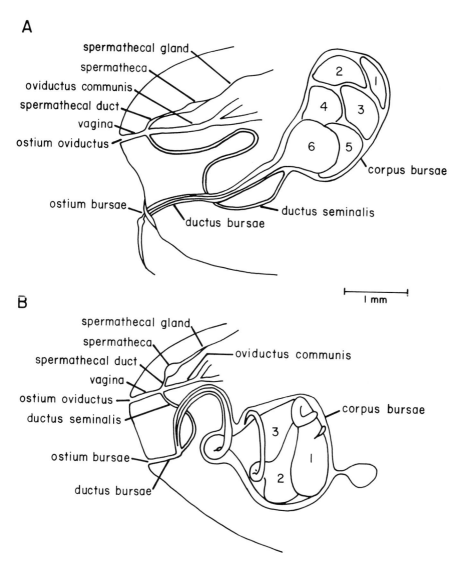

Fig. 9. Spermatophore placement in Ithomiinae (Nymphalidae) from Limoncocha, Ecuador. A. *Methona confusa psamathe* Godman and Salvin. Female with sixth spermatophore in place and five others compacted in upper part of bursa copulatrix. Ductus seminalis leads from base of corpus bursae to oviduct. Collum of spermatophore 6 extends beyond chitinized ostium bursae. B. *Godyris zavaleta amaretta* (Haensch). Female with third spermatophore in place and two others compacted in upper part of bursa copulatrix. Ductus seminalis leads from ductus bursae near the endpoint of the collum of spermatophore 3 (not from corpus bursae, *cf.* A) to oviduct.

The trend demonstrated was from spheroid sperm storage organs in single-mating species to elongate organs in multiply-mating species. More elongate organs presumably cause sequential packing of successive ejaculates such that sperm from the most recent ejaculate, nearest the spermathecal duct, fertilize most or all eggs (in Lepidoptera; Clark and Sheppard 1962, Brower 1975; see also Parker 1970). Elongate sperm storage organs are interpreted as an adaptation of females to facilitate last male sperm precedence. In Lepidoptera, the majority of species examined have elongate spermathecae and exhibit last male sperm precedence (Walker 1980). For species in which females mate multiply to collect nutritional benefits, last male sperm precedence would insure continued interest of males in courting mated females.

Walker (1980) cautions that spermathecal shape as a mechanism to insure last male sperm precedence assumes a low order of sperm mixing within the spermatheca and a sperm storage capacity greater than one insemination. The spermatheca is elastic and distends considerably during sperm storage (Outram 1971, Yang and Chow 1978); however, it is said to be unable to accommodate all sperm received from a single mating in some moths (Hinton 1964). The low order of sperm mixing could result from the forcing of previously stored sperm into the narrow spermathecal gland located cephalad of the lumen of the spermatheca. Here, physical constriction, lack of nutrients, or accumulation of CO_2 could result in inactivity of sperm.

Females could enhance the effectiveness of last male sperm precedence actively by removing or incapacitating stored sperm at the time of each additional mating. Etman and Hooper (1979a) provide circumstantial evidence that mating induces the female to expel stored sperm from the spermatheca in *Spodoptera litura* (Noctuidae). Expansion of the corpus bursae by the arrival of a new spermatophore appears to trigger rhythmic contractions of most (or all?) muscular parts of the female reproductive tract, including the spermatheca (Srivastava and Srivastava 1957, Yang and Chow 1978, Chao 1981). Contractions of the corpus bursae, ductus bursae, ductus seminalis, and bulla seminalis are believed to propel the sperm bundles toward the vestibulum (Norris 1932, Srivastava and Srivastava 1957, Callahan and Cascio 1963, Thibout 1977) while contractions of the spermathecal gland and spermatheca are thought to draw sperm bundles up the spermathecal duct (Norris 1932, Callahan and Cascio 1963). Lum *et al.* (1981) have shown that the flexible main canal of the spermathecal duct is capable of considerable distention during peristaltic contractions of the spermatheca, which can force clusters of sperm down the spermathecal duct. Lum *et al.* (1981) interpreted the flexibility and sperm-carrying capacity of the main canal in *Plodia interpunctella* (Pyralidae) as a mechanism to insure rapid sperm tranfer to the vestibulum for fertilization, although such a system could account for the apparent flushing of sperm from the spermatheca postulated by Etman and Hooper (1979a) for *Spodoptera litura.* If the sperm storage capacity of the spermatheca of *S. litura* is limited, then flushing of stored sperm would insure room in the spermatheca for newly-arriving sperm from the next mating. Sperm flushed from the spermatheca would enter the vestibulum and presumably disintegrate or exit the vagina through the vulva.

Sperm flushing, however, appears to be a risky strategy for *S. litura* because it begins prior to termination of copulation (Etman and Hooper (1979a) and thus prior to the successful exit of sperm from the spermatophore into the ductus seminalis (see section II.B.). Given the levels of mating failure associated with collum misalignment of later matings in some Lepidoptera (Taylor 1967), the possibility exists that a female who began flushing the spermatheca during mating could end up with no new sperm to replace expelled sperm.

The flushing of sperm described for *S. litura* may not occur in some other noctuids, such as *Heliothis zea*, which seems to protect against the movement of large sperm masses out of the spermatheca by having a chitinous area of the spermathecal duct, just below the spermatheca, that bears spines pointing toward the lumen of the spermatheca. No evidence of sperm flushing has been found in *Heliothis* (Callahan and Cascio 1963). Females of both *H. zea* and *S. litura* multiply mate (up to 5-6 times), but the proportion of females in a population engaging in multiple matings (Table I) is greater for *S. litura* (42-47%) than for *H. zea* (22-34%). Perhaps sperm flushing in *S. litura* evolved as a mechanism to permit effective use of sperm from a greater number of matings.

The possibility that sperm stored in the spermatheca are somehow inactivated or rendered impotent by females is a speculation with no evidence to support or refute it. The functions of the secretions of the spermathecal gland are unknown, although it is thought that they provide either nourishment to stored sperm (Norris 1932, Davey 1965) or a medium for sperm traveling up the spermathecal duct (Stockel 1973), or both. It seems unlikely that the secretions could inhibit sperm activity, although withholding the secretions might adversely affect sperm viability.

Last male sperm precedence could be effected by males if a male could somehow displace or inactivate sperm stored from a previous mating. If sperm stored in an elongate spermatheca could be forced to the upper end of the spermatheca or into the spermathecal gland, the second male's sperm would lie closer to the opening of the spermathecal duct leading to the vestibulum where the eggs are fertilized. Although this mechanism has been mentioned as a passive factor imposed by spermathecal shape, the lack of sperm mixing required for it to be successful (Walker 1980) suggests the importance of the integrity of the incoming ejaculate. If sperm from the second mating were to arrive gradually, it is difficult to see how mixing of newly-arrived and stored sperm could be avoided. Furthermore, the eupyrene sperm bundles passed to the female break down before entering the ductus seminalis (Holt and North 1970, Riemann and Thorson 1971, Etman and Hooper 1979b) and sperm migration to the spermatheca is a gradual process requiring several hours (see Appendix). Sperm mixing could be reduced or eliminated if newly arriving sperm could in some way inactivate any stored sperm encountered.

Pair *et al.* (1977) report that apyrene sperm from sterilized males enter the spermatheca of mated, fertile females of *Heliothis virescens* and displace much of the stored eupyrene sperm, thereby reducing female fertility. These authors speculate that the excess of apyrene sperm or some deficiency of the accompanying seminal secretions transferred by sterilized males disrupts the biochemical medium of the stored eupyrene sperm. Under natural conditions, could apyrene sperm from a second mating somehow incapacitate the eupyrene sperm stored from a previous mating? Silberglied *et al.* (1984) have recently summarized the biology of apyrene sperm and reviewed the various hypotheses advanced to account for the substantial numbers of apyrene sperm (usually over 50% of the total sperm complement) produced by lepidopteran males. One of their suggestions is that apyrene sperm remove, inactivate, or destroy eupyrene sperm stored from previous matings, although no mechanisms are proposed. Apyrene sperm are thus envisioned as an inexpensive specialized cell type, preprogrammed to negotiate the tortuous pathway through the female reproductive tract (section II), whose function is to ensure the precedence of their genetically endowed brothers over sperm from previous matings.

3. Sperm Precedence and Reproductive Success

Within the constraints imposed by lepidopteran anatomy and physiology, how should natural selection act on males and females to influence the outcome of sperm use in multiply-mated females? For females in which the disadvantages to multiple-mating outweigh the advantages (see section III.D.), selection on females should favor first male sperm precedence. Complete sperm precedence by the first male to mate would eliminate any advantage to males who mate with mated females, thereby selecting for males who quickly abandon courtship attempts with females who indicate nonreceptivity. Conversely, for females in which the advantages of repeat mating outweigh the disadvantages, selection should favor last male sperm precedence. In such females, sperm precedence of the last male to mate provides an incentive to males to court and attempt to copulate with mated females, and it replenishes the female's sperm supply by providing her with a greater number of fresher (more viable, more competent) sperm. In this case, a balance must be struck between the requirements of the female for sperm renewal or male-contributed nutrients and the persistence of courting males. Efficient courtship rejection behaviors in mated females should be selected for in direct proportion to the persistence of courtship attention that males confer on adequately inseminated females.

In both cases discussed above, selection on females is expected to favor complete sperm precedence (see also Knowlton and Greenwell, this volume). Lack of precedence, or sperm mixing, could be favored if females experienced strong selection to maximize genetic diversity among their offspring. I predict that such selection, leading to a mixed sperm utilization strategy, would be stronger in species that oviposit in clusters over a short reproductive life span than in species that oviposit singly over a longer reproductive life span. In other words, for species in which increased genetic variability of offspring is important, concentration of female reproductive effort in space and time should select against sperm precedence.

As Parker (1970) has pointed out, multiple mating by females that results in sperm from more than one male being present in the female reproductive tract leads to postcopulatory (sperm) competition among males for fertilizations. Opposing selective forces favor adaptations that (1) allow a male to achieve insemination and fertilization at the expense of previously stored sperm (last male precedence), and (2) prevent a male's stored sperm from being replaced by sperm from subsequent matings (first male precedence).

Evidence has been presented establishing the greater frequency of last male precedence in Lepidoptera, but it is important that the predominance of all or none sperm precedence patterns in Lepidoptera is consistent with the predicted selection pressures acting on both males and females to control sperm utilization patterns. That is, selection has favored a strong degree of sperm precedence and a low incidence of sperm mixing, resulting in competition among ejaculates (contest competition) rather than competition among individual sperm (scramble competition).

B. Multiple Mating, Female Receptivity, and Protection of Male Reproductive Investment

It is clear that copulation involving the transfer of a spermatophore, accessory fluids, and sperm results in several changes in the subsequent reproductive physiology and behavior of lepidopteran females (LaChance et al. 1978). These changes include (Hinton 1974, Leopold 1976):

1. stimulation of contractions of the female genital ducts necessary to translocate sperm to the spermatheca (see section II.B. and Appendix);

2. activation of sperm in the reproductive tract of the female (Omura 1936, 1938; Shepherd 1974a, b, 1975; Herman and Peng 1976);

3. acceleration of the maturation of oocytes (Benz 1969, Herman and Barker 1977);

4. stimulation of oviposition (Norris 1933, Benz 1969, North and Holt 1968, Deseo 1972, Stockel 1972, Thibout and Rahn 1972, Karpenko and North 1973, LaChance et al. 1973, Robinson 1973, MacFarlane and Tsao 1974, Truman and Riddiford 1974, Yamaoka and Hirao 1976);

5. incorporation of transferred nutrients into somatic and reproductive tissues of the female (see section III.D.2.);

6. modification of female courtship behavior or obstruction of female genital orifices to delay or prevent remating (see below).

Because the materials passed to a female during copulation represent a sizable energetic investment by the male, especially in those species that contribute

nutrients to enhance female reproduction or survival, males should experience selection to develop mechanisms that reduce the probability of additional copulations
by their mates that result in usurpation of invested materials. The intensity of this
selection increases with (1) the degree of which males invest in nonpromiscuous ME
and PE (Gwynne, this volume), and (2) the propensity of females to engage in
additional matings.

1. Control of Female Receptivity to Courtship

A female refractory period after mating, during which a female stops releasing
male-attractant pheromone (moths) or rejects sexual advances of courting males
(butterflies) is probably universal in Lepidoptera but variable in duration. It may be
as short as 1 (Pliske 1973) to 4 (Taylor 1967) days, but more usually is about
8-9 days (David and Gardiner 1961, Suzuki 1979). Two factors are thought to
stimulate this behavioral switch in the female: (1) the mechanical pressure of the
inflated spermatophore in the bursa copulatrix, detected by stretch receptors
in the corpus bursae that relay the information either by afferent nerve impulses
(Labine 1964, Obara et al. 1975, Sugawara 1979) or by release of a hormone into
the blood (Riddiford and Ashenhurst 1973, Obara 1982), or (2) the presence of
sperm in the spermatheca (Norris 1932, Taylor 1967, Benz 1969, Deseo 1971,
Lum and Brady 1973, Thibout 1975), which also could affect receptivity by either
neural (Lum and Arbogast 1980) or hormonal (Taylor 1967) control. Taylor
(1967) and Leopold (1976) suggest that the initial stimulus inducing nonreceptivity
may be mechanical (neural), whereas long-term nonreceptivity is maintained by a
humoral factor. In this view, presence of the spermatophore initiates nonreceptivity
near the end of copulation, thereby protecting the female (and the genetic and
nutritional contribution of her mate) until the delayed blood-borne hormone can
take control.

In most species the body of the spermatophore collapses to some degree soon
after the sperm exit and, in some species, the spermatophore is dissolved and absorbed in the corpus bursae. Both of these events argue in favor of the spermathecal
stimulus as the more reliable of the two mechanisms for inducing long-term nonreceptivity. Female nonreceptivity during the time required for the sperm to exit
the spermatophore is crucial to male mating success, however, because if the female
remates before this happens, the first spermatophore will be displaced and its
sperm will be unable to reach the spermatheca (sections II, IV.A.). Assuming that
the humoral factor would not begin to operate before the sperm exit the spermatophore (this would certainly be true if the hormone is released because of spermathecal stimulation by the sperm), the early mechanical stimulation of nonreceptivity
by the inflated spermatophore would be very important to males. This may explain
why chitin occurs in spermatophores of higher lepidopteran groups such as the
Noctuidae (Callahan 1958). Perhaps the addition of chitin to the walls of the

spermatophore evolved to prevent the early collapse or dissolution of the spermato-phore, whose inflated state is crucial to inducing nonreceptivity during the vulner-able (to the male) period of sperm migration within the female.

Long-lasting spermatophores are common in the higher Lepidoptera (section III. A.), but, for the female, mere presence of the spermatophore in the bursa copulatrix is a much less reliable indication of successful mating than is detection of sperm in the spermatheca (section III.D.) For females whose first mating is in-effective for reasons other than failure to pass a spermatophore (up to 30% of all matings in *Atteva punctella*, Taylor 1967), reliance on mechanical stimulation of the bursa to induce courtship nonreceptivity would delay the onset of fertile ovi-position considerably in species with long-lasting spermatophores. Detection of viable sperm in the spermatheca is clearly a more efficient system of modulating receptivity and is likely to be widespread, especially in the generally short-lived species of Lepidoptera (Taylor 1967).

Ehrlich and Ehrlich (1978) found a strong association between virginity and immaturity of eggs in females of many groups of butterflies. This suggests either that a freshly eclosed female requires a period of a few hours or days to begin matu-ring eggs, during which time she is unreceptive to courtship and mating, or that mating is required to stimulate egg maturation. In the latter case, insemination may stimulate females to mobilize stored reserves for vitellogenesis, as in *Danaus plex-ippus* (Herman and Barker 1977), or mating may provide nutrients required for egg production. Because they found high proportions of virgins with no mature eggs in species that appear to absorb spermatophores, Ehrlich and Ehrlich (1978) sug-gested that these species are likely to collect nutrients from males to finance egg production. Females of *Zygaena* moths typically mate two or three times before oviposition, even though each mating lasts about 24 h (Hewer 1934). Such females have active sperm crammed into every part of the reproductive tract, suggesting that collection of nutrients rather than accumulation of sufficient sperm accounts for this behavior.

Many female Lepidoptera eclose with a full complement of eggs in various stages of maturity and oviposition proceeds rapidly after mating (Ehrlich and Ehrlich 1978). If one mating supplies sufficient sperm to fertilize all eggs, there would be little incentive for such females to remate (section III.D.), and the loss of time through superfluous matings could be a disadvantage to species with short repro-ductive life spans (*e.g.*, many temperate species). As argued earlier, selection on such species should favor females who become unreceptive after mating and males who transfer a sufficient amount of sperm and who provide an appropriate long-lasting stimulus to maintain female nonreceptivity. Because female nonreceptivity in such cases is an advantage to both partners, it should be considered a cooperative agreement between a mated pair.

For species in which females eclose with no or few matured eggs and who re-quire nutritional input as adults to finance egg production (Dunlap-Pianka *et al.*

1977), there exists a conflict of interests between males and females if females use repeat matings as a means of obtaining the required nutrients. For examples, some groups of tropical butterflies are distasteful to predators as a result of larval or adult feeding on plants containing substances toxic to vertebrates (Ehrlich and Raven 1964; Brown, in press). Most of these species are long-lived as adults because of the protection afforded by distastefulness, which is advertised by aposematic coloration of a cooperative nature (Mullerian mimicry, Wickler 1968). Such adult protection from predators permits a long reproductive life span, measured in months, instead of days and weeks. By contrast, immature stages, whose rapid development in warm tropical climates is favored by high levels of predation and parasitism (Drummond 1976), are relatively short-lived and may be unable to accumulate nutrients for reproduction. In many of these groups (*e.g.*, Ithomiinae, Heliconiini), specialized feeding on protein-rich food sources, such as nectar rich in amino acids (Baker and Baker 1973), pollen (Gilbert 1972, Dunlap-Pianka *et al.* 1977), and bird droppings or detritus (Drummond 1976, Ray and Andrews 1980), provides nutrients needed for reproduction that were not stored during the larval stage. The nutrients needed by adult females for reproduction could also come from males via copulation. It is easy to imagine that long-lived females might have become receptive to repeat matings at first to replenish deteriorating sperm supplies, and later encouraged such matings at more frequent intervals to obtain nutritional benefits from males. Females could select among males according to their potential for providing nutrients; males would be under selection to produce large spermatophores and to advertise this ability to females (see section III.D.3.).

The result of selection in groups with long reproductive life spans and adult-financed egg production should be a mating system that favors increased frequency of mating by females, but that reduces the number of matings per male because of the increased cost of mating. As Gwynne (this volume) has shown, increased investment in reproduction by males leads to selection for increased confidence of paternity. Because females who benefit from multiple matings should influence sperm precedence patterns to favor the last male, selection is strong on males to protect their investment by delaying or preventing additonal matings by females. Selection on males to increase confidence of paternity has modified their behavior and physiology to increase the probability that their matings are successful (*i.e.*, lead to insemination and fertilization) and are likely to remain so (*i.e.*, delay or prevent remating by their female partners). In Lepidoptera, male strategies that enhance the success of copulation include chitinized spermatophores, prolonged copulation, and male-determined mechanisms that promote last-male sperm precedence. Male strategies that prevent loss of mating investment include various mechanisms that delay or prevent females from remating by repelling other males.

2. Mechanisms to Increase Confidence of Paternity

The mechanisms discussed here are limited to those that are solely the result of intrasexual selection on males. They include male attempts to deter other males from mating with mated females, and not with attempts of males to reduce female receptivity, which, if successful, must be viewed as a cooperative contract between male and female (true monogamy).

Parker (1970) has reviewed the variety of sexually selected adaptations in insects that enhance confidence of paternity by reducing the probability of female re-mating. In Lepidoptera these are limited to (1) substances transferred to the female, such as mating plugs or pheromones, that discourage subsequent courting males; and (2) prolonged copulation, when it functions to prevent sexual access to the female by other males.

Most discussions of lepidopteran mating plugs focus on the elaborate external structures known as sphraga that block the ostium bursae. Sphraga attracted attention of naturalists early in this century and the extensive and scattered litera-ture on sphraga is largely descriptive and often speculative. A comprehensive review of early papers is given by Bryk (1919) and the functional significance of the sphragis is discussed by Bryk (1918, 1930), Eltringham (1925), Petersen (1928), and Tykac (1951).

The sphragis is produced by accessory glands of the male (Eltringham 1925) and represents a significant investment in time and materials. During copulation the sphragal fluid is guided by the complex claspers (valvulae) of the male to the sterigma of the female, where it hardens upon contact with air and blocks the ostium bursae. In certain species (*Parnassius, Kailasius*) it may be large enough to cover a significant portion of the underside of the female's abdomen (Fig. 8). The size of the sphragis seems out of proportion to the amount of material needed to block the small ostium bursae, which is penetrated by a part of the sphragal sub-stance (spermatophragma) that extends up the ductus bursae and fuses to the collum of the spermatophore (Petersen 1928). However, Bryk (1918, 1930) suggests that the larger the sphragis, the greater its effectiveness in deterring other males. The sphragis may function secondarily as an intrasexual signalling device; such a function would help explain the diversity of flaglike appendages found on the sphraga of many species (Fig. 8).

The sphragis is soluble in water, so that high humidity and dew could lead to its gradual dissolution (Tykac 1951), presumably decreasing its effectiveness as a mating deterrent. Occasional double plugging has been reported in *Acraea* (Eltring-ham 1916), *Parnassius* (Petersen 1907), and several Papilionidae (Bryk 1930), although it is unclear whether the two plugs represent separate matings or were deposited by the same male during prolonged copulation (Petersen 1928). If the reproductive life span of the insect exceeds the effective life span of the sphragis, remating could occur. Double matings also might occur if a just-mated female was immediately captured by a second male who copulated with her while the first sphragis was still soft. Prolonged copulation may allow the sphragis to harden before the pair separates, thereby reducing the probability of a second mating.

The susceptibility of sphraga to high humidity may explain their apparent restriction to species inhabiting dryer areas (*e.g.*, montane habitats in the circumboreal genus *Parnassius* and scrub forests and grasslands in the paleotropical genus *Acraea*). In wetter areas, smaller copulatory plugs (spermatophraga) are formed in relatives of *Parnassius* (*e.g.*, *Papilio*) and *Acraea* (*e.g.*, *Actinote*, Fig. 8G), but these probably function like those of *Methona* (Fig. 9A), *Argynnis* (Fig. 8F), and *Dircenna* (Fig. 8H) to delay remating rather than to prevent it.

Bryk (1930) suggested that the sphragis serves to hold the pair together during copulation, which would reduce the chance of takeover by other males, who commonly jostle the pair in an attempt to gain access to the female. The genital armature of males is complex and heavily chitinized in *Parnassius, Acraea,* and most other sphragis-producing groups (pers. obs.), and is much more likely to account for *in copula* security than a slowly-hardening sphragis. In most species that produce sphraga (*e.g.*, *Parnassius, Acraea*) there is a pronounced absence of courtship display by males (Eltringham 1912, Scott 1972), who pounce on females in flight and tumble to the ground where copulation ensues ("mating by capture"). Similar mating by capture occurs in *Methona* (Ithomiinae, Pliske 1975c) and after copulation the collum of the spermatophore extends beyond the ostium bursae (Fig. 9A) to which it is sealed by an opaque substance. As Fig. 9A shows, however, multiple mating occurs in *Methona,* and the collum-plug combination probably serves only to delay remating in this long-lived (4-6 months) species.

Although sphraga are restricted to a few taxonomic groups, copulatory plugs (spermatophraga) or spermatophore caps (Holt and North 1970) are probably widespread in the Lepidoptera. In a preliminary survey of 300 species of butterflies, Ehrlich and Ehrlich (1978) found some form of plugging in all subfamilies except Dismorphinae (Pieridae). The incidence of plugging was low in Coliadinae (Pieridae), Riodininae, and Lycaeninae (Lycaenidae), but was common to universal in all other subfamilies. Ehrlich and Ehrlich (1978) found no significant relationship between the incidence of plugging and mean spermatophore counts and concluded that the role of plugs in reducing the incidence of multiple mating is not a simple one. Because of their small size and temporary nature, the copulatory plugs of most species are not effective in preventing remating in the female (Scott 1972), and probably serve to anchor the collum of the spermatophore in the ductus bursae or prevent leakage of sperm through the ostium bursae during sperm migration into the ductus seminalis. It is likely that these were the original functions of the widespread spermatophore caps, which were, in a few species, enlarged into spermatophraga or sphraga to prevent females from remating. If a plug does discourage other males or physically stimulates the female to nonreceptivity, it is probably most effective during the period of sperm transfer to the spermatheca before refractory behavior begins. In mated females of the Lycaeninae, the collum of the spermatophore usually blocks the extruded tube-shaped structure (the henia) that houses the ductus bursae, and the effectiveness of this arrangement in deterring subsequent matings may explain the low incidence (22%) of plugging in this subfamily (Ehrlich and Ehrlich 1978). Whatever the role of a spermatophore cap, copulatory plug, spermatophragis, or sphragis in a species, it is clear that their existence (and persistence) is made possible by the evolution of a separate genital opening for copulation. The vulva is free to function as an ovipore regardless of the size of the structure transferred by the male.

Males of several species of moths (Hirai *et al.* 1978) and butterflies (Pliske 1975c, Gilbert 1976) are known to produce pheromones that are repellent to other males of the same and related species. These function to terminate male-male sexual encounters or to reduce sexual aggressiveness among competing males. In *Heliconius erato*, males transfer a substance to females during mating that imparts a strong long-lasting odor that repels other courting males, a postmating antiaphrodisiac (Gilbert 1976). This cannot be only the result of intramale sexual selection, however, because the females of *H. erato* have evolved unique structures for receiving, storing, and disseminating the pheromone during subsequent courtship encounters. Such cooperation by the female leads to the same conclusion reached for copulation-induced female nonreceptivity: Because both male and female benefit from the arrangement, and have evolved adaptations to facilitate it, the postmating antiaphrodisiac in *H. erato* is a manifestation of a monogamous "agreement."

Postcopulatory guarding of females by their mates as described for Odonata and Diptera (see Parker 1970, for review) does not occur in the Lepidoptera but males may guard their mates from the advances of other males while still *in copula* (Drummond 1976). According to this hypothesis, males should experience selection to extend the duration of copulation beyond that required for transfer of the spermatophore, accessory fluids, and sperm so that most or all of the sperm transferred to the female have entered the ductus seminalis before she is released. This would prevent the first male's spermatophore from being displaced before the exit of his sperm, even if the female were to remate immediately. As indicated earlier (section III), duration of copulation in Lepidoptera is relatively long, averaging 1-3 h or longer in most groups (see reviews in Scott 1972, and Shields and Emmel 1973) and up to 24 h in some moths (Hewer 1934). Although there are few supporting data, it is likely that in many of these species, copulation continues considerably longer than required to complete sperm transfer to the spermatophore. For example, mated pairs of *Leucinodes orbonalis* (Pyralidae) stay *in copula* for 15-30 min after sperm transfer and retraction of the aedeagus (Srivastava and Srivastava 1957). The 24-h copulation period in *Zygaena* (Zygaenidae) allows sufficient time not only for sperm to leave the spermatophore but also to travel to and arrive at the spermatheca, which usually occurs 12-18 h after the start of copulation (Hewer 1934). If this is a widespread phenomenon, I predict that it will be found primarily in species in which mechanically-stimulated stretch receptors are absent from the bursa copulatrix and in species in which first male sperm precedence predominates. This hypothesis assumes that the male determines the length of copulation by controlling its termination, a condition that has been demonstrated for at least one species, *Papilio zelicaon* (Shields 1967).

V. SUMMARY

Female ditrysian Lepidoptera have separate genital openings for copulation and oviposition, with the result that sperm must negotiate a complicated passageway within the female reproductive tract after mating to reach the spermatheca, where they are stored until needed for fertilization. The spatial and temporal dynamics of this sperm migration provide opportunity for natural selection to affect sperm utilization patterns in multiply-mated females by (1) influencing the degree of control that females have over sperm movements and storage; and (2) affecting the incentive that males have to alter the postcopulatory behavior of their mates or to discourage courtship and mating attempts of other males. Considerable variation exists in the degree of complexity of the genital morphology in the Lepidoptera, and increasing complexity in the mechanics of copulation appears to be positively correlated with increased rates of mating failure.

Both males and females of most lepidopteran species are capable of multiple copulations, as verified by field and laboratory experiments. Spermatophore counts of field-collected females show that multiple mating is widespread in natural populations and is correlated with female age. Much of the variation in mating frequencies observed within and among taxa can be explained by variable characteristics of the populations (age structure, sex ratio, density, dispersion, territoriality), by environmental influences that affect mating behavior (temperature, humidity, light intensity), by seasonal effects on adult emergence and activity, and by the relative availability and nutritional quality of larval and adult foodplants.

Mating frequency in male Lepidoptera is limited by the the length of the refractory period between matings, which increases with male age, and by a finite capacity for producing (by apocrine secretion) accessory fluids, which are passed to the female with each successful copulation. Theoretically, multiple matings by female Lepidoptera should be selected against because (1) sperm received from a single copulation can be stored and the quantity is thought to be adequate to fertilize all of a female's eggs; and (2) copulation time in Lepidoptera is relatively long (often measured in hours), so that considerable disadvantages (in the form of reduced time for foraging and oviposition and in increased risk of predation or accident) accrue to females who engage in multiple matings. Courtship refusal behaviors of mated females are usually short and effective, and their costs in time and energy are unlikely to exceed the usually greater costs of lost time or increased bodily risk incurred during copulation, except perhaps under conditions of extremely low or extremely high population densities or where sex ratios are heavily biased in favor of males.

Several factors may contribute to the relatively high mating frequencies observed for lepidopteran females. Failure of the first mating appears to occur at low but consistent rates in most populations and could, perhaps, account for as much as one-quarter or more of the repeat matings by females of some species. Increased within-progeny genetic diversity could be a potentially important factor promoting multiple mating in populations in marginal habitats, exposed to unpredictable environmental conditions, or experiencing frequent colonization events, but the high levels of sperm precedence documented for Lepidoptera minimize the importance of this possibility. Furthermore, the number of factors that affect the importance of genetic diversity in populations makes this hypothesis the most difficult to test. By contrast, techniques are now available to quantify the importance of male nutrient contributions to female survivorship and reproduction, and the recent discoveries of indirect male parental investment in several species suggest that this is one of the most promising avenues of research leading to an understanding of mating frequencies in natural populations of Lepidoptera.

Sperm precedence patterns in Lepidoptera tend to be "all or none" phenomena, with few cases of complete mixing and proportional utilization of sperm from more than one mating. In most cases, sperm from the last male to mate takes preference over sperm from the first male, although the order of sperm precedence may depend on the length of the intermating interval. These results are in accordance with sexual selection theory, which predicts a high degree of sperm precedence and a low incidence of sperm mixing, resulting in competition among ejaculates (contest competition) rather than competition among individual sperm (scramble competition).

Reduction in female courtship receptivity as a result of successful copulation depends on the interaction of a male-produced mechanical (inflated spermatophore) or chemical (constituent of seminal fluid) stimulus and the appropriate neurological (stretch receptors) or physiological (hormone) response of the female. Female nonreceptivity after mating thus reflects a cooperative agreement between the sexes that benefits both.

Selection on males to prevent usurpation by other males of their considerable nutrient investment in copulation has resulted in a bizarre array of substances transferred to the female that serve to repel the advances of other males, including mating plugs, sphraga, and antiaphrodisiac pheromones. The more elaborate sphragal structures are found in groups with truncated courtship sequences that involve mating by capture. The long copulation times in Lepidoptera may be the result of intrasexual selection on males to prevent remating of females until after the transferred sperm have migrated out of the spermatophore. Such a delay would reduce the possibility of last male sperm precedence brought about the displacement of the unemptied first spermatophore.

ACKNOWLEDGMENTS

D. T. Gwynne, J. G. Shepherd, and K. S. Brown, Jr., made available unpublished manuscripts and T. C. Emmel provided unpublished information on butterfly spermatogenesis. This research was funded in part by grants from Sigma Xi, the University of Florida, and Illinois State University. N. J. Price and C. L. Martin aided in the collection of specimens. For all this support I am grateful. I thank R. L. Rutowski for carefully criticizing the manuscript and for several helpful suggestions, although I take full responsibility for any remaining errors or omissions in the selection and presentation of the data. I give special thanks to my wife, Claudia, for unflagging support under extremely trying circumstances, and to R. L. Smith, whose encouragement and patience made this paper possible, for his steadfast commitment to editorial excellence.

APPENDIX

Copulation and Sperm Dynamics in Lepidoptera

Copulation and Sperm Transfer. Copulation begins with the grasping of the tip of the female abdomen by the claspers (valvulae) of the male, and continues from several minutes to several hours with the male and female usually facing in opposite directions. The union requires considerable movement and flexibility of the abdomens. The sclerotized appendages arising from the tegumen and valvulae of the male permit a firm grasp of corresponding structures of the female, who uses some 20 separate muscles to expose her ostium bursae and to dilate the bursa copulatrix (Arnold and Fischer 1977). Some species require several minutes in copula to achieve a secure coupling before insertion of the aedeagus occurs (Norris 1932, Srivastava and Srivastava 1957).

The aedeagus is inserted through the ostium bursae, after which the vesica (endophallus) is everted into the corpus bursae, usually within 5 min of aedeagal penetration (Holt and North 1970, Ferro and Akre 1975). The vesica may be locked into place at the cervix (entry) of the corpus bursae by cuticular caeca (Callahan 1958) or by the coupling of the cornuti of the male with sclerotized plates in the cervix bursae (Allman 1930, Davis 1968, Ferro and Akre 1975). The first material transferred usually enters the female within 15-20 min of the start of copulation.

Preceding the spermatophore into the corpus bursae is a gelatinous, opalescent (sometimes colored) mass (the pearly body of Omura 1938) secreted by the first (lower) secretory area of the primary simplex of the male (Norris 1932, Khalifa 1950, Srivastava and Srivastava 1957, Callahan 1958, Wellso and Adkisson 1962, Callahan and Cascio 1963). The spermatophore is formed from a precursor secreted by the primary simplex into the cuticular simplex, where it is molded into a resilient tube and from which it is later forced into the bursa copulatrix via the vesica. The corpus bursae may expand before the spermatophore enters, possibly by air forced ahead by the advancing aedeagus (Khalifa 1950, Srivastava and Srivastava 1975) or by muscular dilation of the bursal walls by the female (Arnold and Fischer 1977).

As it is forced into the corpus bursae, the spermatophore gradually expands from pressure exerted by a clear fluid secreted by the male about 20 (Srivastava and Srivastava 1957, Holt and North 1970) to 30 (Norris 1932) min after copulation begins. As the body of the spermatophore becomes bulbous, the collum is molded in the aedeagus and is retained there until the sperm are transferred. The expanding bulb of the spermatophore forms around the signa or lamina dentatae, whose spines generally point cephalad (Drecktrah 1978) and thus lock the spermatophore in place. After the spermatophore is almost completely formed, the male ejaculates a milky seminal fluid containing mostly apyrene sperm into the spermatophore, and this is followed immediately by a more compact substance containing the eupyrene sperm bundles (Holt and North 1970). Ejaculation usually begins about 30-40 min after the start of copulation (Norris 1932, Holt and North 1970, Takeuchi and Miyashita 1974) and is completed within 50-70 min from the initiation of pairing (Srivastava and Srivastava 1957, Takeuchi and Miyashita 1974, Ferro and Akre 1975). In species with short ($<$30 min) copulation durations, sperm transfer can occur 3-5 min after pairing (Khalifa 1950, Yang and Chow 1978). In all cases, sperm are the last items to enter the spermatophore.

Before copulation is terminated, the male inserts the collum of the spermatophore into the corpus bursae (Holt and North 1970) or positions it in the ductus bursae (Callahan and Chapin 1960) by eversion and movements of the vesica and its attached cornuti. Such positioning is important because the sperm exit the spermatophore through an aperture, usually located at the tip of the collum or at the base of the frenum. If present, the frenum is positioned at the base of the ductus seminalis (Callahan and Chapin 1960) so that the aperture of the collum lies at the level of the opening of the ductus seminalis. In species without a frenum, the apposition of the aperture with the opening of the seminal duct is maintained by springlike loops of the collum in the ductus bursae (Outram 1971b, Drummond 1976) or corpus bursae (Callahan and Chapin 1960, Holt and North 1970), or by the securing action of the signa or lamina dentatae on the body of the spermatophore (Srivastava and Srivastava 1957, Stern and Smith

1960, Davis 1968). The importance of precise alignment of the aperture and the opening of the ductus seminalis is reflected by the strong intraspecific correlation between the shape and length of the spermatophore collum and the position of the junction of the ductus seminalis with the ductus bursae in most Lepidoptera (Petersen 1907, Williams 1941a, Callahan 1960). In a few groups, the ductus seminalis joins the ductus bursae where the latter leaves the corpus bursae or else joins the corpus bursae directly (Williams 1941a). In these species the spermatophore either lacks a collum and the sperm escape from an aperture in the spherical body of the spermatophore (Yang and Chow 1978), or the collum protrudes through the ostium bursae, such that the sperm leave the collum of the spermatophore at some point other than the tip.

In *Trichoplusia ni,* Holt and North (1970) have described the deposition of a viscous substance around the frenum that eventually forms a semisolid spermatophore cap. This cap may help anchor the spermatophore in place, prevent sperm leakage through the ostium bursae, or discourage further matings by acting as a mating plug (spermatosphragis). In most groups of butterflies, mated females have small plugs in the ductus bursae (Ehrlich and Ehrlich 1978), indicating that the spermatophore caps described by Holt and North (1970) are widespread in the Lepidoptera.

Following ejaculation and the positioning of the spermatophore collum, the vesica is retracted and the male and female separate.

Sperm Migration and Storage. Upon entrance to the lumen of the spermatophore, the apyrene sperm become highly motile, but the bundles of eupyrene sperm remain inactive (Holt and North 1970, Ferro and Akre 1975). The apyrene sperm appear to be activated by a secretion of the lower portion of the primary simplex of the male (the glandula prostatica, Omura 1938) or by a neuro-humeral factor controlled by the cephalic nervous system of the female (Stockel 1973). The eupyrene sperm bundles begin to break down within 1 (Holt and North 1970) or 2 (Ferro and Akre 1975) h after entering the spermatophore. At this time, the sperm mass, which consists of partial bundles of eupyrene sperm surrounded by numerous highly motile apyrene sperm (Srivastava and Srivastava 1957, Holt and North 1970, Thibout 1971, Ferro and Akre 1975), begins to move out of the body of the spermatophore and into the collum. Contractions of the muscles in the wall of the corpus bursae are believed to force the sperm mass out (Norris 1932, Stockel 1973, Thibout 1977), although Proshold *et al.* (1975) suggested that sperm are forced out of the spermatophore by internal pressure generated by a chemical reaction of the seminal fluids or by coiling of the spermatophore. Because much of the accessory gland secretion remains in the spermatophore, Thibout (1977) argued that the sperm must leave partly by their own movement. The intense activity of the apyrene sperm and their orientation around the eupyrene sperm bundles suggest a transport role for apyrene sperm, although in some species the apyrene sperm precede the eupyrene sperm out of the spermatophore (Proshold *et al.* 1975). Gelatinous substances between the wall of the corpus bursae and the body of the spermatophore are thought to aid in the transmission of the bursal contractions to the spermatophore (Norris 1932, Hinton 1964). The gelatinous material may be the pearly body passed by the male (Norris 1932, Omura 1938) or a bursal secretion (from the appendix bursae?) of the female (Weidner 1934, Khalifa 1950).

The sperm mass leaves the spermatophore through the aperture of the collum and immediately enters the ductus seminalis, a migration that underscores the critical importance of collum position to successful sperm transfer in most Lepidoptera. In a few species, the sperm can exit the spermatophore only if the tip of the collum is broken, either by the male during disengagement of the aedeagus (Hewer 1934) or by the contractions of the ductus bursae of the female (Srivastava and Srivastava 1957). After the sperm leave the collum, the spermatophore cap, if present, either disintegrates within a few hours (Callahan and Cascio 1963) or, more often, hardens into a permanent covering over the aperture (Callahan 1960, Holt and North 1970, Proshold *et al.* 1975).

At the time the sperm enter the ductus seminalis, almost all parts of the female reproductive tract undergo rhythmic contractions, which help to propel the sperm mass (Norris 1932, Srivastava and Srivastava 1957, Callahan and Cascio 1963, Thibout 1977, Yang and Chow 1978, Chao 1981). These contractions are thought to be induced by the accessory gland secretions of the male (Callahan and Cascio 1963), which also serve as a medium for the passage of the sperm. The function of the bulla seminalis is unclear. It may serve as a suction pump to draw sperm through the nonmuscular part of the ductus seminalis (Norris 1932) and it may serve as a site of temporary sperm storage (White unpubl., cited in Ferro and Akre 1975). The entrance to the bulla seminalis is guarded by spines (Hewer 1934, Outram 1971b) but Norris (1932) occasionally found one or two eggs in the lumen of the bulla.

Sperm that complete the passage through the ductus seminalis must cross the vestibulum to the opening of the spermathecal duct, up which they move to reach the spermatheca, drawn there by the pumping action of the spermathecal gland and the spermatheca itself (Norris 1932, Callahan and Cascio 1963). The passage across the vestibulum may be aided by secretions of the female accessory glands (Callahan and Cascio 1963) although traditionally these are thought to be the source of egg cement (Common 1970). The lower end of the spermathecal duct is guarded by downward-pointing spines (Hewer 1934) to prevent the entry of eggs or debris (Srivastava and Srivastava 1957) or to funnel the sperm into the duct and toward the spermatheca (Stockel 1973). Secretions of the spermathecal gland fill and lubricate the spermathecal duct, facilitating the migration of the sperm to the spermatheca (Stockel 1973).

The primary sperm storage organ in the female is the spermatheca, although both apyrene and eupyrene sperm have been found in the spermathecal gland (White unpubl., cited in Ferro and Akre 1975; Lum 1976). Sperm transfer to the spermatheca is usually completed within a few hours of copulation, but the time required can vary from as little as 2-3 h (Khalifa 1950, Holt and North 1970) up to 24 h (Norris 1932).

In the Noctuidae, the spermatheca has two lobes that differ in structure and function, the lagena and utriculus. In *Heliothis zea*, sperm arrive first in the lagena of the spermatheca, but within 4 h are transferred to the utriculus (Callahan and Cascio 1963). In *Trichoplusia ni* (Holt and North 1970) the utriculus first receives all of the sperm, but then contracts to force the apyrene sperm into the lagena, where they gradually disintegrate over a 3-day period. The eupyrene sperm remain in the utriculus where they become activated. The spermathecal gland produces secretions that lubricate the spermathecal duct (Stockel 1973) and are believed to provide nutrients to the sperm stored in the spermatheca (Norris 1932, Davey 1965). The recent discovery of mechanoreceptor setae in the spermathecal gland in *Plodia* (Pyralidae, Lum and Arbogast 1980) suggests that the gland may have a role in determining presence of motile sperm in a mated female.

Estimates of the number of sperm transferred to the spermatophore during mating in Lepidoptera range from 22,000 to 46,000 in *Heliothis virescens* (Raulston and Graham 1974, Proshold *et al.* 1975, Pair *et al.* 1977) to 96,000 in *Trichoplusia ni* (Holt and North 1970) with other estimates (Callahan and Cascio 1963, Sims 1979) falling near the center of this range. The number of sperm needed to fertilize all of a female's eggs may be greatly exceeded by the number of sperm transferred to the spermatheca, where excess sperm eventually become immotile (Riemann and Thorson 1971) or die (Khalifa 1950, Yang and Chow 1978). In some groups (*Cerura, Sphinx, Celeris*), more sperm are passed to the female than can be accommodated in the spermatheca, so that the spermatophore is emptied gradually as room is made in the spermatheca (Hinton 1964).

Sperm are stored in the spermatheca until needed to fertilize the eggs, at which time they descend the spermathecal duct to the vestibulum. The spermathecal duct generally consists of two parts, a main canal and a subsidiary spiral or "fertilization" canal (Hewer 1934, Weidner 1934, Omura 1938, Callahan and Cascio 1963, Joubert 1964, Outram 1971b, Hudson 1973, Drektrah 1978). This double structure of the spermathecal duct has long been interpreted as a mechanism to separate incoming sperm traveling up the larger, thin-walled main canal from outgoing sperm moving down the heavily chitinized spiral canal (Wigglesworth 1972, Callahan and Cascio 1963). Recent experimental evidence (Lum *et al.* 1981), however, reveals that in *Plodia interpunctella*, the main and spiral canals are not separated but form a double lumen. The flexible main canal expands during the strong contractions of the circular muscles surrounding the spermathecal duct (Drektrah 1978), which retains its length and shape during contractions from the rigidity of the chitinized spiral canal (Hewer 1934). In many species, eggs are fertilized too rapidly (*e.g.*, every 6.7 sec in *Bombyx mori,* Omura 1938) for the slow (several minutes) movement down the rigid spiral canal to be effective, and Lum *et al.* (1981) argue that the elastic nature of the main canal permits a much more rapid transfer of sperm to the vestibulum.

REFERENCES

Ae, S. A. 1962. Some problems in hybrids between *Papilio bianor* and *P. maakii. Academia* **33**:21-28. (Publ. by Nanzan Univ.)

Ae, S. A. 1966. A study of hybrids between Japanese and Himalayan *Papilio* butterflies. *Spec. Bull. Lepid. Soc. Jpn.* **2**:75-94.

Ahmed, M. S. H., Z. S. Al-Hakkak, and A. M. Al-Sagur. 1972. Inherited sterility in the fig moth, *Cadra (Ephestia) cautella* Walker. In *Peaceful Uses of Atomic Energy, Proc. Conf. Geneva,* 1971, IAEA, Vienna, pp. 383-389.

Alcock, J., E. M. Barrows, G. Gordh, L. J. Hubbard, L. Kirkendall, D. W. Pyle, T. L. Ponder, and F. G. Zalom. 1978. The ecology and evolution of male reproductive behaviour in the bees and wasps. *Zool. J. Linn. Soc. Lond.* **64**:293-326.

Allman, S. L. 1930. Studies of the anatomy and histology of the reproductive system of the female codling moth, *Carpocapsa pomonella* (Linn.). *Univ. Calif. Publ. Entomol.* **5**:135-164.

Amoaka-Atta, B. 1976. Gamma radiation effects on mating frequency and delayed mating of male *Cadra cautella* (Walker) (Lepidoptera:Pyralidae). *J. Kans. Entomol. Soc.* **49**:579 (abstract).

Anwar, M., and M. Feron. 1971. Mating behaviour of gamma irradiated and nonirradiated European corn borer, *Ostrinia nubilialis* (Hübner). *Int. J. Appl. Radiat. Isot.* **22**:89-93.

Anwar, M., M. Ashraf, and M. D. Arif. 1973. Mating, oviposition and gamma sterilization of the spotted bollworm of cotton, *Earias insulana. Entomol. Exp. Appl.* **16**:478-482.

Arnold, R. A., and R. L. Fischer. 1977. Operational mechanisms of copulation and oviposition in *Speyeria* (Lepidoptera:Nymphalidae). *Ann. Entomol. Soc. Am.* **70**:455-468.

Ashraf, M., M. Anwar, N. Chatha, and N. Chatha. 1974. Reproductive biology and radiation sterilization of the sugarcane top borer, *Tryporyza nivella. Entomol. Exp. Appl.* **17**:61-66.

Baker, H. G., and I. Baker. 1973. Amino acids in nectar and their evolutionary significance. *Nature* **241**:543-545.

Benz, G. 1969. Influence of mating, insemination and other factors on oogenesis and oviposition in the moth *Zeiraphera diniana. J. Insect Physiol.* **15**:55-71.

Birket-Smith, S. J. R. 1974a. Morphology of the male genitalia of Lepidoptera. I. Ditrysia. *Entomol. Scand.* **5**:1-22.

Birket-Smith, S. J. R. 1974b. Morphology of the male genitalia of Lepidoptera. II. Monotrysia, Zeugloptera, and discussion. *Entomol. Scand.* **5**:161-183.

Boggs, C. L. 1979. Resource allocation and reproductive strategies in several heliconiine butterfly species. Ph.D. dissertation, Univ. Texas, Austin, TX.

Boggs, C. L. 1981a. Selection pressures affecting male nutrient investment at mating in Heliconiine butterflies. *Evolution* **35**:931-940.

Boggs, C. L. 1981b. Nutritional and life-history determinants of resource allocation in holometabolous insects. *Am. Nat.* **117**:692-709.

Boggs, C. L., and L. E. Gilbert. 1979. Male contribution to egg production in butterflies: Evidence for transfer of nutrients at mating. *Science* **206**:83-84.

Boggs, C. L., and W. B. Watt. 1981. Population structure of pierid butterflies IV. Genetic and physiological investment in offspring by male *Colias. Oecologia (Berl.)* **50**:320-324.

Boorman, E., and G. A. Parker. 1976. Sperm (ejaculate) competition in *Drosophila melanogaster,* and the reproductive value of females to males in relation to female age and mating status. *Ecol. Entomol.* **1**:145-155.

Brower, J. H. 1975. Sperm precedence in the Indian meal moth, *Plodia interpunctella. Ann. Entomol. Soc. Am.* **68**:78-80.

Brower, J. H. 1979. Mating competitiveness in the laboratory of irradiated males and females of *Ephestia cautella. Fla. Entomol.* **62**:41-47.

Brower, L. P., J. V. Z. Brower, and R. P. Cranston. 1965. Courtship behavior of the queen butterfly, *Danaus gilippus berenice* (Cramer). *Zoologica (NY)* **50**:1-39.

Brown, K. S., Jr. 1984. Chemical ecology of dehydropyrrolizidine alkaloids in adult Ithomiinae (Lepidoptera:Nymphalidae). *Rev. Bras. Biol.* **44**(3).

Bryk, F. 1918. Grundzuge der Sphragidologie. *Ark. Zool.* 11:1-38.

Bryk, F. 1919. Bibliotheca Sphragidologica. *Arch. Naturgeschichte* 85:102-183.

Bryk, F. 1930. Monogame Einrichtungen bei Schmetterlingsweibchen. *Arch. Fraunekunde* 16: 308-313.

Burns, J. M. 1966. Preferential mating versus mimicry: Disruptive selection and sex-limited dimorphism in *Papilio glaucus. Science* 153:551-553.

Burns, J. M. 1968. Mating frequency in natural populations of skippers and butterflies as determined by spermatophore counts. *Proc. Nat. Acad. Sci.* 61:852-859.

Burns, J. M. 1970. Duration of copulation in *Poanes hobomok* (Lepidoptera:Hesperiidae) and some broader speculations. *Psyche* 77:127-130.

Burton, R. L., and J. W. Snow. 1970. A marker dye for the corn earworm. *J. Econ. Entomol.* 63:1976-1977.

Byers, J. R. 1978. Biosystematics of the genus *Euxoa* (Lepidoptera:Noctuidae). X. Incidence and level of multiple mating in natural and laboratory populations. *Can. Entomol.* 110: 193-200.

Calderon, M., and M. Gonen. 1971. Effects of gamma radiation on *Ephestia cautella* (Wlk.) (Lepidoptera, Phycitidae). I. Effects on adults. *J. Stored Prod. Res.* 7:85-90.

Callahan, P. S. 1958. Serial morphology as a technique for determination of reproductive patterns in the corn earworm, *Heliothis zea* (Boddie). *Ann. Entomol. Soc. Am.* 51:413-428.

Callahan, P. S. 1960. A morphological study of spermatophore placement and mating in the subfamily Plusiinae (Noctuidae, Lepidoptera). *Proc. XI Int. Congr. Entomol. (Vienna)* 1:339-345.

Callahan, P. S., and T. Cascio. 1963. Histology of the reproductive tracts and transmission of sperm in the corn earworm, *Heliothis zea. Ann. Entomol. Soc. Am.* 56:535-556.

Callahan, P. S., and J. B. Chapin. 1960. Morphology of the reproductive systems and mating in two representative members of the family Noctuidae, *Pseudaletia unipuncta* and *Peridroma margaritosa*, with comparison to *Heliothis zea. Ann. Entomol. Soc. Am.* 53: 763-782.

Campbell, I. M. 1961. Polygyny in *Choristoneura* Led. (Lepidoptera:Tortricidae). *Can. Entomol.* 93:1160-1162.

Campbell, I. M. 1962. Reproductive capacity in the genus *Choristoneura* Led. (Lepidoptera: Tortricidae). I: Quantitative inheritance and genes as controllers of rates. *Can. J. Genet. Cytol.* 4:272-288.

Cantelo, W. W. 1973. Dye markers for moths of the tobacco hornworm. *Environ. Entomol.* 2:393-396.

Chanter, D. O., and D. F. Owen. 1972. The inheritance and population genetics of sex ratio in the butterfly, *Acraea encedon. J. Zool.* 166:363-383.

Chao, W. 1981. The distribution, translocation and function of semen in the female reproductive systems of the armyworm moth (*Leucania separata*). *Acta Entomol. Sin.* 24:135-141.

Clarke, C. A., and P. M. Sheppard. 1962. Offspring from double matings in swallowtail butterflies. *Entomologist (Lond.)* 95:199-203.

Cogburn, R. R., E. W. Tilton, and J. H. Brower. 1973. Almond moth: Gamma radiation effects on the life stages. *J. Econ. Entomol.* 66:745-751.

Common, I. F. B. 1970. Lepidoptera (moths and butterflies). In *The Insects of Australia, a Textbook for Students and Research Workers*, D. F. Waterhouse (ed.), pp. 765-866. Melbourne Univ. Press, Carleton.

Common, I. F. B. 1975. Evolution and classification of the Lepidoptera. *Ann. Rev. Entomol.* 20:183-203.

Cotter, W. B. 1963. Population genetic studies of alleles at the a^+ locus in *Ephestia kuhniella. Evolution* 17:233-248.

Davey, K. G. 1959. Spermatophore production in *Rhodnius prolixus. Q. J. Microscop. Sci.* 100:221-230.

Davey, K. G. 1965. *Reproduction in the Insects*. Oliver and Boyd, Edinburgh and London.

David, W. A. L., and B. O. C. Gardiner. 1961. The mating behaviour of *Pieris brassicae* (L.) in a laboratory culture. *Bull. Entomol. Res.* 52:263-280.

Davis, F. M. 1968. Morphology of the reproductive systems of the southwestern corn borer, *Diatraea grandiosella. Ann. Entomol. Soc. Am.* 61:1143-1147.

Deseö, K. V. 1971. Study of factors influencing the fecundity and fertility of codling moth (*Laspeyresia pomonella* L., Lepid.; Tortr.). *Acta Phytopathol. Acad. Sci. Hung.* 6:243-252.

Deseö, K. V. 1972. The role of farnesylmethyl-ether applied on the male influencing the oviposition of the female codling moth (*Laspeyresia pomonella* L., Lepidopt.; Tortricidae). *Acta Phytopathol. Acad. Sci. Hung.* 7:257-266.

Drecktrah, H. G. 1978. Morphology of the internal reproductive system of the adult female army cutworm, *Euxoa auxiliaris. Ann. Entomol. Soc. Am.* 71:923-927.

Drummond, B. A. III. 1976. Comparative ecology and mimetic relationships of ithomiine butterflies in eastern Ecuador. Ph.D. dissertation, University of Florida, Gainesville, FL.

Dugdale, J. S. 1974. Female genital configuration in the classification of Lepidoptera. *N. Z. J. Zool.* 1:127-146.

Dunlap-Pianka, H., C. L. Boggs, and L. E. Gilbert. 1977. Ovarian dynamics in heliconiine butterflies: Programmed senescence versus eternal youth. *Science* 197:487-490.

Dustan, G. G. 1964. Mating behaviour of the Oriental Fruit Moth, *Grapholitha molesta* (Busck) (Lepidoptera:Olethreutidae). *Can. Entomol.* 96:1087-1093.

Edgar, J. A., C. C. J. Culvenor, and T. E. Pliske. 1976. Isolation of a lactone, structurally related to the esterifying acids of pyrrolizidine alkaloids from the costal fringes of male Ithomiinae. *J. Chem. Ecol.* 2:263-270.

Ehrlich, A. H., and P. R. Ehrlich. 1978. Reproductive strategies in the butterflies. I. Mating frequency, plugging, and egg number. *J. Kans. Entomol. Soc.* 51:666-697.

Ehrlich, P. R., and P. H. Raven. 1964. Butterflies and plants: A study in coevolution. *Evolution* 18:586-608.

Ehrlich, P. R., and R. R. White. 1980. Colorado checkerspot butterflies: Isolation, neutrality, and the biospecies. *Am. Nat.* 115:328-341.

Elliott, W. M., and V. A. Dirks. 1979. Postmating age estimates for female European corn borer moths, *Ostrinia nubilialis* (Lepidoptera:Pyralidae), using time-related changes in spermatophores. *Can Entomol.* 111:1325-1335.

Elthringham, H. 1912. A monograph of the African species of the genus *Acraea,* Fab., with a supplement on those of the Oriental Region. *Trans. R. Entomol. Soc. Lond.* 1912:1-374.

Eltringham, H. 1916. On certain forms of the genus *Acraea. Trans. R. Entomol. Soc. Lond.* 1916:289-296.

Eltringham, H. 1925. On the source of the sphragidal fluid in *Parnassius apollo* (Lepidoptera). *Trans. R. Entomol. Soc. Lond.* 1925:11-15.

Engebretson, J. A., and W. H. Mason. 1980. Transfer of [65]Zn at mating in *Heliothis virescens. Environ. Entomol.* 9:119-121.

Engelmann, F. 1970. *The Physiology of Insect Reproduction.* Pergamon Press, Oxford.

Etman, A. A. M., and G. H. S. Hooper. 1979a. Sperm precedence of the last mating in *Spodoptera litura. Ann. Entomol. Soc. Am.* 72:119-120.

Etman, A. A. M., and G. H. S. Hooper. 1979b. Developmental and reproductive biology *Spodoptera litura* (Lepidoptera:Noctuidae). *J. Aust. Entomol. Soc.* 18:363-372.

Fagerström, T., and C. Wiklund. 1982. Why do males emerge before females? Protandry as a mating strategy in male and female butterflies. *Oecologia (Berl.)* 52:164-166.

Ferro, D. N., and R. D. Akre. 1975. Reproductive morphology and mechanics of mating of the codling moth, *Laspeyresia pomonella. Ann. Entomol. Soc. Am.* 68:417-424.

Flint, H. M., and E. L. Kressin. 1968. Gamma irradiation of the tobacco budworm: Sterilization, competitiveness, and observations on reproductive biology. *J. Econ. Entomol.* 61:477-483.

Flint, H. M., and J. R. Merkle. 1980. Pink bollworm: Irradiation of laboratory and native males. *J. Econ. Entomol.* 73:764-767.

Friedländer, M., and G. Benz. 1981. The eupyrene-apyrene dichotomous spermatogenesis of Lepidoptera. Organ culture study on the timing of apyrene commitment in the codling moth. *Int. J. Invertebr. Reprod.* 3:113-120.

Friedländer, M., and G. Benz. 1982. Control of spermatogenesis resumption in post-diapausing larvae of the codling moth. *J. Insect Physiol.* 28:349-355.

Friedländer, M., and H. Gitay. 1972. The fate of the normal-anucleated spermatozoa in inseminated females of the silkworm *Bombyx mori. J. Morphol.* 138:121-129.

Gehring, R. D., and H. F. Madsen. 1963. Some aspects of the mating and oviposition behavior of the codling moth, *Carpocapsa pomonella. J. Econ. Entomol.* **56**:140-143.

George, J. A., and M. G. Howard. 1968. Insemination without spermatophores in the oriental fruit moth, *Grapholitha molesta* (Lepidoptera:Tortricidae). *Can. Entomol.* **100**:190-192.

Gerber, G. H. 1970. Evolution of the methods of spermatophore formation in pterygotan insects. *Can. Entomol.* **102**:358-362.

Gilbert, L. E. 1972. Pollen feeding and reproductive biology of *Heliconius* butterflies. *Proc. Natl. Acad. Sci. U.S.A.* **69**:1403-1407.

Gilbert, L. E. 1976. Postmating female odor in *Heliconius* butterflies: A male-contributed antiaphrodisiac? *Science* **193**:419-420.

Gilbert, L. E., and M. C. Singer. 1975. Butterfly ecology. *Annu. Rev. Ecol. Syst.* **6**:365-397.

Goldschmidt, R. 1916. The function of the apyrene spermatozoa. *Science* **44**:544-546.

Goodwin, J. A., and H. F. Madsen. 1964. Mating and oviposition behavior of the navel orangeworm *Paramyelois transitella* (Walker). *Hilgardia* **35**:507-525.

Goss, G. J. 1977. The interaction between moths and pyrrolizidine alkaloid-containing plants including nutrient transfer via the spermatophore in *Lymire edwardsii* (Ctenuchidae). Ph.D. dissertation, University of Miami, Coral Gables, FL.

Graham, H. M., P. A. Glick, M. T. Ouye, and D. F. Martin. 1965. Mating frequency of female pink bollworms collected from light traps. *Ann. Entomol. Soc. Am.* **58**:595-596.

Graham, S., W. B. Watt, and L. F. Gall. 1980. Metabolic resource allocation vs. mating attractiveness: Adaptive pressures on the "alba" polymorphism of *Colias* butterflies. *Proc. Natl. Acad. Sci. U.S.A.* **71**:3615-3619.

Hendricks, D. E., H. M. Graham, and A. T. Fernandez. 1970. Mating of female tobacco budworms and bollworms collected from light traps. *J. Econ. Entomol.* **63**:1228-1231.

Henneberry, T. J., and M. P. Leal. 1979. Pink bollworm: Effects of temperature, photoperiod and light intensity, moth age and mating frequency on oviposition and egg viability. *J. Econ. Entomol.* **72**:489-492.

Herman, W. S., and J. F. Barker. 1977. Effect of mating on Monarch butterfly oogenesis. *Experientia* **33**:688-689.

Herman, W. S., and P. Peng. 1976. Juvenile hormone stimulation of sperm activator production in male Monarch butterflies. *J. Insect Physiol.* **22**:579-581.

Hewer, H. R. 1932. Studies in *Zygaena* (Lepidoptera) Part I. (A) The female genitalia; (B) The male genitalia. *Proc. Zool. Soc. Lond.* **1932**:33-75.

Hewer, H. R. 1934. Studies in *Zygaena* (Lepidoptera) Part II. The mechanism of copulation and the passage of sperm in the female. *Proc. Zool. Soc. Lond.* **2**:513-527.

Hill, Jr., H. F., A. M. Wenner, and P. H. Wells. 1976. Reproductive behavior in an overwintering aggregation of monarch butterflies. *Am. Midl. Nat.* **95**:10-19.

Hinton, H. E. 1964. Sperm transfer in insects and the evolution of haemocoelic insemination. In *Insect Reproduction,* K. C. Highnam (ed), pp. 95-107. Symp. R. Entomol. Soc. Lond.

Hinton, H. E. 1974. Symposium on reproduction of arthropods of medical and veterinary importance. III. Accessory functions of seminal fluid. *J. Med. Entomol.* **11**:19-25.

Hirai, K., H. H. Shorey, and L. K. Gaston. 1978. Competition among courting male moths: Male-to-male inhibitory pheromone. *Science* **202**:644-645.

Holt, G. G., and D. T. North. 1970. Effects of gamma irradiation on the mechanisms of sperm transfer in *Trichoplusia ni. J. Insect Physiol.* **16**:2211-2222.

Howell, J. R., R. B. Hutt, and W. B. Hill. 1978. Codling moth: Mating behavior in the laboratory. *Ann. Entomol. Soc. Am.* **71**:891-895.

Hudson, A. 1973. Biosystematics in the genus *Euxoa* (Lepidoptera:Noctuidae). *Can. Entomol.* **105**:1199-1209.

Husseiny, M. M., and H. F. Madsen. 1964. Sterilization of the navel orangeworm *Paramyelois transitella* (Walker), by gamma irradiation (Lepidoptera:Phycitidae). *Hilgardia* **36**:113-137.

Iriki, S. 1941. The two sperm types in the silkworm and their functions. *Zool. Mag. (Tokyo)* **53**:123-124.

Joubert, P. C. 1964. The reproductive system of *Sitotroga cerealella* Oliver (Lepidoptera: Gelechiidae). II. Structure and physiology of the female system. *S. Afr. J. Agric. Sci.* **7**:251-264.

Karpenko, C. P., and D. T. North. 1973. Ovipositional response elicited by normal, irradiated, F_1 male progeny, or castrated male *Trichoplusia ni. Ann. Entomol. Soc. Am.* 66:1278-1280.

Katsuno, S. 1977. Studies on eupyrene and apyrene spermatozoa in the silkworm, *Bombyx mori* L. (Lepidoptera:Bombycidae). IV. The behavior of the spermatozoa in the internal reproductive organs of female adults. *Appl. Entomol. Zool.* 12:352-359.

Kaufmann, T. 1968. Observations on the biology and behavior of the evergreen bagworm moth, *Thyridopteryx ephemeraeformis* (Lepidoptera:Psychidae). *Ann. Entomol. Soc. Am.* 61: 38-44.

Kehat, M., and D. Gordon. 1977. Mating ability, longevity and fecundity of the spiny bollworm, *Earias insulana* (Lepidoptera:Noctuidae). *Entomol. Exp. Appl.* 22:267-273.

Kellen, W. R., D. F. Hoffmann, and R. A. Kwock. 1981. *Wolbachia* sp. (Rickettsiales: Rickettsiaceae) a symbiont of the almond moth, *Ephestia cautella:* Ultrastructure and influence on host fertility. *J. Invertebr. Pathol.* 37:273-283.

Khalifa, A. 1950. Spermatophore production in *Galleria mellonella* L. (Lepidoptera). *Proc. R. Entomol. Soc. Lond. (A)* 25:33-42.

Klots, A. B. 1970. Lepidoptera. In *Taxonomist's Glossary of Genitalia in Insects, 2nd ed.,* S. L. Tuxen (ed.), pp. 115-130. Munksgaard, Copenhagen.

Labine, P. A. 1964. Population biology of the butterfly, *Euphydryas editha.* I. Barriers to multiple inseminations. *Evolution* 18:335-336.

Labine, P. A. 1966. The population biology of the butterfly, *Euphydryas editha.* IV. Sperm precedence—A preliminary report. *Evolution* 20:580-586.

LaChance, L. E., R. A. Bell, and R. D. Richard. 1973. Effects of low dosages of gamma radiation on reproduction of male pink bollworms and their F_1 progeny. *Environ. Entomol.* 2:653-658.

LaChance, L. E., R. I. Proshold, and R. L. Ruud. 1978. Pink bollworm: Effects of male irradiation and ejaculate sequence on female ovipositional response and sperm ratio sensitivity. *J. Econ. Entomol.* 71:361-365.

Lai-Fook, J. 1982. Testicular development and spermatogenesis in *Calpodes ethlius* Stoll (Hesperiidae, Lepidoptera). *Can. J. Zool.* 60:1161-1171.

Lederhouse, R. C. 1981. The effect of female mating frequency on egg fertility in the black swallowtail, *Papilio polyxenes asterius* (Papilionidae). *J. Lepid. Soc.* 35:266-277.

Lederhouse, R. C. 1982. Territorial defense and lek behavior of the black swallowtail butterfly, *Papilio polyxenes. Behav. Ecol. Sociobiol.* 10:109-118.

Lederhouse, R. C., M. D. Finke, and J. M. Scriber. 1982. The contributions of larval growth and pupal duration to protandry in the black swallowtail butterfly, *Papilio polyxenes. Oecologia (Berl.)* 53:296-300.

Leopold, R. A. 1976. The role of male accessory glands in insect reproduction. *Annu. Rev. Entomol.* 21:199-221.

Leppla, N. C., and W. K. Turner. 1975. Carbon dioxide output and mating in adult cabbage looper moths exposed to discrete light regimes. *J. Insect Physiol.* 21:1233-1236.

Levin, M. P. 1973. Preferential mating and the maintenance of the sex-limited dimorphism in *Papilio glaucus:* Evidence from laboratory matings. *Evolution* 27:257-264.

Leviatan, R., and M. Friedländer 1979. The eupyrene-apyrene dichotomous spermatogenesis of Lepidoptera. I. The relationship with postembryonic development and the role of the decline in juvenile hormone titer toward pupation. *Dev. Biol.* 68:515-524.

Lopez, J. D., Jr., J. A. Witz, A. W. Hartstack, Jr., and J. P. Hollingsworth. 1978. Reproductive condition of bollworm moths caught in blacklight traps in corn, sorghum, and cotton. *J. Econ. Entomol.* 71:961-966.

Lum, P. T. M. 1976. Sensillum-like setae to the lumen of the spermathecal gland of *Plodia interpunctella. J. Ga. Entomol. Soc.* 11:247-251.

Lum, P. T. M., and R. T. Arbogast. 1980. Ultrastructure of setae in the spermathecal gland of *Plodia interpunctella* (Hübner) (Lepidoptera:Pyralidae). *Int. J. Insect Morphol. Embryol.* 9:251-253.

Lum, P. T. M., and U. E. Brady. 1973. Levels of pheromone in female *Plodia interpunctella* mating with males reared in different light regimens. *Ann. Entomol. Soc. Am.* 66:821-823.

Lum, P. T. M., J. G. Riemann, and J. E. Baker. 1981. The fertilization canal in *Plodia interpunctella* (Hübner) (Pyralidae:Lepidoptera). *Int. J. Invertebr. Reprod.* **3**:283-289.

Macfarlane, J. H., and C. H. Tsao. 1974. The neural control of spermatophore formation and sperm transfer in the silkworm, *Bombyx mori* L. *Ann. Entomol. Soc. Am.* **67**:759-761.

Makielski, S. K. 1972. Polymorphism in *Papilio glaucus* L. (Papilionidae): Maintenance of the female ancestral form. *J. Lepid. Soc.* **26**:109-111.

Marshall, L. D. 1982. Male nutrient investment in the Lepidoptera: What nutrients should males invest? *Am. Nat.* **120**:273-279.

Matsuda, R. 1976. *Morphology and Evolution of the Insect Abdomen*. Pergamon Press, Oxford.

Mehta, D. R. 1933. Comparative morphology of the male genitalia in Lepidoptera. *Rec. Indian Mus. (Calcutta)* **35**:197-266.

Meves, F. 1902. Über oligopyrene und apyrene Spermien und über ihre Entstehung, nach Beobachtungen an *Paludina* und *Pygaera*. *Arch. Mikrosk. Anat.* **61**:1-84.

Miller, L. D., and H. K. Clench. 1968. Some aspects of mating behavior in butterflies. *J. Lepid. Soc.* **22**:125-132.

Miyashita, K., and M. Fuwa. 1972. The occurrence time, reiterative ability, and duration of mating in *Spodoptera litura* F. (Lepidoptera:Noctuidae). *Appl. Entomol. Zool.* **7**:171-173.

Musgrave, A. J. 1937. The histology of the male and female reproductive organs of *Ephestia kuhniella* Zeller (Lepidoptera). I. The young imagines. *Proc. Zool. Soc. Lond. (B)* **107**: 337-364.

Mutuura, A. 1972. Morphology of the female terminalia in Lepidoptera, and its taxonomic significance. *Can. Entomol.* **104**:1055-1071.

Navon, A., and R. Marcus. 1982. D-Isoascorbic acid fed to *Spodoptera littoralis* moths, induces sterility due to spermatophore malformation. *J. Insect Physiol.* **28**:823-828.

Noordink, J. P. W., and A. K. Minks. 1970. Autoradiography: A sensitive method in dispersal studies with *Adoxophyes orana* (Lepidoptera, Tortricidae). *Entomol. Exp. Appl.* **13**:448-454.

Norris, M. J. 1932. Contributions towards the study of insect fertility. I. The structure and operation of the reproductive organs of the genera *Ephestia* and *Plodia* (Lepidoptera, Phycitidae). *Proc. Zool. Soc. Lond.* **1932**:595-611.

Norris, M. J. 1933. Contributions towards the study of insect fertility. II. Experiments on the factors influencing fertility in *Ephestia kuhniella* Z. (Lepidoptera, Phycitidae). *Proc. Zool. Soc. Lond.* **1933**:903-934.

North, D. T., and G. G. Holt. 1968. Genetic and cytogenetic basis of radiation-induced sterility in the adult male cabbage looper *Trichoplusia ni. I.A.E.A./F.A.O. Symp. on Use of Isotopes and Radiation in Entomology, Vienna* **1967**:391-403.

Obara, Y. 1964. Mating behaviour of the cabbage white, *Pieris rapae crucivora*. II. The 'mate-refusal posture' of the female. *Zool. Mag. (Tokyo)* **73**:175-178.

Obara, Y. 1982. Mate refusal hormone in the cabbage white butterfly? *Naturwissenschaften* **69**:551-552.

Obara, Y., H. Tateda, and M. Kuwabara. 1975. Mating behavior of the cabbage white butterfly, *Pieris rapae crucivora* Boisduval. V. Copulatory stimuli inducing changes of female response patterns. *Zool. Mag. (Tokyo)* **84**:71-76.

Omura, S. 1936. Artificial insemination of *Bombyx mori. J. Fac. Agric. Hokkaido Univ. Ser. Entomol.* **38**:135-150.

Omura, S. 1938. Studies on the reproductive system of the male of *Bombyx mori*. II. Post-testicular organs and post-testicular behaviour of the spermatozoa. *J. Fac. Agric. Hokkaido Univ. Ser. Entomol.* **40**:129-170.

Omura, S. 1939. Selective fertilization in *Bombyx mori. Jpn. J. Genet.* **15**:29-35. (English summary.)

Onukogu, F. A., W. D. Guthrie, W. H. Awadallah, and J. C. Robbins. 1980. Hatchability of eggs and mating success of European corn borer cultures reared continuously on a meridic diet. *Iowa State J. Res.* **54**:347-355.

Outram, I. 1968. Polyandry in spruce budworm. *Can. Dep. For. Rural Dev. Bi-mon. Res. Notes* **24**:6-7.

Outram, I. 1970. Morphology and histology of the reproductive system of the male spruce budworm, *Choristoneura fumiferana. Can. Entomol.* **102**:404-414.

Outram, I. 1971a. Aspects of mating in the spruce budworm, *Choristoneura fumiferana* (Lepidoptera:Tortricidae). *Can. Entomol.* **103**:1121-1128.

Outram, I. 1971b. Morphology and histology of the reproductive system of the female spruce budworm, *Choristoneura fumiferana* (Lepidoptera:Tortricidae). *Can. Entomol.* **103**:32-43.

Ouye, M. T., R. S. Garcia, H. M. Graham, and D. F. Martin. 1965. Mating studies on the pink bollworm, *Pectinophora gossypiella* (Lepidoptera:Gelechiidae), based on presence of spermatophores. *Ann. Entomol. Soc. Am.* **58**:880-882.

Owen, D. F., J. Owen, and D. O. Chanter. 1973a. Low mating frequencies in an African butterfly. *Nature* **244**:116-117.

Owen, D. F., J. Owen, and D. O. Chanter. 1973b. . Low mating frequency in predominately female populations of the butterfly, *Acraea encedon* (L.) (Lep.). *Entomol. Scand.* **4**:155-160.

Page, R. E., Jr., and R. A. Metcalf. 1982. Multiple mating, sperm utilization, and social evolution. *Am. Nat.* **119**:263-281.

Pair, S. D., M. L. Laster, and D. F. Martin. 1977. Hybrid sterility of the tobacco budworm: Effects of alternate sterile and normal matings on fecundity and fertility. *Ann. Entomol. Soc. Am.* **70**:952-954.

Parker, G. A. 1970. Sperm competition and its evolutionary consequences in the insects. *Biol. Rev.* **45**:525-567.

Pease, R. W. 1968. The evolutionary and biological significance of multiple pairing in Lepidoptera. *J. Lepid. Soc.* **22**:197-209.

Perez, R., and W. H. Long. 1964. Sex attractant and mating behavior in the sugarcane borer. *J. Econ. Entomol.* **57**:688-691.

Petersen, W. 1907. Über die Spermatophoren der Schmetterlinge. *Z. Wiss. Zool.* **88**:117-130.

Petersen, W. 1928. Über die Sphragis und das Spermatophragma der Tagfaltergattung *Parnassius* (Lep.). *Dtsch. Entomol. Z.* **1928**:407-413.

Pliske, T. E. 1972. Sexual selection and dimorphism in female tiger swallowtails, *Papilio glaucus* L. (Lepidoptera:Papilionidae): A reappraisal. *Ann. Entomol. Soc. Am.* **65**:1267-1270.

Pliske, T. E. 1973. Factors determining mating frequencies in some New World butterflies and skippers. *Ann. Entomol. Soc. Am.* **66**:164-169.

Pliske, T. E. 1975a. Attraction of Lepidoptera to plants containing pyrrolizidine alkaloids. *Environ. Entomol.* **4**:455-473.

Pliske, T. E. 1975b. Pollination of pyrrolizidine alkaloid-containing plants by male Lepidoptera. *Environ. Entomol.* **4**:474-479.

Pliske, T. E. 1975c. Courtship behavior and use of chemical communication by males of certain species of Ithomiine butterflies (Nymphalidae:Lepidoptera). *Ann. Entomol. Soc. Am.* **68**:935-942.

Pliske, T. E. 1975d. Courtship behavior of the monarch butterfly, *Danaus plexippus* L. *Ann. Entomol. Soc. Am.* **68**:143-151.

Pliske, T. E., J. A. Edgar, and C. C. J. Culvenor. 1976. The chemical basis of attraction of Ithomiine butterflies to plants containing pyrrolizidine alkaloids. *J. Chem. Ecol.* **2**:255-262.

Pointing, P. J. 1961. The biology and behaviour of the European pine shoot moth, *Rhyacionia buoliana* (Schiff.), in southern Ontario. I. Adult. *Can. Entomol.* **93**:1098-1112.

Poulton, E. B. 1931. The gregarious sleeping habits of *Heliconius charithonia* L. *Proc. R. Soc. Entomol. Lond.* **6**:4-10.

Powell, J. A. 1968. A study of area occupation and mating behavior in *Incisalia iroides* (Lepidoptera, Lycaenidae). *J. N. Y. Entomol. Soc.* **76**:47-57.

Proshold, F. I., and L. E. LaChance. 1974. Analysis of sterility in hybrids from interspecific crosses between *Heliothis virescens* and *H. subflexa. Ann. Entomol. Soc. Am.* **67**:445-449.

Proshold, F. I., L. E. LaChance, and R. D. Richard. 1975. Sperm production and transfer by *Heliothis virescens, H. subflexa,* and the sterile hybrid males. *Ann. Entomol. Soc. Am.* **68**:31-34.

Prout, T. 1967. Selective forces in *Papilio glaucus. Science* **156**:534.

Proverbs, M. D., and J. R. Newton. 1962a. Some effects of gamma radiation on the reproductive potential of the codling moth, *Carpocapsa pomonella* (L.) (Lepidoptera:Olethreutidae). *Can. Entomol.* **94**:1162-1170.

Proverbs, M. D., and J. R. Newton. 1962b. Effect of heat on the fertility of the codling moth, *Carpocapsa pomonella* (L.) (Lepidoptera:Olethreutidae). *Can. Entomol.* 94:225-233.

Pruess, K. P. 1967. Migration of the army cutworm, *Chorizagrotis auxiliaris* (Lepidoptera: Noctuidae). I. Evidence for a migration. *Ann. Entomol. Soc. Am.* 60:910-920.

Rakshpal, R. 1944. On the structure and development of the male reproductive organs in the Lepidoptera. *Indian J. Entomol.* 6:87-93.

Raulston, J. R., and H. M. Graham. 1974. Determination of quantitative sperm transfer by male tobacco budworms irradiated at different ages. *J. Econ. Entomol.* 67:463-464.

Raulston, J. R., J. W. Snow, H. M. Graham, and P. D. Lingren. 1975. Tobacco budworm: Effect of prior mating and sperm content on the mating behavior of females. *Ann. Entomol. Soc. Am.* 68:701-704.

Ray, T. S., and C. C. Andrews. 1980. Antbutterflies: Butterflies that follow army ants to feed on antbird droppings. *Science* 210:1147-1148.

Retnakaran, A. 1971a. A method for determining sperm precedence in insects. *J. Econ. Entomol.* 64:578-580.

Retnakaran, A. 1971b. Thiotepa as an effective agent for mass sterilizing the spruce budworm, *Choristoneura fumiferana* (Lepidoptera:Tortricidae). *Can. Entomol.* 103:1753-1756.

Retnakaran, A. 1974. The mechanisms of sperm precedence in the spruce budworm, *Choristoneura fumiferana* (Lepidoptera:Tortricidae). *Can. Entomol.* 106:1189-1194.

Richards, O. W. 1927. Sexual selection and allied problems in the insects. *Biol. Rev.* 2:298-364.

Richerson, J. V., E. A. Cameron, and E. A. Brown. 1976. Sexual activity of the gypsy moth. *Am. Midl. Nat.* 95:299-312.

Riddiford, L. M., and J. B. Ashenhurst. 1973. The switchover from virgin to mated behavior in female cecropia moths: The role of the bursa copulatrix. *Biol. Bull. (Woods Hole)* 144:162-171.

Riemann, J. G. 1970. Metamorphosis of sperm of the cabbage looper, *Trichoplusia ni*, during passage from the testes to the female spermatheca. In *Comparative Spermatology*, B. Baccetti (ed.), pp. 321-331. Academic Press, New York.

Riemann, J. G., and B. J. Thorson. 1971. Sperm maturation in the male and female genital tracts of *Anagasta kuhniella* (Lepidoptera:Pyralidae). *Int. J. Insect Morphol. Embryol.* 1:11-19.

Riemann, J. G., and B. J. Thorson. 1976. Ultrastructure of the ductus ejaculatorius duplex of the Mediterranean flour moth, *Anagasta kuhniella* (Zeller) (Lepidoptera:Pyralidae). *Int. J. Insect Morphol. Embryol.* 5:227-240.

Robin, J. C. 1975. Use of microradiographic techniques instead of dissection for study of the mating status of female armyworm moths: *Mythimna unipuncta,* Haw. (Lepidoptera: Noctuidae) captured in light traps. *Rev. Zool. Agric. Pathol. Veg.* 74:154-157.

Robinson, A. S. 1973. Increase in fertility, with repeated mating, of gamma irradiated male codling moths, *Laspeyresia pomonella* (L) (Lepidoptera:Olethreutidae). *Can. J. Zool.* 51:427-430.

Robinson, A. S. 1974. Gamma irradiation and insemination in the codling moth, *Laspeyresia pomonella* (Lepidoptera, Olethreutidae). *Entomol. Exp. Appl.* 17:425-432.

Roelofs, W. L., and R. T. Carde. 1977. Responses of Lepidoptera to synthetic sex pheromone chemicals and their analogues. *Ann. Rev. Entomol.* 22:377-405.

Roome, R. E. 1975. Activity of adult *Heliothis armigera* (Hb.) (Lepidoptera, Noctuidae) with reference to the flowering of sorghum and maize in Botswana. *Bull. Entomol. Res.* 65:523-530.

Rutowski, R. L. 1978a. The courtship behaviour of the small sulphur butterfly *Eurema lisa* (Lepidoptera, Pieridae). *Anim. Behav.* 26:892-903.

Rutowski, R. L. 1978b. The form and function of ascending flights in *Colias* butterflies. *Behav. Ecol. Sociobiol.* 3:163-172.

Rutowski, R. L. 1979a. The butterfly as an honest salesman. *Anim. Behav.* 27:1269-1270.

Rutowski, R. L. 1979b. Courtship behavior of the checkered white, *Pieris protodice* (Pieridae). *J. Lepid. Soc.* 33:42-49.

Rutowski, R. L. 1982. Mate choice and Lepidopteran mating behavior. *Fla. Entomol.* 65:72-82.

Santorini, A. P., and P. Vassilaina-Alexopoulou. 1976(1977). Morphology of the internal reproductive system in male and female *Palpita unionalis* Hbn. (Lep., Pyralidae). *Entomol. Mon. Mag.* 112:105-108.

Schenck, J. L., Jr., and F. L. Poston. 1979. Adjusting light-trap samples of the southwestern corn borer for age-class bias. *Ann. Entomol. Soc. Am.* **72**:746-748.

Schneider, D., M. Boppré, H. Schneider, W. R. Thompson, C. J. Boriack, R. L. Petty, and J. Meinwald. 1975. A pheromone precursor and its uptake in male *Danaus* butterflies. *J. Comp. Physiol.* **97**:245-256.

Scott, J. A. 1972(1973). Mating of butterflies. *J. Res. Lepid.* **11**:99-127.

Scott, J. A. 1973. Population biology and adult behavior of the circumpolar butterfly *Parnassius phoebus* F. (Papilionidae, Lep.). *Entomol. Scand.* **4**:161-168.

Scott, J. A. 1974. Adult behavior and population biology of *Poladryas minuta,* and the relationship of the Texas and Colorado populations (Lepidoptera:Nymphalidae). *Pan-Pac. Entomol.* **50**:9-22.

Shapiro, A. 1970. The role of sexual behavior in density-related dispersal of pierid butterflies. *Am. Nat.* **104**:367-372.

Shapiro, A. M. 1982. Survival of refrigerated *Tatochila* butterflies (Lepidoptera:Pieridae) as an indicator of male nutrient investment in reproduction. *Oecologia (Berl.)* **53**:139-140.

Shapiro, I. D. 1977. Interaction of population biology and mating behavior of the Fiery Skipper, *Hylephila phylaeus* (Hesperiidae). *Am. Midl. Nat.* **98**:85-94.

Shepherd, J. G. 1974a. Activation of saturniid moth sperm by a secretion of the male reproductive tract. *J. Insect Physiol.* **20**:2107-2122.

Shepherd, J. G. 1974b. Sperm activation in saturniid moths: Some aspects of the mechanism of activation. *J. Insect Physiol.* **20**:2321-2328.

Shepherd, J. G. 1975. A polypeptide sperm activator from male saturniid moths. *J. Insect Physiol.* **21**:9-22.

Shields, O. 1967(1968). Hilltopping: An ecological study of summit congregation behavior of butterflies on a southern California hill. *J. Res. Lepid.* **6**:69-178.

Shields, O., and J. F. Emmel. 1973. A review of carrying pair behavior and mating times in butterflies. *J. Res. Lepid.* **12**:25-64.

Showers, W. B., G. L. Reed, and H. Oloumi-Sadeghi. 1974. Mating studies of female European corn borers: Relationship between deposition of egg masses on corn and captures in light traps. *J. Econ. Entomol.* **67**:616-619.

Sidhu, N. S. 1972(1975). Polymegaly in spermatophores in *Bombyx mori. Indian J. Hered.* **4**:47-50.

Sierra, J. R., W. D. Woggon, and H. Schmid. 1976. Transfer of cantharidin (1) during copulation from the adult female *Lytta vesicatoria* ("Spanish flies"). *Experientia* **32**:142-144.

Silbergleid, R. E. 1977. Communication in the Lepidoptera. In *How Animals Communicate,* T. A. Sebeok (ed.), pp. 362-402. Indiana Univ. Press, Bloomington, IN.

Silberglied, R. E., J. G. Shepherd, and J. L. Dickinson. 1984. Eunuchs: The role of apyrene sperm in Lepidoptera. *Am. Nat.* **123**:255-265.

Sims, S. R. 1978. A red-eyed mutant of *Papilio zelicaon* (Lepidoptera:Papilionidae). *Ann. Entomol. Soc. Am.* **71**:771-772.

Sims, S. R. 1979. Aspects of mating frequency and reproductive maturity in *Papilio zelicaon. Am. Midl. Nat.* **102**:36-50.

Snow, J. W., and T. C. Carlysle. 1967. A characteristic indicating the mating status of male fall armyworm moths. *Ann. Entomol. Soc. Am.* **60**:1071-1074.

Snow, J. W., J. R. Young, and R. L. Jones. 1970. Competitiveness of sperm in female fall armyworms mating with normal and chemosterilized males. *J. Econ. Entomol.* **63**:1799-1802.

Sokolowski, R. J., and Z. W. Suski. 1980. Mating ability and sperm competitiveness of radiosterilized codling moths, *Laspeyresia pomonella* L. (Lepidoptera, Olethreutidae). *Pol. Pismo Entomol.* **50**:463-471.

Sparks, M. R., and J. S. Cheatham. 1973. Tobacco hornworm: Marking the spermatophore with water-soluble stains. *J. Econ. Entomol.* **66**:179-721.

Srivastava, U. S., and B. P. Srivastava. 1957. Notes on the spermatophore formation and transference of sperms in the female reproductive organs of *Leucinodes orbonalis* Guen. (Lepidoptera, Pyraustidae). *Zool. Anz.* **158**:258-266.

Stadelbacher, E. A., M. L. Laster, and T. R. Pfrimmer. 1972. Seasonal occurrence of populations of bollworm and tobacco budworm moths in the central delta of Mississippi. *Environ. Entomol.* **1**:318-323.

Stadelbacher, E. A., and T. R. Pfrimmer. 1973. Bollworms and tobacco budworms: Mating of adults at three locations in the Mississippi Delta. *J. Econ. Entomol.* 66:356-358.

Steklonikov, A. A. 1965. Functional morphology of the copulatory apparatus in some Lepidoptera. *Entomol. Rev. (Engl. Trans. Entomol. Obozr.)* 44:143-149.

Stern, V. M., and R. F. Smith. 1960. Factors affecting egg production and oviposition in populations of *Colias philodice eurytheme* Boisduval (Lepidoptera:Pieridae). *Hilgardia* 29:411-454.

Stockel, J. 1971. Variation de la taille du spermatophore en fonction de son rang d'émission par le mâle de *Sitotroga cerealella* Oliv. (Lepidoptera, Gelechiidae). *C. R. Hebd. Seances Acad. Sci. Ser. D Sci. Nat.* 272:2713-2716.

Stockel, J. 1972. Stimulus inducteur de la ponte chez les femelles de *Sitotroga cerealella* Oliv. (Lepidoptera, Gelechiidae). *C. R. Hebd. Seances Acad. Sci. Ser. D Sci. Nat.* 275:385-387.

Stockel, J. 1973. Fonctionnement de l'appareil reproducteur de la femelle de *Sitotroga cerealella* (Lepidoptera:Gelechiidae). *Ann. Soc. Entomol. Fr.* 9:627-645.

Stride, G. O. 1958. Further studies on the courtship behaviour of African mimetic butterflies. *Anim. Behav.* 6:224-230.

Sugawara, Takashi. 1979. Stretch reception in the bursa copulatrix of the butterfly, *Pieris rapae crucivora*, and its role in behaviour. *J. Comp. Physiol. A (Sens. Neural Behav. Physiol.)* 130:191-199.

Suzuki, Y. 1979. Mating frequency in females of the small cabbage white, *Pieris rapae crucivora* Boisduval (Lepidoptera:Pieridae). *Kontyu* 47:335-339.

Suzuki, Y., A. Nakanishi, H. Shima, O. Yata, and T. Saigusa. 1977. Mating behaviour in four Japanese species of the genus *Pieris* (Lepidoptera, Pieridae). *Kontyu* 45:300-313.

Takeuchi, S., and K. Miyashita. 1975. The process of spermatophore transfer during the mating of *Spodoptera litura* F. *Jpn. J. Appl. Entomol. Zool.* 19:41-46.

Taylor, O. R., Jr. 1967. Relationship of multiple mating to fertility in *Atteva punctella* (Lepidoptera:Yponomeutidae). *Ann. Entomol. Soc. Am.* 60:583-590.

Tedders, W. L., Jr., and V. R. Calcote. 1967. Male and female reproductive systems of *Laypeyresia caryana*, the hickory shuckworm moth (Lepidoptera:Olethreutidae). *Ann. Entomol. Soc. Am.* 60:280-282.

Thibout, E. 1971. Description de l'appareil génital mâle et formation du spermatophore chez *Acrolepia assectella* (Lépidoptère, Plutellidae). *C. R. Hebd. Seances Acad. Sci. Ser. D Sci. Nat.* 273:2546-2549.

Thibout, E. 1975. Analyse des causes de l'inhibition de la réceptivité sexuelle et de l'influence d'une éventuelle seconde copulation sur la reproduction chez la Teigne du poireau, *Acrolepia assectella* (Lepidoptera:Plutellidae). *Entomol. Exp. Appl.* 18:105-116.

Thibout, E. 1977. La migration spermatique chez *Acrolepiopsis (Acrolepia) assectella* Zell. (Lep., Plutellidae): Role de la motilité des spermatozoides et de la musculature de l'appareil génital femelle. *Ann. Soc. Entomol. Fr.* 13:381-389.

Thibout, E. 1981. Evolution and role of apyrene sperm cells of Lepidopterans: Their activation and denaturation in the leek moth, *Acrolepiopsis assectella* (Hypomeutoidea). In *Advances in Invertebrate Reproduction*, W. H. Clark, Jr., and T. S. Adams (eds.), pp. 231-242. Elsevier/North Holland, New York.

Thibout, E., and R. Rahn. 1972. Étude de la variabilité du volume et du pouvoir fécondant des spermatophores successifs d'*Acrolepia assectella* (Lepidoptera:Plutellidae). *Entomol. Exp. Appl.* 15:443-454.

Thornhill, R. 1976. Sexual selection and paternal investment in insects. *Am. Nat.* 110:153-163.

Trivers, R. L. 1972. Parental investment and sexual selection. In *Sexual Selection and the Descent of Man, 1871-1971*, B. Campbell (ed), pp. 136-179. Aldine-Atherton, Chicago.

Truman, J. W., and L. M. Riddiford. 1974. Hormones and behavior. *Adv. Insect Physiol.* 10:297-352.

Turner, J. R. G. 1981. Adaptation and evolution in *Heliconius:* A defense of NeoDarwinism. *Ann. Rev. Ecol. Syst.* 12:99-121.

Turner, W. K., N. C. Leppla. V. Chew, and F. L. Lee. 1977. Light quality influences on carbon dioxide output and mating of cabbage looper moths. *Ann. Entomol. Soc. Am.* 70:259-263.

Tykac, J. 1951. Sphragis Ci Sphragidoid U Motylu. *Acta Soc. Entomol. Cech.* 48:94-98.

Urquhart, F. A. 1960. *The Monarch Butterfly.* Univ. Toronto Press, Toronto.

Wada, T., M. Kobayashi, and M. Shimazu. 1980. Seasonal changes of the proportions of mated females in the field population of the rice leaf roller, *Cnaphalocrocis medinalis* Guenée (Lepidoptera:Pyralidae). *Appl. Entomol. Zool.* **15**:81-89.

Wago, H. 1977a. An artificial control of mating activity in *Callopistria repleta* Walker. *Zool. Mag. (Tokyo)* **86**:77-81.

Wago, H. 1977b. Studies on the mating behavior of the Pale Grass Blue, *Zizeeria maha argia* (Lepidoptera, Lycaenidae). II. Recognition of the proper mate by the male. *Kontyu* **45**:92-96.

Wakamura, S. 1979. Mating behavior of the common cutworm moth, *Agrotis fucosa* Butler (Lepidoptera, Noctuidae). *Jpn. J. Appl. Entomol. Zool.* **23**:251-256.

Walker, T. J. 1978. Migration and re-migration of butterflies through North Peninsular Florida: Quantification with malaise traps. *J. Lepid. Soc.* **32**:178-190.

Walker, W. F. 1980. Sperm utilization strategies in nonsocial insects. *Am. Nat.* **115**:780-799.

Weidner, H. 1934. Beiträge zur Morphologie und Physiologie de Genitalapparates der Weiblichen Lepidopteren. *Z. Angew. Entomol.* **21**:239-290.

Wellington, W. G. 1965. Some maternal influences on progeny quality in the western tent caterpillar, *Malacosoma pluviale* (Dyar). *Can. Entomol.* **97**:1-14.

Wellso, S. G., and P. L. Adkisson. 1962. The morphology of the reproductive system of the female pink bollworm moth, *Pectinophora gossypiella* (Saund.). *J. Kans. Entomol. Soc.* **35**:233-235.

Wickler, W. 1968. *Mimicry in Plants and Animals.* McGraw-Hill (World Library), New York.

Wigglesworth, V. B. 1972. *The Principles of Insect Physiology, 7th ed.* Chapman and Hall, London.

Wiklund, C. 1977. Courtship behaviour in relation to female monogamy in *Leptidea sinapis* (Lepidoptera). *Oikos* **29**:275-283.

Wiklund, C., and T. Fagerström. 1977. Why do males emerge before females? A hypothesis to explain the incidence of protandry in butterflies. *Oecologia (Berl.)* **31**:153-158.

Williams, C. B., G. F. Cockbill, M. E. Gibbs, and J. A. Downes. 1942. Studies in the migration of Lepidoptera. *Trans. R. Entomol. Soc. Lond.* **92**:101-282.

Williams, G. C. 1975. *Sex and Evolution.* Princeton Univ. Press, Princeton, NJ.

Williams, J. L. 1940. The anatomy of the internal genitalia and the mating behaviour of some lasiocampid moths. *J. Morphol.* **67**:411-437.

Williams, J. L. 1941a. The relations of the spermatophore to the female reproductive ducts in Lepidoptera. *Entomol. News* **52**:61-65.

Williams, J. L. 1941b. The internal genitalia of the evergreen bagworm and the relation of the female genital ducts to the alimentary canal. *Proc. Pa. Acad. Sci.* **15**:53-58.

Williams, J. L. 1941c. The internal genitalia of yucca moths, and their connection with the alimentary canal. *J. Morphol.* **69**:217-223.

Williams, J. L. 1947. The comparative anatomy of the internal genitalia of some Tineoidea (Lepidoptera:Gracillariidae, Tischeriidae). *Proc. R. Entomol. Soc. Lond. Ser. A* **22**:77-84.

Yakhontov, V. V. 1960. Notes on the role of the oviductus communis in the fertilization of insects' eggs. *Proc. XI Int. Congr. Entomol. (Vienna)* **1**:346-347.

Yamaoka, K., and T. Hirao. 1976. Stimulation of virginal oviposition by male factor and its effect on spontaneous nervous activity in *Bombyx mori. J. Insect Physiol.* **23**:57-63.

Yang, L., and Y. Chow. 1978. Spermatophore formation and the morphology of the reproductive system of the diamondback moth, *Plutella xylostella* (L) (Lepidoptera:Plutellidae). *Bull. Inst. Zool. Acad. Sin. (Taipei)* **17**:109-115.

10

Sperm Transfer and Use in the Multiple Mating System of Drosophila

MARK H. GROMKO

DONALD G. GILBERT

ROLLIN C. RICHMOND

Sperm Competition and the Evolution
of Animal Mating Systems

I. INTRODUCTION

When females of a particular species not only store sperm from single matings but also remate before sperm from a previous mating is exhausted, there is an opportunity for selection to occur. The coexistence of sperm from two different males within the same storage organs of a female provides the selection pressure of sperm competition, and this can lead to two opposing types of adaptation (Parker 1970). Species of the genus *Drosophila*—particularly *Drosophila melanogaster*—have been used extensively to investigate both types of adaptation. Thus, in *Drosophila melanogaster* we find that: (1) males transfer substances to females during copulation that reduce the female's receptivity to remating, and consequently increase the probability that their sperm is used to fertilize eggs; and (2) when a female does remate, some of the sperm which had been stored is not used to fertilize the female's eggs, while the sperm of the most recent male is used for fertilization. The first kind of adaptation is what we will call a "first male" adaptation, since it tends to maximize the reproductive success of the first male; the latter will be called a "remator male" adaptation. The two kinds of adaptation lead to diametrically opposed outcomes in terms of the reproductive success of the males involved, but they both reduce the degree to which different males' sperm will be used concurrently by a female.

The focus of this review is an investigation of the mechanisms by which "first males" and "remator males" achieve their reproductive success in the context of actual or potential competition from sperm of other males. In section II we review the fundamentals of the reproductive biology of *Drosophila melanogaster* and consider sperm transfer to the female, movement of sperm to the female's sperm storage organs, and utilization of stored sperm by singly mated females. We present evidence that challenges some conventional wisdom about *Drosophila melanogaster* (*i.e.*, that sperm storage is elastic, with no fixed maximum capacity; sperm use in *Drosophila* apparently involves random release of sperm and a large amount of wasted sperm as a consequence). Also, a consideration of interactions between females and ejaculates of different types of males leads us to raise the possibility that the numerical superiority of second male sperm among the progeny of a remated female is due not to a selected response to sperm competition but is due to interaction between the male's ejaculate and the female.

In section III, we establish that repeated mating is a common occurrence in *Drosophila*. Remating time is not arbitrary, however; receptivity of nonvirgin females is apparently controlled by the number of sperm stored.

We introduce two new terms in the discussion of the effect of remating on reproductive success. "P'" is an estimator of the proportion of a first male's sperm that is used by a female before she remates. The theory of sperm competition leads us to expect that selection will increase P' as well as the quantity

P_2, the proportion of second male offspring after remating. We also suggest that the numerical superiority of second male sperm, and the apparent loss or nonuse of previously stored sperm following a repeat mating by a female, should be referred to as "sperm predominance." We suggest this to avoid implying any particular mechanism by which the numerical superiority, or predominance, takes place. Later, in section IV, we present evidence which suggests that actual sperm displacement does not occur in *Drosophila melanogaster.*

Section III concludes with a review of the effects males have on reducing the receptivity of females with which they have just mated. Section IV takes up the converse problem: mechanisms by which sperm predominance occurs. A comparison of the similarities of intra- and inter-ejaculate sperm predominance leads to the suggestion that sperm motility or viability differences are fundamentally involved in this apparent interaction between sperm types.

In section V, we reconsider the evolutionary significance of sperm transfer and use in *Drosophila* in light of information on the mechanisms by which the various adaptations occur.

II. SPERM STORAGE AND UTILIZATION

It is impossible to achieve an understanding of the mechanisms and significance of sperm transfer and utilization in a repeat mating system without basic knowledge of reproductive anatomy and physiology. This section reviews male and female anatomy, physiology of sperm transfer and storage, and utilization of sperm by singly mated females.

A. Reproductive Anatomy

1. The Male

The most detailed and comprehensive review of the structure and ultrastructure of the male reproductive system of *Drosophila melanogaster* is that of Bairati (1967, 1968). Fowler (1973) reviews Bairati's contributions and the earlier light microscopic studies of Miller (1950).

The male reproductive system (Fig. 1) consists of the paired testes each connected to a seminal vesicle by the testiculo-deferential valve. The seminal vesicles narrow at their posterior ends into paired vasa deferentia which join in a short, unpaired deferent duct. This duct joins the ampullary end of the anterior ejaculatory duct as do the outlets of the paired accessory glands or paragonia. At its

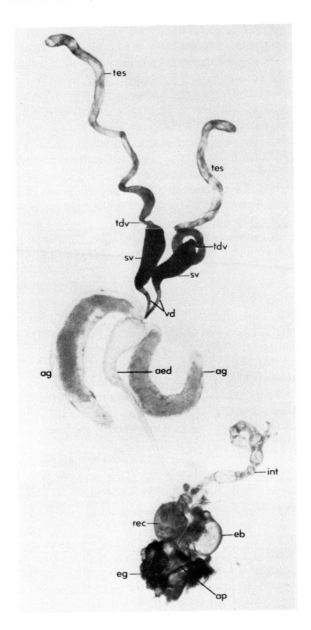

Fig. 1. Photomicrograph of the reproductive system of a male *Drosophila melanogaster.* The organs have been displaced from their normal positions for ease in illustration. tes = testis; tdv = testiculo-deferential valve; sv = seminal vesicle; vd = vas deferens; ag = accessory gland or paragonia; aed = anterior ejaculatory duct; int = anterior intestine; rec = rectum; eb = ejaculatory bulb; eg = external genitalia; ap = basal apodeme of aedeagus.

posterior end, the anterior ejaculatory duct narrows into a tubular region which is connected to the ejaculatory bulb or sperm pump. The ejaculatory bulb is connected to the external genitalia by the tubular, posterior ejaculatory duct.

Meiosis and sperm maturation occur in the main body of the testes in membrane enclosed cysts (Lindsley and Tokuyasu 1980). Individual sperm move from the testes into the seminal vesicles by way of the testiculo-deferential valves. The primary function of the seminal vesicles is to act as reservoirs for mature sperm. The ampullar region of the anterior ejaculatory duct acts as a chamber in which secretion from the accessory glands and the anterior ejaculatory duct itself are combined with sperm emerging from the vasa deferentia. Muscular contractions of the anterior ejaculatory duct draw sperm and secretions from the vasa deferentia and accessory glands and propel them into the ejaculatory bulb. The ejaculatory bulb appears to have at least two functions: production of a viscous secretion and propulsion of this secretion and sperm into the posterior ejaculatory duct and female genital tract. The role of the various male accessory secretions will be considered below.

2. The Female

The anatomy of the female reproductive system (Fig. 2) has been reviewed in detail by Miller (1950), Fowler (1973), and in overview by King (1970). The ovaries are paired structures each consisting of a number (10-20) of ovarioles or egg tubes in which the processes of oogenesis occur (Mahowald and Kambysellis 1980). The ovarioles empty into paired, lateral oviducts which fuse posteriorly into a common oviduct. The common oviduct is connected to the genital chamber which is divided into the anterior uterus and posterior vagina.

There are three sperm storage organs in *Drosophila*: the dorsal, paired spermathecae and the ventral seminal receptacle. Each spermatheca consists of a hemispherical cap resembling an inverted, doubled-walled bowl. The walls of these structures are formed of a pigmented cuticular layer and are surrounded by a single layer of epithelial cells and a second incomplete covering of glandular cells (Filosi and Perotti 1975). The glandular cells are highly specialized and produce a laminar product (possibly of lipoproteic composition) which is secreted directly into the lumen of the spermatheca following insemination. Presumably, this substance acts as a nutrient for stored sperm, although direct evidence of this is lacking. The whole spermatheca is imbedded in a matrix of fat body cells (Fowler 1973). The spermathecae are connected by ducts to the anterior, dorsal end of the uterus (Fig. 2). Wild type strains have been described in which a significant proportion of females have three spermathecae (Sturtevant 1925). Mutant genes are known which modify the shape of spermathecae or result in their elimination (Dobzhansky and Holz 1943, Anderson 1945).

The seminal receptacle is a long tube tightly coiled against the anterior end of the uterus and is composed of cuboidal epithelial cells. It opens into the uterus

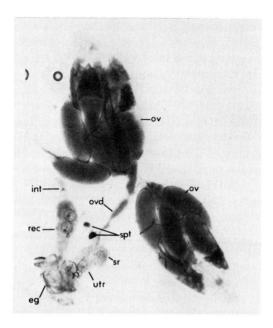

Fig. 2. Photomicrograph of the reproductive system of a female *Drosophila melanogaster*. The organs have been displaced from their normal positions for ease in illustration. ov = ovary; int = anterior intestine; ovd = common oviduct; rec = rectum; spt = spermatheca; sr = seminal receptacle; utr = uterus; eg = external genitalia.

at a point ventral to the common oviduct. Fowler (1973) notes that the lumen of the receptacle widens in the distal half of the tube. Blaney (1970) has demonstrated the existence of a layer of visceral muscle surrounding the tube, and Miller (1950) notes the presence of nerve fibers between the coils of the ventral (=seminal) receptacle.

Slightly posterior to the spermathecae are found the paired parovaria or accessory glands. These structures are connected to the dorsal surface of the uterus by fine ducts. There is no evidence that the parovaria are sperm storage organs, and Riley and Forgash (1967) have demonstrated that they are probably the source of an adhesive mucoprotein which is applied to the egg as it leaves the genital chamber. This substance serves to stick eggs to the oviposition substrate.

B. Sperm Transfer

The sequence of physiological events which is responsible for the transfer of sperm into the female's uterus is poorly understood although considerable progress has been made since Fowler's (1973) review of *Drosophila* reproductive biology.

Since male accessory secretions comprising the seminal fluid have been repeatedly invoked as critical in determining male and possibly female fertility (Leopold 1976), we consider first the known chemical constituents of the seminal fluid.

1. Constituents of the Seminal Fluid

A variety of substances has been detected in insect seminal fluid (see reviews by Mann 1964, and Leopold 1976). Bairati (1967) described a waxy material which is derived from the male ejaculatory bulb of *D. melanogaster* and can be detected in the female genital chamber 5-7 minutes after the beginning of copulation. Bairati postulates that the waxy plug which forms in the uterus acts as a substrate against which sperm can move to the storage organs or simply as a barrier to prevent loss of sperm through the vagina.

The nature of the waxy material described by Bairati (1967) may have been discovered by Butterworth (1969) and Brieger and Butterworth (1970) who showed that male *D. melanogaster* possess a lipid (*cis*-11-octadecen-1-ol acetate) absent in virgin females. It is likely the male lipid forms most or part of the waxy plug identified by Bairati. This lipid material is highly concentrated in the male ejaculatory bulb and can be detected in mated females. It has been demonstrated to act as an anti-aphrodisiac pheromone in *D. melanogaster*, causing an inhibition of male courtship (Jallon *et al.* 1981).

A number of enzymes have been localized to the male reproductive system and identified as components of the seminal fluid. Sheehan *et al.* (1979) and Richmond *et al.* (1980) showed that a non-specific carboxylesterase of *D. melanogaster*, esterase 6 (EST 6), is highly concentrated in the male anterior ejaculatory duct and is transferred to the female within the first 3 minutes after the initiation of copulation (Richmond and Senior 1981). In a series of investigations, the reproductive system function of EST 6 has been explored. In brief, the enzyme has been shown to influence the timing of remating in females, the rate of sperm use by females, productivity at 18°C, copula duration, and mating speed (Richmond *et al.* 1980; Gilbert *et al.* 1981a; Gilbert 1981b; Gilbert and Richmond 1982a, b). The presence or absence of EST 6 in the male ejaculate has no effect on female productivity at 25°C or on interactions among sperm from two males in the female storage organs (Gilbert *et al.* 1981b, Gilbert and Richmond 1981). An *in vivo* substrate for EST 6 has been found to be *cis*-vaccenylacetate, the lipid component of the male's seminal fluid. The EST 6 transferred to females during mating hydrolyzes *cis*-vaccenyl-acetate to produce an alcohol, *cis*-vaccenyl alcohol, which has inhibitory effects on male courtship similar to those of the substrate itself. Purified preparations of the common fast and slow allozymes of EST 6 have been shown to have different specificity for artificial substrates (S. D. Mane *et al.*, unpubl. data), although whether the allozymes differ in their *in vivo* functioning is not yet known.

Many species of *Drosophila* have carboxylesterases localized within the male reproductive system. Johnson and Bealle (1968) surveyed 90 *Drosophila* species for the presence of ejaculatory bulb β-esterases. Korochkin *et al.* (1976) showed that for three species of the *D. virilis* group, the ejaculatory bulb esterases were concentrated within the waxy material produced by the bulb and transferred to the female during copulation. Johnson and Bealle found a significant correlation between the presence of a bulb esterase and the likelihood of successful interspecific crosses. In 50% of interspecific matings involving β-esterase positive males, motile sperm were found in the female receptacle compared with only 5% of mating with β-esterase negative males. The observations reported above that EST 6 influences sperm storage and usage may provide an explanation for these results. This possibility is discussed in more detail below.

A second enzyme which has recently been localized to the male anterior ejaculatory duct is glucose oxidase (GO, Cavener 1980), which catalyzes the oxidation of glucose to gluconate. Cavener has shown that GO is a component of the seminal fluid (pers. comm.) and suggests that it may function as a bactericide and fungicide since GO produces hydrogen peroxide as a by-product of its reaction.

A series of other enzymes have also been localized to the male reproductive system, but it is unknown at present whether these enzymes are secreted into the seminal fluid. Murray and Ball (1967) and Cavener (1980) localized two hexokinases (HEX-B, HEX-t) to the testes. Bischoff (1978) has shown that NAD- and NADP-dependent sorbitol dehydrogenases are found in all male reproductive organs and are likely associated with the midpiece of the sperm tail. The hexokinases, GO, and the sorbitol dehydrogenases are all involved in hexose metabolism and may function both in the male and the female reproductive systems. Beta-glucuronidase has also been found in the male reproductive system (D. L. Hartl, pers. comm.).

The paragonia or accessory glands have long been considered to be the primary source of seminal fluids in *Drosophila* (*e.g.,* Lefevre and Jonsson 1962a). Several components of the paragonial secretions have been identified and their effects on the reproductive physiology of the female explored. The observation that *Drosophila* males which lacked testes or were genetically sterile would mate with females, stimulate oviposition, and cause temporary refractoriness to remating suggested that a chemical factor or factors were involved (Maynard Smith 1956; Kummer 1960; Manning 1962, 1967). This supposition was confirmed by Garcia-Bellido (1964) by transplanting accessory glands into virgin females. Leahy and Lowe (1967) attempted to purify this factor from *D. melanogaster* and succeeded in showing that it was a small peptide which would dramatically stimulate oviposition when injected into the female. Chen *et al.* (1977) have also characterized this factor and believe it not to be a peptide but an ethanolamine-containing galactoside. The contradiction between this result, that of Leahy and Lowe (1967), and Chen's previous work (Chen and Buhler 1970) is striking and deserves further consideration.

Baumann (1974a, b) has studied paragonial substances in *D. funebris.* He was able to isolate two substances which have different effects on the female. One,

PS-1, is a peptide composed of 27 amino acids which reduces the receptivity of virgin females following injection. Another, PS-2, is a low molecular weight, carbohydrate-containing substance that stimulates oviposition in females. Baumann's results suggest the confusion surrounding work on accessory gland substances of *D. melanogaster* may be due to the presence of two substances (Leopold 1976). Clearly, more work is needed, but the conclusion is inescapable that male *Drosophila* produce paragonial substances which affect the behavior and reproductive physiology of their mates.

Filamentous structures with a size and morphology identical to microtubules are another component of the seminal fluid synthesized by the accessory glands (Bairati 1966, Perotti 1971). These microtubules are ejaculated with sperm and can be found in the female's uterus and seminal receptacle packed between sperm. If a female mates a sterile (XO) male, microtubules are deposited in the genital chamber but do not move into the sperm storage organs. Extracellular microtubules such as these are rarely observed, but they have also been found in association with sperm of reptiles and isopods (see references in Perotti 1971). No direct information exists concerning the function of these seminal fluid microtubules, but Bairati (1968) has suggested that they might act as mechanical supports for sperm or assist sperm to move into the storage organs. Perotti (1971) hypothesizes that they may represent a substrate used in sperm metabolism while in storage. It is conceivable that these microtubules mediate the behavioral and physiological effects of the paragonial secretions reviewed above, but the small size of the sex peptide compared with tubulin, the subunit of microtubules (Burnside 1975), makes this hypothesis unlikely.

Table I provides a summary of the substances known to be included in *Drosophila* seminal fluid. A fundamental understanding of the mechanisms involved in sperm storage, utilization, and competition will rely heavily on knowledge of the composition and functions of the seminal fluid. Although a number of components have been identified, the function of most of these substances is poorly perceived.

2. Time of Sperm Transfer

The duration of copulation in the genus *Drosophila* varies enormously from about 30 sec (*e.g., D. mulleri*) to more than 1.5 hr in *D. acanthoptera* (Spieth 1952). *D. melanogaster* copulate for about 20 min, but this varies somewhat between strains (MacBean and Parsons 1966). In *D. melanogaster,* the time during copulation when sperm transfer begins is about 10 min, but some sperm may be transferred as early as 4 min after the initiation of copulation in a small percentage of matings (Gilbert *et al.* 1981a). Contrary to earlier findings (Hinton 1964), there is no evidence for a spermatophore (in the strict sense) in *Drosophila* (Leopold 1976).

TABLE I

Identified Components of the Seminal Fluid of *Drosophila*

Substance	Species	Reference
Lipids		
Waxy plug	*D. melanogaster*	Bairati 1968
Male lipid	*D. melanogaster*	Brieger and Butterworth 1970
Enzymes		
Esterase 6	*D. melanogaster*	Richmond *et al.* 1980
β-esterase	Many	Johnson and Bealle 1968
		Korochkin *et al.* 1976
Glucose oxidase	*D. melanogaster*	Cavener 1980
Other proteins		
Sex peptide	*D. melanogaster*	Leahy and Lowe 1967
PS-1	*D. funebris*	Baumann 1974a, b
Protein aggregates		
Microtubules	*D. melanogaster,*	Bairati 1966
	D. paulistorum	Perotti 1971
Carbohydrate-containing		
Galactoside	*D. melanogaster*	Chen *et al.* 1977
PS-2	*D. funebris*	Baumann 1974a, b

3. Number of Sperm Transferred

Kaufmann and Demerec (1942) and Gilbert (1981b) provide the only direct counts of sperm transferred to the female genital chamber. The earlier investigators found that an average of 3,619 sperm were transferred by four *D. melanogaster* males on their first mating. In subsequent matings on the same day, the numbers transferred decreased to about 1,000-2,000 sperm by the third or fourth consecutive mating. Gilbert (1981b) found that *Oregon R D. melanogaster* males transferred a maximum of 4,690 sperm, while two inbred lines transferred 3,800 and 2,400 sperm. Ejaculated sperm are found in the anterior part of the uterus proximal to the waxy plug described by Bairati (1968, Bairati and Perotti 1970).

Several factors are known to influence the numbers of sperm transferred to females. Kaufmann and Demerec (1942) documented the reduction in transferred sperm numbers with sequential mating of a single male. Lefevre and Jonsson (1962a) and others have attributed this decline in the number of sperm transferred to a reduction in the amounts of seminal fluid available. Richmond *et al.* (1980) provide direct evidence for a reduction in the amount of one component of the seminal fluid (EST 6) transferred to females for virgin vs. once-mated males. Richmond and Senior (1981) found that 24-48 h are required for mated males to recover their virginal levels of EST 6 activity. This result is in accord with the findings of Lefevre and Jonsson (1962a) who noted that an "exhausted" male is fertile if kept isolated from females overnight. Bairati (1968) attributes the reduction in transferred sperm number in sequentially mated males to a reduction in seminal fluid components produced by the accessory glands, the anterior ejaculatory duct, and the ejaculatory bulb, and to a scarcity of mature sperm in the seminal vesicles. Bairati also notes that the senescent decline in male fertility is not due to loss of function in accessory glands or testes, but to early senescence of the secretory epithelium of the anterior ejaculatory duct. Richmond and Sheehan (unpubl. data) have shown than EST 6 levels in aging males fall precipitously as would be expected from the fact that EST 6 apparently is produced by the anterior ejaculatory duct (Sheehan *et al.* 1979).

As might be expected, temperature has also been found to have an influence on the number of sperm a male is able to transfer. Iyenger and Baker (1962) found that a 15 min treatment at $36°C$ increased male fertility while treatment at $-8°C$ decreased fertility. They attributed this difference to changes in sperm motility, but their data are confounded by a behavioral difference as well. Gilbert and Richmond (1982a) have found that female productivity at $18°C$ is substantially reduced over that at $25°C$. This result might be due to reduced numbers of sperm transferred and/or to a reduction in sperm viability and motility. The only general conclusions one can draw from these studies are that under normal conditions, the number of sperm a male transfers to a female is large (2,000-4,000) and that a variety of genetic and environmental factors influence this parameter.

C. Sperm Storage

The mechanistic bases of sperm storage in *Drosophila* are only dimly perceived at present, but as this process is likely to be intimately involved in the phenomena of sperm competition, we review in some detail the existing data.

1. Movement Into Storage Organs

Upon ejaculation, sperm appear to be deposited in the anterior portion of the uterus. This is anatomically advantageous since the openings of the ducts leading to the storage organs are also located here (see Fig. 2). When *Drosophila* males are dissected in saline solution, their sperm rarely exhibit forward movement, presumably because of the enormous length of the sperm tail ($\cong 2$ mm in *D. melanogaster*; Lefevre and Jonsson 1962a). However, observations of sperm in the female genital chamber reveal that sperm appear to be moving in "channels" which Fowler (1973) describes as "physical extensions of the seminal receptacles." The muscular nature of the genital chamber has led some authors to suggest that contractions of the uterine wall aid in the movement of sperm into the storage organs (DeVries 1964).

Lefevre and Jonsson (1962a) and others have pointed to the seminal fluid as an important component in sperm storage. The presence of microtubules in the seminal fluid was mentioned above, and these elements may play a role in sperm storage. Several enzymes have been detected in *Drosophila* seminal fluid (section II, B), and at least one of these (EST 6) appears to affect sperm utilization. It is reasonable to suppose that some components of the seminal fluid participate in the process of storage, and the results of Kiefer (1969, discussed below) reveal that the process of sperm storage may be amenable to genetic investigation.

Once sperm have been deposited in the female, storage apparently begins immediately (Lefevre and Jonsson 1962a, Yanders 1963, DeVries 1964, Lefevre and Moore 1967, Gilbert 1981b). The length of time required for maximum sperm storage reported in the literature is variable. In a brief note lacking quantitative data, Lefevre and Moore (1967) state that storage is complete 15-20 min after copulation. Peacock and Erickson (1965) report that 1-2 h are required for sperm storage to be completed. Gilbert (1981b) has found that maximum sperm storage in the receptacle requires 0.9-9.4 h, while spermathecal storage takes 3.1-7.5 h. The range in values found reflects, at least in part, the use of three different strains of males. It is clear from these results that while sperm storage begins immediately after ejaculation of sperm, several hours may be required for maximum storage to be achieved. Nonidez (1920) noted that sperm fill the seminal receptacle first and then enter the spermathecae. This observation was partially confirmed by Lefevre and Moore (1967) and Gilbert (1981b) who showed that sperm exhibit a preference for the receptacle, but do begin entering the spermathecae before the receptacle is filled. Patterson (1954), examining several other species, found sperm in all storage organs within a few minutes of the termination of copulation. Exceptions to this generalization are *D. pseudoobscura* which do not store sperm in either type of storage organ for more than one hour after mating and species of the *mulleri* subgroup which do not store sperm in their spermathecae. In this subgroup, spermathecae appear to function only as secretory glands.

2. Number of Sperm Stored

Few references report direct counts of total stored sperm in *Drosophila*. Kaplan *et al.* (1962) used Feulgen staining of dissected storage organs of *D. melanogaster* and recorded a maximum of 650 sperm from eight females. Fowler *et al.* (1968) and Zimmering and Fowler (1968) found a maximum storage of 500-600 sperm when females were dissected 3 hr after insemination. Gilbert (1981b), using a modification of the procedures of Fowler *et al.* (1968), has determined the kinetics of sperm storage in *D. melanogaster* females. This information is critical if maximum sperm storage is to be accurately assayed. Gilbert counted uteral, receptacle, and spermathecal sperm in 114 females at various times after the initiation of copulation. The maximum number of sperm stored in all storage organs combined ranged from 954 to 1,032 for three different strains of males. Regardless of the discrepancy in the maximum numbers of sperm stored, these data all show that males transfer to females from two to four times the number of sperm stored by the female. The reason for this large excess of sperm is unclear; however, it may be related to the limited time period for sperm storage before the initiation of oviposition (Gilbert 1981b). A male may be able to increase the number of sperm stored by transferring a larger number.

3. Factors Affecting Sperm Storage

Temperature strongly influences sperm storage. Iyenger and Baker (1962) reported that males treated for 15 min at -8°C or 36°C just before mating produced 33% and 133% as many progeny, respectively, compared with untreated controls. Trosko and Yanders (1963) report that inseminated females (D. melanogaster) stored at 10°C for 2 weeks had full sperm storage organs on dissection but produced fewer progeny than untreated females. Gilbert and Richmond (1982a) have found that the duration of copulation is substantially longer and the number of progeny produced by females at 18°C is substantially reduced compared with production at 25°C. Unfortunately, all these studies confound the process of sperm storage with sperm viability during storage. However, Richards (1963) has found the motility of cockroach sperm increases in direct proportion to temperature, suggesting that more sperm are stored at elevated temperatures.

Perhaps the most interesting and potentially most significant factor affecting sperm storage is an interaction effect between the male ejaculate and the female. This possibility was initially suggested by the preliminary results of Yanders (1963) who found that the number of sperm stored by females of four strains of D. melanogaster varied as a function of the male used. With one exception, Yanders found that females stored the fewest sperm when inseminated by males of the same strain. This result was pursued by DeVries (1964) who found evidence for interactions between the sexes in sperm storage. DeVries suggested that "Certain types of sperm may be inactivated or adversely affected by some factor in the female reproductive tract so that they do not reach the storage organs." Zimmering et al. (1970a) found direct evidence for such an effect when males heterozygous for segregation distorter were mated to yellow or wild-type females. Wild-type females stored about 16% more sperm than yellow females (quoted in Fowler 1973). It is certainly conceivable that some examples of sperm predominance in Drosophila result from an interaction between sperm and the female reproductive system and do not involve competitive interactions among sperm.

Kiefer (1969) has provided direct evidence that male (and sperm) genotype can affect sperm storage. D. melanogaster males which carry a Y chromosome having a deletion in the left arm produce sperm which are transferred to females and are motile in the uterus but do not enter the storage organs. This fascinating finding is not simply a result of genetically defective sperm, as Muller and Settles (1927), Hartl (1973), and others have shown that defective sperm can be stored and may fertilize eggs.

Lefevre and Moore (1967) noted an effect of male age on sperm storage which has not been replicated but deserves attention. They found that sperm from 1-day-old males were transferred and stored more slowly than sperm from 3-day-old males. Our observation (Sheehan *et al.* 1979; and see section II, B, 1) that EST 6 is not actively synthesized by males until they are 24-36 hr old, suggests that Lefevre and Moore's results may have been due to a deficiency of the seminal fluid.

Drosophila reproductive biologists have only begun to understand the factors involved in the processes of sperm storage and have almost no knowledge of the physiological mechanisms involved. The observations of differential sperm storage reported by Fowler (1973) and Beatty and Sidhu (1971) suggest that gamete selection during sperm storage may represent a largely unexplored mechanism of natural or sexual selection. Direct experimental verification of this possibility will be difficult since detection of differences in sperm storage is confounded by the production of dysfunctional sperm during spermatogenesis (Hartl and Hiraizumi 1976) and possible viability differences among zygotes.

4. Differential Sperm Viability During Storage

Myszewski and Yanders (1963) found that the recovery of *D. melanogaster* sperm carrying the recessive lethal but dominant visible mutation *Cy* decreased in comparison with recovery of sperm carrying an X-ray induced recessive lethal gene. DeVries (1964) showed that the number of sperm maintained in storage by *D. melanogaster* females varied as a function of the genotype of the male. These results suggest that interactions which occur between the female reproductive system and the male ejaculate affect not only the process of sperm storage but also the maintenance of sperm viability during storage, providing a little-appreciated mechanism for gametic selection in *Drosophila* (see also section IV).

D. Sperm Utilization

The opportunities for post-mating gametic selection in *Drosophila* do not, of course, end at the moment sperm storage is completed. There are a number of examples of differential sperm use or sperm preference (Childress and Hartl 1972) by females. We now consider the processes and factors involved in the utilization of sperm by females.

1. Patterns of Sperm Usage

The most widely used means of assessing sperm usage by *Drosophila* females is to measure female productivity (*i.e.*, number of progeny produced per female). If simultaneous determination of egg production and egg fertility is made (Pyle and

Gromko 1978), it is found that the proportion of fertile eggs laid begins to fall about 4 days after the female's initial mating. At 25°C, by 14 days after the first mating, few fertile eggs are laid. If females are allowed to remate, the portion of fertile eggs remains high.

Kaufmann and Demerec (1942) suggested that the efficiency of sperm usage in *Drosophila* was low, due at least in part to the use of multiple sperm to fertilize an egg (polyspermy). This hypothesis was tested by Hildreth and Lucchesi (1963), using radioactive sperm; they were unable to find evidence for polyspermy. Lefevre and Jonsson (1962a) concluded that sperm use was highly efficient, approaching 100%. They based their inference on limited data showing that maximum stored sperm numbers and progeny produced by females were about equal. They did, however, express some puzzlement as to how such high efficiency could be obtained since the female storage organs do not have morphological structures which would appear to allow regulation of sperm use. Efficient use of sperm is questioned by more recent findings of Zimmering and Fowler (1968) and others who showed that progeny/sperm ratios (number of progeny per female/number of sperm stored) are generally about 0.50 and may be affected substantially by interactions between the male ejaculate and the female reproductive tract.

Gilbert *et al.* (1981b) studied the relationship between initial sperm storage and female productivity (number of progeny per female) and elaborated a mathematical model to predict productivity curves. This model assumes that the release of sperm from storage organs is a random process which will result in substantial sperm wastage. The hypothesis of inefficient sperm use has been verified by direct counts of sperm storage at various times after mating (Gilbert 1981b). These data show that more than twice the number of sperm needed to fertilize eggs at peak productivity (100 eggs/day) may be released daily and are consistent with the low progeny/sperm ratios frequently reported for *D. melanogaster* (Peacock and Erickson 1965, Zimmering and Fowler 1968, Gilbert *et al.* 1981b).

The pattern of sperm utilization determined from direct counts of sperm stored in the ventral receptacle and spermathecae of *D. melanogaster* is shown in Fig. 3. The number of sperm remaining 4-6 days after mating is below 100, and the rate of sperm loss depends upon a variety of factors. Gilbert (1981b) finds a rapid initial loss of sperm at 24-48 h postmating, and at this time sperm can often be found in the uterus. In *D. melanogaster,* the high proportion of fertile eggs produced for the first 3-4 days after mating is not influenced by the number of sperm transferred to females (Kaufmann and Demerec 1942, Garcia-Bellido 1964) or the presence or absence of spermathecae (Anderson 1945, Bouletreau-Merle 1977). These observations suggest that sperm from the ventral receptacle are initially used for fertilization, and at 3-4 days postmating there is a transition to sperm stored in the spermathecae (Nonidez 1920, Fowler 1973). The lower rate of fertile egg production during this period may result from loss of sperm in the longer route from the spermathecae to the egg micropyle (Gilbert 1981b). The proximity of the opening

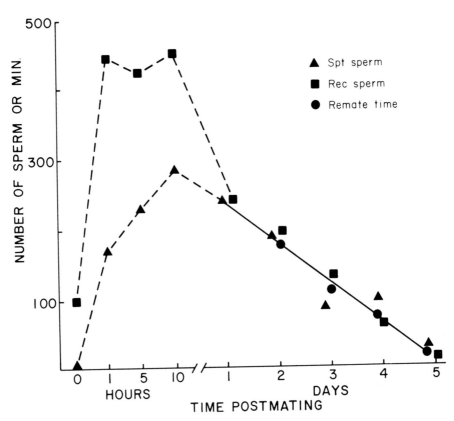

Fig. 3. Average number of sperm stored in the ventral receptacle (Rec) and spermathecae (Spt) and the average number of minutes to remating for *Drosophila melanogaster* females as a function of time from first mating. The solid line represents a highly significant identical regression of all three variables on day postmating.

of the ventral receptacle to the egg micropyle and the apparent release of rather larger numbers of sperm from the receptacle within the first 2-3 days postmating seems to ensure a high degree of egg fertility. The spermathecae appear to be specialized for long-term storage of sperm and are surrounded by an epithelium that produces a laminar material transferred into the organ's lumen (Filosi and Perotti 1975). Females genetically (Anderson 1945) or surgically (Bouletreau-Merle 1977) deprived of spermathecae lose fertility rapidly 3-4 days after mating, even though (nonmotile) sperm remain in the receptacle (Anderson 1945). Given an opportunity, normal females remate when their productivity begins to decrease because of production of infertile eggs (Pyle and Gromko 1978).

Patterson's (1954) examination of storage in other *Drosophila* species reveals a pattern of sperm use similar to that of *D. melanogaster*. In several species, sperm are lost from the ventral receptacle before they are lost from the spermathecae. *D. virilis, D. americana,* and *D. novamexicana* females dissected 6-10 days postmating were still fertilizing about 30% of the eggs laid but had sperm remaining only in the spermathecae.

2. Factors Affecting Sperm Use

Morphological studies of the sperm storage organs have revealed no structures which might allow the female to directly control sperm use. The ventral receptacle is surrounded by a layer of muscle and is innervated (Blaney 1970), but there is no evidence of a sphincter. There are no sphincters or other muscles associated with the spermathecae.

Several investigations have reported that sperm loss occurs when females are prevented from laying eggs by cold treatment or by the use of a non-protein medium (Lefevre and Jonsson 1962b; Trosko and Yanders 1963; Olivieri *et al.* 1970; Gilbert and Richmond, unpubl. data). These results suggest that the exit of sperm from storage organs may be at least partly independent of egg laying as suggested by Gilbert *et al.* (1981b).

The first indication of the physiological basis of sperm usage in *Drosophila* has been provided by analysis of the function of the seminal fluid enzyme, EST 6, in *D. melanogaster.* Richmond *et al.* (1980) and Gilbert *et al.* (1981a) showed that females inseminated by males having active EST 6 remated sooner than females inseminated by males carrying a null allele of *Est 6*. Gilbert *et al.* (1981b) and Gilbert (1981b) have shown that the rate of sperm loss from the ventral receptacle is dependent upon the EST 6 type of a female's mate. Females that receive an active EST 6 from their mates lose sperm at a significantly higher rate than females that do not receive EST 6. There is a strong correlation between the timing of female remating and the number of sperm remaining in the female's ventral receptacle (see section III). These observations support the hypothesis that EST 6 affects the timing of female remating by influencing the rate of sperm loss from the ventral receptacle.

III. REPEATED MATING

Selection arising from physical interactions between the ejaculates of two different males will occur if females remate before previously stored sperm are exhausted. When this happens, sperm from the two males are said to be in competition, and selection will favor adaptations of males which increase the probability that their own sperm will be used to fertilize the female's eggs (Parker 1970).

In this section, we review evidence for the occurrence of repeated mating, its effects on the reproductive success of both sexes, mechanisms regulating the sexual receptivity of mated females, and the genetic basis for repeated mating.

A. Occurrence

1. Natural Populations

That females remate in the wild and that they do so before sperm from previous matings is exhausted has been demonstrated in several species of *Drosophila*. The evidence comes from comparing karyotypes or electrophoretic genotypes of females with the genotypes of their offspring. When the number of alleles or the number of gametes determined to be of paternal origin exceeds two, the inference of multiple paternity is inescapable. Richmond (1976) suggested that this sort of evidence for concurrent multiple paternity was not strong since field studies rely on the use of banana baits, that might create artificially high densities of flies. However, several studies have explicitly controlled for this (Milkman and Zeitler 1974, pers. comm.; Cobbs 1977; Gromko *et al.* 1980; Levine *et al.* 1980). Thus, concurrent multiple paternity has been documented in field populations of *Drosophila melanogaster* (Milkman and Zeitler 1974, Gromko *et al.* 1980, Griffiths *et al.* 1982), *D. pseudoobscura* (Dobzhansky *et al.* 1963, Anderson 1974, Cobbs 1977, Levine *et al.* 1980), *D. euronotus* (Stalker 1976b), *D. silvestris* (Craddock and Johnson 1978), *D. athabasca, D. affinis* (Gromko *et al.* 1980), and *D. subobscura* (Loukas *et al* 1981).

It is tempting to conclude that concurrent multiple paternity is a general phenomenon in *Drosophila*. However, there is considerable variation within the genus for this. It had been throught that *D. subobscura* females which were adequately inseminated at the first copulation would not remate even after the female's sperm storage organs were empty (Maynard Smith 1956); however, convincing evidence of concurrent multiple paternity in this species has since been provided (Loukas *et al.* 1981). At the other extreme, females of *D. euronotus* remate frequently, the sperm storage organs apparently unfilled by single matings (Stalker 1976b).

There is also compelling evidence of variation in the frequency of concurrent multiple paternity among populations of a single species (Levine *et al.* 1980). Populations of *D. pseudoobscura* have been characterized for remating frequencies by several workers. Cobbs' (1977) study of a California population yielded an estimate of 55.4% of females carrying sperm from more than one male, while Levine *et al.* (1980) estimate this parameter to be 93% in their Mexican population of the same species. Both values represent estimates made from a large number of wild females and an estimation procedure which corrected for detectability problems. Both studies also took precautions against the influence of baited-capture on remating frequency. Interestingly, the proportion of multiples that involved three males was much larger in the California than in the Mexican population. The variables contributing to such within-species differences are unknown but could certainly include temperature, population size and density, distribution of resources, season, and possibly genetic factors (Levine *et al.* 1980).

Although the difference between the two studies is so large as to be indicative of real interpopulation differences, there are some sampling problems that should be considered when evaluating concurrent multiple paternity estimates (Gromko *et al.* 1980). Of all the progeny produced by a female following a second mating, the proportion sired by the second male (P_2) is large, having been estimated at 0.84 for *D. athabasca* (Gromko *et al.* 1980), and between 0.86 and 0.96 for *D. melanogaster* (see section IV, C, 1). If a small number of progeny per female is sampled, there is a reasonably high probability that they will all be from one male even when the female has mated twice. The probability of sampling at least one progeny sired by each male is one minus the probability that all are sired by only one of the two males, or $1 - [P_2{}^n + (1 - P_2)^n]$, where n is the number of progeny sampled. In Fig. 4 the relationship between progeny sample size and detectability is shown for a range of values of P_2. The detection of concurrent multiple paternity requires a large sample of progeny per wild-caught female, particularly when P_2 is large.

A second problem arises because multiple inseminations can be detected only when the genotypes of the two (or more) males are different. Estimates of concurrent multiple paternity have dealt with these problems in a variety of ways. Milkman and Zeitler (1974) analyzed 20 progeny per female and corrected for detectability by estimating the frequency of multiple paternity by same-type males under the assumption of random mating. Cobbs (1977) used a maximum likelihood estimation procedure, and Levine *et al.* (1980) used a multilinear least squares criterion to correct for both these sampling problems. An additional sampling problem in the form of a small mean and a large variance in the number of progeny produced per wild female was encountered by Gromko *et al.* (1980). The estimates of concurrent multiple paternity in that case were high but had large confidence limits as well.

2. Laboratory Populations

Concurrent multiple paternity has also been found in laboratory populations of *D. melanogaster* (see section IV, C, 1), *D. pseudoobscura* (Dobzhansky and Pavlovsky 1967, Pruzan 1976, Dejianne *et al.* 1978), *D. paulistorum* (Richmond and Ehrman 1974), *D. mercatorum* (Ikeda 1974), *D. virilis* (Prout and Bundgaard 1977), *D. mojavensis,* and *D. nigrospiracula* (T. Markow, pers. comm.).

Laboratory studies allow observation of the timing of the second mating. Two studies on *D. melanogaster* have shown that remating by females is detectable

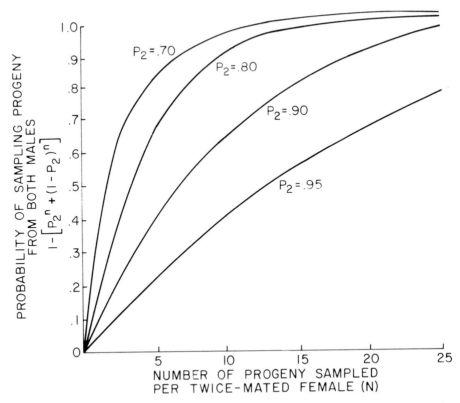

Fig. 4. The relationship between the probability of detecting progeny of both mates of a doubly inseminated female as a function of the total number of progeny sampled. A family of curves is given for several values of P_2.

as early as 2 h following the first mating (Fuerst *et al.* 1973, Bundgaard and Christiansen 1972). Furthermore, both studies demonstrate that increasing lengths of exposure of females to males leads to increased remating, with Fuerst *et al.* having recorded a maximum of 34.2% of females remating within 24 hr. Lefevre and Jonsson (1962a) report additionally that a female's probability of remating is related to the adequacy of the first male's insemination. Females first mated to males with partially depleted accessory glands remated sooner than females mated first to males with full accessory glands.

At first glance, the details of the literature on repeated mating and consequent sperm interactions in *D. melanogaster* are so variable that no simple pattern emerges. Thus, reports vary from remating by females only after sperm exhaustion (and, hence, complete sperm predominance by the second male; Manning 1962, 1967), to rapid remating by females accompanied by a lower degree of sperm predominance (Boorman and Parker 1976, Prout and Bundgaard 1977, Lefevre and

Jonsson 1962a). These results lose their apparent inconsistency when one considers the length of time males and females were allowed to interact each day in order to remate. Across studies, we find that duration of daily male-female interaction is positively correlated with frequency of remating by females. Three studies have shown this result directly (Boorman and Parker 1976, Bundgaard and Christiansen 1972, Fuerst *et al.* 1973). Pyle and Gromko (1981) speculated that remating frequency varies in this way because the female's most effective rejection response (decamping) is rendered ineffective by continuous confinement.

Pyle and Gromko (1981) also present several arguments in support of the idea that a period of male-female daily interaction of intermediate duration (about 2 h) provides a much better laboratory model of the temporal aspects of *Drosophila* courtship than does either of the more extreme period lengths (one-half hour and continuous interaction) that have been frequently used. Hardeland (1972) has shown that courtship in *D. melanogaster* is circadian, most intense in the few hours before dawn. Also, the pattern of sperm use and overlap obtained from the 2-hr daily interaction period more nearly resembles the pattern of sperm use and overlap observed in wild-caught flies. And finally, the intermediate length interaction period allows observation of the effect of first male on female receptivity, expected on theoretical grounds (Parker 1970).

Laboratory studies of enviromental factors that may affect remating are limited. The type of anesthesia used to collect virgins may affect when they first mate, and may also affect remating. Heed (1957) found that ether anesthesia delayed mating times of *D. mulleri* substantially more than carbon dioxide anesthetization. Gilbert (1981a) found the opposite effect for *D. melanogaster;* virgin females collected with carbon dioxide were significantly more reluctant to mate 3 days later than females collected with ether. Gilbert and Richmond (1982b) have found that remating is greatly delayed at 18°C compared with 25°C. While 50% of females from some laboratory strains remate in 1-2 days at 25°C (Gilbert *et al.* 1981a), at 18°C about 30 days are required before 50% of the females remate.

B. Effects on Reproductive Success

1. Effects on the Female

Much of the literature on repeated mating and sperm competition has treated the female as a passive vessel and has supposed that selection on the competitive characteristics of males and their sperm will far outweigh any selective effects of remating on females, should such exist (but see Boorman and Parker 1976). In contrast to this perspective, Walker (1980) has suggested that sperm precedence characteristics, previously interpreted in terms of intrasexual competition among males, may in fact be due primarily to sperm utilization strategies of the female.

Although there is little direct evidence to support this thesis, it seems at least plausible that factors such as genetic variability among offspring, time constraints, exposure to predation, nutritional value of ejaculate, and effects of sperm load on fertility might provide selective pressures that would influence the female's receptivity to remating. The small amount of data that does exist supports this point of view.

If a single insemination is not sufficient to fertilize all the eggs a female might normally lay, then it would seem reasonable to expect that a female's reproductive success would be increased by remating as her supply of stored sperm became low. Pyle and Gromko (1978) found such an effect in a comparison of singly and multiply mated females of *D. melanogaster*. Females given a daily, 2-h opportunity to remate did so up to a maximum of four times in 2 weeks. Remating appeared to be a regularly repeated event in the lives of females. Females that were allowed to remate produced significantly more offspring and maintained their egg production and fertility at a higher level than females not given the opportunity to remate. Furthermore, the rapid increase in second matings corresponded in time to the decrease in productivity, fecundity, and fertility of once-mated females. Hence, there is an advantage to the female associated with remating. By refilling her sperm storage organs when the number of stored sperm gets low, productivity, fecundity, and fertility can be maintained at high levels. Pruzan-Hotchkiss *et al.* (1981) showed a similar productivity increase in remated *D. pseudoobscura* females.

Other studies (Boorman and Parker 1976, Prout and Bundgaard 1977, Pulvermacher and Timner (1977) on *D. melanogaster* have shown no differences in reproductive success of singly and doubly mated females. However, these studies all used an experimental design in which females were continuously paired with males and remating occurred rapidly. Females that did not remate during 24-48 h were not considered further. Rapid remating with the consequent maximization of opportunities for sperm predominance would likely obviate any increases in reproductive success for the female.

Ikeda (1974) has described a strain of *D. mercatorum* that shows rapid rematings, a pattern apparently unusual for this species. Repeated mating reduced longevity, fecundity, and productivity of females. Ikeda postulated that the deleterious effects of remating were due to damage caused by the insemination reaction. The possibility is raised that intraspecific insemination reactions, common in the genus (Patterson and Stone 1952), could have deleterious effects on the females of other species as well, effectively selecting for reduced receptivity to remating.

2. Effects on the Male

The effects of remating by the female on the reproductive success of males has been frequently investigated. Several studies have shown that a male has higher reproductive success when he is the first and only male to mate with a female. This appears well demonstrated in *D. melanogaster;* studies showing this have made use of a variety of strains, a variety of remating procedures, and a variety of techniques for identification of first and second male progeny (Lefevre and Jonsson 1962a, Boorman and Parker 1976, Prout and Bundgaard 1977, Pulvermacher and Timner 1977, Gromko and Pyle 1978). However, these results have been interpreted as due to sperm displacement: the physical removal of previously stored sperm, and replacement by sperm of the copulating male. As discussed in section IV, C, we now feel there is reason to doubt the first male's sperm is displaced from the storage organs. In an effort to accurately describe the effect of remating on reproductive success without implying anything about the mechanism by which it occurs, we refer to this effect as "sperm predominance." Sperm predominance is intended to refer to two effects related to a repeat mating: (1) that following remating, the second male's sperm are more frequently used to fertilize eggs than are sperm from the first male; and (2) that the reproductive success of the first male is reduced. Predominance of second male sperm following a repeat mating by a female could occur by displacement at time of remating, precedence or ordered use, differential sperm viability, or some other mechanism. A precedent for adoption of this nomenclature is found in Boorman and Parker (1976).

Estimates of the degree of sperm predominance in *D. melanogaster* vary among strains and techniques. Boorman and Parker (1976) showed that predominance varied with the time between first and second matings; second males sired about 90% of progeny when females remated 4 days after the first mating, 99% when remating took place more than 4 days later.

Several studies report differences in sperm predominance between strains or marker stocks of *D. melanogaster* (Lefevre and Jonsson 1962a, Prout and Bundgaard 1977; see section IV, C). Such stock differences are also reported for *D. pseudoobscura* (Dejianne *et al.* 1978). Although only a few species have been examined in this regard, we are already led to expect a good deal of variation within the genus. In *D. euronotus,* repeated matings are frequent and without sperm predominance (Stalker 1976b).

Using a 2-h daily period of male-female interaction, Gromko and Pyle (1978) studied the effects of repeated mating on the reproductive success of both males. The degree of predominance was within the range found for different strains of this species (*i.e.,* following a repeat mating, 91% of the progeny were sired by the second male). However, the daily 2-h period of male-female interaction (which emphasizes the female's influence on the timing of repeated matings) revealed an important additional parameter: the proportion of a sperm load used by a female before she remates. We identify this parameter as P' in analogy with the parameter P_2. In the Gromko and Pyle study, females did not remate until about 78% of the sperm from the previous mating had been used. Although the second male's sperm did predominate following the repeat mating ($P_2 = .91$), that fact alone overlooks an equally important adaptation to sperm competition: the effect of the first male on the receptivity of the female (P' = .78). As we have argued, it is probably to a male's advantage to delay or prevent female remating. Selection should tend to increase P' as well as P_2. We now consider the mechanisms by which a male might delay a female's remating until a substantial proportion of his sperm is used.

C. Control of Female Receptivity to Remating

Sperm competition should result in sexual selection for characteristics that promote the use of a male's own sperm (Parker 1970). Any behavioral, morphological, or physiological adaptation that protects a male's sperm against competitive loss due to female remating will be favored by natural selection. In many insect species, such adaptations involve pre- and postcopulatory guarding behaviors or vaginal plugs. In *D. melanogaster,* this effect is achieved apparently through changes in female receptivity to remating.

1. Onset of Receptivity in Females

Factors controlling the sexual receptivity of *D. melanogaster* are numerous, and different mechanisms are involved in the onset of receptivity in young, virgin females and in the return of receptivity in mated females. Manning (1967) provided evidence that an increase in the titer of juvenile hormone released by the corpus allatum results in receptivity in young, virgin females. Originally, Manning proposed the existence of a switch-on and an independent courtship summation process controlling receptivity; either a female was available or not (switched on or off) to the cumulative effects of persistent courtship. Cook's (1973) detailed analysis of receptivity led to a revision of this hypothesis. He showed that the switch-on and the courtship summation were different manifestations of the same underlying process. Summation of male courtship stimuli to a variable threshold (without an abrupt switch mechanism) is important in the control of receptivity in nonvirgin females as well.

After copulation, changes are observed in female *D. melanogaster.* First, egg production and oviposition are stimulated. By implanting various components of the male reproductive system into virgin females and measuring oviposition rate, it has been shown that a substance produced by the male's paragonia is involved in stimulating egg production. In contrast, implants of testes, ejaculatory bulb, or seminal vesicles do not stimulate oviposition (Kummer 1960, Garcia-Bellido 1964, Merle 1968, Burnet *et al.* 1973).

A second change that occurs in females following copulation is loss of receptivity. This loss has a complicated causality and involves at least three components: a sperm factor, paragonial factors, and pheromonal factors.

2. Sperm Factors

Manning (1962, 1967) elaborated the distinction between "sperm" and "copulation" effects on remating by comparing the timing of remating in females that mated either with XO males which do not transfer sperm or with normal fertile males. Females mated to XO males began to remate 24 h after their initial mating, but females mated to normal males did not remate until 5-8 days following their initial mating. Manning found that remating time in the latter group of females was related to the number of sperm remaining in their ventral receptacles. Early remators had fewer sperm stored. Manning (1967) proposed that the delay in female remating which followed copulation with an XO male was due to factors (physiological or behavioral) associated with copulation, whereas the remating delay after copulation with a normal male was due to the presence of sperm in the female's storage organs.

Recently we have extended our studies on the effect of stored sperm on remating to determine the means by which EST 6 causes earlier remating (Gilbert 1981c). When females are paired with males for short periods daily, they usually require several days to remate. With continuous pairing, many females remate within several hours, suggesting that the timing of female remating is determined by the cumulative effects of male courtship. A female's reluctance to remate can be measured simply by recording the interval between pairing and remating. Remating time, or latency to remating with constant pairing, was recorded for four groups of females 2-5 days after their first mating. When females did remate, the copulation was interrupted within 2 min, well before sperm were transferred. Females were then dissected and sperm remaining from the first mating were counted. Thus any causal effect that stored sperm have on female reluctance to remate should appear as a significant regression of remating times on remaining sperm numbers.

Average remating times decline linearly with time from mating in the same manner as sperm number in both the receptacle and spermathecae (see Fig. 3). However, examining the relation between remating time and sperm numbers for individual females rather than averages, we find that remating time is closely related to sperm in the receptacle but is unrelated to sperm in the spermathecae. Fig. 5 shows the regression of remating times on remaining sperm for the group of females tested 2 days after their initial mating. There is a significant regression on receptacle sperm but none on spermathecal sperm. Path analysis of data for days 2-5 reveals that receptacle sperm is a highly significant cause of remating time. However, number of spermathecal sperm, day after first mating, first mating latency, and duration of copulation for the first mating are all insignificant causes (Gilbert 1981c).

The relationship between receptacle sperm number and latency to remating suggests an interaction between the receptacle and stored sperm. The receptacle is an innervated but otherwise simple sacklike tube (Miller 1950), and sperm remain

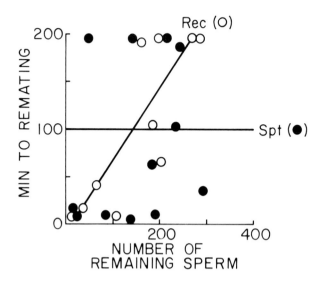

Fig. 5. The relationship between latency to remating and number of sperm in storage at remating for *Drosophila melanogaster* females initially inseminated 2 days earlier. Rec = seminal receptacle sperm; Spt = spermathecal sperm.

motile within it (Patterson 1954, Fowler 1973). If the nerves associated with the receptacle respond to the presence of sperm, this neural response may mediate the relationship between latency to remating and number of stored sperm. A chemical mediator (Manning 1967) seems less likely to exist unless the substance is a component of sperm or its level changes directly with numbers of sperm in the receptacle. We suggest that females assess their stored sperm load by directly sensing stimuli from motile sperm. This mechanism of remating control is a sufficient explanation for the earlier remating by females that lose sperm rapidly because of insemination with ejaculates containing EST 6 (Richmond *et al.* 1980; Gilbert *et al.* 1981a, b; Gilbert 1981b).

3. Paragonial Factors

Implants of paragonia from males are effective in lowering receptivity of virgin females (Merle 1968, Burnet *et al.* 1973). The loss of receptivity induced by paragonial implants is transient, resembling the copulation effect in this regard (Burnet *et al.* 1973).

Analysis of rejection responses of the female has led to a better understanding of the control of receptivity. The rejection behaviors employed by young, virgin females are qualitatively different from those employed by nonvirgins. Virgin females reject by kicking and fending (wing flicking) whereas fertilized females use ovipositor extrusion predominantly (Connolly and Cook 1973). Virgin females implanted with testes still rely on kicking and fending, whereas virgin females implanted with paragonia resemble fertilized females in that these behaviors are less frequent. However, ovipositor extrusion was not significantly different in testes- vs. paragonia-implanted virgins. It may be inferred that some factor other than that produced by the paragonia is important in ovipositor extrusion. Evidence from the behavior of female sterile mutants suggests that this second receptivity control pathway is dependent on feedback from the ovaries and may be part of the sperm effect (Burnet *et al.* 1973).

4. Pheromonal Factors

Ovipositor extrusion, per se, is not necessarily effective as a rejection behavior. Extrusion by older females, in particular, does not discourage male courtship, whereas extrusion by inseminated females does (Connolly and Cook 1973). Furthermore, extrusion is not necessary for successful rejection by fertilized females. Cook (1975) and Cook and Cook (1975) have quantified the courtship of males toward decapitated, inseminated females which do not extrude, although they kick and provide many female stimuli (Spieth 1966). Males court decapitated, inseminated females less often and do so with less persistence than they court decapitated virgin females, despite the lack of ovipositor extrusion by the decapitated females. Markow and Richmond (unpubl. data) have shown that *D. melanogaster* males initially court a virgin female when given a choice between a virgin and a recently inseminated female. These results imply the existence of a pheromone.

A courtship-stimulating pheromone produced by virgin females and young males has been identified recently (Tompkins *et al.* 1980, Venard and Jallon 1980). Using gas chromatography assays, it was shown that mated females produce less of the courtship-eliciting pheromone than do virgins (Tompkins and Hall 1981). Additionally, mated females emit a compound that inhibits courtship by males. The switch to production of an aversive pheromone occurs in females after the first 3-4 min of copulation (prior to sperm transfer). In contrast, ovipositor extrusion is not induced unless copulation has lasted 9-13 min (Tompkins and Hall 1981). Once again, we see evidence of two effects on female receptivity produced by the male during copulation. There is an early (copulation) effect resulting in some changes in female behavior, the cessation of production of a courtship-eliciting pheromone, and possibly the appearance of a volatile courtship-inhibiting substance. The second effect, appearing later in copulation, causes an increase in the frequency of extrusion by the female. The second effect may be mediated by sperm or an associated factor.

Although these are the principal data relevant to the discussion of sperm competition and the effect of the first male in decreasing the probability of remating by the female, the complete story of receptivity is certainly more complicated. For instance, there are the problems of selectivity of females (females are differentially receptive to different types of males; *e.g.,* Spiess 1970, Ehrman 1970) and the effects of conditioning or past experience on the behavior of both sexes (Pruzan 1976, Seigel and Hall 1979).

D. Genetics

The selection pressure of sperm competition can lead to two different avenues of adaptation, one which increases P', and the other which increases P_2. The evolutionary argument that describes these relationships requires genetic variability for traits which affect the magnitude of P' and of P_2. In this section, we discuss a selection experiment for remating time in *D. melanogaster,* a behavior pattern directly related to the magnitude of P'.

Pyle and Gromko (1981) successfully selected, in *D. melanogaster,* for a decrease in time between a female's first and second matings, and thereby demonstrated that the timing of remating is under genetic control in this species. The selection scheme involved exposure of once-mated females to males for 2 h each day, and selection of the first females to remate to start the next generation. The selection scheme imposed selection pressure on both sexes, and both sexes responded. Individuals of each sex from the selected line tested with the opposite sex from the control line remated sooner than did control line males and females together.

There were several indications that the response to selection was not typically polygenic. First, hybridization analysis revealed that genes for rapid remating were acting recessively. Second, genes for rapid remating were localized to the X chromosome rather than distributed throughout the genome (this conclusion has assumed that maternal effects would be very small in comparison with X chromosome effects). Third, existence of a small number of active genes was inferred from the observed rapid response to selection followed by a plateau. Most recently, we have found that the selected line has not reverted to the preselection phenotype despite relaxation of selection for over 2 years (more than double the time required for the original selection). This confirms our original notion that selection involved a small number of genes that have been fixed by selection. It is interesting in this regard that the frequent remating found in a strain of *D. mercatorum* was thought to be due to a small number of genes (Ikeda 1974).

A number of lines of evidence have led us to a rough understanding of the mechanism whereby remating time is decreased in this strain. EST 6 activity is significantly higher in the selected line than in its inbred control (Richmond *et al.* 1980). Two pieces of information indicate the increase in EST 6 activity might be accomplished through modifier or regulatory genes located on the X chromosome: (1) the X chromosome was shown to be involved in the selection response (Pyle and Gromko 1981); and (2) modifiers of EST 6 activity have been found on the X chromosome (Tepper *et al.,* pers. comm; Richmond and Tepper 1983). Comparisons of EST 6 active and null lines have shown that this enzyme increases sperm loss (Gilbert *et al.* 1981b, Gilbert 1981b). Greater sperm loss is a probable cause of the reduction in productivity and more rapid remating characteristic of the selected line (Pyle and Gromko 1981). Thus we are able to trace a probable chain of cause and effect in this strain from X-linked modifiers through increased EST 6 activity and increased sperm loss to decreased productivity and the behavioral outcome of rapid remating for which this line was originally selected.

Male sexual vigor also was increased by selection for rapid remating. Males from the selected line were quicker to begin courtship and more persistent at courting nonvirgin females than were unselected control males. This response, too, may be related to EST 6, although the details of that relationship are not yet clear (Gilbert and Richmond 1982a). Since the first report of the selection experiment, we have confirmed that this increased sexual vigor is also effective in producing more rapid second matings. Control line females mated to control line males and paired daily with selected males remated sooner than did control females mated to selected males and paired daily with control males.

First male properties may have been selected against. EST 6 activity, which was increased in the selected line, promotes more rapid sperm release and earlier rematings (Richmond *et al.* 1980, Gilbert *et al.* 1981a). Another first male property that may have been selected against is the effect the first male has on the appearance of female sex pheromones identified by Tompkins and Hall (1981). The selected line males did not respond to ovipositor extrusion by inseminated females in a typical fashion. Instead of turning away, they frequently approached the extruding female, made physical contact with the female's genital region, and shook, vibrating both flies.

Time to first mating by virgin females was not changed in the line selected for rapid rematings. From this, we inferred that receptivity of virgin females was under control of a different genetic system than the one controlling receptivity of nonvirgin females. It is interesting to note in this regard that Eastwood and Burnet (1977) found that response of males to fertilized females has a different genetic basis than the response to virgins. These findings support the thesis that the development of receptivity in virgin females is controlled by a mechanism different than that which influences the return of receptivity in nonvirgin females.

IV. MECHANISMS OF SPERM PREDOMINANCE

Following a repeat mating by a *Drosophila melanogaster* female, stored sperm do not fertilize as many eggs as they would have if the female had not remated;

sperm of the most recent male predominate. Several mechanisms have been postulated to account for sperm predominance. In this section, we consider sperm predominance not only between ejaculates but also within ejaculates, inasmuch as these two types may be mechanistically related. Most of these mechanisms have been inferred from laboratory experiments employing an array of techniques. Methods for assessing the zygotic contribution of sperm from different males or of different genotypes from one male are often complex. Accordingly, we begin with a review of methods used in the detection of sperm predominance.

A. Methods of Detection

1. Genetic Markers

Experimental studies of sperm predominance in *Drosophila* have relied primarily on readily available morphological mutants of *D. melanogaster* (Lindsley and Grell 1968). Sperm carrying suitable morphological markers allow simple and unambiguous paternity determination by producing a visible offspring phenotype for each sperm type. Use of morphological markers has two drawbacks. The first involves differences in the viabilities of zygotes associated with the marker. In their examination of ejaculate competition, Kaufmann and Demerec (1942) noted that females inseminated by one of three male types produced quite different numbers of adult offspring. These investigations showed that the differences were due to larval competition and differential zygote mortality as well as to differences in the proportion of eggs fertilized. In *Drosophila,* fertilized eggs that die before hatching can be distinguished from unfertilized eggs by their color (Stalker 1976a, Jefferson 1977). This determination can be used to separate sperm competition effects and the zygotic effects of markers.

A second complication to the use of morphological markers is that some have been shown to have pleiotropic effects on female reproductive physiology (Dobzhansky and Holz 1943, Anderson 1945); consequently, caution should be used in designing experiments and interpreting results. For instance, Johnsen and Zarrow (1971) report a brooding effect for attached XY and O sperm that is opposite in direction from that reported by Olivieri *et al.* (1970). This could be due to the influence of different markers carried by these sperm rather than the particular chromosomal constitution. A similar situation is seen in the opposite brooding effects found for *forked* and *carnation* marked sperm (see Fig. 7).

2. Labeling Sperm by Irradiation and Radioisotopes

While nonvisible genetic markers, such as chromosomal inversions or allozymes, can be used for unambiguous paternity determination, they require substantially more effort and are liable to the same zygotic and sperm effects as are visible markers. Irradiation and radioactive labeling of sperm offer attractive alternatives for studies of between-ejaculate competition since each male type can be alternately labeled and unlabeled to control for label effects without confusion with genetic effects. Irradition of adult males with large doses of X or gamma radiation (3-15 kiloroentgens) induces dominant lethal mutations in mature sperm that cause the abortion of eggs. The percentage of eggs that fail is a direct function of the level of radiation used (Riemann and Thorson 1974, Boorman and Parker 1976). Low radiation doses do not greatly affect sperm function. Demerec and Kaufmann (1941) found sperm irradiated with 5 krad in 98 of 99 eggs examined. Larger doses can affect sperm motility; consequently, sperm predominating abilities may be particularly sensitive to radiation treatment (Riemann and Thorson 1974).

The observed percentage of eggs hatching after a double mating with irradiated and normal males will give an estimate of the proportion of eggs fertilized by each male type. For this estimate, the percentage of eggs hatching from single matings with unlabeled males (n), and with labeled males (r), need to be measured independently. The observed percentage hatching from double matings (x) is then transformed into an estimate of the proportion fertilized by the labeled male (P_r) according to the formula (equivalent to Eq. 1 of Boorman and Parker 1976): $P_r = (n - x)/(n - r)$. Thus, the proportion fertilized by the labeled male is equivalent to the observed decrease in normal hatching relative to the maximum decrease if all fertilized eggs receive an irradiated sperm. The main problem with this method is the proper estimation of the hatching percentages n and r. Both decline with time from mating (Kaufmann and Demerec 1942, Boorman and Parker 1976), and n at least declines with female age (David *et al.* 1975).

Labeling sperm with radioactive isotopes is a potentially more accurate method. (Hildreth and Luchessi 1963, Villavaso 1975, Geer *et al.* 1979, Boggs and Gilbert 1979). Fertilization by radioisotope-labeled sperm can be determined relatively by scintillation counting methods or absolutely for individual eggs by autoradiographic methods (Hildreth and Lucchesi 1963, Geer *et al.* 1979). This method allows reciprocal labeling controls as with the irradiation technique and also provides unambiguous paternity determination as with genetic markers.

3. Statistical Estimation of P_2

Estimation of paternity depends also on statistical methods employed. The most concise statistic of paternity for doubly inseminated females is P_2, the proportion of the total progeny after female remating that are sired by the female's second mate (Boorman and Parker 1976). The total number of offspring produced by a female can vary substantially with environmental and genetic effects on female physiology (David *et al.* 1971, 1975; Fitz-Earle 1972; David and Bocquet 1975), and this variation can possibly affect the value of P_2, depending on treatment of the data. The usual method of calculating P_2, or other proportions of sperm recovery (*e.g., k* values for segregation distorter males; Hartl and Hiraizumi 1976) is to sum the number of progeny from different sperm types over all replicate females and divide by total progeny. This is the weighted mean proportion for two sperm types, equivalent to the mean of progeny proportions per female weighted by the female's total productivity. Another method for determining the mean is to average the proportions for each female. This gives a result independent of total productivity.

In the case where the distribution of proportions is highly skewed toward 0 or 1, as it is for P_2 values in Fig. 6, an angular mean (mean of arcsine transformed proportions) gives a better estimate of the central tendency than does either the mean of untransformed proportions or the weighted mean. The weighted and untransformed means are strongly influenced by a small number of observations at the lower tail of the distribution. The angular mean also provides a clearer separation of P_2 from female effects on total productivity. Table II lists the relative amounts of variance in weighted and angular P_2 values contributed by female, first male, and second male genotypes. Female type, which has the major influence on total productivity, contributes significantly to the weighted P_2 variance. In contrast, the angular P_2 values are independent of female influence in this experiment.

B. Intra-ejaculate Sperm Predominance

Intra-ejaculate sperm predominance is identified primarily through the observation of brooding effects—changes in the recovery of one sperm type relative to another with time after mating. While any deviation from expected segregation ratios, such as deviation from a 1/1 zygotic sex ratio, may be due to sperm predominance, it may also be due to the differential production of X- and Y-carrying sperm during spermatogenesis. A brooding effect indicates a change in the relative use of sperm that have already been transferred to and stored by the female.

Several mechanisms have been proposed to account for observed nonrandom sperm use within an ejaculate. These include differences in sperm motility (Olivieri *et al.* 1970), preference (Childress and Hartl 1972), dysfunction (Zimmering 1976), positional precedence of sperm types in the storage organs, differences in sperm viability, storage or release from storage, penetration of the egg, and functioning within the egg.

Some of the hypotheses that have been proposed to account for intra-ejaculate and inter-ejaculate predominance in *Drosophila* have been falsified. One of these is the control of sperm function by the haploid genome of mature sperm. *D. melanogaster* sperm lacking each of the chromosomes, separately or in combination, up to and including sperm lacking all but the minor fourth chromosome, can fertilize eggs (Muller and Settles 1927, McCloskey 1966, Lindsley and Grell 1969), although the normal chromosome complement is required up to the second meiotic division in

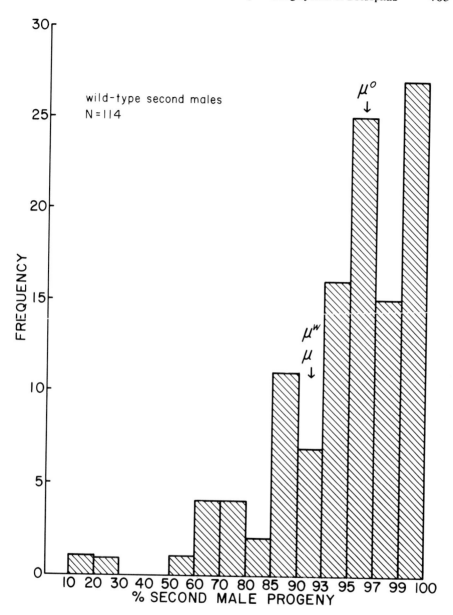

Fig. 6. The distribution of % second male progeny, P_2, for *Drosophila melanogaster* females using marked males as first and wild-type males as second mates. The means of the distribution are: μ^W (weighted mean) = 0.927 ± 0.004; μ (arithmetic mean) = 0.926 ± 0.012; and μ° (angular mean) = 0.956 ± 0.008. Data are from Gilbert and Richmond (1981) and unpublished observations. Because the distribution of P_2 is highly skewed toward 100%, the abscissa is expanded at the upper end to show the detail in that range of the distribution.

TABLE II

Relative Variance Contribution of Female, First Male,
and Second Male Genotypes to Total Progeny Production
and P_2 Values of Doubly Inseminated *D. melanogaster* Females.
Variance components are given as % of total mean square (% V)
and mean square significance level (p); from Gilbert and Richmond (1981)

| | Total Productivity | | Weighted P_2 | | Angular P_2 | |
Source	% V	p	% V	p	% V	p
Female	60.9	<.001	27.4	<.025	1.0	ns
First male	18.1	<.025	24.9	<.005	30.4	<.005
Second male	15.9	<.050	42.9	<.001	62.7	<.001
Error	5.0		4.8		5.9	

spermatogenesis. Olivieri and Olivieri (1965) and Hennig (1967) were unable to detect tritiated uridine incorporated at any stage of spermatogenesis beyond the primary spermatocyte stage, thus corroborating the absence of postmeiotic gene activity in sperm.

1. Evidence from Naturally Occurring Genetic Variants

Evidence that sperm genotypes functioning during early spermatogenesis lead to later differences between sperm phenotypes has been found through the investigation of unequal sperm recovery ratios in phenomena such as meiotic drive (Sandler and Novitski 1957). Unequal sperm recovery ratios from males heterozygous for meiotic drive chromosomes are due to differential function of the heterogametes these males produce. Deviations from the expected 1/1 zygotic sex ratio caused by differential sperm function also depend on male strain (Hanks 1965, 1969), male age (Yanders 1965, Mange 1970), female age (Hannah 1955), and sperm storage time in the female (Mange 1970). This storage time, or brooding effect, indicates that X- and Y-bearing sperm function differently in the female or are differentially used by the female.

Natural populations of several *Drosophila* species are polymorphic for alleles at specific loci that lead to highly deviant sex ratios in the progeny of carrier males (Zimmering *et al.* 1970b, Zimmering 1976). In the case of sex ratio alleles in *D. pseudoobscura* (Policansky and Ellison 1970, and "recovery disrupter" in *D. melanogaster* (Zimmering 1976), the cause of deviant sex ratios is reduced production of Y-bearing sperm. For *sxr* in *D. simulans* (Faulhaber 1967) and *sr* and *msr* in *D. affinis* (Novitski 1947), there is no apparent loss in sperm production but rather a deficiency in fertilizing ability of Y (for *Sxr sr*) or X (for *msr*) sperm, although they are transferred to and stored in the female ventral receptacle.

Faulhaber (1967) found cytological abnormalities of the sperm head in *sxr* males, although such sperm were normally transferred to and stored by females.

Segregation distorter *(SD)* is another naturally occurring variant in many *D. melanogaster* populations that leads to loss in production of, and poorer function of *SD+* sperm in heterozygous males (Hartl and Hiraizumi 1976, Hauschteck-Jungen and Hartl 1978). The recovery ratio of *SD* and *SD+* sperm changes with many factors including storage time in the female, male age, male background genotype, female genotype, temperature during spermatogenesis, and temperature during mature sperm activity (Hartl 1973, Hartl and Hiraizumi 1976). While part of the mechanism of *SD* action is differential production of sperm, the existence of brooding effects, temperature effects on mature sperm, and effects of female storage indicate that differential sperm function is also part of *SD* action. The brooding effect, an increased recovery of *SD* sperm with storage time, occurs for females that have protracted preoviposition as well as those that oviposit normally (Hartl 1973).

2. Evidence from Artificial Variants

Several artifically induced genetic variants in *D. melanogaster* also cause differential sperm function. Denell and Judd (1969) tested the segregation ratios of males heterozygous for the markers *cn* and *bw* when mated to females of different genotypes. They found a small but significant effect of female genotype on the male segregation ratio. This result could arise because of differential sperm storage and/or utilization.

Several investigators have shown that results similar to those of Denell and Judd are probably due to non-random utilization rather than differential storage of sperm. In all these cases, changes in the utilization of two genetically different kinds of sperm have been observed by inseminating females with males heterozygous for various mutants and examining the ratio of progeny produced by these females at various periods after mating. Olivieri *et al.* (1970) mated normal females to males that produce equal proportions of sperm carrying an attached XY chromosome or are nullisomic for the sex chromosomes. These two kinds of sperm are not used in equal proportion through egg laying. Rather, there is a gradual increase in the proportion of nullisomic sperm with time after insemination. Johnsen and Zarrow (1971) conducted a similar experiment, but found an increase in the propagation of attached XY-carrying sperm. The difference between these two experiments may be due to the use of different markers. The effect noted in these experiments is probably not due to differential viability of stored sperm, because females forced to store sperm for 5 days and then allowed to oviposit were not different in their brooding effects from females not forced to store sperm. Mange (1970) also found that the sex ratio in the progeny of females changes as a function of time from mating. The results of all these investigations are best interpreted in terms of subtle functional differences between genetically different sperm. In these experiments, the female appears to be a passive carrier of sperm.

It is also possible that the female discriminates against stored sperm types, and such a situation has been discovered by Childress and Hartl (1972). They used males heterozygous for a genetically marked reciprocal translocation and found that progeny ratios shifted with time after mating in one of four strains of females used. The hypothesis that these females are capable of recognizing sperm types was suggested by showing that females previously exposed to translocation-carrying sperm show an enhanced ability to discriminate against that sperm in a second mating.

All of the cases of non-random sperm utilization depend upon genetically based differences between haploid sperm. A significant body of work (reviewed by Lindsley and Grell 1969) argues against the functioning of genes in *Drosophila* sperm, but several others have suggested that the unequal partitioning of non-chromosomal materials at meiosis might result in differences among the products of a meiotic event (Olivieri *et al.* 1970). Childress and Hartl (1972) suggest that their results with translocation-bearing sperm are reminiscent of the mammalian immune response. This fascinating possibility had been envisioned previously by DeVries (1964).

These several examples indicate that intra-ejaculate predominance occurs between a variety of sperm genotypes. The brooding effects which occur with or without enforced delay of oviposition could result from at least three mechanisms: differences in sperm viability, motility, or storage location. Olivieri *et al.* (1970) suggested that sperm recovered more frequently in early broods are more motile and, as a consequence, are released from storage earliest, fertilizing predominantly the first-laid eggs. Subsequent eggs are fertilized by the less motile sperm, which remain in storage longer. This hypothesis is supported by evidence that sperm release from storage in *D. melanogaster* may be affected by sperm motility (section II, D) and occurs even in the absence of oviposition. A second, non-exclusive mechanism that may produce brooding effects is preferential storage of sperm types in the ventral receptacle or spermathecae. There is evidence that sperm stored in the receptacle are used first after mating, and spermathecally stored sperm are used later (section II, C, D; Gilbert 1981b).

Differential storage locations for sperm types could also account for female genotype "preference" effects as there are distinct genetic differences in storage organ morphology (Dobzhansky and Holz 1943), including differences in number of spermathecae (Sturtevant 1925, Anderson 1945). Although this evidence is indirect, it tends to support the differential storage hypothesis. Another possible mechanism of intra-ejaculate predominance is differential fertilizing ability as indicated by the storage of, but no fertilization by sperm of *D. simulans sxr* males and *D. affinis sr:msr* males. A deficit in late sperm function was identified in a male-sterile mutant studied by Geer *et al.* (1979). Sperm from these males are normally stored in the female and even penetrate eggs, but syngamy within the egg nucleus does not take place. Effects on late sperm function also may be involved in predominance.

C. Inter-ejaculate Sperm Predominance

Drosophila melanogaster and other species (Stone and Patterson 1954) often remate before sperm from their first mate has been exhausted. Consequently, a proportion of the sperm of a female's first mate never fertilizes eggs. This phenomenon has been termed sperm displacement (Lefevre and Jonsson 1962a, Prout and Bundgaard 1977), although direct evidence is lacking of the physical removal of sperm from storage organs. Genetically different males often differ in the proportion of the total number of progeny, P_2, produced by previously mated females. This observation supports the hypothesis that genetic factors affect the utilization of sperm by females.

The mechanisms which might give rise to inter-ejaculate sperm predominance include those presumed to be responsible for intra-ejaculate sperm predominance as well as sperm displacement with a penile structure (Waage 1979), a seminal fluid component (Gilbert et al. 1981b), or stimulation of female storage organ muscles (Villavaso 1975). Sperm predominance might also arise as a result of sperm incapacitation by seminal fluids. In this section, we explore the effects of genetic and environmental variables on inter-ejaculate sperm predominance and examine the evidence for the major mechanisms hypothesized to produce sperm predominance.

1. Evidence for Sperm Predominance

Fig. 6 shows a distribution of P_2 values for wild-type males; the distribution is highly skewed toward 100%, with an angular mean of 95%. A summary of literature on *D. melanogaster* reporting P_2 values is given in Table III. P_2 is listed according to the first and second male genotypes, ordered from highest to lowest P_2, and divided into four qualitative categories. The slope of P_2 on time from remating, *i.e.*, brooding effect, as well as the interval between first and second matings, is given. Several general aspects of sperm predominance are apparent from this table, despite variation in the methods used by various investigators. Four arbitrarily delimited categories of P_2 values occur: I, supernormal (1.00-0.97); II, normal (0.96-0.86); III, subnormal (0.84-0.55); and IV, abnormal (<0.50). Category II is divided into sections that result in part from different methods of calculating P_2; most of the IIa values are angular means whereas most of the IIb values are weighted means. Categories I and II contain all wild-type, wild-type double matings (ref. 5), and 7 of the 12 other matings where wild-type males are second mates.

TABLE III

Summary of Second Male Paternity (P_2) in *Drosophila melanogaster* According to First and Second Male Genotypes

Category Number	Male Genotype[a] First	Second	P_2 Mean[b]	P_2 Slope[c]	Remating Interval[d]	Reference
I						
1	+R	bw:st	1.000	0	1.5-2.5	6
2	+R	bw	1.000	0	1.5-2.5	6
3	bw	bw:st	.998	0	1.5-2.5	6
4	+L	+L	.989	0	14	5
5	cn sp	bw sp	.979	0	4-5	3
IIa						
6	f	+E6	.957	0	1-2	9
7	f	+E6	.954	0	4.6A	9
8	+L	+L	.952	0	8	5
9	car	+E6	.952	0	1.2	9
10	f	+MMS	.951	0	1.1A	9
IIb						
11	+L	+L	.939	0	10	5
12	+L	+L	.918	0	4	5
13	bw:st	bw	.909	0	1.5-2.5	6
14	w	+B	.900	0	<0.8	1
15	+H	f	.896	n.a.	6.0A	8
16	+L	+L	.881	0	6	5
17	sc v f car	+F	.870	0	<0.8	1
18	px bw sp	Cy/Gla	.861	n.a.	3-6	2
19	px bw sp	+R	.860	n.a.	3-6	2
III						
20	+R	Cy/Gla	.835	n.a.	3-6	2
21	bw	+R	.820	+	1.5-2.5	6
22	v f3n car	+M	.813	n.a.	<1	4
23	car	f	.809	+	1-2	9
24	+MMS	f	.774	+	2.1A	9
25	Cy/Gla	px bw sp	.736	n.a.	3-6	2

No.	Wild-type[a]	Genotype	P_2[b]	Days[d]	Sign[c]	Ref.
26	+E6	f	.703	3.0A	+	9
27	+B	w	.653	<0.8	0	1
28	f	+K	.648	<1	−	7
29	+CON	f	.591	2.1A	+	9
30	+E6	car	.585	1.2	−	1
31	+B	sc v f car	.583	<0.8	+	7
32	+K	f	.583	<1	+	9
33	+E6	f	.564	1-2	+	9
34	+M	v f3n car	.561	<1	n.a.	4
IV						
35	+R	px bw sp	.463	3-6	n.a.	2
36	+S	sc v f car	.458	<0.8	+	1
37	+F	sc v f car	.436	<0.8	+	1
38	Cy/Gla	+R	.358	3.6	n.a.	2
39	bw:st	+R	.339	1.5-2.5	+	6
40	f	car	.312	1-2	0	9

[a]Wild-type (+), followed by strain designation; standard abbreviations are used for morphological mutants (Lindsley and Grell 1968).

[b]Angular mean for ref. 6 and 9, weighted mean for others; values for ref. 2 and 4 contain offspring from the pairing interval, i.e.. probably some pre-remating offspring.

[c]+ or − = sign of significant (p<.05) regression slope of P_2 on brooding interval after remating; 0 = no significant slope; n.a. = data not available.

[d]Days between matings; A = average remating time in female choice design; other values indicate maximum range in constant pairing design.

References:

1. Lobashov 1939. P_2 values were taken from Table 1; wild-types are Samara (S), Bukhara (B), and Florida (F).
2. Kaufmann and Demerec 1942. P_2 values calculated from Tables 3-8 using only females producing progeny of two male types, and progeny numbers are corrected for viability differences reported (p. 462); wild-type is Oregon R (R).
3. Meyer and Meyer 1961. P_2 values calculated from Table 1.
4. Lefevre and Jonsson 1962a. P_2 values calculated from Table 1 using only females mated to virgin males; wild-type is M56i (M).
5. Boorman and Parker 1976. Mean P_2 values taken from Table 2; P_2 slopes calculated from hatching percentages in Table 1; wild-type is L48 (L).
6. Prout and Bundgaard 1977. P_2 values estimated by averaging P_2 values for individual females on days 1-7 after remating given in Fig. 1; wild-type is Oregon R (R).
7. Pulvermacher and Timmer 1977. P_2 values calculated from Table 1 for days 2-10 after first copulation; wild-type is Berlin K (K).
8. Gromko and Pyle 1978. P_2 values calculated from Table 1; wild-type is from a population cage of heterogeneous origin (H).
9. Gilbert and Richmond 1981 and unpubl. data; wild-types are EST 6 (E6), CON, and MMS.

Griffiths *et al.* (1982) have estimated P_2 from mother-offspring data on Adh genotypes in natural populations of *D. melanogaster*. Their estimate of 0.83 for P_2 falls at the top of category III, but 95% C.I. range from 0.73 to 0.92. Categories III and IV contain 16 of the 23 matings where marker males are second mates; these subnormal P_2 values generally are not confounded with subnormal zygote viabilities. Three of the marker second males fall into the category of supernormal P_2 values, indicating that ejaculate competitive abilities can be increased as well as decreased from that of wild-type second males.

The variance components contributed by first and second males (Table II) indicate a major influence of second males on P_2. However, there are two instances in Table III where first male type has a strong effect on P_2. In these cases, the P_2 value is much lower for the second male type when competing with first male sperm of the better competitor than with sperm of the poorer one (lines 2 vs. 13, 19 vs. 38). There is also an instance of interaction between first and second male genotypes: *px bs sp* second males are less successful in competition with +R males (line 35) than with *Cy/Gla* first males (line 25), even though +R males are less successful than *Cy/Gla* males when these two types are in reciprocal competition (line 38 vs. 20).

Categories I and II also differ from categories III and IV in brooding effects. Categories I and II show no changes over time, while categories III and IV show several significant increases or decreases in P_2. The lack of slopes in the first two categories indicates that the "first in, last out" phenomenon, or increasing appearance of first male sperm after use of second male sperm, does not normally occur in *D. melanogaster*. The effect of second males in the normal categories is a constant, nearly complete elimination of first male fertilizations. This constancy is illustrated in Fig. 7 for wild-type second males (groups A-E; see also Prout and Bundgard 1977). In contrast, *forked* second males have a consistent, positive increase in P_2 values (Fig. 7, groups F-J). This *forked* male effect was also found by Pulvermacher and Timner (1977) and by Lobashov (1939) for a multiple marker stock incorporating the *forked* allele. A decline in P_2 is produced by *carnation* second males (Fig. 7, group K); this has also been found for a wild-type second male (line 28, Table III). These distinct brooding effects, found in subnormal P_2 categories, indicate greater or lesser use of second male sperm with time in storage. The lack of significant changes in P_2 for wild-type second males in categories I and II (but appearance of such changes with wild-type second males in categories III and IV) suggests that changes in relative sperm function depend on the competitive relation with first male sperm remaining in storage. This is consistent with the dependence of mean P_2 on first and second male genotype.

2. Effect of Remating Interval

Early considerations of sperm competition in *Drosophila* focused on whether second male sperm predominated completely or whether some mixing occurred.

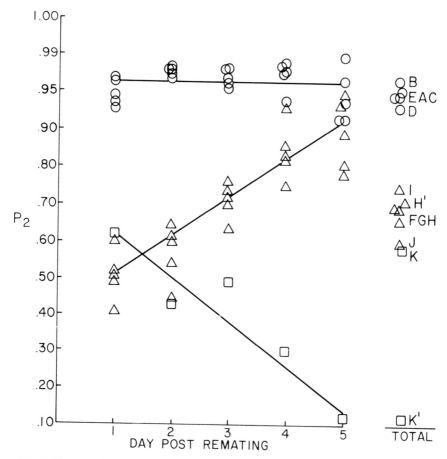

Fig. 7. Mean angular proportions of second male progeny (arcsine or angular P_2) as a function of time after mating for five wild-type second-male groups (A-E), five *forked* second-male groups (F-J), and one *carnation* second-male group ($\bigcirc$ = wild-type second male; $\triangle$ = *forked* second male; $\square$ = *carnation* second male). Lines indicate significant regressions on time for *forked* and *carnation* males, and no significant regression for wild-type males. Groups H vs. H' and K vs K' are comparisons of P_2 for homozygous (H and K) vs. heterozygous (H' and K') offspring. Data are from Gilbert and Richmond (1981) and unpublished observations.

Bateman (1948) concluded that intervals between matings could have an important influence on P_2. Assuming that other factors influencing sperm use remain constant as remating intervals increase, first male sperm would be lost (Fig. 3) and P_2 values would be expected to increase proportionally. Boorman and Parker (1976), using irradiation technique, report a small but significant increase in P_2 with increased remating intervals (0.85% per day). This corresponds to a change in P_2 from 0.92 for rematings after 4 days to 0.99 for rematings after 14 days (Table III, ref. 5). Gilbert and Richmond (1981), using morphological markers, did not find such an effect.

Whether or not there is an effect of remating interval for normally competitive ejaculates, it is clear that some genotypes are capable of very high sperm predominance even for rapid second matings. The sperm of *brown:scarlet* and *brown* second males predominate over essentially all Oregon R first male sperm remaining only 1.5-2.5 days after remating (lines 1 and 2, Table III). Judging from Fig. 3, approximately 400 first male sperm that otherwise might have fertilized eggs were unable to do so. The examples in Table III of high P_2 values with short remating intervals suggest that sperm predominance mechanisms might operate efficiently on large as well as small numbers of first male sperm. However, an alternative explanation for these data exists (Gromko, unpubl. data). The instances of high P_2 and short remating intervals reported in Table III were produced by leaving mated females and males of different strains together for 24-48 h and determining by examination of progeny which pairs remated. Females not remated were excluded from further consideration. Since remating is closely associated with the number of sperm in storage (section III, C, 2), the first females to remate are expected to be those that received fewest sperm at the first mating. In this view, rapidly remating females are not a random sample of the population, and the estimate of "400 first male sperm that otherwise might have fertilized eggs" does not apply. The high P_2 in this (nonrandom) sample of females would be produced because of small numbers of first male sperm. With the data presently available it is not possible to determine which of these hypotheses applies, and we can't say if predominance of second male sperm can occur in the presence of large numbers of first male sperm.

3. Sperm Displacement

Lefevre and Jonsson (1962a) proposed that the highly efficient mechanism of sperm predominance results from physical displacement of first male sperm from the female storage organs. This hypothesis was based on limited evidence (Kaplan *et al.* 1962) suggesting that *D. melanogaster* females could only store enough sperm to account for offspring produced by the second male. More recent information suggests an alternative hypothesis: *D. melanogaster* wild-type females may have the capacity to store enough second male sperm without displacement of first male sperm, to account for all second male offspring. Gugler *et al.* (1965) have shown that radioactively labeled second male sperm enter the ventral receptacle normally after remating. Analysis of *Drosophila* reproductive anatomy (Fig. 1) does not indicate possible mechanical displacement with a penile sperm "scoop" as is found for damselfly males (Waage 1979, this volume). The penis of *D. melanogaster* cannot reach to the anterior uterus area of the storage organs (Miller 1950). There are no spermathecal muscles controlling sperm entry and exit (Miller 1950, Filosi and Perotti 1975) that the male could stimulate to cause displacement, as is found for boll weevils (Villavaso 1975). Although the ventral receptacle is surrounded by a muscular sheath (Blaney 1970), there is no information on whether it could function in the control of sperm release.

The most direct experiment to test the hypothesis of physical displacement of stored sperm in *Drosophila* has not been performed, to our knowledge. This experiment would employ radioactively labeled first male sperm and involve the determination of labeled sperm numbers in singly and doubly mated females. However, Gilbert (1981c) has compared stored sperm numbers in (1) singly mated females immediately following insemination, (2) singly inseminated females immediately following a 1-2 min interrupted remating, and (3) doubly inseminated females following a complete second copulation. Females in the last two groups were remated 2 days after their initial mating. Under the complete sperm displacement hypothesis, average stored sperm number in the first group of females should be equal to that in the third group. However, if no displacement occurs, the mean number of sperm stored by the third group of females should be given by the sum of the means for groups (1) and (2).

The results of this experiment are inconclusive primarily due to the large individual variation in sperm storage (Gilbert 1981b). Total sperm numbers in groups (1) and (3) are not statistically different; however, some differences between the groups are apparent if receptacle sperm are considered separately and the subsets of females taking the longest time to remate are compared. This method of analysis yields a significantly larger number of sperm (375 ± 48 vs. 465 ± 39) in doubly inseminated females, as would be expected if displacement does not occur.

Another method of comparing stored sperm numbers in doubly and singly inseminated females makes use of differences in the timing of female remating to reduce the variance in stored sperm numbers. We have demonstrated previously (section III, B, 2; Fig. 5) that there is a strong, positive relationship between latency to female remating and the number of first male sperm stored in the receptacle at the time of remating. Following a second mating, this relationship should no longer hold if first mating sperm are completely displaced by the second male sperm. However, if sperm are not displaced, the relationship between latency to remating and numbers of sperm stored after remating should remain unchanged. Fig. 8 shows the relationship between remating times and numbers of stored sperm in the receptacle at remating but before transfer of second male sperm (regression line A) and after second male sperm have been transferred (regression line C). Line B gives the expected relationship following second male sperm transfer if displacement were complete. The slopes of lines A and C are not statistically different and they are different from the infinite slope of line B. These results are consistent with the hypothesis that stored receptacle sperm are not displaced at remating.

The method of analysis used in Fig. 8 assumes that differences in stored sperm numbers among females do not arise because of differences in female storage capacities. If such differences were involved, the slopes of lines A and C would be expected to be identical simply because female capacity determines stored sperm numbers. While we cannot rule out this explanation, two pieces of evidence suggest

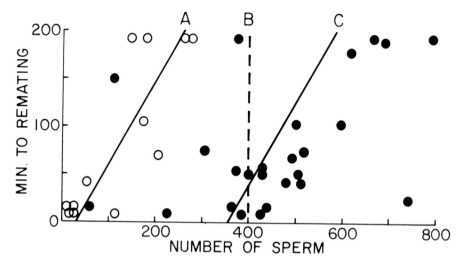

Fig. 8. The relationship between latency to remating and numbers of sperm stored in the seminal receptacle for females initially inseminated 2 days previously and analyzed for sperm storage (A) after the initiation of a second mating but prior to sperm transfer, and (C) following a complete second mating. Line B is the average storage for singly mated females.

that it may not be responsible for the results given in Fig. 8. Several investigators (cited in Section II, D, and II, E) have shown that males may have a marked effect on the number of sperm stored. Secondly, comparison of wild-type and mutant *forked* females which are known to show genetic differences in the morphology of the spermathecae (Dobzhansky and Holz 1943) does not reveal large differences in sperm storage capacities (Gilbert 1981b).

A resolution of this problem awaits the results of further experiments. The data available at present suggest that physical displacement either does not occur or at most results in only a partial loss of previously stored sperm.

4. Sperm Incapacitation

Another potential mechanism of sperm predominance in *Drosophila* is incapacitation of first male sperm by seminal fluids. There is little information on the possible occurrence of this mechanism in *Drosophila*, although the "insemination reaction" noted for many inter- and intra-specific matings may have some relation to sperm incapacitation (Wheeler 1947, Patterson and Stone 1952). Gromko and Elbin (unpubl. data) have conducted three experiments designed to test for incapacitation of stored sperm. Experiments involving first and second matings with the same male, second matings to XO males, and interrupted second matings all failed to give evidence of sperm incapacitation. Furthermore, the bulk of evidence re-reviewed here does not support a sperm incapacitation hypothesis. Daily changes in P_2 are not expected with incapacitation or any other mechanism operating at or just after mating. In particular, declining P_2 values that indicate increases in the relative function of first male sperm are hard to reconcile with such a mechanism. Also, the sometimes pronounced effect of first male genotype and interaction with second male type is more indicative of sperm interactions other than incapacitation, displacement, or other second male mechanisms.

5. Similarities Between Inter- and Intra-ejaculate Mechanisms of Sperm Predominance

There are several important features which inter- and intra-ejaculate sperm predominance have in common, suggesting that the same mechanisms may be operating. Recovery ratios for sperm carrying *sex ratio* and *SD* alleles are dependent on genotypes of the competing sperm as well as genotypes of the female storing the sperm (Hartl and Hiraizumi 1976, Zimmering 1976), as P_2 values are dependent on first as well as second male genotypes. Examples of intra- and inter-ejaculate predominance often include brooding effects which indicate changes in the relative functioning of sperm types within the female. While *sex ratio* and *SD* alleles affect the normal development of the other sperm type, second and first male sperm do not necessarily affect each other's functions. Brooding effects can result from differential motility, differential storage location, or differences in the rate of aging or loss of viability by sperm. Thus, sperm predominance in *Drosophila*, both within and between ejaculates, appears to be influenced by differential sperm function in the female. Functional identity of intra- and inter-ejaculate predominance mechanisms can be tested directly and simply by experiment. The null hypothesis predicts similarity of effects for particular mutant bearing sperm both within and between ejaculates. The existence of male and female genetic effects on both forms of ejaculate competition and the ready availability of a wide range of genetic variants in *D. melanogaster* make this species particularly suitable to genetic and physiological analysis of mechanisms of sperm predominance. The importance of sperm functions in competition for paternity shown by *Drosophila* suggests that such analysis may provide useful insights into the genetic bases of sperm competition in animals that do not store sperm.

V. EVOLUTIONARY SIGNIFICANCE

We now consider the possible impact of sperm predominance and components of the ejaculate that affect female receptivity on genetic variability within populations. We also reconsider the explanation of how these adaptations have evolved in light of the existing information on mechanisms.

A. Impact on Genetic Variability

Although sperm predominance and male effects on female receptivity are consistently treated from the evolutionary point of view, there are only a few studies that have investigated the evolutionary or population consequences of sperm competition through simulation or model building. Johnson (1977) has shown that repeat mating by females, given random encounters with males, does not lead to deviation from Hardy-Weinberg expectations. Prout and Bundgaard (1977) did more extensive modeling with the primary intent of discovering if genetically based differences in sperm predominating ability could be maintained in a population in stable equilibrium. However, they based their model on two assumptions that have since been demonstrated to be rather unrealistic: (1) females die before using their entire supply of stored sperm, and (2) females either mate once or they mate twice in rapid succession. Under these conditions, only if there is overdominance for predominating ability will genetic variability be maintained by this behavior. However, Prout and Bundgaard do suggest that directional selection for predominating ability countered by selection for something else could lead to a stable polymorphism.

The effect of the first male in promoting the resistance of the female to remating could clearly oppose selection for remating and predominating ability. Prout and Bundgaard (1977) present evidence of lines differing in the male's effect on female receptivity to remating. Although there are no published studies that treat this possibility numerically, it seems worthwhile to consider these behaviors from the vantage point of evolutionarily stable strategies (ESS). If we begin by assuming directional selection for remating and predominating ability, after time one would expect a population of males that are all effective at mating with nonvirgin females and at sperm predominance following such matings. However, these males have not been selected for their ability to cause the female to resist mating, and they would therefore be vulnerable to predominance by other males when the female remates. Hence, the population could be invaded by males that do not court nonvirgin females themselves, but mate only with virgins and then effectively resist predominance through a lowering of female receptivity. Similarly, a population of "resistors" could be invaded if the ability to resist remating wanes as the female's supply of stored sperm is used. We expect, then, that the equilibrium condition will be for some level of both strategies, which we might think of as first male, or "resistor," and subsequent male, or "remator." The remator strategy is to devote energy to persistent courtship and sperm predominance so as to mate with females irrespective of the amount of sperm they have stored, whereas the resistor males insure that sperm transferred is largely used before the female's receptivity returns. Boorman and Parker (1976) also suggested a balance of this type. If there were a genetic basis for these two patterns of behavior, one would expect that the equilibrium condition would be characterized by genetic variability. This variability would be the selected outcome of sperm competition. However, there are presently no studies that support this speculation.

B. The Evolutionary Hypothesis of Sperm Competition

One chapter (Thornhill, this volume) considers alternative hypotheses for the explanation of traits believed to have evolved in the context of sperm competition. One noteworthy problem in evaluating the appropriateness of any particular evolutionary explanation is the difficulty of constructing critical tests of such models. In particular, consistency of an observation with a model is rather weak support of that model, given the ease with which consistent evolutionary arguments can be conceived. Gould and Lewontin (1979) have suggested in this context that particular "adaptations" of organisms may not, in fact, have been directly selected. Rather, they suggest that many features of an organism owe their existence to historical, developmental, or genetical restraints on the whole organism. Gould and Lewontin suggest that it is appropriate to first determine if a particular attribute is indeed a selected character or if it is present only because of its association with other features.

The mechanisms of sperm transfer and use seen in the multiple mating system of *Drosophila* are indeed consistent with the sperm competition hypothesis. Furthermore, the substances contained in a male's ejaculate that reduce female receptivity to remating have an obvious effect on his reproductive success. There is no indication that these substances have other functions or that they are there as a result of physiological or genetical restraints on the developing system. What is striking is the effectiveness of at least two overlapping biological mechanisms originating in the male that reduce female receptivity. In constrast, the phenomenon of sperm predominance does not seem to be dependent on any of the mechanisms one might expect to have evolved in the context of sperm competition. We have reason to believe that sperm displacement does not occur; also, haploid sperm are not genetically active, and there is no evidence of sperm inactivation by components of foreign ejaculates. There is evidence that interactions between the female and the sperm she receives may produce at least a part of the phenomenon of sperm predominance. If this process, or differences in sperm viability or motility are the causes of the reproductive outcome observed following a female's repeat mating, it is possible that a high value of P_2 is not a selected response; that is, that sperm predominance has not evolved because of sperm competition. However, such statements must be taken as speculative until the mechanism by which predominance occurs is morely clearly understood.

VI. SUMMARY

Males of *Drosophila melanogaster* transfer large numbers of sperm to females during copulation, but only a small proportion of transferred sperm is stored. Storage of transferred sperm begins immediately and continues for up to several hours, with sperm stored in the ventral receptacle earlier than in the spermathecae. Sperm is first used from the ventral receptacle and later from the spermathecae. The switch to use of spermathecally stored sperm is accompanied by a drop in the percentage of fertilized eggs. Species differences in sperm storage are noted. Many of the other topics treated in this paper also show considerable variation among species of the genus *Drosophila*.

Release of sperm from storage organs is consistent with a model of random use, and this explains the observation of sperm wastage frequently observed in *D. melanogaster*. There is an interaction between the female and male's sperm, both in the process of storage and in the viability of sperm while in storage. Such male-female interactions may account for at least some observations of sperm predominance.

A male contributes sperm and a large number of other substances to the female during copulation. These substances include microtubules, a lipid, one or more peptides, and a variety of enzymes. Among the ejaculatory enzymes is EST 6, an enzyme with an apparent role in the rate of sperm release. This role may be achieved via mediation of sperm motility. As a consequence of its effect on rate of sperm release, EST 6 also affects productivity and the timing of female remating.

Females of many species of *Drosophila* remate on a regular basis throughout their reproductive lives. Such remating keeps the female's egg-laying efficiency at a high level. Some of the substances contained in a male's ejaculate have the effect of lowering female receptivity to remating and consequently increasing P', the proportion of a male's sperm used before the female remates. There are at least three separate sets of these male-mediated effects on female receptivity: a paragonial effect, a pheromonal effect, and a sperm effect. Products of the paragonia are transferred to the female within the first 3-4 min of copulation (before sperm are transferred). The paragonial secretions cause an increase in oviposition and an increase in the frequency of some of the female's courtship-rejecting behaviors. Another effect associated with events early in copulation involves pheromonal changes that result in a reduction in courtship behavior by males. The sperm effect is not seen until sperm are transferred to the female, and is characterized behaviorally by an increase in the frequency of ovipositor extrusion in response to male courtship. The sperm effect diminishes with a decrease in the supply of stored sperm. The number of sperm stored in the ventral receptacle appears to be the primary determinant of receptivity of nonvirgin females.

A hybridization analysis of a line selected for rapid remating revealed that the response to selection was not typically polygenic, involving a relatively small number of genes. An aspect of the behavioral changes in the selected line seems related to the discrete changes found in the genetic analysis, in that an extruded ovipositor was not treated as an aversive stimulus by males in the selected line. This was perhaps due to nonoccurrence of pheromonal changes.

When females of *Drosophila melanogaster* do remate, the reproductive success of all males involved is affected. A consistent observation is that the male whose sperm was in storage has lower reproductive success than if the female had not remated. Also, following the remating, the proportion of progeny sired by the most recent male (P_2) is commonly greater than 0.85 for wild-type flies. These observations have been interpreted as being due to sperm displacement. However, there seems to be reason to doubt that displacement occurs in *D. melanogaster,* and we have suggested that the observations of nonuse of stored sperm and numerical superiority of the remating male's sperm be referred to as "sperm predominance" to avoid any implication of a mechanism responsible for the observation. In addition to the hypothesis of displacement, several other hypotheses for mechanisms of sperm predominance seem unlikely in *D. melanogaster.* Haploid genomes of sperm are not genetically active, and there is no evidence of any substance in the ejaculate that incapacitates sperm of other males.

Similarities between inter- and intra-ejaculate predominance suggest that differences in sperm viability or sperm motility may result in the phenomenon of predominance. This hypothesis is also consistent with the dependence of P_2 on first as well as second male type.

If ejaculate-mediated effects on female receptivity and sperm predominane were both selected traits, it seems possible that one selected outcome of sperm competition would be the maintenance of genetic variability within populations. Although the ejaculate-mediated effects on female receptivity do seem to be traits selected in the context of sperm competition, it seems possible that sperm predominance in *Drosophila* is not. It is possible that sperm predominance results from sperm motility or viability differences brought about because of their different interactions with the female, and not because of direct interactions between sperm types themselves. The importance of sperm competition and selection among reproductive strategies is just beginning to be appreciated by evolutionary biologists. Evidence reviewed here from the genus *Drosophila* demonstrates a large and complex potential for sperm competition as a factor in sexual selection.

ACKNOWLEDGMENTS

The preparation of this paper was made immeasurably easier by the availability of G. L. Fowler's earlier review (1973) of many aspects of the reproductive biology of *Drosophila.* We are also indebted to our colleagues who generously provided copies of unpublished manuscripts. Some of the unpublished research presented here was supported by grants from the National Science Foundation and the National Institutes of Health.

420 Mark H. Gromko, Donald G. Gilbert, and Rollin C. Richmond

REFERENCES

Anderson, R. C. 1945. A study of the factors affecting fertility of lozenge females of *Drosophila melanogaster. Genetics* 30:280-296.

Anderson, W. W. 1974. Frequent multiple insemination in a natural population of *D. pseudoobscura. Am. Nat.* 108:709-711.

Bairati, A. 1966. Filamentous structures in spermatic fluid of *Drosophila melanogaster* Meig. *J. Miscrosc.* 5:265-268.

Bairati, A. 1967. Struttura ed ultrastruttura dell'apparato genitale maschile di *Drosophila melanogaster* Meig. I. Il testicolo. *Z. Zellforsch.* 76:56-99.

Bairati, A. 1968. Structure and ultrastructure of the male reproductive system in *Drosophila melanogaster* Meig. II. The genital duct and accessory glands. *Monit. Zool. Ital.* 2:105-182.

Bairati, A., and M. E. Perotti. 1970. Occurrence of a compact plug in the genital duct of *Drosophila* females after mating. *Drosophila Inf. Serv.* 45:67-68.

Bateman, A. J. 1948. Intra-sexual selection in *Drosophila. Heredity* 2:349-368.

Baumann, H. 1974a. The isolation, partial characterization and biosynthesis of the paragonial substances, PS-1 and PS-2, of *Drosophila funebris. J. Insect Physiol.* 20:2181-2194.

Baumann, H. 1974b. Biological effects of paragonial substances PS-1 and PS-2, in females of *Drosophila funebris. J. Insect Physiol.* 20:2347-2362.

Beatty, R. A., and N. S. Sidhu. 1971. Polymegaly of spermatozoan length and its genetic control in *Drosophila* species. *Proc. R. Soc. Edinb. Sect. B (Biol. Sci.)* 71:14-28.

Bischoff, W. L. 1978. Ontogeny of sorbitol dehydrogenases in *Drosophila melanogaster. Biochem. Genet.* 16:485-507.

Blaney, W. M. 1970. Some observations on the sperm tail of *D. melanogaster. Drosophila Inf. Serv.* 45:125-127.

Boggs, C. L., and L. E. Gilbert. 1979. Male contribution to egg production in butterflies: Evidence for transfer of nutrients at mating. *Science* 206:83-84.

Boorman, E., and G. A. Parker. 1976. Sperm (ejaculate) competition in *Drosophila melanogaster,* and the reproductive value of females to males in relation to female age and mating status. *Ecol. Entomol.* 1:145-155.

Bouletreau-Merle, J. 1977. Role des spermatheques dans l'utilisation du sperme et al stimulation de l'ovogenese chez *Drosophila melanogaster. J. Insect Physiol.* 23:1099-1104.

Brieger, G., and F. M. Butterworth. 1970. *Drosophila melanogaster:* Identity of male lipid in reproductive system. *Science* 167:1262.

Bundgaard, J., and F. B. Christiansen. 1972. Dynamics of polymorphisms: I. Selection components in an experimental population of *Drosophila melanogaster. Genetics* 71:439-460.

Burnet, B., K. Connolly, M. Kearney, and R. Cook. 1973. Effects of male paragonial gland secretion on sexual receptivity and courtship behaviour of female *Drosophila melanogaster. J. Insect Physiol.* 19:2421-2431.

Burnside, B. 1975. The form and arrangement of microtubules: An historical, primarily morphological review. *Ann. N.Y. Acad. Sci.* 253:14-26.

Butterworth, F. M. 1969. Lipids of *Drosophila*: A newly detected lipid in the male. *Science* 163:1356-1357.

Cavener, D. R. 1980. The genetics of male specific glucose oxidase and the identification of other unusual hexose enzymes in *Drosophila melanogaster. Biochem. Genet.* 18:929-937.

Chen, P. S., and R. Buhler. 1970. Paragonial substance (sex peptide) and other free ninhydrin-positive components in male and female adults of *Drosophila melanogaster. J. Insect Physiol.* 16:615-627.

Chen, P. S., H. M. Fales, L. Levenbook, E. A. Sokoloski, and H. J. C. Yeh. 1977. Isolation and characterization of a unique galactoside from male *Drosophila melanogaster. Biochemistry* 16:4080-4085.

Childress, D., and D. L. Hartl. 1972. Sperm preference in *Drosophila melanogaster. Genetics* 71:417-427.

Cobbs, G. 1977. Multiple insemination and male sexual selection in natural populations of *Drosophila pseudoobscura. Am. Nat.* 111:641-656.

Connolly, K. J., and R. Cook. 1973. Rejection responses by female *Drosophila melanogaster:* Their ontogeny, causality, and effects upon the behaviour of the courting male. *Behaviour* 44:142-166.

Cook, R. M. 1973. Physiological factors in the courtship processing of *Drosophila melanogaster. J. Insect Physiol.* 19:397-406.

Cook, R. M. 1975. Courtship of *Drosophila melanogaster:* Rejection without extrusion. *Behaviour* 52:155-171.

Cook, R. M., and A. Cook. 1975. The attractiveness to males of female *Drosophila melanogaster:* Effects of mating, age and diet. *Anim. Behav.* 23:521-526.

Craddock, E. M., and W. E. Johnson. 1978. Multiple insemination in natural populations of *Drosophila silvestris. Drosophila Inf. Serv.* 53:138-139.

David, J. R., and C. Bocquet. 1975. Evolution in a cosmopolitan species: Genetic latitudinal clines in *Drosophila melanogaster* wild populations. *Experientia* 31:164-166.

David, J., Y. Cohet, and P. Fouillet. 1975. The variability between individuals as a measure of senescence: A study of the number of eggs laid and the percentage of hatched eggs in the case of *Drosophila melanogaster. Exp. Gerontol.* 10:17-25.

David, J., J. van Herrewege, and P. Fouillet. 1971. Quantitative underfeeding of *Drosophila:* Effects on adult longevity and fecundity. *Exp. Gerontol.* 6:249-257.

Dejianne, D., A. Pruzan, and S. H. Faro. 1978. Sperm competition in *Drosophila pseudoobscura. Behav. Genet.* 8:544.

Demerec, M., and B. P. Kaufmann. 1941. Time required for *Drosophila* males to exhaust the supply of mature sperm. *Am. Nat.* 75:366-379.

Denell, R. E., and B. H. Judd. 1969. Segregation distorter in *Drosophila* males: An effect of female genotype on recovery. *Mol. Gen. Genet.* 105:262-274.

DeVries, J. K. 1964. Insemination and sperm storage in *Drosophila melanogaster. Evolution* 18:271-282.

Dobzhansky, Th., and A. M. Holz. 1943. A re-examination of the problem of manifold effects of genes in *Drosophila melanogaster. Genetics* 28:295-303.

Dobzhansky, Th., and O. Pavlovsky. 1967. Repeated mating and sperm mixing in *Drosophila pseudoobscura. Am. Nat.* 101:527-533.

Dobzhansky, T., B. Spassky, and T. Tidwell. 1963. Genetics of natural populations. XXXII. Inbreeding and the mutational and balanced genetic loads in natural populations of *Drosophila pseudoobscura. Genetics* 48:361-373.

Eastwood, L., and B. Burnet. 1977. Courtship latency in male *Drosophila melanogaster. Behav. Genet.* 7:359-372.

Ehrman, L. 1970. The mating advantage of rare males in *Drosophila. Proc. Natl. Acad. Sci. U.S.A.* 65:345-348.

Faulhaber, S. H. 1967. An abnormal sex ratio in *Drosophila simulans. Genetics* 56:189-213.

Filosi, M., and M. E. Perotti. 1975. Fine structure of the spermatheca of *Drosophila melanogaster* Meig. *J. Submicrosc. Cytol.* 7:259-270.

Fitz-Earle, M. 1972. Quantitative genetics of fertility. IV. Chromosomes affecting egg production in *Drosophila melanogaster. Can. J. Genet. Cytol.* 14:147-156.

Fowler, G. L. 1973. Some aspects of the reproductive biology of *Drosophila:* Sperm transfer, sperm storage, and sperm utilization. *Adv. Genet.* 17:293-360.

Fowler, G. L., K. E. Eroshevich, and S. Zimmering. 1968. Distribution of sperm in the storage organs of the *Drosophila melanogaster* female at various levels of insemination. *Mol. Gen. Genet.* 101:120-122.

Fuerst, P. A., W. W. Pendlebury, and J. F. Kidwell. 1973. Propensity for multiple mating in *Drosophila melanogaster* females. *Evolution* 27:265-268.

Garcia-Bellido, A. 1964. Das Sekret der Paragonien als Stimulus der Fekunditat bei Weibchen von *Drosophila melanogaster. Z. Naturforsch. Teil. B* 19:491-495.

Geer, B. W., J. T. Bowman, and B. T. Tyl. 1979. A gene necessary for late sperm function in *Drosophila melanogaster. J. Exp. Zool.* 209:387-393.

Gilbert, D. G. 1981a. Effects of CO_2 vs. ether on two mating behavior components of *D. melanogaster. Drosophila Inf. Serv.* 56:45-46.

Gilbert, D. G. 1981b. Ejaculate esterase 6 and initial sperm use by female *Drosophila melanogaster. J. Insect Physiol.* 27:641-650.

Gilbert, D. G. 1981c. Function and adaptive significance of esterase 6 allozymes in *Drosophila melanogaster* reproduction. Ph.D. Dissertation, Indiana University, Bloomington, IN.

Gilbert, D. G., and R. C. Richmond. 1981. Studies of esterase 6 in *Drosophila melanogaster*. VI. Ejaculate competitive abilities of males having null or active alleles. *Genetics* 97:85-94.

Gilbert, D. G., and R. C. Richmond. 1982a. Esterase 6 in *Drosophila melanogaster*: Reproductive function of active and null males at low temperature. *Proc. Natl. Acad. Sci. U.S.A.* 79: 2962-2966.

Gilbert, D. G., and R. C. Richmond. 1982b. Studies of esterase 6 in *Drosophila melanogaster*. XII. Evidence for temperature selection of *Est 6* and *Adh* alleles. *Genetica* 58:109-119.

Gilbert, D. G., R. C. Richmond, and K. B. Sheehan. 1981a. Studies of esterase 6 in *Drosophila melanogaster*. VII. The timing of remating in females inseminated by males having active or null alleles. *Behav. Genet.* 11:195-208.

Gilbert, D. G., R. C. Richmond, and K. B. Sheehan. 1981b. Studies of esterase 6 in *Drosophila melanogaster*. V. Progeny production and sperm use in females inseminated by males having active or null alleles. *Evolution* 35:21-37.

Gould, S. J., and R. C. Lewontin. 1979. The spandrels of San Marco and the Panglossian paradigm: A critique of the adaptationist programme. *Proc. R. Soc. Lond B Biol. Sci.* 205:581-598.

Griffiths, R. C., S. W. McKechnie, and J. A. McKenzie. 1981. Multiple mating and sperm displacement in a natural population of *Drosophila melanogaster*. *Theor. Appl. Genet.* 62: 89-96.

Gromko, M. H., and D. W. Pyle. 1978. Sperm competition, male fitness, and repeated mating by female *Drosophila melanogaster*. *Evolution* 32:588-593.

Gromko, M. H., K. Sheehan, and R. C. Richmond. 1980. Random mating in two species of *Drosophila*. *Am. Nat.* 115:467-479.

Gugler, H. D., W. D. Kaplan, and K. Kidd. 1965. The displacement of first-mating by second-mating sperm in the storage organs of the female. *Drosophila Inf. Serv.* 40:65.

Hanks, G. D. 1965. Are deviant sex ratios in normal strains of *Drosophila* caused by aberrant segregation? *Genetics* 52:259-266.

Hanks, G. D. 1969. A deviant sex ratio in *Drosophila melanogaster*. *Genetics* 61:595-606.

Hannah, A. 1955. The effect of aging the maternal parent upon the sex ratio in *Drosophila melanogaster*. *Z. Indukt. Abstammungs. Vererbungsl.* 86:574-599.

Hardeland, R. 1972. Species differences in the diurnal rhythmicity of courtship behavior within the *melanogaster* group of the genus *Drosophila*. *Anim. Behav.* 20:170-174.

Hartl, D. L. 1973. The mechanism of a brooding effect associated with segregation distortion in *Drosophila melanogaster*. *Genetics* 74:619-631.

Hartl, D. L., and Y. Hiraizumi. 1976. Segregation distortion. In *Genetics and Biology of Drosophila, Vol 1b*, M. Ashburner and E. Novitski (eds.), pp. 615-666. Academic Press, New York.

Hauschteck-Jungen, E., and D. L. Hartl. 1978. DNA distribution in spermatid nuclei of normal and segregation distorter males of *Drosophila melanogaster*. *Genetics* 89:15-35.

Heed, W. B. 1957. XIII. An attempt to detect hybrid matings between *D. mulleri* and *D. aldrichi* under natural conditions. *Univ. Texas Publ. Genet.* 5721:182-185.

Hennig, W. 1967. Untersuchungen zur struktur und Funktion les Lampbursten-Y-chromosoms in der Spermatogenese von *Drosophila*. *Chromosoma (Berl.)* 22:294-357.

Hildreth, P. E., and J. C. Lucchesi. 1963. Fertilization in *Drosophila*. I. Evidence for the regular occurrence of monospermy. *Dev. Biol.* 6:262-278.

Hinton, H. E. 1964. Sperm transfer in insects and the evolution of haemocoelic insemination. *Symp. R. Entomol. Soc. Lond* 2:95-107.

Ikeda, H. 1974. Multiple copulation: An abnormal mating behaviour which deleteriously affects fitness in *Drosophila mercatorum*. *Mem. Ehime Univ., Sci., Ser. B (Biol.)* 3:18-28.

Iyenger, S. V., and R. M. Baker. 1962. The influence of temperature on the pattern of insemination by *Drosophila* males. *Genetics* 47:963-964.

Jallon, J.-M., C. Anthoy, and O. Benamar. 1981. Un anti-aphrodisiaque produit par les males de *Drosophila melanogaster* et transfere aux femelles los de la copulation. *C. R. Acad. Sci. Paris* 292:1147-1149.

Jefferson, M. C. 1977. Breeding biology of *Drosophila pachea* and its relatives. Ph.D. Dissertation, University of Arizona, Tucson, AZ.

Johnsen, R. C., and S. Zarrow. 1971. Sperm competition in the *Drosophila* female. *Mol. Gen. Genet.* 110:36-39.

Johnson, C. 1977. The use of gene frequencies in estimating the mean number of mates in a multiple-mate and stored-sperm system of mating. *Theor. Appl. Genet.* 49:181-185.

Johnson, F. M., and S. Bealle. 1968. Isozyme variability in species of the genus *Drosophila*. V. Ejaculatory bulb esterases in *Drosophila* phylogeny. *Biochem. Genet.* 2:1-18.

Kaplan, W. D., V. E. Tinderholt, and H. D. Gugler. 1962. The number of sperm present in the reproductive tracts of *D. melanogaster* females. *Drosophila Inf. Serv.* 36:82.

Kaufmann, B. P., and M. Demerec. 1942. Utilization of sperm by the female *Drosophila melanogaster*. *Am. Nat.* 76:445-469.

Kiefer, B. I. 1969. Phenotypic effects of Y chromosome mutations in *Drosophila melanogaster*. I. Spermiogenesis and sterility of *KL-1⁻* males. *Genetics* 61:157-166.

King, R. C. 1970. *Ovarian Development in Drosophila melanogaster*. Academic Press, New York.

Korochkin, L. I., E. S. Belyaeva, N. M. Matveeva, B. A. Kuzin, and O. L. Serov. 1976. Genetics of esterases in *Drosophila*. IV. Slow-migrating S-esterase in *Drosophila* of the *virilis* group. *Biochem. Genet.* 14:161-182.

Kummer, H. 1960. Experimentelle Untersuchungen zur Wirkung von Fortpflanzungs-Faktoren auf die Lebensdauer von *Drosophila melanogaster*-Weibchen. *Z. V. Gl. Physiol.* 43:642-679.

Leahy, M. G., and M. L. Lowe. 1967. Purification of the male factor increasing egg deposition in *D. melanogaster*. *Life Sci.* 6:151-156.

Lefevre, G., and L. Moore. 1967. Sperm transfer and storage. *Drosophila Inf. Serv.* 42:77.

Lefevre, G., and U. B. Jonsson. 1962a. Sperm transfer, storage, displacement, and utilization in *Drosophila melanogaster*. *Genetics* 47:1719-1736.

Lefevre, G. B., and U. B. Jonsson. 1962b. The effect of cold shock on *D. melanogaster* sperm. *Drosophila Inf. Serv.* 36:86-87.

Leopold, R. A. 1976. The role of male accessory glands in insect reproduction. *Annu. Rev. Entomol.* 21:199-221.

Levine, L., M. Asmussen, O. Olvera, J. R. Powell, M. E. de la Rosa, V. M. Salceda, M. I. Gaso, J. Guzman, and W. W. Anderson. 1980. Population genetics of Mexican *Drosophila*. V. A high rate of multiple insemination in a natural population of *Drosophila pseudoobscura*. *Am. Nat.* 116:493-503.

Lindsley, D. L., and K. T. Tokuyasu. 1980. Spermatogenesis. In *The Genetics and Biology of Drosophila, Vol. 2d*, M. Ashburner and T. R. F. Wright (eds.), pp. 225-294. Academic Press, New York.

Lindsley, D. L., and E. H. Grell. 1968. Genetic variations of *Drosophila melanogaster*. *Carnegie Inst. Wash. Publ.* 627:1-471.

Lindsley, D. L., and E. H. Grell. 1969. Spermiogenesis without chromosomes in *Drosophila melanogaster*. *Genetics Suppl.* 61:69-78.

Lobashov, M. E. 1939. Mixture of sperm in case of polyandry in *Drosophila melanogaster*. *C. R. (Doklady) Acad. Sci. U.R.S.S.* 23:827-830.

Loukas, M., Y. Vergini, and C. B. Krimbas. 1981. The genetics of *Drosophila subobscura* populations. XVIII. Multiple insemination and sperm displacement in *Drosophila subobscura*. *Genetica* 57:29-37.

MacBean, I. T., and P. A. Parsons. 1966. The genotypic control of the duration of copulation in *Drosophila melanogaster*. *Experientia* 22:101-102.

McCloskey, J. D. 1966. The problem of gene activity in the sperm of *Drosophila melanogaster*. *Am. Nat.* 100:211-218.

Mahowald, A. P., and M. P. Kambysellis. 1980. Oogenesis. In *The Genetics and Biology of Drosophila, Vol 2d*, M. Ashburner and T. R. F. Wright (eds.), pp. 141-224. Academic Press, New York.

Mange, A. P. 1970. Possible nonrandom utilization of X- and Y-bearing sperm in *Drosophila melanogaster*. *Genetics* 65:95-106.

Mann, T. 1964. *The Biochemistry of Sperm and of the Male Reproductive Tract*. Wiley, New York.

Manning, A. 1962. A sperm factor affecting the receptivity of *Drosophila melanogaster* females. *Nature* 194:252-253.

Manning, A. 1967. The control of sexual receptivity in female *Drosophila*. *Anim. Behav.* 15: 239-250.

Maynard Smith, J. 1956. Fertility, mating behavior, and sexual selection in *Drosophila subobscura*. *J. Genet.* 54:261-279.

Merle, J. 1968. Fonctionement ovarien et receptivite sexuelle de *Drosophila melanogaster* apres implantation de fragments de l'appareil genital male. *J. Insect Physiol.* 14:1159-1168.

Meyer, H. U., and E. R. Meyer. 1961. Sperm utilization from successive copulation in females of *Drosophila melanogaster*. *Drosophila Inf. Serv.* 35:90-92.

Milkman, R., and R. R. Zeitler, 1974. Concurrent multiple paternity in natural and laboratory populations of *Drosophila melanogaster*. *Genetics* 78:1191-1193.

Miller, A. 1950. The internal anatomy and histology of the imago of *Drosophila melanogaster*. In *Biology of Drosophila*, M. Demerec (ed.), Wiley, New York.

Muller, H. J., and F. Settles. 1927. The nonfunctioning of genes in spermatozoa. *Z. Indukt. Amstammungs. Vererbungsl.* 43:285-312.

Murray, R. F., Jr., and J. A. Ball. 1967. Testis-specific and sex-associated hexokinases in *Drosophila melanogaster*. *Science* 156:81-82.

Myszewski, M. E., and A. F. Yanders. 1963. The effect of storage upon the differential survival among sperm. *Drosophila Inf. Serv.* 38:35-36.

Nonidez, J. F. 1920. The internal phenomena of reproduction in *Drosophila*. *Biol. Bull.* 39: 207-230.

Novitski, E. 1947. Genetic analysis of an anomalous sex ratio condition in *Drosophila affinis*. *Genetics* 32:526-534.

Olivieri, G., G. Avallone, and L. Pica. 1970. Sperm competition and sperm loss in *Drosophila melanogaster* females fertilized by $Y^SX\cdot Y^L/O$ males. *Genetics* 64:323-335.

Olivieri, G., and A. Olivieri. 1965. Autoradiographic study of nucleic acid synthesis during spermatogenesis in *Drosophila melanogaster*. *Mutat. Res.* 2:366-380.

Parker, G. A. 1970. Sperm competition and its evolutionary consequences in the insects. *Biol. Rev.* 45:525-567.

Patterson, J. T. 1954. II. Fate of the sperm in the reproductive tract of the *Drosophila* female in homogamic matings. *Univ. Texas Publ. Genet.* 5422:19-37.

Patterson, J. T., and W. S. Stone. 1952. *Evolution in the Genus Drosophila*. Macmillan, New York.

Peacock, W. J., and J. Erickson. 1965. Segregation-distortion and regularly nonfunctional products of spermatogenesis in *Drosophila melanogaster*. *Genetics* 51:313-328.

Perotti, M. E. 1971. Microtubules as components of *Drosophila* male paragonia secretion: An electron microscope study with enzymatic tests. *J. Submicrosc. Cytol.* 3:255-282.

Policansky, D., and J. Ellison. 1970. "Sex ratio" in *D. pseudoobscura*: Spermiogenic failure. *Science* 169:888-889.

Prout, T., and J. Bundgaard. 1977. The population genetics of sperm displacement. *Genetics* 85:95-121.

Pruzan, A. 1976. Effects of age, rearing and mating experiences on frequency dependent sexual selection in *Drosophila pseudoobscura*. *Evolution* 30:130-145.

Pruzan-Hotchkiss, A., D. DeJianne, and S. H. Faro. 1981. Sperm utilization in once- and twice-mated *Drosophila pseudoobscura* females. *Am. Nat.* 188:37-45.

Pulvermacher, C., and K. Timner. 1977. Influence of double matings on the offspring of *Drosophila melanogaster*. *Drosophila Inf. Serv.* 52:149-150.

Pyle, D. W., and M. H. Gromko. 1978. Repeated mating by female *Drosophila melanogaster*: The adaptive importance. *Experientia* 34:449-450.

Pyle, D. W., and M. H. Gromko. 1981. Genetic basis for repeated mating in *Drosophila melanogaster*. *Am. Nat.* 117:133-146.

Richards, A. G. 1963. The rate of sperm locomotion in the cockroach as a function of temperature. *J. Insect Physiol.* 9:545-549.

Richmond, R. C. 1976. Frequency of multiple insemination in natural populations of *Drosophila*. *Am. Nat.* 110:485-486.

Richmond, R. C., and L. Ehrman. 1974. The incidence of repeated mating in the superspecies, *Drosophila paulistorum. Experientia* 30:489-490.

Richmond, R. C., D. G. Gilbert, K. B. Sheehan, M. H. Gromko, and F. M. Butterworth. 1980. Esterase 6 and reproduction in *Drosophila melanogaster. Science* 207:1483-1485.

Richmond, R. C., and A. Senior. 1981. Esterase 6 of *Drosophila melanogaster:* Kinetics of transfer to females, decay in females and male recovery. *J. Insect Physiol.* 27:849-853.

Richmond R. C., and C. S. Tepper. 1983. Genetic and hormonal regulation of esterase 6 activity in male *Drosophila melanogaster*. In *Isozymes: Current Topics in Biological and Medical Research.*

Riemann, J. G., and B. J. Thorson. 1974. Viability and use of sperm after irradiation of the large milkweed bug. *Ann. Entomol. Soc. Am.* 67:871-876.

Riley, R. C., and A. J. Forgash. 1967. *Drosophila melanogaster* eggshell adhesive. *J. Insect Physiol.* 13:509-517.

Sandler, L., and E. Novitski. 1957. Meiotic drive as an evolutionary force. *Am. Nat.* 91:105-110.

Sheehan, K., R. C. Richmond, and B. J. Cochrane. 1979. Studies of esterase 6 in *Drosophila melanogaster*. III. The developmental pattern and tissue distribution. *Insect Biochem.* 9: 443-450.

Siegel, R. W., and J. C. Hall. 1979. Conditioned responses in courtship behavior of normal and mutant *Drosophila. Proc. Natl. Acad. Sci. U.S.A.* 76:3430-3434.

Spiess, E. B. 1970. Mating propensity and its genetic basis in *Drosophila*. In *Essays in Evolution and Genetics in Honor of Theodosius Dobzhansky,* M. K. Hecht and W. C. Steere (eds.), pp. 315-379. Appleton-Century-Crofts, New York.

Spieth, H. T. 1952. Mating behavior within the genus *Drosophila* (Diptera). *Bull. Am. Mus. Nat. Hist.* 99:399-474.

Spieth, H. T. 1966. Drosophilid mating behaviour: The behaviour of decapitated females. *Anim. Behav.* 14:226-235.

Stalker, H. D. 1976a. Chromosome studies in wild populations of *D. melanogaster. Genetics* 82:323-347.

Stalker, H. D. 1976b. Enzymes and reproduction in natural populations of *Drosophila euronotus. Genetics* 84:375-384.

Stone, W. S., and J. T. Patterson. 1954. III. Fertilizations in multiple matings of the *virilis* group. *Univ. Texas Publ. Genet.* 5422:38-45.

Sturtevant, A. H. 1925. The seminal receptacles and accessory glands of the Diptera, with special reference to the Acalypterae. *J. N.Y. Entomol. Soc.* 33:195-215.

Tompkins, L., and J. C. Hall. 1981. The different effects on courtship of volatile compounds from mated and virgin *Drosophila* females. *J. Insect Physiol.* 27:17-21.

Tompkins, L., J. C. Hall, and L. M. Hall. 1980. Courtship-stimulating volatile compounds from normal and mutant *Drosophila. J. Insect Physiol.* 26:689-697.

Trosko, J. E., and A. F. Yanders. 1963. Cold storage effect on irradiated *Drosophila* sperm. *Drosophila Inf. Serv.* 38:36-37.

Venard, R., and J.-M. Jallon. 1980. Evidence for an aphrodisiac pheromone of female *Drosophila. Experientia* 36:211-213.

Villavaso, E. J. 1975. Functions of the spermathecal muscle of the boll weevil, *Anthonomus grandis. J. Insect Physiol.* 21:1275-1278.

Waage, J. K. 1979. Dual function of the damselfly penis: Sperm removal and transfer. *Science* 203:916-918.

Walker, W. F. 1980. Sperm utilization strategies in nonsocial insects. *Am. Nat.* 115:780-799.

Wheeler, M. R. 1947. IV. The insemination reaction in intraspecific matings of *Drosophila. Univ. Texas. Publ. Genet.* 4720:78-115.

Yanders, A. F. 1963. The rate of *D. melanogaster* sperm migration in inter- and intra-strain matings. *Drosophila Inf. Serv.* 38:33-34.

Yanders, A. F. 1965. A relationship between sex ratio and paternal age in *Drosophila. Genetics* 51:481-486.

Zimmering, S. 1976. Genetic and cytogenetic aspects of altered segregation phenomena in *Drosophila*. In *The Genetics and Biology of Drosophila, Vol. 1b,* M. Ashburner and E. Novitski (eds.), pp. 569-613. Academic Press, New York.

Zimmering, S., and G. L. Fowler. 1968. Progeny:sperm ratios and nonfunctional sperm in *Drosophila melanogaster. Genet. Res.* 12:359-363.

Zimmering, S., J. M. Barnabo, J. Femino, and G. L. Fowler. 1970a. Progeny:sperm ratios and segregation-distorter in *Drosophila melanogaster. Genetica* 41:61-64.

Zimmering, S., L. Sandler, and B. Nicoletti. 1970b. Mechanisms of meiotic drive. *Ann. Rev. Genet.* 4:409-436.

11

Sperm Competition, Kinship, and Sociality in the Aculeate Hymenoptera

CHRISTOPHER K. STARR

Sperm Competition and the Evolution
of Animal Mating Systems

427

I. INTRODUCTION

The Aculeata are a subgroup of the insect order Hymenoptera. About 60,000 species are known (Brown *et al.* 1983), comprising the bees, ants, and most of the insects recognized as "wasps." The subgroup name derives from an evolutionary modification of the ovipositor into a venom-injecting device, the stinger. This structure is secondarily lost in some species. Two characteristics of the Aculeata cause the group to be of special interest: (1) Sex-determination is of the haplo-diploid type, so that diploid females arise from fertilized eggs and haploid males from unfertilized eggs (arrhenotoky); and (2) species within this group show great variation in social organization such that the full array of insect sociality is manifest in the Aculeata.[1]

This chapter examines the interplay of these two features in the context of sperm competition. I review the applicable literature, but it will quickly be seen that data on sperm competition itself are sparse and usually inexact. Hence, the chapter's principal value will be in identification of problems, formulation of hypotheses, and suggestions for needed research.

II. HYMENOPTERAN REPRODUCTIVE BIOLOGY

A. Reproductive Anatomy

In the Hymenoptera, the spermatheca and its ductwork function to determine sex by regulating release of sperm to eggs. With the exception of this and modification of the ovipositor into a stinger, aculeate reproductive anatomy can be characterized, without offense to specialists, as a generalized pterygote system. It receives little special attention in reviews of insect reproduction (*e.g.,* Engelmann 1970, Scudder 1971, Leopold 1976, Schaller 1979). Pouvreau's (1963) characterization of the intromittent organ of male *Bombus hypnorum* as powerful and complex suits the Aculeata generally, but the male genitalia rarely show any strong departure from antecedents or closely related orders in the way that those of Odonata or spiders do (Scudder 1971). A well-studied exception is found in the honey bee *Apis mellifera* (Bishop 1920, Laidlaw 1944, Fyg 1952) in which the complex intromittent organ is everted into the vagina through internal pressure and

---

[1]Appendix A provides a synthetic summary of insect social classification to acquaint those unfamiliar with the subject with the variation in social organization among the Aculeata and some terms associated with it.

cannot be recovered. Consequently, males mate only once. A similar situation is known from stingless bees (Michener 1974).

B. Sperm Structure

Structure of spermatozoa has been described for only two species, *Apis mellifera* (Rothschild 1955, Hoage and Kessel 1968, Cruz Höfling *et al.* 1970) and the fire ant *Solenopsis saevissima* (Thompson and Blum 1967). Each of these has its peculiarities, but neither differs substantially from the "representative insect spermatozoan" of Breland *et al.* (1968) or Baccetti's (1972, 1979) "typical pterygote sperm." Such sparse data permit only tentative conclusions. Significantly, sperm of other social aculeate species examined in the course of sperm-counting studies (Table I) have prompted no special comment on structure. For this reason, it seems unlikely that any of these is strikingly unique in the manner of gigantic, multi-flagellated sperm of the termite *Mastotermes darwiniensis* (Baccetti and Dallai 1977; see Sivinski, this volume, for other examples and a review of factors in the evolution of sperm form).

C. Sexual Dimorphism

While the comparative study of caste polymorphism in social insects has led to such large treatments as those of Schmidt (1974) and Oster and Wilson (1978), I find no analysis of sexual dimorphism in the Aculeata. There is certainly a great deal of variation among species. While Richards (1978), in his descriptions of New World polistine wasps, repeatedly describes the male as "very like the female," such a phrase is unlikely to be found in any treatment of ants. Indeed, sexual dimorphism is so extreme in army ants that for a great many years the males were not even recognized as ants (Schneirla 1971). This subject seems worth investigating, especially the relationship between sexual dimorphism and social features. I confine myself here to making six limited statements about secondary sexual characteristics in aculeates.

1. Males have not developed bizarre sexual ornaments or weaponry, such as are found among butterflies (Darwin 1874), beetles (Eberhard 1980), or a great many vertebrates. Striking sexual dimorphism is not characteristic of most solitary species (exceptions in the Tiphiidae; Given 1954). Males of many sphecid wasps and megachilid bees have the forelegs expanded into a disk for grasping the female (Richards 1927), and smaller leg structures of similar presumed function are known in various families.
2. Males lack the stinger and tend to have longer antennae. In some few species, such as the wasps *Monobia quadridens* (Rau 1934) and *Eustenogaster luzonensis* (unpubl. data), males can make effective defensive use of sharp structures on the intromittent organ, though the effect is purely physical and venom injection is unknown. Hypothetical functions of differences in male and female antennae have not been demonstrated.

TABLE I

Numbers of Matings by Females in Aculeata[a]

Species	Family	No. of Matings	Method	Reference
SOLITARY				
Trypoxylon politum	SP	often 1	O	Brockmann 1980
Hemithynnus hyalinatus	TI	>1	O	Ridsdill Smith 1970
Andrena vaga	AD	1	O	Vleugel 1947
Nomadopsis puellae	AD	>1	O	Alcock 1980
PRIMITIVELY EUSOCIAL				
Polistes apachus and	VE	~1	A	Lester and Selander 1981
Polistes bellicosus	VE	~2-3	A	Lester and Selander 1981
Polistes exclamans	VE	1	D[b]	Gobbi 1975
Polistes versicolor				
ADVANCED EUSOCIAL				
Vespula atropilosa	VE	>1	O,D	MacDonald *et al.* 1974
Vespula germanica	VE	1	O	Marchal 1896, Thomas 1960
Vespula maculifrons	VE	>1	O	Ross 1984
Vespula pensylvanica	VE	1	O,D	MacDonald *et al.* 1974
Vespula vulgaris	VE	>1	O	Marchal 1896
Dolichovespula saxonica	VE	1	O	Marchal 1896
Apis mellifera adansoni	AP	mean 7.5	V	Kerr and Bueno 1970
Apis mellifera ligustica	AP	mean 5.3	V	Kerr and Bueno 1970

Species	Method		Marker	Reference
Scaptotrigona postica	AP	1	D	Kerr 1974
Formica bradleyi	FO(F)	>1	O	Halverson *et al.* 1976
Formica dakotensis	FO(F)	>1	O	Talbot 1971
Formica exsecta (some populations)	FO(F)	usually 1	A	Pamilo and Rosengren, 1984
Formica pergandei	FO(F)	>1	O	Kannowski and Johnson 1969
Formica polyctena	FO(F)	>1	O	R. Rosengren (unpubl. data), cited from Pamilo 1982
Formica pressilabris	FO(F)	1-2	A	Pamilo 1982
Formica sanguinea	FO(F)	often >1[c]	A	Pamilo 1982
Formica transkaucasica	FO(F)	usually 1	A	Pamilo 1982
Formica yessensis	FO(F)	>1	O	Itô and Imamura 1974
Lasius flavus	FO(F)	>1	O	Forel 1874, cited from Donisthorpe 1915
Lasius niger	FO(F)	>1	O	Imai 1965
Polyergus lucidus	FO(F)	>1	O	Marlin 1971
Mycocepurus goeldii	FO(M)	>1	O	W. D. Hamilton, pers. comm.
Myrmica rubra	FO(M)	>1	O	Donisthorpe 1915
Pogonomyrmex badius	FO(M)	>1	O	Turner 1909
Tranopelta sp.	FO(M)	1	O,D	W. D. Hamilton, pers. comm.

[a]This is a supplement to Page and Metcalf's (1982) review of data. Family and **Method** notation follow Page and Metcalf. AD = Andrenidae, AP = Apidae, FO(F) = Formicidae:Formicinae, FO(M) = Formicidae:Myrmicinae, SP = Sphecidae, TI = Tiphiidae, VE = Vespidae; A = electrophoretic studies, D = dissection (including sperm counts), O = observation, V = genetic (phenotypic) marker.

[b]Gobbi's method was somewhat less direct than other dissection methods. He compared the number of sperm in spermathecae of wild-caught females and those mated once in the laboratory and found no significant difference.

[c]Pamilo and Varvio-Aho (1979) concluded that *Formica sanguinea* in a different part of Finland is singly-mated. On the basis of the result cited here, Pamilo (1982) called the earlier result into question. An alternative explanation, consistent with the general attitude presented in this chapter, is that the difference between populations is real.

Hypothetical functions of differences in male and female antennae have not been demonstrated.

3. Males lack derived structures for nesting and provisioning. Specialized pollen-collecting structures are absent in all male bees, and in many male wasps the mandibles are smaller and simpler (unpubl. data) in a way that presumably makes them less suitable for manipulating prey and nesting materials.

4. The sexes can rarely be distinguished on the basis of gross color pattern in advanced eusocial species, but this can often be done with solitary and primitively eusocial wasps and bees (Starr 1981). Most prominent is the tendency for males to have strikingly lighter yellow or silverish faces. I often find this an easy way to sex live individuals in the field, and presumably it serves to identify sex among the animals themselves. It seems likely that face color dimorphism is involved in epigamic (intersexual) selection. The initial mating position in aculeates has the male above, with his venter to her dorsum, partners facing the same direction (Barrows 1976; J. Alcock, pers. comm.), and in this position the female should be able to see the male's face and little else of him. This is Lamb's (1922) "male ventral position," considered generalized in insects. Position changes are common after copulation has begun (J. Alcock, pers. comm.).

5. Males with wings reduced or absent are not common, though they are known from some ants (Wheeler 1910, Donisthorpe 1915). Flightless queens are common in ants (Wheeler 1910, Donisthorpe 1915, Kusnezov 1962). All mutillids and many tiphiids have wingless females (Riek 1970).

6. The general rule that females are larger than males is well established in solitary species, and in mass-provisioning solitary species an egg can usually be sexed by the size of its mass food provision (this led Fabre [1886: Chapter 3] to undertake experiments that first showed a female "knows" the sex of the egg she lays). For eusocial species the pattern is much less clear, though females appear usually to be larger. A check of 31 species of New World polistine wasps (primitively eusocial) from nine genera in Richards' (1978) revision gave a range in male/male wing length ratios of 0.89 to 1.32, with the female just slightly larger in most species. Overall body form is very similar in polistines, so wing length is a good size dimension. Differences are somewhat more marked in 22 species of vespine wasps (advanced eusocial) from three genera I have examined, with queen/male wing length ratios between 1.05 and 1.40. These figures fail to indicate the queen's greater overall robustness in vespines.

D. Mating Systems

1. Solitary Species

The study of mating in solitary aculeates has recently yielded a rich harvest of observation and theory (Alcock et al. 1978; Alcock 1979, 1980; Eickwort and Ginsberg 1980). Because intrasexual competition will usually have greater consequences for males, the focus has tended to be on male mate-procurement strategies and their relationship to the distribution of food sources, nesting sites, and potential mates, in the approach of Emlen and Oring (1977). This has most especially produced testable predictions of the presence or absence of particular territorial and searching strategies. It has not given rise to description of a general aculeate strategy, though, and is unlikely to do so. The factors and options are very diverse, which is just what makes the subject so interesting. This is not to say, though, that all options within a classification of male insect strategies (e.g., Borgia 1979) are expected to have significant currency. A consideration of basic bionomics indicates two constraints:

1. A female must hunt for brood-food and, in most species, prepare a nest. As a result, she requires unrestrained freedom of movement and should resist any male strategy based on sequestration. Sequestration would also work against his interests. It is not rare for a male to seek and mate an emerging female prior to the nesting cycle. In some cases a male may even physically assist a potential mate in emerging (Alcock *et al.* 1978). If she is to collect the resources necessary to rear their offspring, though, he must subsequently release her to possible contact with other males. Some males minimize this exposure for a time by flying in tandem with their mates (see Plateaux-Quénu 1971:166). This must exact a cost in nesting efficiency, and in no species does this habit extend throughout the nesting cycle.

2. Males are not equipped physically or behaviorally for foraging or brood-care. This is most clearly shown where nest-provisioning is concerned, as male wasps lack a stinger and robust mandibles with which to subdue prey and male bees lack pollen-collecting structures. Consequently, aculeate males cannot provide a substantial pro-teinaceous food offering, as in some Mecoptera (see Thornhill, this volume). The thyn-nine tiphiids are exceptional in that the males often gather food for their flightless mates (Given 1954, Ridsdill Smith 1970, Alcock 1981). Hymenopteran males also have no opportunity to share or take over brood care, as in belostomatine water bugs (Smith 1976a, b).

Low (1978) provides the useful equation:

reproductive effort (RE) = mating effort (ME) + parental effort (PE).

The aforementioned constraints are sufficient to show why there is virtually no male PE in solitary aculeates, as in most other insect taxa (see Smith 1980). On the one hand, males are ill equipped for PE (constraint 2). On the other hand, as emphasized by Thornhill (1979), Werren *et al.* (1980), and Gwynne (this volume), male PE is not expected where there is low paternity assurance (PA). The general ineffectiveness or high cost of efforts to isolate a female from other males (constraint 1) must be an effective barrier to PA where females remain receptive.

In some few cases these constraints have been sidestepped. Males of the sphecid wasps *Oxybelus sublatus* (Peckham *et al.* 1978) and *Trypoxylon* (*Trypargilum*) *politum* (Cross *et al.* 1975, Brockmann 1980) show strong nest-constancy. They effectively repel not only rival conspecific males but also nest parasites, and they attempt copulation each time the female returns to the nest with prey. Observa-tions by Peckham and Peckham (1905:181) on two other species of *Trypargilum* indicate that this behavior may extend throughout the subgenus. Masuda's (1939) observations on *Pison strandi* suggest a similar habit. In *T. politum,* there is some indication that males make a modest contribution to nest-building and provisioning cells (Brockmann 1980), but the value of this contribution cannot be very great. From the male's point of view, the key to this strategy must be effective insemina-tion of the female just before each egg is laid. Logically, increased PA leads to increased return from investment in any of the female's offspring (PE), while the nest-constancy necessary for such high PA at the same time facilitates PE. The interconnectedness of the two factors suggests that this particular departure from reproductive business-as-usual is close to a one-step process and little constrained by phylogenetic inertia.

2. Ants and Advanced Eusocial Bees

Mating in social wasps (Evans and West-Eberhard 1970, Guiglia 1972, Edwards 1980) and bumble bees (Pouvreau 1963, Alford 1975, Lloyd 1981) appears not to differ strikingly from that of solitary species. In the ants and the most highly social bees (stingless bees and honey bees), though, as well as the termites, mating is quite different. Pairing takes place by way of a limited number of mass swarms, often involving the new sexual individuals from many colonies over an extended area. Pairs may be formed in the air and either mate on the wing (e.g., *Apis mellifera; Atta sexdens,* Autuori 1949) or alight together and mate on the ground (most ants, Donisthorpe 1915). Termites usually land again before pairing (Nutting 1969).

In some few nonparasitic ant species (*e.g., Monomorium pharaonis,* Peacock and Baxter 1950, Peacock *et al.* 1950), flights are omitted, and mating takes place in the nest. The significance of mating flights for the great majority of species is indicated, though, by the fact that while wingless females and males are known from ants, in no species are both sexes wingless (Wheeler 1910, Donisthorpe 1915). In the extreme, mating swarms are formed just once a year. Especially in this case, they are key events in colony life, with the exit of sexuals from the nest closely controlled by the workers (Donisthorpe 1915, Kusnezov 1962). Mating options of males are consequently severely time-limited, and as a rule males die shortly after the mating flights, whether they succeed in copulation or not (Kusnezov 1962).

In addition to reducing the chance that a queen will fail to mate, synchrony of mating swarms should increase queen survivorship by swamping potential predators. In spite of the fact that ants are weak fliers and queens lose their wings soon after mating, the physical dimension of their flights can be impressive. While males of some species (*e.g., Mycocepurus goeldii,* Kerr 1961) are not known to fly far from the nest or to pair high in the air, *Solenopsis saevissima* males may be airborne for hours at altitudes of 90-150 m (Markin *et al.* 1971). Mating flights of honey bees, which fly well, are similarly rigorous. They may fly more than a kilometer to the mating site, and queens may fly swiftly for some distance high in the air, pursued by a large group of males. Levenets (1954) records a radius of 3 km for male mate-search. Drone honey bees are strong fliers, capable of carrying the queen along in flight, and they struggle aggressively with each other in the air (Gary 1963). As previously mentioned, a drone is incapable of multiply mating, so there is no time-budget choice between guarding and new mate acquisition strategies. In this situation, it makes sense for him to leave his genitalia obstructively attached to the female, and this is one explanation for the "mating sign."

Mating in *Apis mellifera* requires that the queen submit cooperatively by opening her sting chamber (Gary 1963). Her mate-choice criteria are unknown, but the strenuousness of the mating flight contains an element of "Whoever can catch me can have me" or what we might call the "Dark Dolores strategy" (Runyon 1929), in which the female sets a vigorous course criterion and makes no distinction among finishers. These physical tests led Snell (1932) to the hypothesis that aculeate mating flights are a eugenic system that discriminates against weak hemizygous males, whose weak genes are always phenotypically exposed due to their haploidy. It is interesting to note that Cohen (1969, 1975), in trying to account for the typically very large ratio of sperm/egg numbers in animals, hypothesizes an analogous obstacle-course function for the female genital tract.

E. Sexual Conflict

Males in the Aculeata are often maligned in science and fable as drones, characterized as "greedy, cowardly, and stupid" (Schmitt 1920). In fact, they are probably no more self-serving and shiftless than members of their gender in other taxa, but female aculeates tend to have such clearly defined and apparently purposeful activities that males suffer by comparison. Males do behave quite differently, and since the classic paper of Trivers and Willard (1973), it is a central tenet of sociobiology that members of a breeding pair have only a partial community of reproductive interests and that the reproductive behavior of each is shaped by conflict. This is hinted at above, in the discussion of male RE. How does **sexual conflict** (Parker 1979) affect the number of matings a female undertakes and therefore the degree of sperm competition?

Selection on males is expected to favor those that maximally inseminate, while for females complete insemination (*i.e.,* obtaining a full spermatheca) should rarely be a problem. In the primitive state, a virgin male aculeate starts with enough sperm to completely inseminate more than one female (Garófalo 1980), and the typical male presumably lives long enough to replenish his semen stores a number of times, if spermatogenesis continues. In the extreme he may never need to relinquish a mating for lack of sperm and may generally disregard the cost of semen (but see Dewsbury 1982).

A completely inseminated female has very different interests. Alcock *et al.* (1978) evaluated possible benefits a female might derive from additional matings and concluded that in solitary aculeates there is no obvious advantage (see also Parker, this volume; Knowlton and Greenwell, this volume; but see discussion of this question for advanced eusocial species below). On the other hand, there are at least two serious disadvantages to multiple mating. These are among Daly's (1978) six costs of mating (some similar points were made by Richards 1927).

1. **Courting and mating pairs have increased vulnerability to predation.** This must be almost universal among animals, due to the near impossibility of reaching only conspecifics with sexual advertisement. An example of the risks due to courtship is found in digger wasps: Territory-holding males (species) suffer greater predation from beewolves, *Philanthus* (Gwynne and O'Neill 1980). Mating increases the vulnerability of the pair because of reduced mobility, and the stinger is unavailable for defense during copulation.

2. **Persistent courtship by males may directly lower female fitness.** Females of solitary species are most subject to harassment while searching for nest sites, or while collecting nest materials and brood food. These are exactly the times when freedom of movement is most valuable. Loss of a host spider by a pompilid female, for example, represents forfeiture of the one offspring that would have used that host for its entire development.

Mated females of some species successfully resist pursuant males with minimal time expenditure. Where this occurs, male net reward from persistence may be so low that selection has favored alternative strategies (Alcock *et al.* 1978). Walker (1980) finds a low probability of sperm displacement in species where females typically mate only once and suggests that it is in this way she discourages subsequent males.

In some cases harassment by males may be so intense or the female's ability to resist so low that her best tactic is to submit and get it over with. This is the cause adduced by Alcock *et al.* (1977) for the tendency of females of the bee *Anthidium maculosum* to mate repeatedly. *Anthidium* is one of few genera with males larger than females. Females of other species apparently employ this tactic as well (*e.g.,* Alcock *et al.* 1978). In the "arms race" analogy of sexual conflict over mating, *A. maculosum* males appear to be winning the race. In most eusocial species, on the other hand, females have a distinct advantage in the arms race, as reproductive females (queens) remain on or in the nest, removed from activities leading to sexual harassment. Even where queens found the colony alone and must undertake all construction and brood-care until the first workers emerge, there are sometimes ways to minimize harassment.

In very seasonal habitats the mating and nesting seasons may be separated. All temperate social wasps (Evans and West-Eberhard 1970) and bumble bees (Heinrich 1979) overwinter as adults. In all but one of these (*Polistes annularis;* Brimley 1908, Hermann *et al.* 1974), mating takes place in the fall, while nesting is in the spring. In tropical wasps this factor is probably very rare.

In the great majority of nonparasitic ant species, colony-founding is by a single queen, working alone. Among these, most practice **claustral** founding (*i.e.,* the queen sequesters herself and rears the first worker cohort from breakdown products of her wing muscles; Wheeler 1910). In the primitive groups, queens forage feed first brood, but usually leave the nest only at night and thus are unlikely to be exposed to males. In any event, universal flightlessness (*i.e.,* loss of wings) by ant queens after nest-initiation eliminates airborne encounters with males.

III. SPERM COMPETITION IN THE ACULEATA

Having reviewed the factors that affect multiple mating in the Aculeata, I now turn to the question of its actual incidence. This is facilitated by a recent review (Page and Metcalf 1982). Table I supplements that of Page and Metcalf. The extreme unevenness of the data must be emphasized. They were obtained with methods of varying reliability and, more importantly, comprise measurements taken at different stages in the mating process: copulation, insemination, and fertilization. As noted by Page and Metcalf (1982), observation (O) may show that a female mates many times, but the belief that she is therefore multiply inseminated may be incorrect. At the opposite end, isozyme (A) and genetic marker (V) data are not direct measures of sperm competition, if it occurs, but rather measures of its outcome; where one male's sperm, by whatever mechanism, preempts all others, there will be no indication that competition has occurred. In addition, while observation and dissection (D) can be used to test the hypothesis of single copulation or insemination, they are less effective against the opposite hypothesis. As an example, Alcock's (1981) observations on Australian thynnine wasps indicate that in at least two species, females sometimes mate more than once. My interpretation of Alcock's results is that females are usually unreceptive after mating and behave such that subsequent matings are unlikely. This is necessarily not clearly shown, though. Because of this contradiction, I have not included these results in Table I, though Ridsdill Smith's (1970) data on a different thynnine are included.

Inferences of female unreceptivity (some are included in Table I) represent a special difficulty. At least in solitary species, females appear to have nothing to gain from multiple mating and assume a cost in sexual harassment. Female resistance is expected where it usually succeeds. Otherwise, a submissive tactic would be expected. However, successful resistance is not necessarily even similar to monogamy. To a female, the difference between five and 100 matings, for example, is likely to be very large while the difference between one and five matings may be insignificant, but in terms of sperm competition (including offspring relatedness; see discussion below) the difference between one and five matings is much greater.

In spite of these limitations, the data are extensive, cover diverse taxa, and demonstrate that both sperm competition and its absence are well represented in aculeates. Field biologists generally conclude that in most species, females mate only once. Alford (1975) has said this of bumble bees and Eickwort and Ginsberg (1980) of bees in general, and the tentative conclusion of Alcock *et al.* (1978) from the rarity of observed copulations in bees and wasps and the frequency of observed refusals is that in most species a once-mated female repulses all subsequent suitors.

There are no apparent phylogenetic or sociobehavioral trends in the number of times females mate. Perhaps this is not just an artifact of the unsystematic nature of the data, and further studies should not be expected to reveal any bold trends. When sperm do compete, what is the outcome of the contests? On this question our ignorance is nearly complete. Prout and Bundgaard (1977) cite data on sperm displacement from a variety of insect taxa, but these did not include Hymenoptera. There is no *a priori* reason to expect that aculeates will deviate from the general rule that the last male to inseminate a female achieves the best fertilization (Parker 1970, this volume), but there is only modest evidence that this is in fact the case:

> 1. Martinho's (1979) artificial multiple-insemination results are consistent with a last-come first-serve arrangement of sperm in the spermatheca. It should be noted, though, that the competitive effects of this will be blunted by the fact that a long-lived and fertile queen can use up the ejaculate of several males (see below).
> 2. The multiple-mating scheme in *Trypoxylon politum* is likewise consistent with most fertilization attributable to the most recent ejaculate (Brockmann 1980).

Walker (1980) indicates a correlation between spheroid spermathecae and single insemination or precedence of first ejaculates, and elongate spermathecae are purported to favor the last ejaculate. The Aculeata usually have spheroid spermathecae, predicting first-ejaculate precedence. The generality of Walker's rule is far from established (see Gwynne, this volume; Thomas and Zeh, this volume). Nothing now known suggests unique patterns of sperm competition in the Aculeata. If extraordinary patterns do in fact exist, their origin should not be sought in the structure of the spermatheca or the male genitalia or the behavior of either sex, but rather in male haploidy.

Parker's (1970) statement that there is "every reason to expect that selection acts on individual sperm" within the same ejaculate is central to the subject matter of this volume and to sociobiology (*i.e.,* the biology of conflicts of interest). Where gametes, like organisms, are not genetically identical, they have a degree of disunity of interests. And where one type achieves greater success, the underlying advantageous mechanism will be selected (see Sivinski, this volume, for discussion of sperm evolution). In the Hymenoptera, though, unlike the vast majority of taxa, sperm from a single male **are** identical and therefore have a complete identity of interests. By close analogy with whole organisms, all bases for parent-offspring conflict (*i.e.,* between a male and his sperm) and for resistance to parental manipulation are abolished. Where it serves the interests of the male (*i.e.,* the ejaculate as a whole), selection for sperm altruism may occur. Some manifestations of this might include:

> 1. An ejaculate adopting a unitary form (see Sivinski, this volume). This would be advantageous when it occupies the duct position or can compete to do so, in which case the unanimous ejaculate should attempt to obtain and keep exclusive access to the duct and to seal its border against alien sperm.

2. Suicidal interference with other ejaculates. If n ejaculates occupy a spermatheca, a sperm might be selected to self-destruct if it can destroy more than n foreign sperm with no harm to its fellows.

Each of these requires ejaculate-mate recognition. One result of recent research in insect social discrimination is that genetically based chemical (odor) differences are much more important than had been previously thought. Hölldobler and Michener (1980) and Holmes and Sherman (1983) have reviewed these findings and discussed their adaptive significance for social organisms. Conceivably, analogous biochemical discrimination among spermatozoa might occur, especially where these are genetically identical.

These tactics can be rendered more effective by sperm specialization. Where this is well advanced we should find distinct physical types within an ejaculate. Lee and Wilkes (1965) report at least five such physical types from the parasitic wasp *Dahlbominus fuscipennis,* and Sivinski (1980) suggests "a complex competitive environment" in the vagina and spermatheca. Haploidy provides the ideal condition for evolution of specialized self-sacrificial individuals. Significantly, Lee and Wilkes (1965) observed that three of the five sperm types in *Dahlbominus* rarely reach the spermatheca. *Dahlbominus* females multiply mate (Wilkes (1966).

A peculiar analog to sperm altruism happens in a place where it is very difficult ot explain, in the process of aggregation and formation of the spore-capsule in the cellular slime mold *Dictyostelium discoideum.* The majority of the aggregating diploid cells go to form reproducing spores, but about 15% sacrifice their reproductive potential by forming the stalk of the capsule (Loomis 1975). Unless cells making up the spore capsule are almost always a clone, such altruism is a problem to explain, while anything similar within a hymenopteran ejaculate would very easily explained.

The existence of such effects is an unexplored possibility. It must be acknowledged that present evidence is against the widespread existence of sperm polymorphisms. They are not found in the honey bee (Rothschild 1955) or imported fire ant (Thompson and Blum 1967). And as with the existence of remarkable monomorphic sperm, it seems that it would have caught the attention of the several researchers who have looked at spermatozoa in many species for the purpose of counting them. The other effects mentioned are unlikely to be seen by anyone not specifically looking for them, so that it tells us nothing that they have not been reported. While **haploid effects**, of which these are examples, are common in plants, they are rare or unknown in animals (Beatty 1975; Sivinski, this volume). If they do in fact exist in animals, multiple-mating Hymenoptera would be a good place to look for them.

IV. SOCIAL CONSEQUENCES OF SPERM COMPETITION

A. Sperm Requirements of Queens

Social life, especially advanced eusociality, introduces qualitatively new effects of sperm competition. Because she produces workers, a queen lays more eggs and utilizes more sperm than a solitary female. Especially in advanced eusocial species, the increase is likely to be in orders of magnitude. Engelmann (1970) cites some examples. A European yellowjacket, *Vespula vulgaris,* queen may lay 22,000 eggs (most of them fertilized) in her one year lifetime. A honey bee queen, *Apis mellifera,* can lay 120,000 eggs/year (1,000-2,000/day at peaks) in the temperate zone, 200,000 in the subtropics or tropics. Ant queens are often even more prolific. As part of the cyclical colony economy, army ant eggs are produced in discrete broods. The neotropical *Eciton burchelli* queen lays about 120,000 eggs in a 36-day brood cycle, the African *Anomma wilverthi* about 3-4 million in a 25-day cycle. This amounts to 1.2 million and 44-58 million/year, respectively. Predictably, this requires some specialization of the reproductive system. Social insect queens often have many more ovarioles than solitary females and may have greatly distended abdomens, and/or individual ovarioles may operate more efficiently (Iwata 1964, Iwata and Sakagami 1966). Virtually all solitary and primitively eusocial aculeates have three or four ovarioles/ovary, vespines have six or seven, *Apis mellifera* queens have 75-175, and ants show a great range from very few to more than 1,000 (in *Eciton burchelli*) (Iwata 1955, 1960, 1964, 1965; Iwata and Sakagami 1966; Kugler *et al.* 1976).

There are two fundamental ways the sperm requirements of such queens may be satisfied. First, they may mate repeatedly as sperm supply is exhausted; second, they may take on a very large sperm supply at the beginning of their reproductive lives. The first option is characteristic of the termites. Réaumur (1722:345) may have had the original insight that some social Hymenoptera employ the second. He doubted that males were a necessary continuing feature of the colony in two European yellowjackets, but believed that the fall mating provided sperm sufficient for all of the next year's egg production. With few exceptions, aculeate queens mate only within a span of one or very few days. There are no proven exceptions to this rule, but it would not be surprising if in some ants the queen's lifetime sperm demands simply exceed her storage capacity. Rettenmeyer (1963) has suggested that *Eciton burchelli* queens mate as often as every year, and Raignier and Van Boven (1955) suggest that it is even more frequent in *Anomma wilverthi.* In addition, if the considerable hazards of the mating flight are an obstacle to selection for repeated mating, we might look for a change of habit where this obstacle is weakened or removed, as in *Monomorius pharaonis* (the ant cited above which has no mating flights and mates in the nest). Büchner (1881) advanced the imaginative idea that male yellowjackets are regularly produced as a simple result of the exhaustion of the sperm supply. If this were the case, it could cause fluctuations in the sex ratio. A male-rich year would give rise to well inseminated new queens, that will in turn produce queen-rich broods of sexuals the next year. Aside from the anarchic nature of such a method of primary sex-ratio determination, the longevity of sperm (or ability of the spermatheca to maintain sperm) appears not to be a factor in limiting a queen's reproductivity. The extended viability of sperm in the spermatheca over years is well known in honey bees and some ants (Taber and Blum 1960), and Parker's (1970) prediction that the longest-lived sperm should be found in the Hymenoptera is almost certainly correct.

The requirements of acquiring great numbers of sperm during a relatively brief period of time can be met by mating with more males or through increased sperm delivery per male. The question of which of these two strategies is employed has recently been addressed by Garófalo (1980) and it appears that both take place. Iwata and Sakagami (1966) showed that in species where the female has a greater number of ovarioles, the male likewise has an increased number of testioles. Garófalo measured sperm from males and mated female bees of a range of social types and reviewed other data from the literature. In three advanced eusocial species, females contained between 67-134% of a male's full (standing crop) sperm complement. When she contains more than 100%, a female is necessarily multiply mated. In 10 of his 14 other species, Garófalo found that females contained only 2-4% of a full male complement, and in none of the 14 was it higher than 29%.

Significantly, two primitively eusocial (Garófalo's "mesosocial") species in this latter group (both bumble bees) had male/female stored sperm ratios indistinguishable from those of solitary species. This result is corroborated by other ratios from bumble bees (Table II). The conclusion to be taken from available data is that the transition from primitive to advanced eusocial involves sharply reduced male/female stored sperm ratios. It cannot be inferred that multiple mating may not occur where males contain enough sperm to inseminate many females and the sex ratio is balanced, but in that situation if multiple mating evolves it will not be in response to increased female sperm requirements.

Even at this introductory level, the subject of sperm ratios is far from exhausted. It would be very useful to have more facts about male sperm load, female sperm capacity, and the proportion of total mass taken up by the reproductive system in each sex in socially diverse wasps and bees.

B. Contest to Make Queens

Parker (this volume) has noted that the extended longevity of insect sperm and their provisioning in the spermatheca provide special opportunity for overlapping ejaculates and competition for ova. This is especially applicable to the social aculeates where sperm competition has become a more complex affair. The primitive

TABLE II

Male and Queen Sperm Counts in Six Species of Bumble Bees[a]

Species	Mean Sperm in Male (Range)	n	Mean Sperm in Female (Range)	n	Mean Sperm in Female/Male
Bombus lucorum	2.4×10^5 ($1.6 - 3.3 \times 10^5$)	3	2.4×10^4 ($4.2 \times 10^3 - 3.7 \times 10^4$)	10	0.10
B. lapidarius	1.9×10^5	1	2.6×10^4 ($1.5 - 4.4 \times 10^4$)	6	0.14
B. terrestris	3.3×10^5 ($1.5 \times 10^4 - 1.1 \times 10^6$)	8	4.5×10^4 ($3.1 - 6.5 \times 10^4$)	5	0.14
B. hortorum	1.6×10^4 ($3.9 \times 10^3 - 2.9 \times 10^4$)	2	4.2×10^{3b}	1	0.26
B. agrorum	2.6×10^5 ($3.5 \times 10^4 - 5.5 \times 10^5$)	10	1.8×10^4	1	0.07
Psithyrus campestris	8.5×10^4	1	3.2×10^4	1	0.38

[a]All data are from W. D. Hamilton (unpubl.). Röseler (1973) gave sperm counts for females only for the same five species of *Bombus*. Sperm counts are taken from wild-caught individuals. The last column represents the estimated number of males needed to completely inseminate a queen. *Psithyrus* is the genus of obligate social-parasitic bumble bees.
[b]Röseler's mean count from five females of this species was 1.5×10^4, the only case in which his counts differ greatly from Hamilton's for the same species.

contest to fertilize any and all eggs persists, but there is now the added struggle to specifically fertilize **queen-producing** eggs. The overwhelming majority of fertilized (female) eggs in a eusocial species will become workers. These may achieve some small level of classical fitness through the deposition of unfertilized (male) eggs, but in any event it may be discounted that selection should ever favor males with a bias toward fertilizing worker-producing eggs. Given that worker-producing and queen-producing eggs are in themselves identical, though, can males make such a manipulation? This would require that there be information leakage from the colony's reproductive state to the sperm in the spermatheca and that sperm rise up to compete at the decisive moment. Such competition is possible only if queen production is a nonrandom occurrence. If, for example, it is a late season event (as in vespines and bumble bees), it may be in the male's best interest that his sperm not take advantage of early season opportunities to fertilize (worker) eggs. As discussed below, though, it may be in the queen's interest to introduce a different type of bias. We would not expect her interests to be overruled within her own spermatheca.

C. Kinship and Within-colony Conflicts

The swift general acceptance of kinship as a key factor in animal behavior is in large part due to a theoretical triumph right at the outset: Hamilton's (1964) explanation of why most eusocial insects and origins of eusociality are found in the Aculeata. He showed that haplodiploid sex-determination in the Hymenoptera produces asymmetries in relatedness that correlate well with certain major asymmetries in social behavior. The principles involved have been analyzed and discussed at all levels by many authors (for a succinct review of the background, see Charnov 1978b). Recently, a number of authors have suggested that aculeate eusociality may be a consequence of other peculiarities of the group (Scudo and Ghiselin 1975; Craig 1980b). Starr (1981) argues that the stinger is the key preadaptation, while others focus on resource utilization and the nest as the paramount factors. At present, both kinship theory and alternatives account for the same major features of social life in the Aculeata, so it is difficult to isolate one of them as a sole explanation.

While kinship theory's headline triumph is now contested by some, its very foundation, Hamilton's central concept of **inclusive fitness**, is not, and it has rather effortlessly entered the mainstream of evolutionary thought. This concept includes in the fitness gain from an individual's actions not only their effect on this individual's classical fitness, measured in personal offspring, but also on the classical fitness of relatives, weighted by the degree of their relatedness.

Hamilton's formulations neatly accounted for the key stage in social evolution in which a newly emerged female has a choice of nesting solitarily or remaining in the natal nest and assisting her mother, at the expense of some or all of her own classical fitness. In most eusocial insects, though, workers no longer have this choice, so questions of maximizing inclusive fitness seem irrelevant. It is in this context that Trivers and Hare (1976) presented the most important addition to kinship theory to date. They pointed out that workers and queens have a conflict over (1) who lays the male eggs and (2) the sex-ratio of investment in brood, that workers can compete with queens in these two questions, and that the outcome of this struggle can be known. It is a measure of the extent to which kinship theory had been assimilated since 1964, as well as of the importance of Trivers and Hare's paper, that after minimal publishing lag time it provoked a substantial literature of discussion, criticism, and extension (Alexander and Sherman 1977, Benford 1978, Charnov 1978a, Oster and Wilson 1978, Craig 1979, Crozier 1979, Starr 1979). Some authors have turned these ideas back to the question of the origin of eusociality (Charnov 1978b, Craig 1980b). This is something that Trivers and Hare touched on in passing, but was not central to their thesis and seems unlikely to become as important as the analysis of competitive equilibira in eusocial colonies. Appendix B recapitulates the main part of Trivers and Hare's argument.

The idea that workers maintain a measure of fitness antedates kinship theory and has often been called upon to explain why workers are willing to be workers at the origin of eusociality. If Kropotkin (1902) had been aware of worker fecundity it seems certain he would have emphasized it. The most complete modern expression of this idea is perhaps Lin and Michener's (1972) statement that "eusocial colonies without altruism are possible if male production by workers is important enough." How important it is varies greatly throughout the social Aculeata, and values of $1 - p$ from 0 to almost 1 are known (Oster and Wilson 1978:Chapter 3). It is highly likely that such classical fitness will be unevenly distributed within the caste so that not all workers will have identical interests. Trivers and Hare indicate this by reference to the extreme case, in which a single worker lays all male eggs not laid by the queen, so that q, the fraction of male eggs from a given worker, is equal either to 0 or $1 - p$. Variation in q can have a moderate effect where both p and P_e are small, so that if $p = 0.1$ and $P_e = 1.5$, for example, the preferred investment ratio of the single laying worker is 10% greater than that of a full-sister worker. The greatest possible difference between the two x values is 16% (Fig. 1A).

Whether this can have serious social consequences in small colonies is unknown. It might do so if a worker can recognize her own offspring. However, even in social wasps and those social bees having larvae in fixed cells, there is no evidence that females can distinguish sons from brothers and nephews.[1] Where x values differ sufficiently, selection should favor workers that remember how much of the male brood is theirs and are not entirely ignorant of what fraction of the remainder will typically come from the queen. In large colonies, laying workers should be such a small part of the colony and have such a small opportunity to bias the investment

---

1. Note added in proof. Klahn and Gamboa's (1983) results with *Polistes fuscatus* make it a little more plausible that this might be done.

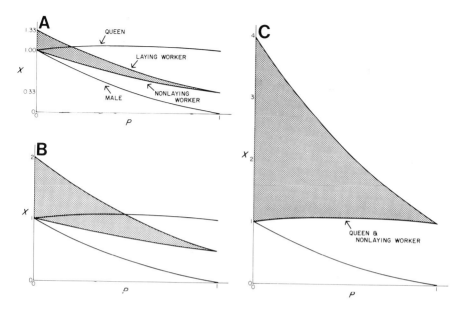

Fig. 1. Preferred investment ratios x of queen, laying worker (lays all male eggs not laid by queen), nonlaying worker, and male (queen's mate) as a function of p. Explanation in Appendix B. The shaded area shows the total possible range of worker x. A. $P_e = 1$. B. $P_e = 2$. C. $P_e = \infty$.

ratio in their favor that they can conveniently be ignored (Oster and Wilson 1978) and all workers can be treated as identical, with $q = 0$.

Trivers and Hare calculated x values for queens and workers of 1.0 and 0.33, respectively. They assumed that the queen mates just once and lays all of the male eggs. Using data from 21 species of ants believed to fit these assumptions, they obtained a scatter of y values, most < 1 and many not very different from 0.33. Various authors (*e.g.*, Benford 1978) have interpreted this to indicate that y is largely equivalent to worker x in ants, a decisive victory for offspring in this parent-offspring conflict. Trivers and Hare were properly more reserved in their conclusion: "Instead of supporting a general principle predicting who shall dominate in situations of conflict, our work supports the notion that there is no inherent tendency for evolution to favor any particular party in situations of conflict." For an even stronger view on why this must be so in mound-building ants, see Herbers (1979). Thus, findings such as those of Noonan (1978) that y is not significantly different from 1 in *Polistes fuscatus* should not be seen as a test of Trivers and Hare's imaginary general conclusion, but as an indication of the outcome in a particular wasp species.

The first published criticism of Trivers and Hare's paper came from Alexander and Sherman (1977), who raised an array of objections. The most prominent of these was their contention that Trivers and Hare had attached an incorrect interpretation to their own data, which in fact showed the effects of local mate competition (Hamilton 1967). I concur with Oster and Wilson (1978) and Crozier (1980) that most of these ant species, with their large explosive mating swarms (see Kusnezov 1962), seem poor candidates for local mate competition. On the other hand, Trivers and Hare's data collection methods are clearly inadequate (Alexander and Sherman 1977). Among others, their use of dry weights can be questioned. If their data are adjusted to live weight, the results would be substantially altered. Noonan (1978) used live weights as her index of investment production costs.

Models of queen-worker conflict in eusocial insects have often included the assumption that the queen is singly mated (*e.g.*, Craig 1980a). This is one of Trivers and Hare's assumptions, and they believed it to be justified for the ant species they examined. Without taking up the question of whether this belief is justified, I will examine some consequences when the queen is multiply mated. Where the effects of multiple mating have been treated, it has been usual to use the term n to designate the number of males, assuming that they each contribute equally to paternity (*e.g.*, Page and Metcalf 1982). This assumption is not realistic. Where relative contribution has been evaluated, large variance among ejaculates has been found. Substitution of Starr's (1979) index of **effective promiscuity**, P_e, for n solves the problem without complicating the mathematics. **Promiscuity** is widely used as a measure of the number of males mating with a female (Gwynne, this volume), and P_e is equal to:

$$\frac{1}{\sum\limits_{i=1}^{n} f_i^2}$$

where f_i is the fractional contribution to daughters by the ith male. As an example, if a *Polistes metricus* queen on average uses sperm from two males in a 9:1 ratio (Metcalf and Whitt 1977), $P_e = 1.22$. Any given value of P_e has the same effect on progeny relatedness as would an equivalent value of n if all males contribute equally, so that $1 \leqslant P_e \leqslant n$. In the absence of inbreeding, then, relatedness between siblings is equal to $\frac{1}{4} + 1/(2P_e)$. Effective promiscuity in diploid organisms, such as termites, is calculated in exactly the same way, but relatedness between siblings is equal to $\frac{1}{4} + 1/(4P_e)$. The effective promiscuity index has one unfortunate limitation in that a mean of several P_e values gives rise to underestimated mean relatedness. This can be a large difference when variance in P_e is large. This must be taken into account when dealing with population-wide average P_e.

D. Why Multiple Mating? Possible Answers From Honey Bees

Of what value is multiple mating to the queen? A semiproximal answer is that in some species she requires more than one male to fill her spermatheca. This does not settle the question, but rather presents another: Why have males not evolved to carry enough sperm to inseminate completely?

Let us examine this question in relation to the domestic honey bee, *Apis mellifera*. Apiculture is a sophisticated agricultural enterprise, comparable to dairy farming, and the subject of a great deal of genetic research. Consequently, honey bees are the best known social insect species. Given the extremely male-biased sex ratio in this species, it is not surprising that males are designed somewhat like kamikaze aircraft, with features directed to effective one-time delivery and without provision for a return engagement. The reproductive system fills most of the abdominal cavity; in copulation the eversible copulatory organ is irretrievably forced into the vagina by haemostatic pressure resulting from a full contraction of the abdominal skeletal muscles. Death of the drone unavoidably follows mating (Laidlaw 1944). It is odd, then, that he delivers so little. Woyke (1962) calculated that an average drone ejaculate is only 1/8 to 1/9 of the female's spermathecal capacity. Other authors have reached comparable conclusions (see Page and Metcalf 1982, and Table I).

Is this due to phylogenetic inertia, the impossibility of a full-delivery drone even under selection for it? It is very unlikely. Drones are strong fliers that travel distances (to mating areas) similar to foraging workers. *Apis mellifera ligustica* workers are known to carry nectar loads weighing as much as 70% of the unloaded body (Park 1922), and *A. m. adansoni* workers carry loads heavier than their unloaded body (Neves *et al.* 1978). There is no obvious reason that drones could not be similarly loaded with fluid and still fly well (this is open to experiment). Even more compelling is the aforementioned fact that drones sometimes carry the (larger) queen along in flight while in copula. From the colony perspective, it would seem a better economy of scale could be achieved; that is, production of drones with eight or nine times the sperm storage capacities should require much less than eight to nine times as much investment.

Parker (this volume) suggests (and I concur) that the male sperm load in *A. mellifera* is small because of intense intersexual selection imposed on drones by queens. The swift mating flights of queens allows them to elude all but the fastest flying males. In this circumstance, an additional volume and mass of stored sperm could impose a severe handicap on the male, such that an extraordinarily encumbered male could never catch a queen. Heavily-laden workers are noticeably slowed in flight (K. G. Ross, pers. comm.), as males would presumably be.

By setting a high pursuit-pace and altitude, virgin queens provide the selection which mandates multiple mating. To return to the question, of what value is multiple mating to queens? I suggest three possibilities:

1. The queen is merely channeling the intense male-male competition into a form that selects physically strong partners, and therefore good genes. This is simply Snell's (1932) eugenic hypothesis, with multiple mating as an unimportant side effect. It may in fact contain just this advantage for the queen, but this says nothing special about social insects. If additional important benefits can be discerned, then the eugenic explanation is at least incomplete.

2. Offspring genetic diversity is favored in an exceptional way. It may be suggested that genetic heterogeneity among workers is beneficial to the colony (and therefore the queen). This argument, similar to Williams' (1975) general selective advantage for sex, is given by Pamilo et al. (1978) and Pamilo (1982) to account for genetic diversity within ground-nesting ant colonies. Page (1980) hypothesizes a special reason why multiple mating should be adaptive in the honey bee. It has long been known that diploid males are occasionally produced and that these are inviable. Sex is determined by a single locus with multiple alleles, and diploid males result from homozygosity at this locus, while females result from zygotes heterozygous at this locus. While the average expected proportion of diploid males does not vary with P_e under outbreeding, the variance does. Page's model addresses the question of why it should be adaptive (in honey bees) to decrease the variance through multiple mating. This model has more recently been extended to the Aculeata in general (Page and Metcalf 1982).

3. Interests of nonlaying workers are shifted to the queen's advantage. It is generally agreed that as colony size increases the queen is less able to directly affect the ratio of investment in brood. At the same time, the number of workers laying eggs is expected to remain small or be depressed. Under these conditions, an increase in effective promiscuity and decrease in average worker-worker relatedness will have two effects (Fig. 1A, B, C). First, the preferred investment ratio of nonlaying workers moves closer to that of the queen, until these are identical at infinite P_e. Second, the conflict of interests between nonlaying and laying workers broadens, until at infinite P_e there is a fourfold difference in x values (Fig. 1C).

As shown in the figures, nonlaying workers have no conflict of interest with the queen when workers lay all the male eggs, and they have none among themselves when the queen lays all the male eggs. It is the middle range of p that is of interest here. The interest shifts of nonlaying workers with increased P_e should turn their efforts to the queen's purposes because they control the allocation of investment, so that she benefits from bringing their x closer to hers. In addition, workers are related to the three possible classes of male offspring in the following way: full-sisters' sons > brothers > half-sisters' sons. They are on average less related to nephews than brothers whenever $P_e > 2$ (Fig. 2) and should prefer that the queen lay all the male eggs. Workers would therefore be expected to interfere with each others' reproduction.

Of these three possible queen advantages from multiple mating, the third is the most compelling, as it has the greatest universality and explanatory value. It gives rise to two testable predictions:

1. In large colonies with single queens that are multiply mated, workers should never lay a substantial fraction of the male eggs.
2. Workers in orphaned colonies of Aculeata should work less diligently and lay more male eggs than in the presence of a functioning queen.

There are no data on the first of these predictions. However, it has often been observed that workers in orphaned colonies are predictably less industrious and lay more male eggs (Wilson 1971:Chapter 15). As worker-worker conflict of interest will lead them to police each other's reproduction in queenright colonies, so in

	QUEEN	WORKER		MALE
G_M	$\dfrac{1 + p}{4}$	$\dfrac{1 + p + 3q}{8}$	$+ \; \dfrac{1 - p - q}{4 P_e}$	$\dfrac{1 - p}{2 P_e}$
G_F	$1/2$	$\dfrac{2 + P_e}{4 P_e}$		$1/2 P_e$

Fig. 2. Average relatedness to each of the queen, workers, and males (*i.e.,* queen's mates) of male (G_M) and female (G_F) brood in monogynous colonies under outbreeding. p = fraction of male brood coming from the queen; q = fraction of male brood coming from the worker in question; P_e = the index of effective promiscuity (see text). If inbreeding and local mate competition are significant, then G_F to workers = $\frac{1}{2}P_e + (1 + 5F + 2r')/(4 + 4F)$, in which r' = the correlation coefficient of sperm drawn randomly from two males, and F = the inbreeding coefficient. By a different route, Hamilton (1972) has arrived at the expression: $(1 + 5F + 2r_s)/(4[1 + F])$ for average worker-sister relatedness, in which r_s is the correlation of sperm drawn randomly from the spermatheca. If each of these is correct, and if $r' \approx F$, then $r_s \approx (1 + [P_e - 1]F)/(P_e)$. This seems intuitively a good approximation of r_s to each of us, though Dr. Hamilton (pers. comm.) doubts the possibility of deriving a general expression of r_s in the absence of exact knowledge of a species' mating system. Note that where inbreeding and local mate competition are discounted, r_s is simply the reciprocal of P_e.

orphaned colonies we expect the most intense reproductive competition in species where policing was most complete.

It is unfortunate that this analysis does not lead to an unequivocal prediction about the relationship between multiple mating and the retention or loss of functional ovaries in workers. As long as workers effectively interfere with each others' egg-laying, the queen should be indifferent to whether they **could** lay eggs in her absence. Alexander (1974) has explained this as being a factor in her interests; she should benefit by having the orphaned colony produce her grandsons rather than nothing. A logical extension of prediction 2 is that worker ovaries will be **most** functional in species with queenright colonies having the **least** worker egg-laying. This would represent preparation for reproductive competition in the event of queen loss. It assumes that orphaning is a sufficiently common event to select for preparations to compete. It appears to be uncommon in species with very large colonies. In species with smaller colonies such as some *Polistes* (Strassmann 1981), it is more common, so that the reproductive potential of workers is very relevant. A larger question is whether worker ovaries are selected by regular low level egg-laying opportunity in queenright colonies or rare bursts of egg-laying in orphaned colonies. The answer will probably vary among taxa and levels of social organization..

E. Consequences of Sperm Clumping

It has often been pointed out that if the ejaculates of different males remain separate in the spermatheca, sequentially used sperm will have a high probability of coming from the same ejaculate. Thus short-term effective promiscuity could be close to 1 in multiply mated females (Trivers and Hare 1976; Charnov 1978a, b; Crozier and Brückner 1981; among others). Sperm clumping would assure that daughters of a given cohort will tend to be much more closely related than if there were a high level of ejaculate mixing. Clumping is assumed by Trivers and Hare (1976) and Charnov (1978a, b). The mainstay of this assumption has been Taber's (1955) results from artificially twice inseminated honey bee queens, which showed no appreciable sperm mixing, as inferred from the sequence of offspring types. This was consistent with Koehler's (1962) report of distinct ejaculate clumps in the spermatheca, and Jordan's (1967) similar though less direct observations. Kerr (1961) has reported physical clumping in two leafcutter ant species, such that each ejaculate was not easily broken apart by the experimenter. More recently Martinho (1979), using both artificial insemination and free flights with two races of drones, concluded not only that sperm clumped but also that ejaculates rotate about every 20 days within the spermatheca. That is to say, the queen draws sperm mostly from one ejaculate for a time and then somehow switches to using another.

The function of this sperm rotation is obscure. One possibility is that, in the absence of the queen's ability to break up each ejaculate and thereby increase short-term effective promiscuity, she achieves the same results by rotating them frequently. It is suggested by Kerr et al. (1980) that exactly the opposite effect results and that rotation ensures that daughters of a given age group tend to be full-sisters. Inasmuch as workers are short-lived, this would by no means be an insignificant effect. But why a period of about 20 days? This makes Martinho's conclusion and the hypothesis of Kerr et al. intriguing. Queens require 16 days to develop from eggs, and brood-care is the domain of young bees (3 to 10 days old; Lindauer 1961). If queen eggs are laid in the second half of a 20-day period, then as larvae they will be cared for by full sisters. This may increase the return to these workers for efficient brood-care.

Perhaps more importantly, by the time queens emerge, their sisters will constitute most (if not the total) in-hive workforce (Lindauer 1971). Workers are not passive in treatment of newly emerged virgin queens. It is usual for more than one of these to be produced simultaneously and usually only one survives beyond the first few days of adult life. It is often observed (Gary and Morse 1962, Hölldobler and Michener 1980) that workers kill one or more virgin queens before they permit one to mate and establish herself as the new queen. In addition, they may physically impede the first emerging queen in her attempts to kill others. At this critical stage in colony reproduction the cooperation of the workers is essential to the reigning mother-queen. While she is the one to designate queen eggs, by laying them in distinct queen cells, workers can exercise their veto power over any of her queen candidates, at a number of levels. If the 20-day sperm period is in fact a real phenomenon, it may represent a mechanism by which the queen gains cooperation from workers.

But why is the sperm rotation period not of much greater length? Consistent with the scheme of Kerr *et al.* is the hypothesis that the shortness of the cycle serves to divide the workforce into two genetic groups at the time a new queen is produced, and thereby facilitates the decision of one group to swarm with the old queen. If this could be demonstrated, it would provide a wonderful example of **queen control** (Michener and Brothers 1974) or **parental manipulation** (Alexander 1974). In fact, Getz *et al.* (1982) show a tendency for the two groups to assort along kinship lines when the workers' fathers are of different genetic strains. They did not find, though, that the groups could be divided by age and concluded that direct recognition of kinship is the cue used. In recent years, it is increasingly accepted that animals can do just this in many cases (Hölldobler and Michener 1980, Holmes and Sherman 1983, Klahn and Gamboa 1983).

My interpretation of Martinho's data is that the sperm rotation period is much less regular than suggested by Kerr *et al.* More critically, there is room for doubt that the rotation itself really takes place, for two main reasons:

1. There are several methodological points that remain in doubt, especially with regard to accurately determining the paternity of offspring. As a result of these, alternate interpretations of the data can be entertained. This is a difficult problem, so it is not surprising that Martinho's study has not entirely settled it.

2. Page (1980) was unable to confirm the presence of ejaculate segregation in *Apis mellifera*. Over 11 weeks he found no major changes in the phenotypic frequencies of different offspring types from two naturally mated queens. This is a repetition of Taber's (1955) much-cited study, with improved experimental methodology. Page's results are contrary to Taber's and call the latter's into question. Alber *et al.* (1955) likewise found sperm from different males to be used approximately equally in the short term. Getz *et al.* (1982) found some change over time in use of sperm from two males, but far from strict rotation.

Most likely it is the interpretations more than the data from different studies that are contradictory. As Crozier and Brückner (1981) have shown, complete sperm clumping does not occur, and even Martinho's results show significant mixing. I suggest that randomness should be viewed as the null hypothesis and to show, as Page has done, that a female uses sperm from more than one male at one time does not really go to the heart of the matter, which is whether there is significant temporal clumping of full sisters. Given the popularity of honey bees as experimental animals, we can hope for continuing progress in this question.

F. Male Interests in Queen/Worker Conflicts

Until now I have not considered the question of the preferred ratio of investment of the queen's mate(s) in advanced eusocial Aculeata. This was explicitly discounted by Trivers and Hare (1976), on the grounds that such males are rarely present in the colony and in any case have few behavioral means of intervening. Most subsequent authors on the subject have explicitly or implicitly concurred; Craig (1979) suggested obliquely that male interests might be examined as a sort of exercise. Benford (1978) showed male relatedness to offspring along with those of queens and workers, though without further comment. If male interests are at all relevant, it is a question of the limits of parental manipulation. Is it possible in sexually reproducing organisms for a parent to genetically program an offspring to act in the parent's interest at a cost to its own? It is generally doubted (with Craig among the doubters) that this is possible. In Charlesworth's (1978) models no genes persist that cause daughters to behave in their fathers' interests if such behavior is detrimental to the daughters. Nonetheless, the subject of this chapter revolves around male interests, so we might wonder how they work into the calculations.

A male's relatedness to the offspring of his mate's colony is shown in Fig. 2. His preferred investment ratio at all values of P_e, then, is $2(1 - p)(3 - p)/3 + p$. This is included with the preferred ratios of queens and workers in Fig. 1A-C. The purported conditions for Trivers and Hare's data are shown in Fig. 1A at $p = 1$. If only queen and worker interests are taken into account, a realized investment ratio in the middle range between 0 and 1 would seem to represent a victory for the workers, and Trivers and Hare give this interpretation to many of their results. With males in the picture, though, it could also be interpreted as a compromise between the two classes of parents, a case of undecided sexual conflict coming to resemble offspring victory.

If sexual conflict is irrelevant to investment ratio in advanced eusocial colonies, y should never be lower than nonlaying worker x. If parent-offspring conflict is irrelevant and sexual conflict all-important, y should never be higher than queen x. What I want to point out here is that available data leave each of these contradictory testable propositions unscathed.

V. CONCLUSION: WHAT IS TO BE DONE?

Part of the tone of this chapter is set by doubt. I doubt that many of the large patterns which are assumed or suspected to apply across social types or taxa really exist. While it may be fruitful to seek general trends in the number of matings per female, the relationship between n and P_e, the degree of sperm clumping in the spermatheca and the outcome of queen-worker conflict, we should not be surprised when they prove very elusive. We must also be vigilant against the temptation to see general trends where the data are insufficient to support them. Perhaps nowhere more than in behavioral ecology is the impulse to seek universal, simplifying principles—so persuasively decried by Bondi (1977)—likely to be mistaken. After all, this is not physics, and species are not subatomic particles. Even as the large patterns are sought, they should not be expected. And if found they should not be wholeheartedly accepted.

We know a reasonable amount about the external aspect of courtship and mating in the Aculeata, and a substantial body of theory is developing to deal with the consequences of sperm competition. Events inside the spermatheca, standing between these areas of relative knowledge, though, remain unknown. They form the bottleneck in our understanding of this sequence of processes. To widen this bottleneck we need more **facts**. Facts which bear on the following questions should be welcomed:

1. What fraction of body mass does each sex devote to the reproductive system?
2. Are sperm from haploid males unusual? Are they usually polymorphic? Do they have special features which suggest adaptation for coordinated action by members of an ejaculate?
3. Do members of an ejaculate clump together in the spermatheca?
4. Is there sperm precedence?
5. How many sperm does a virgin male have? How many does a completely inseminated female have?
6. What is the relationship between number of matings per female (n) and diversity of the sperm she takes in (P_e)? What is the variance in sperm contribution (f_i) of different males? What is the female's sperm-usage pattern?
7. What are the population sex-ratios of number and weight?

VI. SUMMARY

The Aculeata are a large and familiar group, comprising the wasps, bees, and ants. In sperm structure and the reproductive system they appear to be generalized. They show some interesting features of sexual dimorphism, which are poorly understood. The two characteristics of aculeates which are of special interest here are (1) their haplodiploid sex-determination, and (2) their diversity of social organization. The group contains the full range of major insect social types.

Mating systems likewise show great diversity. In the ants and advanced eusocial bees, mating is usually by way of occasional mass mating flights. In the wasps, solitary bees and primitively eusocial bees, mating is more closely tied with the daily events of the nesting cycle.

There is a diverse body of data on the number of times females mate. At present they show no patterns among taxa or social types, but indicate that both single- and multiple-mating are widespread. Almost nothing is known about the relationship between number of matings and number of inseminations, so that the distribution of sperm competition in the group is little understood.

Unlike diploid organisms, a (haploid) male aculeate produces genetically identical sperm, so that there is no conflict of interests within his ejaculate. This raises the possibility for ejaculates to behave very differently than in diploids, especially in competing with each other for inseminations.

Social life raises the sperm requirements of reproductive females, because of the production of workers. In species with very large colonies queens require the full sperm complement of several males, so that sperm competition necessarily takes place. Sperm competition in social species is at two levels: (1) the ordinary competition to maximize fertilizations, and (2) to specifically fertilize queen-producing eggs.

The earlier idea that multiple mating is uncommon in social aculeates, at least in primitively social species, is no longer tenable. The effects of multiple mating on social evolution must be taken into account. Multiple mating introduces an increased disunity of interests between nestmate workers, especially between those which lay some male-producing eggs and those which do not. At the same time, it makes no change in the overlap of interests between queen and workers. Multiple mating, then, may evolve through selection on the queen to monopolize male-production by creating disunity among workers, so that they prevent each other from laying male eggs.

It has generally been accepted that in honey bees the members of an ejaculate clump together. Sperm cannot then be used randomly, so that age-mates are likely to be full sisters, even if their mother mates with many males. While sperm clumping now seems less complete than earlier, sperm-usage is not random. It may be a way that the queen can partly regulate worker-worker relatedness to her own advantage.

The interests of the queen's mate(s) in within-colony conflicts have received almost no attention. A male's preferred sex-ratio of investment in brood is unaffected by multiple mating, unlike with workers. Multiple mating increases the conflict of interest between workers and their fathers, while bringing them relatively closer to the queen. If males can influence the ratio of investment in any way, then, multiple mating can serve the queen's interests by dividing her competitors.

We have an increasing fund of information about events preceding sperm competition. Likewise, the conseqences of sperm competition have received theoretical treatment. The magnitude and mechanisms of sperm competition itself, though, are almost completely unknown for aculeates. New facts about ejaculate-interactions inside the spermatheca are what are most needed in this field.

ACKNOWLEDGMENTS

As is proper in such a volume, communication with others of the authors has helped me on many points. I am also grateful to a number of other persons for generously answering my questions: John Alcock, Robin Edwards, John MacDonald, Rick Duffield, M. R. Martinho, Bill Hamilton, Bob Matthews, Rob Page, Tom Seeley, and J. van der Vecht. Ross Crozier and Pekka Pamilo read an earlier version and made many valuable suggestions; their efforts went beyond ordinary colleaguely review and I am most grateful to them. I appreciate also the comments of Robert Abugov, Rick Michod, and Ken Ross. Rick Duffield helped to compile Table I and Roy Snelling identifed two ant species included in Table I. Permission to use unpublished data from Bill Hamilton is gratefully acknowledged. I began this chapter while still a student at the University of Georgia and enjoying the concomitant benefits.

APPENDIX A

Social Organization in the Aculeate Hymenoptera

The following is an outline of three distinct social systems, modified from Michener (1969). This classification, based on just three social characteristics, is in general use for insects, though it necessarily conceals immense variation (Michener 1974:38).

SOLITARY. In the simplest scheme, a female wasp stings a host (insect or spider) into temporary paralysis and lays an egg on it. The hatched larva develops on the revived host as an ectoparasitoid. More commonly, the female prepares a nest cell by excavation or construction. She then provisions each cell with permanently paralyzed hosts (= prey), lays an egg in the cell, and closes it. This is all done unassisted. Through changes in the number and sequence of these elements, more complex nesting cycles are elaborated. Phyletic sequences of this elaboration have been analyzed by Evans and West-Eberhard (1970). Most nonparasitic aculeates have this sort of nesting cycle. For present purposes, a variety of types of simple organization are lumped under **solitary**. See Michener (1969) or Wilson (1971) for a more detailed breakdown. The bees and most masarine wasps utilize pollen rather than prey as their protein source, but otherwise will be treated here as wasps.

PRIMITIVELY EUSOCIAL. The transition from the most complex solitary to the least complex eusocial system represents a quantum leap in social behavior which is not always made clear in evolutionary schemata. For at least part of its life cycle, the eusocial species has each of (1) an overlap of adult generations; (2) cooperative brood-care; and (3) a differentiation of reproductive roles (castes) among adult females (Michener 1969, Wilson 1971:4). Females of primitively eusocial species cannot reliably be separated into egg-laying **queens** and nonreproductive **workers** on the basis of physical structure alone, except in some cases by size. Most social wasps, including the large, widespread genus of paper wasps, *Polistes,* are primitively eusocial, as are some bees. The term **primitively eusocial** is accepted usage, but it is rather unfortunate, as the implied meaning doesn't necessarily fit the definition. As R. H. Crozier points out, in some ants (*e.g.,* most *Rhytidoponera,* all known *Diacamma*) the fully-differentiated queen form has been lost and replaced by mated workers. Dr. Crozier's question: Are these species "secondarily primitively eusocial"? By definition, yes, though it makes semantic nonsense.

ADVANCED EUSOCIAL. A useful distinction is drawn between primitively eusocial species and those in which workers are physically distinct from queens. The importance of this physical differentiation is that it severely limits the worker's potential to be queenlike and therefore her options in the struggle to maximize individual fitness. In all known cases, workers retain a functional reproductive system and some potential for laying eggs for at least part of adult life, but this is much less than in the queen, and they can rarely be inseminated. In many species physical differentiation extends into the worker caste itself, especially where **soldiers** differ markedly from the workers proper (Wilson 1971, Schmidt 1974:various chapters). The advanced eusocial aculeates comprise the vespine wasps, all ants, all stingless bees, and the five or six species of honey bees, *Apis.* All termites (Isoptera) are also of this social type.

In some species of *Polybia* social wasps there are small, statistically significant physical differences between queens and workers (Richards 1978) that could be transitional from primitive to advanced eusocial. Size differences are much more pronounced in the bumble bees, *Bombus,* so that the castes are usually readily distinguished by size (Michener 1974). Transitions from primitive to advanced eusocial tends to accompany increasing group size, role differentiation, and social complexity.

APPENDIX B

Queen-worker Conflict in Eusocial Aculeates
(Adapted from Trivers and Hare 1976)

1. In most species, workers retain functional ovaries and can sometimes lay eggs, though they are rarely or never inseminated (Lin and Michener 1972, Oster and Wilson 1978). It is possible, then, that some male eggs will come from workers. A queen differs in her relatedness to a son and a grandson, and a worker in her relatedness to a son, a brother, and a nephew. These inequalities of relatedness produce conflict over who should lay male eggs.

2. Workers differ in their relatedness to female and male brood. They handle most of the direct resource-investment in brood, so that there is potential for selfish manipulation if workers: (a) recognize the sex of brood individuals; and (b) can bias the ratio of investment toward the sex to which they are more closely related.

3. Let x equal the ratio of investment in male/female brood that most benefits the inclusive fitness of a given class of individuals in the colony population, *i.e.*, that class's **preferred investment ratio**. By "class" is meant any subgroup with relatedness to brood distinct from that of other subgroups, *e.g.*, queens, laying workers, nonlaying workers, males (queens' mates). The ratio of investment in the sexes usually takes into account only males and potential queens, as investment in worker brood should be seen strictly as a step in investment in these other types. The reason for this is that only sexuals leave the colony as propagules. Where this is not the case, *e.g.*, in honey bees, stingless bees, army ants, and swarm-founding social wasps, x is more complex (Oster and Wilson 1978, Macevicz 1979).

4. Let y equal the actual investment ratio found in the population. Wherever $y = x$, members of that class should be indifferent as to which sex receives the next unit of investment. In all other cases, they should attempt to bias investment in all brood over which they have influence, such that y approaches x. Comparison of x and y shows who is winning the queen-worker conflict (*i.e.*, whose interests are more nearly satisfied). It is not expected that an individual should prefer that her own colony's investment ratio be x, where $y \neq x$. Rather if she perceives or expects that the population invests **disproportionately** in one sex (from her point of view), she should prefer that her colony invest **exclusively** in the other sex. For this reason, if $y = 0.5$, $x_A = 0.3$, $x_B = 0.4$, and $x_C = 0.6$, for example, no conflict exists between classes A and B. Each wants the colony to produce queens. On the other hand, each of these is in sharp conflict with class C, which wants the colony to produce males. From C's point of view, females are devalued by this low value of y relative to x. The important point, not sufficiently stressed in the past, is that workers and queens are not unavoidably in conflict, even if their x values differ widely. It is only when y falls between their x values that there is a basis for conflict. At the same time, it is this same conflict that should perpetuate itself by holding y at an intermediate value. Colonies that produce one sex exclusively in a given season are well known in ants (Wheeler 1910, Oster and Wilson 1978, Herbers 1979) and illustrate the need for large samples to determine y. Wheeler interpreted this as a device for outbreeding, analogous to dioecy or self-incompatibility in flowering plants.

The preferred ratio of investment of any class in a population at equilibrium is equal to:

$$x = \frac{G_M}{G_F} \cdot \frac{RS_M}{RS_F}, \text{ where}$$

G_M is the average relatedness of that class to the colony's male brood, G_F is its average relatedness to queen brood, RS_M is the expected reproductive success of a male, and RS_F that of the queen. Like most authors, Trivers and Hare use the coefficient r rather than G. See Pamilo and Crozier (1982) for a discussion of the various coefficients and reasons for preferring G. Trivers and Hare use dry weight as an index of individual cost, so that the population's ratio of investment is equal to:

$$y = \frac{\text{total dry weight of males produced}}{\text{total dry weight of queens produced}}$$

Under the assumptions that mating is not significantly nonrandom and inbreeding is insignificant, G values for queens and workers are given in Fig. 2.

$$RS_M = \frac{3-p}{4} \text{ and } RS_F = \frac{3+p}{8},$$

so that the reproductive success ratio is:

$$\frac{RS_M}{RS_F} = \frac{2(3-p)}{3+p}$$

where p is the fraction of male eggs which comes from the queen, and P_e is a measure of the number of males with which the queen has mated. This ratio does not vary for different classes within a colony.

REFERENCES

Alber, M., R. Jordan, F. Ruttner, and H. Ruttner. 1955. Von der Paarung der Honigbiene. *Z. Bienenforsch.* **3**:1-28.

Alcock, J. 1979. The evolution of intraspecific diversity in male reproductive strategies in some bees and wasps. In *Sexual Selection and Reproductive Competition in Insects*, M. S. Blum and N. A. Blum (eds.), pp. 381-402. Academic Press, New York.

Alcock, J. 1980. Natural selection and the mating systems of solitary bees. *Am. Sci.* **68**:146-153.

Alcock, J. 1981. Notes on the reproductive behavior of some Australian thynnine wasps (Hymenoptera:Tiphiidae). *J. Kans. Entomol. Soc.* **54**:681-693.

Alcock, J., E. M. Barrows, G. Gordh, L. J. Hubbard, L. Kirkendall, D. W. Pyle, T. L. Ponder, and F. G. Zalom. 1978. The ecology and evolution of male reproductive behaviour in the bees and wasps. *Zool. J. Linn. Soc.* **64**:293-326.

Alcock, J., G. C. Eickwort, and K. R. Eickwort. 1977. The reproductive behavior of *Anthidium maculosum* (Hymenoptera:Megachilidae) and the evolutionary significance of multiple copulations by females. *Behav. Ecol. Sociobiol.* **2**:385-396.

Alexander, R. D. 1974. The evolution of social behavior. *Annu. Rev. Ecol. Syst.* **5**:325-383.

Alexander, R. D., and P. W. Sherman. 1977. Local mate competition and parental investment in social insects. *Science* **196**:494-500.

Alford, D. V. 1975. *Bumblebees.* Davis-Poynter, London.

Autuori, M. 1949. Investigações sôbre a biologia de saúva. *Ciênc. Cult. (São Paulo)* **1**:4-12.

Baccetti, B. 1972. Insect sperm cells. *Adv. Insect Physiol.* **9**:315-395.

Baccetti, B. 1979. Ultrastructure of sperm and its bearing on arthropod phylogeny. In *Arthropod Phylogeny*, A. P. Gupta (ed.), pp. 609-644. Van Nostrand Reinhold, New York.

Baccetti, B., and R. Dallai. 1977. The spermatozoon of Arthropoda. III. The first multi-flagellate spermatozoon in *Mastotermes darwiniensis*. *J. Cell Biol.* **76**:569-576.

Barrows, E. M. 1976. Sweat bees. *Insect World Digest* **3**:12-16.

Beatty, R. A. 1975. Genetics of animal spermatozoa. In *Gametic Competition in Plants and Animals*, D. L. Mulcahy (ed.), pp. 61-69. North Holland, Amsterdam.

Benford, F. A. 1978. Fisher's theory of the sex ratio applied to the social Hymenoptera. *J. Theor. Biol.* **72**:701-728.

Bishop, G. H. 1920. Fertilization in the honey-bee. I. the male sexual organs: Their structure and physiological functioning. II. Disposal of the sexual fluids in the organs of the female. *J. Exp. Biol.* **31**:225-265, 265-286.

Bondi, H. 1977. The lure of completeness. In *The Encyclopaedia of Ignorance*, R. Duncan and M. Weston Smith (eds.), pp. 5-8. Pocket Books, New York.

Borgia, G. 1979. Sexual selection and the evolution of animal mating systems. In *Sexual Selection and Reproductive Competition in Insects*, M. S. Blum and N. A. Blum (eds.), pp. 19-80.

Breland, O. P., C. D. Eddleman, and J. J. Biesele. 1968. Studies of insect spermatozoa. I. *Entomol. News* **79**:197-216.

Brimley, C. S. 1908. Male *Polistes annularis* survive the winter. *Entomol. News.* **19**:109.

Brockmann, H. J. 1980. Diversity in nesting behavior of mud-daubers (*Trypoxylon politum* Say; Sphecidae). *Fla. Entomol.* **63**:53-64.

Brown, W. L., *et al.* 1982. Hymenoptera, Aculeata. In *Taxonomy and Classification of Living Organisms*, S. Parker (ed.), pp. 670-680. McGraw-Hill, New York.

Büchner, L. 1881. *La vie psychique des bêtes.* C. Reinwald, Paris. A translation of *Aus dem Geistesleben der Tiere*, 1876.

Charlesworth, B. 1978. Some models of the evolution of altruistic behaviour between siblings. *J. Theor. Biol.* **72**:297-320.

Charnov, E. L. 1978a. Sex-ratio selection in eusocial Hymenoptera. *Am. Nat.* **112**:317-326.

Charnov, E. L. 1978b. Evolution of eusocial behavior: Offspring choice or parental parasitism? *J. Theor. Biol.* **75**:451-465.

Cohen, J. 1969. Why so many sperms? An essay on the arithmetic of reproduction. *Sci. Prog. Oxford* **57**:23-41.

Cohen, J. 1975. Gamete redundancy—waste or selection? In *Gamete Competition in Plants and Animals*, D. L. Mulcahy (ed.), pp. 99-112. North Holland, Amsterdam.

Craig, R. 1979. Parental manipulation, kin selection, and the evolution of altruism. *Evolution* 33:319-324.

Craig, R. 1980a. Sex investment ratios in social Hymenoptera. *Am. Nat.* 116:311-323.

Craig, R. 1980b. Sex ratio changes and the evolution of eusociality in the Hymenoptera: Simulation and games theory studies. *J. Theor. Biol.* 87:55-70.

Cross, E. A., M. G. Stith, and T. R. Bauman. 1975. Bionomics of the organ-pipe mud-dauber, *Trypoxylon politum* (Hymenoptera:Sphecoidea). *Ann. Entomol. Soc. Am.* 68:901-916.

Crozier, R. H. 1979. Genetics of sociality. In *Social Insects, Vol. 1*, H. R. Hermann (ed.), pp. 223-287. Academic Press, New York.

Crozier, R. H. 1980. Genetical structure of social insect populations. In *Evolution of Social Behavior: Hypotheses and Empirical Tests*, H. Markl (ed.), pp. 129-146. Verlag Chemie, Weinheim, West Germany.

Crozier, R. H., and D. Brückner. 1981. Sperm clumping and the population genetics of Hymenoptera. *Am. Nat.* 117:561-563.

Cruz Höfling, M. A. da, C. da Cruz Landim, and E. W. Kitajima. 1970. The fine structure of spermatozoa from the honeybee. *An. Acad. Bras. Cienc.* 42:69-78.

Daly, M. 1978. The cost of mating. *Am. Nat.* 112:771-774.

Darwin, C. 1874. *The Descent of Man and Selection in Relation to Sex. Second Edition.* John Murray, London.

Dewsbury, D. A. 1982. Ejaculate cost and mate choice. *Am. Nat.* 119:601-610.

Donisthorpe, H. St. J. K. 1915. *British Ants: Their Life History and Classification.* Wm. Bredon & Son, Plymouth.

Eberhard, W. G. 1980. Horned beetles. *Sci. Am.* 242:166-182.

Edwards, R. 1980. *Social Wasps: Their Biology and Control.* Rentokil, East Grinstead, U.K.

Eickwort, G. C., and H. S. Ginsberg. 1980. Foraging and mating behavior of Apoidea. *Annu. Rev. Entomol.* 25:421-446.

Emlen, S. T., and L. W. Oring. 1977. Ecology, sexual selection and the evolution of mating systems. *Science* 197:215-223.

Engelmann, F. 1970. *The Physiology of Insect Reproduction.* Pergamon, Oxford.

Evans, H. E., and M. J. West-Eberhard. 1970. *The Wasps.* Univ. Michigan, Ann Arbor, MI.

Fabre, J.-H. 1886. *Souvenirs Entomologiques. Vol. 3.* Delagrave, Paris.

Fyg, W. 1952. The process of natural mating in the honeybee. *Bee World* 33:129-139.

Garófalo, C. A. 1980. Reproductive aspects and evolution of social behavior in bees (Hymenoptera, Apoidea). *Rev. Bras. Genet.* 3:139-152.

Gary, N. E. 1963. Observations of mating behavior in the honey bee. *J. Apic. Res.* 213-219.

Gary, N. E., and R. A. Morse. 1962. The events following queen cell construction in honey bee colonies. *J. Apic. Res.* 1:3-5.

Getz, W. M., D. Brückner, and T. R. Parisian. 1982. Kin structure and the swarming behavior of the honey bee *Apis mellifera*. *Behav. Ecol. Sociobiol.* 10:265-270.

Given, B. B. 1954. Evolutionary trends in the Thynninae with special reference to feeding habits of Australian species. *Trans. R. Entomol. Soc. Lond.* 105:1-10.

Gobbi, N. 1975. Aspectos evolutivos de bionomia das vespas, vizualizados atraves de estudos de reprodução (Hym. Aculeata). M.S. Thesis, Univ. Sao Paulo, Ribeirao Preto, Brasil. (Abstract.)

Guiglia, D. 1972. *Les Guêpes Sociales (Hymenoptera, Vespidae) d'Europe Occidentale et Septentrionale.* Masson, Paris.

Gwynne, D. T., and K. M. O'Neill. 1980. Territoriality in digger wasps results in sex biased predation on males (Hymenoptera:Sphecidae:*Philanthus*). *J. Kans. Entomol. Soc.* 53: 220:224.

Halverson, D. D., J. Wheeler, and G. C. Wheeler. 1976. Natural history of the sandhill ant, *Formica bradleyi* (Hymenoptera:Formicidae). *J. Kans. Entomol. Soc.* 49:280-303.

Hamilton, W. D. 1964. The genetical evolution of social behavior. II. *J. Theor. Biol.* 7:17-52.

Hamilton, W. D. 1967. Extraordinary sex ratios. *Science* 156:477-488.

Heinrich, B. 1979. *Bumblebee Economics.* Harvard Univ. Press, Cambridge, MA.

Herbers, J. M. 1979. The evolution of sex-ratio strategies in Hymenopteran societies. *Am. Nat.* 114:818-834.

Hermann, H. R., D. Gerling, and T. F. Dirks. 1974. The cohibernation and mating activity of five polistine wasp species (Hymenoptera:Vespidae:Polistinae). *J. Ga. Entomol. Soc.* 9:203-204.

Hoage, T. R., and R. G. Kessel. 1968. An electron microscope study of the process of differentiation during spermatogenesis in the drone honey bee (*Apis mellifera* L.) with special reference to centriole replication and elimination. *J. Ultrastruct. Res.* 24:6-32.

Hölldobler, B., and C. D. Michener. 1980. Mechanisms of identification and discrimination in social Hymenoptera. In *Evolution of Social Behavior: Hypotheses and Empirical Tests,* H. Markl (ed.), pp. 35-58. Verlag Chemie, Weinheim, West Germany.

Holmes, W. G., and P. W. Sherman 1983. Kin recognition in animals. *Am. Sci.* 7:46-55.

Imai, H. T. 1965. Nuptial flight and multiple mating observed in the formicine ant, *Lasius niger. Annu. Rep. Nat. Inst. Genet. Jpn.* 16:54-55.

Itô, M., and S. Imamura. 1974. Observations on the nuptial flight and internidal relationship in a polydomous ant, *Formica (Formica) yessensis* Forel. *J. Fac. Sci. Hokkaido Univ. Ser. VI Zool.* 19:681-694.

Iwata, K. 1955. The comparative anatomy of the ovary in Hymenoptera. Part I. Aculeata. *Mushi* 29:17-34.

Iwata, K. 1960. The comparative anatomy of the ovary in Hymenoptera. Supplement on Aculeata with descriptions of ovarian eggs of certain species. *Acta Hymenopterol.* 1:205-211.

Iwata, K. 1964. Egg giantism in subsocial Hymenoptera, with ethological discussion on tropical bamboo carpenter bees. *Nat. Life Southeast Asia, Kyoto* 3:399-434.

Iwata, K. 1965. The comparative anatomy of the ovary in Hymenoptera. (Records on 64 species of Aculeata in Thailand, with descriptions of ovarian eggs.) *Mushi* 38:101-109.

Iwata, K., and S. F. Sakagami. 1966. Gigantism and dwarfism in bee eggs in relation to the modes of life, with notes on the number of ovarioles. *Jpn. J. Ecol.* 16:4-16.

Jordan, R. 1967. Über die Ursachen, die sanftmütige carnica stechlustig zu machen vermögen. *Bienenvater* 3:72-76.

Kannowski, P. B., and R. L. Johnson. 1969. Male patrolling behavior and sex attraction in ants of the genus *Formica. Anim. Behav.* 17:425-429.

Kerr, W. E. 1961. Acasalmento de rainhas com varios machos em duas especies da tribu Attini (Hym., Formicoidea). *Rev. Bras. Biol.* 21:45-48.

Kerr, W. E. 1974. Advances in cytology and genetics of bees. *Annu. Rev. Entomol.* 19:253-268.

Kerr, W. E., and D. Bueno. 1970. Natural crossing between *Apis mellifera adansonii* and *Apis mellifera ligustica. Evolution* 24:145-145.

Kerr, W. E., M. R. Martinho, and L. S. Goncalves. 1980. Kinship selection in bees. *Rev. Bras. Genet.* 111:339-344.

Klahn, J. E., and G. J. Gamboa. 1983. Social wasps: Discrimination between kin and nonkin brood. *Science* 221:482-484.

Koehler, F. 1962. Recherches expérimentales sur la fertilité du receptaculum seminis à l'aide de la fécondation artificielle. *Bull. Apic. Inform.* 411:222-223.

Kropotkin, P. 1902. *Mutual Aid: A Factor in Evolution.* Heinemann, London.

Kugler, J., T. Orion, and J. Ishay. 1976. The number of ovarioles in the Vespinae (Hymenoptera). *Insectes Soc.* 23:525-533.

Kusnezov, N. 1962. El vuelo nupcial de las hormigas. *Acta Zool. Lilloana* 18:385-344.

Laidlaw, H. H. 1944. Artificial insemination of the queen bee (*Apis mellifera* L.): Morphological basis and results. *J. Morphol.* 74:429-465.

Lamb, C. G. 1922. The geometry of insect pairing. *Proc. R. Entomol. Soc. Lond. (B) Taxon.* 94:1-11.

Lee, P. E., and A. Wilkes. 1965. Polymorphic spermatozoa in the hymenopterous wasp *Dahlbominus. Science* 147:1445-1446.

Leopold, R. A. 1976. The role of male accessory glands in insect reproduction. *Annu. Rev. Entomol.* 21:199-221.

Lester, L. J., and R. K. Selander. 1981. Genetic relatedness and the social organization of *Polistes* colonies. *Am. Nat.* 117:147-166.

Levenets, I. P. 1954. The flight range of drones. *Apic. Abstr.* 14:58.

Lin, N., and C. D. Michener. 1972. Evolution of sociality in insects. *Q. Rev. Biol.* 47:131-159.

Lindauer, M. 1961. *Communication Among Social Bees.* Harvard Univ. Press. Cambridge, MA.

Lloyd, J. E. 1981. Sexual selection: Individuality, identification, and recognition in a bumblebee and other insects. *Fla. Entomol.* 64:89-118.

Loomis, W. F. 1975. *Dictyostelium discoideum. A Developmental System.* Academic Press, New York.

Low, B. S. 1978. Environmental uncertainty and the parental strategies of marsupials and placentals. *Am. Nat.* 112:197-213.

Macevicz, S. 1979. Some consequences of Fisher's sex ratio principle for social Hymenoptera that reproduce by colony fission. *Am. Nat.* 113:363-371.

MacDonald, J. F., R. D. Akre, and W. B. Hill. 1974. Comparative biology and behavior of *Vespula atropilosa* and *V. pensylvanica* (Hymenoptera:Vespidae). *Melanderia* 18:1-66.

Marchal, P. 1896. La reproduction et l'évolution des guêpes sociales. *Arch. Zool. Exp. Gén.* 4:1-100.

Markin, G. P., J. H. Dillier, S. O. Hill, M. S. Blum, and H. R. Hermann. 1971. Nuptial flight and flight ranges of the imported fire ant, *Solenopsis saevissima richteri* (Hymenoptera: Formicidae). *J. Ga. Entomol. Soc.* 6:145-156.

Marlin, J. C. 1971. The mating, nesting and ant enemies of *Polyergus lucidus* Mayr (Hymenoptera:Formicidae). *Am. Midl. Nat.* 86:181-189.

Martinho, M. R. 1979. Competição reproductiva entre machos de *Apis mellifera* L. e migração de espermatozoides para a espermateca de rainhas. D.Sc. Thesis, Univ. São Paulo, Riberão Preto, Brasil.

Masuda, H. 1939. Biological notes on *Pison iwatai* Yasumatsu. *Mushi* 12:114-146.

Metcalf, R. A., and G. S. Whitt. 1977. Intra-nest relatedness in the social wasp *Polistes metricus.* A genetic analysis. *Behav. Ecol. Sociobiol.* 2:339-351.

Michener, C. D. 1969. Comparative social behavior of bees. *Annu. Rev. Entomol.* 14:299-342.

Michener, C. D. 1974. *The Social Behavior of the Bees: A Comparative Study.* Harvard Univ. Press, Cambridge.

Michener, C. D., and D. J. Brothers. 1974. Were workers of eusocial Hymenoptera initially altruistic or oppressed? *Proc. Nat. Acad. Sci. U.S.A.* 71:671-674.

Neves, L. H. M., A. C. Stort, and J. Chaud-Neto. 1978. Observações comparativas entre abelhas africanizadas (*Apis mellifera adansonii*) e italianas (*Apis mellifera ligustica*) en relação ao compartamento de coleta de alimento. *An. II. Congr. Lat. Iber. Am. Mar del Plata.*

Noonan, K. M. 1978. Sex ratio of parental investment in colonies of the social wasp *Polistes fuscatus. Science* 199:1354-1356.

Nutting, W. L. 1969. Flight and colony foundation. In *Biology of Termites, Vol. I,* K. Krishna and F. M. Weesner (eds.), pp. 233-282. Academic Press, New York.

Oster, G. F., and E. O. Wilson. 1978. *Caste and Ecology in the Social Insects.* Princeton Univ. Press, Princeton, NJ.

Page, R. E. 1980. The evolution of multiple mating behavior by honey bee queens (*Apis mellifera* L.). *Genetics* 96:263-273.

Page, R. E., and R. A. Metcalf. 1982. Multiple mating, sperm utilization, and social evolution. *Am. Nat.* 119:263-281.

Pamilo, P. 1982. Multiple mating in *Formica* ants. *Hereditas* 97:37-45.

Pamilo, P., and R. H. Crozier. 1982. Measuring genetic relatedness in natural populations: Metholology. *Theor. Popul. Biol.* 21:171-193.

Pamilo, P., and R. Rosengren. 1984. Evolution of nesting strategies of ants: Genetic evidence from different population types of *Formica* ants. *Biol. J. Linn. Soc.* 21:331-348.

Pamilo, P., R. Rosengren, K. Vepsäläinen, S.-L. Varvio-Aho, and B. Pisarski. 1978. Population genetics of *Formica* ants. I. Patterns of enzyme gene variation. *Hereditas.* 89:233-248.

Pamilo, P., and S.-A. Varvio-Aho. 1979. Genetic structure of nests in the ant *Formica sanguinea. Behav. Ecol. Sociobiol.* 6:91-98.

Park, O. W. 1922. Time and labor factors in honey and pollen gathering. *Am. Bee J.* 62:254-255.

Parker, G. A. 1970. Sperm competition and its evolutionary consequences in the insects. *Biol. Rev.* 45:525-567.

Parker, G. A. 1979. Sexual selection and sexual conflict. In *Sexual Selection and Reproductive Competition in Insects*, M. S. Blum and N. A. Blum (eds.), pp. 123-166. Academic Press, New York.

Peacock, A. D., and A. T. Baxter. 1950. Studies in Pharaoh's ant, *Monomorium pharaonis* (L). 3. Life history. *Entomol. Mon. Mag.* 86:171-178.

Peacock, A. D., D. W. Hall, I. C. Smith, and A. Goodfellow. 1950. The biology and control of the ant pest *Monomorium pharaonis* (L). *Misc. Publ. Dept. Agric. Scotland* 17:1-51.

Peckham, D. J., F. E. Kurczewski, and D. B. Peckham. 1973. Nesting behavior of nearctic species of *Oxybelus* (Hymenoptera:Sphecidae). *Ann. Entomol. Soc. Am.* 66:647-661.

Peckham, G. W., and E. G. Peckham. 1905. *Wasps, Social and Solitary*. Houghton Mifflin, New York.

Plateaux-Quenú, C. 1972. *La Biologie des Abeilles Primitives*. Masson, Paris.

Pouvreau, A. 1963. Observations sur l'accouplement de *Bombus hypnorum* L. (Hymenoptera, Apidae) en serre. *Insectes Soc.* 10:111-118.

Prout, T., and J. Bundgaard. 1977. The population genetics of sperm displacement. *Genetics* 85:95-124.

Raignier, A., and J. Van Boven. 1955. Etude taxonomique, biologique et biométrique des *Dorylus* du sous-genre *Anomma* (Hymenoptera, Formicidae). *Ann. Mus. Congo Belge* 2:1-359.

Rau, P. 1934. The sting of the male wasp, *Monobia quadridens. Psyche* 41:245.

Réaumur, F. A. F. de. 1722. Histoire des guêpes. *Mém. Acad. R. Sci., Paris* 21(1719):302-364.

Rettenmeyer, C. W. 1963. Behavioral studies of army ants. *Univ. Kans. Sci. Bull.* 44:281-465.

Richards, O. W. 1927. Sexual selection and allied problems in the insects. *Biol. Rev.* 2:298-364.

Richards, O. W. 1978. *The Social Wasps of the Americas, Excluding the Vespine*. British Museum (Natural History), London.

Ridsdill Smith, T. J. 1970. The biology of *Hemithynnus hyalinatus* (Hymenoptera: Tiphiidae), a parasite on scarabaeid larvae. *J. Aust. Entomol. Soc.* 9:183-195.

Riek, E. F. 1970. Hymenoptera (wasps, bees, ants). In *The Insects of Australia*, D. F. Waterhouse (ed.), pp. 867-943. Melbourne Univ. Press, Carlton, Victoria.

Röseler, P.-F. 1973. Die Anzahl der Spermien im Receptaculum Seminis von Hummelköniginnen (Hym., Apoidea, Bombinae). *Apidologie* 4:267-274.

Ross, K. G. 1983. Laboratory studies of the mating biology of the Eastern Yellowjacket, *Vespula maculifrons* (Hymenoptera:Vespidae). *J. Kans. Entomol. Soc.* 56:523-537.

Rothschild, L. 1955. The spermatozoa of the honey-bee. *Trans. R. Entomol. Soc. Lond.* 107:289-294.

Runyon, D. 1929. Dark Dolores. *Cosmopolitan.*

Schaller, F. 1979. Significance of sperm transfer and formation of spermatophores in arthropod phylogeny. In *Arthropod Phylogeny*, A. P. Gupta (ed.), pp. 587-608. Van Nostrand Reinhold, New York.

Schmidt, G. H. (ed.). 1974. *Sozialpolymorphismus bei Insekten. Probleme der Kastenbildung im Tierreiche.* Wiss. Verlagsgesellschaft, Stuttgart.

Schmitt, C. 1920. Beiträge zur Biologie der Feldwespe (*Polistes gallicus* L.). *Z. Wiss. Insektenbiol.* 16:221-230. (Last of three parts.)

Schneirla, T. C. 1971. *Army Ants: A Study in Social Organization.* Freeman, San Francisco.

Scudder, G. G. E. 1971. Comparative morphology of insect genitalia. *Annu. Rev. Entomol.* 16:379-406.

Scudo, F. M., and M. T. Ghiselin. 1975. Familial selection and the evolution of social behavior. *J. Genet.* 62:1-31.

Sivinski, J. 1980. Sexual selection and insect sperm. *Fla. Entomol.* 63:99-111.

Smith, R. L. 1976a. Brooding behavior of a male water bug *Belostoma flumineum. J. Kans. Entomol. Soc.* 49:333-343.

Smith, R. L. 1976b. Male brooding behavior of the water bug *Abedus herberti* (Hemiptera: Belostomatidae). *Ann. Entomol. Soc. Am.* **69**:740-747.

Smith, R. L. 1980. Evolution of exclusive postcopulatory paternal care in the insects. *Fla. Entomol.* **63**:63-78.

Snell, G. D. 1932. The role of male parthenogenesis in the evolution of the social Hymenoptera. *Am. Nat.* **66**:381-384.

Starr, C. K. 1979. Origin and evolution of insect sociality: A review of modern theory. In *Social Insects, Vol. I*, H. R. Hermann (ed.), pp. 35-79. Academic Press, New York.

Starr, C. K. 1981. Defensive tactics of social wasps. Ph.D. Thesis, Univ. Georgia, Athens, GA.

Strassmann, J. E. 1981. Evolutionary implications of early male and satellite nest production in *Polistes exclamans* colony cycles. *Behav. Ecol. Sociobiol.* **8**:55-64.

Taber, S. 1955. Sperm distribution in the spermatheca of multiple-mated queen honey bees. *J. Econ. Entomol.* **48**:522-525.

Taber, S., and M. S. Blum. 1960. The preservation of honey bee semen. *Science* **131**:1734-1735.

Talbot, M. 1971. Flights of the ant *Formica dakotensis* Emery. *Psyche* **78**:169-179.

Thomas, C. R. 1960. The European wasp (*Vespula germanica* Fabr.) in New Zealand. *Inf. Serv. Dept. Sci. Ind. Res. New Zealand* **17**:1-74.

Thompson, T. E., and M. S. Blum. 1967. Structure and behavior of spermatozoa of the fire ant *Solenopsis saevissima* (Hymenoptera:Formicidae). *Ann. Entomol. Soc. Am.* **60**:632-642.

Thornhill, R. 1979. Male and female sexual selection and the evolution of mating strategies in insects. In *Sexual Selection and Reproductive Competition in Insects*, M. S. Blum and N. A. Blum (eds.), pp. 81-121. Academic Press, New York.

Trivers, R. L., and H. Hare. 1976. Haplodiploidy and the evolution of the social insects. *Science* **191**:249-263.

Trivers, R. L., and D. E. Willard. 1973. Natural selection of parental ability to vary the sex ratio of offspring. *Science.* **179**:90-92.

Turner, C. H. 1909. The mound of *Pogonomyrmex badius* Latr. and its relation to the breeding habits of the species. *Biol. Bull.* **17**:161-169.

Vleugel, D. A. 1947. Waarnemingen an het gedrag van de Grijze Graafke (*Andrena vaga* Panz.) (Hym.). *Entomol. Berichten* **12**:185-192.

Walker, W. F. 1980. Sperm utilization strategies in nonsocial insects. *Am. Nat.* **115**:780-799.

Werren, J. H., M. R. Gross, and R. Shine. 1980. Paternity and the evolution of male parental care. *J. Theor. Biol.* **82**:619-631.

Wheeler, W. M. 1910. *Ants: Their Structure, Development and Behavior.* Colombia Univ. Press, New York.

Wilkes, A. 1966. Sperm utilization following multiple insemination in the wasp *Dahlbominus fuscipennis. Can. J. Genet. Cytol.* **8**:541-461.

Williams, G. C. 1975. *Sex and Evolution.* Princeton Univ. Press, Princeton, NJ.

Wilson, E. O. 1971. *The Insect Societies.* Harvard Univ. Press, Cambridge, MA.

Woyke, J. 1962. Natural and artificial insemination of queen honeybees. *Bee World* **42**:21-25.

12

Sperm Competition in Poeciliid Fishes

GEORGE D. CONSTANTZ

I. INTRODUCTION

Sperm competition is the process and consequences of the ejaculates from two or more males occurring simultaneously within a single female. Among male insects, where the phenomenon has been best studied (Parker 1970), sperm competition has apparently contributed to the evolution of male traits such as a penis which flushes out of females the sperm of other males, prolonged postcopulatory guarding of females to repel additional suitors, and the deposition of copulatory plugs within the reproductive tract of females to block ejaculates of subsequent males. Male insects abound in such traits because they possess two critical preadaptations: limbs for grasping females, and genitalia which effect tight coupling.

Sperm Competition and the Evolution
of Animal Mating Systems

Livebearing fishes of the family Poeciliidae (Order Cyprinodontiformes) display an equally fascinating set of reproductive traits. Females are inseminated by several males and produce broods of mixed paternity. Relative to the males of other fishes, poeciliids exhibit intense sexual pursuit, determinate growth, alternate mating behaviors, and intromittent organs with complex hooks and spines. However, in contrast to insects, with perhaps only a single exception, male poeciliids do not physically manipulate females or their reproductive tracts. Further, it appears that male poeciliids control few resources (*e.g.*, spawning sites) vital to female fitness. Emlen and Oring (1977) have emphasized the importance of resource availability in constraining the evolution of mating systems. Relative to females, male poeciliids would appear to be in an evolutionarily weak position. What is the nature of sperm competition when males do not control females by direct manipulation or indirectly via resources?

The answer to this question is developed in four steps. First, I review the relevant biology of poeciliid fishes. Second, I describe some of their curious reproductive adaptations. Third, I demonstrate the occurrence of sperm competition by showing that multiple insemination and sperm replacement occur. Lastly, I integrate previous information to elucidate how sperm competition could have contributed to the evolution of poeciliid reproductive features.

II. BIOLOGY OF POECILIID FISHES

The Poeciliidae are endemic to the New World where they are a dominant fish group of tropical fresh and brackish waters (Rosen and Bailey 1963). The family includes familiar aquarium species such as the guppy (*Poecilia reticulata*), swordtail (*Xiphophorus helleri*), platyfish (*X. maculatus*), and the widespread mosquitofish (*Gambusia affinis*) which is used to control mosquito larvae.

A. Morphology and Physiology

Poeciliids include some of the smallest vertebrates, with most species less than 100 mm long, and males about two-thirds the length of females (Fig. 1). All but one species bear living young, and all have internal fertilization. Although the males of different species vary in size, color, and morphology, females are often gray-brown without bright colors.

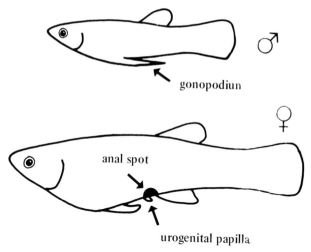

gonopodiun

anal spot

urogenital papilla

Fig. 1. Sketch of a generalized poeciliid fish.

1. The Female

Female poeciliids undergo definite ovarian cycles (Borowsky and Kallman 1976, Hopper 1943, Rosenthal 1952, Scrimshaw 1944, Siciliano 1972, Thibault and Schultz 1978, Turner, 1937, Wolf 1931). After embryos complete their gestation within the ovarian follicles (20-30 days), they are evacuated to the ovarian cavity. From there, they descend the short oviduct to the exterior. As the brood develops in the ovary, another group of ova accumulates yolk. A few days after the brood is born, ova are fertilized in the follicles. A third group of ova begins to grow and the cycle is continued. Ovarian cycles vary from 30-35 days in length.

Females carry mature, fertilizable eggs for only 1-7 days during each cycle (Hildemann and Wagner 1954, Rosenthal 1952). Poeciliid males apparently perceive the female's fertile phase through two sensory modes, visual and gustatory. Females of some poeciliids have dark spots at their urogenital region (Fig. 1). Peden (1973) demonstrated that such an anal spot facilitates orientation of the male's gonopodium. In some species (*e.g.*, *Gambusia affinis*), the anal spot varies in size during the ovarian cycle. Peden suggested that in such species, females with eggs ready for fertilization advertise their condition to males.

A pheromone produced periodically by females stimulates sexual activity in males (Amouriq 1964, 1965; Crow and Liley 1979; Gandolfi 1969; Liley 1966; Parzefall 1973). Opening of the oviduct coincides with periods of female receptivity and pheromone secretion (Weishaupt 1925). I suggest that the pheromone may function to incite male-male competition (*sensu* Cox and LeBoeuf 1977), improving the female's ability to discriminate among males.

Females with depleted sperm stores exhibit degeneration and absorption of unfertilized eggs (McKay 1971, Tavolga 1949). Consequently, sperm insufficiency results in decreasing brood size (Vallowe 1953), in contrast to the growing brood size typical of reinseminated females (Rosenthal 1952).

2. The Male

Among male poeciliids, traits exhibiting sexual modifications are the anal, pelvic, and pectoral fins, body size, and color patterns. The anal fin, modified in males for the transport of sperm, is called the gonopodium (Fig. 1). The poeciliid gonopodium is located more anteriorly than the anal fin in oviparous cyprinodontiforms, possibly because this places the gonopodium at a body region that travels forward without oscillating (Rosen and Tucker 1961). In most species, as the gonopodium is swung forward, its uppermost ray rotates and folds against the lowermost (Rosen and Tucker 1961), resulting in a temporary groove along one side of the fin (Peden 1972a). Spermatophores are presumably transmitted through this channel, although the mechanism of their transport to the gonopodial tip is a mystery.

In some species with long gonopodia (*e.g., Girardinus*), the tip of the erected gonopodium reaches to or beyond the eye (Rosen and Tucker 1961) and such fishes may orient their gonopodia visually. In contrast, species with short gonopodia have modified pectoral and pelvic fins for assisting copulatory aim (Clark and Kamrin 1951, Rosen and Bailey 1963).

There is enormous variation among species in the structure of hooks and serrae on gonopodial tips (Fig. 2). *Xiphophorus maculatus* has a small, complex gonopodial tip (Clark *et al.* 1949). Although the gonopodium of *Poecilia reticulata* has less prominent skeletal structures, it has a thin cecum with an extensive nerve plexus which serves to direct gonopodial thrusts (Clark and Aronson 1951). The gonopodium has no apparent mechanism for effecting a seal with the female genitalium or for generating hydraulic pressure within the female's body.

To my knowledge, *Xenodexia ctenolepis* is the only poeciliid in which males grasp conspecific females (Hubbs 1950). By examining preserved specimens, Hubbs inferred that insemination occurs when the male holds the female in a recess behind the modified base of his right pectoral fin. It is consistent that muscles at the breast of the right pectoral fin are twice as thick as the corresponding muscles on the left. In the context of sperm competition, the functions of this "clasper" are unknown; however, it is possible that *Xenodexia* males employ it to inhibit subsequent copulation by females or to force tight genital coupling.

Although not rigidly substantiated, there is the strong belief among students of poeciliid fishes that males grow little, if any, after they mature. For example, *Gambusia affinis* males grow no more than 1 mm after their gonopodia mature (N=50; S. Stearns, unpubl. data). Such determinate growth could have two con-

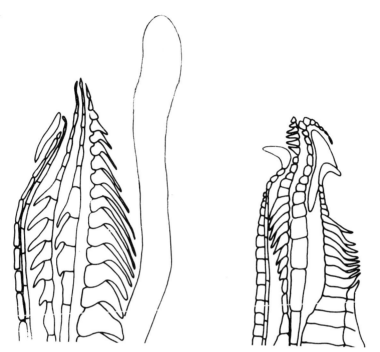

Fig. 2. Bones of gonopodial tips of *Poecilia reticulata* (left) and *Xiphophorus helleri* (right). Note elongate hood on right of *Poecilia*'s gonopodium. (Redrawn from Rosen and Bailey 1963).

sequences for males. First, individual males may be locked into specific positions in local size-dependent social hierarchies. Second, because females continue growing after maturing, males are generally smaller than females.

Within species, adult males occur at a variety of body sizes. This variation may result from, among other causes, genetic variation for age at maturity and lack of post-maturation growth. Schreibman and Kallman (1978) identified a sex-linked gene in *Xiphophorus maculatus* which controls the age at which the gonadotropic zone of the pituitary gland becomes active. The alternate alleles "early" and "late" account for the following mean ages of maturation: 12.5 weeks for individuals homozygous for "early," 20 weeks for heterozygotes, and 26.5 weeks for homozygous "late." Apparently because males grow little after maturing, early maturing males are significantly smaller than late maturing males. In addition, small, early maturing "low" males and large, late-maturing "high" males have been documented in *X. helleri* (Peters 1964), *X. montezumae* (Zander 1965), *X. pygmaeus* (Rosen and Kallman 1969), and *X. maculatus* (Kallman, unpubl. data). In some species, the variation may be discontinuous, whereas in others (*e.g., Poeciliopsis occidentalis*; Constantz 1975) the frequencies of body sizes fit a continuous, normal distribution.

Males of several poeciliids are polymorphic for body color. *Poecilia reticulata* males are so polychromic that seldom are similar individuals encountered within the same population (Baerends *et al.* 1955, Endler 1978, Haskins *et al.* 1961). Another genetic polymorphism determines gonopodial color in *Xiphophorus maculatus, X. variatus,* and *X. milleri* (Kallman and Borowsky 1972), in which some males have conspicuously black gonopodia. Other color polymorphisms may reflect social influences. For example, relative to subordinates, dominant males may be light colored (*Gambusia heterochir*; Warburton *et al.* 1957), yellow-red (*Xiphophorus variatus*; Borowsky 1973), or black (*Poeciliopsis occidentalis*; Constantz 1975).

3. Coevolution of Male and Female Genitalia

There apparently has been coevolution between the morphologies of male gonopodia and female papillae. The female's urogenital opening is located on the tip of the papilla (Fig. 1). Peden (1972b) observed that within *Gambusia*, pointed gonopodia occur in species with large urogenital papillae; whereas papillae are small or absent in species with blunt or rounded gonopodia. Peden suggested that this was an adaptation for efficient sperm transfer. I suggest an alternative hypothesis: long, sharp gonopodia are more successfully parried, and the male thwarted, by long papillae; males with blunt gonopodia may be less of a threat in this regard. Thus, I propose that a function of the papillae of female poeciliids is to deflect gonopodia, thereby improving the female's control of paternity.

4. Insemination

Sperm are massed in tiny, unencapsulated spermatophores which dissolve in ovarian fluid (Collier 1936, Philippi 1909) within 10-15 min after their introduction into the female's body (Kadow 1954). Sperm may be stored for 3-10 months (Gordon 1947; Haskins *et. al.* 1961; Hogarth and Sursham 1972; McKay 1971; Schmidt 1919; Turner 1947; Winge 1927, 1937) in epithelial folds of the ovary and oviduct (Fraser and Renton 1940, Gordon 1947, Kadow 1954, Philippi 1909, Stepanek 1928, Winge 1922). Philippi (1909) reported that in two species of *Glaridichthys*, sperm are ingested by amoeboid cells in the ovary.

B. Ecology and Life Histories

Poeciliids inhabit a variety of shallow, relatively predator-free, non-marine waters. For example, *Poecilia reticulata* occurs in springs (Liley and Seghers 1975), streams (Haskins *et al.* 1961), and drainage ditches (Liley 1966); *Xiphophorus maculatus* inhabits flooded pastures (Kallman 1975); and *Brachyraphis rhabdophora* teems in isolated pools of rainforest headwaters (Constantz 1981). Because such shallow waters are influenced by aquatic, terrestrial, and meteorologic processes, I suggest that their fish inhabitants are adapted to fluctuating environments.

A general poeciliid life history can be distilled from the literature (Constantz 1979; Haskins *et al.* 1961; Kallman 1975; Krumholz 1948, 1963; McPhail 1978; Reznick 1980; Stearns 1975; Thibault and Schultz 1978). Breeding may occur year-round in warm springs and in the tropics, or within a clearly limited season in temperate areas. Brood sizes vary from one to several hundred embryos. Fry are born 1-9 mm in length, with complete fins, fully motile, and competent to find food and avoid predators. Poeciliids become sexually mature at ages from one to 11 months, and live for 6 months to 3 years.

III. BEHAVIOR OF POECILIID FISHES

A. Mating Sequence

Although each poeciliid displays a unique ethogram, generalized patterns of sexual behavior can be inferred from the literature (Baerends *et al.* 1955; Breder and Coates 1935; Clark and Aronson 1951; Constantz 1975; Farr 1975; Farr and Herrnkind 1974; Gordon 1948; Haskins and Haskins 1949; Itzkowitz 1971; Kadow 1954; Liley 1966; Martin 1975; McPhail 1978; Parzefall 1968; Peden 1972a, 1975; Rosen and Tucker 1961). Males search for females in a rapid, sometimes frantic fashion. For example, *Gambusia affinis* swim swiftly among females and spend only 3 sec with each female (Itzkowitz 1971). Typically females do not cooperate with approaching males; however, if the female does not withdraw, males proceed with additional motor patterns which may fall into two broad categories: (1) slow movements in front of the female for the apparent purpose of advertising male traits; or (2) copulatory approaches performed outside of the female's field of vision. Male *Gambusia* and tiny *Poeciliopsis* approach the female from behind, adopt a tense, sigmoid position, and make periodic lunges at the female (Peden 1975 and Constantz 1975, respectively). These clandestine approaches may deprive females of mate choice opportunity and may reduce aggressive interactions with males higher in the dominance hierarchy.

Relative to other fishes, courtship by male poeciliids seems to be of moderate complexity. *Poecilia reticulata* give a sigmoid display (Baerends *et al.* 1955, Clark and Aronson 1951) and *P. sphenops* erects its dorsal fin (Baake 1932, Parzefall 1968). Peden (1972a) suggested that *Gambusia* courtship falls into two categories:

(1) frontal display, in which the male remains in front of the female and orients his body at 90° to the female's body, with median fins folded and body quivering, in a sigmoid posture before arcing to a position behind the female for gonopodial thrusting; and (2) lateral display, in which, with the male and female facing the same direction, the male drifts stiffly towards the female, then suddenly swims to a position below the female and thrusts. Color changes and prolonged interactions usually associated with pair bonding are absent. Male poeciliids attempt to copulate by swinging their gonopodia forward, and thrusting them at the female's urogenital opening. In some species, for example *Gambusia heterochir,* males steady the erect gonopodium in a notch on the upper margin of the pectoral fin (Peden 1972a, Warburton *et al.* 1957). Rosen and Tucker (1961) observed that species with long gonopodia and nonspecialized pectoral and pelvic fins (*e.g., Tomeurus, Heterandria, Poeciliopsis*) exhibit little display or physical contact between the sexes before thrusting. In contrast, species with shorter gonopodia and sexually dimorphic fins show more physical contact and display before copulation (*e.g., Belonesox, Poecilia, Xiphophorus*).

Among poeciliids, sperm transfer is achieved in two ways. First, executed without prior display, "gonopodial thrust" is a momentary contact in which the gonopodial tip appears to bounce off the female. Depending upon the species, thrusts either occasionally transfer sperm (*Poecilia picta,* Liley 1966) or apparently never result in insemination (*Xiphophorus helleri* and *X. maculatus,* Clark *et. al.* 1948; *Poecilia reticulata,* Stepanek 1928, Kadow 1954). Second, "copulation" is a rare, comparatively prolonged act with a receptive female. Stepanek (1928) was apparently the first to observe female cooperation during sperm transfer in a poeciliid. Subsequently, receptive responses have been reported for several species (Clark and Aronson 1951, Clark *et al.* 1948, Haskins and Haskins 1949, Liley 1966). Female *Gambusia affinis* facilitate gonopodial contact by "tilting" their bodies toward males (Martin 1975). In *Poecilia reticulata,* females cooperate when mature eggs are ready for fertilization (Kadow 1954).

In *Xiphophorus helleri* and *X. maculatus,* copulation may last 5.6 sec, during which the male and female appear to be hooked together (Clark *et al.* 1948). Physical coupling also occurs occasionally in *Poecilia reticulata* (J. Endler, pers. comm.). Species with large gonopodial holdfasts (*e.g., Xiphophorus helleri, Gambusia affinis*) have long contacts, appear to have their gonopodia locked in the female, and may terminate copulations with a violent break (Clark *et al.* 1954, Peden 1972a). Peden (1972a) observed that male *Gambusia affinis* make one or more vigorous thrusts of their body to free themselves. Upon separation, the gonopodial hooks and spines of *Poecilia reticulata* may injure female genital tissue, and sometimes possibly cause bleeding (Kadow 1954). Clark *et al.* (1954) reported that some *Xiphophorus helleri* females bled from their genital openings following withdrawal of the gonopodial tip. These observations suggest that in some poeciliids, the gonopodial tip catches and tears female genital tissue.

After copulation, the male may swim in erratic spurts. Such "post-copulatory jerking" (Baerends *et al.* 1955, Warburton *et al.* 1957) indicates that sperm have been ejaculated (Kadow 1954) and may serve to rearm the gonopodial tip with spermatophores.

B. Rate of Sexual Activity

The males of many poeciliids exhibit nearly continuous and vigorous sexual behavior (Breder and Coates 1935; Chesser *et al.* 1984; Clark and Aronson 1951; Constantz 1975; Dempster 1947; Dulzetto 1934, 1935; Farr 1980; Hildemann and Wagner 1954; Kadow 1954; Nelson 1976; Schlosberg *et al.* 1949). For example, male *Poecilia reticulata* display to females at rates of 1.4-1.6 times/min and thrust 0.1-0.7 times/min (Farr 1975); male *Xiphophorus maculatus* perform 3.2 thrusts/min and 0.8 gonopodial swings/min (Kamrin and Aronson 1954); and male *Gambusia affinis* perform an average of 0.9 sexual acts/min (Martin 1975). In many species, this level of activity persists throughout the day.

Male poeciliids often attempt to copulate with females of other species (Balsano *et al.* 1981). Male *Poecilia reticulata* court nearly anything (Breder and Coates 1935). Liley (1966) observed that males of *Poecilia reticulata, P. parae, P. picta,* and *P. vivipara* are relatively unselective and promiscuous. Subordinate male *Poeciliopsis* inseminate heterospecific females (McKay 1971). Thus, intense and promiscuous sexual behavior appears to be characteristic of male poeciliids.

C. Alternate Mating Behaviors

The phrase "alternate mating behaviors" identifies the attempts by conspecific males to fertilize eggs employing different behavior patterns (Alcock 1979). Although in some fishes, individuals may vary their mating behavior (*e.g.,* sneaking and territorial behavior by individual male *Gasterosteus aculeatus,* Van den Assem 1967), in the case of poeciliids, each individual male seems to display only one of several possible mating patterns.

Two cases of alternate mating behaviors have been described for poeciliids. In *Poecilia latipinna,* Parzefall (1968) noticed a distinction between early and late maturing males on the basis of body growth and the time of sexual differentiation. Only large *P. latipinna* males display in front of the female. Small *P. latipinna* males seem to compensate for their size disadvantage by "ambushing" females (Parzefall 1979). Male *Poeciliopsis occidentalis,* at extremes of a continuous body size distribution, exhibit contrasting mating behaviors (Constantz 1975). The largest males aggressively defend territories, display black body color, and court visiting females. Their lengthy interactions with females include head and anal nibbling, and repeated gonopodial thrusting. In contrast, tiny males flee from other males, are light colored, and roam through the population. Upon encountering a female, they remain at a distance behind her before darting in for a single gonopodial thrust. Such "sneak" copulatory attempts seem to be enhanced by a relatively long gonopodium. Other than genital contact, sneak males do not physically contact females.

IV. MULTIPLE INSEMINATION

Multiple insemination has been demonstrated to occur in six species of poeciliid fishes (Table I). Although the frequency of multiple insemination has not been estimated for *Poecilia reticulata,* broods of mixed paternity have been reported by six authors in both wild and laboratory populations (Table I). At least 56% of female *Gambusia affinis* were multiply inseminated, as determined by the frequencies of enzyme phenotypes in mother-offspring combinations (Chesser *et al.* 1984). Using similar methods, Leslie and Vrijenhoek (1977) estimated that at least 23% of *Poeciliopsis monacha* females were multiply inseminated. For *Xiphophorus maculatus* and *X. variatus,* the frequency of multiple insemination, inferred from the ratio of color patterns among progeny, was 66% (Borowsky and Kallman 1976) and 42% (Borowsky and Khouri 1976), respectively. In *Neoheterandria tridentiger,* two of six broods were sired by two males (Hall 1980).

Thus, multiple insemination has been demonstrated to occur in the following five genera: (1) *Gambusia,* (2) *Neoheterandria,* (3) *Poecilia,* (4) *Poeciliopsis,* and (5) *Xiphophorus.* Because these genera represent three of the six tribes within the Poeciliidae (Rosen and Bailey 1963), and because all species studied exhibit it, I propose that multiple insemination may be a general characteristic of the family.

Three hypotheses have been proposed to explain the adaptive significance of multiple insemination in female animals. First, Chesser *et al.* (1984) suggested that multiple insemination contributes to a larger effective population size than if females were monogamous. For example, should a few females become trapped in an isolated pool, the number of effective founders would be greater if females carried several ejaculates. This is a population level argument with overtones of group selection. An alternative hypothesis, framed at the individual level, is that females in isolated pools minimize inbreeding depression among their grandchildren by producing offspring of mixed paternity. Second, mating restores sperm supply. I previously described how sperm depletion leads to unfertilized eggs, egg resorption, and lower brood size. Boorman and Parker (1976) suggested that multiple insemination is an adaptation which allows inadequately inseminated females to refill their sperm stores. This may suffice as a proximal explanation, but fails to

Table I

Evidence of Multiple Insemination in Poeciliid Fishes

Species	Frequency of Multiple Insemination	Method of Determination	References
Gambusia affinis	≥56% of females, and probably near 100%	analyzed frequencies of enzyme phenotypes in mother-offspring combinations	Chesser *et al.* 1984
Neoheterandria tridentiger	2 of 6 broods were sired by 2 males	observed enzyme phenotypes among progeny	Hall 1980
Poecilia reticulata	The majority of broods of wild females were sired by 2 or 3 males	observed color patterns among progeny	Haskins *et al.* 1961
"	?	observed color patterns among progeny	Gandolfi 1971, Hildemann and Wagner 1954, Rosenthal 1952 Schmidt 1979, Winge 1937
Poeciliopsis monacha	≥23% of females, and probably greater	statistical analysis of enzyme phenotypes of broods	Leslie and Vrijenhoek 1977
Xiphophorus maculatus	66% of females had been fertilized by 2 males: 48% by 2 males, 24% by 3 males, and 4% by 4 males	observed color patterns among progeny	Borowsky and Kallman 1976
"	?	observed color patterns among progeny	Gordon 1947, Kallman 1965, 1970, Vallowe 1953
Xiphophorus variatus	42% of females had been fertilized by two or more males	observed color patterns among progeny	Borowsky and Khouri 1976

explain why sperm stores become depleted. And third, selection may favor females that produce heterogeneous offspring in a fluctuating environment. In this
case, I use the word "environment" to include both physical and social factors.
Shallow, seasonally variable habitats may represent an unpredictable environment
to a short-lived fish. Similarly, when a previously mated female encounters a new
more desirable male, her fitness may be increased if future offspring are fertilized
by his sperm.

The first and third hypotheses seem evolutionarily plausible. Therefore, it is
possible that female poeciliids seek multiple inseminations to achieve phenotypic
variation among their offspring and that this variation functions to minimize inbreeding depression and to track seasonal changes.

V. SPERM REPLACEMENT

The process of paternal turn-over has been studied in two poeciliid fishes,
Poecilia reticulata and *Xiphophorus maculatus*. When female *P. reticulata* were
paired successively with males of different morphs, they produced a series of
broods which alternated in their dominant phenotype (Schmidt 1919, Winge
1937). The general consequence was that the last male to mate prior to zygote
formation sired the greatest proportion of the next brood (Winge 1937).

However, sperm from such "last" males only gain most (Borowsky and
Kallman 1976, Schmidt 1919, Vallowe 1953, Winge 1937), not necessarily all
(Hildemann and Wagner 1954), fertilizations. Broods of mixed paternity result
from the following process. The female's eggs are fertilizable for only one (Rosenthal 1952) to 7 days (Hildemann and Wagner 1954). Although the eighth day
post-partum is too late for sperm replacement to occur, remating on the sixth
day yields replacement in some females (Rosenthal 1952). Thus, if a female is
paired simultaneously with males of different genetic morphs during her fertile
period, broods of mixed phenotypes may result (Borowsky and Kallman 1976,
Schmidt 1919, Winge 1937).

What is the physiological cause of such turn-over in paternity? Sperm replacement in poeciliid fishes cannot result from physical displacement of previously
deposited sperm by a subsequent male, as is the case for some insects, because
male poeciliids can neither grasp onto females nor apply hydraulic pressure to the
female urogenital tract. Therefore, paternal replacement in poeciliid fishes may
be controlled entirely by the female.

Clark and Aronson (1951) suggested a physiological mechanism consistent
with the previous conclusion. Even though a single copulation may transfer hundreds of spermatophores, sperm become undetectable in the female's reproductive

system in accordance with the following approximate schedule. A few hours after insemination, milky fluid in the oviduct teems with sperm. After 2 days, the ovarian fluid is clear but still contains a low density of sperm. At 7 days post-copulation, females contain considerably fewer sperm, and some are devoid of sperm. By 11-14 days after insemination, few, if any, sperm can be recovered from the female's oviduct. Clark and Aronson suggested that such reduction of sperm could be due to two factors: (1) sperm ingestion by amoeboid cells in the ovary, as found in *Glaridichthys* (Philippi 1909), and (2) the accumulation of sperm in folds of the oviduct (*Glaridichthys*, Philippi 1909) and ovary (*Poecilia*, Stepanek 1928, Winge 1922). Regardless of how the sperm in a female's reproductive tract are diminished in number, it is clear that they are no longer available to fertilize ova, thus allowing sperm of a subsequent male to take precedence.

What is the adaptive significance of sperm replacement? Borowsky and Kallman (1976) suggested that sperm replacement allows females to optimize fitness through the simple strategy of multiple insemination. If a female is inseminated and has sufficient sperm stores it would still be advantageous to mate with a new male if he were clearly superior to the last, and that advantage would be increased if sperm replacement were obtained. In the next section, I conjecture further on the selective advantage of sperm replacement.

The preceding discussion points to differences between poeciliid fishes and insects in the mechanisms of sperm replacement. In most insects, sperm replacement appears to be controlled by males that "smother" or hydraulically displace each other's sperm (Parker, this volume). In contrast, female poeciliid fishes seem to control sperm replacement in what may be an adaptation for increasing variation among their offspring.

VI. THE ROLE OF SPERM COMPETITION IN THE EVOLUTION OF POECILIID REPRODUCTIVE TRAITS

Poeciliids are small, soft, armorless fishes that typically occur in shallow habitats, where they are minimally exposed to predation by larger fishes. These habitats seem to be temporally variable because they are influenced by aquatic, terrestrial, and meteorologic factors. In fluctuating environments, natural selection may favor females that produce offspring of several different phenotypes. I propose that female poeciliids effect sperm replacement and allow multiple insemination as a means of achieving greater heterogeneity among offspring than would result from monogamy. This is a behavioral analog to bet-hedging in life history theory (Stearns 1976), and as such may point to the general priority of producing offspring of different phenotypes in fluctuating environments. Other theoretical considerations

(Parker 1970, this volume) suggest that selection on males should favor mechanisms to minimize the probability of their sperm being replaced or diluted. If successful, such male adaptations would tend to homogenize broods. This places males and females in evolutionary conflict. Determinants of the outcome may include the control by each sex of resources vital to reproduction, and their ability to physically manipulate members of the opposite sex.

In this context, it may be profitable to evaluate constraints on members of each sex. Poeciliid females give birth to living young with the presumed advantages that offspring are immediately able to feed and escape predators. Another benefit of livebearing may be that it precludes male control of resources vital to reproduction by females (*e.g.*, optimal oviposition substrate, or defense of eggs). Further, livebearing may obviate selection on males to provide post-partum care owing to the advanced state of neonates and because of their uncertain genetic relatedness to the male (Blumer 1979). Consequently, male poeciliids neither monopolize resources vital to female reproduction nor contribute materially to offspring survival. Female poeciliids provide eggs and nurture to embryos, both of which limit male reproduction.

With one possible exception, male poeciliids are unable to physically manipulate females. This contrasts with male insects that physically control females in two ways not available to poeciliids. First, some male insects clasp the female's body to prevent subsequent copulation by another male and possible displacement of their own sperm. Second, positive genital coupling may allow male insects to displace previously deposited sperm (see Parker, Waage, and Gromko *et al.*, this volume). The absence of clasping limbs and genitalic locking mechanisms precludes male manipulation of female sperm stores in poeciliids.

In summary, male poeciliids control neither females nor resources vital for female reproduction. Females provide eggs and nurture, both resources which are vital to male fitness. I submit that this disparity may account for many of the reproductive traits of male poeciliids.

Female poeciliids have asynchronous reproductive cycles (but see Borowsky and Diffley 1981) through a protracted breeding season. Thus only a few females are receptive at any given time (Balsano *et al.* 1981). The female's body does not appear to change throughout the ovarian cycle in any manner that would serve as a long-distance cue for males. Thus, males must closely approach or actually touch females to determine their breeding condition (*i.e.*, if gravid with mature eggs). The optimal male reproductive strategy in such a system may be for males to constantly move among females, approach each closely to determine their reproductive status, then quickly mate and/or resume the search.

Individuals allocate energy to the basic functions of growth, maintenance, and reproduction; disbursement to one function necessarily decreases resources available for other processes (Williams 1966). In the case of male poeciliids, if maintenance costs are determined by ambient temperatures and body mass, energy available for growth and reproduction must vary inversely through tradeoff. I suggest natural selection may favor male poeciliids that allocate to sexual behavior all energy above that required for somatic maintenance, and this may account for the fact that male poeciliids grow little after maturing sexually.

This thesis may be testable. It predicts a negative correlation between the intensity of male sexual activity and the degree of post-maturation growth. Thus, males of species in which sexual behavior is frantic should grow less after maturing than males of species with more leisurely courtship. At present, data for evaluating this prediction are not available. Because the growth of adult male poeciliids is very slow or even determinate, males that mature at small body sizes may be disadvantaged. If female choice is based on male size, small males would have little chance of ever being chosen. This disadvantage could select small males that are able to copulate by circumventing female choice. Thus, programmed plasticity in the ontogeny of reproductive behavior may produce demonstrative large males and sneaky small males. If so, alternate mating behaviors should be most prevalent in species where mature males grow the least.

Relative to many other fishes, poeciliids exhibit intense, but relatively uncomplicated courtship. In general, courtship by male animals functions to advertise genetic attributes, ability to control resources important for reproduction by females, and/or a willingness and skill to provide parental care. Since male poeciliids contribute neither resources nor parental care, their courtship must simply advertise genetic quality.

Poeciliid gonopodia are structurally complex, and highly variable among species in the size and shape of their bony projections. I propose that the function of some of these skeletal elements in some species is to tear, irritate, or otherwise traumatize female genital tissue. As previously noted, forceful copulatory termination results in vaginal breeding, oviducal blood clots, and females subsequently unreceptive to copulation—all consistent with my hypothesis. This minimizes sperm replacement in two ways. First, an irritated female may be sexually unreceptive until her brood has been fertilized, and second, a blood clot produced after such an injury may act as a copulatory plug (*sensu* Murie and McLean 1980, Ross and Crews 1977), inhibiting the penetration of sperm from subsequent copulations. I consider this possibility of enforced chastity to be one of the most exciting questions awaiting students of poeciliid fishes.

I close this paper by extending present inferences drawn from data on poeciliids to other livebearing fishes. I have proposed that some characteristics of male poeciliids are evolutionary consequences of the fact that males neither monopolize resources nor dominate females. If these are fundamental evolutionary constraints, my reasoning should apply to other fishes with internal fertilization. Let us briefly consider the correlates of each type of control by the males of some other species.

Among fishes, one of the most significant resources for female reproduction, and therefore the object of male defense, is a spawning substrate (*e.g.*, gravel depressions, nests made of plant material, spawning rocks). However, because such objects are irrelevant to livebearing females, one of the most powerful forces selecting for territorial defense is absent. Similarly, possibly because food is spatially diffuse, defending a food patch is precluded. Thus, relative to other fishes, males of livebearing species have little opportunity to monopolize resources critical to females.

In contrast to the uniformity in the low level of territorial behavior, livebearing fishes vary in the degree to which males dominate females. This diversity should result in differences in the nature of sperm competition. For example, because male goodeids have no means of grasping females and therefore may exercise little control over them, they should display a constellation of traits similar to that reported for poeciliids. It is consistent, for example, that male goodeids compete aggressively for females (Kingston 1980). In contrast, male elasmobranchs (sharks and rays) latch onto females with teeth and genital claspers, possibly to inhibit subsequent copulation by the female or as a means to flush out sperm deposited by another male. Consequently, I would expect elasmobranchs to exhibit a relaxed and prolonged courtship, lengthy copulation, continued growth after sexual maturity, and only infrequent broods of mixed paternity. Although little is known of the mating behavior of sharks and rays, preliminary sketches (Klimley 1980, McCourt and Kerstitch 1980, Tricas 1980) are consistent with the first two expectations, *i.e.*, they display lengthy bouts of courtship and copulation. I will consider this review to be successful if it stimulates research on sperm competition in other livebearing fishes, thereby forcing a revision of present hypotheses.

VII. SUMMARY

Mated females of the fish family Poeciliidae effect sperm replacement when they are inseminated by a new male. The adaptive significance of this mating system may be its facilitation of change in the phenotypes of offspring in the shallow, variable habitats typical of poeciliids. In contrast, males experience higher fitness if they are able to inhibit subsequent copulation by females; if successful, this produces lower phenotypic variance among a female's offspring. The outcome of this intersexual evolutionary conflict may be determined by the ability of members of each sex to physically manipulate individuals of the opposite sex and to control resources vital to reproduction. Male poeciliids neither physically dominate (*e.g.*, hold on to females) nor control resources critical to female fitness (*e.g.*, spawning sites). In contrast, females provide eggs and nurture, both limited resources for males.

In response to such disparity, natural selection may have favored males that expend enormous time and energy on reproductive activities. Typically, male poeciliids search for females at a rapid pace. Possibly because such enormous output leaves little energy for somatic growth or maintenance of additional soma, males grow little, if any, after attaining sexual maturity. In some species, males that mature at a small body size, and therefore may be less successful in courtship, attempt to override female choice by quick, sneak copulations. There is an indication in the literature that males are able to manipulate females in one way: using their intromittent organs to traumatize female genitalia and thereby induce temporary post-copulatory chastity in females.

ACKNOWLEDGMENTS

Research funds from the Division of Limnology and Ecology, Academy of Natural Sciences, supported the preparation of this review. Robin Vannote, Director of the Stroud Water Research Center, and Harold and Shirley Parsons graciously provided office space. The criticisms of John Endler, Tim Halliday, Astrid Kodric-Brown, and Steve Stearns significantly improved the manuscript. For unpublished data, I thank Klaus Kallman and Ron Chesser. I want to thank Charlie Reimer for his patience in translating German papers into English. Renée Askins helped search the literature. Thanks to all.

REFERENCES

Alcock, J. 1979. *Animal Behavior: An Evolutionary Approach, 2nd Ed.* Sinauer, Sunderland, MA.

Amouriq, L. 1964. L'activite et le phenomene social chez *Lebistes reticulatus* (Poeciliidae, Cyprinodontiformes). *C. R. Acad. Sci. Paris* 259:2701-2702.

Amouriq, L. 1965. Origine de la substance dynamogene emise par *Lebistes reticulatus* femelle (Poisson Poeciliidae, Cyprinodontiformes). *C. R. Acad. Sci. Paris* 160:2334-2335.

Baake, K. 1932. *Mollienesia velifera* (Regan). *Bl. Aquar. Terrarienk.* 43:69-71.

Baerends, G. P., B. Brouwer, and H. Tj. Waterbolk. 1955. Ethological studies on *Lebistes reticulatus* (Peters): 1. An analysis of the male courtship. *Behaviour* 8:249-334.

Balsano, J. S., K. Kucharski, E. J. Randle, E. J. Rasch, and P. J. Monaco. 1981. Reduction of competition between bisexual and unisexual females of *Poecilia* in northeastern Mexico. *Environ. Biol. Fishes* 6:39-48.

Blumer, L. S. 1979. Male parental care in the bony fishes. *Q. Rev. Biol.* 54:179-161.

Boorman, E., and G. A. Parker. 1976. Sperm (ejaculate) competition in *Drosophila melanogaster* and the reproductive value of females in relation to female age and mating status. *Ecol. Entomol.* 1:145-155.

Borowsky, R. 1973. Relative size and the development of fin coloration in *Xiphophorus variatus*. *Physiol. Zool.* 46:22-28.

Borowsky, R., and J. Diffley. 1981. Synchronized maturation and breeding in natural populations of *Xiphophorus variatus* (Poeciliidae). *Environ. Biol. Fishes* 6:49-58.

Borowsky, R., and K. D. Kallman. 1976. Patterns of mating in natural populations of *Xiphophorus* from Belize and Mexico. *Evolution* 30:693-706.

Borowsky, R. L., and J. Khouri. 1976. Patterns of mating in natural populations of *Xiphophorus*. II. *X. variatus* from Tamaulipas, Mexico. *Copeia* 1976:727-734.

Breder, C. M., and C. W. Coates. 1935. Sex recognition in the guppy, *Lebistes reticulatus* Peters. *Zoologica* 19:187-107.

Chesser, R. K., M. W. Smith, and M. H. Smith. 1984. Biochemical genetics of mosquitofish populations. I. incidence and significance of multiple insemination. *Genetica,* in press.

Clark, E., and L. R. Aronson. 1951. Sexual behaviour in the guppy, *Lebistes reticulatus. Zoologica* 36:49-66.

Clark, E., and R. P. Kamrin. 1951. The role of the pelvic fins in the copulatory act of certain poeciliid fishes. *Am. Mus. Novit.* 159:1-14.

Clark, E., L. R. Aronson, and M. Gordon. 1948. An analysis of the sexual behavior of two sympatric species of poeciliid fishes and their laboratory induced hybrids. *Anat. Rec.* 101:42.

Clark, E., L. R. Aronson, and M. Gordon. 1949. The role of the distal tip of the gonopodium during the copulatory act of the viviparous teleost *Platypoecilus maculatus. Anat. Rec.* 105:26-27.

Clark, E., L. R. Aronson, and M. Gordon. 1954. Mating behavior patterns in two sympatric species of xiphophorin fishes: Their inheritance and significance in sexual isolation. *Bull. Am. Mus. Nat. Hist.* 103:139-115.

Collier, A. 1936. The mechanism of internal fertilization in *Gambusia. Copeia* 1936:45-53.

Constantz, G. D. 1975. Behavioral ecology of mating in the male Gila topminnow, *Poeciliopsis occidentalis* (Cyprinodontiformes:Poeciliidae). *Ecology* 56:966-973.

Constantz, G. D. 1979. Life history patterns of a livebearing fish in contrasting environments. *Oecologia* 40:189-201.

Constantz, G. D. 1981. Freshwater fishes of Corcovado National Park, Costa Rica. *Proc. Acad. Nat. Sci. Phila.* 133:15-19.

Cox, C. R., and R. J. LeBoeuf. 1977. Female incitation of male competition: A mechanism in sexual selection. *Am. Nat.* 111:317-335.

Crow, R. T., and N. R. Liley. 1979. A sexual pheromone in the guppy, *Poecilia reticulata* (Peters). *Can. J. Zool.* 57:184-188.

Dempster, R. P. 1947. Guppies. *Aquar. J.* 18:4-6.

Dulzetto, F. 1934. Osservationi sulla vita e sul rapporto sessuale dei nati *Gambusia holbrookii* (Grd.). *Arch. Zool. Ital.* 10:45-65.

Dulzetto, F. 1935. Nuove osservationi sulla vita e sul rapporto sessuale di "*Gambusia holbrookii* Grd." *Rend. R. Accad. Naz. Lincei, Rome* 21:524-532.

Emlen, S. T., and L. W. Oring. 1977. Ecology, sexual selection, and the evolution of mating systems. *Science* 197:215-223.

Endler, J. A. 1978. A predator's view of animal color patterns. In *Evolutionary Biology, Vol. II,* M. K. Hecht, W. E. Steere, and B. Wallace (eds.), pp. 319-364. Plenum, New York.

Farr, J. A. 1975. The role of predation in the evolution of social behavior of natural populations of the guppy. *Evolution* 19:151-158.

Farr, J. A. 1980. Social behavior patterns as determinants of reproductive success in the guppy, *Poecilia reticulata* (Pisces:Poeciliidae). *Behaviour* 74:38-91.

Farr, J. A., and W. F. A. Hermkind. 1974. A quantitative analysis of social interaction of the guppy, *Poecilia reticulata* (Pisces:Poeciliidae) as a function of population density. *Anim. Behav.* 22:582-592.

Fraser, E. A., and R. M. Renton. 1940. Observations on the breeding and development of the viviparous fish *Heterandria formosa. Q. J. Microsc. Sci.* 81:479-516.

Gandolfi, G. 1969. A chemical sex attractant in the guppy *Poecilia reticulata* Peters (Pisces, Poeciliidae). *Monit. Zool. Ital.* 3:89-98.

Gandolfi, G. 1971. Sexual selection in relation to the social status of males in *Poecilia reticulata* (Teleostei:Poeciliidae). *Boll. Zool.* 38:35-48.

Gordon, M. 1947. Speciation in fishes: Distribution in time and space of seven dominant multiple alleles in *Platypoecilus maculatus. Adv. Genet.* 1:95-132.

Gordon, M. 1948. Guppy's antics analyzed by scientists. *Aquarium* 17:243-246.

Hall, E. 1980. Size dependent reproductive success in *Neoheterandria tridentiger*. B.A. Thesis, Reed College, Portland, OR.

Haskins, C. P., and E. F. Haskins. 1949. The role of sexual selection as an isolating mechanism in three species of poeciliid fishes. *Evolution* 3:160-169.

Haskins, C. P., E. F. Haskins, J. J. A. McLaughlin, and R. E. Hewitt. 1961. Polymorphism and population structure in *Lebistes reticulatus* (an ecological study). In *Vertebrate Speciation*, R. W. Blair (ed.), pp. 320-395. Univ. Texas Press, Austin.

Hildemann, W. H., and E. D. Wagner. 1954. Intraspecific sperm competition in *Lebistes*. *Am. Nat.* 88:87-91.

Hogarth, P. J., and C. M. Sursham. 1972. Antigenicity of *Poecilia* sperm. *Experientia* 28:463.

Hopper, A. F., Jr. 1943. The early embryology of *Platypoecilus maculatus*. *Copeia* 1943: 218-224.

Hubbs, L. 1950. Studies of cyprinodont fishes. XX. A new subfamily from Guatemala with ctenoid scales and a unilateral pectoral clasper. *Misc. Publ. Mus. Zool. Univ. Mich.* 78.

Itzkowitz, M. 1971. Preliminary study of the social behavior of male *Gambusia affinis* (Baird & Girard) (Pisces:Poeciliidae) in aquaria. *Chesapeake Sci.* 12:219-224.

Kadow, P. E. 1954. An analysis of sexual behavior and reproductive physiology in the guppy, *Lebistes reticulatus* (Peters). Ph.D. Dissertation, New York University.

Kallman, K. D. 1965. Genetics and geography of sex determination in the poeciliid fish, *Xiphophorus maculatus*. *Zoologica* 50:151-190.

Kallman, K. D. 1970. Sex determiantion and the restricton of sex-linked pigment patterns to the X and Y chromosomes in populations of a poeciliid fish, *Xiphophorus maculatus*, from the Belize and Sibun rivers of British Honduras. *Zoologica* 55:1-16.

Kallman, K. D. 1975. The platyfish, *Xiphophorus maculatus*. In *Handbook of Genetics, Vol. IV*, R. C. King (ed.), pp. 81-132. Plenum, New York.

Kallman, K. D., and R. Borowsky. 1972. The genetics of gonopodial polymorphism in two species of poeciliid fish. *Heredity* 28:197-310.

Kamrin, R. P., and L. R. Aronson. 1954. The effects of forebrain lesions on mating behavior in the male platyfish, *Xiphophorus maculatus*. *Zoologica* 39:133-140.

Kingston, D. I. 1980. Eye-color changes during aggressive displays in the goodeid fishes. *Copeia* 1980:169-171.

Klimley, A. P. 1980. Observations of courtship and copulation in the nurse shark, *Ginglymostoma cirratum*. *Copeia* 1980:878-882.

Krumholz, L. A. 1948. Reproduction in the western mosquitofish, *Gambusia affinis affinis* (Baird and Girard) and its use in mosquito control. *Ecol. Monogr.* 18:1-43.

Krumholz, L. A. 1963. Relationships between fertility, sex ratio, and exposure to predation in populations of the mosquitofish *Gambusia manni* Hubbs at Bimini, Bahamas. *Int. Rev. Gesamten Hydrobiol.* 48:201-256.

Leslie, J. F., and R. C. Vrijenhoek. 1977. Genetic analysis of natural populations of *Poeciliopsis monacha*: Allozyme inheritance and pattern of mating. *J. Hered.* 68:301-306.

Liley, N. R. 1966. Ethological isolating mechanisms in four sympatric species of poeciliid fishes. *Behav. Suppl. 13*.

Liley, R. N., and B. H. Seghers. 1975. Factors affecting the morphology and behavior of guppies in Trinidad. In *Function and Evolution in Behaviour*, G. P. Baerends, C. Beer, and A. Manning (eds.), pp. 92-118. Oxford University Press.

Martin, R. 1975. Sexual and aggressive behavior, density and social structure in a natural population of mosquitofish, *Gambusia affinis holbrooki*. *Copeia* 1975:445-454.

McCourt, R. M., and A. N. Kerstitch. 1980. Mating behavior and sexual dimorphism in dentition in the stingray *Urolophus concentricus* from the Gulf of California. *Copeia* 1980: 900-901.

McKay, F. E. 1971. Behavioral aspects of population dynamics in unisexual-bisexual *Poeciliopsis* (Pisces:Poeciliidae). *Ecology* 52:778-790.

McPhail, J. D. 1978. Sons and lovers: The functional significance of sexual dichromisms in a fish, *Neoheterandria tridentiger* (Garman). *Behaviour* 64:329-339.

Murie, J. O., and I. G. McLean. 1980. Copulatory plugs in ground squirrels. *J. Mammal.* 62: 355-356.

Nelson, J. L. 1976. Sexual selection and the swordtail fish, *Xiphophorus helleri*. Ph.D. Dissertation, Uniiversity of California, Santa Cruz.

Parker, G. A. 1970. Sperm competition and its evolutionary consequences in the insects. *Biol. Rev.* 45:525-568.

Parzefall, J. 1968. Comparative ethology of several varieties of *Mollienesia* including a cave form of *M. sphenops*. *Behaviour* 33:1-37.

Parzefall, J. 1973. Attraction and sexual cycle of poeciliids. In *Genetics and Mutagenesis of Fish*, J. H. Schroder (ed.), pp. 177-183. Springer-Verlag, Berlin.

Parzefall, J. 1979. Zur genetik und biologischen bedeutung des aggressionsuerhaltens von *Poecilia sphenops* (Pisces, Poeciliidae). *Z. Tierpsychol.* 50:399-422.

Peden, A. E. 1972a. The function of gonopodial parts and behavior pattern during copulation by *Gambusia* (Poeciliidae). *Can J. Zool.* 50:955-968.

Peden, A. E. 1972b. Differences in the external genitalia of female gambusiin fishes. *Southwest Nat.* 17:265-272.

Peden, A. E. 1973. Variation in anal spot expression of gambusiin females and its effect on male courtship. *Copeia* 1973:159-263.

Peden, A. E. 1975. Differences in copulatory behavior as partial isolating mechanisms in the poeciliid fish *Gambusia*. *Can. J. Zool.* 53:1290-1296.

Peters, G. 1964. Vergleichende untersuchungen an drei subspecies von *Xiphophorus helleri* (Heckel) (Pisces). *Z. Zool. Syst. Evolutionforsch.* 2:185-271.

Philippi, E. 1909. Fortpflanzugsgeschicte der viviparen teleosteer *Glaridichthys januarius* und *G. decemmaculatus* in ihrem ein fluse auf lebensweise makrosckopische und microskopische anatomie. *Zool. Jahrb.* 27:1-94.

Reznick, D. N. 1980. Life history evolution in the guppy (*Poecilia reticulata*). Ph.D. Dissertation, University of Pennsylvania, Philadelphia.

Rosen, D. E., and R. M. Bailey. 1963. The poeciliid fishes (Cyprinodontiformes), their structure, zoogeography and systematics. *Bull. Am. Mus. Nat. Hist. 126*.

Rosen, D. E., and K. D. Kallman. 1969. A new fish of the genus *Xiphophorus* from Guatemala, with remarks on the taxonomy of endemic forms. *Am. Mus. Novit. 2379*.

Rosen, D. E., and A. Tucker, 1961. Evolution of secondary sexual characters and sexual behavior patterns in a family of viviparous fishes (Cyprinodontiformes:Poeciliidae). *Copeia* 1961:102-212.

Rosenthal, H. L. 1952. Observations on reproduction of the poeciliid *Lebistes reticulatus* (Peters). *Biol. Bull.* 102:30-38.

Ross, P., Jr., and D. Crews. 1977. Influence of the seminal plug on mating behaviour in the garter snake. *Nature* 267:344-345.

Schlosberg, H., M. C. Duncan, and B. Daitch. 1949. Mating behavior of two live-bearing fish *Xiphophorus helleri* and *Platypoecilus maculatus*. *Physiol. Zool.* 22:148-161.

Schmidt, J. 1919. Racial investigations. III. Experiments with *Lebistes reticulatus* (Peters) Regan. *C. R. Trav. Lab. Carlsberg, 14*.

Schreibman, M. P., and K. D. Kallman. 1978. The genetic control of sexual maturation in the teleost, *Xiphophorus maculatus* (Poeciliidae); a review. *Ann. Biol. Anim. Biochim. Biophys.* 18:957-962.

Scrimshaw, N. S. 1944. Superfetation in poeciliid fishes. *Copeia* 1944:180-183.

Siciliano, M. J. 1972. Evidence for a spontaneous ovarian cycle in fish of the genus *Xiphophorus*. *Biol. Bull.* 142:480-488.

Stearns, S. C. 1975. A comparison of the evolution and expression of the life history in stable and fluctuating environments: *Gambusia affinis* in Hawaii. Ph.D. Dissertation, University of British Columbia, Vancouver.

Stearns, S. C. 1976. Life-history tactics: A review of the ideas. *Q. Rev. Biol.* 51:3-47.

Stepanek, O. 1928. Morfologie a biologie genitalnich organu u *Lebistes reticulatus* Ptrs. (Cyprinodontidae viviparae). *Publ. Fac. Sci. Univ. Charles* 79:1-30.

Tavolga, W. N. 1949. Embryonic development of the platyfish (*Platypoecilus*), the swordtail (*Xiphophorus*), and their hybrids. *Bull. Am. Mus. Nat. Hist.* 94:163-229.

Thibault, R. E., and R. J. Schultz. 1978. Reproductive adaptations among viviparous fishes (Cyprinodontiformes:Poeciliidae). *Evolution* 32:320-333.

Tricas, T. C. 1980. Courtship and mating-related behavior in myliobatid rays. *Copeia* **1980**: 553-556.

Turner, C. L. 1937. Reproductive cycles and superfoetation in poeciliid fishes. *Biol. Bull.* **72**:145-164.

Turner, C. L. 1947. Viviparity in teleost fishes. *Sci. Monthly* **74**:508-518.

Vallowe, H. H. 1953. Some physiological aspects of reproduction in *Xiphophorus maculatus*. *Biol. Bull.* **104**:240-249.

van den Assem, J. 1967. Territory in the three-spined stickleback, *Gasterosteus aculeatus* L.: An experimental study in intraspecific competition. *Behav. Suppl. 16.*

Warburton, B., C. Hubbs, and D. W. Hagen. 1957. Reproductive behavior of *Gambusia heterochir. Copeia* **1957**:299-300.

Weishaupt, E. 1925. Die ontogenie der genitalorgane von *Girardinus reticulatus. Z. Wiss. Zool.* **126**:571-611.

Williams, G. C. 1966. *Adaptation and Natural Selection.* Princeton Univ. Press, Princeton, NJ.

Winge, O. 1922. A peculiar mode of inheritance and its cytological explanation. *J. Genet.* **12**:137-144.

Winge, O. 1927. The location of eighteen genes in *Lebistes reticulatus. J. Genet.* **18**:1-43.

Winge, O. 1937. Succession of broods in *Lebistes. Nature* **140**:467.

Wolf, L. E. 1931. The history of the germ cells in the viviparous teleost, *Platypoecilus maculatus. J. Morphol. Physiol.* **52**:115-153.

Zander, C. D. 1965. Die geschlechtsbestimmung bei *Xiphophorus montezumae cortezi* Rosen (Pisces). *Z. Vererbungsl.* **96**:128-141.

13

Sperm Competition in Amphibians

T. R. HALLIDAY

P. A. VERRELL

I. INTRODUCTION

Since sperm competition has not yet been demonstrated in any species of amphibian, this chapter is essentially speculative. Our aim is to examine the reproductive biology of amphibians and identify situations in which sperm competition may be an important factor. In some amphibians there appears to be considerable scope for sperm competition to occur and this possibility makes it appropriate to formulate new hypotheses about certain aspects of their sexual behavior. We consider the three orders of living amphibians, the anurans (Salientia;

frogs and toads), the apodans (Gymnophonia; caecilians), and the urodeles (Caudata; salamanders and newts), separately, because of their very different modes of sperm transfer. In most anurans fertilization is external. In apodans fertilization is internal, with sperm being transferred by means of an intromittent organ. In most urodeles there is internal fertilization, but sperm is transferred indirectly by means of a spermatophore. Clearly, the potential for sperm competition of the kind found in insects and other groups of animals is much greater in the two amphibian orders having internal fertilization.

II. ANURANS

In frogs and toads the typical mating pattern is for the male to grasp the female dorsally in amplexus which is maintained until the female spawns. As the eggs emerge, the male sheds sperm onto them. In most species the sex ratio within mating populations is skewed towards an excess of males (Wells 1977a) and males in amplexus commonly have to defend females against single males who attempt to displace them. A typical example of such behavior is provided by the European toad (*Bufo bufo*; Davies and Halliday 1977, 1978, 1979).

Sperm competition could occur in anurans if single males shed sperm onto the eggs of a pair, either during oviposition or very shortly after. Such behavior would be analogous to the opportunistic fertilizations achieved by some males in a number of fishes that employ external fertilization. Examples include the three-spined stickleback (*Gasterosteus aculeatus;* van den Assem 1967), the longear sunfish (*Lepomis megalotis*; Keenleyside 1972), the bluegill sunfish (*Lepomis macrochirus;* Gross 1979. Gross and Charnov 1980, Dominey 1980), the fourspine stickleback (*Apeltes quadracus*; Rowland 1979), and the blenny (*Tripterygion tripteronotus*; Wirtz 1978). In these species, small males rush towards a spawning pair and shed sperm in the vicinity of the newly-laid eggs.

There is only one detailed report of comparable behavior for an anuran, that described by Pyburn (1970) in the leaf frogs *Phyllomedusa callidryas* and *Pachymedusa dacnicolor* in Mexico. Males call from vegetation near pools and are approached by females. A male grasps a female and the pair moves down from the male's calling perch to a pool where the female fills her bladder with water. The pair then leaves the pool and climbs back into the vegetation, where eggs are laid in batches on the upper surfaces of leaves. The female releases water from her bladder onto each batch, returning to the pond, with the male on her back, to refill her bladder between each bout of egg-laying. During these movements, pairs frequently pass unpaired males who try to climb onto the backs of females. Paired males resist these attempts, but if an amplectant male fails to dislodge an intruder, both males may shed sperm onto the female's eggs.

Multiple paternity is also likely to occur in the African treefrogs (*Chiromantis*), in which a single female may be grasped by up to three males as she constructs a foam nest and lays her eggs in it (Coe 1974). This behavior has been interpreted as an example of cooperative breeding, possibly involving closely related animals (Wilson 1975), but there is no evidence for this apart from the observation that all the males contribute to the process of whipping up the foam with their hindlimbs.

The position of a male anuran during amplexus can make it difficult for him to repel rival males quickly. He is limited to kicking out with his hindlegs, since he must use his forelimbs to retain a firm grip on the female. In toads (*Bufo bufo*) a single strong kick by an amplectant male will sometimes knock a rival clear, but in many fights an intruder gains a grip on the female and the ensuing struggle may last for several hours (Davies and Halliday 1979). Fights sometimes occur while the female is spawning, a process which takes several hours in this species. Toad fights clearly seem to be directed towards establishing exclusive possession of a female which, during oviposition, involves attempts by an amplectant male to prevent intruders from fertilizing some of his female's eggs.

In some species males prevent or reduce intrusion by rivals during mating by establishing a territory around the spawning site. Anurans vary considerably in the extent to which they show territorial behavior (Wells 1977a). In some species, such as the North American bullfrog (*Rana catesbeiana*; Howard 1978a, 1978b), the green frog (*R. clamitans*; Wells 1977b, 1978), and the gladiator frog (*Hyla rosenbergi*; Kluge 1981), males defend a territory from which they call to attract females and in which spawning occurs. In others, such as the European natterjack toad (*Bufo calamita*), males defend calling sites, but spawning occurs elsewhere in the pond at a site that is not defended (Arak 1983). Territorial behavior occurs only in those species that have prolonged breeding seasons (Wells 1977a). In species in which the breeding season is very short, described as explosive breeders, there tends to be a pattern of "scramble competition" (Wells 1977a), in which males attempt to clasp any individual they meet and struggle with one another for the possession of females.

In some explosive breeders, spawning is communal and takes place at a very localized site within a breeding pond. In the frogs *Rana temporaria* (unpubl. data) and *R. sylvatica* (Howard 1980, Berven 1981), individual spawn masses are laid on top of one another to produce a single, very large spawn mass. The functional significance of such behavior probably has at least two aspects. First, a dense aggregation of eggs and, when they hatch, of tadpoles, will generate a "selfish herd" effect whereby individual progeny are less likely to be predated (Hamilton 1971). Secondly, the temperature within a large spawn mass may be significantly higher than that of the surrounding water for at least part of the day, promoting faster development (Howard 1980, Waldman 1982). In both European frogs (*Rana temporaria*) in which spawning is highly gregarious, and toads (*Bufo bufo*) in which spawning pairs are rather less densely aggregated, single males gather in considerable numbers among the spawn, often close to spawning pairs. These single males may include those that have already mated, as well as those that have failed to find a mate, and it is possible that they shed sperm indiscriminately onto the spawn mass. Such behavior could be adaptive if amplectant males fail to fertilize all of the eggs laid by the females with whom they have paired.

Laboratory experiments with single pairs of both *Bufo bufo* and *Rana temporaria* suggest that there may be considerable variance in the proportion of a female's eggs which an amplectant male can successfully fertilize (Halliday and Davies, unpubl.). In *Bufo bufo* the fertilization rate of captive pairs ranged from 0 to 90.4%, with a mean of 38.0% and a standard deviation of 24.8 (N=41). For *Rana temporaria* comparable figures are: range 0 to 98.2%, mean 64.9%, standard deviation 28.5 (N=55). These figures contrast markedly with what we have observed in nature where, in both species, virtually 100% of eggs appear to be fertilized. However, it is very unlikely that the disparity between the high fertilization rate observed in the wild and the often low values obtained in the laboratory can be attributed solely to the activities of single males in the natural situation. The experimental spawnings were carried out under very artificial conditions, with pairs confined in plastic containers (Davies and Halliday 1977), and this is probably the major cause of the variance in fertilization rate that we observed. However, it seems reasonable to suggest that single males may move about over spawn masses and shed sperm that may fertilize eggs not fertilized when they were laid. If single males do behave in this way, then it is also possible that their sperm may come into competition with that of amplectant males to fertilize eggs as females lay them. There is evidence, however, that in at least one anuran, *Rana sylvatica,* a male's sperm supply is quite quickly depleted (Smith-Gill and Berven 1980); this would limit the scale of any sperm competition effect.

In a few anurans fertilization is internal. In the tailed frog (*Ascaphus truei*) the male possesses a penis-like cloacal protuberance with which he inseminates the female (Slater 1931, Wernz 1969). This appears to be an adaptation to a fast-flowing mountain stream habitat, where external fertilization would be highly ineffective. In *Nectophrynoides occidentalis,* a viviparous, terrestrial species, male and female directly oppose their cloacae during sperm transfer (Noble 1931, Boisseau and Joly 1975). Internal fertilization has also been reported for *Nectophrynoides malcolmi* (Wake 1980). Townsend *et al.* (1981) have found evidence of internal fertilization in the Puerto Rican frog *Eleutherodactylus coqui,* and they suggest that it may be more common than is generally supposed in species with terrestrial breeding habits. The ovoviviparity described for *Eleutherodactylus jasperi* by Wake (1978) indicates that fertilization is internal in this species also. It is possible that sperm competition could occur within the female's body in these species but, at present, very little is known of their breeding biology.

III. APODANS

The reproductive biology of apodans has been described by Wake (1977), who also has made detailed studies of the urogenital morphology of these creatures (Wake 1968, 1970a, 1970b, 1972). Fertilization is internal, sperm being transferred by means of an erectile intromittent organ, called the phallodeum, consisting of the posterior part of the cloaca. Many caecilians give birth to live young. The Mullerian glands of the male produce a secretion that provides both a fluid medium and metabolites for the sperm (Wake 1981), but in contrast to urodeles, the female possesses no special structure for the storage of sperm (Wake 1972). Very little is known of mating behavior, in particular whether females are inseminated by more than one male, and so it is impossible to assess the potential for sperm competition in this group.

IV. URODELES

Urodeles, because of a number of features of their reproductive biology, are the most likely of the three amphibian orders in which sperm competition may be important. They also are a group for which there is now a considerable amount of relevant data.

In all but the primitive cryptobranchids and hynobiids, which together account for only about 10% of living urodele species, fertilization in urodeles is internal. Sperm transfer is indirect, by means of a spermatophore deposited on the substrate (Salthe 1967; Arnold 1972, 1977). In some genera, such as *Ambystoma* (Arnold 1976) and *Triturus* (Halliday 1974, 1977), mating takes place in water; in others, such as *Plethodon* (Arnold 1976) and *Salamandra* (Joly 1966, Himstedt 1965), it occurs on land. There is much variation between genera, both in the manner in which males stimulate females, and in the extent to which males physically restrict female movements (Arnold 1972, 1977; Halliday 1977). In *Salamandra* the female is firmly held by the male throughout the entire courtship and mating sequence (Joly 1966), in *Taricha* (Davis and Twitty 1964) and *Notophthalmus* (Arnold 1972, Verrell 1982) she is released just before spermatophore transfer, and in *Triturus* she is not restrained at any stage during courtship (Halliday 1974, 1977). As discussed below, the extent to which males restrain females may be an important factor in determining the extent to which females may be inseminated by several males.

Before discussing urodele reproductive biology in detail, it is useful to recall the four conditions, listed by Parker (1970, this volume), which must exist if sperm competition is to occur. These are: (1) individual females are inseminated by more than one male; (2) females can store sperm, at least for the duration of the period over which matings with different males occur; (3) sperm remain viable during the storage period; and (4) sperm are stored and used efficiently by females. Any loss of sperm from one male, through inefficient storage or use, will reduce the competitive potential of that male.

A. Multiple Mating

In many, and perhaps most urodeles, there is considerable potential for females to engage in matings with several males because sex ratios in breeding ponds are often biased towards an excess of males; males commonly produce, and females may accept, many more spermatophores than are required to fertilize all available eggs; and the courtship period for many terrestrial species is extended over many months, during which a female can mate repeatedly and can store sperm until oviposition.

What is important in relation to mating dynamics is not simply the overall sex ratio within a breeding population, but that which exists at the time when mating actually occurs, called the operational sex ratio (Emlen 1976). The most detailed study of sex ratio changes in a urodele is that by Douglas (1979) of *Ambystoma jeffersonianum* in which, over the course of a breeding season, there are more than two males to every female visiting a breeding pond. However, at any one time the operational sex ratio is considerably more skewed, because females spend significantly less time in a breeding pond than males. Over the course of two seasons, Douglas obtained daily operational sex ratio values varying between 2.7 and 10.1 males per female.

Table I summarizes the data available for overall sex ratios in the urodeles that migrate to water to breed. In most species there is an excess of males in breeding populations. The tendency for males to stay longer than females in breeding populations will produce an even more skewed operational sex ratio in many species, depending on how great is the disparity in visit duration.

Sex ratios are very much more difficult to assess in species that mate on land, such as *Salamandra salamandra* and most plethodontids, simply because the breeding population is not spatially confined. Approximately even sex ratios have been reported for *Salamandra salamandra* (Degani and Warburg 1978), *Desmognathus ochrophaeus* (Tilley 1980), *Bolitoglossa rostrata* (Houck 1977), and *Plethodon glutinosus* (Wells 1980). An excess of males has been reported for *Plethodon jordani* (Highton 1962, Arnold 1976). Some male plethodontids show territorial behavior (*e.g.,* Cupp 1980, Jaeger *et al.* 1982, Wrobel *et al.* 1980). If this limits the number of males that can gain access to females, it will tend to produce a biased operational sex ratio with females commoner than males in localities where males take up residence.

TABLE I

Data on the Overall Sex Ratio in Breeding Populations of Aquatic-breeding Urodeles

Species	Sex Ratio	Males Stay Longer Than Females	References
Ambystoma jeffersonianum	2.3, 2.7[a]	+	Douglas 1979
Ambystoma talpoideum	ca. 1.0	+	Hardy and Raymond 1980
Ambystoma maculatum	1.87	+	Husting 1965
Ambystoma annulatum	>1.0	+	Spotila and Beumer 1970
Taricha torosa	>1.0	+	Smith 1941
Taricha rivularis	ca. 5.0[b]	+	Hedgecock 1978
Triturus vulgaris	⌈ 0.78-1.52[c] ⌊ <1.0	?	Hagstrom 1979 Bell 1977
Triturus cristatus	⌈ 1.28, 1.72[d] ⌊ ca. 1.0	?	Hagstrom 1979 Bielinski, pers. comm.
Notophthalmus viridescens	⌈ >1.0 >1.0 ca. 2.0 ⌊ 4.7-1.6[e]	+	Hurlbert 1969 Bellis 1968 Gill 1978a, b Healy 1974

All ratios are expressed as number of males/number of females.
[a]Data for two different years.
[b]Data for 1955. Data for subsequent years show increasing ratio, due to experimental removal of females.
[c]Data from several populations.
[d]First value = estimated for population, second value = ratio for recorded animals.
[e]First value = spring breeding season, second value = fall breeding season.

Data on male spermatophore production are limited. Experiments in which males have been allowed to court females repeatedly over successive days have been carried out with *Triturus vulgaris* (Halliday 1976) and *Ambystoma maculatum* (Arnold 1976). In both studies, sample sizes were small (only five males) and the variance high. The five *Triturus vulgaris* males tested produced 27 to 73 spermatophores (mean = 47.8) over 40 days; the five *Ambystoma maculatum* males produced 28 to 137 over three days. These are both likely to be underestimates of total spermatophore production, for *A. maculatum* because of the limited duration of the experiment, and for *T. vulgaris* because males were collected from the wild where they may already have deposited spermatophores. These data do indicate that, at least in these two species, males are capable of producing large numbers of spermatophores. L. D. Houck (pers. comm.) has shown that female *Ambystoma tigrinum* and *Desmognathus ochrophaeus* that have picked up only one spermatophore can lay complete clutches of eggs. This suggests that each spermatophore contains sufficient sperm to fertilize a full clutch of eggs and that the potential for males to inseminate many females is therefore considerable. The sperm content of a spermatophore is clearly a crucial unknown in need of investigation.

Males differ from species to species in the way that they distribute their spermatophore production over the course of a breeding season (Arnold 1977). In *Ambystoma*, the breeding season is very short and males deposit large numbers of spermatophores per night. In some *Ambystoma* species, males deposit spermatophores in an apparently indiscriminate way, producing a "field" of spermatophores through which females move (Garton 1972; Arnold 1972, 1976). Under such circumstances, the chances that females will pick up spermatophores from more than one male could be very high. In *Triturus*, breeding activity lasts for about three months and, from laboratory studies, it seems that males produce a few spermatophores each day for several weeks (Halliday 1976). In *Plethodon*, males deposit only about one spermatophore per week for several weeks (Arnold 1977). The data available for these three genera, though incomplete, suggest a positive relationship between the amount of reproductive effort that a male invests in each spermatophore and the probability of its being accepted by a female (Arnold 1977). Thus, in *Plethodon*, males produce relatively few spermatophores, each associated with lengthy and elaborate courtship behavior; consequently, each has a high probability of being picked up. By contrast, *Ambystoma maculatum* males produce many spermatophores during a somewhat perfunctory courtship, and each has a low chance of being picked up. *Triturus* falls somewhere between these two extremes.

Direct evidence that females pick up spermatophores from more than one male was obtained by P. A. Verrell for *Triturus vulgaris* (Table II) and for *Notophthalmus viridescens* (Table III). The data for *T. vulgaris* indicate that, although only 8% of the females picked up sperm from two males, females remained sexually responsive to new males for at least 20 days after mating. The data for *N. virides-*

TABLE II

The Responses of a Group of 25 Inseminated Female Smooth Newts
(*Triturus vulgaris*) to Further Male Courtship

| Time Since Last Spermatophore Picked Up | Number of Females That | | | |
	Showed No Response	Showed Weak Response	Elicited Spermatophore Deposition	Picked Up a Spermatophore
1 hour	17	4	1	0
12 hours	17	2	2	1
2 days	22	1	0	1
4 days	21	0	1	0
8 days	24	0	1	0
16 days	21	1	2	0
10 days	22	1	2	0

Note: Not all of the 25 females were tested at all time intervals.

TABLE III

The Responses of Inseminated Female Red-spotted Newts
(*Notophthalmus viridescens*) to Further Male Courtship

Time Since Last Spermatophore Picked Up	Number of Females That	
	Were Tested	Picked Up a Spermatophore
1 day	8	4
5 days	7	2
8 days	4	1

cens indicate that multiple matings are more common in this species and that females remain responsive for at least eight days. These results suggest that in these two species, a female could build up a stock of sperm contributed by several males.

The only conclusive evidence that a female's eggs have been fertilized by more than one male is that obtained by testing the paternity of her progeny. Only one published study, that by Tilley and Hausman (1976) on *Desmognathus ochrophaeus*, has adopted this approach. They investigated seven populations in North Carolina and used electrophoretic techniques to examine variations at loci controlling esterase and lactate dehydrogenase. By comparing females and their progeny, they established that at least eight of 100 broods had two fathers. They estimate that the frequency of multiple inseminations is at least 7%. Houck *et al.* (unpubl. data) have obtained preliminary results from multiple mating experiments in this salamander, suggesting that sperm derived from earlier matings may enjoy greater fertilization success.

Taken together, the various lines of evidence described in this section, though fragmentary and gathered from different species, indicate that multiple mating certainly occurs in some urodeles and that it is likely to occur in many others.

B. Sexual Interference

The spermatophore method of sperm transfer shown by the great majority of urodeles is open to a particular form of sexual competition between males, in which they interfere with one another's efforts to inseminate females. Such behavior has been called sexual interference by Arnold (1976), who described it in three species, *Plethodon jordani*, *Ambystoma tigrinum*, and *A. maculatum*. Recent work by P. A. Verrell (unpubl.) has investigated similar behavior in *Triturus vulgaris* and *Notophthalmus viridescens*.

Sexual interference takes a variety of forms. For example, a *Triturus vulgaris* male is clearly stimulated by the sight of another male displaying to a female and

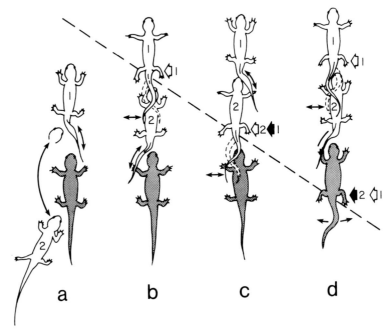

Fig. 1. Sexual interference in *Ambystoma tigrinum*. a. As male 1 leads the female (stippled) in a "tail-nudging walk", male 2 moves between them. b. Male 2 nudges male 1's tail, mimicking female behavior; in response, male 1 deposits a spermatophore (arrows). c. Male 2 deposits a spermatophore on top of male 1's. d. The female picks up only male 2's sperm, that of male 1 being inaccessible to her. The dashed line joins equivalent points on the substrate. From Arnold (1976).

will approach her and start to display to her himself. This is the simplest form of sexual interference. In this discussion, we confine ourselves to those forms which relate to spermatophore transfer, because it is these that are relevant to the phenomenon of multiple insemination of females and consequent sperm competition.

The general pattern of spermatophore transfer among salamandrids is as follows. Following a period in which the male stimulates the female, he turns away from her and deposits a spermatophore, usually in response to a tactile stimulus, such as a nudge of the female's snout against the base of his tail (Halliday 1975). He then moves away from the spermatophore and the female walks over it with her cloaca extended. Depending on her exact path, her cloaca may contact the sperm mass on top of the spermatophore, which is then drawn up into her body. One form of sexual interference, in *Ambystoma tigrinum*, *Triturus vulgaris,* and *Plethodon jordani*, is mimicry of female behavior by males. An interfering male moves between a courting male and a female and nudges the courting male's tail to elicit spermatophore deposition (Fig. 1). He then either deposits his own spermatophore on top of that of the first male, as in *Ambystoma tigrinum,* or leads the female away to initiate the deposition and transfer of his own spermatophore, as in *Triturus vulgaris.*

Spermatophore stacking, in which a male deposits his spermatophore on top of a rival's, can be regarded as a form of sperm competition, since the first male's sperm is rendered quite inaccessible to the female (Fig. 2), that occurs outside the female's body. The form of sexual interference shown by *Triturus vulgaris* males may lead to internal sperm competition, since it can result in a female picking up spermatophores from more than one male in rapid succession. Table IV summarizes the results of several courtship encounters in which two males courted a single female. These data show that it is quite possible for two males to inseminate the same female within the same bout of courtship, although this appears to be a rare event.

C. Sperm Storage

Female urodeles typically have a spermatheca opening into the cloaca; this is homologous to the pelvic glands of the male. Spermathecal morphology has been reviewed by Kingsbury (1895) and Boisseau and Joly (1975); both reviews distinguish two basic types of spermathecae. *Necturus,* a proteid, and salamandrids, such as *Notophthalmus,* have numerous short tubules that open into the roof of the female's cloaca (Fig. 3a). Most plethodontids, such as *Desmognathus, Plethodon,* and *Hydromantes,* have a few short tubules, or diverticulae, opening into a common duct that links them to the cloaca (Fig. 3c). *Ambystoma* has an intermediate structure, consisting of many tubules opening into a short duct (Fig. 3b; Noble 1931).

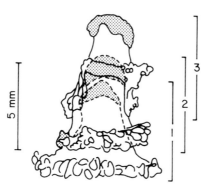

Fig. 2. Stacked spermatophores of *Ambystoma maculatum.* Three spermatophores have been deposited, one on top of another, so that the sperm masses (stippled) of the first two are covered and are thus inaccessible to a female. From Arnold (1976).

TABLE IV

The Incidence of Insemination of Female Smooth Newts (*Triturus vulgaris*)
by One or Two Males as a Result of Sexual Interference Between Two Males

Total number of triadic courtship episodes observed	39
Number in which female picked up spermatophore(s) deposited by first male only	13 (33%)
Number in which female picked up spermatophore(s) deposited by intruding male only	3 (8%)
Number in which female picked up spermatophores deposited by both males	2 (5%)
Number in which no spermatophores picked up	21 (54%)

In a review of sperm storage and competition in insects, Walker (1980) pointed out that the shape of the sperm storage organ may have an important influence on the nature and extent of sperm competition. In species in which there is only one insemination, and in those in which the first of several inseminations has precedence in fertilizing eggs, the spermatheca tends to be spherical. In contrast, in insects in which the last insemination has precedence, females tend to have elongate, tubular spermathecae. It is possible that the same generalizations apply to the urodeles, in which case last male precedence will be most likely in proteids, salamandrids, and ambystomatids. However, the situation is complicated by the absence, in proteids and salamandrids, and by the presence, in ambystomatids and plethodontids, of a common spermathecal duct. Possible relationships between spermathecal anatomy and sperm competition in urodeles will require empirical investigation.

The microscopic structure of the urodele spermatheca appears to be more uniform than the gross anatomy. Dent (1970) described the ultrastructure of the spermathecal tubules of *Notophthalmus viridescens,* which have an inner lining of epithelial cells and an outer covering of myoepithelial cells. He suggests that the epithelial cells provide nourishment for the stored sperm, and contractions of the myoepithelial cells bring about the expulsion of sperm from the tubules. Pool and Hoage (1973) have described an essentially similar structure for the tubules of *Eurycea quadridigitatus,* a plethodontid. In this species, the tissue surrounding the inner, secretory cells is elastic, allowing the tubules to expand as they become filled with sperm.

The spatial distribution of stored sperm within the spermathecal tubules varies among species. Sperm appear to be randomly distributed in *Eurycea quadridigitatus* (Pool and Hoage 1973), in close contact with the epithelium in *Notophthalmus viridescens* (Dent 1970) and *Desmognathus fuscus* (Marynick 1971), and buried in the epithelium in *Cynops pyrrhogaster* (Tsutsui 1931).

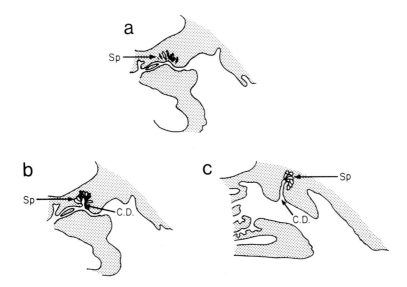

Fig. 3. Diagrammatic sagittal sections of the cloacae of three female urodeles. The anterior end is to the left. a. *Necturus*, with numerous spermathecal tubules opening independently into the root of the cloaca. b. *Ambystoma*, with rather fewer tubules opening into a common duct. c. *Desmognathus*, with very few tubules and a common duct. CD = common duct, Sp = spermathecal tubules. From Noble (1931).

It has been suggested that entry of sperm into the spermatheca may involve chemotaxis by the sperm, the female secreting some kind of sperm attractant from her spermathecal tubules (Kingsbury 1895, Boisseau and Joly 1975). In plethodontids, sperm may be carried passively up the spermathecal duct by cilia (Pool and Hoage 1973). Chemical attraction of sperm from the spermatophore to the spermatheca may provide a mechanism by which defective sperm are selected out by the female, since they may not react appropriately to the female's attractant secretion (D. M. Sever, pers. comm.).

D. Viability of Sperm

There is considerable variation among urodeles in the interval between mating and oviposition, when fertilization occurs. In axolotls (*Ambystoma mexicanum*), sperm do not survive for more than 12 days (Humphrey 1977). In the European newts (*Triturus*), females start to lay eggs within a few days of mating, but in many plethodontids there is an interval of up to 8 months between mating in the spring or autumn and egg-laying during the summer or subsequent spring (Marynick 1971; D. M. Sever, pers. comm.). Sperm remain viable for about 8 months in

Eurycea quadridigitatus (Pool and Hoage 1973). In *Salamandra salamandra,* eggs are commonly not fertilized until a year has elapsed since mating (Boisseau and Joly 1975).

There is evidence that urodele sperm can remain viable for considerably longer periods under certain circumstances. For example, Tsutsui (1931) reported a female *Cynops pyrrhogaster* laying eggs 190 days after insemination, and Boisseau and Joly (1975) suggest that sperm may survive for as long as 2 years in *Salamandra salamandra*. Such observations suggest that the epithelial cells lining the female's spermathecal tubules provide nourishment for stored sperm for a considerable period of time.

D. M. Sever (pers. comm.) reports that, from a comparison of the cloacae and associated structures of female *Plethodon cinereus* examined in the March mating season and at egg-laying time in June, there appears to be a reduction in the quantity of sperm stored by females. He suggests that females resorb some sperm and that one of the functions of a long storage period may be that less viable sperm are somehow selected out. Dent (1970) suggests that spermathecal epithelial cells ingest some sperm by phagocytosis in *Notophthalmus viridescens*.

E. Utilization of Stored Sperm

As Parker (1970, this volume) has pointed out, the competitive potential of an individual male may be strongly influenced by the efficiency with which a female stores and uses his sperm (see also Knowlton and Greenwell, this volume). This is particularly true of insects, in which males discharge relatively few spermatozoa in each ejaculate. Nothing is known about how efficiently female urodeles use sperm when fertilizing eggs. Most urodeles are oviparous, producing eggs singly or in batches and, presumably, a female expels a small quantity of sperm onto each egg or batch of eggs as it passes through her cloaca, though chemotactic attraction between eggs and sperm may be involved (Boisseau and Joly 1975). In viviparous species, sperm make their own way from the cloaca to the eggs. In *Salamandra atra* fertilization occurs close to the cloaca, in the uterus; in *S. salamandra* sperm have to swim further, up into the oviduct (Boisseau and Joly 1975). In both oviparous and viviparous species, it is possible that sperm stored close to the openings of the spermathecal tubules may have a competitive advantage, since they may be the first to emerge. In the oviparous species, the strength of any such competitive effect will depend on how many sperm females extrude onto each egg or batch of eggs. If a large number are extruded onto each egg, competition between sperm to penetrate an egg may be of greater significance than any effect due to position in the storage organ.

F. Possible Counter-measures

In many animals in which sperm competition has been demonstrated, there are associated patterns of behavior that can be interpreted as adaptations on the part of males that counter its effects. Such adaptations include guarding females between insemination and egg-laying and the use of mating plugs to prevent subsequent inseminations by other males. If comparable behavior patterns exist in urodeles, they may be construed as possible evidence for sperm competition.

There is no evidence that males of any urodele species attempt to prevent other males gaining access to females with whom they have just mated. Indeed, it is generally the case that pairs separate and lose contact with one another as soon as mating is finished. Aggressive behavior and territoriality between males have been reported for some plethodontids (Cupp 1980, Jaeger *et al.* 1982, Wrobel *et al.* 1980). The adaptive significance of such behavior may be that it reduces the chances that a male's courtship of those females that enter his territory will be subject to sexual interference by rival males. Wells (1980) sampled slimy salamanders (*Plethodon glutinosus*) hiding under logs, and found that males tended not to be spatially associated with one another, whereas males and females did tend to be closely associated, especially in the spring. If females do remain close to particular males for long periods, then male spacing, whether it is brought about by aggression or by mutual avoidance between males, may tend to increase the probability that a male will have exclusive paternity of the eggs of those females with whom he mates.

As described above, sexual interference during courtship sometimes results in more than one male inseminating a female, thus generating potential sperm competition. In some species males show behavior patterns that are clearly adapted to counter sexual interference; Arnold (1976) labeled these sexual defense. Males of *Ambystoma tigrinum* grasp females and carry them away from rival males just before spermatophore transfer, and male *Plethodon jordani* bite and chase males that approach them when they are courting females (Arnold 1976). Territoriality in plethodontids may be regarded as a form of sexual defense (Arnold 1976).

The high spermatophore productivity shown by some urodeles, notably *Ambystoma* species, may also be an adaptive response to sperm competition, although its primary significance is more likely to be related to the probability with which spermatophores are picked up by females. In species in which courtship is relatively inefficient in terms of spermatophore transfer, male productivity tends to be high (Arnold 1977).

In some insects, the spermatophore has acquired the secondary function of a sperm plug, defending a male's paternity of a female's eggs by preventing other males inseminating her (Davey 1960, Schaller 1979). It is possible that the spermatophore of some urodeles may fulfill a similar function. Typically, the urodele spermatophore consists of a conical base which adheres to the substrate, surmounted by a sperm cap. The female removes the sperm cap with her cloaca, leaving the base behind. The sperm cap of plethodontids is unusual in being surrounded by a number of tough concentric membranes (Organ and Lowenthal 1963). Noble and Weber (1929) reported that, in many plethodontids, the sperm cap is large enough to occlude the female's cloacal opening and that fragments of the spermatophore base sometimes adhere to it. It takes some days for the sperm cap to disintegrate and, until it does so, it may effectively prevent further insemination. There is no evidence that spermatophores can act as sperm plugs in aquatic-breeding urodeles such as *Triturus*. Indeed, in *T. vulgaris*, a female sometimes picks up two, or even three spermatophores during a single courtship encounter with a male (Halliday 1974). This clearly would be impossible if the first spermatophore to be picked up acted as a plug.

V. MATING IN HYNOBIIDS

The spermatophore method of sperm transfer shown by the majority of urodeles is generally believed to have evolved from a form of external fertilization like that shown by the hynobiids (Arnold 1977, Salthe 1967). Females produce batches of eggs enclosed in sacs and males shed sperm onto them. Two or more males may participate in the egg-laying activity of a single female and may show aggression towards one another (Kusano 1980). This aggression is apparently directed towards the defense of egg sacs, rather than females, and probably serves as a form of paternity assurance. It thus appears to be analogous to the defense of eggs shown by some anurans, as discussed above.

VI. TESTING FOR SPERM COMPETITION

The combination of multiple mating and sperm storage shown by female urodeles makes it very likely that some form of sperm competition occurs in them. There is an urgent need to establish what form this might take, whether there is first or last-male precedence or whether a male's fertilization success is simply a function of the number of sperm he passes to a female, relative to those contributed by other males. Methods which have been used in the study of sperm competition fall into two categories:

1. Those using experimentally induced differences between males. For example, a female is mated with two males, one of which has been irradiated to render his sperm incapable of normal fertilization (Parker 1970). By allowing "labeled" and normal males to mate with a female in different orders, the effects of sperm competition can be deduced from the failure of eggs fertilized by irradiated sperm to develop normally.

2. Those exploiting naturally occurring differences between males. Females are mated with a number of males which carry genetic markers that can be detected in the progeny. Such studies require detailed understanding of the genetics of any markers used. Heritable characters that have been used include coat color in rodents (e.g., Dewsbury and Baumgardner 1981), insect cuticle color or pattern (e.g., Retnakaran 1974, Smith 1979, McLain 1980), and allozymes (e.g., Hanken and Sherman 1981). This last technique is currently being used by Houck et al. in a study of sperm competition in the salamander Desmognathus ochrophaeus. Populations of this species differ at several allozyme loci that can be analyzed by electrophoresis. This method appears to be particularly appropriate to urodeles because their demic population structure is likely to be associated with a high level of interpopulation genetic variability (Hedgecock 1978).

VII. SUMMARY

The information brought together in this review is fragmentary and has been gathered from a variety of sources. We stress that none of the published studies referred to involved investigations concerned primarily with sperm competition. Consequently, our conclusions are highly speculative. Despite this reservation, we feel that there are good grounds for believing that sperm competition may be an important factor in the reproductive biology of many amphibians, particularly of urodeles, and that it promises to be a fruitful area for future research.

Among anurans, the predominant pattern of external fertilization involves a potential for external sperm competition between the ejaculates of more than one male attending a spawning female. Aggression between males may, among other functions, serve to minimize multiple paternity of a female's eggs. Townsend et al. (1981) suggested that internal fertilization, generally considered to be extremely rare among anurans, may prove to be quite widespread, especially among terrestrial-breeding species. Whether sperm competition within the female's body is important in these species is a question that awaits the collection of basic information about their sexual behavior.

Although apodans have internal fertilization, their potential for sperm competition is probably low, because of the lack of any sperm storage organ in the female.

In contrast, the combination of internal fertilization and a well-developed sperm storage capacity found in many urodeles suggests a strong potential for sperm competition. There is an urgent need for investigation into the form that this might take. There is good evidence from some species that females acquire sperm from more than one male. A crucial question is whether or not ejaculates from different males maintain their integrity while they are being stored by the female, or whether they become mixed. If they remain discrete, then sperm competition may be of the type in which either the first or the last male to mate is at a competitive advantage. If they are mixed, then the advantage may simply go to the male who transfers most sperm to the female. The paucity of examples of behavior patterns that might be adapted to ensure that a male is either the first or the last to inseminate a female, combined with the fact that many urodeles are capable of maintaining high levels of sperm production, suggests that sperm competition may be a question of quantity of sperm, rather than order of insemination. However, those observations that indicate that the amount of sperm stored by a female becomes attenuated with time raise the possibility that females are able to impose some kind of direct selection on sperm. If this is so, then a male's competitive advantage may be determined, neither by the order in which he inseminates a female, nor by the amount of sperm he transfers, but rather by the quality of his sperm. Cohen and McNaughton (1974) have suggested that sperm may be selectively destroyed in the reproductive tract of female rabbits. Such an effect, were it to be demonstrated conclusively, would be of profound importance as a form of sexual selection. As Walker (1980) has suggested, it is probably misleading to regard sperm competition simply as a form of inter-male competition. Since sperm competition can occur only if females, through their anatomy, physiology, and behavior, provide appropriate conditions, it is important that we also consider possible advantages that females may derive from its occurrence. The capacity of female urodeles to store and, perhaps, to nourish or destroy sperm selectively, is therefore a phenomenon that deserves intensive study.

ACKNOWLEDGMENTS

We are very grateful to David M. Sever for drawing our attention to a number of phenomena discussed in this chapter, and for allowing us to quote his unpublished observations. We also thank him, Lynne D. Houck, Kentwood D. Wells, and Stevan J. Arnold, for reading and commenting on the manuscript, and Andrew Bielinski, for allowing us to refer to his unpublished data.

REFERENCES

Arak, A. 1983. Male-male competition and mate choice. In *Mate Choice,* P. P. G. Bateson (ed.), pp. 181-210. Cambridge Univ. Press.
Arnold, S. J. 1972. The evolution of courtship behavior in salamanders. Ph.D. Dissertation, Univ. Michigan.

Arnold, S. J. 1976. Sexual behavior, sexual interference and sexual defense in the salamanders *Ambystoma maculatum, Ambystoma tigrinum* and *Plethodon jordani. Z. Tierpsychol.* 42: 247-300.

Arnold, S. J. 1977. The evolution of courtship behavior in New World salamanders with some comments on Old World salamanders. In *The Reproductive Biology of Amphibians,* D. H. Taylor and S. I. Guttman (eds.), pp. 141-183. Plenum, New York.

Bell, G. 1977. The life of the smooth newt (*Triturus vulgaris*) after metamorphosis. *Ecol. Monogr.* 47:279-299.

Bellis, E. D. 1968. Summer movement of red-spotted newts in a small pond. *J. Herpetol.* 1: 86-91.

Berven, K. A. 1981. Mate choice in the wood frog *Rana sylvatica. Evolution* 35:707-722.

Boisseau, C., and J. Joly. 1975. Transport and survival of spermatozoa in female Amphibia. In *The Biology of Spermatozoa,* E. S. E. Hafez and C. G. Thibault (eds.), pp. 94-104. Karger, Basel.

Coe, M. 1974. Observations on the ecology and breeding biology of the genus *Chiromantis* (Amphibia:Rhacophoridae). *J. Zool. Lond.* 172:13-34.

Cohen, J., and D. C. McNaughton. 1974. Spermatozoa: The probable selection of a small population by the genital tract of the female rabbit. *J. Reprod. Fertil.* 39:297-310.

Cupp, P. V. 1980. Territoriality in the green salamander, *Aneides aeneus. Copeia* 1980: 463-468.

Davey, K. G. 1960. The evolution of spermatophores in insects. *Proc. Roy. Entomol. Soc. Lond. (A)* 35:107-113.

Davies, N. B., and T. R. Halliday. 1977. Optimal mate choice in the toad *Bufo bufo. Nature* 269:56-58.

Davies, N. B., and T. R. Halliday. 1979. Competitive mate searching in male common toads, *Bufo bufo. Anim. Behav.* 27:1253-1267.

Davis, J. W. F., and V. C. Twitty. 1964. Courtship behavior and reproductive isolation in the species of *Taricha* (Amphibia, Caudata). *Copeia* 1964:601-610.

Degani, G., and M. R. Warburg. 1978. Population structure and seasonal activity of the adult *Salamandra salamandra* L. (Amphibia, Urodela, Salamandridae) in Israel. *J. Herpetol.* 12: 437-444.

Dent, J. N. 1970. The ultrastructure of the spermatheca in the red-spotted newt. *J. Morphol.* 132:397-424.

Dewsbury, D. A., and D. J. Baumgardner. 1981. Studies of sperm competition in two species of muroid rodents. *Behav. Ecol. Sociobiol.* 9:121-133.

Dominey, W. J. 1980. Female mimicry in male bluegill sunfish—a genetic polymorphism? *Nature* 284:546-548.

Douglas, M. E. 1979. Migration and sexual selection in *Ambystoma jeffersonianum. Can. J. Zool.* 57:2303-2310.

Emlen, S. T. 1976. Lek organization and mating strategies in the bullfrog. *Behav. Ecol. Sociobiol.* 1:283-313.

Garton, J. S. 1972. Courtship of the small-mouthed salamander, *Ambystoma texanum,* in southern Illinois. *Herpetologica* 28:41-45.

Gill, D. E. 1978a. Effective population size and interdemic migration rates in a metapopulation of the red-spotted newt, *Notophthalmus viridescens* (Rafinesque). *Evolution* 32: 839-849.

Gill, D. E. 1978b. The metapopulation ecology of the red-spotted newt, *Notophthalmus viridescens* (Rafinesque). *Ecol. Monogr.* 48:145-166.

Gross, M. R. 1979. Cuckoldry in sunfishes (*Lepomis;* Centrarchidae). *Can. J. Zool.* 57:1507-1509.

Gross, M. R., and Charnov, E. L. 1980. Alternative male life histories in bluegill sunfish. *Proc. Nat. Acad. Sci. U.S.A.* 77:6937-6940.

Hagstrom, T. 1979. Population ecology of *Triturus cristatus* and *T. vulgaris* (Urodela) in S.W. Sweden. *Holarctic Ecol.* 2:108-114.

Halliday, T. R. 1974. Sexual behavior of the smooth newt, *Triturus vulgaris* (Urodela:Salamandridae). *J. Herpetol.* 8:277-292.

Halliday, T. R. 1975. An observational and experimental study of sexual behaviour in the smooth newt, *Triturus vulgaris* (Amphibia:Salamandridae). *Anim. Behav.* 23:291-322.

Halliday, T. R. 1976. The libidinous newt: An analysis of variations in the sexual behaviour of the male smooth newt, *Triturus vulgaris*. *Anim. Behav.* 24:398-414.

Halliday, T. R. 1977. The courtship of European newts: An evolutionary perspective. In *Reproductive Biology of Amphibians*, D. H. Taylor and S. I. Guttman (eds.), pp. 185-232. Plenum, New York.

Hamilton, W. D. 1971. Geometry for the selfish herd. *J. Theor. Biol.* 31:295-311.

Hanken, J., and P. W. Sherman. 1981. Multiple patenity in Belding's ground squirrel litters. *Science* 212:351-353.

Hardy, L. M., and L. R. Raymond. 1980. The breeding migration of the mole salamander, *Ambystoma talpoideum,* in Louisiana. *J. Herpetol.* 14:327-335.

Healy, W. R. 1974. Population consequences of alternative life histories in *Notophthalmus v. viridescens. Copeia* 1974:221-229.

Hedgecock, D. 1978. Population subdivision and genetic divergence in the red-bellied newt, *Taricha rivularis. Evolution* 32:271-286.

Highton, R. 1962. Geographic variation in the life history of the slimy salamander. *Copeia* 1962:597-613.

Himstedt, W. 1965. Beobachtungen zum Paarungsverhalten des Feuersalamanders. *Zool. Anz.* 175:295-300.

Houck, L. D. 1977. Reproductive biology of a Neotropical salamander, *Bolitoglossa rostrata. Copeia* 1977:70-83.

Howard, R. D. 1978a. The influence of male-defended oviposition sites on early embryo mortality in bullfrogs. *Ecology* 59:789-798.

Howard, R. D. 1978b. The evolution of mating strategies in bullfrogs, *Rana catesbeiana. Evolution* 32:850-871.

Howard, R. D. 1980. Mating behaviour and mating success in woodfrogs, *Rana sylvatica. Anim. Behav.* 28:705-716.

Humphrey, R. R. 1977. Factors influencing ovulation in the Mexican axolotl as revealed by induced spawning. *J. Exp. Zool.* 199:209-214.

Hurlbert, S. H. 1969. The breeding migrations and inter-habitat wandering of the vermilion-spotted newt, *Notophthalmus viridescens* (Rafinesque). *Ecol. Monogr.* 39:465-488.

Husting, E. L. 1965. Survival and breeding structure in a population of *Ambystoma maculatum. Copeia* 1965:352-362.

Jaeger, R. G., D. Kalvarsky, and N. Shimizu. 1982. Territorial behaviour of the red-backed salamander: Expulsion of intruders. *Anim. Behav.* 30:490-496.

Joly, J. 1966. Sur l'ethologie sexuelle de *Salamandra salamandra* L. *Z. Tierpsychol.* 23:8-27.

Keenleyside, M. H. A. 1972. Intraspecific intrusions into nests of spawning longear sunfish (Pisces:Centrarchidae). *Copeia* 1972:272-278.

Kingsbury, B. F. 1895. The spermatheca and methods of fertilization in some American newts and salamanders. *Proc. Am. Microscopical Soc.* 17:261-305.

Kluge, A. G. 1981. The life history, social organization and parental behavior of *Hyla rosenbergi* (Boulenger), a nest-building gladiator frog. *Misc. Publ. Mus. Zool. Univ. Mich.* 160: 1-170.

Kusano, T. 1980. Breeding and egg survival of a population of a salamander, *Hynobius nebulosus tokyoensis* Tago. *Res. Popul. Ecol.* 21:181-196.

Marynick, S. P. 1971. Long term storage of sperm in *Desmognathus fuscus* from Louisiana. *Copeia* 1971:345-347.

McLain, D. K. 1980. Female choice and the adaptive significance of prolonged copulation in *Nezara viridula* (Hemiptera:Pentatomidae). *Psyche* 87:325-336.

Noble, G. K. 1931. *The Biology of the Amphibia.* McGraw-Hill, New York.

Noble, G. K., and J. A. Weber. 1929. The spermatophores of *Desmognathus* and other plethodontid salamanders. *Am. Mus. Novit.* 351:1-15.

Organ, J. A., and L. A. Lowenthal. 1963. Comparative studies of macroscopic and microscopic features of spermatophores of some plethodontid salamanders. *Copeia* 1963:659-669.

Parker, G. A. 1970. Sperm competition and its evolutionary consequences in the insects. *Biol. Rev.* 45:525-568.

Pool, T. B., and T. R. Hoage. 1973. The ultrastructure of secretion in the spermatheca of the salamander *Manculus quadridigitatus* (Holbrook). *Tissue & Cell* 5:303-313.

Pyburn, W. F. 1970. Breeding behavior of the leaf frogs *Phyllomedusa callidryas* and *Pachymedusa dacnicolor* in Mexico. *Copeia* 1970:209-218.

Retnakaran, A. 1974. The mechanism of sperm precedence in the spruce budworm *Choristoneura fumiferana* (Lepidoptera:Tortricidae). *Can. Entomol.* 106:1189-1194.

Rowland, W. J. 1979. Stealing fertilizations in the fourspine stickleback, *Apeltes quadracus*. *Am. Nat.* 114:602-604.

Salthe, S. N. 1967. Courtship patterns and phylogeny of the urodeles. *Copeia* 1967:100-117.

Schaller, F. 1979. Significance of sperm transfer and formation of spermatophores in arthropod phylogeny. In *Arthropod Phylogeny*, A. P. Gupta (ed.), pp. 587-608. Van Nostrand Holland, New York.

Slater, J. R. 1931. The mating behavior of *Ascaphus truei* Stijneger. *Copeia* 1931:62-63.

Smith, R. E. 1941. Mating behavior in *Triturus torosus* and related newts. *Copeia* 1941:255-262.

Smith, R. L. 1979. Repeated copulation and sperm precedence: Paternity assurance for a male brooding water bug. *Science* 205:1029-1031.

Smith-Gill, S., and K. A. Berven. 1980. In vitro fertilization and assessment of male reproductive potential using mammalian gonadotropin-releasing hormone to induce spermiation in *Rana sylvatica. Copeia* 1980:723-728.

Spotila, J. R., and R. J. Beumer. 1970. The breeding habits of the ringed salamander, *Ambystoma macuiatum* (Cope), in north-western Arkansas. *Am. Midl. Nat.* 84:77-89.

Townsend, D. S., M. M. Stewart, F. H. Pough, and P. F. Brussard. 1981. Internal fertilization in an oviparous frog. *Science* 212:469-471.

Tilley, S. G. 1980. Life histories and comparative demography of two salamander populations. *Copeia* 1980:806-821.

Tilley, S. G., and J. S. Hausman. 1976. Allozymic variation and occurrence of multiple inseminations in populations of the salamander *Desmognathus ochrophaeus. Copeia* 1976:734-741.

Tsutsui, Y. 1931. Notes on the behavior of the common Japanese newt, *Diemyctylus pyrrhogaster* Boie. I. Breeding habit. *Mem. Fac. Sci. Kyoto Univ. Ser. Biol.* 7:159-179.

van den Assem, J. 1967. Territory in the three-spined stickleback *Gasterosteus aculeatus. Behaviour Suppl.* 16:1-164.

Verrell, P. A. 1982. The sexual behaviour of the red-spotted newt, *Notophthalmus viridescens* (Amphibia, Urodela, Salamandridae). *Anim. Behav.* 30:1224-1236.

Wake, M. H. 1968. Evolutionary morphology of the caecilian urogenital system. I. The gonads and the fat bodies. *J. Morphol.* 126:291-332.

Wake, M. H. 1970a. Evolutionary morphology of the caecilian urogenital system. II. The kidneys and urogenital ducts. *Acta Anat.* 75:321-358.

Wake, M. H. 1970b. Evolutionary morphology of the caecilian urogenital system. III. The bladder. *Herpetologica* 26:120-128.

Wake, M. H. 1972. Evolutionary morphology of the caecilian urogenital system. IV. The cloaca. *J. Morphol.* 136:353-366.

Wake, M. H. 1977. The reproductive biology of caecilians. In *Reproductive Biology of Amphibians,* D. H. Taylor and S. I. Guttman (eds.), pp. 73-101. Plenum, New York.

Wake, M. H. 1978. The reproductive biology of *Eleutherodactylus jasperi* (Amphibia, Anura, Leptodactylidae), with comments on the evolution of live-bearing systems. *J. Herpetol.* 12:121-133.

Wake, M. H. 1980. The reproductive biology of *Nectophrynoides malcolmi* (Amphibia, Bufonidae), with comments on the evolution of reproductive modes in the genus *Nectophrynoides. Copeia* 1980:193-209.

Wake, M. H. 1981. Structure and function of the male Mullerian gland in caecilians, with comments on its evolutionary significance. *J. Herpetol.* 15:17-22.

Waldman, B. 1982. Adaptive significance of communal oviposition in wood frogs (*Rana sylvatica). Behav. Ecol. Sociobiol.* 10:169-174.

Walker, W. F. 1980. Sperm utilization strategies in nonsocial insects. *Am. Nat.* 115:780-799.
Wells, K. D. 1977a. The social behaviour of anuran amphibians. *Anim. Behav.* 15:666-693.
Wells, K. D. 1977b. Territoriality and male mating success in the green frog (*Rana clamitans*). *Ecology* 58:750-762.
Wells, K. D. 1978. Territoriality in the green frog (*Rana clamitans*): Vocalizations and agonistic behaviour. *Anim. Behav.* 26:1051-1063.
Wells, K. D. 1980. Spatial associations among individuals in a population of slimy salamanders (*Plethodon glutinosus*). *Herpetologica* 36:271-275.
Wernz, J. G. 1969. Spring mating in *Ascaphus. J. Herpetol.* 3:167-169.
Wilson, E. O. 1975. *Sociobiology: The New Synthesis.* Harvard Univ. Press, Cambridge, MA.
Wirtz, P. 1978. The behavior of the Mediterranean *Tripterygion* species (Pisces, Blennioidei). *Z. Tierpsychol.* 48:142-174.
Wrobel, D. J., W. F. Gergits, and R. G. Jaeger. 1980. An experimental study of interference competition among terrestrial salamanders. *Ecology* 61:1034-1039.

14

Potential for Sperm Competition in Reptiles: Behavioral and Physiological Consequences

MICHAEL C. DEVINE

I. INTRODUCTION

The literature directly addressing the topic of sperm competition and its consequences in reptiles is quite limited. However, indirect implications from the extensive literature on reptilian reproductive behavior, anatomy, and physiology strongly suggest possible sperm competition in numerous reptiles. Presently, this subject area is open to considerable speculation, but empirical studies in the reptiles should prove fruitful. I hope the following treatment, while speculative in part, will stimulate future research.

Sperm Competition and the Evolution
of Animal Mating Systems

II. PREREQUISITES FOR SPERM COMPETITION IN REPTILES

The potential for sperm competition in reptiles is high because the important prerequisites are widespread through this class. Sperm storage is well documented, with records reviewed by Fox (1956), Cuellar (1966a), and Fox (1977). The reported cases (Table I) include at least three species of turtles from two families, eight species of lizards from four families, and 29 species of snakes from six families. Fox and Dessauer (1962) ventured the opinion that storage receptacles probably are characteristic of all female snakes. This opinion is made reasonable by the fact that snake mating seasons and ovulatory seasons are seldom synchronous (Saint Girons 1982). Sperm may be viable in female reproductive tracts for up to 2½ months in lizards (Cuellar 1966b), 4 years in turtles (Barney 1922), and 5 years (Haines 1940) to perhaps 7 years (Magnusson 1979) in snakes.

TABLE I

Species of Reptiles Reported to Store Sperm,
Show Delayed Fertilization, or Have Seminal Receptacles[a]

TURTLES	
Malaclemys terrapin	*Testudo graeca*
Terrapene carolina	
LIZARDS	
Anolis carolinensis	*Chamaeleo jacksoni*
Uta stansburiana	*Microsaura pumila*
Coleonyx variegatus	*Eumeces egregius*
Phyllodactylus homolepidurus	*Hemiergis peronii*
SNAKES	
Agkistrodon contortrix	*Natrix natrix*
Bothrops lanceolatus	*Natrix vittata*
Causus rhombeatus	*Rhabdophis submineata*
Crotalus viridis	*Storeria dekayi*
Trimeresurus popeorum	*Thamnophis brachystoma*
Vipera aspis	*Thamnophis couchi*
Bungaris fasciatus	*Thamnophis elegans*
Naja naja	*Thamnophis sirtalis*
Acrochordus javanicus	*Tropidoclonion lineatum*
Boiga multimaculata	*Xenodon merremi*
Coronella austriaca	*Leptotyphlops dulcis*
Drymarchon corias	*Leptotyphlops humilis*
Leimadophis viridis	*Typhlops angolensis*
Leptodeira albofusca	*Typhlops braminus*
Leptodeira annulata	

[a]Compiled from Fox (1956), Fox and Dessauer (1962), Clark (1963), Smyth and Smith (1968), Schaefer and Roeding (1973), Fox (1977), Magnusson (1979), and Saint Girons (1982).

The potential for a female to mate with more than one male before she ovulates is high in species with long-term sperm storage because ovulation need not immediately follow mating. Some reptiles certainly do mate long before they ovulate. Fox (1956) found sperm in female gartersnakes (*Thamnophis elegans* and *T. sirtalis*) months before ovulation, and Blanchard and Blanchard (1941) noted fall-mated gartersnakes that produced young without additional mating in the following spring. Ludwig and Rahn (1943) found sperm in female rattlesnakes (*Crotalus viridis*) whose ovarian condition indicated that they would not have ovulated until almost a year later.

However, evidence for actual multiple paternity in nature is very limited, partly because few genetic markers are known since reptiles are not as easy to breed as some other animals. Blanchard and Blanchard (1941) bred gartersnakes (*T. sirtalis*) over 10 years to compile evidence that the melanistic morph from a population near Lake Erie was determined by a homozygous recessive state at a single locus. Gibson and Falls (1975) analyzed the frequencies of melanic vs. normal littermates in broods of naturally mated Lake Erie gartersnakes. They found ratios incompatible with a hypothesis of single paternity and the genetics postulated by the Blanchards. If the genetics of melanism is the same for all Lake Erie gartersnake populations, this is probably the best documented case of natural multiple mating in reptiles to date. Other techniques, such as electrophoresis, might usefully add to this type of evidence in the future.

In territorial lizards, females sometimes mate repeatedly but usually with the same male (Rand 1967). Multiple matings with the same male may be common in the context of territoriality, and they might be thought of as "forced copulations." The female who submits to repeated matings may benefit from sharing the resources in the male's territory. The female who is unreceptive, or who is likely to have been fertilized by another male, may be excluded from the territory by the male (Milstead 1970). Multiple matings by the female in these species may be of little consequence for sperm competition because the same male usually mates with her. But multiple mating by the male in this context might be evolved as a strategy to dilute the sperm of infrequent competitors, and so be a consequence of sperm competition, in part.

Direct observations of multiple mating of a female snake exist. They show that females will mate more than once, but they usually fall short of providing evidence that sperm mixing is a natural event. Frequently, these reports are incidental to a captive breeding program, and only one male performs the matings. Lewke's (1979) observations on kingsnakes (*Lampropeltis getulus*) are typical. The confinement of captivity raises doubts about naturalness when more than one male does mate, as with the pythons (*Python molurus*) observed by Barker *et al.* (1979). In a seminatural enclosure, Hammerson (1978) saw one female whipsnake (*Masticophis lateralis*) mate twice with one male, and another female mate with one male and accept courtship from another male later. Early reports of supposed multiple mating (Perry 1920, Brennan 1924) were rightly dismissed by Noble (1937) as being misinterpretations of mere courtship.

Multiple mating would probably be much more of a female option in the snakes. Unlike most lizards, snakes are thought to be nonterritorial as a rule, and females cannot improve their access to resources by submitting to multiple copulations. Moreover, the hemipenes of snakes are not designed for forcing penetration by thrusting. Each is a tubular sac, inverted into the base of the male's tail, which everts to fill the cavity of the female's cloaca (Edgrin 1953) when she actively opens her vent to allow intromission (Noble 1937). The anal scale that covers the vent normally forms such a snug, smooth, sturdy shield that forced copulation (with a hemipenis) is difficult to imagine. Similar arguments against forced copulation might hold true for nonterritorial lizards and perhaps other reptiles with analogous characteristics.

Although a significant potential for multiple matings by female reptiles is a key prerequisite for the discussion that follows, it must be considered more as an assumption than as an established fact. I do not believe it is an unreasonable assumption. Sperm storage is extensive, and the paucity of documentation of multiple matings may be more a consequence of observational difficulties than a reflection of true rarity. Mating is seldom seen at all in reptiles, and single observations still warrant publication for many species.

III. SPERM COMPETITION IN FEMALES
AND ITS EVOLUTIONARY CONSEQUENCES FOR MALES

I know of no studies that are directly concerned with what happens within a female reptile that has mated with more than one male. Sperm of different genotypes may be differentially effective in fertilization. Alternatively, their effectiveness may be equal and, if their semen is completely mixed, each male's success would merely be diluted proportionately to the contribution from other males. Seminal mixing might take place in the oviducal lumen type of storage site that Ludwig and Rahn (1943) described for prairie rattlesnakes. But complete mixing would be less likely in gartersnakes with their numerous small discrete sperm receptacles in the walls of the infundibulum (Fox 1956). According to Fox, sperm cells tend to fill these receptacles as they encounter them, and the most cephalic receptacles are usually empty. A form of sperm precedence could take place if sperm from the latest mating filled the anteriormost receptacles and, so, were in the best position to encounter eggs first. Such hypotheses remain to be tested.

Sperm displacement (as a consequence of the male's delivery system rather than sperm competition *per se*) is not likely to occur in reptiles because the known sperm storage sites are beyond the known reach of the males' intromittent organs. In squamates that have been studied, penetration of the hemipenis is limited to the cloaca (Pope 1941, Inger and Marx 1962, Conner and Crews 1980) or, at most, the vaginal pouches of the oviducts (Ludwig and Rhan 1943). Anterior to this is a coiled tubular part of the oviduct (misleadingly called the vagina) where stored sperm have been found in iguanid lizards (Cuellar 1966a) and skinks (Schaefer and Roeding 1973). Similarly, the turtle, *Terrapene carolina,* stores sperm in the caudal part of the oviduct (Hattan and Gist 1975). Sperm storage is in the more anterior vagina and uterus of crotalid snakes (Ludwig and Rahn 1943), between the uterus and infundibulum of gekkonid lizards (Cuellar 1966a), and in the infundibulum of colubrid snakes (Fox 1956).

Guarding behavior before and after copulation may be widespread. Territoriality is certainly common in lizards, and it may occur among crocodilians, turtles, and snakes (Porter 1972). The functional context in which territorial behavior evolved may be complex with no one factor acting exclusively. But one consequence of territoriality that probably played a role in its evolution is that the resident male reduces the chances of other males mating with females in his territory. Furthermore, some male lizards ride the backs of females for extended periods following copulation (Greenberg and Noble 1944, Vitt and Lacher 1981). Male-to-male combat with the appearance of a ritualized wrestling match has been reported in many snake species (reviewed by Shine 1978) and some lizards (Murphy and Mitchell 1974). Some turtles also display combat (Weaver 1970, Booth and Peters 1972, Barzilay 1980). Most of these observations have been fortuitous or staged and insufficiently detailed to reveal the basis for the males' aggression. The more detailed reports suggest that this is most likely guarding behavior or the defense of a "floating" territory in which a female is the only resource (Bennion and Parker 1976, Shine *et al.* 1981). Perhaps the best example comes from Prestt's (1971) study of the viper (*Vipera berus*) in which males established and defended stations in the immediate vicinity of isolated females.

Prestt's study also revealed what may be an important prerequisite to the evolution of guarding behavior in snakes. Females and, subsequently, males moved away from their crowded hibernation sites before any sexual activity began. Since the females moved first, this is not a male strategy of moving to a low risk of take-over area. Once the snakes have dispersed enough so that a defending male would sustain threats usually from no more than one competitior at a time, guarding behavior could be effective and advantageous. Effective guarding would be much more difficult if the snakes mated at the crowded hibernaculum, and different male mechanisms might be more advantageous.

Mating plugs (Fig. 1) have been recognized in only a few species of snakes so far: *Nerodia taxispilota, Thamnophis butleri, T. sauritus, T. sirtalis, Virginia striatula* (unpubl. data), *T. radix* (Ross and Crews 1977), and *Coluber constrictor priapus* (L. Vitt, pers. comm.). The hard gelatinous plug is formed from male kidney

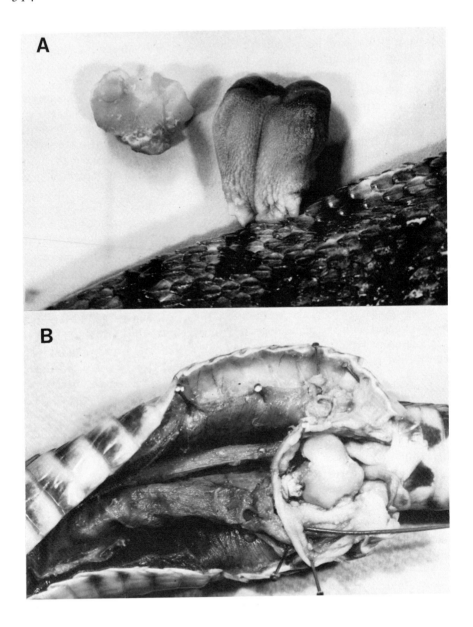

Fig. 1. A mating plug of *Nerodia taxispilota*. A. Next to the hemipenis of a male. B. Replaced in the dissected cloaca of the female that expelled it. Photos by Paul E. Feaver.

secretions deposited in the female's anterior cloaca at the end of copulation and after sperm transfer (Devine 1975). The plug adheres to the cloacal walls obstructing the mouths of the oviducts and forming a barrier to the passage of sperm while it is in place. Its duration in females is temperature dependent in these poikilotherms, and natural temperatures may fluctuate unpredictably during their early spring mating season. I have recorded durations as brief as 2 days at 21°C and as long as 2 weeks at 4°C under constant laboratory conditions.

The snake mating plug probably evolved, like insect plugs (Parker 1970), to reduce sperm competition by forming a temporary barrier to the effective insemination of a female by other males. This could be considered to be an alternative to guarding behavior which was evolved in those species in which guarding is ineffectual because females tend to mate in areas of high male density. The most intense period of mating activity for gartersnakes, for example, takes place at the hibernacula (Aleksiuk and Gregory 1974, Gregory 1974, Devine 1977). Females emerge one by one after the males have emerged *en masse,* and so the effective male to unmated female ratio is usually very high. Competition among males does not take the form of combat or territoriality, but it resembles more a scramble for position to transmit tactile courtship cues along the length of a female (Noble 1937). I have seen entwined clusters of up to six males courting a female, and Gregory (1974) reported courting aggregates with as many as 30 or more males per female. Guarding behavior by a male under these conditions would probably be futile and counterproductive. From the limited available data, it seems that species that form plugs also have been seen courting in aggregates (Devine 1975), whereas at least some species that guard and do not form courting aggregates also do not form plugs (*e.g.,* rattlesnakes, Ludwig and Rahn 1943; kingsnakes, Murphy *et al.* 1978).

IV. FACTORS INFLUENCING THE INCLINATION TO MATE MORE THAN ONCE

Why should female reptiles tend to mate with more than one male? This seldom addressed question is crucial to the foregoing discussion because, without multiple mating, hypotheses involving sperm competition and its consequences in reptiles are meaningless. Multiple mating is easily explained in mating systems that favor forced copulation, and I have already alluded to the importance of territoriality in this regard. But the answers are much less obvious in species in which forced copulation is improbable, and the snakes are the outstanding example of these. Consequently, unless otherwise specified, the following discussion is intended to apply to snakes rather than reptiles in general.

Multiple mating must have disadvantages as well as advantages to female snakes. Excessive courtship and mating must decrease the time and energy available to a female for other important activities. But the most obvious disadvantage is an increased risk from predation because male courtship tends to make the pair or group more conspicuous. Gibson and Falls (1975) suggested that successive matings are simply less disadvantageous for female gartersnakes than their submitting to endless courtship. This argument assumes that females have no control over their attractivity and that competitive males would endlessly court unreceptive females when receptive females might be nearby. I doubt the validity of either of these assumptions, and experiments by Garstka *et al.* (1982) suggest that neither is true.

If females "deliberately" maintain attractivity and receptivity after one mating, we must consider other possible advantages. Male snakes contribute no known benefits to females or their progeny other than sperm and, since I have seen thousands of sperm in a small sample of a single ejaculate, sperm quantity is probably not limiting. Perhaps females mate again as insurance against the slight possibility that the first male was sterile. This is merely an extreme case of the more general hypothesis that females should promote sperm competition *per se* because, with any degree of difference in sperm effectiveness, the most effective sperm will give rise to more sons with effective sperm thus favoring the passage of the female's genes.

Female choice might be important. A female snake might benefit from mating again if a subsequent male was significantly "better" than the first in the same sense as the term is used in other discussions of female choice. A key assumption here is that memory would be more important than we usually assume it to be because females would be making sequential rather than simultaneous comparisons between males.

An even more extraordinary kind of female choice is at least imaginable in the form of postmating, nonrandom use of sperm by females. Since females provide sperm storage organs that are multiple and disjunct, they would seem to have the potential for controlling the use or lives of specific populations of sperm. If a later male was in some way "better" than a previous mate, a female could conceivably disfavor the storage or use of the earlier set of sperm and favor the new set. If such female adaptations exist, their effects would be difficult to distinguish from the effects of sperm precedence mediated by male adaptations.

Insemination in itself should not impose constraints on a female that are contrary to her own reproductive best interests. Chemical or physical aspects of insemination might trigger a reduction in the female's receptivity and attractivity, but only if the disadvantages of multiple mating already predominate or if the male also effectively enforces the female's ineligibility to other males. Territoriality, guarding, and plugging tend to enforce female chastity. Thus, the coition-induced inhibition of receptivity in female *Anolis* lizards reported by Crews (1973) might have its effects only because male territoriality restricts the mating options of females.

To the extent that mating with several males confers net benefits to females, there should be evolved mechanisms in females to counteract any weakly enforced male strategies that would restrict female mating options. The mating plugs of gartersnakes have a maximum known duration of only 2 weeks. Since most mating occurs in March and April, but ovulation occurs in May (Fox 1956), females regain the option of remating before any sperm are used. I predict that antiplug enzymes will be found to be a part of the female's cloacal secretions. I imagine female gartersnakes have evolved progressively better enzymes to dissolve mating plugs while males were evolving plug proteins more resistant to dissolution.

The mating behavior of gartersnakes provides the best available evidence that snake mating plugs effectively enforce female ineligibility while they are in place. I have observed intense competition in Michigan among numerous male *Thamnophis sirtalis* and *T. butleri* in seeking out and courting the limited number of available females. Most courtship scenes involved more than one male. I found many plugged females (19 *T. sirtalis* and 4 *T. butleri*) that had not left the arena of activity, but none of these was ever being courted (Devine 1977). Field experiments, in which I presented hand-held plugged and unplugged females to unrestrained males, showed that males can and do discriminate between these types of females, preferentially courting females without plugs. In independent laboratory studies by Ross and Crews (1977, 1978), *T. radix* males showed similar preferences. The behavioral change is rapid. The unsuccessful males in a courting aggregation tend to diperse soon after their successful rival achieves intromission (Noble 1937; Devine 1975, 1977). Furthermore, when Ross and Crews artificially expressed plugs from recently mated females, the snakes regained attractivity for courtship (but none remated within 48 h). These behavior patterns strongly suggest that the plug is an effective barrier to insemination, and selection has favored males that waste no time and energy on plugged females but court females without plugs.

Whether or not females exercise their option to remate after naturally expelling the plugs remains in question. Ross and Crews believed that the females from whom they removed plugs were attractive but unreceptive. This seems maladaptive given the disadvantages of courtship without mating. Ross and Crews may have abnormally affected the receptivity of their snakes by artifically expressing the plugs or by inducing sexual behavior out of season with exogenous hormones, or females may simply be more selective having mated once.

The source and function of the cues that mediate the inhibition of courtship are matters of debate. Ross and Crews established that the gartersnake cues are chemical. Moreover, they intimated that a pheromone emanating from the plug itself renders the female unattractive, and exposure to the odor temporarily depresses courtship activity in all males to all (even unmated) females while the mated male is sexually refractory (24 to 72 h in their report). However, their experiments in transferring cloacal wash materials from mated to unmated females and in mating females to vasectomized males only show that a chemical cue comes from successfully mated females, not specifically from the plug alone. Apparently, they took no precautions to remove the odor of mated females from the testing cages of males whose courting of unmated females purportedly was repressed. The lingering, abnormally confined odor probably confused and affected the subsequent behavior of males to unmated females. Males certainly are not repressed similarly in the field (Devine 1977). Finally, the lengthy male refractory period that they reported may reflect an abnormal sexual physiology induced by exogenous hormones, a treatment having questionable value (Garstka *et al.* 1982). Normal males can resume sexual activity at least within one to a few hours (Blanchard and Blanchard 1941, Devine 1977).

An evolutionary argument must consider all affected individuals. While recognizing that curtailment of unproductive courtship benefits the female, Ross and Crews emphasized how highly advantageous it would be to the mating male to produce an odor that caused other males to disregard females and removed competitors from the breeding population temporarily. I have argued (Devine 1977) that such an odor would be most advantageous to a producing male at the initiation of courtship (not mating, as Ross and Crews [1978] misquoted), but this is not observed. If a male odor alone had these effects, energetics would favor minimal as well as early secretions of plug material, and the effectiveness of the plug as an enforced barrier would be less important. Ross and Crews failed to recognize that their proposed male pheromone would exert selection against the responding males, removing their type from the breeding population permanently. But males that disregarded a potentially unenforced male cue and tested the female's responsiveness to their courtship anyway would be favored by selection. Mere signals have too high a probability of being used as a bluff by males. The behavior of rival males must be influenced by an effective physical barrier, the plug. Only the female in this mating system could unambiguously verify the futility of continued courtship because the plug has formed an effective barrier to further inseminations. Therefore, the pheromone that inhibits male courtship of a specific female probably comes from the female that is plugged rather than from the plug itself.

V. SUMMARY

The potential for sperm competition in reptiles is high. Sperm storage in females is widespread, and females with this ability have ample time to mate with more than one male before they ovulate and use the stored sperm. Reported observations suggest that females can tend to mate more than once, but the evidence is limited. Females of territorial species may be constrained to mate with the same male usually. Forced copulation is improbable in nonterritorial species. Multiple mating and sperm competition can be assumed to occur with significant frequency in reptiles. The interactions of different sperm populations within a female reptile have not been studied, but female anatomy provides a basis for some speculation. Sperm displacement by mixing and dilution or sperm precedence may occur in different species. Sperm displacement directed by the male's copulatory organ is unlikely. Aspects of male behavior would be adaptive for reducing the potential for sperm competition. Territoriality and male combat are widespread. Alternnatively, some snakes form mating plugs. If females mate more than once, some benefit must outweigh the disadvantages of waste and risk. Possible advantages include promoting sperm competition and female choice, exercised before or after mating. Chemical properties of the male's ejaculate should not constrain the reproductive strategies of females or other males, but physical barriers to insemination can do so. The behavior of snakes indicates that mating plugs are effective barriers. Female snakes with mating plugs become unattractive to males, and a pheromone signals this condition. The suggestion that males produce this pheromone that inhibits other males is discounted because males sould not respond to a cue that might well be a bluff. Mated females probably produce the signal of their unavailability.

REFERENCES

Aleksiuk, M., and P. T. Gregory. 1974. Regulation of seasonal mating behavior in *Thamnophis sirtalis parietalis. Copeia* 1974:681-689.

Barker, D. G., J. B. Murphy, and K. W. Smith. 1979. Social behavior in a captive group of Indian pythons, *Python molurus* (Serpentes, Boidae), with formation of a linear social hierarchy. *Copeia* 1979:466-471.

Barney, R. L. 1922. Further notes on the natural history and artificial propagation of the diamond-back terrapin. *Bull. U.S. Bur. Fish.* 38:91-111.

Barzilay, S. 1980. Aggressive behavior in the wood turtle, *Clemmys insculpta. J. Herpetol.* 14:89-91.

Bennion, R. S., and W. S. Parker. 1976. Field observations on courtship and aggressive behavior in desert striped whipsnakes, *Masticophis t. taeniatus. Herpetologica* 32:30-35.

Blanchard, F. N., and F. C. Blanchard. 1941. The inheritance of melanism in the garter snake, *Thamnophis sirtalis sirtalis* (L.), and some evidence of effective autumn mating. *Pap. Mich. Acad. Sci.* 26:177-193.

Booth, J., and J. A. Peters. 1972. Behavioural studies on the green turtle (*Chelonia mydas*) in the sea. *Anim. Behav.* 20:808-812.

Brennan, G. A. 1924. A case of simultaneous polyandry in snakes. *Copeia* 1924:52.

Clark, R. J. 1963. On the possibility of an autumnal mating in the tortoise (*Testudo graeca ibera*). *Br. J. Herpetol.* 3:85-86.

Conner, J., and D. Crews. 1980. Sperm transfer and storage in the lizard, *Anolis carolinensis. J. Morphol.* 163:331-348.

Crews, D. 1973. Coition-induced inhibition of sexual receptivity in female *Anolis carolinensis. Physiol. Behav.* 11:463-468.

Cuellar, O. 1966a. Oviducal anatomy and sperm storage structures in lizards. *J. Morphol.* 119:7-20.

Cuellar, O. 1966b. Delayed fertilization in the lizard, *Uta stansburiana. Copeia* 1966:549-552.

Devine, M. C. 1975. Copulatory plugs in snakes: Enforced chastity. *Science* 187:844-845.

Devine, M. C. 1977. Copulatory plugs, restricted mating opportunities, and reproductive competition among male garter snakes. *Nature* 267:345-346.

Edgren, R. A. 1953. Copulatory adjustment in snakes and its evolutionary implications. *Copeia* 1953:162-164.

Fox, H. 1977. The urogenital system of reptiles. In *Biology of the Reptilia, Vol. 6,* C. Gans and T. S. Parsons (eds.), pp. 1-157. Academic Press, New York.

Fox, W. 1956. Seminal receptacles of snakes. *Anat. Rec.* 124:519-540.

Fox, W., and H. C. Dessauer. 1962. The single right oviduct and other urogenital structures of female *Typhlops* and *Leptotyphlops. Copeia* 1962:590-597.

Garstka, W. R., B. Camazine, and D. Crews. 1982. Interactions of behavior and physiology during the annual reproductive cycle of the red-sided garter snake (*Thamnophis sirtalis parietalis*). *Herpetologica* 38:104-123.

Gibson, A. R., and J. B. Falls. 1975. Evidence for multiple insemination in the common garter snake *Thamnophis sirtalis. Can. J. Zool.* 53:1362-1368.

Greenberg, B., and G. K. Noble. 1944. Social behavior of the American chameleon (*Anolis carolinensis*). *Physiol. Zool.* 17:392-439.

Gregory, P. T. 1974. Patterns of spring emergence of the red-sided garter snake (*Thamnophis sirtalis parietalis*) in the Interlake region of Manitoba. *Can. J. Zool.* 52:1063-1069.

Haines, T. P. 1940. Delayed fertilization in *Leptodeira annulata polysticta. Copeia* 1940:116-118.

Hammerson, G. A. 1978. Observations on the reproduction, courtship, and aggressive behavior of the striped racer, *Masticophis lateralis euryxanthus* (Reptilia, Serpentes, Colubridae). *J. Herpetol.* 12:253-255.

Hattan, L. R., and D. H. Gist. 1975. Seasonal receptacles in the eastern box turtle, *Terrapene carolina. Copeia* 1975:505-510.

Inger, R. F., and H. Marx. 1962. Variation of hemipenis and cloaca in the colubrid snake *Calamaria lumbricoidea. Syst. Zool.* 11:32-38.

Lewke, R. E. 1979. Neck-biting and other aspects of reproductive biology of the Yuma kingsnake (*Lampropeltis getulus*). *Herpetologica* 35:154-157.

Ludwig, M., and H. Rahn. 1943. Sperm storage and copulatory adjustment in the prairie rattlesnake. *Copeia* 1943:15-18.

Magnusson, W. E. 1979. Production of an embryo by an *Acrochordus javanicus* isolated for seven years. *Copeia* 1979:744-745.

Milstead, W. W. 1970. Late summer behavior of the lizards *Sceloporus merriami* and *Urosaurus ornatus* in the field. *Herpetologica* 26:343-354.

Murphy, J. B., B. W. Tryon, and B. J. Brecke. 1978. An inventory of reproduction and social behavior in captive gray-banded kingsnakes, *Lampropeltis mexicana alterna* (Brown). *Herpetologica* 34:84-93.

Murphy, J. B., and L. A. Mitchell. 1974. Ritualized combat behavior of the pygmy mulga monitor lizard, *Varanus gilleni* (Sauria:Varanidae). *Herpetologica* 30:90-97.

Noble, G. K. 1937. The sense organs involved in th courtship of *Storeria, Thamnophis* and other snakes. *Bull. Am. Mus. Nat. Hist.* 73:673-725.

Parker, G. A. 1970. Sperm competition and its evolutionary consequences in the insects. *Biol. Rev.* 45:525-567.

Perry, A. 1920. The mating of watersnakes. *Copeia* 1920:49-50.

Pope, C. H. 1941. Copulatory adjustment in snakes. *Field Mus. Nat. Hist. Publ. Zool. Ser.* 24:249-252.

Porter, K. R. 1972. *Herpetology.* Saunders, Philadelphia, PA.

Prestt, I. 1971. An ecological study of the viper, *Vipera berus,* in southern Britain. *J. Zool. (Lond.)* 164:373-418.

Rand, A. S. 1967. Ecology and social organization in the iguanid lizard *Anolis lineatopus. Proc. U.S. Nat. Mus.* 122:1-79.

Ross, P., and D. Crews. 1977. Influence of the seminal plug on mating behavior in the garter snake. *Nature* 267:344-345.

Ross, P., and D. Crews. 1978. Stimuli influencing mating behavior in the garter snake, *Thamnophis radix. Behav. Ecol. Sociobiol.* 4:133-142.

Saint Girons, H. 1975. Sperm survival and transport in the female genital tract of reptiles. In *Biology of Spermatozoa,* S. Karger, Basel.

Saint Girons, H. 1982. Reproductive cycles of male snakes and their relationship with climate and female reproductive cycles. *Herpetologica* **38**:5-16.

Schaefer, G. C., and C. E. Roeding. 1973. Evidence for vaginal sperm storage in the mole skink, *Eumeces egregius. Copeia* **1973**:346-347.

Shine, R. 1978. Sexual size dimorphism and male combat in snakes. *Oecologia* **33**:169-278.

Shine, R., G. C. Grigg, T. G. Shine, and P. Harlow. 1981. Mating and male combat in Australian blacksnakes, *Pseudechis porphyiacus. J. Herpetol.* **15**:101-107.

Smyth, M., and M. J. Smith. 1968. Obligatory sperm storage in the skink *Hemiergis peronii. Science* **161**-575-576.

Vitt, L. J., and T. E. Lacher, Jr. 1981. Behavior, habitat, diet, and reproduction of the iguanid lizard *Polychrus acutirostris* in the Caatinga of Northeastern Brazil. *Herpetologica* **37**: 53-63.

Weaver, W. G., Jr. 1970. Courtship and combat behavior in *Gopherus berlandieri. Bull. Fla. State Mus. Biol. Sci.* **15**:1-43.

15

Sperm Competition
in Apparently Monogamous Birds

FRANK McKINNEY

KIMBERLY M. CHENG

DAVID J. BRUGGERS

I. INTRODUCTION

Students of avian mating systems generally agree that most bird species are monogamous. Usually the generalization is made either to contrast birds with mammals, in which pair-bonding is rare, or to emphasize that the spectacular avian examples of polygyny, polyandry, and promiscuity really are exceptional phenomena in birds (*e.g.*, Lack 1968, Selander 1972, Emlen and Oring 1977). Monogamy is a "prolonged association and essentially exclusive mating relationship between one male and one female" (Wittenberger and Tilson 1980). Although this definition states that mates normally copulate only with one another, the literature contains many reports of extrapair copulations in monogamous birds (Table I).

Sperm Competition and the Evolution
of Animal Mating Systems

TABLE I

The Number of Apparently Monogamous Avian Species
in Which Extrapair Copulation Has Been Recorded

Family	Forced Extrapair Copulations[a]	Unforced Extrapair Copulations[b]	Number of Species[c]
Diomedeidae	3	1	4
Procellariidae		1	1
Spheniscidae		1	1
Sulidae	1	1	1
Pelecanidae	1	1	2
Phalacrocoracidae		1	1
Anhingidae	1		1
Ardeidae	6	2	7
Threskiornithidae		1	1
Phoenicopteridae		1	1
Anatidae	39		39
Gruidae		1	1
Haematopodidae	1	1	1
Recurvirostridae		1	1
Laridae	5	4	9
Alcidae	1	2	2
Columbidae	2	2	2
Meropidae		1	1
Hirundinidae	4	3	5
Corvidae	5		5
Cracticidae	1	1	1
Muscicapidae			
Turdinae[d]		1	1
Emberizidae			
Parulinae		2	2
Emberizinae	1	2	3
Fringillidae	2	1	2
Passeridae		1	1
Estrildidae	8		8
Totals	81	33	104

[a]This category includes both apparently successful FEPC's and FEPC attempts.

[b]Extrapair courtship and/or ambiguous copulation attempts (see Corvidae, Parulinae, Fringillidae, Estrildidae in part V) are not included in this category.

[c]A single species may practice more than one form of extrapair copulation. Thus the number of species may be less than the number obtained by summing the number of species in which FEPC's and UEPC's have been observed.

[d]Electrophoretic evidence indicating EPC.

Such incidents have usually been regarded as exceptions to the species-typical pattern of behavior, and until very recently the possibility that they can result in the fertilization of eggs has not been seriously considered. The purpose of this paper is to review the evidence relating to this possibility and to evaluate the potential for sperm competition (as defined by Parker 1970) in species generally considered to be monogamous.

Experimental and analytical methods that provide direct evidence on sperm competition and multiple paternity of clutches in wild birds (*e.g.,* genetic markers, sterilization of males, biochemical polymorphisms) have been used rarely by ornithologists (*e.g.,* Bray *et al.* 1975, Roberts and Kennelly 1980), but their importance is now widely recognized (Sherman 1981). In future studies, indirect behavioral evidence of the kind reviewed here may be useful in suggesting which species to test for the incidence of sperm competition by these direct methods.

In theory, at least four circumstances could result in sperm competition in basically monogamous birds. These entail either (a) possible fitness gains for paired males, unpaired males, or females that engage in extrapair copulations, or (b) incidental consequences of mate-switching. If the birds are individually marked and of known status, it may be possible to distinguish between these circumstances by observing the behavior of the participants. In the case of paired males, the main benefit is likely to be an increase in the total number of offspring sired, but there is also the possibility of insurance against the chance that the male's mate is sterile or genetically inferior. If females are scarce, unpaired males may have little chance of acquiring mates of their own and they may benefit from attempting to sire offspring by inseminating already paired females. Paired females may benefit by receiving sperm from more than one male if there is a chance that their mate is sterile or genetically inferior, or if males contribute parental care to the young of females with whom they have copulated. If the benefits of extrapair copulation are to males only, we might expect to see reluctance or rejection on the part of females; if females benefit, they can be expected to solicit or at least tolerate these copulations.

Sperm competition might occur also in situations where birds benefit by switching mates or remating quickly. In both instances, sperm from the previous mate might survive in the female's sperm storage organs and compete with sperm from the new mate. If this is a significant hazard for newly paired males, then males should behave in ways that reduce the risk of sperm competition. Two possibilities are that males may behave aggressively toward new mates and so delay fertilization of the next clutch (Erickson and Zenone 1976), or they may copulate repeatedly to overwhelm competing sperm.

In spite of the theoretical advantages (to females, to unpaired males, to paired males), there may be conflicting costs and risks involved in attempting extrapair copulations. The possibilities of desertion by the mate, disruption of vital breeding activities, and risk of physical damage could be especially serious. Because of their fundamentally different roles in reproduction, the cost-benefit ratio is sure to be different for males and females. Also, depending on which sex plays the major role in shaping the mating system, there could be subtle factors that tip the balance between costs and benefits in different directions even among closely related species.

There is one general hypothesis, however, on how natural selection may be expected to act on the sexes in regard to extrapair copulations. Trivers (1972) argued that when strict monogamy is not enforced by ecological or social circumstances, males may be expected to copulate with and abandon as many females as possible. In species where selection is strong for male parental care, he proposed that a "mixed" strategy is likely to be advantageous for males. In such species, a male should be expected to "help a single female raise young, while not passing up opportunities to mate with other females whom he will not aid." Under Trivers' hypothesis, such mixed male strategies are likely to be widespread in monogamous animals.

Some of the literature relevant to this topic was assembled by Gladstone (1979) for a discussion of extrapair copulations in monogamous colonial birds in relation to energetic investments of the sexes. Our review covers a broader survey of avian groups and our focus is on behavioral aspects of extrapair copulations in relation to sperm competition.

II. PHYSIOLOGICAL CONSIDERATIONS

Many experiments involving sperm competition have been carried out on domestic chickens, turkeys, and ducks, and much has been learned through the use of artificial insemination techniques (Payne and Kahrs 1961, Allen and Champion 1955, Lake 1975). By removing males from a captive flock of laying females, Elder and Weller (1965) showed that female mallard ducks (*Anas platyrhynchos*) can store viable spermatozoa for up to 17 days, but viability drops rapidly after about 5 days. This is similar to ring doves (*Streptopelia risoria*) in which viable sperm can be stored for nearly 6 days (Erickson and Zenone 1976). Studies on chickens have shown a similar decline in viability with the age of the sperm (although the longevity record is 35 days), and in general the most recent of competing inseminations is likely to be the most effective. When hens were caught immediately after copulation and artificially inseminated with semen from a rooster of a different breed, half of the progeny were sired by each male (Warren and Gish 1943). If competing inseminations were 4 hours apart, however, 80% of the progeny resulted from the second insemination (Compton *et al.* 1978). Evidently semen from the two inseminations remains in separate layers in the sperm storage glands of the female with the most recently inseminated semen staying on top and being used first to fertilize eggs. Recent experiments with mallard ducks have shown similar results suggesting that the same storage mechanism is operating (Cheng *et al.* 1982). Comparable information on order effects is lacking for most birds and, judging from the striking differences reported by Dewsbury (this volume) in rodents, we should be prepared to find interspecific variations.

In birds that lay multi-egg clutches, one egg is typically laid each day. In some species, however, the interval between successive eggs is longer. In chickens, turkeys, quail, and ducks (all of which lay large clutches), ovulation occurs normally within 15-75 minutes after laying of the previous egg (Sturkie 1976). The ovulated ovum is fertilizable for only about 15 minutes before albumen is deposited around the yolk. An insemination occurring while the egg is passing down the oviduct is likely to be ineffective (Bobr *et al.* 1964). Thus the timing of inseminations in relation to time of egg-laying each day could be very important in species such as this.

A further factor to be taken into account is the time required for sperm to reach the ovum after insemination. Most birds do not have a male intromittent organ and it seems unlikely that sperm can be inserted very far into the oviduct at copulation. In chickens, sperm reached the ovum 26 minutes after insemination (Mimura 1939) and in turkeys sperm traveled from the vagina to the infundibulum within 15 minutes (Howarth 1971), but it is difficult to say whether these experimental results approximate the times required under natural conditions of copulation. Tinamous (Tinamidae), curassows (Cracidae), waterfowl (Anatidae), ratites, and a few other birds do have intromittent organs (King 1981) and presumably sperm has a shorter distance to travel in such species. We have not found any studies that attempt to relate cloacal morphology to sperm competition but research along these lines could be rewarding. There seems to be no evidence that birds have copulatory plugs comparable to those of insects and certain mammals.

III. COPULATORY BEHAVIOR IN BIRDS

In pair copulations, performed by mutual agreement between mates, the male mounts the female's back, maintains balance by shuffling with his feet or flapping his wings (gulls) or by grasping the female's crown feathers in his bill (waterfowl), presses his tail around one side of the female's tail and ejaculates when the cloacae of both birds are in apposition. Females signal their readiness by performing special movements or adopting a receptive posture. The act of copulation may be very brief, entailing a single thrust, or it can be prolonged (as in gulls) with much tail-waggling by the mounted male. Reversed mounts occur in some species (*e.g.*, pukeko, *Porphyrio p. melanotus*; Craig 1980).

In some birds, copulations occur at times when they can have nothing to do with the fertilization of eggs. In the mallard, for example, copulations are frequent during the period of pair-formation which begins 6 months before egg-laying starts. At this time, gonads are regressed and sperm are not being produced (Höhn 1947). Perhaps these mountings are part of the mate-assessment process whereby females test male competence in copulating.

Males of many kinds of birds have been reported attempting to force copulation on unwilling females, and in waterfowl (a group in which males have an intromittent organ) some of these attempts definitely succeed (see below). Among passerine birds (which do not have intromittent organs), some observers (*e.g.,* Rohwer 1978) believe that insemination cannot be achieved without the female's cooperation. However, Morris (1957) described "highly efficient rape" by male bronze mannikins (*Lonchura cucullata*), involving a female with a damaged wing which spent much time on the ground. Morris concluded that it is almost impossible for a male estrildine finch to force copulation when the female is perched, but it is apparently comparatively easy on the ground. Similarly, forced copulation attempts in breeding colonies of birds such as herons, gulls, and geese, are often directed at females on their nests. In summary, very little is known about the ability of birds to achieve insemination by force and for many species this topic is likely to be very difficult to study in the field.

IV. FORCED COPULATION AND SPERM COMPETITION IN WATERFOWL

Almost all waterfowl (ducks, geese, and swans of the family Anatidae) form pair-bonds and are generally regarded as monogamous. In many species of ducks, however, while males escort and copulate with their own mates they also force copulation (FC) with other females (Fig. 1). ("FC" has been used in recent papers on waterfowl as shorthand for "FEPC" [=forced extrapair copulation; Gladstone 1979].) This behavior has been reported in the literature for at least 70 years, and more is known about FC in waterfowl than in any other bird group. Here we summarize various theories and briefly review the current state of knowledge. A detailed review is presented elsewhere (McKinney *et al.* 1983).

Huxley (1912) thought that FC provides an outlet for the sex drive of mallard drakes after females begin incubating and are no longer available for copulation. This idea seemed plausible because FC attempts are frequent at the time of year when many females are incubating, but recent studies of marked birds have shown that the temporal association is misleading. For example Smith (1968) showed that pursuits are directed mainly at pre-laying or laying females thus indicating

Fig. 1. Forced copulation attempt in captive northern shovelers (*Anas clypeata*). The female's mate (bird with open wings) is attacking the assaulting male. These birds carry numbered plastic nasal discs for individual identification.

that fertilization of eggs may be involved rather than simply sublimation of male sex drive.

Some authors have argued that females encourage males to chase them and that females benefit from FC's. Christoleit (1929a, b) thought that the competition between males for the opportunity to copulate ensures that the eggs are fertilized by the strongest male, and Milstein (1979) has suggested that after losing early clutches to predators female mallards ensure fertilization of renest clutches through FC "timed by the female." Although female incitement has been reported in certain mammals (Cox and LeBoeuf 1977), there is no evidence that it occurs in waterfowl. Female ducks can be exhausted, wounded, or even killed during FC attempts (Huxley 1912); they try to avoid FC, and apparently pair-bonds reform for renest attempts.

In some species of dabbling duck (genus *Anas*), paired males defend a breeding territory from which they drive away other pairs. Weidmann (1956) suggested that, in the mallard, FC is a strong deterrent to intruding females and it discourages other pairs from settling on the territory. However, the interpretation that this is the primary function of FC is unsatisfactory because (a) male mallards will leave their territories to make FC attempts elsewhere (Lebret 1961) and (b) FC is highly developed in certain species that do not defend territories (*e.g.,* northern pintail; Smith 1968).

Most studies of mallard behavior have been made on urban populations living in parks or zoos, and sometimes FC has been regarded as a by-product of crowding, artificial feeding, and contamination with domesticated breeds. This explanation is inadequate, however, because FC occurs also in wild mallards (*e.g.,* Dzubin 1969) and in wild populations of some 29 additional species of waterfowl (McKinney *et al.* 1983).

Barash (1977a) suggested that unpaired males are responsible for FC in ducks because they have "little to lose and much to gain" by trying to inseminate paired females "as opposed to mated males whose options are exactly reversed." To date, studies of marked birds do not support this hypothesis. Paired males are known to engage in FC in all species of *Anas* studied to date and unpaired males appear to be mainly concerned with courtship and pair-formation (*e.g.,* Seymour and Titman 1979.)

The interpretation of FC that fits best with current views on how natural selection operates was outlined by Heinroth (1911) but it did not become generally accepted. Heinroth considered FC to be an integral component of mallard reproductive behavior. He thought that males combine pair-bonding with promiscuity via FC and that females do not benefit by being inseminated by males. He noted similar male promiscuity in a number of other duck species but stressed that it does not occur in species in which males play a major role in cooperative brood-care (swans, geese, and shelducks). Essentially this view agrees with Trivers' (1972) theory of mixed male reproductive strategies.

Support for the hypothesis that FC is an insemination strategy of paired males is strongest for several species of dabbling duck (*e.g.,* mallard, northern pintail, green-winged teal). The following are the key findings. Eggs can be fertilized by insemination during FC. This was shown for captive mallards in flight pens by using a genetic plumage marker (Burns *et al.* 1980). FC attempts are made by paired males during the months when eggs are being fertilized and they are directed primarily at pre-laying and laying females (Lebret 1961, Dzubin 1969, McKinney and Stolen 1982, Cheng *et al.* 1982). When pursued by males attempting FC, females try to escape by flying, diving, or hiding in vegetation, and paired males usually defend their mates by attacking males that attempt FC (*e.g.,* Bailey *et al.* 1978). After an FC attempt, the female's mate may try to force copulation himself. The occurrence of these "forced pair copulations" (FPC's; Barrett 1973,

Barash 1977b) suggests that sperm from the mate is being introduced to compete with sperm from other males.

Evidence pointing to the existence of a mixed male insemination strategy in the lesser snow goose (*Anser caerulescens caerulescens*) has been presented by Mineau and Cooke (1979a, b). Egg-laying is highly synchronous and there is no renesting in this arctic-breeding species. Paired males defend their mates closely until the clutch is complete and then they make FC attempts on neighboring females, mostly while they are on the nest. Only 10 of the 53 FC attempts observed were directed at females that were still fertilizable (the others were on incubating females), indicating a low potential payoff. However, even if only a few eggs could be fertilized each year by FC, this might still be of selective advantage to males in a long-lived species.

Apart from the lesser snow goose and closely related Ross's goose (*Anser rossii*), FC appears to be absent from the swan, goose, and shelduck groups. As Heinroth (1911) pointed out, these are species in which pairs hold exclusive territories and males are fully occupied with territorial defense and brood-care. There are records of FC in certain species belonging to most other waterfowl tribes (whistling ducks, perching ducks, pochards, sea ducks, and stiff-tails), and it is especially widely distributed among dabbling ducks. Most of these duck species are dispersed nesters but their social systems differ greatly, and the factors favoring FC in one more than another are not yet clear.

V. EXTRAPAIR COPULATION IN OTHER MONOGAMOUS BIRDS

With few exceptions the behavioral literature relevant to sperm competition in groups of birds other than waterfowl is fragmentary and difficult to evaluate. Until a few years ago, students of monogamous birds were more interested in cooperative aspects of pair-bonding than in the possibility that mates have conflicting interests. The "pair" or the "species" was often viewed as the unit that natural selection acts to preserve, and extrapair copulations were usually dismissed as rare "mistakes" having no biological significance. Thus valuable observations were probably omitted from many publications, and the descriptions that remain are often sketchy.

Authors have used a variety of terms to refer to extrapair copulations (*e.g.*, rape, pouncing, sexual chase, promiscuous mating, stolen mating) and usually the characteristics used to identify them have not been given. In particular, the criteria used to decide on two difficult questions were omitted: Did the female resist, and was insemination achieved? In the following summaries of observations reported in the literature we use the term "extrapair copulation" (EPC) for all copulations between individuals that are not paired to each other. The distinction between EPC in which the female resisted ("forced extrapair copulation," FEPC; Gladstone 1979), and EPC in which the female solicited or appeared to acquiesce ("unforced extrapair copulation," UEFC) is made only when authors make explicit statements on the female's behavior. Some authors have noted when EPC attempts were "incomplete" or "unsuccessful," but when they have not, we generally report "attempt" to indicate the uncertainty. We have followed the AOU Checklist (1982) and Peters (1931-79) for scientific and vernacular names.

Diomedeidae (Albatrosses).—FEPC occurs in Buller's albatross (*Diomedea bulleri*) when females visit the breeding colony before laying (Richdale 1949). Males tend to arrive first and then spend much time on the nest site where their mates visit them and pair copulations take place. Female visits, each lasting several hours, occur at about 6-day intervals expecially during the period 6-10 days before egg-laying. During these visits, apparently successful FEPC's by "intruder males" were recorded. Usually the female protested vigorously but was overpowered and mounted forcibly. In one instance the intruder male was known to be paired.

Similar behavior has been reported during the pre-egg stage in the wandering albatross (*D. exulans*; Tickell 1968) and Laysan albatross (*D. immutabilis*; Fisher 1971). In the former species, both paired and unpaired birds were involved, but in the latter species, almost all FEPC attempts were by paired, experienced breeders. UEPC's have been reported in the grey-headed albatross (*D. chrysostoma*; Tickell and Pinder 1966).

Procellariidae (Petrels).—In great-winged petrels (*Pterodroma macroptera*), copulation occurs during a brief mating period, following which females forage at sea for about 60 days prior to laying their eggs. Imber (1976) reported that in about 10% of all pairs observed, females had copulated with a male other than their mate.

Spheniscidae (Penguins).—Mated male jackass penguins (*Spheniscus demersus*) performed UEPC's with prospecting females (who later formed pair-bonds with other males) during the pre-laying and near fledging stages, but mated females were not observed copulating with strange males (Eggleton and Siefgried 1979).

Sulidae (Gannets).—Prospecting female northern gannets (*Sula bassanus*) frequently copulate with many males before pairing with one (Nelson 1965). Both UEPC and FEPC appear to be involved.

Pelecanidae (Pelicans).—Male pink-backed pelicans (*Pelecanus rufescens*) were observed copulating with females from neighboring nests when the female's mate was absent (Din and Eltringham 1974). FEPC's have been observed in the great white pelican (*Pelecanus onocrotalus*) but may occur too early in the breeding season to be successful in fertilizing eggs (Brown and Urban 1969).

Phalacrocoracidae (Cormorants).—UEPC's involving unpaired males and paired females have been reported in the Brandt's cormorant (*Phalacrocorax pencillatus*; Williams 1942).

Anhingidae (Anhingas).—In a study of nesting *Anhinga anhinga* in Mexico (Burger *et al.* 1978), males attempted FEPC with nearby females up to 15 days after the start of incubation.

Ardeidae (Herons).—FEPCs were reported in colonies of cattle egrets (*Bubulcus ibis*) in South Africa (Blaker 1969) and Colombia (Lancaster 1970), but a recent study in Japan (Fujioka and Yamagishi 1981) provides the most detailed analysis. In the latter study, 10 neighboring pairs were watched, three of which were followed through all stages of the breeding cycle. Ninety-two pair copulations and 147 FEPC attempts (25% apparently complete) were recorded. Failure of FEPC attempts was caused by mate defense, interference by other males attempting FEPC, or refusal by the female. Mate-defense was very effective when the mate was present, and all 38 apparently complete FEPC's occurred when he was absent. FEPC attempts were frequent during pre-laying and egg-laying stages, and infrequent during incubation.

In the reddish egret (*Egretta rufescens*), Allen (1955) noted that the female readily submits to copulation with other males (even up to four in rapid succession) when her mate is away from the nest prior to egg-laying. Both FEPC's and UEPC's have been reported in colonies of little blue herons (*Egretta caerulea*; Meanley 1955, Rodgers 1980, Werschkul 1982). Paired males usually copulated with a neighboring, egg-laying female when her mate was absent. Werschkul (1982) has suggested that inter-colony variation in the rate of EPC's is related to how much time males spend away from their mates gathering nest material. Verwey (1930) observed 10 FEPC's on four female grey herons (*Ardea cinerea*). Three of the females were incubating, and one was rebuilding her nest. Two of the FEPC's were performed by an unpaired male. The rest were performed by mated males. Verwey also observed a paired female copulate with a male that was not her mate. A single FEPC attempt was recorded in black-headed herons (*Ardea melanocephala*) just prior to incubation (Taylor 1984). In some studies of great blue herons (*A. herodias;* Cottrille and Cottrille 1958) and great egrets (*Casmerodius albus;* Wiese 1976, Gladstone 1979) FEPC's have been reported but in other studies they have not (Mock 1976, 1978).

Threskiornithidae (Ibises).—Kushlan (1973) and Rudegeair (1975) observed apparent UEPC's in the white ibis (*Eudocimus albus*) with males leaving their nests to copulate with nearby incubating females. Males fought so vigorously that they bloodied each other (Rudegeair 1975).

Phoenicopteridae (Flamingos).—Female lesser flamingos (*Phoeniconaias minor*) occasionally permit more than one male to mount them in quick succession (Brown and Root 1971).

Gruidae (Cranes).—Littlefield (1981) observed UEPC in the sandhill crane (*Grus canadensis*).

Haematopodidae (Oystercatchers).—Makkink (1942) reported occasional FEPC's and more common UEPC's throughout the breeding period of the European oystercatcher (*Haematopus ostralegus*).

Recurvirostridae (Avocets).—Three marked males and one marked female American avocet (*Recurvirostra americana*) copulated with unmarked birds, which were not their mates, during the incubation period of each marked bird (Gibson 1971).

Laridae (Gulls and Terns).—FEPC attempts have been reported in herring gulls (*Larus argentatus*; Goethe 1937, MacRoberts 1973), black-headed gulls (*L. ridibundus*; Moynihan 1955, Tinbergen 1956), lesser black-backed gulls (*L. fuscus*; MacRoberts 1973), laughing gulls (*L. atricilla*; Burger 1976, Burger and Beer 1976), and apparently successful FEPC's have been reported in glaucous-winged gulls (*L. glaucescens*; Vermeer 1963). The FEPC's and FEPC attempts were most frequent during egg-laying and were made by paired males on females at neighboring nests while her mate was absent. UEPC's have been recorded or inferred in western gulls (*L. delawarensis*; Ryder and Somppi 1979), California gulls (*L. californicus*; Conover *et al.* 1979), and Caspian terns (*Sterna caspia*; Cuthbert 1981).

Alcidae (Murres).—Many FEPC attempts by paired males on paired females were recorded in common murres (*Uria aalge*) by Birkhead (1978), mainly during the pre-laying period. UEPC's have been recorded for this species and for thick-billed murres (*U. lomvia*; Tuck 1961).

Columbidae (Pigeons).—The Heinroths (1958) reported UEPC's in rock doves (*Columba livia*) and Goodwin (1967) described copulation interference in which one or more sexually active males struggled to dislodge the mounted male and then mounted the female themselves. Apparently a male rock dove is unable to forcibly restrain a female for copulation, but copulations can be stolen when a female has assumed the receptive posture in response to courtship by her mate (Samelson 1982). Once a female has crouched, she will remain receptive for several seconds regardless of what goes on around her, and pigeon fanciers can achieve desired inseminations at that moment by removing her mate and substituting a different male. Males constantly "drive" their mates to keep them away from other males (Goodwin 1967), and copulation interference may prove to be an extrapair insemination strategy in this species.

Mate-guarding and other anti-cuckoldry behavior have also been documented in ring doves (Erickson and Zenone 1976, Zenone *et al.* 1979, Lumpkin *et al.* 1982). Attempted EPCs have been reported for mourning doves (*Zenaida macroura*; Irby 1964), and we have one observation of a vigorous FEPC attempt that was foiled by a mate-guarding male in this species (D. J. Bruggers, unpubl. data).

Meropidae (Bee-eaters).—In red-throated bee-eaters (*Merops bulocki*) UEPCs occasionally occur between a male helper and the female of the breeding pair he is helping (Fry 1972).

Hirundinidae (Swallows).—FEPC attempts are frequent during the pre-laying and laying stages in cliff swallows (*Hirundo pyrrhonota*; Emlen 1954), cave swallows (*H. fulva*; Martin 1980), bank swallows (*Riparia riparia*; Petersen 1955, Hoogland and Sherman 1976, Beecher and Beecher 1979), and purple martins (*Progne subis*;

Brown 1978). Close mate-guarding is characteristic of the period when females are fertilizable in bank swallows, and Beecher and Beecher suggest that laying females are pursued because they have a "heavy" flight. Emlen saw male cliff swallows pounce on nesting or mud-gathering birds of either sex, and Brown suggests that male purple martins pursue females carrying nesting material. Deposition of semen was documented by Hoogland and Sherman when male bank swallows made FEPC attempts on a dead swallow decoy, and successful insemination has been documented by the presence of cave x barn (*Hirundo rustica*) swallow hybrid young in nests attended by barn swallow adults (Martin 1980). In addition to FEPC's, UEPC's were recorded in cliff swallows (Emlen 1954) and bank swallows (Petersen 1955).

   Corvidae (Crows).—FEPC's are frequent in the rook (*Corvus frugilegus*). Goodwin (1955) and Coombs (1960) describe many males struggling over females on the nest, usually during the incubation period. Goodwin recorded one instance of a forced copulation by a male on his own mate after he dislodged three males attempting FEPC. FEPC attempts have also been recorded in carrion crows (*C. corone*) by Wittenberg (1968) and in common ravens (*C. corax*) by Kramer (1932). In both species, paired males from neighboring territories attempted FEPC with females sitting on nests prior to incubation.

   Three FEPC attempts were recorded in black-billed magpies (*Pica pica*) by Birkhead (1979). All were on paired females during egg-laying and were unsuccessful. In one case, the male of a neighboring pair sneaked into the nest while the guarding male dozed. The other two incidents took place during a gathering and involved males that were probably unpaired. Mate-guarding and extrapair courtship have also been described for this species by Buitron (1983). In the yellow-billed magpie (*P. nuttalli*), Verbeek (1973) has recorded at least eight attempts by paired males to copulate with females of other pairs. Seven were in response to the begging calls of the females, and one was clearly an FEPC attempt in response to an observed copulation.

   Balda and Balda (1978) observed dominant male Pinyon jays (*Gymnorhinus cyanocephalus*) attempting to court and copulate with females other than their own mates. Males were also observed feeding young in other nests.

   Cracticidae (Bellmagpies).—Western Australian magpies (*Gymnorhina tibicen dorsalis*) live in groups in which the dominant male copulates with the dominant female. Robinson (1956) reported some UEPC's within groups but more between groups. FEPC attempts were common between groups and were made by subordinate males. Such a male sneaks into another group's territory, pounces on a female, who tries to break away, and is then attacked and driven off by the group. Neither FEPC attempts nor mate-guarding have been observed in *G. t. tibicen* (C. Veltman, pers. comm.).

   Turdinae (Thrushes).—Mate-guarding and mate-defense have been reported in mountain bluebirds (*Sialia currucoides*; Power and Doner 1980). P. A. Gowaty and A. A. Karlin (pers. comm.) used electrophoresis and found mixed paternity in broods of eastern bluebirds (*S. sialis*).

Sturnidae (Starlings).—In order to guard their mate and presumably to ensure paternity, male European starlings (*Sturnus vulgaris*) follow their mates about closely during egg-laying, leaving their nests unguarded and exposed to parasitism, usurpation, and predation (Power *et al.* 1981).

Nectariniidae (Sunbirds).—Malachite sunbirds (*Nectarina famosa*) are typically monogamous, and close mate-guarding and mate-defense were observed before and during egg-laying (Wolf and Wolf 1976).

Parulinae (Wood Warblers).—UEPC's have been reported in ovenbirds (*Seiurus aurocapillus;* Hann 1937), and American redstarts (*Setophaga ruticilla;* Ficken 1962). Male prairie warblers (*Dendroica discolor*) often courted and attempted to copulate with their neighbors' mates, and at least one paired female was observed soliciting copulations from a male other than her mate (Nolan 1978). N. L. Ford (pers. comm.) reports that paired male yellow warblers (*Dendroica petechia*) frequently intrude on territories where the resident female is nest-building. Typically they follow females closely and chase them in flight. Twice males were seen to grapple with flying females, but the birds tumbled out of sight into vegetation and, although suspected, EPC was not observed.

Emberizinae (Buntings).—Howard (1929) observed many FEPC attempts on paired females by neighboring males in the yellow bunting (*Emberiza citrinella*). Male Lapland longspurs (*Calcarius lapponicus*), whose females were incubating eggs, pursued unpaired females and paired females that were laying eggs (Seastedt and MacLean 1979). UEPC's have been reported for American tree sparrows (*Spizella arborea;* Weeden 1965) and savannah sparrows (*Passerculus sandwichensis;* Welsh 1975).

Fringillidae (Finches).—FEPC attempts are common in the hawfinch (*Coccothraustes coccothraustes;* Mountfort 1957), while both FEPC's and UEPC's are reported in the chaffinch (*Fringilla coelebs;* Marler 1956). Persistent extrapair courtship by mated and unmated male twites (*Acanthis flavirostris*) and close mate-guarding occur during egg-laying, but no EPC's were reported by Marler and Mundinger (1975).

Passeridae (House Sparrows).—Summers-Smith (1963) observed many multiple male chases of females but only one EPC in the house sparrow (*Passer domesticus*).

Estrildidae (Waxbills).—FEPC attempts have been seen in captive cut-throat (*Amadina fasciata;* Goodwin 1982), red-billed firefinch (*Lagonosticta senegala;* Kunkel 1959), dark firefinch (*L. rubricata;* Kunkel 1965), black-rumped waxbill (*Estrilda troglodytes;* Kunkel 1965), bronze mannikin (*Lonchura cucullata;* Güttinger (1970) and spice finch (*L. punctulata;* Moynihan and Hall 1954). Also FEPC attempts on sick or damaged females have been recorded in cordon-bleu (*Uraeginthus angolensis;* Goodwin 1965), as Morris (1957) noted in bronze mannikins. Copulation interference, similar to that seen in rock doves (p. 534), has been recorded in bronze mannikins (Morris 1957), and red-cheeked cordon-bleu (*U. bengalis;* Goodwin 1982). In the latter species, well developed mate-guarding tactics, including driving behavior similar to that of rock doves, was noted. In addition, male long-tailed finches (*Poephila acuticauda*) and black-throated finches (*P. cincta*) frequently attempted EPC's with paired females, but the females refused to cooperate (Zann 1977).

VI. EXTRAPAIR COPULATION IN PARTIALLY MONOGAMOUS BIRDS

Many species are difficult to classify as being typically monogamous, polygamous, or promiscuous (see Verner and Willson 1959, Wittenberger 1978, Wittenberger and Tilson 1980). Even in those species that have been classified as being typically polygynous (*e.g.*, house wren *Troglodytes aedon*, red-winged blackbird *Agelaius phoeniceus*; Verner and Willson 1969), there are many instances of apparent monogamy. Sperm competition obviously occurs in birds that are simultaneously polyandrous or promiscuous, but this problem requires separate treatment (see Faaborg and Patterson 1981). We are interested here in situations where sperm competition could result from copulations that occur outside the established mating bonds.

Charadriidae (Plovers).—Male and female northern lapwings (*Vanellus vanellus*) are reported to solicit and achieve copulations with birds other than their mates throughout the breeding season (Bannerman 1961).

Phalaropodinae (Phalaropes).—In red phalaropes (*Phalaropus fulicaria*), Ridley (1980) observed a paired male copulating with a female of unknown status, while Kistchinski (1975) observed frequent UEPC's.

Picidae (Woodpeckers).—Stacey (1979) reported that UEPC's were common between the dominant female and subordinate males in acorn woodpeckers (*Melanerpes formicivorus*) and suggested that in many areas promiscuity rather than monogamy may be the rule in this communally breeding species.

Troglodytidae (Wrens).—Kendeigh (1941) reported UEPC's between a paired female house wren (*Troglodytes aedon*) and the male from a neighboring territory during the late-nestling stage.

Muscicapinae (Old World Flycatchers).—Female pied flycatchers (*Ficedula hypoleuca*) have been observed copulating with strange males when their mates were absent, and the aggressiveness of paired males is closely correlated with pre-laying and early egg-laying stages (von Haartman 1956).

Icteridae (Icterids).—The red-winged blackbird (*Agelaius phoeniceus*), which is typically polygynous, is noteworthy for the experiment involving surgical sterilization of males to demonstrate fertilization of eggs by presumed UEPC's. Females mated to vasectomized males produced fertile clutches (Bray *et al.* 1975, Roberts and Kennelly 1980). FEPC's have also been observed in this species (Nero 1956).

VII. DISCUSSION

The occurrence of EPC's in monogamous species belonging to 26 avian families suggests that this behavior is widespread in this class (Table I). Although we have found published records for only a few species in each of these families (39 waterfowl, 65 other kinds of birds) this is undoubtedly a fragmentary sample of the species that exhibit this behavior. Records of forced copulation in dabbling duck species have increased rapidly in recent years because we have been looking for it. There are enough records for large families such as the albatrosses, herons, gulls, and swallows to indicate that FEPC will probably be discovered in many other species in these groups at least. We predict that EPC will prove to be a common phenomenon in birds.

Coincidence of EPC's with the period when eggs are being fertilized, and guarding of females by their mates at this time have been noted in a number of species and are strong indications that sperm competition is involved. Observations of groups of males struggling to copulate with a female (in the Laysan albatross, rook, and bank swallow) are reminiscent of the melees that occur in dabbling ducks and lesser snow geese, leaving no doubt that males are competing intensely. The occurrence of FEPC attempts when the female's mate is temporarily absent as noted in herons, and the "stealthy" behavior described for males attempting FEPC in black-billed magpies and Australian magpies suggest that tactics may be highly developed in these species.

The reluctance of females to accept EPC's is obvious from the descriptions of many authors and the word "forced" seems appropriate for most species. Reports of UEPC's with apparent acceptance by the female have been given for a number of species but they are difficult to evaluate. In some cases, they could have been involved in mate-assessment (jackass penguins, gannets), but it is also possible that the female's willingness was apparent rather than real. In colonial species where EPC occurs as the female sits on the nest (*e.g.,* herons, gulls), it is especially difficult to distinguish between female acceptance of copulation from a neighboring male and her reluctance to leave her nest. In such situations, females must be especially dependent on defense by the mate and, when he is not present, often they must remain in full view of neighbors. There may be risk of egg breakage when females struggle on the nest (*e.g.,* lesser snow goose, rook) and perhaps they are better off to submit readily to males to avoid loss of eggs.

Most behavioral evidence suggests that EPC's benefit males but not females. Paired males have been seen making forced copulation attempts on females paired to other males in many kinds of birds (albatrosses, herons, ibises, ducks, geese, gulls, guillemots, rooks, magpies, buntings, finches). At least in cattle egrets and several dabbling ducks, mixed male reproductive strategies of the kind postulated by Trivers (1972) appear to be operating. It seems likely that this type of male mating system occurs in other species currently considered to be monogamous.

There are relatively few reports of unpaired males engaging in EPC's with paired females. In part this may reflect lack of information about the activities of unpaired males because their status is difficult to establish for certain in the field. On the other hand, it may be that unpaired males tend to be more concerned with courtship and pair-formation than with attempts to steal copulations as seems to be the case in dabbling ducks.

The reports of EPC's involving subordinate males and the female of breeding pairs in red-throated bee-eaters, acorn woodpeckers, and Australian magpies suggest that unpaired males behave differently in some species with helper systems (references in Emlen 1982). A recent study of the pukeko by Craig (1980) illustrates how complex copulatory activities may prove to be in communal breeders. In New Zealand, this gallinule lives in pairs or in communal groups of three to six birds. In groups, Craig found that all adult members may engage in copulations and, although male-female copulations were most frequent, he concluded that these activities must have social functions in addition to achieving insemination. In pair territories, juvenile males tried to copulate with their mothers and sisters but successful copulations were seen only between mates. Such findings emphasize the need for caution in generalizing about species-characteristic mating systems.

Gladstone (1979) drew attention to the occurrence of EPC's in monogamous birds that breed in colonies. While it seems logical that EPC is more likely to occur when neighboring pairs are close by, it should be remembered that a great deal of research attention has been focused on colonial species. Intensive observations on individually marked pairs are necessary to detect EPC's, and in dispersed nesters this takes a great deal of time. The occurrence of forced copulation in the lesser snow goose may well be attributable to its colonial-nesting habit, which is unusual in geese, but this behavior is highly developed in many dabbling ducks which are usually dispersed nesters. In view of the fragmentary evidence available it is probably too early to generalize about the distribution of EPC in birds or about factors that favor or prevent its occurrence in particular species.

VIII. SUMMARY

Little is known about the mechanics of sperm competition in birds except in domestic breeds for which artificial insemination techniques have been developed. Experiments on chickens, turkeys, and ducks have shown that sperm from separate inseminations can compete. Females can store sperm for more than 2 weeks, but viability drops rapidly after a few days. Semen from inseminations several hours apart is stored in separate layers, the most recent remaining on top and being most likely to fertilize the next egg. Each egg in a clutch is fertilized separately a few hours after the previous egg is laid.

Most species of birds are thought to be monogamous and paired males contribute in various ways to the care of eggs and young. Observations of paired females copulating with other males as well as the mate are widespread but usually they have been considered exceptional or aberrant. There is evidence in certain waterfowl and one heron for mixed male reproductive strategies in which males combine pair-bonding with forced copulation on other paired females. Forced copulations occur in certain albatrosses, herons, and swallows during the period when eggs are being fertilized, but in gulls and anhingas they can involve incubating females. Male rock doves interrupt pairbond copulations and, if the female remains in the receptive posture after her mate has been dislodged, a stolen copulation (apparently accepted by the female) may follow. Extrapair copulations that appear to be tolerated by the female have been reported for a number of species, but their significance is not clear. Some could be components of mate-assessment, non-functional in fertilization; others may have been forced rather than tolerated if the female was strongly motivated to remain on the nest. Sperm competition and mixed male reproductive strategies may prove to be characteristic of many birds currently considered to be strictly monogamous, including dispersed as well as colonial nesting species.

ACKNOWLEDGMENTS

Our research on mating systems and sperm competition in waterfowl was supported by the National Science Foundation (grants BNS-7602233 and BNS-7924692) and the Graduate School, University of Minnesota. Flight pen studies were carried out at the Cedar Creek Natural History Area and we acknowledge the support of D. F. Parmelee,, Field Biology Program, University of Minnesota. We thank N. L. Ford, D. Mock, and P. Sherman for helpful comments on the manuscript.

REFERENCES

Allen, R. P. 1955. The reddish egret. *Audubon* 57:24-27.
Allen, C. F., and L. R. Champion. 1955. Competitive fertilization in the fowl. *Poult. Sci.* 34: 1332-1342.
American Ornithologists' Union. 1982. Thirty-fourth supplement to the A.O.U. Check-list of North American birds. *Suppl. to Auk* 99, No. 3.
Bailey, R. O., N. R. Seymour, and G. R. Stewart. 1978. Rape behavior in blue-winged teal. *Auk* 95:188-190.
Balda, R. P., and J. H. Balda. 1978. The care of young Piñon Jays (*Gymnorhinus cyanocephalus*) and their integration into the flock. *J. Ornithol.* 119:146-171.
Bannerman, D. A. 1961. *The Birds of the British Isles.* Vol. 10. Oliver and Boyd, Edinburgh.
Barash, D. P. 1977a. *Sociobiology and Behavior.* Elsevier North-Holland, New York.
Barash, D. P. 1977b. Sociobiology of rape in mallards (*Anas platyrhynchos*): Response of the mated male. *Science* 197:788-789.

Barrett, J. 1973. Breeding behavior of captive mallards. M.S. Thesis, Univ. of Minnesota, Minneapolis, MN.

Beecher, M. D., and I. M. Beecher. 1975. Sociobiology of Bank Swallows: Reproductive strategy of the male. *Science* 205:1282-1285.

Birkhead, T. R. 1978. Behavioural adaptations to high density nesting in the Common Guillemot *Uria aalge. Anim. Behav.* 26:321-331.

Birkhead, T. R. 1979. Mate guarding in the Magpie *Pica pica. Anim. Behav.* 27:866-874.

Blaker, D. 1969. Behaviour of the Cattle Egret, *Ardeola ibis. Ostrich* 40:75-129.

Bobr, L. W., F. X. Ogasawara, and F. W. Lorenz. 1964. Distribution of spermatozoa in the oviduct and fertility of domestic birds. II. Transport of spermatozoa in the fowl oviduct. *J. Reprod. Fertil.* 8:49-58.

Bray, O. E., J. J. Kennelly, and J. L. Guarino. 1975. Fertility of eggs produced on territories of vasectomized Red-winged Blackbirds. *Wilson Bull.* 87:187-195.

Brown, C. R. 1978. Sexual chase in Purple Martins. *Auk* 95:588-590.

Brown, L. H., and E. K. Urban. 1969. The breeding biology of the Great White Pelican *Pelecanus onocrotalus roseus* at Lake Shala, Ethiopia. *Ibis* 111:199-237.

Brown, L. H., and A. Root. 1971. The breeding behaviour of the Lesser Flamingo *Phoeiconaias minor. Ibis* 133:147-172.

Buitron, D. 1983. Extra-pair courtship in black-billed magpies. *Anim. Behav.* 31:211-220.

Burger, J. 1976. Daily and seasonal activity patterns in breeding laughing gulls. *Auk* 93:308-323.

Burger, J., and C. G. Beer. 1976. Territoriality in the Laughing Gull. *Behaviour* 55:301-320.

Burger, J., L. M. Miller, and D. C. Hahn. 1978. Behavior and sex roles of nesting Anhingas at San Blas, Mexico. *Wilson Bull.* 90:359-375.

Burns, J. T., K. M. Cheng, and F. McKinney. 1980. Forced copulation in captive mallards. I. Fertilization of eggs. *Auk* 97:875-879.

Cheng, K. M., J. T. Burns, and F. McKinney. 1982. Forced copulation in captive mallards. II. Temporal factors. *Anim. Behav.* 30:695-699.

Cheng, K. M., J. T. Burns, and F. McKinney. 1983. Forced copulation in captive mallards. III. Sperm competition. *Auk,* 100:302-310.

Christoleit, E. 1929a. Ueber das Reihen der Enten. *Beitr. Fortpfl. Vögel* 5:45-53.

Christoleit, E. 1929b. Nochmals das Reihen der Enten. *Beitr. Fortpfl. Vögel* 5:212-216.

Compton, M. M., H. P. Van Krey, and P. B. Siegel. 1978. The filling and emptying of the uterovaginal sperm-host glands in the domestic hen. *Poult. Sci.* 57:1696-1700.

Conover, M. R., D. E. Miller, and G. L. Hunt, Jr. 1979. Female-female pairs and other unusual reproductive associations in Ring-billed and California Gulls. *Auk* 96:6-9.

Coombs, C. J. F. 1960. Observations on the Rook *Corvus frugilegus* in southwest Cornwall. *Ibis* 102-:394-419.

Cottrille, W. P., and B. C. Cottrille. 1958. Great Blue Heron: Behavior at the nest. *Misc. Publ. Mus. Zool. Univ. Mich.* 102:3-15.

Cox, C. R., and LeBoeuf, J. J. 1977. Female incitation of male competition: A mechanism in sexual selection. *Am. Nat.* 111:317-335.

Craig, J. L. 1980. Pair and group breeding behaviour of a communal gallinule, the Pukeko, *Porphyrio p. melanotus. Anim. Behav.* 28:593-603.

Cuthbert, F. J. 1981. Caspian tern colonies in the Great Lakes: Responses to an unpredictable environment. Ph.D. Dissertation, Univ. of Minnesota, Minneapolis, MN.

Din, N. A., and S. K. Eltringham. 1974. Breeding of the Pink-backed Pelican *Pelecanus rufescens* in Rwenzori National Park, Uganda. *Ibis* 116:477-493.

Dzubin, A. 1969. Comments on carrying capacity of small ponds for ducks and possible effects of density on mallard production. *Saskatoon Wetlands Seminar, Can. Wildl. Serv. Rept. Ser.* 6:138-160.

Eggleton, P., and W. R. Siegfried. 1979. Displays of the Jackass Penguin. *Ostrich* 50:139-167.

Elder, W. H., and M. W. Weller. 1954. Duration of fertility in the domestic mallard hen after isolation from the drake. *J. Wildl. Manage.* 18:495-502.

Emlen, J. T. 1954. Territory, nest building and pair formation in the Cliff Swallow. *Auk* 71:16-35.

Emlen, S. T. 1982. The evolution of helping. II. The role of behavioral conflict. *Am. Nat.* **119**: 40-53.

Emlen, S. T., and L. W. Oring. 1977. Ecology, sexual selection and the evolution of mating systems. *Science* 197:215-223.

Erickson, C. J., and P. G. Zenone. 1976. Courtship differences in male ring doves: Avoidance of cuckoldry? *Science* 192:1353-1354.

Faaborg, J., and C. B. Patterson. 1981. The characteristics and occurrence of cooperative polyandry. *Ibis* 123:477-484.

Ficken, M. S. 1962. Agonistic behavior and territory in the American Redstart. *Auk* 79:607-632.

Fisher, H. E. 1971. The laysan albatross: Its incubation, hatching and associated behaviors. *Living Bird* 10:19-78.

Fry, C. H. 1972. The social organization of bee-eaters (Meropidae) and co-operative breeding in hot-climate birds. *Ibis* 114:1-14.

Fujioka, M., and S. Yamagishi. 1981. Extramarital and pair copulations in the Cattle Egret. *Auk* 98:134-144.

Gibson, F. 1971. The breeding biology of the American Avocet (*Recurvirostra americana*) in central Oregon. *Condor* 73:444-454.

Gladstone, D. E. 1979. Promiscuity in monogamous colonial birds. *Am. Nat.* 114:545-577.

Goethe, F. 1937. Beobachtung und Untersuchungen zur Biologie der Silbermöwe auf der Bogelinsel Memmerstand. *J. Ornithol.* 85:1-119.

Goodwin, D. 1955. Some observations on the reproductive behaviour of rooks. *Br. Birds* 48:97-107.

Goodwin, D. 1965. A comparative study of blue waxbills (Estrildidae). *Ibis* 107:285-315.

Goodwin, D. 1967. *Pigeons and Doves of the World.* The British Museum (Natural History), London.

Goodwin, D. 1982. *Estrildid Finches of the World.* The British Museum (Natural History), London.

Güttinger, H. R. 1970. Zur Evolution von Verhaltensweisen und Lautäusserungen bei Prachtfinken (Estrildidae). *Z. Tierpsychol.* 27:1011-1075.

Hann, H. W. 1937. Life history of the Oven-bird in southern Michigan. *Wilson Bull.* 49:145-237.

Heinroth, O. 1911. Beiträge zur Biologie, namentlich Ethologie und Psychologie der Anatiden. *Verh. V. Int. Ornithol. Kongr.* 589-702.

Heinroth, O., and K. Heinroth. 1958. *The Birds.* (Trans. from German.) Univ. Mich. Press, Ann Arbor, MI.

Höhn, E. O. 1947. Sexual behaviour and seasonal changes in the gonads and adrenals of the mallard. *Proc. Zool. Soc. Lond.* 117:281-304.

Hoogland, J. L., and P. W. Sherman. 1976. Advantages and disadvantages of Bank Swallow coloniality. *Ecol. Monogr.* 46:33-58.

Howard, H. E. 1929. *An Introduction to the Study of Bird Behaviour.* Cambridge Univ. Press, Cambridge.

Howarth, B., Jr. 1971. Transport of spermatozoa in the reproductive tract of turkey hens. *Poult. Sci.* 50:84.

Hunt, G. L., Jr., and M. W. Hunt. 1977. Female-female pairing in Western Gulls (*Larus occidentalis*) in southern California. *Science* 196:1466-1467.

Huxley, J. S. 1912. A "disharmony" in the reproductive habits of the wild duck (*Anas boschas* L.). *Biol. Zentralbl.* 32:621-623.

Imber, M. J. 1976. Breeding biology of the Grey-faced Petrel *Pterodroma macroptera* gouldi. *Ibis* 118:51-64.

Irby, H. D. 1964. The relationship of calling behavior to Mourning Dove populations and production in southern Arizona. Ph.D. Dissertation, Univ. of Arizona, Tucson, AZ.

Kendeigh, S. C. 1941. Territorial and mating behavior of the House Wren. *Ill. Biol. Monogr.* 18:1-120.

King, A. S. 1981. Phallus. In *Form and Function in Birds,* A. S. King and J. McLelland (eds.), pp. 107-147. Academic Press, London.

Kistchinski, A. A. 1975. Breeding biology and behaviour of the Grey Phalarope *Phalaropus fulicarius* in East Siberia. *Ibis* 117:285-301.

Kramer, G. 1932. Beobachtungen und Fragen zur Biologie des Kolkraben. *J. Ornithol.* 80: 329-342.

Kunkel, P. 1959. Zum Verhalten einiger Prachtfinken. *Z. Tierpsychol.* 16:302-350.

Kunkel, P. 1967. Displays facilitating sociability in waxbills of the genera *Estrilda* and *Lagonosticta* (family Estrildidae). *Behaviour* 29:237-261.

Kushlan, J. A. 1973. Promiscuous mating behavior in the White Ibis. *Wilson Bull.* 85:331-332.

Lack, D. 1968. *Ecological Adaptations for Breeding in Birds*. Methuen, London.

Lake, P. E. 1975. Gamete production and the fertile period with particular reference to domesticated birds. *Symp. Zool. Soc. Lond.* 35:225-244.

Lancanster, D. A. 1970. Breeding behavior of the Cattle Egret in Colombia. *Living Bird* 9:167-194.

Lebret, T. 1961. The pair formation in the annual cycle of the mallard, *Anas platyrhynchos* L. *Ardea* 49:97-158.

Littlefield, C. D. 1981. Mate-swapping of sandhill cranes. *J. Field Ornithol.* 52:244-245.

Lumpkin, S., K. Kessel, P. G. Zenone, and C. J. Erickson. 1982. Proximity between the sexes in ring doves: Social bonds or surveillance? *Anim. Behav.* 30:506-513.

McKinney, F., S. R. Derrickson, and P. Mineau. 1983. Forced copulation in waterfowl. *Behaviour* 86:250-294.

McKinney, F., and P. Stolen. 1982. Extra-pair-bond courtship and forced copulation among captive green-winged teal (*Anas crecca carolinensis*). *Anim. Behav.* 30:461-474.

MacRoberts, M. H. 1973. Extramarital courting in Lesser Black-backed and Herring Gulls. *Z. Tierpsychol.* 32:62-74.

Makkink, G. F. 1942. Contribution to the knowledge of the behaviour of the Oyster-Catcher (*Haematopus ostralegus* L.). *Ardea* 31:23-74.

Marler, P. 1956. Behaviour of the chaffinch *Fringilla coelebs*. *Behaviour Suppl.* 5:1-184.

Marler, P., and P. C. Mundinger. 1975. Vocalizations, social organization and breeding biology of the Twite *Acanthus flavirostris*. *Ibis* 117:1-17.

Martin, R. F. 1980. Analysis of hybridization between the hirundinid genera *Hirundo* and *Petrochelidon* in Texas. *Auk* 97:148-159.

Meanley, B. 1955. A nesting study of the Little Blue Heron in eastern Arkansas. *Wilson Bull.* 67:84-99.

Milstein, P. le S. 1979. The evolutionary significance of wild hybridization in South African Highveld ducks. *Ostrich Suppl.* 13:1-48.

Mimura, H. 1939. On the mechanism of travel of spermatozoa through the oviduct in the domestic fowl, with special reference to artificial insemination. *Sonderabdruck aus Okjimas Folia Anatomica Japonica* 17(5). (Cited by Sturkie 1976.)

Mineau, P., and F. Cooke. 1979a. Rape in the lesser snow goose. *Behaviour* 70:280-291.

Mineau, P., and F. Cooke. 1979b. Territoriality in Snow Geese or the protection of parenthood—Ryder's and Inglis's hypotheses re-assessed. *Wildfowl* 30:16-19.

Mock, D. W. 1976. Pair-formation displays of the Great Blue Heron. *Wilson Bull.* 88:185-230.

Mock, D. W. 1978. Pair-formation displays of the Great Egret. *Condor* 80:159-172.

Morris, D. 1957. The reproductive behaviour of the Bronze Mannikin, *Lonchura cucullata*. *Behaviour* 11:156-201.

Mountfort, G. 1957. *The Hawfinch*. Collins, London.

Moynihan, M. 1955. Some aspects of reproductive behaviour in the Black-headed Gull (*Larus ridibundus* L.) and related species. *Behav. Suppl.* 4:1-201.

Moynihan, M., and M. F. Hall. 1954. Hostile, sexual and other social behaviour patterns of the Spice Finch (*Lonchura punctulata*) in captivity. *Behaviour* 7:33-76.

Nelson, J. B. 1965. The behaviour of the Gannet. *Br. Birds* 58:233-288, 313-336.

Nero, R. W. 1956. A behavior study of the Red-winged Blackbird. I. Mating and nesting activities. *Wilson Bull.* 68:5-37.

Nolan, V., Jr. 1978. The ecology and behavior of the Prairie Warbler *Dendroica discolor*. *A.O.U. Monogr.* 26:1-595.

Parker, G. A. 1970. Sperm competition and its evolutionary consequences in the insects. *Biol. Rev.* 45:525-567.

Payne, L. F., and A. J. Kahrs. 1961. Competitive efficiency of turkey sperm. *Poult. Sci.* **40**: 1598-1604.

Peters, J. L. 1931-1979. *Check-list of Birds of the World.* Vol. 1-15. Harvard Univ. Press, Cambridge, MA.

Petersen, A. J. 1955. The breeding cycle in the Bank Swallow. *Wilson Bull.* **67**:235-286.

Power, H. W., and C. G. P. Doner. 1980. Experiments on cuckoldry in the Mountain Bluebird. *Am. Nat.* **116**:689-704.

Power, H. W., E. Litovich, and M. P. Lombardo. 1981. Male starlings delay incubation to avoid being cuckolded. *Auk* **98**:386-389.

Richdale, L. E. 1949. The pre-egg stage in Buller's Mollymawk. *Biol. Monogr. 2.* Dunedin, New Zealand.

Ridley, M. W. 1980. The breeding behaviour and feeding ecology of Grey Phalaropes *Phalaropus fulicarius* in Svalbard. *Ibis* **122**:210-226.

Roberts, T. A., and J. J. Kennelly. 1980. Variation in promiscuity among Red-winged Blackbirds. *Wilson Bull.* **92**:110-112.

Robinson, A. 1956. The annual reproductive cycle of the Magpie *Gymnorhina dorsalis,* in South Western Australia. *Emu* **56**:233-336.

Rodgers, J. A., Jr. 1980. Little Blue Heron breeding behavior. *Auk* **97**:371-384.

Rohwer, S. 1978. Passerine subadult plumages and the deceptive acquisition of resources: Test of a critical assumption. *Condor* **80**:173-179.

Rudegeair, T. J. 1975. The reproductive behavior and ecology of the White Ibis. Ph.D. Dissertation, Univ. of Florida, Gainesville, FL.

Ryder, J. P., and P. L. Somppi. 1979. Female-female pairing in Ring-billed Gulls. *Auk* **96**:1-5.

Samelson, D. A. 1982. Copulation interference in *Columba livia*: Controlling proximate factors and ultimate adaptive significance. M.A. Thesis, Univ. of Minnesota, Minneapolis, MN.

Seastedt, T. R., and S. F. MacLean. 1979. Territory size and composition in relation to resource abundance in Lapland Longspurs breeding in Arctic Alaska. *Auk* **96**:131-142.

Selander, R. K. 1972. Sexual selection and dimorphism in birds. In *Sexual Selection and the Descent of Man 1871-1971,* B. Campbell (ed.), pp. 180-230. Aldine, Chicago.

Seymour, N. R., and R. D. Titman. 1979. Behaviour of unpaired male black ducks (*Anas rubripes*) during the breeding season in a Nova Scotia tidal marsh. *Can. J. Zool.* **57**:2421-2428.

Sherman, P. W. 1981. Electrophoresis and avian genealogical analyses. *Auk* **98**:419-422.

Smith, R. I. 1968. The social aspects of reproductive behavior in the pintail. *Auk* **85**:381-396.

Stacey, P. B. 1979. Kinship, promiscuity and communal breeding in the Acorn Woodpecker. *Behav. Ecol. Sociobiol.* **6**:53-66.

Sturkie, P. E. 1976. *Avian Physiology, 3rd Ed.* Springer-Verlag, New York.

Summers-Smith, D. 1963. *The House Sparrow.* Collins, London.

Taylor, J. S. 1948. Notes on the nesting and feeding habits of the Black-headed Heron, *Ardea melanocephala. Ostrich* **19**:203-210.

Tickell, W. L. N. 1968. Biology of the Great Albatrosses, *Diomedea exulans* and *Diomedea epomophora.* In *Antarctic Bird Studies,* O. L. Austin, Jr. (ed.), pp. 1-56. Antarctic Research Series Vol. 12.

Tickell, W. L. N., and R. Pinder. 1966. Two-egg clutches in albatrosses. *Ibis* **108**:126-129.

Tinbergen, N. 1956. On the functions of territory in gulls. *Ibis* **98**:401-411.

Trivers, R. L. 1972. Parental investment and sexual selection. In *Sexual Selection and the Descent of Man 1871-1971,* B. Campbell (ed.), pp. 136-179. Aldine, Chicago.

Tuck, L. M. 1961. *The Murres.* Vol. 1. Can. Wildl. Serv., Ottawa.

Verbeek, N. A. M. 1973. The exploitation system of the yellow-billed magpie. *Univ. Calif. Publ. Zool.* **99**:1-58.

Vermeer, K. 1963. The breeding ecology of the glaucous-winged gull (*Larus glaucescens*), on Mandarte Island, B.C. *Occas. Pap. B.C. Prov. Mus.* **13**:1-104.

Verner, J., and M. F. Willson. 1969. Mating systems, sexual dimorphism, and the role of male North American passerine birds in the nesting cycle. *Ornithol. Monogr.* **9**:1-76.

Verwey, J. 1930. Die Paarungsbiologie des Fischreihers. *Zool. Jahrb. Abt. Allg. Zool. Physiol. Tiere.* **48**:1-120.

von Haartman, L. 1956. Territory in the Pied Flycatcher *Muscicapa hypoleuca. Ibis* **98**:460-475.

Warren, D. C. and, C. L. Gish. 1943. The value of artificial insemination in poultry breeding work. *Poult. Sci.* **22**:108-117.

Weeden, J. S. 1965. Territorial behavior of the tree sparrow. *Condor* **67**:193-209.

Weidmann, U. 1956. Verhaltensstudien an der Stockente (*Anas platyrhynchos* L.). I. Das Aktionssystem. *Z. Tierpsychol.* **13**:108-271.

Welsh, D. A. 1975. Savannah sparrow breeding and territoriality on a Nova Scotia dune beach. *Auk* **92**:235-251.

Werschkul, D. F. 1982. Nesting ecology of the Little Blue Heron: Promiscuous behavior. *Condor* **84**:381-384.

Wiese, J. H. 1976. Courtship and pair formation in the Great Egret. *Auk* **93**:709-724.

Williams, L. 1942. Display and sexual behavior of the Brandt Cormorant. *Condor* **44**:85-104.

Wittenberg, J. 1968. Freilanduntersuchungen zu Brutbiologie und Verhalten der Rabenkrähe (*Corvus c. corone*). *Zool. Jahrb. Abt. Syst. Oekol. Geogr. Tiere.* **11**:197-232.

Wittenberger, J. F. 1979. The evolution of mating systems in birds and mammals. In *Handbook of Behavioral Neurobiology, Vol. 3. Social Behavior and Communication,* P. Marler and J. Vandenbergh (eds.), pp. 271-349. Plenum, New York.

Wittenberger, J. F., and R. L. Tilson. 1980. The evolution of monogamy: Hypotheses and evidence. *Annu. Rev. Ecol. Syst.* **11**:197-232.

Wolf, L. L., and J. S. Wolf. 1976. Mating system and reproductive biology of Malachite Sunbirds. *Condor* **78**:27-39.

Zann, R. 1977. Pair-bond and bonding behaviour in three species of grassfinches of the genus *Poephila* (Gould). *Emu* **77**:97-106.

Zenone, P. G., M. E. Sims, and C. J. Erickson. 1979. Male ring dove behavior and the defense of genetic paternity. *Am. Nat.* **114**:615-626.

16

Sperm Competition in Muroid Rodents

DONALD A. DEWSBURY

I. INTRODUCTION

The phenomenon of sperm competition concerns the competition or potential competition among spermatozoa from different males for fertilization of ova within a single female. This has been well documented in insects (Parker 1970, Boorman and Parker 1976). However, there has been comparatively little interest in the phenomenon of sperm competition in mammals. In a classical paper that played a pivotal role in stimulating the study of sperm competition, Parker wrote, "A relatively low level of sperm competition might be expected, for example, in most mammals, since the sperm usually survive for only a short time and there are no special sperm stores" (Parker 1970). More recent evidence suggests that sperm competition may be extremely important in the evolution of reproductive behavior in mammals. The purposes of this paper are to present some of the evidence

Sperm Competition and the Evolution of Animal Mating Systems

547

favoring re-evaluation of the significance of sperm competition in mammals and to explore some of its implications.

The critical consideration in determining if the conditions for sperm competition prevail is to establish that more than one male mates a single female within a short enough interval to promote fertilization contests. It is now apparent that in mammalian species as diverse as sheep (Allison 1977) and lions (Bertram 1976) several males may mate a single female during a given estrus. Among rodents it has been repeatedly observed that in both field and semi-natural situations several male Norway rats (*Rattus norvegicus*) may mate with a single female (Calhoun 1962, Robitaille and Bovet 1976, Telle 1966). Electrophoretic methods have been used on field-conceived litters of deer mice, (*Peromyscus maniculatus*) and Belding's ground squirrels (*Citellus beldingi*) to show that multiple paternity is relatively common (Birdsall and Nash 1973, Hanken and Sherman 1981, Merrit and Wu 1975). Fitch (1957) described an apparent instance of males fighting over a trapped female prairie vole, *Microtus ochrogaster*. Madison (1980) has provided evidence of multi-male competition at the time of mating in meadow voles, *M. pennsylvanicus*). These observations suggest that multi-male mating may be more common in mammals than previously thought. In addition, sperm longevity in mammals can be protracted (Thibault 1973). Vespertilionid bats use stored sperm from autumn matings in the spring (Fenton, this volume), and fertilization 7-9 days after coitus is possible in humans (see Smith, this volume). It is, therefore, apparent that conditions for sperm competition are prevalent in mammals, but the dynamics of the phenomenon are poorly known.

II. REPRODUCTIVE SYSTEMS IN RODENTS

The research to be reported here was conducted on four species of muroid rodents: Norway rats, Syrian golden hamsters (*Mesocricetus auratus*), deer mice, and prairie voles. Before this research can be presented properly, it is necessary to make a brief digression in order to discuss the reproductive systems in these species. Some characteristics for each are summarized in Table I. The pattern of copulatory behavior in laboratory rats is typical of rodent species (see Dewsbury 1975). Three events are distinguished: mounts, intromissions, and ejaculations. On intromissions the male mounts the female from behind, displays shallow pelvic thrusting, achieves a single deep intravaginal thrust lasting less than a second, and rapidly dismounts. There are no sperm transferred. Mounts are scored when the male mounts the female but fails to effect vaginal penetration. Mounts and intromissions can be distinguished using behavioral criteria. After a cluster of intromissions, with or without mounts, the male displays a longer vaginal insertion that is coupled with a distinctive pattern of dismounting, which is indicative of ejaculation. Ejaculations only oc-

Table I

Characteristics of Four Species Studied in Experiments on Sperm Competition

Common Name	Laboratory Rats	Syrian Golden Hamsters	Deer Mice	Prairie Voles
Latin Name	*Rattus norvegicus*	*Mesocricetus auratus*	*Peromyscus maniculatus*	*Microtus ochrogaster*
Copulatory Behavior	no lock no thrust mult. intro. mult. ejac.	no lock no thrust* mult. intro. mult. ejac.	no lock no thrust mult. intro. mult. ejac.	no lock thrust mult. intro. mult. ejac.
Male Ejaculation Frequency	7.0	10.1	3.6	2.0
Ovulation	spontaneous	spontaneous	spontaneous	induced
Luteal Phase	induced	induced	induced	spontaneous
Mating System	promiscuous	solitary?	promiscuous?	monogamous or promiscuous?

cur after one or more intromissions, and are followed by a refractory period of no copulation. An "ejaculatory series" includes the mounts and intromissions associated with a given ejaculation. Male rats typically attain a mean of approximately seven series prior to reaching sexual satiation, operationally defined as 30 min with no intromissions. The pattern of copulatory behavior in rats is characterized by no lock (mechanical tie between penis and vagina), no intravaginal thrusting, multiple intromissions requisite to ejaculations, and multiple ejaculations per episode. This is pattern no. 13 as described by Dewsbury (1972) and is typical of many rodent species. The pattern of deer mice is quite similar, although males typically achieve fewer ejaculations and copulate in an incomplete series near the end of the test. Golden hamsters display a similar pattern, with copulations slightly longer in duration and more closely spaced than the other two species. In addition, after about ten ejaculations, male hamsters show a pattern of intravaginal thrusting. Prairie voles display intravaginal thrusting on most intromissions and a mean of just two ejaculations per episode.

The adaptive significance of these behavioral patterns is puzzling. One wonders why several intromissions might precede ejaculation and why multiple ejaculations are attained with a single female. It is now known that a female rat receiving an ejaculation after relatively few intromissions is less likely to become pregnant than one receiving the normal complement of pre-ejaculatory intromissions (Adler 1969, Wilson *et al.* 1965). There are at least two reasons for this. First, sperm are

transported up the female reproductive tract more effectively if the full complement of intromissions precedes ejaculation (see Matthews and Adler 1978). The
second reason relates to the pattern of reproductive physiology in laboratory
rats. Although ovulation is spontaneous, in that it occurs in every estrous cycle
(whether or not the female copulates), a functional luteal phase (preparation of the
uterus for implantation) occurs only if the female mates, and is thus "induced."
The multiple intromissions are therefore critical in triggering a functional luteal
phase in rats. Deer mice and hamsters likewise also have spontaneous ovulation
and induced luteal phases, and prairie voles are induced ovulators. Consequently,
pre-ejaculatory intromissions appear to be important for pregnancy initiation
in all four species (see Dewsbury 1978).

Because young female rats in cycling estrus become pregnant after one complete
ejaculatory series, the function of multiple ejaculations has not been clear. It
has been noted that the activity seems "wasted" and that male rats would seem
to be best served by delivering just one ejaculatory series to each of many females
(Adler 1978, Zucker and Wade 1968). However, it is now clear that multiple
ejaculations may maximize the probability of pregnancy. Such results have been
reported for old, multiparous female rats, female rats in postpartum estrus, and
for a variety of species including hamsters and deer mice (see Dewsbury 1978).
Although multiple ejaculations clearly function in pregnancy initiation, males
of species such as hamsters and deer mice copulate longer than appears necessary
for maximal pregnancy initiation. It will be suggested below that such prolongation
of copulatory activity may be explained in the context of sperm competition.

According to the evidence summarized above it appears that the mating systems
in the field often result in copulation by several males with a single female during
a single period of receptivity. The mating system in prairie voles is controversial,
with the literature containing proposals of both monogamy and polygamy (Fitch
1957, Getz 1978, Getz and Carter 1980, Thomas and Birney 1979). Golden hamsters have not been well-studied in their native Syria. It appears that they may be
relatively solitary, with females dominant over males (Murphy 1979). These mating
systems seem correlated with various aspects of copulatory patterns (Dewsbury
1982c).

III. SPERM COMPETITION, ORDER EFFECTS, AND DIFFERENTIAL FERTILIZING CAPACITY

With this background on reproductive behavior and mating systems, it is now
possible to address the issue of sperm competition in muroid rodents. To study the
dynamics of sperm competition it is necessary that the paternity of offspring from

multiply-mated females be discernible. This has generally been accomplished by use of genetic markers. If a female homozygous for the recessive genotype is mated to males homozygous for the dominant and recessive genotypes, paternity will be unambiguous. This basic method was used with each of the four species.

A fundamental question in the study of sperm competition relates to order effects. Is there a differential advantage to the male mating first or to the male mating last, or is there no order effect? In the insects, the second male to mate has an advantage in many species (Boorman and Parker 1976; Parker 1976, this volume). However, there are exceptions to this (Holmes 1974, Linley 1975). A first step in the study of sperm competition is the determination of mating order effects.

A model experiment on mammals was conducted by Levine (1967) who mated female house mice (*Mus musculus*) of the ST strain sequentially to ST males and to agouti males of the CBA strain. A female was first permitted to copulate for one ejaculatory series with a male of one strain, then removed. At ejaculation male house mice, like males of all other rodent species discussed herein, deposit a copulatory plug. Levine removed the copulatory plug from the female before the second mating. The female was then permitted to mate with the second male for one series. Half of the tests were conducted in each mating order. When the albino males mated first they sired 95% of the offspring; when CBA males mated first, only 70% of the offspring were albino. There are two important aspects to Levine's results. First, males were more effective at gaining representation in litters when they mated first, rather than last. Over and above this effect, however, was a strain difference. Regardless of the order of mating, significantly more of the offspring were albino than agouti.

Differential representation of one of two strains in litters as found in Levine's experiment is a common result in experiments of this type. We have termed this phenomenon "differential fertilizing capacity" (Lanier *et al.* 1979). Differential fertilizing capacity (DFC) is the relative ability of sperm from males of different genotypes to gain representation in litters that are the result of competitive matings, when order of mating, number of ejaculates, and timing of matings are controlled. Although males may differ with respect to number of sperm ejaculated, this appears not to provide a full explanation of DFC in most cases. In species as diverse as chickens, swine, rabbits, and cattle, differences between individual males or lines have been found even when approximately equivalent numbers of sperm were introduced into the female's tract via artificial insemination (Beatty 1960, Beatty *et al.* 1969, Edwards 1955, Martin and Dziuk 1977). Information on the DFC's of the genotypes under study was necessary before the results of more complex studies of sperm competition could be properly interpreted.

In order to obtain basic information on order effects and DFC, we have conducted experiments similar to those of Levine (1967) in our four species of muroid rodents. Both the normal agouti genotype and a cream-colored line were used in hamsters (Oglesby *et al*. 1982). Females were homozygous for cream, the recessive allele. Because of the relatively high stimulus requirements for pregnancy initiation in hamsters (Lanier *et al*. 1975), the female was permitted to receive five ejaculations from each of two males. When the agouti male mated first, 31% of the offspring were agouti; when the cream male mated first, 72% of the offspring were agouti. This strain difference was not significant; however, there was a significant order effect, with the advantage accruing to the second male (Fig. 1).

Three different combinations of strains were studied in laboratory rats (Lanier *et al*. 1979). The females and males of one strain in each study were of the albino F344 genotype. Each of two males mated with the female for one ejaculatory series (Fig. 1). In none of the three experiments was there a significant order effect; sperm apparently mixed without respect to order of mating. In the studies with BN and MAXX males, albino males significantly outcompeted pigmented males, and were therefore not used in further studies. The distribution of offspring in the Long Evans/F344 combination did not deviate significantly from chance, and so these latter two strains were selected for use in further research.

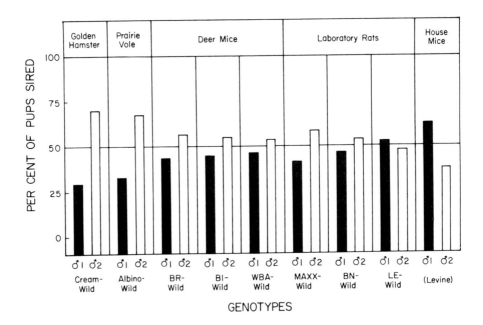

Fig. 1. Litter composition as a function of order of mating in five species of muroid rodents. In each study two males copulated with a single female with each male securing the same number of ejaculations. In studies of rats and deer mice three combinations of strains were used. There is a statistically significant advantage for males ejaculating first in house mice and last in golden hamsters and prairie voles. There are no order effects in deer mice and laboratory rats.

In three studies with deer mice, each male attained two ejaculations (Dewsbury and Baumgardner 1981). As the brown recessive and blonde mutants are recessive to the wild-type coat color, females in those studies were homozygous for the mutant allele. In the study with wide-band agouti males, females were of the wild genotype. The results were parallel to those for rats (Fig. 1). There was consistent agreement across the three experiments in that there were no significant order effects. When brown recessive and wide-band agouti males competed with wild-type males there was a disproportionate number of wild-type offspring regardless of which male mated first. Although blonde males outcompeted wild males when wild males mated first, there were no significant differences when the blonde males mated first. The blonde/wild-type combination was selected as most appropriate for use in further study.

Albino and wild-type prairie vole males were each permitted to mate for one ejaculation with albino females. A significant advantage accrued to the last male to mate (Fig. 1). There was a significant departure in litter composition from chance only when the wild males mated first; progeny were evenly distributed when albino males mated first.

To summarize results with respect to effects of order, a significant advantage accrued to the first male in house mice, a significant advantage accrued to the last male in prairie voles and hamsters, and there were no significant order effects in laboratory rats or deer mice (Fig. 1). In a study with swine, sows were mated with five different boars in succession; no order effects were observed (Sumption and Adams 1961). Obviously, order effects vary dramatically in different mammalian species. Both the effects of mating order and of DFC need be assessed in each combination of strains of each species selected for study in research on sperm competition.

IV. IMPLICATIONS AND EXTENSIONS OF STUDIES
OF MATING ORDER
AND DIFFERENTIAL FERTILIZING CAPACITY

There are a number of implications, extensions, and additional aspects of the research on mating order and DFC that require some elaboration as they relate to the evolution of mating systems. The first matter concerns the search for a generalization concerning species differences in the effects of mating order. When substantial species differences of the sort just described are discovered, it seems appropriate to seek some general principle from which the specific results can be derived. There is no obvious principle from which all results might be predicted; however, one relevant factor may be paternal investment. It is generally agreed that in situations in which a male will make substantial investment, as in caring for young, mechanisms should evolve to ensure paternity of young for whom he cares. Frequently the male is adapted to ensure that he is the last male to copulate with a female where there is precedence for sperm from the last male. Such a system has been described for giant water bugs, *Abedus herberti,* by Smith (1979). If, as proposed by Getz (1978) and Thomas and Birney (1979), prairie voles are monogamous, the last-male advantage would be consistent with this hypothesis. On the other hand, if Norway rats and deer mice are as promiscuous as available data would suggest, mixing of sperm without regard to mating order appears likely. This hypothesis leaves unexplained the first-male advantage of house mice and the last-male advantage found in hamsters.

A second matter meriting discussion is the role of copulatory plugs as adaptations to sperm competition in muroid rodents. Like house mice, all four species under study deliver copulatory plugs at the time of ejaculation (see Hartung and Dewsbury 1978). It is generally believed that such plugs function to prevent subsequent inseminations. Thus Voss (1979) has suggested that the copulatory plugs of these rodents have evolved in the context of "chastity enforcement." There is convincing evidence that the plugs of male guinea pigs block subsequent matings (Martan and Shepherd 1976). Plugs of garter snakes (Devine 1975, 1977) appear functionally similar. It is obvious that the copulatory plugs of muroid rodents do not exclude subsequent inseminations. Indeed, in the only species studied that shows an advantage to the first male, the plug was removed between the first and second matings. One reason that plugs may not prevent subsequent inseminations is that they are frequently displaced. For example, Mosig and Dewsbury (1970) found that of 162 plugs delivered by male rats, 69% were subsequently dislodged; 82% of these were dislodged within three or fewer intromissions. Males ate 65% of the plugs, females ate none. It is not clear whether the entire plug is dislodged or whether a fragment of adequate size to be functional may remain in the female's tract.

As the damselfly penis functions to remove sperm (Waage 1979, this volume), Milligan (1979) has proposed that the spines on the glans penis of muroid rodents have evolved in the context of plug removal. Further, the multiple-intromission copulatory pattern of rodents may have evolved in part to cleanse the female reproductive tract of plugs, sperm, vaginal casts, and plug fragments prior to ejaculation and deposition of a new plug. Thus both genital morphology and reproductive behavior in muroid rodents may be highly relevant to the context of sperm competition (Dewsbury 1981a).

Various authors have used electrophoretic methods on field-conceived, laboratory born animals to make inferences regarding the multiple paternity of litters (*e.g.,* Birdsall and Nash 1973, Foltz 1981, Hanken and Sherman 1981). One factor that must be remembered in such studies is that multiple matings do not always produce multiply-sired litters. Further, there may be appreciable species differences with respect to the probability of multiple paternity given multiple mating. For example, in the experiments discussed above in which two males copulated for equal numbers of ejaculations, pups of both genotypes were born in about 50% of litters in deer mice and less than 10% of prairie vole litters (Fig. 2). The incidence of multiple paternity following multiple mating is likely to be even more reduced in situations where two males differ in the number of ejaculates delivered to the female or where the sperm stores of one male are depleted. These factors must be considered when making inferences about mating systems from data on multiple paternity.

In all studies in which two successive males mate one female there is inevitably some delay between the last ejaculation by the first male and the beginning of mating by the second male. This is particularly a problem with deer mice, in that males typically require approximately 20 min to initiate copulation (Dewsbury 1979a). On some occasions a male of the second strain failed to mate and was replaced by a different male of the same strain. This variability with respect to delay between the matings of males of the first and second strains permits a correlational analysis of the effect of delay on litter composition. In none of the three experiments on deer mice was there a significant correlation between the length of the delay and litter composition.

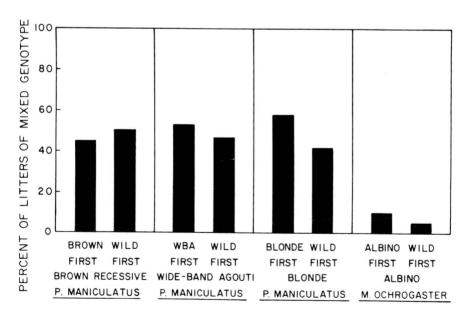

Fig. 2. Percentage of litters of mixed paternity in three studies of sperm competition in deer mice, *Peromyscus maniculatus,* and one study of prairie voles, *Microtus ochrogaster.* Whereas an average of about 50% of the deer mouse litters were of mixed paternity, only about 10% of prairie vole litters were even though all females copulated for equal numbers of ejaculations with two different males.

Because the above correlational approach is of limited power, both because of the inherent weakness of the correlational approach as a research design and because of limited range of values, the role of delay was further studied via experimental manipulation. Two groups of female blonde deer mice both mated first with a blonde male and then with a wild-type male. For one group a delay of 2 hr was imposed between the first male's last ejaculation and introduction of the second male; for the other group the second male was introduced immediately. In addition, the number of ejaculations permitted each male was manipulated. In one test for each female the first male attained three ejaculations and the second male one; in the other test the first male achieved one ejaculation and the second male three. This was done in an attempt to demonstrate experimentally that it is the relative number of ejaculations that is critical in determining litter composition in deer mice, rather than the order of mating or the delay between males. Results were consistent with the hypothesis (Fig. 3). The effect of manipulating the relative number of ejaculations was significant for both delay conditions; there was no significant effect of delay. In *Drosophila* the predominance of the second male over the first increases with increased time between the two matings (Boorman and Parker 1976; Gromko *et al.*, this volume). In rabbits, by contrast, spermatozoa that have resided in the genital tract of a doe for an interval of 2 hr have an advantage over sperm deposited more recently (Dziuk 1965). Although no such effects of delay have been detected in deer mice, it is possible they might appear should longer delays be imposed.

The three studies of mating order and DFC in deer mice were difficult to complete because of the low incidence of pregnancy encountered (Dewsbury and Baumgardner 1981). Incidence was much lower than in studies in which females mated for approximately the same number of copulations with just one male. The difference was not just a function of strain (Dewsbury and Baumgardner 1981). In subsequent work (Dewsbury (1982a), three studies of pregnancy initiation with multiple-male copulation or exposure were completed. In the first experiment more females were pregnant after mating with one male, with or without imposition of a delay, than after receiving the same number of ejaculations from two males. In a second experiment, females that had received three ejaculations from one male were placed opposite either the familiar male or a strange blonde or wild-type male. More females were pregnant after exposure to the familiar than the strange males. Finally, mated females were more likely to become pregnant if the 2 hr after mating was spent alone in the familiar male's home cage than in a cage that had contained a strange male. These data reveal the existence of a pregnancy block after multiple-male exposure of mating in this species. The phenomenon differs from the classical Bruce effect (Bruce 1959) in that multiple-male exposure occurs at the time of mating and the duration of exposure to both the familiar and novel male is quite short. It remains to be determined whether underlying mechanisms are the same. This pregnancy block appears related in the field to multiple-male mating rather than to the takeover of a deme as with traditional interpretations of the Bruce effect (*e.g.*, Bronson and Coquelin 1980).

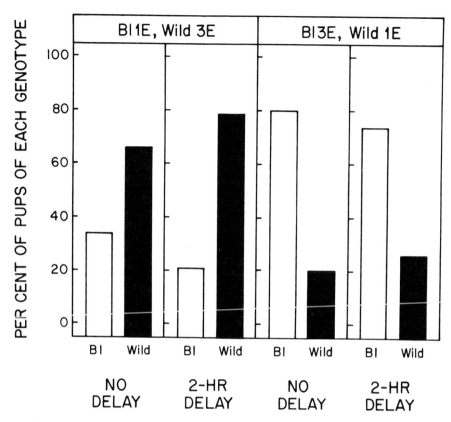

Fig. 3. Genotypes of offspring in a study of sperm competition in deer mice. All females copulated for a total of four ejaculations. They mated first with a blonde male and then with a wild-type male. In half of the tests the blonde male attained one ejaculation and the wild male three ejaculations; in half of the tests the blonde male achieved three ejaculations and the wild male one ejaculation. Tests were conducted with the second male introduced soon after the first male's last ejaculation or after a 2-hr delay. Litter composition was affected by the relative number of ejaculations but not by delay.

Although there was no relationship between amount of stimulation and probability of pregnancy in studies with prairie voles and wide-band agouti deer mice, females that became pregnant or pseudopregnant had received significantly more copulations than those showing no disruption of estrous cycles in the studies with both brown-recessive and blonde deer mice (Dewsbury and Baumgardner 1981). It should be remembered that females of all species under study became pregnant only if they received sufficient vaginal stimulation prior to ejaculation by the male. In fact, a relationship between the number of copulations received and the probability of pregnancy could be detected even in a study in which all females received four ejaculations. Clearly, selective pressures should act to ensure that males deliver stimulation sufficient for pregnancy. Otherwise, their ejaculates, and the associated effort in delivering them, would be wasted.

It should be emphasized that in none of these experiments were significant alterations in either litter size or gestation period detected. The amount of stimulation and the number of ejaculates are important, however, in determining the probability of a pregnancy and the paternity of the pups produced. It is likely that these two aspects of differential reproduction are controlled by two distinct mechanisms.

V. EJACULATION FREQUENCY AND SPERM COMPETITION

It is now possible to return to the problem of the adaptive significance of multiple ejaculatory patterns of copulation in muroid rodents and the role sperm competition may have played in their evolution. Parker (1970, this volume) proposed that in situations of sperm competition two prominent evolutionary forces should prevail. On the one hand, selection should favor males that, when copulating with previously-mated females, can somehow displace or otherwise neutralize sperm already in the female's reproductive tract. On the other hand, mechanisms should be favored that enable a male to avoid or reduce the effects of subsequent matings by other males. There would thus be an evolutionary "push-pull" between forces acting on males to displace previous ejaculates and on early males to prevent such displacement. Among the mechanisms that may have evolved to deter subsequent displacement are mating plugs, prolonged copulation, passive phases, non-contact guarding phases, and avoidance of "take-over" (Parker 1970, this volume). We have proposed that the multiple ejaculatory patterns of muroid rodents may have evolved partially in such a context.

Experiments designed to assess the role of multiple ejaculations in protecting against subsequent displacement have been conducted on laboratory rats and golden hamsters (Lanier *et al.* 1979, Oglesby *et al.* 1981). If, as hypothesized, multiple ejaculations function in reducing the effectiveness of subsequent matings, it ought to be possible to detect differences between litters conceived when two males copulate with a single female and the first male either does or does not attain multiple ejaculations. In our first experiment with rats, albino females mated with two males in succession. The first male achieved either one ejaculation or five ejaculations; the second male was permitted to copulate until satiated. In some tests the first male was of the albino F344 strain and the second male was of the pigmented ACI strain; in the remaining tests the order of mating was reversed. When the albino male mated first, the results were statistically significant and consistent with the hypothesis: 32% of the offspring were sired by the first male when he attained one ejaculation whereas 98% were sired by the first male when he attained five ejaculations. When the albino male mated first, however, there was no difference as a function of the number of ejaculations by the first male; the second male sired 96-100% of the offspring regardless of the number of ejaculations by the first male. Although the data generated when pigmented males mated first were sufficient to demonstrate the validity of the hypothesis, a cleaner demonstration would include reciprocity of the effect (*i.e.*, it would appear regardless of which male mated first). This necessity led us to conduct studies of mating order and DFC in laboratory rats in an effort to find a pair of strains having equivalent DFC's so as to achieve reciprocity (Lanier *et al.* 1979). The multiple ejaculation experiment was repeated using Long-Evans males (rather than ACI males) together with F344 males and females. As shown in Fig. 4, the expected reciprocity obtained regardless of mating order. Males of both types gained significant increases in percentages of pups sired from five ejaculations over one ejaculation.

A parallel experiment was conducted with golden hamsters (Oglesby *et al.* 1981). Animals were of the same agouti and cream-colored strains discussed earlier. The first male was permitted to mate for one ejaculation, five ejaculations, or to satiety; the second male was permitted to mate to satiety. In some tests the cream male mated first; in some the agouti male mated first. Results paralleled

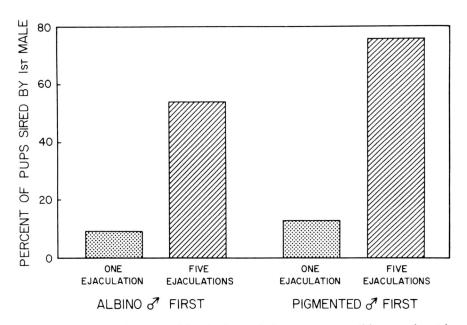

Fig. 4. Percent of pups sired by the first male in a sperm competition experiment in laboratory rats in which either an albino or pigmented male mated first for one or five ejaculations and was followed by a male of the opposite genotype that mated to satiety.

those of the first experiment with rats. When the cream male mated first there was no significant effect on litter composition of the number of ejaculations he attained; the cream male sired from 12-24% of the pups. However, when the agouti male mated first there was a significant effect due to the number of ejaculations attained by the first male. Interpretation of the latter result is complicated slightly because in five of 11 satiety-satiety tests, both cream males failed to mate after an agouti male had mated to satiation. This apparently was attributable to agouti males having induced female non-receptivity by extended mating bouts. At any rate, agouti males sired 30% of the pups when mated for one ejaculation, 58% when mating for five ejaculations, and 86-92% when mating to satiety. We have not located a pair of hamster strains having reciprocally equivalent DFC's.

These studies were followed up in both rats and hamsters with tests designed to determine the mechanism underlying the protective effect of multiple ejaculations. There are at least three possibilities. It may be critical not that males ejaculate, but that they prevent other males from ejaculating for some period of time. Copulating may be simply one sure means of securing that time. In rabbits, for example, sperm that have resided in the doe for 2 hr or more have an advantage over subsequently deposited sperm (Dziuk 1965). Alternatively, vaginal stimulation may be critical in producing the effect. Repeated copulations result in decreased receptivity in both rats and hamsters (Carter 1973; Carter and Schein 1971; Hardy and DeBold 1971, 1972) and in coition-induced hormone release (e.g., Leavitt and Blaha 1970). In either of these ways, vaginal stimulation may be the critical variable. Finally, the full copulatory pattern with ejaculation may be paramount. For example, Martin et al. (1974) found the relative proportion of sperm from males of two strains of chickens to be critical in determining brood composition.

In the experiment with hamsters (Fig. 5), all females copulated first with an intact agouti male for one ejaculatory series, underwent an intermediate treatment, and then were mated with a cream male for five ejaculations. To assess the effect of delay alone, a 10 min interval was imposed between the first and second male in one condition. To assess the effects of stimulation alone, the intermediate treatment was four series from a vasectomized agouti male. In the third condition a second intact agouti male copulated for four series. In the final condition there was no delay. Neither a 10 min delay nor four series with a vasectomized male produced a significant increase in the percentage of offspring sired by males of the first strain over the condition in which the first male attained just one ejaculation and there was no delay (Fig. 5). However, when an intact male mated there was a significant increase in the representation of the agouti genotype in the resulting litters. These data would indicate that the complete ejaculatory series is critical to the protective effect. Thus, it is most likely the relative concentrations of sperm from males of the different lines that is critical.

The parallel experiment in rats was similar except the "no delay" treatment was not included and the time interval imposed was 45 min, reflecting the slower pace

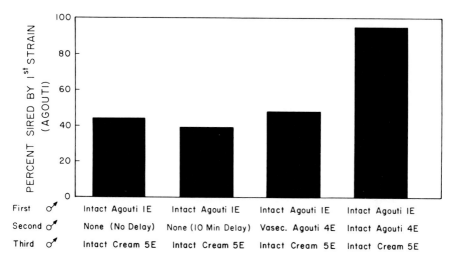

Fig. 5. Results of an experiment designed to determine the mechanisms underlying a protective of multiple ejaculations by golden hamsters in a sperm competition situation. The figure represents the percentage of offspring sired by the first male to mate as a function of the test conditions explained along the abscissa.

of copulation in rats. Twenty-two percent of the pups were sired by males of the first strain when there was delay only (in the presence of a restrained albino male), 17% of the pups were sired by the first male when a vasectomized male was interposed, and 46% were sired by males of the first strain when two males combined for five complete ejaculatory series. These results suggest that the competitive advantage secured with multiple ejaculations in both rats and hamsters is due to the additional sperm deposited, rather than protracted copulatory time alone or vaginal stimulation.

This series of experiments provides substantial evidence that the multiple ejaculatory patterns of muroid rodents functions to advantage in sperm competition and may have evolved under selection for that advantage.

VI. SIMULTANEOUS SPERM COMPETITION

In studies of successive sperm competition just discussed, the experimenter controlled the order and amount of mating by each male. This is an ideal procedure for experimental analysis, but is contrived and unlike the natural situation in which timing and relative frequency of copulations are individual interactions of the males and female involved. In more recent research we have allowed the rodents to "design" the study by permitting two males and one female to copulate *ad lib.* to satiety.

In the study with laboratory rats (Dewsbury and Hartung 1980), 20 triads were formed with a Long-Evans male, a F344 male, and an F344 female. The primary data to be considered along with litter composition relate to which male ejaculated first, which ejaculated last, and which ejaculated the greater number of times. There was no correlation between first ejaculation and either last ejaculation or greater number of ejaculations. The tendencies to ejaculate last and more were seriously confounded because in 17 of the 20 tests the last male to ejaculate also attained more ejaculations. The relationships between these behavioral data and litter composition are portrayed in Fig. 6. Litter composition was independent of whether an albino or pigmented male ejaculated first; in either event approximately

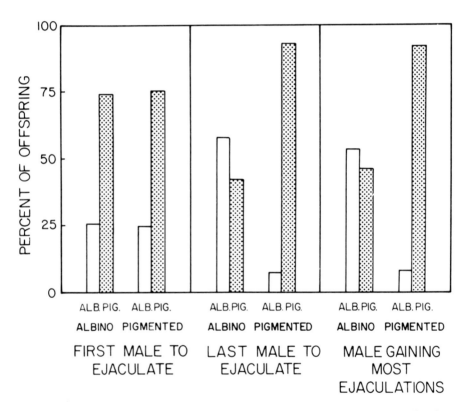

Fig. 6. Litter composition in a study of laboratory rats in which a pigmented and an albino male had simultaneous access to a single receptive albino female and both males mated. Results of 20 tests are portrayed in relation to the first male to ejaculate, the last male to ejaculate, and the male gaining more ejaculations. There was an advantage to ejaculating last and/or more, but not to ejaculating first.

75% of the offspring were pigmented. By contrast, litter composition was greatly affected by which male ejaculated last and/or more. When the pigmented male ejaculated last and/or more, about 92% of the offspring were pigmented. When the albino male ejaculated last and/or more, about 45% of the pups were pigmented. Although earlier data would suggest that it is the number of ejaculations, rather than order of ejaculating, that is critical in affecting litter composition, the effects cannot be disentangled in the present data. The correlation between the percentage of ejaculations delivered by a pigmented male and the percentage of the pups in the resulting litter that were pigmented was +0.67 (N = 17, p < 0.01).

A similar experiment has been completed with blonde and wild-type deer mice (Dewsbury 1981b). Although there were 26 tests with copulation, in only six tests did both males ejaculate and the recipient female produce a litter. There was no significant correlation between the number of ejaculations delivered by the blonde male and the percentage of pups sired by the blonde male (r = -0.31, p > 0.05).

In general, the results in the more natural tests resemble those obtained from the controlled tests. In both situations there is an indication that the relative number of ejaculations attained by the two males is an important determinant of success in sperm competition.

VII. IMPLICATIONS FOR THE EVOLUTION OF MATING STRATEGIES

The data on pregnancy initiation and sperm competition can now be used to permit some preliminary speculations regarding the evolution of mating strategies in these species. As we lack sufficient information, a full explanation is not yet possible. Nevertheless, some of the factors influencing mating strategies are now apparent. The female appears to face several major problems. Given that environmental circumstances are appropriate, it is to her advantage to become pregnant and to rear a litter. In order to do that she must not only copulate with a male and receive sperm, but she must also copulate long enough to ensure that she receives sufficient stimulation for the initiation of sperm transport and for activation of the neuroendocrine reflexes necessary for ovulation or a functional luteal phase and for successful pregnancy. Females may be further selected to ensure that they mate with males bearing genes that will result in offspring maximally able to survive and produce viable, fertile offspring themselves. Should males sometimes contribute other resources to females, and/or paternal care of offspring, females should select males that will act appropriately. By maintaining pregnancies selectively after mating with one male, a female may be acting to ensure stable conditions in which to rear her litter.

Male rodents also face problems. It is generally in the male's best interest that the female he mates becomes pregnant. It would be pointless for a male to mate with a female for one ejaculatory series if the one series did not provide sufficient stimulation to produce a pregnancy. Further, when a female rodent multiply mates with one estrus, males she mates should be selected to maximize their genetic representation in the resulting litter. The solutions to these problems constitute the major elements in muroid rodent mating strategies.

Reproductive characteristics of the four species are summarized in Table II. For simplicity deer mice will be used to exemplify mating strategies; however, many of the same principles apply to the remaining three species. Consider first the problem of pregnancy initiation. The young female deer mouse in cycling estrus requires approximately two ejaculatory series to initiate pregnancy (Dewsbury 1979b). If she is in postpartum estrus, the female will require just one series. Old, multiparous females require additional series. Thus, stimulus requirements for females represent a dynamic system, changing with time and conditions (see Dewsbury 1978). The other factor complicating pregnancy initiation in deer mice is the reduced incidence of pregnancy observed when females mate with or are exposed to two or more males (Dewsbury 1982a). The mechanism underlying this effect is not yet clear. Nevertheless, the implications are that as a male mates with

TABLE II

Pregnancy and Sperm Competition in Four Species of Muroid Rodents

Characteristic	Laboratory Rats	Syrian Golden Hamsters	Deer Mice	Prairie Voles
Mean ejaculation frequency	7.0	10.1	3.6	2.0
Stimulus requirements for reliable (80%) pregnancy				
Cycling estrus	1 series	~5 series	2 series	1 series
Postpartum estrus	<5 series	?	1 series	<2 series
Old, multiparous females	~4 series	?	<2 series	?
Effects of exposure to 2nd male	?	?	decrease	?
Sperm competition order effects	none	advantage last male	none	advantage last male
Sperm competition delay effects	?	?	none	none

a female and permits her to mate with a subsequent male, he suffers a risk not only of sperm competition but of a decreased probability that the female will become pregnant. Further, if a male copulates with a previously mated female he may reduce the probability that the female will become pregnant. It may be significant that male rats prefer unmated females (Krames and Mastromatteo 1973). Successful conception in muroid rodents appears to be a combined function of the condition of the female, the amount of stimulation she receives, and the number of males with which she mates.

The next problem for the male concerns allocation of ejaculates. In much theory relating to sexual selection, males are treated as if they were capable of mating an unlimited number of females. The notion is that sperm are cheap and a male loses little with each ejaculate. According to Dawkins (1976), "the word excess has no meaning for a male." It is only the female that makes a large investment and suffers enormously from an inappropriate mating. Males should mate with as many females as possible (*e.g.*, Williams 1966). Such views neglect some basic facts of male reproductive physiology. A single ejaculate in the rat, for example, may contain 500×10^5 sperm. As noted in Table II, fully rested male rats attained approximately seven ejaculations before satiation; deer mice attained a mean of 3.6. What is critical, however, as was noted by Trivers (1972) is the rate of recovery. In rats, many aspects of reproductive behavior are depressed 3 days after satiation and some are still lowered 6 days following mating to satiety (Beach and Jordan 1956, Jackson and Dewsbury 1979). Prolonged copulation results in an emptying of the seminal vesicles, which are only 1/3 replenished 26 hr later. It is estimated that 3-6 days are required for full restitution (Pessah and Kochva 1975). Together, these data suggest that males have a limited capacity to produce sperm and other constituents of the ejaculate. As the ability to produce ejaculates is limited, males would be expected to be prudent in their allocation (Dewsbury 1982b). The finding that primate testis size is related to mating system (Harcourt *et al.* 1981; Harvey and Harcourt, this volume) is consistent with the notion that selection is acting on the capacity of males to produce ejaculates. The word "excess" does have meaning for males.

Data on sperm competition in deer mice suggest that the order of mating and the intervals between ejaculates are of relatively little importance in determining litter composition. What appears critical in the determination of paternity are the relative numbers of ejaculates the female receives from each male. This is illustrated in Fig. 7. A male is thus faced with a dilemma: The more ejaculates he can deliver to a given female, the higher the proportion of that litter he will sire; however, by mating with one female he depletes his capacity to inseminate other females and sire their offspring. Adler (1978) has suggested that male rats will be "better off inseminating seven different females than concentrating on just one." This will not always be the case. The optimal strategy will vary with the relative number of other active males and receptive females in the vicinity—the operational sex ratio (Emlen and Oring 1977). If females are many and males are few, the optimal

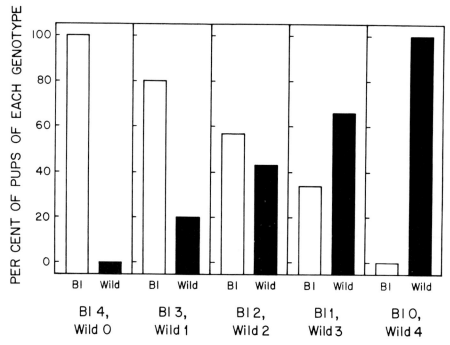

Fig. 7. Litter composition in deer mice as a function of the number of ejaculations attained by two males. Results of several studies are summarized in this figure. In each instance blonde females mated for four ejaculations. When two males mated, the blonde male mated first and the wild-type male second. Litter composition reflects the relative number of ejaculations attained by the two males.

strategy will be to mate with each female just enough to ensure her pregnancy. If males are many and females are few, it would seem optimal to sequester a female and copulate with her repeatedly. The calculation of optimal strategies for all situations will require extensive data on relative costs and benefits resulting from allocation of ejaculates to one or more females. Given the factors already apparent, and perhaps some not yet identified, the solution is sure to be complex. A preliminary effort for deer mice is presented by Dewsbury (1982b). What is already clear is that males have limited ejaculatory capacity and they should be selected to use this limited resource in an efficient way that will maximize fitness. Optimal use will vary with the operational sex ratio. Whether or not males actually alter their copulatory behavior in accordance with the operational sex ratio has yet to be investigated.

In emphasizing sperm competition among males, one should not forget that male strategies can evolve only in conjunction with female strategies (see Knowlton and Greenwell, this volume). There are good reasons that females would be selected both to curtail and to prolong copulatory activity. The optimal solution for females probably lies in a compromise among factors. Excessive prolongation can produce tissue damage in the vaginal and perineal regions, wasted energy, loss of sperm already in the reproductive tract, and risk of predation. Also, it has been suggested that females should mate with just one male so that conflict among individuals in the resulting litter would be reduced through kin selection (Hamilton 1964, Trivers 1972). On the other hand, copulation with more than one male would protect against the possibility of mating with a sterile male and might contribute to greater variability in the litter, with advantage in an unpredictable environment. In addition, females may be selected to promote male-male competition both through overt behavioral interactions (Cox and LeBoeuf 1977) and more directly among spermatozoa (Cohen 1971), thus producing sons with a competitive edge. Under some conditions females might be selected to engage in multiple matings rather than to offer active resistance simply because mating entails fewer risks and less energetic cost than resistance (Gibson and Falls 1975, Parker 1970). Whatever strategy would be favored in males would necessarily coevolve with female strategies.

VIII. CONCLUSIONS AND DIRECTIONS
FOR FURTHER RESEARCH

Field data on Norway rats and deer mice suggest that the conditions necessary for sperm competition may often exist. Laboratory situations designed to permit analysis of the kinds of interactions that are likely to occur in the field indicate that sperm competition can be a major factor in differential reproduction in the four species of muroid rodents studied here. Although sperm competition may be important for all species, the mechanisms underlying such competition may vary across species. For example, the last male to ejaculate has an advantage only in golden hamsters and prairie voles; order effects appear negligible in deer mice and rats. Far from being wasted activity, multiple ejaculations are important not only in pregnancy initiation, but in the context of sperm competition. Males that attain multiple ejaculations with a female gain a reproductive advantage over males attaining just one ejaculation. The strategy that is optimal for a given male will vary with the prevailing conditions, especially the operational sex ratio. The complete mating system will be a product of various factors, including the necessity for pregnancy initiation, the potential for sperm competition, predator pressure, and operational sex ratios, acting on both males and females.

If the evolution of mating strategies in these species is to be better understood, a variety of additional information will be needed. First, it will be necessary to obtain more complete data on ejaculatory capacities for males of the various species, the rates at which capacity declines with repeated matings, and the rates of recovery. Next, it will be necessary to quantify costs and benefits to both male and females of varying numbers of copulations and the timing thereof. In addition, it will be important to determine if male and female reproductive behavioral patterns are optimal to given circumstances and whether patterns are adjusted to changing conditions in ways that make biological sense. All of these factors must be assessed in the context of different social organizational patterns and in relation to factors such as dominance and territoriality. Finally, all are likely to vary across species—especially in species having divergent mating systems (*e.g.*, monogamy vs promiscuity).

IX. SUMMARY

Although much attention has been directed to the study of sperm competition in insects, mammals have stimulated less interest. It is now apparent that the conditions necessary for sperm competition exist in natural populations of mammals. Laboratory studies have revealed some of the dynamics of sperm competition in muroid rodents and their implications for the evolution of mating strategies. Essential to all such studies is an understanding of the differential fertilizing capacities of the individuals and strains under study. Effects of mating order on litter composition vary from species to species. The delay between successive matings by different males appears relatively unimportant in determining litter composition. Multiple ejaculations function in protecting a male's sperm investment; males attaining multiple ejaculations with a given female are better sperm competitors than males mating for just one ejaculation. The underlying mechanism for this effect is apparently sperm concentrations rather than the passage of time alone or vaginal stimulation. In a situation in which two males had simultaneous access to a single receptive female the male rats ejaculating last and/or more often, sired a disproportionate percentage of the resulting litter; there was no advantage to ejaculating first. In the evolution of mating strategies males will be be selected to make maximally efficient use of the limited number of ejaculates they can discharge. Male strategies must evolve in the context of a variety of factors, including female strategies.

ACKNOWLEDGMENTS

This research was supported by Grant BNS78-05173 from the National Science Foundation. The contributions of the various graduate students and other associates affiliated with this research are gratefully acknowledged.

REFERENCES

Adler, N. T. 1969. Effects of the male's copulatory behavior on successful pregnancy of the female rat. *J. Comp. Physiol. Psychol.* 6-613-622.

Adler, N. T. 1978. On the mechanisms of sexual behavior and their evolutionary constraints. In *Biological Determinants of Sexual Behavior*, J. B. Hutchison (ed.), pp. 655-695. Wiley, New York.

Allison, A. J. 1977. Flock mating in sheep. II. Effect of number of ewes per ram on mating behavior and fertility of two-tooth and mixed-age Romney ewes run together. *New Zealand J. Agr. Res.* 20:123-128.

Beach, F. A., and L. Jordan. 1956. Sexual exhaustion and recovery in the male rat. *Q. J. Exp. Psychol.* 8:121-133.

Beatty, R. A. 1960. Fertility of mixed semen from different rabbits. *J. Reprod. Fertil.* 19: 52-60.

Beatty, R. A., G. H. Bennett, J. G. Hall, J. L. Hancock, and D. L. Stewart. 1969. An experiment with heterospermic insemination in cattle. *J. Reprod. Fertil.* 19:491-501.

Bertram, B. C. R. 1976. Kin selection in lions and in evolution. In *Growing Points in Ethology* P. G. Bateson and R. A. Hinde (eds.), pp. 281-301. Cambridge University Press, Cambridge.

Birdsall, D. A., and D. Nash. 1973. Occurrence of successful multiple insemination of females in natural populations of deer mice (*Peromyscus maniculatus*). *Evolution* 17:106-110.

Boorman, E., and G. A. Parker. 1976. Sperm (ejaculate) competition in *Drosophila melanogaster*, and the reproductive value of females to males in relation to female age and mating status. *Ecol. Entomol.* 1:145-155.

Bronson, F. H., and A. Coquelin. 1980. The modulation of reproduction by priming pheromones in house mice: Speculations on adaptive function. In *Chemical Signals: Vertebrates and Aquatic Invertebrates*, D. Müller-Schwarze and R. M. Silverstein (eds.), pp. 243-265. Plenum, New York.

Bruce, H. M. 1959. An exteroceptive block to pregnancy in the mouse. *Nature* 184:105.

Calhoun, J. B. 1962. *The Ecology and Sociology of the Norway rat.* U.S. Dept. Health, Educ., Welfare, Bethesda, MD

Carter, C. S. 1973. Stimuli contributing to the decrement in sexual receptivity of female golden hamsters (*Mesocricetus auratus*). *Anim. Behav.* 21: 827-834.

Carter, C. S., and M. W. Schein. 1971. Sexual receptivity and exhaustion in the female golden hamster. *Horm. Behav.* 2:191-100.

Cohen, J. 1971. The comparative physiology of gamete populations. *Adv. Comp. Physiol. Biochem.* 4:167-380.

Cox, C. R., and B. J. LeBoeuf. 1977. Female incitation of male competition: A mechanism in sexual selection. *Am. Nat.* 111:317-335.

Dawkins, R. 1976. *The Selfish Gene.* Oxford University Press.

Devine, M. C. 1975. Copulatory plugs in snakes: Enforced chastity. *Science* 187:844-845.

Devine, M. C. 1977. Copulatory plugs, restricted mating opportunities and reproductive competition among garter snakes. *Nature* 267:345-346.

Dewsbury, D. A. 1972. Patterns of copulatory behavior in male mammals. *Q. Rev. Biol.* 47: 1-33.

Dewsbury, D. A. 1975. Diversity and adaptation in rodent copulatory behavior. *Science* 190: 947-954.

Dewsbury, D. A. 1978. The comparative method in studies of reproductive behavior. In *Sex and Behavior: Status and Prospectus*, T. E. McGill, D. A. Dewsbury, and B. D. Sachs (eds.), pp. 83-112. Plenum, New York.

Dewsbury, D. A. 1979a. Copulatory behavior of deer mice (*Peromyscus maniculatus*): I. Normative data, subspecific differences, and effects of cross-fostering. J. *Comp. Physiol. Psychol.* 93:151-160.

Dewsbury, D. A. 1979b. Copulatory behavior of deer mice (*Peromyscus maniculatus*): III. Effects on pregnancy initiation. *J. Comp. Physiol. Psychol.* 93:178-188.

Dewsbury, D. A. 1981a. On the function of the multiple-intromission, multiple-ejaculation copulatory patterns of rodents. *Bull. Psychonomic Soc.* 18:221-223.

Dewsbury, D. A. 1981b. Social dominance, copulatory behavior, and differential reproduction in deer mice (*Peromyscus maniculatus*). *J. Comp. Physiol. Psychol.* 95:880-895.

Dewsbury, D. A. 1982a. Pregnancy blockage following multiple-male copulation of exposure at the time of mating. *Behav. Ecol. Sociobiol.*, in press.

Dewsbury, D. A. 1982b. Ejaculate cost and male choice. *Am. Nat.* 119:601-610.

Dewsbury, D. A. 1982c. An exercise in the prediction of monogamy in the field from laboratory data on 42 species of muroid rodents. *Biologist* 63:138-162.

Dewsbury, D. A., and D. J. Baumgardner. 1981. Studies of sperm competition in two species of muroid rodents. *Behav. Ecol. Sociobiol.* 9:121-133.

Dewsbury, D. A. and T. G. Hartung. 1980. Copulatory behavior and differential reproduction of laboratory rats in a two-male, one-female competitive situation. *Anim. Behav.* 28:95-102.

Dziuk, P. J. 1965. Double mating of rabbits to determine capacitation time. *J. Reprod. Fertil.* 10:389-395.

Edwards, R. G. 1955. Selective fertilization following the use of sperm mixtures in the mouse. *Nature* 175:215-223.

Emlen, S. T., and L. W. Oring. 1977. Ecology, sexual selection, and the evolution of mating systems. *Science* 197:215-223.

Fitch, H. S. 1957. Aspects of reproduction and development in the prairie vole (*Microtus ochrogaster*). *Univ. Kans. Publ. Mus. Nat. Hist.* 10:129-161.

Foltz, D. W. 1981. Genetic evidence for long-term monogamy in a small rodent (*Peromyscus polionotus*). *Am. Nat.* 117:665-675.

Getz, L. L. 1978. Speculation on social structure and population cycles of microtine rodents. *Biologist* 60:134-147.

Getz, L. L., and C. S. Carter. 1980. Social organization in *Microtus ochrogaster*. *Biologist* 60:134-147.

Gibson, A. R. and J. B. Falls. 1975. Evidence for multiple insemination in the common garter snake, *Thamnopis sirtalis*. *Can. J. Zool.* 53:1362-1368.

Hamilton, W. D. 1964. The evolution of social behavior. I and II. *J. Theor. Biol.* 7:1-52.

Hanken, J. and P. W. Sherman. 1981. Multiple paternity in Belding's ground squirrel litters. *Science* 212:351-353.

Harcourt, A. H., P. H. Harvey, S. G. Larson, and R. V. Short. 1981. Testis weight, body weight and breeding system in primates. *Nature* 293:55-57.

Hardy, D. F., and J. F. DeBold. 1971. Effects of mounts without intromission upon the behavior of female rats during the onset of estrogen-induced heat. *Physiol. Behav.* 7:643-645.

Hardy, D. F., and J. F. DeBold. 1972. Effects of coital stimulation upon behavior of the female rat. *J. Comp. Physiol. Psychol.* 78:400-408.

Hartung, T. G., and D. A. Dewsbury. 1978. A comparative analysis of copulatory plugs in muroid rodents and their relationship to copulatory behavior. *J. Mammal.* 59:717-723.

Holmes, H. B. 1974. Patterns of sperm competition in *Nasonia vitripennis*. *Can J. Genet. Cytol.* 16:789-795.

Jackson, S. B., and D. A. Dewsbury. 1979. Recovery from sexual satiety in male rats. *Anim. Learn. Behav.* 7:119-124.

Krames, L., and L. A. Mastromatteo. 1973. Role of olfactory stimuli during copulation in male and female rats. *J. Comp. Physiol. Psychol.* 85:528-535.

Lanier, D. L., D. Q. Estep, and D. A. Dewsbury. 1975. Copulatory behavior of golden hamsters: Effects on pregnancy. *Physiol. Behav.* 15:209-212.

Lanier, D. L., D. Q. Estep, and D. A. Dewsbury. 1979. Role of prolonged copulatory behavior in facilitating reproductive success in a competitive mating situation in laboratory rats. *J. Comp. Physiol. Psychol.* 93:781-792.

Leavitt, W. W., and G. C. Blaha. 1970. Circulating progesterone levels in the golden hamster during the estrous cycle, pregnancy, and lactation. *Biol. Reprod.* 3:353-361.

Levine, L. 1976. Sexual selection in mice. IV. Experimental demonstration of selective fertilization. *Am. Nat.* 101:289-294.

Linley, J. R. 1975. Sperm supply and its utilization in doubly inseminated flies, *Culicoides melleus*. *J. Insect Physiol.* 21:1785-1788.

Madison, D. M. 1980. Space use and social structure in meadow voles, *Microtus pennsylvanicus. Behav. Ecol. Sociobiol.* 7:65-71.

Martan, J. and B. A. Shepherd. 1976. The role of the copulatory plug in reproduction of the guinea pig. *J. Exp. Zool.* 196:79-84.

Martin, P. A., and P. J. Dziuk. 1977. Assessment of relative fertility of males (cockerels and boars) by competitive mating. *J. Reprod. Fertil.* 49:323-329.

Martin, P. A., T. J. Reimers, J. R. Lodge, and P. J. Dziuk. 1974. The effect of ratios and numbers of spermatozoa mixed from two males on proportions of offspring. *J. Reprod. Fertil.* 39:251-258.

Matthews, M. K., Jr., and N. T. Adler. 1978. Systematic interrelationship of mating, vaginal plug position, and sperm transport in the rat. *Physiol. Behav.* 20:303-309.

Merritt, R. B., and B. J. Wu. 1975. On the quantification of promiscuity (or *Promyscus maniculatus?*). *Evolution* 29:575-578.

Milligan, S. R. 1979. The copulatory pattern of the bank vole (*Clethrionomys glareolus*) and speculation on the role of penile spines. *J. Zool.* 188:279-300.

Mosig, D. W., and D. A. Dewsbury. 1970. Plug fate in the copulatory behavior of rats. *Psychon. Sci.* 20:315-316.

Murphy, M. R. 1971. Natural history of the Syrian golden hamster—a reconnaissance expedition. *Am. Zool.* 11:632.

Oglesby, J. M., D. L. Lanier, and D. A. Dewsbury. 1981. The role of prolonged copulatory behavior in facilitating reproductive success in male Syrian golden hamsters (*Mesocricetus auratus*) in a competive mating situation. *Behav. Ecol. Sociobiol.* 8:47-54.

Parker, G. A. 1970. Sperm competition and its evolutionary consequences in the insects. *Biol. Rev.* 45:525-567.

Pessah, H., and E. Kochva. 1975. The secretory activity of the seminal vesicles in the rat after copulation. *Biol. Reprod.* 13:557-560.

Robitaille, J. A., and J. Bovet. 1976. Field observations on the social behavior of the Norway rat, *Rattus norvegicus* (Berkenhout). *Biol. Behav.* 1:189-308.

Smith, R. L. 1979. Repeated copulation and sperm precedence: Paternity assurance for a male brooding water bug. *Science* 205:1029-1031.

Sumption, L. J., and J. C. Adams. 1961. Multiple sire mating in swine. III. Factors influencing multiple paternity. *J. Hered.* 52:214-218.

Telle, H. J. 1966. Contribution to the knowledge of behavioral patterns in two species of rats, *Rattus norvegicus* and *Rattus rattus. Z. Angew. Zool.* 53:129-196. (Tech. Trans. 1608, Nat. Res. Counc. Can., 1972, V. N. Nekrassoff, trans.)

Thibault, C. 1973. Sperm transport and storage in vertebrates. *J. Reprod. Fertil. Suppl.* 18:39-53.

Thomas, J. A., and E. C. Birney. 1979. Parental care and mating system of the prairie vole, *Microtus ochrogaster. Behav. Ecol. Sociobiol.* 5:171-186.

Trivers, R. L. 1972. Parental investment and sexual selection. In *Sexual Selection and the Descent of Man 1871-1971*, B. Campbell (ed.), pp. 136-179. Aldine, Chicago.

Voss, R. 1979. Male accessory glands and the evolution of copulatory plugs in rodents. *Occ. Pap. Mus. Zool. Univ. Mich.* 968:1-27.

Waage, J. K. 1979. Dual function of the damselfly penis: Sperm removal and transfer. *Science* 102:916-918.

Williams, G. C. 1966. *Adaptation and Natural Selection: A Critique of some Current Evolutionary Thought.* Princeton Univ. Press.

Wilson, J. R., N. Adler, and B. LeBoeuf. 1965. The effects of intromission frequency on successful pregnancy in the female rat. *Proc. Nat. Acad. Sci. U.S.A.* 53:1392-1395.

Zucker, I., and G. Wade. 1968. Sexual preferences of male rats. *J. Comp. Physiol. Psychol.* 66:816-819.

17

Sperm Competition?
The Case of Vespertilionid
and Rhonolophid Bats

M. BROCK FENTON

I. INTRODUCTION

There is potential for competition among ejaculates of different males in mammals with promiscuous mating systems, particularly species capable of storing sperm before ovulation and fertilization. In some rodents promiscuous mating produces litters of mixed paternity (Birdsall and Nash 1973; Hanken and Sherman 1981; Dewsbury, this volume), but in species where sperm are stored for extended periods, potential for sperm competition could be higher, and in this context, the reproductive patterns of some bats make them likely candidates for the occurrence of sperm competition. The purpose of this chapter is to review the data on mating systems in bats where females are known to store sperm, and to consider the possibility that sperm competition has had a role in shaping certain aspects of bat reproductive biology.

Sperm Competition and the Evolution
of Animal Mating Systems

Storage of sperm by some female bats allows mating at the most propitious time and birth of young in the most favorable season. Analogous adaptations include delayed implantation and/or development, phenomena common in several groups of mammals, including bats (Vaughan 1978). Sperm storage and delayed ovulation were first described in vespertilionid bats by Pagenstecher (1859) who studied *Pipistrellus pipistrellus.* Adaptive sperm storage is demonstrated when stored sperm are capable of fertilizing eggs (Hartman and Guyler 1972; Gates 1936; Wimsatt 1942, 1944). The ability to store sperm has been shown in some species of *Rhinolophus* (Rhinolophidae), and in many Vespertilionidae (Table I), and the fertility of stored sperm has been demonstrated in eight species of vespertilionids (Racey 1979).

Female bats may store sperm for from 16 to almost 200 days (*Pipistrellus ceylonicus* and *Nyctalus noctula,* respectively; Racey 1979), and the sites of storage likewise vary among species. In rhinolophids and some vespertilionids sperm is stored in the oviducts; in other vespertilionids it is stored at the uterotubal junction, or in the uterus (Table I). Although there appears to be consistency within species, the species within a genus (*e.g., Myotis* or *Pipistrellus*) may use different sites (Table I). Variability in the distribution of sperm in the female reproductive tract does not necessarily reflect the sites of sperm storage. This is particularly true during the mating season when the tract may be distended with sperm, but storage in the vagina seems unlikely (Racey 1979). In cases where sperm storage has been demonstrated, there frequently is a close spatial relationship between stored sperm and the epithelial lining of the storage organ (*e.g.,* epithelial microvilli in *P. pipistrellus* [Fig. 1A, B], Racey and Potts 1970; sperm lodged in indentations in oviducal epithelial cells in *Tylonycteris pachypus,* Racey *et al.* 1975). The close association between stored sperm and the lining of the female tract may indicate an important female role in successful sperm storage. Males may also store viable sperm for long periods (Racey 1973).

Since sperm storage was first reported in temperate bats, it seemed to be an adaptation for overwintering, and a specialization related to hibernation. This interpretation was supported by evidence that females of nonhibernating populations of *Myotis austroriparius* did not appear to store sperm (Rice 1957). However, more recently tropical vespertilionids have been found to store viable sperm (*e.g.,* Gopalakrishna and Madhavan 1971, 1978; Wimsatt 1969; Myers 1977; Racey *et al.* 1975), thus clouding the proposed association among seasonal climatic extremes, hibernation, and sperm storage. An ecological significance of sperm storage has yet to be proposed for tropical species; however, the process may be correlated with wet and dry seasons and consequent fluctuations in food abundance.

Actually, a variety of temperate and tropical vespertilionids and at least two temperate rhinolophids mate at one time of the year and females store fertile sperm for some time before ovulation and fertilization. In many species the period of sperm storage corresponds to hibernation, but this is not always the case and the suggestion that nonhibernating females cannot store sperm, usually made with reference to temperate species (*e.g., Pipistrellus hesperus;* Krutzsch 1975), has been disputed (Racey 1975, Thomas *et al.* 1979).

Storage of sperm by females per se does not produce potential competition between sperm of different males, or postcoital processing of sperm from different males by females. For competition to occur, sperm from more than one male (overlapping ejaculates) must be present in the female reproductive tract at the time of fertilization. Our knowledge of mating systems in bats, particularly species known to store sperm, is even less complete than our knowledge of which species store sperm (Table I). At present it is impossible to put sperm storage by bats in behavioral context for most species (Bradbury 1977, Thomas *et al.* 1979).

II. CHASTITY PLUGS?

In some rhinolophids and vespertilionids that store sperm, the female tract may be plugged some time after copulation, and there are at least three categories of blockages, two female-generated, and one male-generated. In some species (*e.g., Nyctalus noctula;* Grosser 1903) the plug is produced by connective tissue hypertrophy of the cervix, while in others the vagina is blocked to varying degrees by the cornified epithelium which characterizes the vagina of estrous mammals (*e.g., Pipistrellus pipistrellus*). In *Rhinolophus hipposideros* the plug is formed by coagulation of accessory gland secretions (Gaisler 1966). Some researchers (*e.g.,* Harrison Mathews 1937, Eckstein and Zuckerman 1956, Gaisler 1966, Kitchener 1975, Bernard 1980) proposed that plugs were sites of sperm storage, but there are few sperm in these plugs and those contained tend to be decapitate. Consequently, Racey (1979) rejects this suggestion. It would seem that vaginal plugs, whether of male or female origin, may function as barriers to further insemination. This is an attractive hypothesis in a context of possible competition among stored sperm, but the presence of a well-developed plug in *Hipposideros caffer,* a species that does not store sperm (Bernard 1980), complicates the picture.

Furthermore, female *Rhinolophus ferrumequinum* can expel plugs after mating (Racey 1975) and do so in nature (Ransome 1980). In this case the male-produced plug appears to be an ineffective barrier as it does not prevent the female from obtaining additional ejaculates from other males. An alternative, and the classical interpretation of mammalian plugs, is that they simply prevent sperm leakage from the female's reproductive tract.

Although vaginal plugs and similar cervical occlusions have been reported from several species of bats (Table I), in only one of these species (*N. noctula*) are there any data on mating behavior (Likhachev 1961, van Heerdt and Sluiter 1965, Sluiter

TABLE I

Species of Bats in Which Sperm Storage by Females is Known or Suggested, With Details of the Duration and Site of Sperm Storage, Vaginal Plugs, and Number of Offspring per Litter

Species	Suggested/Reported	No. of Days Females Can Store Viable Sperm	Site of Storage[a]	Vaginal Plug[b]	Mating System	Litter Size	Sources
RHINOLOPHIDAE							
Rhinolophus ferrumequinum	+	–	O	SF	male harems?	1	Racey 1975, Ransome 1978
Rhinolophus hipposideros	+	–	O	SF	–	1	Racey 1975
VESPERTILIONIDAE							
Myotis lucifugus	+	138	UTJ	NO	random, promiscuous	1	Wimsatt 1944, Racey 1979, Thomas et al. 1979
Myotis sodalis	+	68	–	–	–	1	Gates 1936
Myotis tricolor[c]	+	–	–	–	–	1	Bernard 1980
Myotis albescens[c]	+	–	–	–	–	1	Myers 1977
Myotis nattereri	+	–	U	–	–	1	Racey 1975
Myotis daubentoni	+	–	UTJ	–	–	1	Racey 1975
Pipistrellus pipistrellus	+	151	U	–	promiscuous?	2	Racey 1973, 1975; Ognev 1928

Species							Reference
Pipistrellus abramus	+	175	U	—	—	2	Hiraiwa and Uchida 1956
Pipistrellus ceylonicus[c]	+	16	O	—	—	2	Gopalakrishna and Madhavan 1971, Racey 1979
Nyctalus noctula	+	198	U	CV	single male harems	2	Racey 1973, 1975; van Heerdt and Sluiter 1965
Eptesicus fuscus	+	156	—	—	—	1-2	Wimsatt 1944
Eptesicus pumilus[c]	+	—	—	—	—	2	Green 1965
Eptesicus regulus[c]	+	—	—	VA	—	1	Kitchener and Halse 1978
Eptesicus furinalis[c]	+	—	—	—	—	2	Myers 1977
Tylonycteris pachypus[c]	+	21	O	—	—	2	Racey 1979, Racey *et al.* 1975
Tylonycteris robustula[c]	+	—	—	—	—	2	Racey *et al.* 1975
Scotophilus heathi[c]	+	—	UTJ	—	—	2	Krishna and Dominic 1978, Gopalakrishna and Madhavan 1978
Chalinolobus gouldii[c]	+	—	—	VA	—	2	Kitchener 1975
Lasiurus ega[c]	+	—	—	—	—	2	Myers 1977
Plecotus townsendii	+	—	—	NO	—	1	Pearson *et al.* 1952
Antrozous pallidus	+	—	—	—	—	1-2	Orr 1954

[a]Sperm may be stored in the oviduct (O), the uterotubal junction (UTJ) or in the uterus (U).
[b]The vaginal plug may be formed from male accessory gland secretions (SF), or by either cornified vaginal epithelium (VA) or hypertrophy of cervical connective tissue (CV), or it may be absent (NO).
[c]Tropical or subtropical species.
—Indicates no data.

577

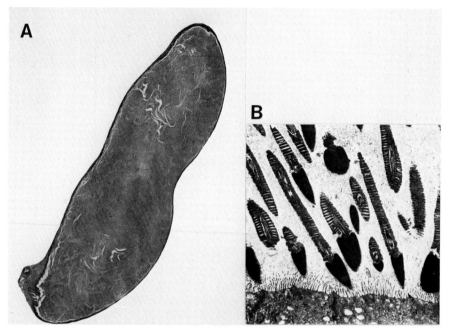

Fig. 1. A. The uterus of an inseminated *Pipistrellus pipistrellus* distended with semen containing spermatozoa in very high density (X32); courtesy of P. A. Racey. B. The uterine epithelium contains potential sperm nutrients, and these spermatozoa in contact with it are frequently oriented toward the epithelium with their heads embedded in microvilli (X6500); electron micrograph courtesy of P. A. Racey.

and van Heerdt 1966). In *N. noctula,* males are aggressive toward one another (Kleiman and Racey 1969) and form single male harems during the mating season. Females may mate with only one male, but the data base is not drawn from individually recognizable animals and since in other bats females may switch harems (*e.g.,* Bradbury 1977, Porter 1979) this inference is not justified. If further work reveals that in some bats, blockages of the female reproductive tract prevent further copulations and that females therefore mate with only one male, then the possibility of sperm competition may be minimal.

It would be instructive to learn how many copulations occur prior to formation of the plug, and whether the plug can be circumvented. Courrier (1927) suggested that the baculum might permit a male to penetrate a vagina occluded by cornified epithelium and, as already noted, some females can expel plugs of male origin apparently at will. So there may be potential for competitive "games" in spite of plugs (see Knowlton and Greenwell, this volume).

III. "DISCO" MATING SYSTEM

Bats that store little sperm are not known to produce vaginal plugs, although the vagina may be partly blocked by cornified epithelium. The lack of a plug presumably permits unimpeded further mating, but the dearth of information about bat mating behavior makes it difficult to generalize. Multiple mating and lack of intermale aggression have been reported in *Pipistrellus pipistrellus* and *Myotis lucifugus* (Ognev 1928, Thomas *et al.* 1979), but most information is available for the latter species, "the bat" for many studies of reproduction (*e.g.*, Guthrie 1933, Guthrie and Jeffers 1938, Wimsatt 1945, Gustafson and Shemesh 1976, Quay 1976). In *M. lucifugus* the mating system appears to be random and promiscuous (Thomas *et al.* 1979).

In Ontario most mating by *M. lucifugus* occurs in August and early September, but some continues later. There are two seasonal periods of mating, an autumn period when both partners are active, and a winter period when males move through hibernacula mating with torpid bats (curiously, both males and females; Thomas *et al.* 1979). The winter mating is commonly observed by bat banders visiting hibernacula (*e.g.*, Barbour and Davis 1969). Torpid partners of either sex may or may not arouse during these encounters, and examination of the perineal region of the accosted individuals has revealed that ejaculation may occur, regardless of whether the torpid bats are male or female (Thomas 1978).

During the autumn period males appear to pacify struggling females with a copulation call (Barclay and Thomas 1979), but these calls do not occur if the female is receptive. Evidence that most copulations occur in August and early September is based on our observations taken from late July to May. Although we have found occasional pairs of coupled bats (Fig. 2) on visits to hibernacula from November to May, the incidence of copulation during the winter period is very low. Even after bats were disturbed during hibernation, the numbers that awoke and copulated never approached levels observed at the same site on any night in August or September. In-hibernaculum mating activity was regularly distributed throughout this period with no peaks prior to the bats' departure for summer quarters (Thomas *et al.* 1979). There is no indication that mating occurs outside hibernacula before or after the onset of in-hibernaculum mating (Thomas *et al.* 1979). Therefore, mating is apparently confined to hibernacula under natural condtions. Most mating occurs before or just after initiation of hibernation, with the strong implication that active females store viable sperm (Thomas *et al.* 1979).

Although male little brown bats display from sites in the hibernacula, usually from those sites most likely to foster successful copulations, Thomas *et al.* (1979) found no evidence of skewed mating success among males or of strong intermale aggression. Tenure at an appropriate site was important for successful copulation, as couples mating on horizontal surfaces (small ledges in caves and drill holes in

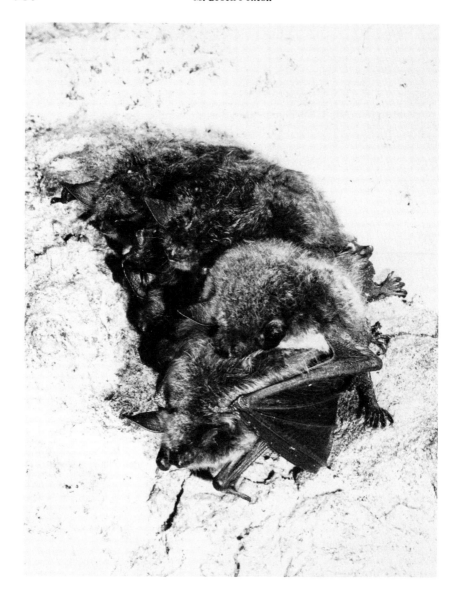

Fig. 2. Mating in *Myotis lucifugus* during the winter period. During copulation the male holds the female with his thumbs and wrists (foreground) and bites at her back. The other bats in this cluster (background) are torpid. The difference in texture of the fur of the male and the torpid bats reflects postarousal grooming. In this case both copulating bats were active; it is more common to find the accosted individual still in torpor.

mines) were less likely to be dislodged by other bats during the fall matings. Copulating pairs of bats attracted considerable attention from conspecifics, and these "curious onlookers" (adult females and subadults) often climbed over and under mating pairs. Bats coupled on vertical surfaces were easily dislodged by spectators, preempting the copulation (Thomas 1978).

Male *M. lucifugus* appeared not to occupy preferred mating sites longer than about 2 h, because temperatures within the hibernaculum (5-10°C) ultimately resulted in the occupants becoming torpid and thus unable to defend optimal positions. Prolonged defense of a site would have required a large expenditure of energy, apparently an unprofitable strategy for these animals about to enter hibernation. Males seemed to mate with as many females as possible (Thomas *et al.* 1979) and on several occasions, individual males mated sequentially with several different females. Because males become torpid as the season progresses, they cannot defend females with whom they have mated.

Torpid female little brown bats are similarly unable to protect whatever investment they have made in mate choice. Before joining a male at a suitable site in a hibernaculum, a female may fly about in a cave or mine and visit several locations occupied by different males. Examination of their reproductive tracts revealed that females mate more than once, perhaps with different males (Thomas *et al.* 1979). However, as winter approaches and temperatures decline, females become torpid and are passively available to winter mating males.

If the peaks observed in the incidence of mating activity are real (*i.e.,* there are few if any copulations outside the hibernacula), then active females must store sperm. Our research has demonstrated this to be the case. It is not known whether sperm from active matings and female passive matings have equal viability. Furthermore, nothing is known concerning patterns of sperm utilization in females that have been multiply mated. Finally, there are no data on the relative contribution to fertilization of sperm from fall and midwinter matings. This has been suggested, however, particularly for females in their first year, because they may not have become fully estrus in their first fall (Guthrie 1933). It is tempting to suggest that matings during hibernation are more energetically expensive than those that occur during the fall period because of the additional costs of arousal from torpor; however, we know relatively little about the bioenergetics and dynamics of bat hibernation, the role of clustering behavior, the frequency of arousal, etc. (*e.g.,* Fenton 1970). Arousal is the most expensive part of hibernation, but if torpid animals must arouse as a part of the hibernation cycle, the cost of arousal could not be assigned to winter mating. If arousal is expensive, its costs might be offset by a competitive advantage obtained by "sneak" copulations with sleeping females.

Myotis lucifugus females produce but a single young from late May to early July, depending upon latitude (Barbour and Davis 1969). They selectively nurse their own young and will actively (sometimes violently) reject attempts by other infants to suckle (Thomson 1980). The female invests considerable energy in her offspring, before it is able to fly (*e.g.,* Kunz 1980), but it is not known if mothers bring insect food to their young during weaning (Buchler 1980, Thomson 1980). Young are reared in nursery colonies whose populations are composed almost entirely of adult females and young, and rarely males. Given the low fecundity and large maternal investment per reproductive unit required in this species, it would seem that females should be under intense selection to choose their mates carefully, and minimize the potential for preemption of that choice. Male mechanisms to provide postcoital defense against other males should likewise be selected. For these reasons, it is curious to find plugs lacking in this species.

IV. SPERM COMPETITION?

Competition among ejaculates from different males, or postcoital processing of sperm by females, or both processes may occur in bats. Males of temperate species that arouse from hibernation to mate with torpid females might obtain a competitive advantage for their sperm over those stored from fall matings. But, no data are available that would elucidate any of these issues. The necessary data must be secured by direct observation of matings, ascertaining what sperm are successfully stored, and employing some system to identify paternity of progeny. Isozyme analysis, useful in determining paternity in some bats (McCracken and Bradbury 1977, 1981) may be applicable, but two important problems intervene. The first involves the physical separation of mating sites and nurseries. In spite of a long period of intensive banding of *M. lucifugus* (*e.g.,* Davis and Hitchcock 1965, Griffin 1970, Fenton 1970, Humphrey and Cope 1976), examples of situations where known females have moved between known hibernacula and known nurseries are rare. The probability of observing mating activities of a female and successfully tracking her to a particular nursery seem very low. An even more difficult task would be to determine how many males had inseminated a given female. Another major obstacle is the time delay between mating, ovulation, fertilization, and parturition, often the period from mid-August until mid-April and mid-June.

A long-standing unresolved problem with *M. lucifugus* is the disappearance of large numbers of females between summer and winter roosts (reviewed by Griffin 1970). In spite of considerable effort we seem no closer to understanding the situation. Studies of swarming behavior at hibernacula in North America have clarified the problem to some extent (Fenton 1969). Swarming involves precopulatory tours through hibernacula, and populations include much higher percentages of females than are known from many hibernating populations, suggesting that differential mortality (Hitchcock 1965) may not account for the summer-winter discrepancy.

The disappearance of females, coupled with their propensity for hibernating in smaller clusters, and often at different locations within hibernacula (*e.g.*, Davis and Hitchcock 1965, Fenton 1970) may indicate dispersal to as yet undiscovered hibernacula, perhaps sites not extensively used by males. McNab (1974) suggested that selection of cooler sites by hibernating females might reduce competition. If females successfully exploit sites not used by males, they could reduce the chances of unwanted matings and achieve some protection of investment in mate choice. This scenario is compatible with available data on structure (sex ratios) of swarming and hibernating populations, and with the data of Thomas *et al.* (1979) on mating behavior, for it is possible females that had mated more than once could have done so with the same male. The relatively few females that hibernate with males in many known hibernacula may be exceptional. However, the incidence of females in hibernacula rises sharply toward the center of the species' range (Fenton 1970, Humphrey and Cope 1976).

The idea of as yet undiscovered hibernacula filled with hibernating females is made less attractive by recent observations that *M. lucifugus* use echolocation calls of conspecifics to locate hibernation sites (Barclay 1982). These bats may not have a rich repertoire of vocalizations (Barclay *et al.* 1979), but they are very gregarious and individuals "eavesdrop" on others to find localized resources.

V. WHAT NEXT?

I suspect that the current state of the art in bat studies is not likely to quickly produce data on sperm competition. Further, a number of complications suggest that *M. lucifugus* may not be the best candidate species for sperm competition studies (see for example Dewsbury, this volume). However, it is clear that the subject species should be one with a promiscuous mating system, and to date *M. lucifugus* and *P. pipistrellus* are obvious choices.

There are other species from temperate areas whose mating systems are unknown but which do not appear to form vaginal plugs. Some, such as *Myotis leibii*, commonly hibernate in narrow crevices that may provide protection from unsolicited copulations. Others, such as *Eptesicus fuscus*, have no obvious means of avoiding unsolicited copulations, are widespread, common, and large enough to carry small radio transmitters (Bradbury *et al.* 1979, Thomas 1979), but *E. fuscus* has problems of temporal and spatial separation of mating and parturition.

A recent study of the small (3 g) *Pipistrellus nanus* in Kenya (O'Shea 1980) suggests that these bats may store sperm, but does not permit clear identification of sperm storage or a promiscuous mating system. However, the fact that sperm storage appears to be relatively common at least among tropical vespertilionids, suggests that those interested in an unequivocal demonstration of sperm competition may find a suitable experimental subject in tropical or subtropical habitats.

VI. SUMMARY

Females of some temperate species of rhinolophids and temperate and tropical vespertilionids store and nurture spermatozoa for periods ranging from 16 to almost 200 days. In temperate species, this sperm storage may be an adaptation to seasonal climatic extremes and hibernation. The adaptive significance of sperm storage in tropical species is unknown. In any case, multiple mating and long-term storage of sperm by females creates an extraordinary potential for sperm competition in bats among the mammals.

In some cases the female reproductive tract is partially or completely blocked after copulation by a vaginal plug of male or female origin. Not all sperm-storing species form plugs, and in *Myotis lucifugus,* which lacks a plug, mating seems to be promiscuous. In fact, promiscuity typifies species that lack plugs, making them especially suitable candidates for studies on dynamic sperm competition.

Myotis lucifugus is extraordinary in that most mating takes place during the fall prior to hibernation, but some males wake up to mate with hibernating bats during the winter. This system suggests the hypothesis that males at competitive disadvantage in the fall mating may obtain competitive advantage for their sperm by arousal and "sneak" matings with hibernating females in the winter.

For a variety of reasons, it is currently difficult to study the phenomenon of sperm competition in bats. *Myotis lucifugus,* though a prime candidate, may be a particularly problematical species for sperm competition studies.

ACKNOWLEDGMENTS

I am particularly grateful to Paul A. Racey who carefully examined an earlier draft of this manuscript and made many helpful suggestions. The comments of W. A. Wimsatt and D. W. Thomas were also valuable and I thank them for their effort on my behalf. My research on bats has been generously supported by grants from the National Sciences and Engineering Research Council of Canada.

REFERENCES

Barclay, R. M. R. 1982. Interindividual use of echolocation calls: Eavesdropping by bats. *Behav. Ecol. Sociobiol.* **10**:271-275.

Barclay, R. M. R., M. B. Fenton, and D. W. Thomas. 1979. Social behavior of the little brown bat, *Myotis lucifugus:* II. Vocal communication. *Behav. Ecol. Sociobiol.* **6**:137-146.

Barclay, R. M. R., and D. W. Thomas. 1979. Copulation call of *Myotis lucifugus:* A discrete, situation-specific communication signal. *J. Mammal.* **60**:632-634.

Barbour, R. W., and W. H. Davis. 1969. *Bats of America.* University of Kentucky Press, Lexington, KY.

Bernard, R. T. F. 1980. Female reproduction in five species of Natal cave-dwelling Microchiroptera. Ph.D. Dissertation, University of Natal, Pietermaritzburg, South Africa.

Birdsall, D. A., and D. Nash. 1973. Occurrence of successful multiple insemination of females in natural populations of deer mice *(Peromyscus maniculatus).* *Evolution* **27**:106-110.

Bradbury, J. W. 1977. Social orgaization and communication. In *Biology of Bats, Vol. 3,* W. A. Wimsatt (ed.), pp. 2-72. Academic Press, New York.

Bradbury, J. W., D. Morrison, E. Stashko, and R. Heithaus. 1979. Radio-tracking methods for bats. *Bat Res. News* **20**:9-17.

Buchler, E. R. 1980. The development of flight, foraging, and echolocation in the little brown bat *(Myotis lucifugus).* *Behav. Ecol. Sociobiol.* **6**:211-218.

Courrier, R. 1927. Etude sur le déterminism des caractères sexuels secondaires chez quelques mammifères à l'activité testiculaires périodique. *Arch. Biol., Paris* **37**:173-334.

Davis, W. H., and H. B. Hitchcock. 1965. Biology and migration of the bat, *Myotis lucifugus* in New England. *J. Mammal.* **46**:296-313.

Eckstein, P., and S. Zuckerman. 1956. The oestrus cycle in the Mammalia. In *Marshall's Physiology of Reproduction I,* A. S. Parkes (ed.), pp. 78-80. Longmans, London.

Fenton, M. B. 1969. Summer activity of *Myotis lucifugus* (Chiroptera:Vespertilionidae) at hibernacula in Ontario and Quebec. *Can. J. Zool.* **47**:596-602.

Fenton, M. B. 1970. Population studies of *Myotis lucifugus* (Chiroptera:Vespertilionidae) in Ontario. *Life Sci. Contrib. R. Ont. Mus.* **77**:1-34.

Gaisler, J. 1966. Reproduction in the lesser horseshoe bat *(Rhinolophus hipposideros,* Bechstein 1800). *Bijdr. Dierkd.* **36**:45-64.

Gates, W. H. 1936. Keeping bats in captivity. *J. Mammal.* **17**:268-273.

Gopalakrishna, A., and A. Madhavan. 1971. Survival of spermatozoa in the female genital tract of the Indian vespertilionid bat, *Pipistrellus ceylonicus chrysothrix* (Wroughton). *Proc. Indian Acad. Sci. (B)* **73**:43-49.

Gopalakrishna, A., and A. Madhavan. 1978. Variability of inseminated spermatozoa in the Indian vespertilionid bat *Scotophilus heathii* (Horesfield). *Indian J. Exp. Biol.* **16**:852-854.

Grosser, O. 1903. Die physiologische bindegewebige Atresie des Genitalkanales von *Vesperugo noctula* nach erfolgter Kohabitation. *Verh. Anat. Ges.* **17**:129-132.

Green, R. A. 1965. Observations on the little brown bat, *Eptesicus pumilus* (Gray) in Tasmania. *Rec. Queen Victoria Mus.* **10**:1-16.

Griffin, D. R. 1970. Migrations and homing of bats. In *Biology of Bats, Vol. 1,* W. A. Wimsatt (ed.), pp. 233-265. Academic Press, New York.

Gustafson, A. W., and M. Shemesh. 1976. Changes in plasma testosterone levels during the annual reproductive cycle of the hibernating bat, *Myotis lucifugus,* with a survey of plasma testosterone levels in adult male vertebrates. *Biol. Reprod.* **5**:9-24.

Guthrie, M. J. 1933. The reproductive cycle of some cave bats. *J. Mammal.* **14**:199-216.

Guthrie, M. J., and K. R. Jeffers. 1938. The ovaries of the bat, *Myotis lucifugus,* after injection of hypophyseal extract. *Anat. Rec.* **72**:11-36.

Hanken, J., and P. W. Sherman. 1981. Multiple paternity in Belding's ground squirrel litters. *Science* **212**:351-353.

Harrison Mathews, L. 1937. The female sexual cycle in British bats *Rhinolophus ferrumequinum insulans* Barrett-Hamilton and *R. hipposideros minutus* Montagu. *Trans. Zool. Soc. Lond.* **23**:224-255.

Hartman, C. G. 1933. On the survival of spermatozoa in the female tract of the bat. *Q. Rev. Biol.* 8:185-193.

Hartman, C. G., and W. K. Guyler. 1927. Is the supposed long life of the bat spermatozoa fact or fable? *Anat. Rec.* 35:39.

Haraiwa, Y. K., and T. Uchida. 1956. Fertilization capacity of spermatozoa stored in the uterus after copulation in the fall. *Sci. Bull. Fac. Agric. Kyushu Univ.* 31:565-574.

Hitchcock, H. B. 1965. Twenty-three years of bat banding in Ontario and Quebec. *Can. Field-Nat.* 79:4-14.

Humphrey, S. R., and J. B. Cope. 1976. *Population Ecology of the Little Brown Bat, Myotis lucifugus, in Indiana and Southcentral Kentucky.* Am. Soc. Mammalogists, Stillwater, OK.

Kitchener, D. J. 1975. Reproduction in female Gould's wattled bat, *Chalinolobus gouldii* (Gray) (Vespertilionidae), in western Australia. *Aust. J. Zool.* 23:257-267.

Kitchener, D. J., and S. A. Halse. 1978. Reproduction in female *Eptesicus regulus* (Thomas) (Vespertilionidae) in south-western Australia. *Aust. J. Zool.* 26:257-267.

Kleiman, D. G., and P. A. Racey. 1969. Observations on noctule bats *(Nyctalus noctula)* breeding in captivity. *Lynx* 10:65-77.

Krishna, A., and C. J. Dominic. 1978. Storage of sperm in the female genital tract of the Indian vespertilionid bat, *Scotophilus heathii* Horesfield. *J. Reprod. Fertil.* 56:319-321.

Krutzsch, P. H. 1975. Reproduction of the canyon bat, *Pipistrellus hesperus,* in south-western United States. *Am. J. Anat.* 143:163-200.

Kunz, T. H. 1980. Daily energy budgets of free-living bats. *Proc. Int. Bat Res. Conf., Lubbock* 5:369-392.

Likhachev, G. N. 1961. Use by bats of bird nesting boxes. *Prioksho-Terranish Gos. Zapovdnika* 3:85-156; *Ref. Zh., Biol.* 19:302 (1962).

McCracken, G. F., and J. W. Bradbury. 1977. Paternity and genetic heterogeneity in the polygynous bat, *Phyllostomus hastatus. Science* 198:303-306.

McCracken, G. F., and J. W. Bradbury. 1981. Social organization and kinship in the polygynous bat, *Phyllostomus discolor. Behav. Ecol. Sociobiol.* 8:11-34.

McNab, B. K. 1974. The behavior of temperate cave bats in a subtropical environment. *Ecology* 55:943-958.

Myers, P. 1977. Patterns of reproduction of four species of vespertilionid bats in Paraguay. *Univ. Calif. Publ. Zool.* 107:1-41.

Ognev, S. I. 1928. *Mammals of Eastern Europe and Northern Asia, Vol. 1.* Trans. Israel Program for Scientific Translations, Jerusalem (1962).

Orr, R. T. 1954. Natural history of the pallid bat, *Antrozous pallidus* (LeConte). *Proc. Calif. Acad. Sci.* 28:165-246.

O'Shea, T. J. 1980. Roosting, social organization and the annual cycle in a Kenya population of the bat, *Pipistrellus nanus. Z. Tierpsychol.* 53:171-195.

Pagenstecher, H. A. 1859. Über die Begattung von *Vesperugo pipistrellus. Verh. Naturh.-Med. Ver. Heideb.* 1:194-195.

Pearson, O. P., M. R. Koford, and A. K. Pearson. 1952. Reproduction of the lump-nosed bat *(Corynorhinus rafinesquei)* in California. *J. Mammal.* 33:273-320.

Porter, F. L. 1979. Social behavior in the leaf-nosed bat, *Carollia perspicillata:* I. Social organization. *Z. Tierpsychol.* 49:406-417.

Quay, W. B. 1976. Seasonal cycle and physiological correlates of pinealocyte nuclear and nucleolar diameters in the bats *Myotis lucifugus* and *Myotis sodalis. Gen. Comp. Endocrinol.* 29:369-375.

Racey, P. A. 1973. The viability of spermatozoa after prolonged storage by male and female European bats. *Period. Biol.* 75:201-205.

Racey, P. A. 1975. The prolonged survival of spermatozoa in bats. In *The Biology of the Male Gamete,* J. G. Duckett and P. A. Racey (eds.), pp. 385-416. Academic Press, London.

Racey, P. A. 1979. The prolonged storage and survival of spermatozoa in Chiroptera. *J. Reprod. Fertil.* 56:403-416.

Racey, P. A., and D. M. Potts. 1970. Relationship between stored spermatozoa and the uterine epithelium in the pipistrelle bat *(Pipistrellus pipistrellus). J. Reprod. Fertil.* 22:57-63.

Racey, P. A., R. Suzuki, and Lord Medway. 1975. The relationship between stored spermatozoa and the oviducal epithelium of the genus *Tylonycteris*. In *The Biology of Spermatozoa: Transport, Survival and Fertilizing Capacity*, E. S. E. Hafex and C. G. Thibault (eds.), pp. 123-133. Karger, Basel.

Ransome, R. D. 1978. Spatial distribution of different age and sex groups in populations of *Rhinolophus ferrumequinum*. *Bat Res. News* **19**:96.

Ransome, R. D. 1980. *The Greater Horseshoe Bat*. Blandford Press, Poole, England.

Rice, D. W. 1957. Life history and ecology of *Myotis austroriparius* in Florida. *J. Mammal.* **38**:15-32.

Sluiter, J. W., and P. F. van Heerdt. 1966. Seasonal habits of the noctule bat (*Nyctalus noctula*). *Arch. Neerl. Zool.* **16**:423-435.

Thomas, D. W. 1978. Aspects of the social and mating behaviour of the bat, *Myotis lucifugus* (Chiroptera:Vespertilionidae) at hibernacula in Ontario. M.S. Thesis, Carleton University, Ottawa, Canada.

Thomas, D. W. 1979. Plans for a lightweight, inexpensive radio transmitter. In *A Handbook on Biotelemetry and Radio Tracking*, C. J. Amlaner, Jr. and D. W. MacDonald (eds.), pp. 175-179. Pergamon Press, Oxford.

Thomas, D. W., M. B. Fenton, and R. M. R. Barclay. 1979. Social behavior of the little brown bat, *Myotis lucifugus*: I. Mating behavior. *Behav. Ecol. Sociobiol.* **6**:129-136.

Thomson, C. E. 1980. Mother-infant interactions in free-living little brown bats, *Myotis lucifugus* (Chiroptera:Vespertilionidae). M.S. Thesis, Carleton University, Ottawa, Canada.

van Heerdt, P. F., and J. W. Sluiter. 1965. Notes on the distribution and behavior of the noctule bat (*Nyctalus noctula*), in the Netherlands. *Mammalia* **29**:463-477.

Vaughan, T. A. 1978. *Mammalogy, 2nd ed.* W. B. Saunders, Philadelphia.

Wimsatt, W. A. 1942. Survival of spermatozoa in the female reproductive tract of the bat. *Anat. Rec.* **83**:299-307.

Wimsatt, W. A. 1944. Further studies on the survival of spermatozoa in the female reproductive tract of the bat. *Anat. Rec.* **88**:193-204.

Wimsatt, W. A. 1945. Notes on breeding behavior, pregnancy, and parturition in some vespertilionid bats of the eastern United States. *J. Mammal.* **26**:23-33.

Wimsatt, W. A. 1969. Some interrelations of reproduction and hibernation in mammals. *Symp. Soc. Exp. Biol.* **23**:511-549.

18

Sperm Competition, Testes Size, and Breeding Systems in Primates

PAUL H. HARVEY

A. H. HARCOURT

I. INTRODUCTION

In *The Descent of Man and Selection in Relation to Sex,* Darwin (1871) developed his ideas about sexual selection. He sought to differentiate between two modes of sexual selection which were later called intrasexual selection and intersexual (or epigamic) selection by Huxley (1938). If sexual selection rests on differential abilities to acquire mates (Wilson 1975), we must also recognize another level of sexual competition: the differential ability to acquire reproductive success after mating.

These different types of selection might be expected to account for most, though not all (Selander 1966, 1972), of the sexual dimorphism seen in nature. But how can we assess the relative importance of each in a particular case? It is usually impractical to test evolutionary hypotheses by experimental manipulation.

Sperm Competition and the Evolution
of Animal Mating Systems

589

However such hypotheses claim generality because, when similar selective forces act on independently evolving lineages, we might expect convergent evolution to produce traits that are correlated in similar ways within different taxonomic groups. Hypotheses which propose that sexual dimorphism results from adaptive evolution can, then, be tested by examining relationships among sets of characters in a variety of taxa. We shall use this comparative method to test hypotheses concerning one aspect of sexual dimorphism, testes size, among primates.

One component of selection in mammals results from competition among spermatozoa within the reproductive tract of a female. Despite occasional exceptions such as decreased motility of diploid sperm (Mortimer 1977, 1979), the lack of haploid gene expression in mammals (Beatty and Gleucksohn-Wallsch 1972) means that selection among the sperm from a single male does not generally result in adaptive evolution. However, competition among the spermatozoa from the ejaculates of different males is likely to have evolutionary significance.

Multiple paternity within litters has often been demonstrated in laboratory animals (e.g., Whitney 1940; Levine 1958, 1967; Beatty 1960; Napier 1961; Dewsbury and Hartung 1980; Lanier et al. 1979; Dewsbury, this volume), but evidence from natural populations is sparse and seems limited to two cases (Birdsall and Nash 1973, Hanken and Sherman 1981). The latter study on Belding's ground squirrel (Spermophilus beldingi) indicates multiple paternity in at least 55%, and probably nearer 80% of litters. Selection among males for success in sperm competition is, therefore, potentially enormous. Apart from intrasexual selection acting to reduce access by competing males to ovulating females, selection may favor a variety of physiological devices that increase the chances of outcompeting the spermatozoa from other males. One simple mechanism would be to increase the probability of successful fertilization by sheer weight of numbers (Parker, this volume). That is, when females are likely to be mated by more than one male during the same estrous cycle, males might be selected to inseminate more sperm.

Among mammals, spermatozoa are produced by the seminiferous tubules in the testes. Since sperm production per spermatogenic cell remains roughly constant across species, we might expect males belonging to species where females are likely to mate with more than one male per estrus to have larger testes than those where mating access is restricted to a single male. But, this would be too simplistic a view (Short 1981). Testes serve two major functions: They produce spermatozoa and they are endocrine organs. Among mammals with the same breeding system, we would expect larger species to have larger testes in order to produce more spermatozoa counteracting the dilution effect of larger reproductive tracts. Also, larger testes would be required to produce more hormones for maintenance of threshold concentrations (Short 1981). Other reproductive and endocrine organs also increase with body size (Brody 1945). However, as with most other allometric (size-dependent) relationships (Gould 1966), the expected quantitative increase of testes size with body size cannot easily be predicted. As a consequence we have to work with empirical relationships (Harvey and Mace 1982).

In a series of papers on the great apes (Pongidae) and man, Short (1977, 1979, 1981) has argued that variation in testes size is partly a consequence of species differences in breeding system. Although he was only dealing with four species, comparisons among them revealed striking differences. For instance male chimpanzees (*Pan troglodytes*) are about one-quarter the weight of male gorillas (*Gorilla gorilla*) and yet their testes are about four times heavier. At least 25% of chimpanzee conceptions occur after an estrus in which the female mated more than one male (Tutin 1980), while gorillas are characterized by a breeding system in which estrous females mate only one male (Harcourt 1981). The differences are in the predicted direction. Short (1979) wrote that "it remains to be seen whether the significance of these simple anatomical clues will be confirmed by examination of a far wider range of species." Within the mammals, primates are a particularly convenient group; body weights and testes weights are known for at least 33 species from 21 genera and 8 families in the order (Table I). For almost all these species, recent field studies not only provide information on group structure (the numbers of adult and subadult males and females), but also on breeding system (whether females are likely to mate with more than one male during estrus) Breeding systems are fairly well distributed among families so that taxonomic effects can be examined. For example, certain families of primates might contain animals with particularly large testes for their body sizes as a consequence of phylogenetic inertia (*sensu* Wilson 1975).

II. METHODS

Data on body weights, testes weights, and breeding systems were obtained from the literature or were unpublished results provided by research workers (see Harcourt *et al.* 1981, and Table I). For weights, only data on mature males are included, and all body weights are from animals whose testes were weighed. When more than one individual of a species was available, the testes weight of the animals with median body weight was used except for the squirrel monkey (*Saimiri sciureus*) and the Hanuman langur (*Presbytis entellus*) where the original sources required use of the mean, and midpoint of extreme values for the two, respectively. We stress that **breeding system and not social system or group structure** is used throughout this chapter. Species such as the gorilla (Harcourt 1981) live in multi-male groups but have single-male breeding systems (*i.e.*, only one male in the group usually mates with estrous females); as a consequence, they are classified as single-male. The three categories of breeding system used are single male multi-female, multi-male multi-female, and monogamous.

Previous analyses of testes size in primates (Schultz 1938, Collins 1978, Short 1979) have used either absolute testes weight and ignored body size effects, or

TABLE I

Measures of Body Weight, Combined Testes Weight (excluding epididymis),
Relative Testes Size (*i.e.,* observed/expected for a given body weight),
and Mating System of Primates. – (w) = wild caught; (c) = captive specimens;
P = monogamous; M = multi-male; S = single male breeding system (see text).
Data sources are given in Harcourt *et al.* (1981).

Families and Species	N	Body Weight (kg)	Testes Weight (g)	Relative Testes Size*	Breeding System
FAMILY LORISIDAE					
Loris tardigradus (w)	114	0.27	1.8	1.39	?
FAMILY CALLITRICHIDAE					
Callithrix jacchus (c)	15	0.32	1.3	0.88	P
Saguinus oedipus (w)	56	0.52	3.4	1.64	P
FAMILY CEBIDAE					
Saimiri sciureus (w)	40	0.78	3.2	1.17	M
Aotus trivirgatus (w)	1	1.02	1.2	0.37	P
Lagothrix lagotricha (c)	1	5.22	11.2	1.17	M
Alouatta palliata (w)	8	7.26	23.0	1.93	M
Ateles geoffroyi (w)	1	7.94	13.4	1.06	M
FAMILY CERCOPITHECIDAE					
SUBFAMILY CERCOPITHECINAE					
Cercopithecus aethiops (w)	6	4.95	13.0	1.41	M
Cercocebus atys (c)	1	8.68	25.1	1.93	?
Macaca fascicularis (w,c)	20	4.42	35.2	3.71	M
Macaca radiata (w,c)	2	8.65	48.2	3.71	M
Macaca mulatta (c)	26	9.20	46.2	3.71	M
Macaca nemestrina (w)	1	9.98	66.7	3.71	M
Macaca arctoidea (c)	20	10.51	48.2	3.71	M
Papio hamadryas (w)	6	20.17	27.1	1.16	S
Papio cynocephalus (c)	21	24.32	52.0	2.54	M
Papio anubis (w)	4	26.40	93.5	2.54	M
Papio ursinus (?)	1	31.75	72.0	2.54	M
Papio papio (w)	1	31.98	88.9	2.54	M
Theropithecus gelada (w)	1	20.40	17.1	0.72	S
SUBFAMILY COLOBINAE					
Presbytis rubicunda (w)	12	6.23	3.4	0.33	?
Presbytis cristata (w)	12	6.58	6.2	0.49	S
Presbytis obscura (w)	14	7.45	4.8	0.49	S
Presbytis entellus (w)	6	17.00	11.1	0.49	S
Colobus polykomos					
(= guereza) (w)	3	10.25	10.7	0.72	S
Nasalis larvatus (w)	8	20.64	11.8	0.50	S
FAMILY HYLOBATIDAE					
Hylobates moloch (w)	7	5.44	6.1	0.59	P
Hylobates lar (c)	1	5.50	5.5	0.59	P
FAMILY PONGIDAE					
Pan troglodytes (c)	3	44.34	118.8	3.00	M
Pongo pygmaeus (w)	2	74.64	35.3	0.63	S
Gorilla gorilla (w)	2	169.00	29.6	0.31	S
FAMILY HOMINIDAE					
Homo sapiens	4	65.65	40.5	0.79	S/P

*"generic" (see text).

testes weight as a proportion of body weight and ignored allometric effects. Most organs increase with an exponential relationship to body weight when viewed across taxa (Huxley 1932, Brody 1945) so that: $y = ax^b$, where y is organ weight, x is body weight, and a and b are constants. For analysis, logarithmic transformations are used to produce a linear relationship: $\log(y) = \log(a) + b\log(x)$, where b is the slope of $\log(y)$ plotted against $\log(x)$. For interspecies studies, distributions of y and x are usually lognormal before transformation and so normality is also achieved when the data are log-transformed (Harvey 1982). When b is greater than one, then the relationship is said to be **positively allometric**, and when b is less than one, it is **negatively allometric**. It is only when b equals one that proportions remain constant with changes in body weight.

A convenient way of removing allometric relationships is to examine deviations from some best-fit line, thus producing measures of relative testes size. A best-fit line is defined as that which minimizes deviations about itself, and two considerations dictate which line to use. First, since both body weight and testes weight are not free of statistical error, and because they are not dependent and independent variables in the the usual sense of the terms (Kermack and Haldane 1950), regression analysis is not appropriate. Major axis analysis is more suitable (Jolicoeur and Mosimann 1968, Harvey and Mace 1982). Second, species with different breeding systems have different ranges of body weights (see Clutton-Brock and Harvey 1979, and Harvey and Mace 1982).

As a consequence, major axis slopes are calculated separately for each breeding system. As long as slopes are not heterogeneous, we can estimate the common "average" slope and draw this through the mean on both axes. Deviations from this line, drawn at right-angles to the body weight axis, estimate testes size with body weight effects removed. These values are approximately normally distributed and are used for statistical analysis in the results section. However, they provide measures of the log of observed testes size/expected testes size. A more easily interpreted number is the antilogarithm of such values, *i.e.*, observed testes size/ expected testes size, and we have quoted these values in Table I and used them to produce Fig. 2 (see below). The values are referred to as relative testes size (RTS).

Analysis of species values produced RTS measures that had body weights removed, but retained taxonomic effects. This was because species in a number of genera have very similar testes and body weights. When the analysis was performed at the generic level both body weight and taxonomic effects were removed as shown below.

Generic values were calculated from the means of the logarithmically transformed species values, with the proviso that congenerics were kept separate when they differed in breeding system. Thus the "generic" classification used did not necessarily coincide with taxonomic status: The Hamadryas baboon (*Papio hamadryas*) was placed in a separate "genus" from the other baboons (*Papio* spp.) because of its single-male breeding system compared with their multi-male systems. We consider only "generic" results for the rest of this chapter.

III. RESULTS

A maximum likelihood test (Harvey and Mace 1982) revealed no significant heterogeneity ($\chi^2_2 = 3.10$) among the slopes of logarithmic plots of testes weight on body weight for the three breeding systems. Thus, use of a common slope was permissible. It had a negatively allometric value of 0.66, and a line with this slope is fitted through the mean of all data points on Fig. 1. Deviations are calculated from this line. Within breeding systems, there is no heterogeneity of deviations from the line among families (monogamous, $F_{2,1} = 2.57$; single-male, $F_{2,4} = 2.43$; multi-male, $F_{2,5} = 3.33$; by one-way analyses of variance. We could not perform an overall two-way analysis of variance for families and breeding systems combined because all families lacked "genera" belonging to at least one breeding system). Therefore, any difference in RTS among "genera" belonging to different

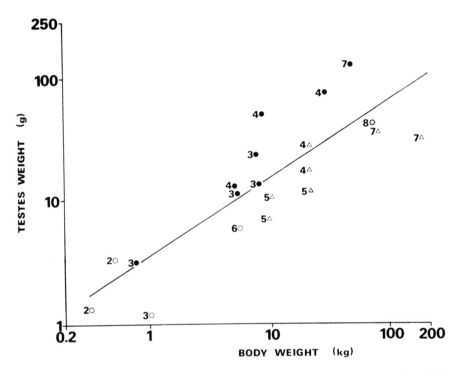

Fig. 1. Combined testes weight (g) plotted against body weight (kg) on a logarithmic scale for different primate "genera" (see text). ● = multi-male, △ = single-male, ○ = monogamous breeding systems; 2 = Callitrichidae, 3 = Cebidae, 4 = Cercopithecinae, 5 = Colobinae, 6 = Hylobatidae, 7 = Pongidae, 8 = Hominidae (non-multi-male). 4 and 5 are amalgamated for statistical analysis.

breeding systems is unlikely to be due to taxonomic effects. Among breeding systems, as predicted, significant heterogeneity of deviations from the common line is evident ($F_{2,16}$ = 10.29, p $<$ 0.05). Both monogamous and single-male "genera" have small RTS compared with multi-male "genera" (t_6 = 2.45, p $<$ 0.05 and t_{15} = 4.76, p $<$ 0.01 respectively), while RTS does not differ significantly between monogamous and single-male "genera" (t_5 = 0.62).

Records were available from more than ten individuals for eight species in this study, and for three of these species there were significant positive correlations between testes weight and body weight (rank correlation, p $<$ 0.05; median N = 17.5). This correlation would be expected if larger individuals within a species have larger reproductive tracts and require a greater absolute amount of hormone to maintain equivalent hormone titer (see above).

IV. DISCUSSION

A. Testes Size

Although we were unable to predict the magnitude of testes size increase, our predictions were qualitatively correct. Testes size increased with body size and the relationship was negatively allometric. For their body size, "genera" in which females are likely to mate with more than one male during estrus have larger testes than those where only one male usually gains access.

The hypothesis that sperm competition is important in the correlation between breeding system and testes size requires that multi-male primates, as well as having larger testes, have higher rates of sperm production than single-male or monogamous species. In other words, we need to show that the larger RTS is not merely due to an increase in non-spermatogenic tissue. If the value of the seminiferous tubules is applied as a measure of potential sperm production, our hypothesis is supported. Chimpanzees, baboons, and macaques (*Macaca* spp.), all with multi-male breeding systems, have ratios of tubules to connective tissue in their testes that range from 2.2:1 to 2.8:1; while man, gibbons (*Hylobates* spp.), and langurs (*Presbytis* spp.), which are not multi-male have ratios of 0.9:1 to 1.3:1. In addition, the sperm production rate of the rhesus macaque (*Macaca mulatta*) is about 23 x 10^6, while the corresponding value for man is only 4.4 x 10^6 (Amann *et al.* 1976, Amann and Howards 1980). These data suggest that multi-male species not only have larger testes but also have a higher sperm production capacity per unit weight of testis tissue. Our measure of RTS may, therefore, underestimate the effects of sperm competition as a selective force increasing sperm production rates.

Nevertheless, sperm production is unlikely to be the only reason for differences in RTS among primates. For example, seasonality of breeding could be an important factor; species with a short breeding season and hence periods of intense copulatory activity may need larger testes. Certainly, the multi-male genus *Macaca* which contains predominantly seasonal breeders has the largest RTS. In this connection, it is interesting to note that the single-male "genera" do not have larger RTS than monogamous primates (on average it is rather less), and sperm production does not, therefore, seem to relate to the number of females that a male serves during the year.

As with all comparative studies, aberrant taxa are revealed. *Saguinus oedipus* has rather larger testes for its body size and a monogamous breeding system; likewise for the Hamadryas baboon with its single-male breeding system. Nevertheless, this study reveals a consistent trend and emphasizes the need for similar studies on other mammalian groups. Among primates, further progress will require studies on the fine structure of primate testes, on sperm production rates and on copulation frequencies and timing in natural populations. Even then, controlled experiments using genetic markers will be necessary to validate correlative evidence.

B. Intrasexual Selection and Sperm Competition

Recent surveys of sexual selection (*e.g.,* Halliday 1978) restrict themselves to the traditional distinction made between intrasexual and epigamic selection for access to mates. This chapter has so far considered one component of **post-copulatory** intrasexual selection, sperm competition (Short 1977). Epigamic selection concerning active female choice on the male phenotype is particularly difficult to identify. For instance, in many primate species females develop sexual swellings when they are in estrus. It is generally accepted that these serve to attract males. These species, with one exception, live in multi-male troops and whether sexual swellings are associated with active choice of male (epigamic selection) or passive "choice" (provoking inter-male conflict so that the dominant male achieves access to the female or where males from neighboring troops are enticed to compete with resident males for access) is not known. However, recent comparative studies on primates have indicated a strong influence of intrasexual selection on both sexual dimorphism in body size (Clutton-Brock *et al.* 1977) and canine tooth size (Harvey *et al.* 1978). In this section we compare the effects of intrasexual selection and sperm competition on primates belonging to "genera" with different breeding systems.

Intrasexual competition for mates favors increased fighting ability among males and, therefore, increased body size and canine size in polygynous species. Monogamous primates are expected to be less sexually dimorphic for both body size and canine size. The outcome of intrasexual competition among polygynous male primates may be access to a single-male or a multi-male breeding group. In the former breeding system, sperm competition is almost non-existent since one male usually monopolizes all the females, while sperm competition is expected among multi-male primates because more than one male commonly has access to each estrous female. Consequently, body size dimorphism and relative canine size (RCS), a similar measure to relative testes size, should be highest among polygynous primates (multi-male and single-male) and lowest in monogamous primates. This contrasts with RTS which is high among multi-male and low in both single-male and monogamous primates.

Fig. 2 uses data from Clutton-Brock and Harvey (1977), and Harvey *et al.* (1978) together with those from more recent studies (Table I, and see Harcourt *et al.* 1981) which correct the breeding system of some species (*e.g., Gorilla gorilla* and *Nasalis larvatus*). Unlike Clutton-Brock *et al.* (1978) and Harvey *et al.* (1978) the analyses are performed at the "generic" level. The figure attests to the different expected outcomes of intrasexual competition for mates and sperm competition.

C. Correlates and Causes of Sexual Selection

Convergent evolution might be expected to produce repeated occurrences of the same functionally related constellations of characters in different taxonomic groups. Such clusters of characters that are usually found occurring together may ultimately be related to environmental variation, for example food dispersion or predator occurrence (Clutton-Brock and Harvey 1978). A simple causal chain of argument is that, among primates, the dispersion of food determines female grouping patterns and males respond to and influence these grouping patterns in their attepts to gain mating access to females (Wrangham 1980). Cross-species statistical comparisons do not reveal either a unitary pattern of female grouping in relation to food sources, or enivronmental correlates of single-male versus multi-male breeding systems (Clutton-Brock and Harvey 1977), and the source(s) of environmental variation that lead ultimately to sperm competition among primates remain a matter for speculation.

V. SUMMARY

Among primates, sperm competition is likely to be important in species where females are routinely mated by more than one male during estrous. These species are defined as having a multi-male breeding system.

Testes size varies among primates and shows a negatively allometric relationship with body size. Because sperm competition is likely to be important, it was predicted that multi-male primates would have larger testes for their body size than harem (*i.e.,* single-male, multi-female) or monogamous primates.

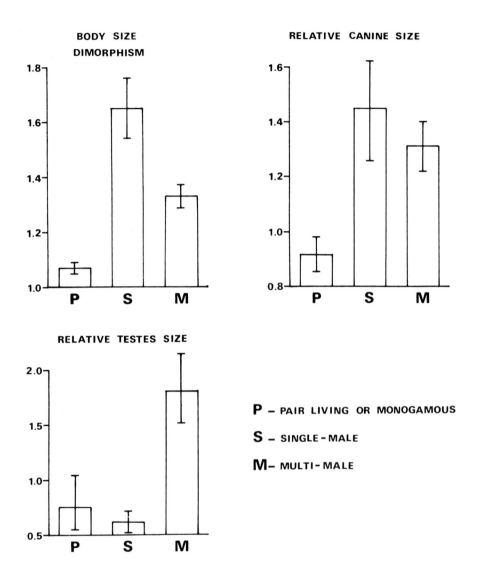

Fig. 2. Body size dimorphism, relative canine size and relative testes size for primates belonging to different breeding systems. All analyses are "generic" (see text). Sample sizes for P, S and M respectively: body size dimorphism 12, 9, 14; relative canine size 4, 7, 9; relative testes size 4, 7, 8. Bars indicate one standard error in each direction from the mean.

Comparative data on body size, testes size, and breeding system confirm the prediction that multi-male primates have large testes for their body size.

These findings are compared with breeding system correlates of body size dimorphism and relative canine size, and the pattern is shown to be different, so emphasizing the different expected outcomes of intrasexual selection for mates and sperm competition.

ACKNOWLEDGMENTS

We thank Professor Roger V. Short F.R.S. for his seminal ideas and discussions.

REFERENCES

Amann, R. P., L. Johnson, D. L. Thompson, and B. W. Pickett. 1976. Daily spermatozoal production, epididymal spermatozoal reserves and transit times of spermatozoa through the epididymis of the rhesus monkey. *Biol. Reprod.* 15:586-592.

Amann, R. P., and S. S. Howards. 1980. Daily spermatozoal production and epididymal spermatozoal reserves of the human male. *J. Urol.* 124:211-215.

Beatty, R. A. 1960. Fertility of mixed semen from different rabbits. *J. Reprod. Fertil.* 1: 52-60.

Beatty, R. A., and S. Gluecksohn-Wallsch. 1972. *Edinburgh Symposium on the Genetics of the Spermatozoon.* R. A. Beatty and S. Gluecksohn-Wallsch (eds.). Edinburgh and New York.

Birdsall, D. A. and D. Nash. 1973. Occurrence of sucessful multiple insemination of females in natural populations of deer mice *(Peromyscus maniculatus). Evolution* 27:106-110.

Brody, S. D. 1945. *Bioenergetics and Growth.* Collier, MacMillan, London.

Clutton-Brock, T. H., and P. H. Harvey. 1977. Primate ecology and social organization. *J. Zool. Lond.* 183:1-39.

Clutton-Brock, T. H., and P. H. Harvey. 1978. Mammals, resources and reproductive strategies. *Nature* 273:191-195.

Clutton-Brock, T. H., and P. H. Harvey. 1979. Comparison and adaptation. *Proc. R. Soc. Lond. B. Biol.* 205:547-565.

Clutton-Brock, T. H., P. H. Harvey, and B. Rudder. 1978. Sexual dimorphism, socionomic sex ratio and body weight in primates. *Nature* 269:797-800.

Collins, A. 1978. Why do some baboons have red bottoms? *New Sci.* 78:12-14.

Darwin, C. 1871. *The Descent of Man, and Selection in Relation to Sex.* John Murray, London.

Dewsbury, D. A., and T. C. Hartung. 1980. Copulatory behaviour and differential reproduction of laboratory rats in a 2-male 1-female competitive situation. *Anim. Behav.* 28:95-102.

Gould, S. J. 1966. Allometry and size in ontogeny and phylogeny. *Biol. Rev. Camb. Phil. Soc.* 41:587-640.

Halliday, T. R. 1978. Sexual selection and mate choice. In *Behavioural Ecology, an Evolutionary Approach*, J. R. Krebs and N. B. Davies (eds.), pp. 180-214. Blackwell, Oxford.

Hanken, J., and P. W. Sherman. 1981. Multiple paternity in Belding's ground squirrel litters. *Science* 212:351-353.

Harcourt, A. H. 1981. Intermale competition and reproductive behavior of the great apes. In *Reproductive Biology of the Great Apes,* C. E. Graham (ed.), pp. 301-318. Academic Press, London.

Harcourt, A. H., P. H. Harvey, S. G. Larson, and R. V. Short. 1981. Testis weight, body weight and breeding system in primates. *Nature* 293:55-57.

Harvey, P. H. 1982. On rethinking allometry. *J. Theor. Biol.* 95:37-41.

Harvey, P. H., M. Kavanagh, and T. H. Clutton-Brock. 1978. Sexual dimorphism in primate teeth. *J. Zool. Lond.* 186:475-485.

Harvey, P. H. and G. M. Mace. 1982. Comparisons between taxa and adaptive trends: problems of methodology. In *Current Problems in Sociobiology*, King's College Research Group (eds.), pp. 343-361. University, Cambridge.

Huxley, J. S. 1932. *Problems of Relative Growth*. Methuen and Co., London.

Huxley, J. S. 1938. The present standing of the theory of sexual selection. In *Evolution*, G. R. de Beer (ed.), pp. 11-42. Clarendon, Oxford.

Jolicoeur, P. and J. E. Mosimann. 1968. Intervalles de confiance pour la pente de l'axe majeur d'une distribution bidimensionelle. *Biometrie-Praximetrie* 9:121-140.

Kermack, K. A., and J. B. S. Haldane. 1950. Organic correlation and allometry. *Biometrika* 37:30-41.

Lanier, D. L., D. Q. Estep, and D. A. Dewsbury. 1979. Role of prolonged copulation behavior in facilitating reproductive success in a competitive mating situtation in laboratory rats. *J. Comp. Physiol. Psychol.* 93:781.

Levine, L. 1958. Studies on sexual selection in mice. I. Reproductive competition between albino and black-agouti males. *Am. Nat.* 92:21-26.

Levine, L. 1967. Sexual selection in mice. IV. Experimental demonstration of selective fertilization. *Am. Nat.* 101:189-294.

Mortimer, D. 1977. The survival and transport to the site of fertilization of diploid rabbit spermatozoa. *J. Reprod. Fertil.* 51:99-1044.

Mortimer, D. 1979. Differential motility of diploid rabbit spermatozoa. *Arch. Androl.* 2: 41-47.

Napier, R. A. N. 1961. Fertility in the male rabbits. III. The estimation of spermatozoan quality by mixed insemination, and the inheritance of spermatozoan characters. *J. Reprod. Fertil.* 2:273-289.

Selander, R. K. 1966. Sexual dimorphism and differential niche utilization in birds. *Condor* 2:113-149.

Selander, R. K. 1972. Sexual selection and dimorphism in birds. In *Sexual Selection and the Descent of Man*, B. Campbell (ed.), pp. 180-229. Aldine-Atherton, Chicago.

Short, R. V. 1977. Sexual selection and the descent of man. In *Proceedings of the Canberra Symposium on Reproduction and Evolution*, pp. 3-19. Australian Academy of Sciences.

Short, R. V. 1979. Sexual selection and its component parts, somatic and genital selection, as illustrated by man and the great apes. *Adv. Study Behav.* 9:131-158

Short, R. V. 1981. Sexual selection in man and the great apes. In *Reproductive Biology of the Great Apes*, C. E. Graham (ed.), pp. 319-341. Academic Press, London.

Schultz, A. H. 1938. The relative weight of the testes in primates. *Anat. Rec.* 72:387-394.

Tutin, C. E. G. 1980. Reproductive behaviour of wild chimpanzees in the Gombe National Park, Tanzania. *J. Reprod. Fertil. Suppl.* 28:43-57.

Whitney, L. F. 1940. The timing of ovulation in the bitch. *Vet. Med. Small Anim. Clin.* 35: 182-200.

Wilson, E. O. 1975. *Sociobiology, the New Synthesis*. Harvard University Press.

Wrangham, R. W. 1980. An ecological model of female-bonded primate groups. *Behaviour* 75:262-300.

19

Human Sperm Competition

ROBERT L. SMITH

I. INTRODUCTION

Alexander (1979) has invited ". . . biologists to contribute to the analysis of human behavior on all legitimate fronts . . ." He considered it ". . . especially relevant that [they] take up the problem of relating human attributes to evolutionary history." The analysis of human sexual behavior surely qualifies as a legitimate topic in evolutionary biology. This chapter represents a contribution to the argument that sperm competition does occur in humans and has been a selective force in the evolution of certain human characteristics.

There has been considerable controversy over what may be the "natural" sexual inclinations (promiscuous, polygynous, serially polygynous, monogamous, or some mixture of these) of human males (*e.g.,* Trivers 1972; Wilson 1975; Alexander 1977; Short 1977, 1979, 1981; Daly and Wilson 1978; Symons 1979; Lovejoy 1981; Barash 1982; Harvey and Harcourt, this volume), but relatively much less debate over the sexual predilections of human females (Hrdy 1981). Females are widely assumed to be monogamous, with little formal recognition of alternative female strategies (but see Hrdy 1981, and Knowlton and Greenwell, this volume). The compromise view of human male mating strategy proposes mixed tactics (Trivers 1972) where males attempt to pair-bond with one or more females by high investment, and opportunistically (more or less promiscuously) mate with other females. All combinations of male tactics from rape (Shields and Shields 1983, Thornhill and Thornhill 1983) to high investment (Trivers 1972) and the environmental and social circumstances that occasion their expression have received analysis in the literature. As Hrdy (1981) observed: "The sociobiological literature stresses the travails of males—their quest for different females, the burdens of intrasexual competition, the entire biological infrastructure for the double standard. No doubt this perspective has led to insights concerning male sexuality. But it has also effectively blocked progress toward understanding female sexuality—defined here as the readiness of a female to engage in sexual activity."

The biological irony of the double standard is that males could not have been selected for promiscuity if historically females had always denied them opportunity for expression of the trait. If strict monogamy were the singular human female mating strategy, then only rape would place ejaculates in position to compete and the potential role of sperm competition as a force in human evolution would be substantially diminished.

Here I shall explore the literature for evidence of the evolutionary significance of sperm competition in humans. I present data on the circumstances that would place ejaculates from different human males together in the reproductive tract of a female during a single reproductive cycle. I summarize the evidence that human sperm competition actually occurs. And, finally, I speculate on how selection within the context of potential or actual sperm competition may have operated in human evolutionary history to shape some aspects of human anatomy, physiology, behavior, and culture.

II. HUMAN SPERM COMPETITION IN CONTEXT

Human sperm are motile for up to 7-9 days in the reproductive tract of the female (Morris 1977, Porter and Finn 1977). Therefore, any circumstance that places ejaculates from two or more males in the vagina or uterus of a female within a 7-9 day period creates the potential for contested fertilization of her ova. The contexts that may cause human sperm to compete are communal sex, rape, prostitution, courtsthip, and facultative polyandry.

A. Communal Sex

Communal sex is used here as a collective term to include all of the consensual sexual arrangements among females and their primary mates that create contexts for sperm competition. Included are "orgies," "wife sharing," "wife swapping," and variations on these themes. This is a relatively insignificant context that seems to have received publicity greatly disproportionate to its historical and modern importance. The cross-cultural index shows 60.9% of indexed societies have no communal sex of any kind, and in the 35.4% that permit some form (Fig. 1H), it is typically highly regulated, reciprocal, involves wife sharing only with close relatives, and/or results in some benefit to the consenting male spouse. In only 3.6% of societies was there a high level of permissiveness for these activities (Broude and Green 1976). A 1970s survey (conducted during a period of sexual experimentation in the United States) on "swinging," *i.e.*, various forms of mate exchange, showed about 1% of respondents had engaged in these activities with any frequency, and only 3.5% had tried them once or twice (Athanasiou 1973). The infamous orgies of classical Greece and Rome apparently always involved prostitutes, and never wives (Bullough and Bullough 1978), so appropriately are included under prostitution rather than communal sex.

B. Rape

Rape is forced copulation of a female by a male (Thornhill and Thornhill 1983). Rape by an individual male (with ejaculation into the victim's reproductive tract) of a pair-bonded female will usually place the rapist's sperm in competition with that of her principal mate. "Gang rape," *i.e.*, sequential copulations forced on a

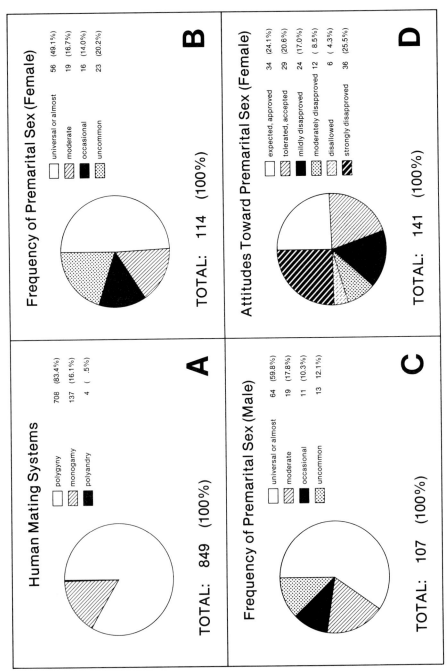

Fig. 1. Cross-cultural indices of human mating systems, and sexual practices and attitudes. A. Data from Murdock (1967). B-F and H-K. Data from Broude and Green (1976). G. Data from Burley and Symanski (1981). L. Data from Gaulin and Schlegel (1980).

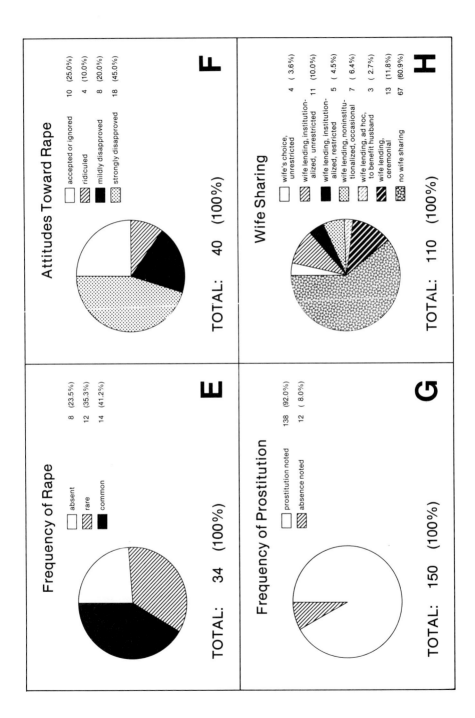

Frequency of Rape

☐ absent	8	(23.5%)
▨ rare	12	(35.3%)
■ common	14	(41.2%)

TOTAL: 34 (100%)

E

Attitudes Toward Rape

☐ accepted or ignored	10	(25.0%)
▨ ridiculed	4	(10.0%)
■ mildly disapproved	8	(20.0%)
▦ strongly disapproved	18	(45.0%)

TOTAL: 40 (100%)

F

Frequency of Prostitution

☐ prostitution noted	138	(92.0%)
▨ absence noted	12	(8.0%)

TOTAL: 150 (100%)

G

Wife Sharing

☐ wife's choice, unrestricted	4	(3.6%)
▨ wife lending, institutionalized, unrestricted	11	(10.0%)
■ wife lending, institutionalized, restricted	5	(4.5%)
▦ wife lending, noninstitutionalized, occasional	7	(6.4%)
▨ wife lending, ad hoc, to benefit husband	3	(2.7%)
▨ wife lending, ceremonial	13	(11.8%)
▦ no wife sharing	67	(60.9%)

TOTAL: 110 (100%)

H

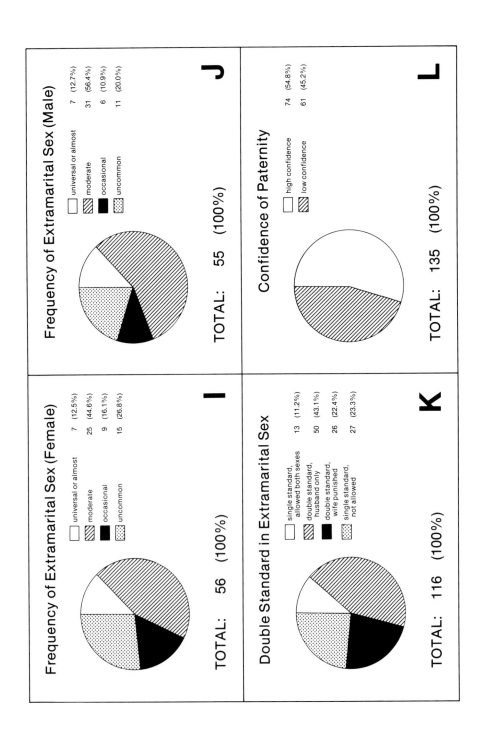

Frequency of Extramarital Sex (Female)

☐ universal or almost	7	(12.5%)
▨ moderate	25	(44.6%)
■ occasional	9	(16.1%)
⊡ uncommon	15	(26.8%)

TOTAL: 56 (100%)

I

Frequency of Extramarital Sex (Male)

☐ universal or almost	7	(12.7%)
▨ moderate	31	(56.4%)
■ occasional	6	(10.9%)
⊡ uncommon	11	(20.0%)

TOTAL: 55 (100%)

J

Double Standard in Extramarital Sex

☐ single standard, allowed both sexes	13	(11.2%)
▨ double standard, husband only	50	(43.1%)
■ double standard, wife punished	26	(22.4%)
⊡ single standard, not allowed	27	(23.3%)

TOTAL: 116 (100%)

K

Confidence of Paternity

☐ high confidence	74	(54.8%)
▨ low confidence	61	(45.2%)

TOTAL: 135 (100%)

L

female by several males in succession, probably creates the highest conceivable levels of ejaculate competition.

Broude and Green (1976) found rape to be common in 41.1%, uncommon in 35.3%, and absent in 23.5% of indexed societies (Fig. 1E). Sanday (1981) broadly classified 95 societies, finding 45 to be "rape-free" and 17 to be "rape-prone." She concluded that rape was a cultural phenomenon; however, her criteria for identifying "rape-free" societies and her conclusions have been challenged (see Shields and Shields 1983). More than 60,000 rapes are **reported** in the United States each year, yet it is estimated that this figure represents only 10% of the actual cases (Green 1980).

Frequency of rape is highest during wars. Rape has been a universal aspect of human conflict from tribal battles to world wars, and has occurred throughout history. Brownmiller (1975) reviews the history of rape and provides an exhaustive assemblage of modern anecdotes as well as some statistics. For example, during the 9-month West Pakistani occupation of Bangladesh in 1971, between 200,000 and 400,000 rapes took place.

Rape is widely distributed and occurs with sufficient frequency to be considered a significant context for human sperm competition. Although it is not obvious that cultural evolution has progressively reduced the incidence of rape through time, I assume (as apparently Alexander and Noonan [1979] do also) that it was probably a more important sperm competition context in early human history, and relatively much more important in human prehistory.

C. Prostitution

Prostitution is indiscriminate sexual activity for profit. Of specific importance here is the *ad hoc* sexual intercourse by females with males in exchange for resources as an agreed-to precondition for the intercourse. Because prostitutes may copulate with many men in a single day, the practice has potential for creating very intense sperm competition.

Prostitution definitely occurs in 142 of ca. 300 societies coded in the Human Relations Area Files (HRAF) and is explicitly excluded in only 12 of the 300 (Fig. 1G). Censuses of prostitutes in various societies are not available generally, but some estimates have been accumulated by Symanski (1981). For the United States in the last decade, estimates ranged from 100,000 to 550,000 (Winick and Kinsie 1972), and Symanski presumed 250,000 to 350,000 full- and part-time prostitutes a credible estimate in 1981. In 1957, Poland allegedly had 230,000 prostitutes, Budapest ca. 10,000, and Tokyo more than 130,000. In the late 1960s, Rome was said to have had 100,000 prostitutes. A recent study reported 50,000 registered prostitutes in West Germany, and estimated 150,000 to work illegally. The estimate of 80,000 working prostitutes in Addis Ababa, Ethiopia, in 1974 represented ca.

25% of the city's adult female population (Dirasse 1978), and illustrates the extremely high levels of sperm competition due to prostitution under certain socioeconomic conditions. Marco Polo found over 20,000 women living as prostitutes in Peking, and "too many to estimate" in Hangchau in the 13th century (Parks 1929). Use by males of female prostitutes was routine and almost universal in ancient Greece and Rome (Bullough and Bullough 1978). Kinsey *et al.* (1948) found that 69% of white males surveyed had had experience with prostitutes, and 15% used prostitutes regularly.

Commercial sex has had a long and colorful history (see Bullough and Bullough 1978), and though it is currently practiced by relatively few females in most societies, I shall assume that some form of prostitution has been an important context for sperm competition in human prehistory.

D. Courtship

In cultures that permit adolescent females to choose their primary mates, the time between puberty and marriage may represent a brief period of relatively dynamic sexual activity, and concomitant potential for sperm competition. This is because the unmarried adolescent girl may not yet be under the control of any one male who has a reproductive interest in her, and hence may be free to sample the population of prospective long-term mates. Unmarried !Kung couples, for example, often leave their village to have intercourse in the bush, and when individuals find partners they like based on these experiences, they may marry (Howell 1979). Adolescent promiscuity (under the guise of courtship by both sexes) is well-known and generally tolerated in Western culture, and the cross-cultural index (Fig. 1D) reveals that 44.7% of indexed societies approve of premarital sex or tolerate it without punishment. Another 25.5% disapprove only mildly and administer token punishment to violators, for a cumulative total of 70.2% lenient societies. Female premarital sex occurs in at least 79.8% of indexed societies (Fig. 1B). Males (in a revealing discrepancy) from at least 89.9% of indexed societies are reported to engage in premarital sex (Fig. 1C).

Some U.S. schoolgirls may use multi-male mating as a primary mate acquisition strategy (N. Chagnon, pers. comm.; see section V.B). An adolescent female may copulate repeatedly with several youths until impregnated, then name as father the favored phenotype (or best potential provider) from among the lot. Malinowski (1929) reported that Trobriand Island females were free to choose their lovers as an adolescent privilege but after marriage, female adultery was an offense that could be punished by death.

I contend that the period of female courtship is a common context for relatively intense human sperm competition, but both its historic and later prehistoric importance may be limited by its short duration in each female's reproductive life.

E. Facultative Polyandry

The typical human female has one principal mate from whom she and her children receive care, protection, and material resources. Marriage (socially recognized bonding of a female to her principal mate) is a cross-cultural universal that applies virtually to all members of society (Daly and Wilson 1978, Hawthorn 1970). The male benefits in this arrangement by having routine sexual access to the female, and thus a consistent competitive advantage over that of any facultative mates, for his sperm in the fertilization of her ova. Overt long-term polyandry is extremely rare (Fig. 1A); it seems to occur only under conditions of limited resources and most usually involves brothers sharing one wife (Aiyappan 1935, Kurland 1979, Beall and Goldstein 1981, Gowaty 1981, Hughes 1982). The universal convention of female monogamous pair-bonding is almost certainly an early hominid adaptation (Lovejoy 1981) that has served to limit the occurrence of sperm competition.

When a female voluntarily copulates with a male or males other than her primary mate, she is practicing facultative polyandry. Facultative polyandry may place ejaculates of extrabond mates at competition with that of the primary mate and/or cause competition among sperm of extrabond mates.

This context for human sperm competition almost certainly exceeds all others in importance. Female extramarital sex occurs in 73.2% of indexed societies and is common in 57.1% (Fig 1I). The cross-cultural occurrence of male and female extramarital sex is surprisingly similar (Fig. 1J). Gaulin and Schlegel (1980) found low levels of paternity assurance in nearly half (N = 61) of the 135 societies they examined. Presumably, sperm competition is especially intense in these societies. Kinsey et al. (1953) found ca. 26% of U.S. adult females had engaged in extramarital intercourse, and over 50% of respondents in the *Cosmopolitan* survey (Wolfe 1981) indicated that they had had extramarital liaisons. Although similar data are not available for other cultures, it seems safe to assume that female extrabond sex occurs in all but the most restrictive societies, and that it is common.

III. POTENTIAL FEMALE BENEFITS FROM FACULTATIVE POLYANDRY

Why should a female mate with males other than a primary mate? Parker, Knowlton and Greenwell, and others in this volume have enumerated the benefits that could accrue to a female of any species by multiple mating. These include: (1) good genes, (2) sons' effects (both somatic and gametic), (3) genetic diversity, (4) fertility backup, (5) material resources, and (6) protection for self and offspring. To this might be added another, perhaps uniquely human, benefit: enhancement of social status (Symons 1979, Barash 1982, Dickeman 1978). Proximate psychophysiological benefits, such as comfort and sexual pleasure, are not considered here except to the extent that they may represent mechanisms selected because they facilitate evolutionarily important results.

The good genes, sons' effect, and genetic diversity benefits have in common that all require extrabond mating to supply genetic material somehow superior to that available exclusively from a female's primary mate. All of these possible benefits are controversial and assailable on theoretical grounds.

A. Good Genes

Most societies support polygyny (Fig. 1A), but there are always constraints on the number of wives that even a "very high quality male" may effectively maintain. Consequently, some females will always have to settle for primary mates that are of apparently "inferior quality." This problem is intensified in cultures that prohibit polygyny.

A female who perceives her primary male to be genetically inferior to other males with whom she can casually mate may employ a mixed strategy, *i.e.,* accept resources from her primary mate and opportunistically copulate with an available superior male or males. The successful mixed-strategist female may deceive (cuckold) her primary mate into supporting a child or children fathered by the genetically superior male or males. This assumes that human females are capable of discriminating genetically-based differences among potential mates and deciding which phenotypes are superior.

The extent to which female mate choice may operate is a topic of current interest and no little controversy (see Bateson 1983; Parker 1983, this volume). Female intersexual selection is challenged on the grounds that if it occurred, it would quickly eliminate all of the variability in the male population. Others doubt that variation in human male status or ability to secure material resources has a genetic component.

The first of these objections would seem more credible if human females based mate-choice decisions on one or a very few simple aspects of the male phenotype such as physiognomy. However, physical appearance seems to rank relatively low among the characteristics females evaluate (Symons 1979). It is difficult to address the second objection, but I submit that there are many good human male phenotypes (and genes) when viewed against the backdrop of human culture and its occupational specializations. I do not doubt human females' abilities to discern especially good gene assemblages, but the task must be intellectually challenging. Processing mate-choice decisions was perhaps an important prehistoric challenge to human female intelligence.

In many cultures, however, choice of the first primary mate for a girl is made entirely by the parents and often exclusively by the daughter's male parent (Lévi-Strauss 1969, Daly and Wilson 1978, Symons 1979). The parents' interests may, but probably will not be coincident with the daughter's. Thus a parent-offspring conflict (see Trivers 1974) is created that may result in extramarital adventures by dissatisfied daughters.

Fickleness (mate choice ambivalence) is a characteristic that young males in Western culture attribute to the objects of their courtship. That human females have difficulty selecting primary mates when allowed the option is not surprising given the importance of the decision and the bewildering array of characteristics that must be comparatively evaluated. Female equivocation after a marriage is equally comprehensible, especially if the wife did not participate in the mate-choice decision or if her choice seems to have been a mistake. Such confusion may lead to facultative polyandry. Symons (1979) points out that polyandrous behavior may provide opportunities for the female to change principal mates.

B. Sons' Effect

Sons' effect assumes that male children fathered by a particularly charming man may inherit paternal charm and thus a competitive advantage in seduction (see Weatherhead and Robertson 1979). Likewise, the sperm production and sperm delivery systems of male children produced by paternal sperm with a competitive edge may deliver sperm exceptionally good at getting eggs in contests with gametes from other males. Recently, some theorists have turned their attention to the possibility of selection for daughters' effect (R. L. Trivers, pers. comm.).

C. Genetic Diversity

Maintenance of genetic diversity is the evolutionary hedge against an unpredictably changing environment, and dynamic environments are responsible for retention of sexual reproduction in most animal populations (Williams 1966, 1975; Maynard Smith 1978). Human environments are dynamic and unpredictable. A small technological or cultural innovation, or a political event may dramatically alter human environment within a single generation, thus causing selection to shift favor from one phenotype to another or to suddenly reward a previously ignored or rejected behavioral morph. A recombinant perfect for computer programming may not have produced a good Pleistocene hunter, yet a modern person endowed with this set may secure large rewards, and possibly enhanced fitness. On a shorter time scale, male aggressivity may be considered in the contexts of war or peace.

The highly aggressive male may be a better warrior than a more pacific type and thus win material rewards and differential reproductive privileges from a society grateful for his services during periods of conflict. In times of peace and social stability, with no acceptable outlet for his aggressivity, the warrior phenotype may commit criminal acts and be incarcerated and thus deprived of the opportunity to reproduce. Human females may benefit by diversifying the paternity of their children, especially during times of social instability when the future is relatively unpredictable.

D. Fertility Backup

Failure to conceive may be the most compelling biological reason for faculta-tive polyandry. It is estimated that 25% of couples in the United States have fertility problems, and in up to 35% of these, fault is with the male alone (Stangel 1979). A variety of factors may affect male fertility, including general health, diet, disease, stress, injury, frequency of ejaculation, genetic abnormalities or incom-patibilities, and specific vaginal antibiosis. Male infertility may be temporary or permanent, depending on causes.

Failure of a female to conceive can be attributed to one of four general faults: (1) female sterility or reduced fertility, (2) male sterility or reduced fertility, (3) mutual sterility or reduced fertility, and (4) gametic incompatibility. In three of these four cases a female could benefit reproductively from extrabond mating. If extrabond mating is practiced, sperm competition will occur in all but the cases of absolute female or primary mate sterility.

Wives of polygynous males in West Africa are less fecund than those in monog-amous marriages (Musham 1956; Dorjahn 1958, 1959; Isaac 1979). This may be attributable to the polygynous husband's rotational patterns among wives, or to reduced active sperm counts per ejaculate due to increased coital frequency. What-ever the cause, it makes facultative polyandry of greater potential benefit to the polygynously-married wife than to the monogamously-married one. It is possible that the effect of polygyny on female fecundity in the subject cultures may in fact be mitigated by facultative polyandry. Dorjahn (1959) notes that polygynously-married wives of the Temne are freer from supervision and are therefore more adulterous than monogamously-married females. Little (1967) implies this situation for the Mende of Sierra Leone as well.

E. Material Resources

Prostitution is one extreme in a continuum of practices that capitalize on sex as a female service (see Burley and Symanski 1981). It probably has its origin in courtship gifting. Malinowski (1929) observed that Trobriand men regularly give their mistresses small presents; if nothing is offered, the mistress may refuse intercourse. It is said by the Bush Negroes: "The men who have the most success with women are always short of money" (Hurault 1961). Rich Hopi men are most successful in their "love affairs," presumably because Hopi women are in the habit of accepting money for providing sexual access (Titiev 1972). Among the Birom of northern Nigeria, it is customary for married women to have "lovers." The lover pays the husband for sexual access to his wife, and has commitments to the married mistress (Smedley 1976, in Daly and Wilson 1978). Pygmy men pay parents for the privilege of having affairs with their daughters (Turnbull 1965). Truk wives may be loaned to other men for material gain by their husbands (Kramer 1932). In a number of cultures, women not classified as prostitutes acknowledge that they use sex to secure supplemental income (Daly and Wilson 1978).

Nuptial gifting is common in many animal groups (Alcock 1984) and is probably a universal aspect of human courtship with its origins in distant prehistory (see section VIII). The first gifts were almost certainly high-quality food, probably meat. A good male hunter could give females excess meat in exchange for sex or sexual obligations. Females could likewise venture sex on a good prospect against the possibility of some future return in material resources. Males are more likely to share resources with females who cultivate investment by providing sex than randomly with all females. Males also may be inclined to provide resources to a married consort's children on the chance that some may be his. Female behavior in this regard represents a hedge against the possibility that a principal mate will be unable or unwilling to supply sufficient resources.

In virtually all societies, females are economically dependent on males, and the extent to which a male is willing to support a female mate is a function of his perception of her reproductive value. Women of low reproductive value and high liability, such as nonvirgin girls, adulteresses, women of "low moral character," women past their reproductive prime, divorced or widowed women with children, and those afflicted with disease or deformity may be shunned by potential support providers, with the result that their only recourse may be to exchange sex directly and indiscriminately for economic return (Burley and Symanski 1981). The benefit from prostitution, for some females in many cultures, may simply be survival.

F. Protection

Human males typically provide protection to their mates and children. Such protection may be delivered against all forms of exploitation (inter- and intraspecific) from rape to predation. This is an essential benefit to females from the pair bond, and female mate-choice decisions must be based in part on a male's perceived ability to protect.

A primary mate cannot always be available to defend his wife and children and, in his absence, it may be advantageous for a female to consort with another male for the protection he may offer. This arrangement could provide a female the advantage of secondary mate choice and simultaneously offer some protection against loss of choice by rape. Absence of the primary mate may create the opportunity and the need for extrabond mating. As in the case of resource sharing, a male may be inclined to protect the children of a married lover on the chance that his genes are represented among them.

Infanticide (see Hrdy 1979) occurs among some primates, including our closest extant relatives, the chimpanzees (*Pan*) and gorillas (*Gorilla*). In some cases, infanticide is apparently a male reproductive strategy that eliminates the dependent offspring of a competitor and reproductively releases the mother from lactational amenorrhea to the perpetrator (Hrdy 1979). Gorilla and chimpanzee females may reduce the probability of having offspring become victims of infanticide by mating with different males (see section VI). Among humans, infanticide is most commonly a parental manipulation tactic based on resource scarcity or sex preferences (Alexander 1974). However, there are occasional examples in several cultures of men killing children fathered by other men (Hrdy 1977), so there is at least the possibility that human females might likewise reduce infanticide under certain circumstances by obscuring paternity with multi-male mating.

G. Status Enhancement

There is little question that human females prefer to marry high-status males (see van den Berghe and Barash 1977, Symons 1979, Trivers and Willard 1973), and in some cultures the tendency has become institutionalized (Dickeman 1979). Symons (1979) asserts that if no material compensation is given, selection can be expected to favor the female desire for high-status sex partners (as distinct from husbands) only if the status has a genetic basis and the female succeeds in conceiving by the high-status mate. I have addressed this possibility in section III.A, but another possible benefit is enhancement of the female's social position by association with a high-status male exclusive of any material or genic benefit he might supply. Liaison with a high-status but noncontributing male may permit the female temporary access to and possible opportunity in a higher social stratum (*i.e.*, hypergamy). It may at least elevate her position in her own stratum. A hint of this effect may be found in attempts at high status imitation by occupants of lower social strata (Symons 1979).

IV. POSSIBLE FEMALE COSTS
OF FACULTATIVE POLYANDRY

My objective in the previous section has been to suggest benefits to females from facultative polyandry. These benefits may be considerable, but they may be offset by risks and costs.

Male sexual jealousy (see section VII.A.7) may result in punishment for female adultery that may include loss of resources, physical injury, or even death. Females must weigh the risks and consequences of discovery against the potential benefits to be derived from extrabond coitus. This model predicts that a female mated to a physically attractive, high status, wealthy, fertile, nurturant primary mate (if such exists) in times of societal stability, is not likely to engage in facultative polyandry. Also, the wife of a particularly brutal man, or of any man in a particularly brutal society, may reject facultative polyandry.

V. EVIDENCE FOR HUMAN SPERM COMPETITION

There are hard data (albeit few) on sperm competition in humans. These come from (1) judicial proceedings of Western countries, supported by forensic genetics studies performed to exclude paternity; and (2) human population genetics studies conducted by geneticists and anthropologists. It is in the nature of these data that sperm competition is revealed only where the genes of contestant sperm other than those of putative fathers are expressed and can be detected. Therefore these data are conservative.

A. Paternity Law and Forensic Genetics

The fact of legally contested paternity itself provides an indirect measure of probable sperm competition in Western culture. For example, courts in West Germany hear over 5,000 paternity cases each year (Spielmann and Kühnl 1980); 4,200 paternity suits were filed in New York in 1959 (Anon. 1982); and 12,000 paternity suits were filed in Cuyahoga County, Ohio, between 1945 and 1959 (Whitlatch and Masters 1962).

A legal infrastructure including statutes, legal theory, legal specialists, and supporting forensic genetics laboratories, has grown of the need to adjudicate thousands of disputed paternity cases in Europe and the United States (Ellman and Kaye, pers. comm.). Use of blood test evidence in paternity actions did not begin in the United States until the 1930s. The first tests were based on the conventional ABO red blood cell antigens, and were relatively insensitive. At best the tests could only exclude ca. 14% of putative fathers who were not the biological fathers (Ellman and Kaye, pers. comm.). Later combination of ABO typing with other blood antigen systems succeeded in excluding about two-thirds of all falsely-accused defendants (Mendelson 1982) and, more recently, human leukocyte antigen (HLA) tests have produced exclusion probabilities of 95-99% (Terasaki 1978, Terasaki *et al.* 1978, Hummel 1979).

Presumably, few putative fathers would protest unless they had reason to suspect they were being cuckolded. It has been estimated, based on post-trial test results, that in ca. 40% of cases heard by the New York courts system, the defendant is innocent of paternity (Anon. 1982). However, in only about 10% of New York's paternity cases have the accused requested blood tests that may possibly exclude them; the reason seems to be the high cost of the tests which must be borne by the defendants. Sussman (in Anon. 1982) applied blood grouping tests *ex post judicum* to 67 cases that had been resolved in favor of the plaintiff. In each case the defendant had conceded paternity, yet Sussman's tests revealed that 18% of these men could not be the fathers of the offspring for whom they had accepted legal responsibility.

The defendant in most paternity cases is the primary or most conspicuous mate of the plaintiff and therefore may not credibly deny coitus with the mother around the time of conception. Defense and justification for paternity testing are therefore based on "multiple access" to (*excepto plurium concubentium*) or "promiscuity" (*excepto plurium generalis*) of the mother (Sass 1977). It is significant that prior to the availability of reliable paternity exclusion testing, the law in at least some judicial systems seems to have recognized the statistical advantage enjoyed by sperm of the principal mate, relative to that of the female's casual sex partners, in deciding paternity cases (Sass 1977).

The fertilization of separate ova in a single female by different males and the subsequent plural birth of half siblings (superfecundation) provides indisputable evidence of human sperm competition. More than a dozen cases of superfecundation have been reported since the early 1800s and several recent ones have attracted coverage by the popular media (Archer 1810, Sorgo 1973, Terasaki *et al.* 1978, Spielmann and Kühnl 1980, Associated Press 1983). The most celebrated and rigorously investigated cases are those discovered by Terasaki *et al.* (1978) and by Spielmann and Kühnl (1980). The latter case, reported in West Germany, was characterized by a conspicuous genetic marker: One of the fathers was a black U.S. soldier, the other a caucasian German businessman (Shearer 1978).

B. Population Studies

Only three population genetics studies are known to illuminate levels of cuckoldry in humans, and none was done specifically to investigate confidence of paternity. Trivers (1972) reported a personal communication from the anthropologist Henry Harpending, whose preliminary analysis of biochemical data on the !Kung indicated that about 2% of the children in that society were not the progeny of men to whom they were attributed. More recently, however, Harpending (pers. comm., and Howell 1979) has indicated that these paternity exclusions were statistically no more frequent than maternity exclusions, and should therefore be attributed to a small amount of error in labeling of samples. It seems likely, however, that the close conformity of actual to putative paternity in the !Kung is not attributable to fidelity, but rather to the fact that affairs usually lead to rapid divorce and remarriage of the participants (Howell 1979).

Neel and Weiss (1975) performed blood typing tests on 132 Yanomama children and their putative parents and found eight paternity exclusions. Considering that the typings would allow about one-third of true exclusions to go undetected, the researchers estimate that there would be four instances of undetectable nonpaternity in the 132 children. This yields an estimate that roughly 9% of children are fathered by males other than putative fathers.

Finally, an extensive study conducted by the University of Michigan Department of Human Genetics on a rural midwestern U.S. community allegedly revealed a ca. 10% discrepancy from putative paternity. This discrepancy may have been confined largely to first births, and was apparently primarily attributable to girls that "got pregnant" and then chose a husband from among the possible fathers (N. Chagnon, pers. comm.).

These meager population data are clearly insufficient to derive any generalizations, but it is remarkable that the discrepancy from presumed paternity in the !Kung and Yanomama were low, given the reported high levels of pre- and extramarital sexual relations that characterize each culture (Howell 1979, Chagnon 1979). The results of these studies suggest that sperm competition adaptations, including marriage, successfully assure paternity for principal mates.

Recent advances in biotechnology may stimulate interest in and facilitate human population studies of this kind. Future investigators, whatever their primary objective, might well consider questions of paternity in their survey designs.

VI. COMPARATIVE HOMINOID REPRODUCTIVE
AND SOCIAL BIOLOGY

Although there is legitimate controversy on the exact relationship of humans and the great apes, it is generally agreed that the balance of evidence places *Pan* as our closest living relative, followed by *Gorilla* and *Pongo* (Pilbeam 1984). Schwartz (1984) offers an alternative, though weakly supported, view that sees the fossil *Sivapithecus, Pongo,* and *Homo sapiens* in the same clade, and envisions an earlier divergence of *Pan,* with *Gorilla* arising from the chimpanzee line.

Of interest here are not disputed relative positions of taxa, but rather the undisputed certainty that we are extremely close to three species having diverse ecologies, social systems, and sexually selected characteristics, and that the study of these relatives offers possibility for insights into the evolution of human characteristics. Short (1976, 1977, 1979, 1981) has contributed significantly by applying the comparative method (see Alcock 1984; Thornhill, this volume) to extant hominoids. Following Short, and in collaboration with him, Harcourt *et al.* (1981a) and Harvey and Harcourt (this volume) have used clever statistics and ingenious design to test predictions about the operation of selection on testis weight in the primates. Results of these studies are generally consistent with predictions based on species' mating systems.

I rely a great deal on the aforementioned work for this discussion. Tables I and II encapsulate anatomical and physiological aspects of male and female hominoid reproductive biology, and Table III assembles reproductive behavioral attributes for each species. Fig. 2 graphically represents sexual dimorphism and relative size of reproductive anatomical features in humans and the apes. These features are discussed below.

A. Orangutan

Apparently because its food is randomly and sparsely distributed in space, the orangutan has evolved a solitary lifestyle. During estrus, orangutans form monogamous consortships, and pairs copulate about every two days over a period of 3-8 days. Females do not advertise estrus until a mate has been selected. Female mate choice is apparently paramount for, despite their much larger size, male orangutans cannot control females in their arboreal habitat. Forced copulation is not uncommon in free-living orangutans, but bonded females are successfully defended by their mates. So sparse is their distribution that a consorting pair may never contact another male during the average 5.4-day period of estrus. So there may be little opportunity for polyandrous mating even if bonded females were so inclined. And, since male orangutans make no parental investment, contribute nothing to the female (except limited defense), and are not reported to kill infants, there is little reason for a bonded female orangutan to mate with other males if her current consort is of high quality. This is a male dominance system, with insignificant incidence of sperm competition. The genital attributes of the species are consistent with expectation for negligible selection by ejaculate competition. The male has an extremely short unspecialized penis and very small nonscrotal testes that contribute low sperm concentration to low volume ejaculates. As is typical of male dominance polygyny, the male orangutan is much larger than the female (Fig. 2).

B. Gorilla

Gorillas live in small bisexual groups that are dominated by an alpha (usually silverbacked) male. Within each group, the dominant male is responsible for all or nearly all of the conceptions (Harcourt *et al.* 1981b). Intragroup conflict is minimal. Female gorillas show a slight swelling of the external genitalia during estrus. Estrous females solicit copulations from the dominant male. The dominant male prevents subordinates from mating with fertile females with little violence, and allows them to mate with subadult and pregnant females. It is significant that immature and pregnant females submit to copulation by subordinate males and that alpha males allow it. This behavior may be adaptive in the context of the potential for adult male-perpetrated infanticide known to occur in gorillas (Fossey 1974, 1976; see section III.F, and Hrdy 1979, 1981). The absolute dominance of a single male within each gorilla group seems to militate against any significant sperm competition. Consequently, male gorillas, though much larger than females, have an unspecialized penis of about the same relative length as that of the orangutan and, like the orangutan, gorilla testes are small and nonscrotal. Mean ejaculate volume and sperm concentration for *Gorilla* are lowest of the hominoids (Table I). Copulation occurs about once every 2.5 hours during daylight for from 1-3 days. A dominant male gorilla may go for months or even as long as a year without mating with an estrous female. Short (1979) characterizes gorilla sexual behavior as ". . . a rare treat for these amiable vegetarians."

C. Chimpanzee

Chimpanzees are reproductively extraordinary among the great apes. They are not strikingly dimorphic for size as are the other two species (Fig. 2), but male chimpanzees have enormous scrotal testes, proportionately about 5 and 10 times larger than *Pongo* and *Gorilla,* respectively, and a specialized penis more than twice as long as that of the much larger gorilla. Short (1981) has calculated that the testes of an average chimpanzee can sustain sperm production at a level that will produce at least four full-strength ejaculates/day, each containing several times the number of sperm in an average gorilla or orangutan ejaculate.

TABLE I

Male Reproductive Anatomy/Physiology in Humans and the Great Apes

Characteristic	Human	Chimpanzee	Gorilla	Orangutan
Testes weight, g % body weight	40.5 0.079	118.8 0.300	29.6 0.031	35.3 0.063
Scrotum[a]	highly developed, pendulous, hairless	moderately developed, pendulous, hairless, unpigmented skin	barely perceptible post-penial bulge, hair-covered	barely perceptible post-penial bulge, hairless skin, but concealed by hair
Penis length and other features[b]	13 cm, thick, exposed when flaccid, conspicuous, no os penis, glans penis present	8 cm, narrow, filiform, pointed at tip, concealed when flaccid, os penis present, no glans penis	3 cm, inconspicuous, conical, concealed when flaccid, os penis present, glans penis present	4 cm inconspicuous, concealed when flaccid, os penis present, glans penis present
Seminal vesicle[b]	moderately large	extremely large	small	moderately large
Mean ejaculate volume[d] ml (range)	2.5 (0.1-12.0)	1.1 (0.1-2.5)	0.4 (0.2-0.8)	1.2 (0.2-3.6)
Mean sperm density[d] no./ml x 10^6 x vol. of ejaculate (range)	70 (0.1-600)	548 (54-2750)	162 (29-375)	76 (10-165)
Mean sperm number/ ejaculate[e] no./ml x 10^6 x vol. of ejaculate (range)	175 (0.01-7,200)	603 (5.4-6,875)	65 (5.8-300)	91 (2-594)
Spermatozoa, qualitative characteristics[d]	pleiomorphic for size, F-body visible on Y chromosome	F-body visible on autosomes	pleiomorphic for size, F-bodies visible on autosome and / chromosome	F-body not present
Mean length, microns Head length microns	58.4 3.5	57.4 2.9	61.4 7.1	66.6 3.8

[a]Harvey and Harcourt (this volume). [b]Short (1979). [c]Graham and Bradley (1972). [d]Martin and Gould (1981). [e]Calculated.

TABLE II

Female Reproductive Anatomy/Physiology of Humans and the Great Apes

Characteristic	Human	Chimpanzee	Gorilla	Orangutan
Ovary size, % body weight[a]	0.014	0.010	0.012	0.006
Vagina length, cm	?	<12.5[b]	ca. 70–80[b]	?
Hymen	present[c]	absent, no evidence of homologue[b]	absent, no evidence of homologue[b]	absent, no evidence of homologue[b]
Breasts[a]	highly developed, pendulous, in nonlactating and lactating females from puberty for life	developed at end of first pregnancy for duration of lactation	developed at end of first pregnancy for duration of lactation	developed at end of first pregnancy for duration of lactation
Sexual sign[a]	no sexual sign of any kind, cryptic ovulation	pronounced swelling of labia and circum-anal region with onset prior to menarche, lasting into pregnany	slight perineal tumescence at time of estrus	perineal swelling absent during estrus, but present during pregnancy
Estrus period, days (range)	mean 28 (24–35)[d] continuously receptive[a]	mean 9.6 (7–17)[e]	median 1 (1–3)[f]	mean 5.4 (3–8)[g]
Menstrual cycle, days (range)	mean 28 (24–35)[d]	mean 37.3 (22–187)[h]	median 49 (36–72)[h]	median 30.5 (26–32)[h]
Lactational amenorrhea, months	49.2[i]	68.4[j]	45.6[j]	?
Orgasm	common[k]	not reported	reported[l]	not reported

[a]Short (1979, 1981). [b]K. G. Gould (pers. comm.) [c]*Gray's Anatomy.* [d]Tortora and Anagnostakos (1984). [e]Tutin and McGinnis (1981). [f]Harcourt *et al.* (1981b). [g]Galdikas (1981). [h]Graham (1981). [i]Short (1976). [j]Nadler *et al.* (1981). [k]Fisher (1973). [l]Mitchell (1979).

621

TABLE III

Social and Sexual Behavior of Humans and Great Apes

Characteristic	Human	Chimpanzee	Gorilla	Orangutan
Social system	highly social, family and larger social units[a]	moderate size, bisexual unit-group[b]	small bisexual groups with a single dominant "silverback" male[b]	solitary[b]
Mating system (male view)	monogamous or polygynous and opportunistically promiscuous[c]	promiscuous and/or sequentially monogamous in consortships[d]	polygynous[e]	polygynous[f]
Mating system (female view)	monogamous and facultatively polyandrous; infrequently promiscuous[c]	promiscuous and/or sequentially monogamous in consortships[d]	monogamous[e]	monogamous[f]
Frequency of copulation	1–3.5 copulations/week, female continuously receptive[g]	mean 0.52 copulations/male/hour, female up to 50 copulations/day[d]	median 0.4 copulations/hour only when female in heat[e]	estimated 0.5/day during consortship lasting 5 days[h]
Courtship	extended courtship usually lasting several days at minimum, each coitus preceded by a bout of "foreplay"[g]	male penile display and invitation directed to estrous females, duration <1 min[d]	estrous female initiates by soliciting dominant (silverback) male[e]	consortship initiated by dominant male or by female, female choice exercised, short courtship[f]
Duration of copulation	variable, but usually extended period of foreplay and median 10 min coitus (range 2– >60 min)[i]	mean 7 sec (range 3–15 sec)[d]	median 96 sec[e]	mean 10.8 min (range 3–28 min)[f]

Copulatory position (female/male)	variable, but primarily ventral/ventral[l]	dorsal/ventral[d]	dorsal/ventral[e]	variable, but usually ventral/ventral[f]
Copulation, place and situation	variable, typically in sleeping place, privacy usually sought[j]	arboreal or on the ground[d]	on the ground[e]	invariably arboreal[f]
Copulation, time of day	variable, but usually nocturnal for pair-bonded individuals[j]	diurnal[d]	diurnal[e]	diurnal[f]
Rape (forced copulation)	common[k]	not reported	not reported	occasional[f]
Male masturbation	common[g]	not reported in wild, common in captivity	not reported in wild	occasionally observed in juveniles[l]
Female masturbation	common[g]	not reported in wild	not reported in wild	occasionally observed in juveniles[m]

aWilson (1975). bHamburg and McGown (1979). cVarious sources (see introduction). dTutin and McGinnis (1981). eHarcourt et al. (1981b). fGaldikas (1981). gKinsey et al. (1948), Hunt (1974), Pietropinto and Simenauer (1977). hHarcourt (1981). iAge-dependent (male); Hunt (1974). jAlexander and Noonan (1979); Cross-cultural Index, Schlegel (pers. comm.). kBrownmiller (1975, Fig. 1E. lMitchell (1979). mMartin and Gould (1981).

624 Robert L. Smith

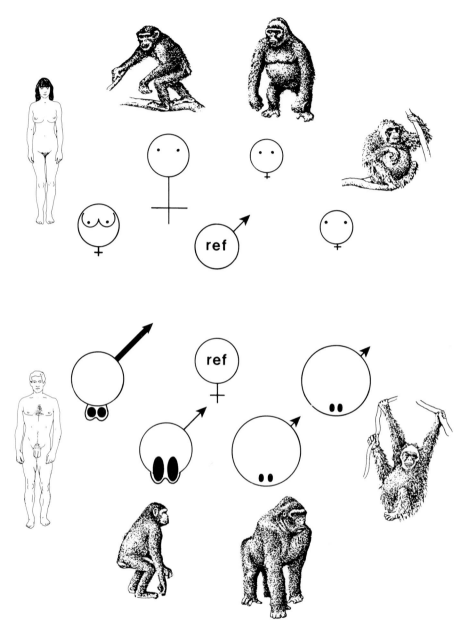

Fig. 2. Diagrammatic representation of size dimorphism, and relative sizes and features of sexual anatomy in humans and the great apes. Clockwise: human female, chimpanzee female, gorilla female, orangutan female, orangutan male, gorilla male, chimpanzee male, and human male. Figures labeled "ref" serve as size references for the figures of opposite sex that surround them. Adapted from Short (1977) and Halliday (1980).

Female chimpanzees in estrus have greatly swollen external genitalia that advertise their receptivity. This display is visually discernible at a distance, and promotes the promiscuous mating typical of these animals. "Opportunistic mating" (Tutin and McGinnis 1981) may result in a female copulating as many as 50 times in one day with a dozen different males (Tutin 1975, in Hrdy 1979). This behavior would result in the highest levels of sperm competition of any hominoid species. Every male member of the group in turn is permitted sexual access by the estrous female, and no fights occur over coital privilege. Opportunistic mating is noncompetitive and even cooperative. It seems to promote group solidarity with probable defensive advantages for all participants. And, as noted for *Gorilla,* multimale mating may result in some future advantage for females in the differential survival of offspring when infanticide by adult males is a possibility (see Hrdy 1979, 1981 for reviews and discussion). Chimpanzee females are in estrus for about 10 days each month for 2-3 years prior to first pregnancy. Sexual swelling and receptivity begin before the menarche, and continue into the early months of pregnancy (Short 1979). These observations support the idea that female chimpanzees may use coitus for purposes other than obtaining sperm. I believe these conditions foreshadow cryptic ovulation and continuous receptivity in human females.

Chimpanzees have two mating strategies in addition to opportunistic mating (Tutin and McGinnis 1981). These are "possessiveness" and "consortship." Possessiveness is a short-term relationship initiated by a male in which he persistently attends to a particular female and restricts her copulations with other subordinate males. Such behavior confers an obvious competitive advantage on the sperm of the possessing male.

Consortship contains the elements of possessiveness, but also involves the consort pair removing itself from the rest of the community. The male initiates consortship, but requires cooperation of the female. The advantage to the male is almost exclusive sexual access to "his" female, and greatly reduced sperm competition during the peak of her estrus. The female advantage is that of mate choice, shared premium food, and perhaps future special defense of her young. There are disadvantages as well. In isolation, both sexes are vulnerable to dangerous encounters with members of other communities. These encounters can result in death of the male and of the female's offspring. Also, both partners can expect hostile reunions with their own group when the consortship is terminated. Finally, the male who becomes involved in consortship gives up opportunistic matings with other females during the consortship.

D. Humans

Fig. 2 shows similarities of *Pan* and *Homo*. Males of both species have large specialized penises and large testes. Human testes, though absolutely and proportionately smaller than those of the chimpanzee, are about twice as large as those of

Gorilla and *Pongo;* chimpanzee and human males both have testes in scrota, in contrast to the peritoneal testes of the other two species. Finally, both human and chimpanzee males are only slightly larger than the females, while gorillas and orangutans are strikingly dimorphic.

Female chimpanzees and humans each have conspicuous, but different, anatomical features that distinguish them from unadorned orangutan and gorilla females. In the chimpanzee, the female feature of genital swelling is not present in the orangutan and is only slightly developed in the gorilla. Genital swelling is absent in human females, but they have pendulous breasts.

Behaviorally, chimpanzees and humans seem to share a preoccupation with sex compared to the relative indifference seen in the gorilla and orangutan. Gorilla males not infrequently reject the advances of an estrous female, while chimpanzee males actively and frequently seek to copulate with any available and willing female. This tendency is also seen in human males, though usually mitigated by society. Although we do not know what apes think about, human males in the age class 12-19 report thinking about sex on average once every 5 minutes (Shanor 1978). Fortunately for the progress of civilization, this rate declines to from once every hour to only several times a day from age 50 on.

VII. HUMAN ATTRIBUTES POSSIBLY EVOLVED IN THE CONTEXT OF SPERM COMPETITION

If sperm competition has been a significant factor in human evolution, we would expect human males to possess attributes to maximize the probability of fertilization by sperm from their every ejaculate, at the expense of those in competing ejaculates. Females, on the other hand, should have evolved characteristics to provide maximum control over the success in fertilization of sperm from competing ejaculates, and mechanisms that resist male control over their reproduction. This section explores human attributes that may have evolved by sperm competition. Where possible, I have tested characteristics by the comparative method, but for many intriguing elements, insufficient or inappropriate data make it impossible to do more than mention possible adaptation.

A. Male Attributes

1. Testes, Accessory Glands, and Ducts

Human males have proportionately large testes relative to the great apes (Table I, Fig. 2). Spermatogenic capacity in the hominoids is a function of testicular weight. Short (1981) suggests that the size of the human testis (like that of the chimpanzee) must be related to high rates of copulation (relative to *Gorilla* and *Pongo;* see Table I). However, he does not mention sperm competition as a possible factor in the evolution of testis size.

Each gorilla ejaculate contains ca. 65 x 10^6 sperm (Table I). Given a testes weight of ca. 30 g, daily sperm production for *Gorilla* is calculated to be 680 x 10^6, which allows about 10 daily ejaculates with undiminished sperm densities (see Short 1981, for assumptions). The same calculations applied to the chimpanzee allow about 4.5 ejaculations per day with constant sperm counts, but each chimpanzee ejaculate contains almost 10 times the sperm found in a gorilla ejaculate. At ejaculation rates of greater than 3.5 per week, human sperm counts begin to decrease (Freund 1962, 1963). At an ejaculation rate of 8.6 per week, human sperm counts are reduced to approximately that of *Gorilla,* but the average human ejaculate (at the 3.5/week rate) contains ca. 2.7 times the gorilla sperm count per ejaculate (see Table III). In other words, a gorilla with relatively small testes can copulate without reducing his sperm count, at over twice the frequency of a chimpanzee, and 20 times more frequently than a human male, although its absolute sperm production is much lower than that of the other two species. If these estimates are correct, selection for frequent copulation per se will not explain testis weight in hominoids. An alternative hypothesis is that increased testis weight evolved by sperm competition to produce ejaculates with larger numbers of spermatozoa (see Parker, this volume, and section VII.A.4).

Most other aspects of the male reproductive system have been studied in man, but not in the great apes. Data on the seminal vesicles of all four species are presented in Table I (from Short 1979). The size of the seminal vesicles follows the pattern for other genital organs, except in humans and the orangutan they are similar in size. Seminal vesicles in man contribute about 60% of semen volume. This secretion is alkaline and contains fructose; it apparently functions to nourish ejaculated sperm and raise the pH of the vagina. It probably affects sperm longevity and vagility. The prostate gland and the bulbourethral glands are well developed in both chimpanzees and humans, but apparently not in *Gorilla* or *Pongo* (Graham and Bradley 1972, Martin and Gould 1981). The epididymis is clearly exposed to selection in the context of sperm competition because of the sperm storage function of its caudal portion (see section VII.A.2), but there are no comparable morphometric data on this organ for the great apes.

Finally, the thickness of the muscle surrounding the lumen of the vas deferens is greater in relation to the lumen diameter than in any other tubular structure in man (Turner and Howards 1977). These muscles control the force of ejaculation and would therefore be subject to selection by sperm competition. It would be instructive to compare the thickness of vas deferens muscular layers among man and the great apes. Humans and chimpanzees should have a greater muscle to lumen ratio for this organ than *Pongo* and *Gorilla* if the theory is correct.

2. The Scrotum

The scrotum, a thin pouch of skin and subcutaneous tissue that contains the testes of sexually mature males, is found in many eutherian mammals. It is a curious structure. Why should males of some mammal species and not others have been selected to move their testes from the relative protection of the abdominal cavity to a position of seemingly extreme vulnerability? This question is especially intriguing when applied to the great apes. Fig. 2 illustrates the relative positions of testes in the male orangutan, gorilla, chimpanzee, and human. The orangutan has a postpenial bulge of bare black skin under which the testes reside. A similar arrangement is found in the gorilla, but the postpenial bulge is covered with hair. The chimpanzee has a true pendulous scrotum that is hairless and unpigmented. It is exceeded in its pendulosity by the scrotum of human males.

Hypotheses advanced to explain the scrotum range from the mechanistic view of Woodland (1903) that the weight of maturing testes wears a hole in the lower abdominal wall, to the suggestion that the scrotum has evolved as an epigamic display (Portmann 1952). Since the scrotum is found only in mammals, and mammals are homeotherms, it has been proposed that the scrotum may function to maintain testicular temperatures lower than that of the body cavity, and this effect has indeed been demonstrated. Scrotal testicular temperature is typically lower (from 2-6°C, depending on species) than the abdominal cavity (Carrick and Setchell 1977).

Bedford (1977), in an insightful paper, takes issue with all previously proposed hypotheses and garners impressive support for an alternative explanation, *i.e.* that testicular migration to the scrotal position has been driven by progressive selective advantage in epididymal sperm storage. Fig. 3A-C depicts a possible evolutionary progression. It shows migration of the sperm-storage region of the epididymis preceding descent of the testes into the scrotum. The reverse is illustrated by a nonexistent arrangement depicted in Fig. 3F. Scrotal topographies in a variety of species over diverse taxa are apparently patterned to ensure cooling of the epididymal sperm-storage area rather than the testes. Typically, the pattern involves abrupt hairlessness at the lower border of the scrotum adjacent to the cauda epididymis where mature spermatozoa are stored. Experiments with rabbits and rats, both scrotal mammals, in which the epididymis was surgically reflected into the abdomen, produced no effect on sperm maturation but dramatically altered the longevity of stored sperm. Rat sperm in the scrotal cauda epididymis remained motile for up to 42 days and viable for 21 days. In the rabbit, viability of scrotal stored sperm was 35 days. Sperm retained by ligatures in **abdominal** epididymis remained viable for only 3-4 days in rats, and 8-10 days in rabbits.

The sperm storage hypothesis to explain the scrotum is intuitively satisfying. Mammalian sperm that compete would be selected to possess metabolic enzymes that operate most efficiently at female deep body temperatures. However, sperm

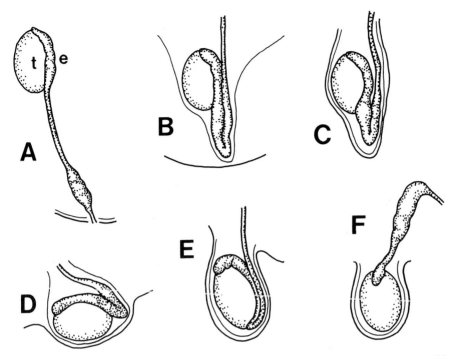

Fig. 3. A-C and D. Diagrammatic examples of eutherian mammal testes showing a possible evolutionary progression to the scrotal condition, led by migration of the epididymis. E. Human. F. Hypothetical arrangement expected if testis function were the only important element in the evolution of the scrotum. Adapted from Bedford (1977).

that contain such enzymes are likely to store much better at lower temperatures because of enzyme deactivation and conservation of metabolic substrate. Conflicts between storage and operational efficiencies of sperm in a system subject to selection by sperm competition could have driven the migration of the sperm storage organ to a scrotal position. Sperm of mammals with negligible sperm competition may not evolve high temperature adapted enzymes, therefore causing no selection for migration of the epididymis into a scrotum.

Bedford (1977) suggests that the scrotal state should be found in mammal species where "the male must be able to produce fertile ejaculates repeatedly over a relatively short period." He emphasizes polygyny as a mating system that would create these demands on the male system, but fails to identify competitive mating and sperm competition as a context. I propose that sperm competition drove the evolution of the scrotum in *Homo* and *Pan,* and note that this hypothesis would also account for the lack of a scrotum in *Pongo* and *Gorilla,* species that are polygynous, but without intense sperm competition.

3. The Penis

The size of the human male penis is extraordinary relative to other hominoids. It is nearly twice as long and over twice the diameter of the chimpanzee penis (Table I, Fig. 2). The mean length of the adult human penis is 13 cm (range = 11-15 cm). Penis lengths measure 3 cm, 4 cm, and 8 cm in the gorilla, orangutan, and chimpanzee, respectively (Table I, and Short 1980).

Short (1980, 1981) and Halliday (1980) review speculation on why so large an intromittent organ has evolved in humans. Speculation as to function include: aggressive display (intrasexual selection), attractive display (epigamic selection), and facilitation of a variety of copulatory positions to enhance female sexual stimulation during coitus. To these proposed functions, I add what would seem to be the most obvious: delivery of ejaculate as close as possible to ova.

The display theses are not compelling. Penis size is inherited independently of other aspects of human male physiognomy (Schonfield 1943). So the idea that the human penis is used in aggression raises the improbable scene of a large muscular man with a small penis cowering at the penile display of a "well-endowed" shrimp. There are no anthropological data to suggest that the human penis is used in aggressive display between males, and no evidence that human males experience erection in agonistic encounters.

It seems equally unlikely that the large human penis evolved by epigamic selection for display, or because larger penises enhance female enjoyment of intercourse. In a survey conducted by the *Village Voice* (Diagram Group 1981), women were asked to select the anatomical features they admired in men. Only 2% indicated an interest in penises (buttocks ranked highest at 39%). An indirect measure of female indifference to penises is the failure of U.S. magazines featuring photographs of nude males to secure a substantial female readership. This contrasts dramatically with the male market for photographs of nude females (featuring genitalia) that supports a dozen or so periodicals in the United States alone (see Symons 1979). The work of Masters and Johnson (1966) and results of recent surveys (Hite 1976, Wolfe 1981) have shown that penis size has little to do with female enjoyment of coitus and the achievement of female orgasm. The idea that the large human penis evolved to permit a variety of copulatory positions is tenuous at best (see Table III). Orangutan males have penises that average 4 cm, and they regularly copulate in the ventral/ventral position. Chimpanzees have 8 cm penises that would probably permit ventral/ventral coitus, but they always adopt the dorsal/ventral position. The ventral/ventral position is cross-cultural in humans and apparently can be achieved by males having penises through the full range of sizes (Masters and Johnson 1966). Finally, Symons (1979) observed that if females had the tendency to be sexually aroused by the sight of male genitalia, men would be able to obtain copulations by genital display, as is apparently the case with chimpanzees; however, human females seem to be singularly unimpressed and decidedly unaroused by "flashers."

An excessively large penis produced by epigamic selection would be no less effective in delivering sperm than one of optimal length, but a substantially shorter (than optimal) penis would obviously place its owner's ejaculates at a disadvantage in competition with those deposited by a longer organ. Therefore it may be possible to test some predictions of the ejaculate delivery/sperm competition hypothesis for human penis length and distinguish this function from one of display.

If human penis length has been selected to deliver sperm as close as possible to ova descending the female tract, then length should not on average exceed the mean depth of the vaginal column (of the sexually aroused female) from labia to cervical os, and there should be racial differences in vaginal column and penis length consistent with this prediction.

If penis length has been selected for display, then the organ may on average be longer or shorter than the vagina, and its length or proportional length perhaps randomly distributed among races, depending on the intensity of epigamic selection from race to race. Epigamic selection intensity might be greatest in the tropics where sparse clothing would permit the community of females to regularly view male genitalia. However, in this regard, Short (1980) notes that penis sheaths were worn out of modesty and decorum rather than for display by the Telefomin of New Guinea. Unfortunately, there are no genitalic morphometric data that would permit interracial comparisons.

Short (1979) also argues that the human penis probably has a display function because the flaccid organ is exposed outside of the body cavity. An alternative hypothesis is that in attaining its present length by sperm competition, it lost the os penis (penis bone) and the protection of an abdominal prepuce.

The chimpanzee penis is pointed at the apex and differs from that of the other apes and man in lacking a glans. The specialized apex may permit the penis tip to penetrate the cervical os; gross anatomy of the female tract suggests that this is possible (Gould and Martin 1981). Although there are no data on chimpanzee vaginal length, K. G. Gould (pers. comm.) uses a 12.5 cm speculum to dilate chimpanzee vaginas for artificial insemination. This is longer than the 8 cm mean penis length in these animals, but if the cervix dips a couple of centimeters into the vagina, and the speculum is not inserted completely, the fit could be made.

In the absence of selection by sperm competition or female preference, penis length should stabilize at that which consistently transfers an effective ejaculate, and would likely be less than optimal length because of compromising selection. *Gorilla* and *Pongo* are candidates for testing this prediction.

4. Semen

Seminal fluid is not essential for effective fertilization. Spermatozoa taken directly from the vas deferens will fertilize ova (Johnson and Everitt 1980). So, although many constituents of human seminal plasma have been identified, and some of the more obvious functions for major constituents established, there is yet much left to be learned. Human reproductive physiologists apparently have not considered the possibility that accessory gland secretions may function in the context of sperm competition. In this regard, prostaglandins, so named because they were originally discovered in extracts from the human prostate gland, are known to cause strong uterine smooth muscle contractions (Eliasson and Posse 1965, Pickles 1967). Much work has been done to identify the prostaglandin content and secretions of the reproductive systems of humans. The concentrations of "E" prostaglandins (PGE) in human semen have been found to be about 100 times greater than found in all other reproductive tissues (Bergström 1973). Goldberg and Ramwell (1977) observe that "PG's in the male tract occupy a somewhat existential niche in biology: although everyone admits that they exist, no one knows why." Bygdeman *et al.* (1970) found that 40% of infertile men whose infertility could not be attributed to another cause had significantly lower PGE levels than fertile men. I propose that these compounds are produced and incorporated into ejaculatory plasma as a means for controlling the female reproductive tract by causing uterine contractions that move the sperm in an ejaculate toward the ovum (see section VII.B.4). This intriguing possibility is reminiscent of factors found in *Drosophila* ejaculates that impose some control on female physiology and behavior (see Gromko *et al.,* this volume). If this is a function, presence and concentration of prostaglandins would be sensitive to selection by sperm competition. In this regard, it would be useful to compare prostaglandin titers among the ejaculates of the apes and man.

A large seminal plasma volume might secure some competitive advantage for the sperm it contains by providing optimal environment and nutrients for spermatozoa, and/or by debilitating alien sperm. Human males have not only the largest mean ejaculate volume, over twice as large as that of any of the great apes, but also volumes that range to 12 ml, more than four times the volume of any ape species.

Martin and Gould (1981) compare the morphometrics and gross anatomical features of spermatozoa among the apes and man (see Table I) and reveal similarities among them. No currently available data suggest the differential operation of selection on the anatomy of any species of hominoid sperm. Future comparative ultrastructural studies may, however, reveal unique adaptations. Acrosomal chemistry may be selected for egg penetration efficiencies in the final context among competing spermatozoa (see section VII.B.5).

We know from artificial insemination technology that a small fraction of sperm contained in an ejaculate can achieve fertilization when accurately placed at the proper time. So why do ejaculates contain so many sperm? Parker (this volume) shows that high sperm densities can evolve by sperm competition (the strategy being to overwhelm opponent sperm by sheer numbers). The comparative work of Short (1979) on humans and the great apes provides results that fit this model. Chimpanzees and humans have much higher levels of sperm competition relative to the other two species and, predictably, produce ejaculates that contain many more sperm than those of *Gorilla* and *Pongo* (Table I).

Foote (1973) notes that the quality (sperm density) of ejaculates obtained from bulls for artificial insemination is dependent on providing proper "teasers" to sexually stimulate the animal prior to ejaculation. He speculates that this factor may also be of "considerable importance" in humans. Is it possible that the human male is able to adjust sperm density levels to meet anticipated competitive challenges to his ejaculate? There are no data on this, but the well-known Coolidge effect (sexual habituation to an old mate, and arousal in response to the presence of a potential new mate; see Symons 1979 for full discussion) certainly demonstrates the psychosexual power of a novel stimulus, and portends the possibility of other physiological effects.

Masturbation seems a bizarre, maladaptive behavior because it wastes sperm, but the practice is almost universal among U.S. males and is particularly common in adolescents (Kinsey *et al.* 1948, Sorensen 1973). Masturbation is also commonly observed in some captive primates. It has been said that "primates masturbate, because they can," but if autoeroticism were maladaptive, the tendency should have been repressed by natural selection. A clue to a possible function is found in the fact that males who do not ejaculate by coitus or masturbation for some period of time experience nocturnal emissions, *i.e.* spontaneous seminal discharge during sleep. This suggests that stored sperm and/or accessory gland products may have a definite "shelf life" after which they are best discarded and replaced with new in order to stay competitive. Another possibility that has not been properly investigated is that frequent ejaculation may activate some feedback mechanism (*e.g.*, testosterone/inhibin) to stimulate higher levels of spermatogenesis and accessory gland secretion.

No studies are available that make comparisons of human sperm motility, longevity, time to capacitation, or other measure of performance that would reveal mixed strategies within an ejaculate, or specializations between ejaculates. Although there has been a great deal of work done on the causes of male infertility, I find no data that compare performance of sperm among ejaculates of fully fertile men. No published observations of sperm in mixed human ejaculates exist that might reveal incapacitation or selective spermaticidal effects as competitive tactics. Finally, none of these parameters has been studied in the great apes, so there is no basis for interspecific comparison.

5. Mating Patterns

Human males employ a variety of highly variable reproductive tactics, each mix depending on the opportunities to place sperm in competition and the need to defend against competing sperm. Symons (1979) devotes a chapter in his book to the human male desire for sexual variety, which he introduces with the following reproductive arithmetic: "A male hunter/gatherer with one wife (who will probably produce four or five children during her lifetime) may increase his reproductive success an enormous 20 to 25% if he sires a single child by another woman during his lifetime." Small wonder that selection has favored sex-crazed human males.

The most common human male reproductive strategy involves attempts to monopolize sexual access to one or more women through pair-bonding. Maintenance of wives requires a regular outlay of resources, and failure of the male to provide adequately may result in his loss of sexual exclusivity and even possible loss of a wife.

As will be discussed (section VIII), marriage will on average produce higher male reproductive returns than any mix of other tactics that could be pursued at the expense of equivalent resources. Thus whatever mix of extramarital tactics a human male employs should be free or cost less than required to support another wife or equivalent (*i.e.*, mistress/concubine). Prostitution, for example, is an institution that provides males the opportunity to make small cost highly speculative reproductive investments. If marriage is a blue chip stock, then intercourse with a prostitute is the reproductive equivalent of purchasing a lottery ticket. A **very few** purchasers win big.

Shields and Shields (1983) and Thornhill and Thornhill (1983) have meticulously searched and analyzed the available data on rape with the purpose of seeking an explanation for its ultimate cause. Both teams found that rape has reproductive value, and is therefore subject to the forces of natural selection. There is no question that rape results in pregnancies. Best estimates are that 25,000 women became pregnant as a result of the hundreds of thousands of rapes that occurred during the West Pakistani occupation of Bangladesh (Brownmiller 1975), and several gynecologists recorded hundreds of conceptions following the many rapes that occurred in Germany in the three postwar years of 1945-1947 (Jöchle 1973).

Shields and Shields (1983) concluded that ultimately, rape should be expected to occur when its potential benefit (in production of extra offspring) exceeds its potential costs (in energy expended and risk taken). They agree with the sociocultural view that as male hostility increases so likewise does the probability of rape. They contend that the external stimulus for rape is a sufficiently vulnerable female.

Shields and Shields (1983) suggest that warfare is a context that may precipitate rape by almost any man because the risks of detection and punishment are low, and hostilities are high. They find substantial anecdotal support (in Brownmiller 1975) for their contention that the extremely high frequency of rape during war is a function of more males raping, rather than simply some males raping more.

Thornhill and Thornhill (1983) propose and supply much data in support of the position that rape is an evolved facultative reproductive strategy employed primarily by men who are unable to compete effectively for the status and resources necessary to attract and mate with desirable females. They further demonstrate that rape may be incorporated into a repertoire of other patterns, including low committal pairbonding and/or directing resources to sister's offspring (the avunculate).

There is apparently some evidence for unusually high rates of conception following rape (Fox *et al.* 1970; Jöchle 1973, 1975). Two theories attempt to explain this phenomenon. Fox *et al.* (1970) suggest that the release of adrenalin caused by fear and anger may induce contractions in the uterus that facilitate uptake of semen. Jöchle (1973, 1975) proposed that rape may induce ovulation in human females. Both the incidence of conception following rape, and these theses bear further investigation, especially in consideration of their apparent maladaptiveness for females.

Another reason rape may succeed as a reproductive strategy is that it often eliminates subsequent competition from sperm of the victim's primary mate. Genitalic and somatic injury and the psychological trauma that usually accompany rape minimize the possibility that the victim will soon be receptive to coitus (see Thornhill and Thornhill 1983). This leaves the rapist's sperm uncontested in the victim's reproductive tract. Furthermore, the rape victim's principal mate will very likely reject her or refuse normal sexual activity with her following sexual assault (Thornhill and Thornhill 1983). A substantial number of cultures take a surprisingly relaxed view toward rape (Fig. 1F). This may be due to low incidence of rape, or an exceptionally low reproductive payoff potential for forced copulation. In any case, the study of these societies may offer significant new insights into cultural causes of rape.

6. Coitus

Aspects of copulation itself (*i.e.*, position, duration, time of day, and frequency) can be male-controlled and may therefore be optimized by natural selection and adjusted for circumstances to achieve maximum probability of fertilization.

Humans are able to copulate in a variety of positions, but most are variations on two basic positions, dorsal/ventral and ventral/ventral. Most primates, and other mammals for that matter, copulate in the dorsal/ventral position, but humans use primarily the ventral/ventral ("missionary") position. It is reported that females prefer this position because it produces maximum clitoral stimulation and probability of orgasm (Masters and Johnson 1966, Beach 1973, Fisher 1973). It is probably in the male's interest to have his mates experience orgasm, but it is not always in the female interest (see section VII.B.4). The ventral/ventral position does not, however, result in ideal placement of the ejaculate. Ejaculation in the ventral/ventral position produces a semen pool below the cervix (Masters and Johnson 1966), and if the female remains supine following coitus, most of the seminal reservoir is unavailable to the uterus. In contrast, the dorsal/ventral position leads to appression of the male glans to the cervix at full penetration such that semen may be ejaculated directly through the cervical os into the uterus, producing an obvious advantage over vaginal ejaculation. In addition, the ejaculate would be aided by gravity in the dorsal/ventral position. Consequently, this position should be the unmitigated preference of males, but I find no useful data on male preferences to test this prediction. In a similar vein, males should prefer to ejaculate during coitus rather than by any other stimulation. Melanesian island couples of "East Bay" routinely engage in extended mutual heterosexual masturbation, but coitus always takes place just before male orgasm (Davenport 1977).

Human females require about 8 minutes of coitus on average to achieve orgasm with a range of from 1 to 30 minutes, and women may achieve multiple orgasms during extended coitus (Masters and Johnson 1966, Fisher 1973). If female orgasm facilitates uptake of semen and fertilization (as suggested below), or if it is important in mate retention, then males should be selected to choose coital positions and sustain copulation for a minimum period that would allow the typical female to achieve orgasm. As noted, humans usually copulate in the ventral/ventral position and average duration of coitus in humans is about 10 minutes, so the prediction is supported (Table III).

The preference of human males to copulate with their mates at night in the sleeping place (Table III) secures a competitive advantage for primary mates in two ways. First, "sleeping together" permits the investing male to passively guard his mate for several hours after coitus. Second, the male's ejaculate is likely to be retained longer in the sleeping female's reproductive tract.

Human males usually initiate coitus and, therefore, generally have control over its frequency. Coital frequency is highly variable for humans, but typically occurs from 1-3.5 times/week (Table III). It might be expected that the frequency of male-initiated copulation should relate to the perceived potential threat of sperm competition in a relationship. For example, copulation frequency should be low in an old and stable relationship (a corollary to the Coolidge effect), and high in a young or threatened relationship. Clinical psychologists may possess some data useful for testing this prediction.

7. Male Sexual Jealousy and Paternity Assurance

Male sexual jealousy is a behavioral/motivational complex that causes the "dogged inclination" of men to possess and control women, and to use violence or threat of violence to achieve sexual exclusivity and control (Daly *et al.* 1982). The emotion, by defending males' exclusive sexual access to their wives, functions to assure paternity in a system that threatens sperm competition. The proximate manifestations of male jealousy include male precautionary attitudes and methods designed to prevent the occurrence of sperm competition, violence directed toward unfaithful wives and their facultative mates, and universal societal rules and laws that buttress and facilitate these male attitudinal and behavioral predispositions.

Male anticuckoldry tactics have been reviewed by Dickeman (1979, 1981). These include the more conventional practices of virginity tests and guarantees, veiling, chaperoning, and claustration. Also of note are the bizarre and brutal practices of footbinding to prevent females from "straying," clitorectomy to preclude female orgasm and thereby diminish feminine interest in extramarital sex, and infibulation (suturing shut the labia majora) of girls to prevent infelicitous or untimely intercourse. Although footbinding is no longer practiced in China, Hosken (1979) estimates that 65 million living African women have undergone some sort of genital mutilation.

Human males are typically unpleasant about their wives' infidelity. Daly *et al.* (1982) point out that most wife beating is the result of the husband's suspicion or knowledge of his wife's sexual infidelity. Also, male jealousy has ranked high as a motive for domestic homicide. Up to one-third of the wife killing in the United States can be attributed to male sexual jealousy (see Daly *et al.* 1982), and this cause is important in other countries and cultures. For example, a Canadian study credited 85.3% of spouse killings to "sexual matters" (Chimbos 1978); Lobban (1972) discovered male jealousy was the leading motive category for all homicides in the Sudan; Tanner (1970) in a Ugandan study found killings due to male sexual jealousy followed only those committed during robbery and property disputes; 45% of all murdered wives in several cultures in British colonial Africa (Bohannan 1960) involved sexual matters (primarily adultery); and studies of homicide among aboriginal people in India revealed a high proportion attributable to adultery and related sexual problems (see Daly *et al.* 1982).

The double standard is cross-cultural (Fig. 1K). Over 65% of indexed societies have rules against female extramarital sex, but allow male affairs. Twenty-three percent prohibit extramarital liaisons by both sexes, but the prohibition on males in most of these seems designed more to create the illusion of equity than to enforce it (*i.e.,* females are punished for their indiscretions and males are not, or if so, less severely). Daly *et al.* (1982) studied adultery law and found a remarkable consistency of concept: "Sexual intercourse between a married woman and a man other than her husband is an offense." Typically, the offense is viewed as a property violation. The husband is the victim and, depending on the culture to which he belongs, is entitled to extract compensation from the offending male and/or violent revenge on his wife and/or her facultative mate. Clearly, male-dominated politics have universally produced laws and social conventions that reflect and support male sexual jealousy and its ultimate function: to prevent sperm competition and thereby protect male investment in wives and offspring.

B. Female Attributes

1. Cryptic Ovulation and Continuous Sexual Receptivity

Reproductive interests of human males and females are often in conflict. Males attempt to achieve exclusive sexual access to wives with minimum expenditure. By contrast, selection favors females who extract maximum contribution from principal mates while maintaining the option of extramarital mating (see section III). Overt female tactics are not likely to have been successful, and because of males' superior physical strength and political power, would result in reduced female fitness. Thus, the evolution of more subtle female stratagems.

Human males could efficiently allocate their resources with relatively low expenditure per mate if they were provided unambiguous information about the probability of a female conceiving at any given time. I propose that natural selection has produced just the opposite effect. I argue (following others) that continuous sexual receptivity, cryptic ovulation, and some other feminine characteristics have evolved to obscure a human female's current reproductive value and confuse males as a countermeasure to male resource allocation and anticuckoldry strategies. These female adaptations enhance opportunities for facultative polyandry and thus promote human sperm competition.

Most animals have regular periods of mating when copulation can result in pregnancy. This periodicity is usually synchronized between the sexes, and is mediated by hormones released in response to a variety of environmental stimuli. Some tropical mammals reproduce continuously; in these species, female receptivity is coincident with cyclical ovulation and males are always fertile.

Heape (1900) observed that primates deviate from the typical mammalian pattern in that males occasionally copulate with infertile females. This behavior is widespread among the primates (Hrdy 1979, 1981) and is most highly developed in humans. Human females do not experience estrus, offer no conspicuous morphological or behavioral evidence of ovulation, and are more or less continuously receptive to coitus from the onset of puberty.

Cryptic ovulation and perennial sexual receptivity in humans has intrigued sociobiologists, and many have attempted to identify the ultimate causation of these phenomena (Etkin 1963, Pfeiffer 1972, Alexander and Noonan 1979, Hrdy 1979, Burley 1979, Benshoof and Thornhill 1979, Strassmann 1981, Daniels 1983). Most theorists suggest that cryptic ovulation and continuous female receptivity facilitate monogamous pairbonding through the mechanism of permanent sexual attractiveness (Etkin 1954, Morris 1967, Washburn and Lancaster 1971, Crook 1972, Campbell 1974, Washburn 1974, Cook 1975, Halliday 1980, Short 1981, Lovejoy 1981). These notions imply that diminution of or loss in signals about a female's reproductive condition and the extension of sexual receptivity have somehow trapped promiscuous males into monogamy, which is maintained by conjugal sexual contentment for the ultimate benefit of altricial offspring.

Comparative primate behavior studies do not support this hypothesis. Cryptic ovulation is unlikely to evolve in a promiscuous mating system. Consider, for example, the chimpanzee (Table III). Male chimpanzees prefer to mate with females who are on the verge of ovulating. High status males are likely to be most successful in realizing this preference. If the majority of females in such a system provide an honest sign of high reproductive value, such as labial and circumanal swelling (Table II), then the best males will attend the most swollen females. Female chimpanzees at the peak of estrus (those most swollen) do get much more attention from males than do less swollen females (Tutin and McGinnis 1981). Therefore any female variant that de-emphasized swelling, as a first move toward concealed ovulation, would have had diminished attractiveness to the best males, and would therefore have been selected against. Another difficulty with the hypothesis comes from the fact that increased infant dependency could only evolve in response to increased parental investment and not the reverse. Greater offspring dependency unmet by nurturant parents would only have increased infant mortality and could not have evolved. Finally, natural selection favors individuals whose behavior patterns tend to maximize their reproductive success and not their sexual contentment, so males could not have been bound by the continuous sexual attractiveness of wives if the binding itself did not contribute to fitness.

Of the several theories on the original function of concealed ovulation and continuous receptivity, Benshoof and Thornhill's (1979) discussion best recognizes and avoids these pitfalls. They propose that unrevealed ovulation and continuous female receptivity evolved after female monogamy because any group-living female with inconspicuous estrus in a monogamous mating system would be in a better position to mate with a superior male without her being detected by her primary mate. Benshoof and Thornhill further suggest that ovulation is apparently concealed from the female because some degree of self-deception facilitates the deceit of mates.

There is some debate on whether ovulation is actually concealed from the female, or if the event is only protected from her conscious detection. *"Mittelschmertz,"* a pain in the lower abdomen occurring midway through a female's intermenstrual interval is thought to be caused by irritation of the pelvic peritoneum by blood or other fluid escaping from the ovary at ovulation (Osol 1972). If *mittelschmertz* occurs regularly, it might somehow trigger complex adaptive behavioral or physiological processes in a female without her conscious involvement. This could permit females' selectivity of sperm without alerting males. Support for this idea comes from studies that reveal females copulate more frequently at midcycle than at other times (Udry and Morris 1968, Adams *et al.* 1978, and see Hrdy 1979).

Heightened female libido during the first few months of pregnancy (see Masters and Johnson 1966) may also contribute to female reproductive inscrutability, and facilitate cuckoldry under conditions of facultative polyandry. In this regard, Howell (1979) found that !Kung women do not reveal or acknowledge their pregnancy until it shows after about the third or fourth month.

2. Perennial Pendulous Breasts

Following parturition, female mammals cease to cycle reproductively for variable periods of time generally coincident with the duration of lactation (Tortora and Anagnostakos 1984). Lactation is maintained by prolactin, a hormone produced by the adenohypophysis in response to suckling stimulation of the breasts. Active mammary gland tissues expand, causing breast enlargment and pendulosity. Non-human primates, including the great apes, develop pendulous breasts only at the onset of lactation, and their breasts remain pendulous only for the duration of lactation.

Human females are unique among the hominoids (and mammals) in that they develop pendulous breasts at puberty and retain them throughout life irrespective of lactation. Human pubertal breast development involves estrogen-induced hypertrophy of the adipose and stromal tissue of the mammary gland, rather than activity of the glandular epithelium itself. The quantity of contained adipose determines breast size, and size bears no relationship to milk production ability (Tortora and Anagnostakos 1984).

Short (1980), Halliday (1980), and others have suggested that human breasts evolved their perennially pendulous condition to maintain constant male sexual interest in the female and therefore facilitate pair bonding and paternal investment. Although males in Western cultures are undeniably fascinated by female breasts, it is not clear this preoccupation is a cross-cultural universal. Male African graduate students have expressed surprise and amusement at this focus of U.S. males. "We look at the whole woman," is their typical reaction.

Although cross-cultural studies on intersexual attraction have been conducted (e.g., Rosenblatt 1974), the question of which specific female features and which qualities of these are admired by males in different cultures apparently has not been investigated. Ford and Beach (1951) noted that appearance is important crossculturally, especially in males' assessment of females, but found that the criteria of physical attractiveness vary widely among cultures. The only regularly recurrent characteristics that could be identified were health and feminine plumpness. This latter characteristic, although alien to members of Western food-rich, diet-conscious cultures, must provide an important indication of female health and reproductive potential in a predominantly food-limited world. Plumpness must also have been an excellent predictor of female fecundity throughout most of our evolutionary history. High investment males may therefore be attracted by females having pendulous breasts because this feature conspicuously displays stored feminine resources, emphasizes the principal organs of maternal contribution, and may predict females' ability to contribute. However, a problem is apparent with all of these proposed "functions." None of the aforementioned seems likely to have **initiated** selection for perennially pendulous breasts in hominids.

Pendulous breasts should initially have been sexually repulsive, at least to high status males, because they would have signaled lactation, amenorrhea, anovulation, and therefore a female of little or no current reproductive value. Astute males certainly would tend to ignore lactating females in order to focus their attention on others with obvious current reproductive value. As explained later (section VIII. B), selection for perennial pendulous breasts prior to the evolution of extended human pair bonding would have been especially problematical. In this regard, Tutin and McGinnis (1981) observed that female chimpanzees who resumed sexual cycles while still lactating mated infrequently with adult males. In contrast, a female who resumed sexual cycling (with flattened breasts) 2 months after the death of her 2.5-year-old son was observed to be extremely attractive to males.

So, if the evolution of perennial breast pendulosity was not initiated by epigamic selection, what forces were involved? I propose the same function as discussed above for the evolution of concealed ovulation and continuous sexual receptivity, *i.e.,* to render additional ambiguity of female reproductive value in an imperfect system of female monogamy imposed by male domination.

After the evolution of concealed ovulation and continuous receptivity by females, males still would have had two conspicuous indicators of low female reproductive value: the swollen abdomen of third trimester pregnancy, and the swollen breasts of lactation. Both signs indicate current low reproductive value, and neither of these functional conditions could be disguised. However, the informational content of enlarged breasts could be progressively obscured by gradually extending the period of enlargement. Presumably, lactating females were not as assiduously guarded by their principal mates, and a mechanism to extend the relaxation of male vigilance into a post-lactational ovulation would enable sexual liaisons with other males. Such facultative polyandry would have been advantageous for the social, material, and genetic benefits discussed previously (section III). Though potential facultative mates would perceive females with enlarged breasts to have low reproductive value, mating would occur and female benefits accrue as is seen in other primates (see section VI, and Hrdy 1979). In a primarily monogamous mating system, promiscuous mating opportunities for males are relatively rare and should therefore almost always be beneficial, even with low value females. So, the evolution of perennial pendulous breasts could have initially evolved to provide females greater opportunity for facultative polyandry. Male epigamic selection may have operated, perhaps on the size and form of breasts, but only after the general trait had been fixed by selection for its original function.

3. The Hymen

Males in most societies place a high premium on virginity of prospective wives, and this preference has been institutionalized in some cultures (Dickeman 1978, 1981). Virginity tests typically involve examination of the female to determine that the hymen is intact, or of the nuptial bedding for blood stains—evidence of hymen rupture by first intercourse.

The hymen is a membranous partition partially blocking the orifice of the vagina. It may take on a variety of forms, but is typically circular or crescentic. Occasionally, it may be multiple, entirely lacking, or imperforate (Osol 1972). Although the structure may be ruptured by activities other than sexual intercourse, usually the first coitus breaks the hymen, causing female pain and bleeding.

The hymen is one of the great unsolved mysteries of human anatomy. I know of no plausible hypothesis for any physiological function it may serve, and I know of no other organ in the animal kingdom evolved inevitably to be injured. However, the trauma caused females of certain bugs (see Thornhill and Alcock 1983) and poeciliid fishes (see Constantz, this volume) by male genitalia during copulation is analogous. If the hymen does have some as yet undiscovered physiological function, why should it cease to be important after the female's first copulation?

Dickeman (1978), among others, apparently believes that the vaginal hymen evolved by selection due to male preference for virgin wives in hypergamous systems. It is true that members of most societies are aware of the hymen and its significance as an indicator of female virginity. Therefore, in some cultures, a mutant female who lacked this "evidence" of virginity could be selected against.

The question here, as with concealed ovulation and perennial pendulous breasts, is how did the characteristic originally increase in frequency? The trait in a new variant would have no instant value in the context of intersexual selection, and would seem disadvantageous since any structure subject to injury should be selected against unless it confers some compensatory advantage. If the hymen were present in females of closely related species, some primitive function might be discerned, but K. G. Gould (pers. comm.) of the Yerkes Primate Research Center finds no evidence for a structure homologous to the hymen in any of the great apes. I assume, then, that the structure originated in the hominid line and perhaps first arose in *Homo sapiens.*

I offer no hypothesis for the initial evolution of the hymen, but concede that the structure may have been maintained and even further developed in human females by intersexual selection in the context of male defense against sperm competition.

4. Female Orgasm

It would be adaptive for facultatively polyandrous or promiscuous females to assist the sperm of preferred mates and handicap those of nonpreferred mates. In this way a female might exercise some control over the paternity of her offspring while pursuing the nonreproductive benefits of facultative polyandry. Orgasm might permit females to achieve both of these objectives.

Female orgasm has been investigated by Masters and Johnson (1966). They found that single or multiple orgasms may be achieved by either direct or indirect clitoral stimulation. Indirect stimulation of the clitoris sufficient to induce female orgasm can be accomplished most effectively during intercourse in the ventral/ventral position (Masters and Johnson 1966, and Beach 1973). At the peak of female orgasm, the outer part of the vagina contracts rhythmically from 3-15 times, the inner blind end of the vagina expands, and the uterus contracts rhythmically for several seconds.

These events are believed to be mediated by the hormone oxytocin released from the posterior pituitary into the circulation as an autonomic response to clitoral stimulation (Fox and Fox 1969; Bishop 1971, 1973). Until recently, this belief was based largely on studies of nonhuman animals and circumstantial evidence; however, recent experiments have revealed a consistent, significant elevation of blood serum oxytocin in female subjects coincident with their orgasms induced by masturbation (J. Davidson, pers. comm.).

Symons (1979) views female orgasm as an essentially nonadaptive, or even "dysfunctional" byproduct of mammalian bisexual potential. He state ". . . orgasm may be possible for female mammals because it is adaptive for males." Symons sees the capacity of human females to experience multiple orgasms as an incidental effect of their inability to ejaculate. These attitudes seem as naive as they are chauvinistic. If there is as much variation in the ability of human females to achieve orgasm as is noted by Symons in support of his thesis, variants must have existed upon which natural selection could have operated to remove this complicated "artifactual, possibly maladaptive trait" from female populations.

An alternative to Symon's view is that human female orgasm may facilitate the transport of semen from the vagina into the uterus, and thereby provide the female with some control over the use of ejaculates. Grafenberg (1950) and Masters and Johnson (1966) conducted experiments on human females that addressed this possibility. However, they failed to demonstrate the entry of radio-opaque fluid from the cervical cap into the uterus after either coitus or clitoral stimulation. The latter experiment has been criticized by Fox and Fox (1967). In addition, Fox et al. (1970) used radio telemetry to demonstrate a positive intrauterine pressure of ca. +40 cm H_2O during coitus and a negative intrauterine pressure of ca. −26 cm H_2O immediately after female orgasm (see also Fox and Fox 1971).

These results confirm the uterus creates suction immediately following orgasm and support the idea that female orgasm may actively transport semen and thereby facilitate fertilization. If this conclusion is indeed correct, orgasm may provide females with some opportunity to control fertilization. For example, a female might achieve orgasm during, or masturbate to orgasm following coitus with a preferred mate. Conversely, she could avoid orgasm during or following coitus with a less desirable potential father.

The Mangaia islanders practice penile subincision, a curious form of male genital mutilization which entails slitting the urethra from its orifice for a variable distance toward the scrotum (Marshall 1971). Aboriginals appropriately refer to this process as *mika,* meaning "terrible rite" (Short 1980). Subincision almost certainly results in semen being shallowly discharged against the vaginal wall with the result that female orgasm may be required in order for fertilization to occur. It is perhaps no coincidence therefore that the Mangaian culture is also known for its elaborate instruction to pubertal males on coital methods designed to cause multiple orgasms in the female and to insure female orgasm simultaneous with male ejaculation (Marshall 1971).

Psychosexual anecdotes suggest that female orgasm during coitus is most easily accomplished if the female feels generally secure and is convinced of at least some degree of future support from her partner (Fisher 1973, Hite 1976). It is significant that call girls, prostitutes who are well paid and well treated by high status clientele, have orgasms as frequently as non-prostitutes. Streetwalkers, by contrast, have difficulty experiencing orgasm (Exner *et al.* 1977). These prostitutes are subject to harassment from both clients and procurers and are unable to exercise selectivity on clients who typically belong to the lowest socioeconomic strata. This pattern is consistent with the thesis that females may choose the fathers of their children after mating, or at least minimize the possibility of impregnation by undesirable males.

Finally, if female orgasm facilitates sperm transport and fertilization, its absence during and following rape might militate against conception by a rapist's sperm. Consistent with this idea, I find no reports of rape victims having experienced orgasm during the sexual assault. However, this issue is confounded by the possible existence of other mechanisms that may enhance the prospects for rapists' sperm (see section VII.A.5).

5. Female Reproductive Tract and Ova

Females may also control sperm by means of the physical, chemical, and immunological characteristics of their reproductive tract secretions. The secretions of the female tract have been studied both to determine causes of infertility and in search of improved contraceptive technologies, but never from the viewpoint of sperm competition. Here, I trace the movement of sperm in the female tract and consider how and where the female system may aid or impede a particular ejaculate, or discriminate among sperm from different ejaculates.

Female emotional factors such as attraction, affection, and trust together with the investment in sexual foreplay rendered by a mate prior to coitus are known to influence the vaginal environment. Specifically, the quality and quantity of vaginal mucus produced by the vestibular glands are affected (Tortora and Anagnostakos 1984). Mucus lubricates the vagina to facilitate intercourse and perhaps the passage of sperm to the cervical os.

Insufficient lubrication prior to coitus, as in rape, may result in trauma to the vaginal wall with the discharge of blood into the vaginal cavity. Blood serum contains the spermatotoxic protein gamma globulin and often sperm antibodies as well (Franklin and Dukes 1964a, b), so females who experience vaginal trauma with bleeding during forced copulation may be defended to some extent against impregnation by constituents of their own blood, and by the lack of vaginal mucus to aid sperm movement.

The human vaginal environment is considered generally hostile toward sperm. Evidence for this is partially circumstantial in that only a very small fraction of ejaculated sperm ever reach the uterine tubes where fertilization usually occurs. Of the hundreds of millions of sperm contained in each ejaculate, only about 2,000 arrive in the vicinity of the descending ovum (Tortora and Anagnostakos 1984). Selection on the female vagina in the context of ejaculate competition may have favored the evolution of secretions that produce extraordinarily rigorous environments for sperm. This possibility might be explored by experiments to test the prediction that vaginal secretions in humans and the chimpanzee will be more hostile toward sperm than those of the gorilla and orangutan.

The cervix may be an important filter of sperm because of its key position in the route to fertilization (Porter and Finn 1977). All ascending sperm must negotiate the cervix and contact cervical mucus. During most of the female reproductive cycle, the high viscosity and elevated protein and leucocyte content of the cervical mucus impede the progress of sperm. At midcycle, there is a 10-fold increase in the volume of cervical mucus produced. The midcycle mucus secreted beginning about 6.2 days prior to ovulation is a low viscosity, low protein, more hydrated product than the earlier cycle secretions. Midcycle mucus facilitates passage of sperm, and is therefore helpful to male gametes contained in ejaculates deposited during this time. Females may use subliminal cues to copulate with preferred mates during this time and avoid coitus with non-preferred mates. The change in cervical mucus is apparently so reliably associated with ovulation and easily recognizable by females, that it has been suggested as a means of predicting ovulation (see Porter and Finn 1977).

Franklin and Dukes (1964a, b) first reported a high incidence of sperm agglutinin in the blood serum of females with sterility of unknown etiology. A sperm agglutination factor believed to be transferred from serum is found in cervical mucus (Isojima 1973). Antibodies in cervical mucus have been reported at higher levels than in the blood (Eyquem *et al.* 1976, Parish *et al.* 1967). The highest levels of sperm agglutinin factor have been found in prostitutes, followed by married females, who in turn have higher levels than unmarried women. Conceivably, antibodies produced by married women may be specific to husbands' sperm. This

raises the discomfiting possibility (for married men) that the gametes of facultative mates may enjoy a competitive advantage over those of husbands' handi-
capped by wives' immune systems. Significantly, "condom therapy" that shields
wives from exposure to husbands' sperm for a period of several months, apparently
halts production of antibodies and causes reduced female titers of sperm agglutinating and immobilizing factors (Kay 1977).

The uterine tube is the final leg in the journey of human sperm. Sperm are
usually capacitated and activated (made biochemically ready for penetrating the
ovum) in the tubal environment. Tubal secretions are known to be involved in these
processes, although the molecular mechanisms are not well understood (Johnson
and Everitt 1980). Oviducal fluid is produced most copiously around the time of
ovulation and its composition at this time stimualtes maximum spermatic metabolism (Blandau et al. 1977). The change in the quantity and characteristics of
oviducal fluid is apparently under hormonal control.

Psychological factors are known to influence reproductive endocrine functions
(Okamura et al. 1973), and psychosomatic sterility is an established syndrome in
human females (Mori et al. 1973, Uemura and Suzuki 1973). Although the details
of this syndrome are vague, it is known that emotional anxiety in psychologically
normal females can influence oviducal motility and the constitution of tubal
fluid (Asaoka et al. 1973). Since both tubal mobility and fluid composition have a
critical influence on the progress and activation of sperm, it seems reasonable to
suggest that females have the capability of exercising control over sperm in the
uterine tubes.

The final barrier to fertilization is the zona pellucida of the ovum. In order for
fertilization to occur, the acrosome of a spermatozoon must release its enzymes
against the egg's gelatinous covering, and the enzymes must succeed in dissolving
a hole so that penetration may occur. The ovum may be surrounded by many
sperm attempting to penetrate, but only one will succeed and this may not necessarily be the sperm first to arrive at the membrane (Chang et al. 1977). Though
no data specifically address the subject, it seems evident from descriptions of in
vitro fertilization that egg membranes have properties that may favor penetration
by some sperm over others within an ejaculate. The much higher variability among
sperm from mixed competitive ejaculates would seem to permit greater opportunity
for discrimination by the egg membrane.

VIII. SPERM COMPETITION AND HUMAN EVOLUTION

Pilbeam (1984) recently provided an update on hominoid historical evolution
that synthesizes the past 5 years' fossil and archaeological discoveries. I have

abstracted this useful paper in the following paragraphs to provide a factual frame-work upon which I attempt to contruct a scenario for the evolution of hominoid mating systems and to fit the elements presented in Section VII.

A. Hominoid Evolution

It is not clear precisely when the hominid line diverged from the Old World monkeys, but current consensus places the event after middle Oligocene and before early Miocene (20-30 million years BP). The molecular "clock" and other evidence has the orangutan line branching about 15 million years BP, followed by the gorilla line 9-10 million BP. Then the chimpanzee line diverged between 6-10 million years ago from the lineage that ultimately gave rise to the hominids. Experiments on DNA hybridization have revealed at least 98% identity in nonrepeated DNA in humans and the chimpanzee, indicating a remarkably close relationship (see Lovejoy 1981).

Bipedal hominids were present in eastern Africa by at least 3.5-4 million years BP. *Australopithecus africanus,* the first species reasonably well represented in the fossil record (primarily from Hadar, Ethiopia), was strongly dimorphic, with males from 50-100 percent larger than females. This species had chimpanzee-like facial features, but the overall skull was more gorilla-like. The canines, however, were low-crowned and without ape-like forward projection. Between 2-2.5 million years BP, another sexually dimorphic species, *A. boisei,* appeared in Africa. This species lived in an area of woodland and savanna, away from the forests. Ecologically, these early hominids shared more in common with modern *Gorilla* than with humans. Their dentition (*i.e.,* large cheek teeth capped with thick enamel) indicates a vegetable diet. Walker and Grime (see Pilbeam 1984) have studied the teeth of *Australopithecus* and concluded that all were apelike vegetarians, eating broadly similar diets that required a great deal of chewing. The *Australopithecus* diet probably varied from that of modern forest-dwelling apes only in that it may have contained more roots and tubers and less fruit. No stone tools have been found associated with fossils, so it is assumed that if tools were constructed, they must have been of perishable materials. Tool use is inferred from reduction of canines, bipedal locomotion, and skeletal evidence of manual dexterity.

Homo habilis was present at the same time as *A. boisei,* and persisted until about 1.75 million years ago. *H. habilis* vanished and was replaced by *H. erectus,* who had a larger brain than the australopithecines, but resembled members of the older genus. The appearance of *H. habilis* was coincident with the first archaeological sites that contained used or altered stone, as well as non-hominoid animal remains.

It is widely acknowledged that this period is marked by a significant shift in diet from plant to animal food. It is hazardous to suppose any particulars in the lifestyle of *H. habilis.* Perhaps it was a hunter-gatherer with home camps and division of labor with shared resources. Alternatively, the species may have been primarily vegetarian with some opportunisitic scavenging of meat and a lower level of social organization. The physical evidence is sufficient to support either scenario, but inadequate to distinguish one from the other. *Homo habilis* survived for a few hundred thousand years and was replaced by *H. erectus.*

Homo erectus was a widely-distributed species that first appeared in Africa. At 1 million years BP, it was present in southeastern and eastern Asia, where it survived until at least 300,000 years ago. This species made flaked stone tools and lived in groups, some of which may have used fire. It resembled early *H. sapiens* in gross body morphology, and seems to have been both morphologically and culturally stable for about 1.5 million years. A gradual transition from *H. erectus* to our own species began about a half million years ago. The transition from "archaic" *H. sapiens* (including Neanderthal man) to "modern" *H. sapiens* apparently represents a continuum.

Neanderthals and their contemporaries were different from us both physically and behaviorally. Skeletal evidence suggests that they were much more robust and muscular than modern humans. Their teeth were larger than those of modern *H. sapiens.* Dental wear patterns indicate that Neanderthals used their teeth for non-feeding activities, probably chewing of hides to soften them, by analogy with the modern Eskimo. This implies the use of clothing.

Homo sapiens sapiens appeared sometime between 50,000 and 100,000 years BP, and was characterized by loss of robustness and changes in the female pelvis. The female skeletal changes are indicative of some change in pregnancy and births. These alterations may have included reduction in the gestation period from 11 months to the present 9 months and a concomitant increase in dependency of infants on parental care. The prediction is made on the basis of morphometric comparisons with the great apes and other mammals (see Pilbeam 1984).

B. Hominid Mating Systems and Sperm Competition

There is no indication that the strongly dimorphic species *Australopithecus africanus* and *A. boisei* had a sexual division of labor that would necessitate the separation of males from females or give males control over high quality food resources not directly available to females. There is likewise no physical evidence that the australopithecines used tools or sophisticated weapons for hunting big game, and in fact nothing to suggest that they regularly ate meat. Consequently, their social organization was probably similar to that of extant gorillas, *i.e.,* small mixed sex groups, each with a dominant male who was in constant contact with and control of females. It may be reasonable to suggest that their mating pattern consisted of male dominance polygyny and female monogamy, as in the system of

Gorilla. If this assessment is correct, sperm competition should not have been a significant evolutionary force in the australopithecines. Two factors, intergroup competition and the sharing of high quality food, probably played a major role in destabilizing the single male dominance mating system of these early ancestors. By analogy, the first of these factors seems to have been important in development of female promiscuity in chimpanzees, and may well have played a role in humans (see Alexander and Noonan 1979). Some fossil evidence exists to mark the historical onset of the second element.

Homo habilis probably hunted meat by scavenging. Hunting for meat first took the form of scavenging, which initiated a sexual division of labor. Scavenging the kills of large predators clearly involved more risk than gathering vegetable material, insect larvae, and the like. Early hominid scavengers would regularly encounter large predators, and no doubt had to compete directly with other species highly skilled in this specialization. The rigors and risks of scavenging would almost certainly have excluded pregnant females and those with infants.

Possession of meat by subordinate males would have been disruptive to the single male dominance system in two ways. First, it would have given subordinate males the power to lure females away from dominant males while giving females a strong incentive to mate with subordinates. Second, if a dominant male began to engage in scavenging, he would be separated from "his" females, thus increasing their opportunities for facultative polyandry. Females at this time were probably selected to deceive the dominant males in order to maximize their harvest of meat from subordinate males. These changes in turn diminished most of the male intersexual advantage for being large, and therefore selection for large male size was relaxed.

Sperm competition would have intensified, triggering a series of reciprocal evolutionary changes in the reproductive organs of males and females. First, males probably rapidly developed both enlarged penises and large scrotal testes. In response, females would have evolved more rigorous vaginal, uterine, and tubal environments. Males countered with the elaboration of protective and manipulative accessory gland secretions, while at the same time increasing the volume of their ejaculates to neutralize vaginal defenses and otherwise mitigate against female control of sperm. Females may now have begun to exert psychological control over their reproductive tracts and secretions.

Males were selected to identify and favor females of the highest reproductive value, in the exchange of meat for sex. As a consequence, females may initially have evolved honest signals that emphasized their actual high reproductive value, and later modified these signals to create the illusion of high value. Selection could have favored extension of female sexual receptivity before and after ovulation (as has occurred in chimpanzees), but it seems unlikely that female reproductive crypsis and perennial pendulous breasts could have been initiated at this stage because any tendency to crypsis in a female would have made her less desirable

than signaling females with the result that she would have received less male attention and, hence, fewer male resources. Similarly, pendulous breasts would have
signaled low reproductive value and would have resulted in reduced male interest.

The sex-for-meat strategy pursued by males may have been the driving force
behind the transition from scavenging to hunting, a far more hazardous, but potentially more rewarding method of obtaining meat. Individual hunting with primitive
weapons may not have been as efficient as scavenging, but cooperative hunting of
big game developed into a hugely successful operation probably beginning with
H. erectus about 1 million years ago.

Though a certain level of cooperation may have evolved earlier by selection
arising from intergroup conflict, the benefits of cooperative hunting certainly
resulted in even greater selection for behavior patterns and cultural adaptations
tending to increase male solidarity and reduce overt intermale competition and
its disruptive consequences. It is reasonable to assume that these adaptations would
have included mechanisms to reduce conflict over mates. The form of these adaptations may have resembled those of the chimpanzee system of female promiscuity.

So, the early social contract would have guaranteed meat and sex for every
contributing male. This obviously would have created very intense sperm competition that drove genital selection to remove all of the heritable variation in the male
population. As a result all males had approximately equal chance to reproduce,
but no male could have any confidence of his paternity for any offspring of any
female. This situation would seem to be unstable, especially if the product of
cooperative hunting was not equally distributed. Good hunters or hunt leaders
probably demanded and were afforded an extra share of meat from each kill. This
inequity would have created the opportunity to translate hunting/leadership
skills into extra matings and hence potential differential reproductive success.
However, variance in male reproductive success would have been reduced by female
promiscuity. Consequently, the reproductive success of superior males with abundant resources to exchange for sex would be limited only by the time to copulate,
and by ejaculate production capability. At this point, as in the chimpanzee analogy,
high quality males would be selected to supplement promiscuous mating with
temporary consortship. The relatively high level of paternity confidence afforded
by consortship would have been of great advantage, especially to top-ranking
males with the most abundant resources at their disposal for investment in offspring. The gradual shift in mating strategies from purely promiscuous to a mix
that emphasized mate defense polygyny would not destabilize the primitive society.
The now obligatory interdependence of males would make for stability and nonconsorting males would continue to have sexual access to nonconsorting females.

Improved weaponry and hunting techniques would have increased the resource
base so that progressively more males could afford to engage in consortships. This
caused competition among females for males able to establish consortships, and
selected for female inclination to consort rather than to mate promiscuously. The

disappearance of female promiscuous mating may have begun selection for forced copulation as a low investment male reproductive option.

Consortship would have gradually given way to long-term pairbonding with progressively higher levels of male investment in mates and offspring. Increased male investment coincided with the increased levels of paternity assurance that resulted from improved male efficiencies in controlling their mates' reproductive activities. Selection strongly favored sexually jealous males who maintained tight control over their mates, and this progressively and substantially reduced the levels of sperm competition. Females evolved cryptic ovulation, perennial pendulous breasts, and behavioral deceptions to militate against male control efficiencies, and to achieve reproductive and other advantages from facultative polyandry. Paradoxically, female opportunity to secure advantages by facultative polyandry was created and propagated by the ability of males to mix high and low investment strategies.

Long-term pairbonding and a willingness of males to make parental investment, as distinguished from mating effort (see Gwynne, this volume) probably enabled reduced gestation and increased infant dependency. As evidenced in the changes of the female pelvis that distinguish *Homo sapiens sapiens* from earlier forms of our species, paternal care and increased infant dependency may be relatively recent events in hominid evolution.

The potential for sperm competition apparently has remained sufficiently strong into recent prehistoric and historical times for there to be no relaxation of selection on what may be very old male genitalic and emotional/behavioral anticuckoldry adaptations. Historically, male-dominated societies have universally evolved politics that reinforce individual anticuckoldry adaptations and have instituted a variety of new practices that serve this function. Very recent social institutions evolved in the context of sperm competition may even have selected a female structure, the hymen, that is unique to humans.

SUMMARY

Sperm competition occurs in humans and apparently has been a force in the evolution of certain human characteristics. It occurs in the contexts of communal sex, rape, prostitution, courtship, and (most commonly) facultative polyandry. Facultative polyandry permits a female to maximize genic benefits for her offspring, and to secure material resources and protection from males. It may also accomplish social benefits. Data on the incidence of human sperm competition are meager. Most come from forensic genetics studies conducted to exclude paternity, and very few from human population genetics studies. These data indicate the occurrence of sperm competition and the threat of cuckoldry, but reveal that long-term pairbonding (*i.e.,* marriage) achieves a high level of paternity assurance.

Among our relatives the great apes, chimpanzees have a social system and some sexual behavior patterns that most closely resemble our own. Relatively high levels of sperm competition occur in humans and chimpanzees, compared with the gorilla and the orangutan. Coincident with these differences are male genitalic similarities between humans and chimpanzees that include large specialized penises and large scrotal testes. Male gorillas and orangutans have small penises and nonscrotal testes. Human and chimpanzee females both have conspicuous sexual ornamentation (breasts in humans and genital swelling in chimpanzees), but female gorillas and orangutans lack conspicuous sexual displays.

A variety of human anatomical, physiological, behavioral, and cultural characteristics may have evolved in the context of sperm competition. Among male characteristics probably influenced by this force are the size and structure of the testes, accessory glands, ejaculatory duct, and penis. The human scrotum probably evolved by sperm competition favoring efficient storage of spermatozoa. Seminal plasma constituents and ejaculate production capability were almost certainly influenced by interejaculate competition. Male masturbation and nocturnal emission in lieu of intercourse may maintain highly competitive ejaculates. A variety of human male mating tactics have evolved to place sperm in competition and to defend against competing sperm. Male controlled aspects of copulation including coital position, time, duration, and frequency may be optimized in the context of sperm competition. Finally, male sexual jealousy encompasses a constellation of masculine attitudes and aggressive behavior patterns evolved to defend against sperm competition. These attitudes and behavior patterns are universally reinforced by cultural institutions.

Human female characteristics, including continuous sexual receptivity, cryptic ovulation, and perennial pendulous breasts, have apparently evolved as countermeasures to male resource allocation and anticuckoldry strategies. These female attributes enhance opportunities for facultative polyandry, promote sperm competition, and may permit females some control over paternity of offspring. The motility and secretions of the female reproductive tract are influenced by female emotional states, which offers females opportunity to control the fertilization of ova in a facultatively polyandrous mating system by selectively aiding sperm from preferred mates or handicapping sperm from undesirable mates. The hymen is a mysterious structure because of its obscure origin and its absence in the great apes, but it may have been maintained and even elaborated in human females by an institutionalized male preference to marry virgins in hypergamous societies.

Early hominids, *i.e., Australopithecus* spp., probably formed mixed groups, each with a large dominant male. These species would then have had a male dominance polygyny/female monogamy mating system, and thus little or no sperm competition. I postulate that this system was destabilized by hunting (scavenging) and possession of meat by subordinate males beginning with *Homo habilis.* Cooperative hunting was probably begun by *H. erectus* and set conditions that led to a promiscuous mating system and high levels of sperm competition. This in turn would have caused intense reciprocal selection on the genital organs of males and females such that all of the heritable variation was quickly removed from populations. Unequal distribution of the product of cooperative hunting probably caused some males to begin consortships which would have reduced sperm competition, improved individual paternity assurance, and begun to destabilize the promiscuous mating system. Improved paternity assurance permitted evolution of longer term pairbonding, paternal investment, and intensifed male sexual jealousy. More efficient male control of females selected female reproductive crypsis in the form of continuous sexual receptivity, concealed ovulation, perennial pendulous breasts, and other deceptive tactics. Paternal investment enabled reduced gestation and increased infant dependency in *H. sapiens sapiens.*

Historically, sperm competition has apparently remained a sufficiently important force to permit no relaxation of selection on very old genitalic and male emotional/behavioral anticuckoldry adaptations. During recorded human history, male attitudes forged by the potential for sperm competition have come to be reflected in political, legal, and social institutions that repress human females.

ACKNOWLEDGMENTS

I am grateful to the following people for assistance by way of useful discussions, papers, and personal communications: J. Alcock, J. Brewer, N. Burley, N. A. Chagnon, J. M. Davidson, M. Dickeman, I. M. Ellman, K. G. Gould, H. Harpending, K. Hummel, D. Kaye, H. P. Krutsch, A. Schlegel, P. I. Terasaki, R. Thornhill, R. L. Trivers, and D. W. Zeh. J. Alcock, D. N. Byrne, T. Myles, W. L. Nutting, F. G. Werner, J. Smith, and D. W. Zeh reviewed the manuscript and all provided useful suggestions. David Zeh deserves special recognition for carefully editing the editor and substantially improving the final product. Finally, I thank Jill Smith for acting as the sounding board for all of my ideas, for producing the figures, and for helping with all aspects of the project.

REFERENCES

Adams, D. B., A. R. Gold, and A. D. Rise. 1978. Rise in female-initiated sexual activity at ovulation and its suppression by oral contraceptives. *New England J. Med.* **229**:1145-1150.
Aiyappan, A. 1935. Fraternal polyandry in Malabar. *Man India* **15**:108-118.
Alcock, J. 1984. *Animal Behaviour: An Evolutionary Approach.* Sinauer, Sunderland, MA.
Alexander, R. D. 1974. The evolution of social behavior. *Annu. Rev. Ecol. Syst.* **5**:325-383.
Alexander, R. D. 1977. Natural selection and the analysis of human sociality. In *Changing Scenes in the Natural Sciences: 1776-1976,* C. E. Goulden (ed.), pp. 283-337. Philadelphia Acad. Nat. Sci. Special Publ. 12.
Alexander, R. D. 1979. *Darwinism and Human Affairs.* Univ. Washington Press, Seattle, WA.

Alexander, R. D., and K. N. Noonan. 1979. Concealment of ovulation, parental care, and human social evolution. In *Evolutionary Biology and Human Social Behavior: An Anthropological Perspective,* N. A. Chagnon and W. G. Irons (eds.), pp. 436-453. Duxbury Press, North Scituate, MA.

Anonymous. 1982. Thirty-nine percent of paternity charges are phony, tests reveal. *Current Med. Attorneys* **29**:10.

Archer, J. 1810. Facts illustrating a disease peculiar to the female children of Negro slaves. *Med. Reposit.* 1:319.

Asaoka, T., S. Iwabuchi, and H. Yamamoto. 1973. Influence of emotional anxiety on tubal factor in infertile women. In *Fertility and Sterility,* T. Hasegawa, M. Hayashi, F. J. G. Ebling, and I. W. Henderson (eds.), p. 970. Excerpta Medica, Amsterdam.

Athanasiou, R. 1973. A review of public attitudes on sexual issues. In *Contemporary Sexual Behavior,* J. Zubin and J. Money (eds.), pp. 361-390. Johns Hopkins Univ. Press, Baltimore, MD.

Barash, D. P. 1982. *Sociobiology and Behavior,* 2nd ed. Elsevier, New York.

Bateson, P. (ed.) 1983. *Mate Choice.* Cambridge Univ. Press, Cambridge.

Beach, F. A. 1973. Human sexuality and evolution. In *Advances in Behavioral Biology, Vol. II Reproductive Behavior,* W. Montagna and W. A. Sadler (eds.), pp. 333-365. Plenum Press, New York.

Beall, C. M., and M. S. Goldstein. 1981. Tibetan fraternal polyandry: A test of sociobiological theory. *Am. Anthropol.* 83:5-12.

Bedford, J. M. 1977. Evolution of the scrotum: The epididymis as the prime mover? In *Reproduction and Evolution,* J. H. Calaby and C. H. Tyndale-Boscoe (eds.), pp. 171-182. Aust. Acad. Sci., Canberra.

Benshoof, L., and R. Thornhill. 1979. The evolution of monogamy and concealed ovulation in humans. *J. Soc. Biol. Struct.* 2:95-106.

Bergström, S. 1973. Prostaglandins and human reproduction. In *Fertility and Sterility,* T. Hasegawa, M. Hayashi, F. J. G. Ebling, and I. W. Henderson (eds.), pp. 40-44. Excerpta Medica, Amsterdam.

Bishop, D. W. 1971. Sperm transport in the fallopian tube. In *Pathways to Conception: The Role of the Cervix and the Oviduct,* A. I. Sherman (ed.), pp. 99-109. Charles Thomas Publ., Springfield, IL.

Bishop, D. W. 1973. Biology of spermatozoa. In *Sex and Internal Secretions, Vol. 2,* W. C. Young and G. W. Corner (eds.), pp. 707-796. Robert E. Krieger Publ., Huntington, NY.

Blandau, R. J., B. Brackett, R. M. Brenner, J. L. Boling, S. H. Broderson, C. Hammer, and L. Mastroianni. 1977. The oviduct. In *Frontiers in Reproduction and Fertility Control,* R. O. Greep and M. A. Koblinski (eds.,), pp. 132-145. MIT Press, Cambridge, MA.

Bohannan, P. 1960. *African Homicide and Suicide.* Princeton Univ. Press, Princeton, NJ.

Broude, G. E., and S. J. Greene. 1976. Crosscultural codes on twenty sexual attitudes and practices. *Ethnology* 15:410-429.

Brownmiller, S. 1975. *Against Our Will: Men, Women, and Rape.* Simon and Shuster, New York.

Bullough, V., and B. Bullough. 1978. *Prostitution, An Illustrated Social History.* Crown Publ., Inc., New York.

Burley, N. 1979. The evolution of concealed ovulation. *Am. Nat.* 114:835-858.

Burley, N., and R. Symanski. 1981. Women without: An evolutionary and cross-cultural perspective on prostitution. In *The Immoral Landscape: Female Prostitution in Western Societies,* pp 239-273. Butterworth, Toronto.

Bygdeman, M., B. Fredricsson, K. Svanborg, and B. Samuelsson. 1970. The relation between fertility and prostaglandin content of seminal fluid in man. *Fertil. Steril.* 21:622-629.

Campbell, B. G. 1974. *Human Evolution,* 2nd ed. Aldine, Chicago.

Carrick, F. N., and B. P. Setchell. 1977. The evolution of the scrotum. In *Reproduction and Evolution,* J. H. Calaby and C. H. Tyndale-Boscoe (eds.), pp. 165-170. Aust. Acad. Sci., Canberra.

Chagnon, N. A. 1979. Mate competition favoring close kin, and village fissioning among the Yanomamo Indians. In *Evolutionary Biology and Human Social Behavior, An Anthropological Perspective,* N. A. Chagnon and W. G. Irons (eds.), pp. 86-132. Duxbury Press, North Scituate, MA.

Chang, M. C., C. R. Austin, J. M. Bedford, B. G. Brackett, R. H. F. Hunter, and R. Yanagimachi. 1977. Capacitation of spermatozoa and fertilization in mammals. In *Frontiers in Reproduction and Fertility Control*, R. O. Greep and M. A. Koblinsky (eds.), pp. 434-451. MIT Press, Cambridge, MA.

Chimbos, P. D. 1978. *Marital Violence: A Study of Interspouse Homicide*. R&E Research Associates, San Francisco.

Cook, J. M. 1975. *In Defense of Homo Sapiens*. Farrar, Straus, and Giroux, New York.

Crook, J. H. 1972. Sexual selection, dimorphism, and social organization in the primates. In *Sexual Selection and the Descent of Man, 1871-1971*, B. Campbell (ed.), pp. 231-281. Aldine-Atherton, Chicago.

Daly, M., and M. Wilson. 1978. *Sex, Evolution, and Behavior*. Duxbury Press, North Scituate, MA.

Daly, M., M. Wilson, and S. J. Weghorst. 1982. Male sexual jealousy. *Ethol. Sociobiol.* 3:11-27.

Daniels, D. 1983. The evolution of concealed ovulation and self deception. *Ethol. Sociobiol.* 4:69-87.

Davenport, W. H. 1977. Sex in cross-cultural perspective. In *Human Sexuality in Four Perspectives*, F. A. Beach (ed.), pp. 115-163. Johns Hopkins Univ. Press, Baltimore, MD.

Diagram Group, The. 1981. *Sex: A User's Manual*. Perigee Books, New York.

Dickeman, M. 1978. Confidence of paternity mechanisms in the human species. (Circulated and cited unpublished manuscript.)

Dickeman, M. 1979. Female infanticide, reproductive strategies, and social stratification: A preliminary model. In *Evolutionary Biology and Human Social Behavior, An Anthropological Perspective*, N. A. Chagnon and W. G. Irons (eds.), pp. 321-367. Duxbury Press, North Scituate, MA.

Dickeman, M. 1981. Paternal confidence and dowry competition: A biocultural analysis of purdah. In *Natural Selection and Social Behavior*, R. Alexander and D. Tinkle (eds.), pp. 417-438. Chiron Press, New York.

Dirasse, L. 1978. The Socioeconomic Position of Women in Addis Ababa: The Case of Prostitution. Ph.D. Dissertation, Boston Univ.

Dorjahn, V. R. 1958. Fertility, polygyny, and their interrelations in Temne society. *Am. Anthropol.* 60:838-860;

Dorjahn, V. R. 1959. The factor of polygyny in African demography. In *Continuity and Change in African Cultures*, W. R. Bascom and M. J. Herskovits (eds.), pp. 87-112. Univ. Chicago Press, Chicago.

Eliasson, R., and N. Posse. 1965. Rubin's test before and after intravaginal application of prostaglandin. *Int. J. Fertil.* 10:373-377.

Etkin, W. 1954. Social behavior and the evolution of man's mental facilities. *Am. Nat.* 88:129-134.

Etkin, W. 1963. Social behavior factors in the emergence of man. *Human Biol.* 35:299-310.

Exner, J. E., J. Wylie, A. Leura, and T. Parrill. 1977. Some psychological characteristics of prostitutes. *J. Personality Assessment* 41:474-485.

Eyquem, A., M. D'Almeida, and J. Rothman. 1976. Local antispermic immunity in the genital tract. In *Biological and Clinical Aspects of Reproduction*, F. J. G. Ebling and I. W. Henderson (eds.), pp. 174-177. Excerpta Medica, Amsterdam.

Fisher, S. 1973. *The Female Orgasm*. Basic Books, New York.

Foote, R. H. 1973. My experience with artificial insemination. In *Fertility and Sterility*, T. Hasegawa, M. Hayashi, F. J. G. Ebling, and I. W. Henderson (eds.), pp. 353-358. Excerpta Medica, Amsterdam.

Ford, C. D., and F. A. Beach. 1951. *Patterns of Sexual Behavior*. Harper & Row, New York.

Fossey, D. 1974. Development of the mountain gorilla (*Gorilla gorilla beringei*) through the first thirty-six months. Paper presented at Berg Wartenstein, symposium no. 62, the behavior of the Great Apes. Wenner-Gren Foundation for Anthropological Research. (See Hrdy 1979).

Fossey, D. 1976. The behavior of the mountain gorilla. Ph.D. Dissertation, Cambridge Univ.

Fox, C. A., and B. Fox. 1967. Uterine suction during orgasm. *Br. Med. J.* 1:300-301.

Fox, C. A., and B. Fox. 1969. Blood pressure and respiratory patterns during human coitus. *J. Reprod. Fertil.* 19:405-415.

Fox, C. A., and B. Fox. 1971. A comparative study of coital physiology, with special reference to the sexual climax. *J. Reprod. Fertil.* **24**:319-336.

Fox, C. A., H. S. Wolff, and J. A. Baker. 1970. Measurement of intra-vaginal and intra-uterine pressures during human coitus by radio-telemetry. *J. Reprod. Fertil.* **22**:243-251.

Franklin, R. R., and C. D. Dukes. 1964a. Antispermatozoal antibody and unexpected infertility. *Am. J. Obstet. Gynecol.* **89**:6-9.

Franklin, R. R., and C. D. Dukes. 1964b. Further studies on sperm agglutinating antibody and unexplained infertility. *J. Am. Med. Assoc.* **190**:682-683.

Freund, M. 1962. Interrelationships among the characteristics of human semen and factors affecting semen specimen quality. *J. Reprod. Fertil.* **4**:143-159.

Freund, M. 1963. Effect of frequency of emission on semen output and an estimate of daily sperm production in man. *J. Reprod. Fertil.* **6**:269-286.

Galdikas, B. M. F. 1981. Orangutan reproduction in the wild. In *Reproductive Biology of the Great Apes: Comparative and Biomedical Perspectives,* C. E. Graham (ed.), pp. 281-300. Academic Press, New York.

Garfenberg, E. 1950. The role of the urethra in female orgasm. *Int. J. Sexol.* **3**:145-148.

Gaulin, S. J. C., and A. Schlegel. 1980. Paternal confidence and paternal investment: A cross-cultural test of a sociobiological hypothesis. *Ethol. Sociobiol.* **1**:301-309.

Goldberg, V. J., and P. W. Ramwell. 1977. The role of prostaglandins in reproduction. In *Frontiers in Reproduction and Fertility Control,* R. O. Greep and M. A. Koblinsky (eds.), pp. 219-235. MIT Press, Cambridge, MA.

Gould, K. G., and D. E. Martin. 1981. The female ape genital tract and its secretions. In *Reproductive Biology of the Great Apes: Comparative and Biomedical Perspectives,* C. E. Graham (ed.), pp. 105-125. Academic Press, New York.

Gowaty, P. A. 1981. An extension of the Orians-Verner-Willison model to account for mating systems besides polygyny. *Am. Nat.* **118**:851-859.

Graham, C. E., and C. F. Bradley. 1972. Microanatomy of the chimpanzee genital system. In *The Chimpanzee, Vol. 5,* G. C. Bourne (ed.), pp. 77-126. Karger, Basel, Switzerland.

Graham, C. G. 1981. Menstrual cycle of the great apes. In *Reproductive Biology of the Great Apes: Comparative and Biomedical Perspectives,* C. E. Graham (ed.), pp. 1-43. Academic Press, New York.

Green, R. 1980. Variant forms of human sexual behavior. In *Reproduction in Mammals, Book 8, Human Sexuality,* C. R. Austin and R. V. Short (eds.), pp. 68-97. Cambridge Univ. Press, Cambridge.

Halliday, T. 1980. *Sexual Strategy.* Oxford Univ. Press, Oxford.

Hamburg, D. A., and E. McGown (eds.). 1979. *The Great Apes.* Benjamin/Cummings Publ., Menlo Park, CA.

Harcourt, A. H. 1981. Intermale competition and the reproductive behavior of the great apes. In *Reproductive Biology of the Great Apes: Comparative and Biomedical Perspectives,* C. E. Graham (ed.), pp. 301-318. Academic Press, New York.

Harcourt, A. H., P. H. Harvey, S. G. Larson, and R. V. Short. 1981a. Testis weight, body weight, and breeding system in primates. *Nature* **293**:55-57.

Harcourt, A. H., K. J. Stewart, and D. Fossey. 1981b. Gorilla reproduction in the wild. In *Reproductive Biology of the Great Apes: Comparative and Biomedical Perspectives,* C. E. Graham (ed.), pp. 265-279. Academic Press, New York.

Hawthorn, G. 1970. *The Sociology of Fertility.* Collier-Macmillan, London.

Heape, W. 1900. The "sexual season" of mammals and the relations of the "pro-oestrum" to menstruation. *Q. J. Microsc. Sci.* **44**:1-70.

Hite, S. 1976. *The Hite Report.* Macmillan, New York.

Hosken, F. P. 1979. *The Hosken Report. Genital and Sexual Mutilation of Females,* 2nd rev. ed. Women's International Network News, Lexington, MA.

Howell, N. 1979. *Demography of the Dobe !Kung.* Academic Press, New York.

Hrdy, S. B. 1977. *The Langurs of Abu.* Harvard Univ. Press, Cambridge, MA.

Hrdy, S. B. 1979. Infanticide among animals. A review, classification, and examination of the implications for the reproductive strategies of females. *Ethol. Sociobiol.* **1**:13-40.

Hrdy, S. B. 1981. *The Woman That Never Evolved.* Harvard Univ. Press, Cambridge, MA.

Hughes, A. L. 1982. Confidence of paternity and wife-sharing in polygynous and polyandrous systems. *Ethol. Sociobiol.* **3**:125-124.

Hummel, K. 1979. Das biostatistische Gutachten als forensisches Beweismittel. *Arztl. Lab.* **25**: 131-137.

Hunt, M. M. 1974. *Sexual Behavior in the 1970s.* Playboy Press, Chicago.

Hurault, J. 1961. *The Boni Refugee Blacks of French Guiana.* Institut Francais d'Afrique Noire, Dakar.

Isaac, B. L. 1979. Female fertility and marital form among the Mende of rural upper Bambara chiefdom, Sierra Leone. *Ethnology* **19**:297-313.

Isojima, S. 1973. The nature of antibodies against spermatozoa found in women with unexplained sterility. In *Fertility and Sterility,* T. Hasegawa, M. Hayashi, F. J. G. Ebling, and I. W. Henderson (eds.), pp. 105-107. Excerpta Medica, Amsterdam.

Jöchle, W. 1973. Coitus-induced ovulation. *Contraception* **7**:523-565.

Jöchle, W. 1975. Current research in coitus-induced ovulation: A review. *J. Reprod. Fertil. Suppl.* **22**:165-207.

Johnson, M. H., and B. J. Everitt. 1980. *Essential Reproduction.* Blackwell Sci. Publ., Oxford.

Kay, D. J. 1977. Clinical significance of antibodies to antigens of the reproductive tract. In *Immunological Influence on Human Fertility,* B. Boettcher (ed.), pp. 119-123. Academic Press, New York.

Kinsey, A. C., W. B. Pomeroy, and C. E. Martin. 1948. *Sexual Behavior of the Human Male.* W. B. Saunders, New York.

Kinsey, A. C., W. B. Pomeroy, C. E. Martin, and P. H. Gebhard. 1953. *Sexual Behavior of the Human Female.* W. B. Saunders, New York.

Kramer, A. 1932. *Truk.* Friederichsen, De Gruyter, Hamburg.

Kurland, J. A. 1979. Paternity, mother's brother and human sociality. In *Evolutionary Biology and Human Social Behavior,* N. A. Chagnon and W. G. Irons (eds.), pp. 145-180. Duxbury Press, North Scituate, MA.

Lévi-Strauss, C. 1969. *The Elementary Structure of Kinship.* Beacon Press, Boston, MA.

Little, K. 1967. *The Mende of Sierra Leone.* Routledge and Kegan Paul Limited, London.

Lobban, C. F. 1972. *Law and Anthropology in the Sudan (An Analysis of Homicide Cases in Sudan).* African Studies Seminar Series No. 13, Sudan Research Unit, Khartoum Univ.

Lovejoy, C. O. 1981. The origin of man. *Science* **211**:341-350.

Malinowski, B. 1929. *The Sexual Life of Savages in North-Western Melanesia.* Routledge, London.

Marshall, D. S. 1971. Sexual behavior on Mangaia. In *Human Sexual Behavior,* D. S. Marshall and R. C. Suggs (eds.), pp. 103-162. Basic Books, New York.

Martin, D. E., and K. G. Gould. 1981. The male ape genital tract and its secretions. In *Reproductive Biology of the Great Apes: Comparative and Biomedical Perspectives,* C. E. Graham (ed.), pp. 127-161. Academic Press, New York.

Masters, W. H., and V. E. Johnson. 1966. *Human Sexual Response.* Little Brown and Co., Boston, MA.

Maynard Smith, J. 1978. *The Evolution of Sex.* Cambridge Univ. Press, Cambridge, MA.

Mendelson, A. R. 1982. From here to paternity. *Barrister* **9**:12-15,55.

Mitchell, G. 1979. *Behavioral Sex Differences in Nonhuman Primates.* VanNostrand Reinhold, New York.

Mori, I., Y. Maki, Y. Ijuin, A. Tukuda, and M. Takeda. 1973. Psychosomatic aspects of sterility. In *Fertility and Sterility,* T. Hasegawa, M. Hayashi, F. J. G. Ebling, and I. W. Henderson (eds.), pp. 960-961. Excerpta Medica, Amsterdam.

Morris, D. 1967. *The Naked Ape.* McGraw-Hill, New York.

Morris, J. M. 1977. The morning-after pill: A report on postcoital contraception and interception. In *Frontiers in Reproduction and Fertility Control,* R. O. Greep and M. A. Koblinsky (eds.), pp. 203-208. MIT Press, Cambridge, MA.

Murdock, G. P. 1967. *Culture and Society.* Univ. Pittsburgh Press, Pittsburgh, PA.

Musham, H. V. 1956. The fertility of polygamous marriages. *Population Studies* **10**:3-16.

Nadler, R. D., C. E. Graham, D. C. Collins, and O. R. Kling. 1981. Post partum amenorrhea and behavior of apes. In *Reproductive Biology of the Great Apes: Comparative and Biomedical Perspectives,* C. E. Graham (ed.), pp. 69-81. Academic Press, New York.

Neel, J. V., and K. M. Weiss. 1975. The genetic structure of a tribal population, the Yanomama Indians. XIII. Biodemographic studies. *Am. J. Phys. Anthropol.* **42**:25-51.

Okamura, Y., M. Kitazia, K. Arakawa, H. Tateyama, M. Nagakawa, T. Goto, A. Kurano, R. Mori, and I. Taki. 1973. Psychological aspects of gynecological endocrine diseases. In *Fertility and Sterility,* T. Hasegawa, M. Hayashi, F. J. G. Ebling, and I. W. Henderson (eds.), pp. 965-967. Excerpta Medica, Amsterdam.

Osol, A. (ed.). 1972. *Blakiston's Gould Medical Dictionary, 3rd ed.* McGraw-Hill, New York.

Parker, G. A. 1983. Mate quality and mating decisions In *Mate Choice,* P. Bateson (ed.), pp. 141-166. Cambridge Univ. Press, Cambridge, England.

Parks, G. R. (ed.). 1929. *Marco Polo, Travels.* Book League of America, New York.

Parish, W. E., J. A. Carron-Brown, and C. B. Richards. 1967. The detection of antibodies to spermatozoa and to blood group antigens in cervical mucus. *J. Reprod. Fertil.* **13**:469-483.

Pfeiffer, J. E. 1972. *The Emergence of Man.* Harper and Row, New York.

Pickles, V. R. 1967. Uterine suction during orgasm. *Br. Med. J.* **1**:427.

Pietropinto, A., and J. Simenauer. 1977. *Beyond the Male Myth.* Times Books, New York.

Pilbeam, D. R. 1984. The descent of hominoids and hominids. *Sci. Am.* **150**:84-96.

Porter, D. G., and C. A. Finn. 1977. The biology of the uterus. In *Frontiers in Reproduction and Fertility Control,* R. O. Greep and M. A. Koblinsky (eds.), pp. 146-156. MIT Press, Cambridge, MA.

Portmann, A. 1952. *Animal Forms and Patterns.* Farber and Farber, London.

Rosenblatt, P. C. 1974. Cross-cultural perspective on attraction. In *Foundations of Interpersonal Attraction,* T. L. Huston (ed.), pp. 79-95. Academic Press, New York.

Sanday, P. R. 1981. The socio-cultural context of rape: A cross-cultural study. *J. Social Issues* **37**:5-27.

Sass, S. L. 1977. The defense of multiple access (*exceptio plurium concubentium*) in paternity suits: A comparative analysis. *Tulane Law Rev.* **51**:468-509.

Schonfeld, W. A. 1943. Primary and secondary sexual characteristics. Study of their development in males from birth through maturity, with biometric study of penis and testes. *Am. J. Dis. Child.* **65**:535-549.

Schwartz, J. H. 1984. The evolutionary relationships of man and orangutans. *Nature* **308**:501-505.

Shanor, K. 1978. *The Shanor Study: The Sexual Sensitivity of the American Male.* Dial Press, New York.

Shearer, L. 1978. Sex sensation. In "Intelligence Report," *Parade Mag.,* Sept. 10.

Shields, W. M., and Shields, L. M. 1983. Forcible rape: An evolutionary perspective. *Ethol. Sociobiol.* **4**:155-136.

Short, R. V. 1976. Definition of the problem: The evolution of human reproduction. *Proc. R. Soc. Lond. B* **195**:3-24.

Short, R. V. 1977. Sexual selection and the descent of man. In *Reproduction and Evolution,* J. H. Calaby and C. H. Tyndale-Briscoe (eds.), pp. 3-19. Aust. Acad. Sci., Canberra.

Short, R. V. 1979. Sexual selection and its component parts, somatic and genital selection, as illustrated by man and the great apes. *Adv. Study Behav.* **9**:131-158.

Short, R. V. 1980. The origins of human sexuality. In *Reproduction in Mammals, Book 8, Human Sexuality,* C. R. Austin and R. V. Short (eds.), pp. 1-33. Cambridge Univ. Press, Cambridge.

Short, R. V. 1981. Sexual selection in man and the great apes. In *Reproductive Biology of the Great Apes,* C. E. Graham (ed.), pp. 319-341. Academic Press, New York.

Sorensen, R. C. 1973. *Adolescent Sexuality in Contemporary America.* World Publ. Co., New York.

Sorgo, G. 1973. Das Problem der Superfecundatio im Vaterschaftsgutachten. *Beitr. Gerichtl. Med.* **30**:415-421.

Spielmann, W., and P. Kuhnl. 1980. The efficacy of modern blood group genetics with regard to a case of probable superfecundation. *Haematologia* **134**:75-85.

Stangle, J. J. 1979. *Fertility and Conception: An Essential Guide for Childless Couples.* Paddington Press, New York.

Strassmann, B. I. 1981. Sexual selection, paternal care, and concealed ovulation in humans. *Ethol. Sociobiol.* **2**:31-40.

Symanski, R. 1981. *The Immoral Landscape: Female Prostitution in Western Societies.* Butterworth, Toronto.

Symons, D. 1979. *The Evolution of Human Sexuality.* Oxford Univ. Press, New York.

Tanner, R. E. S. 1970. *Homicide in Uganda, 1964. Crime in East Africa.* Scandinavian Inst. African Studies, Uppsala.

Terasaki, P. I. 1978. Resolution by HLA testing of 1000 paternity cases not excluded by ABO testing. *J. Fam. Law* 16:543-557.

Terasaki, P. I., D. Gjertson, D. Bernoco, S. Perdue, M. R. Mickey, and J. Bond. 1978. Twins with two different fathers identified by HLA. *New England J. Med.* 299:590-592.

Thornhill, R., and J. Alcock. 1983. *The Evolution of Insect Mating Systems.* Harvard Univ. Press, Cambridge, MA.

Thornhill, R., and N. W. Thornhill. 1983. Human rape: An evolutionary analysis. *Ethol. Sociobiol.* 4:137-173.

Titiev, M. 1972. *The Hopi Indians of Old Oraibi: Change and Continuity.* Univ. Michigan Press, Ann Arbor, MI.

Tortora, G. J., and N. P. Anagnostakos. 1984. *Principles of Anatomy and Physiology.* Harper & Row, New York.

Trivers, R. L. 1972. Parental investment and sexual selection. In *Sexual Selection and the Descent of Man, 1871-1971,* B. Campbell (ed.), pp. 136-179. Aldine-Atherton, Chicago.

Trivers, R. L. 1974. Parent-offspring conflict. *Am. Zool.* 14:249-264.

Trivers, R. L., and D. E. Willard. 1973. Natural selection of parental ability to vary the sex ratio of offspring. *Science* 179:90-92.

Turner, T. L., and S. S. Howards. 1977. Sperm maturation, transport, and capacitation. In *Male Infertility,* A. T. K. Cockett and R. L. Urry (eds.), pp. 29-57. Grune and Stratton, New York.

Turnbull, C. M. 1965. *The Mbuti Pygmies: An Ethnographic Survey.* Am. Mus. Nat. Hist., AMNH Board of Trustees, New York.

Tutin, C. E. G. 1975. Sexual behavior and mating patterns in a community of wild chimpanzees (*Pan troglodytes*). Ph. D. Dissertation, Univ. Edinburgh.

Tutin, C. E. G., and P. R. McGinnis. 1981. Chimpanzee reproduction in the wild. In *Reproductive Biology of the Great Apes: Comparative and Biomedical Perspectives,* C. E. Graham (ed.), pp. 239-264. Academic Press, New York.

Udry, J. R., and N. M. Morris. 1968. Distribution of coitus in the menstrual cycle. *Nature* 220:593-596.

Uemura, T., and N. Suzuki. 1973. Treatment for psychosomatic sterility. In *Fertility and Sterility,* T. Hasegawa, M. Hayashi, F. J. G. Ebling, and I. W. Henderson (eds.), pp. 969-970. Excerpta Medica, Amsterdam.

Van den Berghe, P. L., and D. P. Barash. 1977. Inclusive fitness theory and human family structure. *Am. Anthropol.* 79:809-823.

Washburn, S. L. 1974. *Ape Into Man.* Little, Brown and Co., Boston, MA.

Washburn, S. L., and C. S. Lancaster. 1971. The evolution of hunting. In *Background for Man,* P. Dolhinow and V. M. Sarich (eds.), pp. 386-403. Little, Brown and Co., Boston, MA.

Weatherhead, P. J., and R. J. Robertson. 1979. Offspring quality and the polygyny threshold: "The sexy son hypothesis." *Am. Nat.* 113:201-208.

Whitlatch, W. G., and R. W. Masters. 1962. Comment: Contribution of blood tests in 734 disputed paternity cases: Acceptance by the law of blood tests as scientific evidence. *Western Reserve Law Rev.* 14:114-115.

Williams, G. C. 1966. *Adaptation and Natural Selection.* Princeton Univ. Press, Princeton, NJ.

Williams, G. C. 1975. *Sex and Evolution.* Princeton Univ. Press, Princeton, NJ.

Wilson, E. O. 1975. *Sociobiology: The New Synthesis.* Belknap Press, Cambridge.

Winick, C., and P. M. Kinsie. 1972. *The Lively Commerce: Prostitution in the United States.* Quadrangle, Chicago.

Wolfe, L. 1981. *The Cosmo Report.* Arbor House, New York.

Woodland, W. 1903. On the phylogenetic case of the transposition of the testes in mammalia. *Proc. Zool. Soc. Lond.* 1:319-340.

Index

682   Index